4th International Conference on Manufacturing, Material and Metallurgical Engineering (ICMMME 2019)

IOP Conference Series: Materials Science and Engineering
Volume 538

Chengdu, China
22 – 25 March 2019

ISBN: 978-1-5108-8812-8
ISSN: 1757-8981

Printed from e-media with permission by:

Curran Associates, Inc.
57 Morehouse Lane
Red Hook, NY 12571

Printed by Curran Associates, Inc. (2019)

Additional copies of this publication are available from:

Curran Associates, Inc.
57 Morehouse Lane
Red Hook, NY 12571 USA
Phone: 845-758-0400
Fax: 845-758-2633
Email: curran@proceedings.com
Web: www.proceedings.com

Table of contents

Volume 538

ICMMME 2019, The 4th International Conference on Manufacturing, Material and Metallurgical Engineering

22–25 March 2019, Chengdu, China

Accepted papers received: 17 April 2019

Preface

Papers

Composite Preparation and Performance Testing

Industrial Materials and Manufacturing Technology

Advanced Manufacturing Processes and Applications

Aerospace Dynamics and Safety Engineering

Vehicle Structure and Control Engineering

IOP Conf. Series: Materials Science and Engineering **538** (2019) 011001 doi:10.1088/1757-899X/538/1/011001

PREFACE

As a Conference Chair, I am very pleased to welcome all the participants in the the 4th International Conference on Manufacturing, Materials and Metallurgical Engineering (ICMMME 2019) held in Chengdu, China, 22 - 25 March, 2019. The conference focuses on the highly popular, but challenging in the fields of manufacturing, materials and metallurgical engineering and the outcome of our deliberations will play a crucial role in progress achieved in these areas.

ICMMME 2019 comprehensively presents the exciting results in Manufacturing, Materials and Metallurgical Engineering from our keynote speakers of leading scientists and a variety of authors from around the world. We would like to express sincerest thanks to the honorable keynote speakers: Prof. Ramesh K. Agarwal from Washington University in St. Louis, USA, Prof. Changduk Kong from Chosun University, South Korea, Assoc. Prof. Dong Xiang from Southwest Petroleum University, China and Prof. Seok-Soon Lee from Gyeongsang National University, South Korea. Their speeches motivated the discussions among participants and promoted the idea exchanges multi-disciplines and multi-fields. The conference team truly believes that active participants will find the discussion erudite and informative, and enjoy the opportunity for sharing research results. We hope this success can be evolved into persistent success annually.

We also would like to express our sincerest gratitude to our colleagues in the disciplines who have kindly volunteered in the reviewing process. And special thanks go to the chairs of the conference and the sessions.

The proceedings include a variety of advanced scientific and academic achievements in the field of manufacturing, materials and metallurgical engineering. Full papers submissions were blind peer-reviewed by at least two experts from the International Technical Committee in order to ensure high quality standards of research. We certainly appreciate the efforts and contribution from the conference chairs, organization staff and the members of the technical committee for their work. We sincerely hope that all participants and guests benefit significantly from the published proceedings.

At last, we hope all of you will again get together at the next ICMMME, to have intelligent and exciting, academic and practical discussions which will be rewarding and most enjoyable experience.

Prof. Ramesh K. Agarwal
Washington University in St. Louis, USA
25 March 2019

IOP Conf. Series: Materials Science and Engineering **538** (2019) 011002 doi:10.1088/1757-899X/538/1/011002

Conference Chairs

Ying Zhou, Southwest Petroleum University, China

Ramesh K. Agarwal, Washington University in St. Louis, USA

Changduk Kong, Chosun University, South Korea

Program Chairs

Seong Soo Choi, Sun Moon University, South Korea

S. Lee, Gyeongsang National University, South Korea

Technical Committee

Michel Bellet, Institut Mines-Telecom. Mines ParisTech, France

Pavel Lukac, Univerzita Karlova, Czech Republic

Jiri Kliber, VSB-TU Ostrava, Czech Republic

N. Emre Altun, Middle East Technical University, Turkey

H. Park, Howon University, South Korea

K. Oh, Howon University, South Korea

Abel Ribeiro de Jesus, Federal University of Bahia - Polytechnic School, Brazil

Xianfeng Ma, Sun Yat-sen University, China

Jhewn-Kuang Chen, National Taipei University of Technology, Taiwan

Recai KUS, Selcuk University, Turkey

Huei-Sen Wang, I-Shou University, Taiwan

Chia-Yen Lee, National Pingtung University of Science and Technology, Taiwan

Nakamura Yuzo, Kagoshima University, Japan

Raj Das, RMIT University, Australia

Datin Napsiah Ismail, Universiti Putra Malaysia

Md Mustafizur Rahman, Universiti Malaysia Pahang, Malaysia

Poi Sim Khiew, The University of Nottingham, Malaysia

Megat Mohamad Hamdan bin Megat Ahmad, Universiti Pertahanan Nasional Malaysia

Mellal Mohamed Arezki, M'Hamed Bougara University (UMBB), Algeria

Laichang Zhang, Edith Cowan University, Australia

Lijia Zhao, Colorado School of Mines, USA

Rusdi Nur, Politeknik Negeri Ujung Pandang, Indonesia

Rosli Ahmad, Universiti Tun Hussein Onn Malaysia

Chou Chau Yuan Fong, Universiti Brunei Darussalam, Brunei

Thanakom Soontornchainacksaeng, King Mongkut's University of Technology North Bangkok, Thailand

Quanli Han, Xi'an Technological University, China

Jinqiang Ning, Georgia Institute of Technology, USA

Muhammad Jahan, Miami University, USA

Kelvin Fu, University of Delaware, USA

Zhibin Lin, North Dakota State University, USA

Jose Xavier, New University of Lisbon, Portugal

Elammaran Jayamani, Swinburne University of Technology Sarawak Campus, Malaysia

Amar Rezoug, Superieure National School of Technology (ENST), Algeria

Hoong Thiam Toh, Universiti Teknologi Malaysia

Cristian Paul Chioncel, "Eftimie Murgu" University of Resita, Romania

Nyuk Yoong Voo, Universiti Brunei Darussalam, Brunei Darussalam

Supacharee Roddecha, Kasetsart University, Thailand

Hongqiang Wan, Xi'an Technological University, China

IOP Conf. Series: Materials Science and Engineering **538** (2019) 011003 doi:10.1088/1757-899X/538/1/011003

Peer review statement

All papers published in this volume of *IOP Conference Series: Materials Science and Engineering* have been peer reviewed through processes administered by the proceedings Editors. Reviews were conducted by expert referees to the professional and scientific standards expected of a proceedings journal published by IOP Publishing.

Graphene-polymer Composites for Enhancing the Mechanical Properties

Xiao Zhang[1]*, Jian Zheng[1] and Zilong Ma[2]

[1]Shijiazhuang Campus, Army Engineering University, Shijiazhuang, 050003, China
[2]Huayin Ordnance Test Ctr., Huayin, 714200, China

*Corresponding author: zxleo@foxmail.com

Abstract. Graphene is widely used in the improvement of mechanical properties of composite materials due to its excellent mechanical properties. In order to more fully utilize the enhanced performance of graphene, many preparation methods have been developed and gradually improved. In this paper, the research status of graphene to improve the mechanical properties of polymer composites is reviewed. The preparation process of materials and the effects of different processes on the mechanical properties of the materials are studied.

1. Introduction

Graphene is currently the strongest material, 200 times higher than the best performing steel. The defect-free graphene has a Young's modulus of 1.0TPa and a breaking strength of 130GPa. Due to its excellent mechanical properties, graphene is widely used to enhance the mechanical properties of materials. Graphene is currently mainly used in polymer matrix composites and metal matrix composites. In this paper, the application status of graphene in improving the mechanical properties of polymer composites is reviewed. The main preparation methods and processes for improving the mechanical properties of polymer composites are summarized. The main problems are pointed out.

2. Methods

In polymer/graphene nanocomposites, the key to improving the mechanical reinforcement of graphene is improving the surface compatibility between graphene and polymer and promoting stress transfer. The covalent bond and non-covalent bond modification have improved the interface properties to a certain extent, so it has become the main research direction to improve the mechanical reinforcement of graphene. Lu [1] used molecular mechanics and molecular dynamics simulation to study the effect of chemical functional groups on the surface of graphene on the interfacial bonding properties between graphene and polymer. The results show that both energy and shear stress between graphene and polymer increase with increasing chemical modification, and it has been found that some special functional groups modified on the surface of graphene play an important role in enhancing the shear stress between graphene and polymer interface. Therefore, chemical modification of graphene surface is an effective method to enhance stress transfer between interfaces when preparing graphene/polymer composites. Wu [2] proposed an effective way to synthesize hyperbranched aromatic polyamine functionalized graphene sheets (GS-HBA) with 3,5-diaminobenzoic acid as monomer. The obtained GS-HB exhibits stable solution-like dispersibility and good compatibility with thermoplastic polyurethane. By controlling the content of GS-HBA, the mechanical properties of the composite can be adjusted. Zhang [3] synthesized an epoxy group-containing hyperbranched polysiloxane-grafted graphene oxide (HPE-GO) by hydrolysis condensation method, and prepared HPE-GO by melt blending into DCPDCE resin. The HPE-GO/DCPDCE nanocomposite system, the addition of HPE-GO increases the impact and bending strength of the resin. Wu et al. [4] first synthesized a high fluorine content fluorinated graphene, and then used the nucleophilic substitution reaction activity of the secondary amine group on the aramid III (PA) imidazole ring to replace the fluorine in the fluorinated graphene. The atom forms a chemical bond with the graphene sheet layer, thereby achieving the purpose of fluorinated graphene grafting to enhance the mechanical properties of the PA film. Li et al. [5] prepared ethylene diamine (EDA) covalently functionalized modified graphene oxide sheets (GS-EDA)/acidified multi-walled carbon nanotubes (MWNT-COOH)/polystyrene grafting by solution-composite method. The surface structure of the filler was characterized by infrared spectroscopy and X-ray diffraction. The mechanical properties of the material were tested by tensile, impact and melt flow rate tests. It is shown that the EDA is successfully grafted onto the surface of graphene oxide, and the mechanical properties of the composite are obviously improved. Su et al. [6] prepared a thermotropic liquid crystal polymer/graphene oxide (TLCP/GO) hybrid material by solution mixing of oxidized graphene oxide with a thermotropic liquid crystal polymer, and through roll milling and extrusion. Thermotropic liquid crystal polymer/graphene oxide/phenolic resin (TLCP/GO/PF) composites were prepared by compression molding process. The effects of TLCP/GO hybrid filler content on the mechanical properties of phenolic resin composites were investigated. When the amount is low, the bending strength and flexural modulus of the composite material are significantly improved. Cano et al. [7] grafted PVA molecules onto the surface of graphene oxide by covalent modification and used to improve the mechanical properties of the composite.

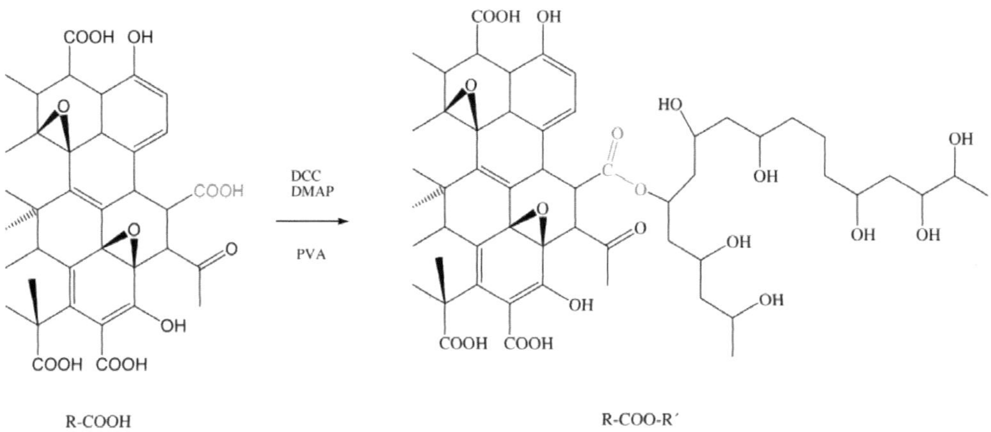

Figure 1. Functionalisation of graphene oxide with poly(vinyl alcohol) by a carbodiimide esterification reaction [7].

Kuila et al. [8] regarded octadecylamine as a surface modifier, and prepared vinyl acetate/graphene composites using functionalized graphene oxide to improve the mechanics of materials. Fang et al. [9] covalently grafted polystyrene chains onto the surface of graphene by adding diazonium salt and atom transfer radical polymerization to modify the surface of graphene.

3. Mechanical Properties

The improvement of mechanical properties of graphene is divided into two aspects: static mechanical properties and dynamic mechanical properties. The research on the mechanical properties of graphene/polymers mainly focuses on these two aspects. On the other hand, the creep property has also been studied.

3.1. Static and Dynamic Mechanical Properties

Numerous studies have shown that the addition of a small amount of graphene or graphene oxide can significantly improve the strength, fracture toughness and fatigue properties of polymer materials. Bao [10] selected polylactic acid (PLA), polyvinyl alcohol (PVA) and polystyrene (PS) as polymers， and studied the different interfacial interactions between graphene and these polymers. Based on GO, graphene and the different polymeric matrices, the essential relationship between the microstructure and properties of GPNC has been revealed. It is proposed that graphene can enhance the mechanical properties of polymer composites on the one hand, but at the same time block the mutual interaction between polymer chains, resulting in the reducing of the mechanical properties of the composite material. The mutual competition between the two leads to the repetition and inflection point of the polymer composite when the graphene content reaches a certain value. Zhang [11] studied the mechanism of mechanical reinforcement of graphene-carbon nanotube hybrid composites. A graphene-carbon nanotube hybrid structure reinforcement model was proposed. The molecular structure simulation method was used to study the hybrid structure/polymer. The mechanical response of composite materials under axial load and the mechanism of interface mechanics enhancement are analyzed. The mechanism of mechanical response of hybrid structures is analyzed. The effects of different structural parameters on the strength and mechanical behavior of the composites are studied. Huang et al. [12] obtained a boron cross-linked graphene oxide-polyvinyl alcohol (B-GO/PVA) film by incorporating a boron cross-linked structure into a graphene oxide-polyethylene structure, and examined it with a dynamic analyzer. Compton et al. [13] studied the stiffness characteristics of graphene oxide film and PVA/graphene oxide composite film by combining experimental, theoretical and numerical calculation methods. The effect of water molecules between interfaces was investigated. The results showed that water between molecular layers Molecules can act as a network of hydrogen bonds to enhance the stiffness of the film. Li et al. [14] established a sandwich structure model of single-layer graphene and ultra-thin polymer layer in order to study the enhancement of interface mechanical properties. The results of this model show that the significantly enhancing effect of graphene occurs in the small strain range and the interfacial shear strength of the structure is given by this method.

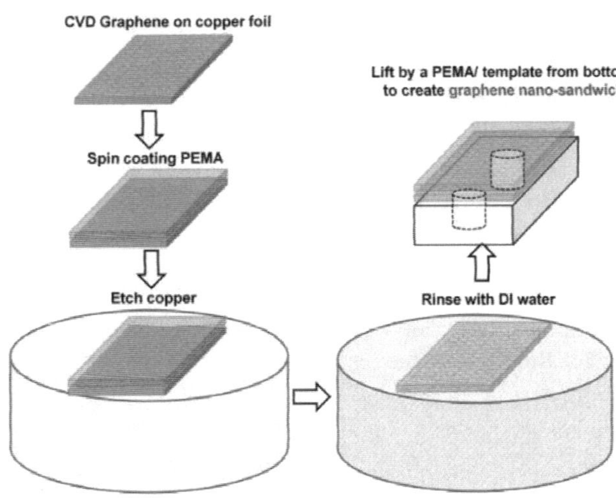

Figure 2. Schematic of the method to create the graphene nano-sandwiches [14].

Shen et al. [15] analyzed the mechanical properties of graphene reinforced epoxy resin and epoxy resin/carbon fiber composite laminates, analyzed the ultimate tensile strength and toughness of graphene/epoxy resin, and studied the tensile fatigue strength of epoxy resin/carbon fiber composite laminate at the graphene filling mass fraction of 0.25%. Yan et al. [16] coated a thin layer of polymer on the surface of CVD-grown graphene to form a polymer/graphene composite film, and analyzed the mechanical properties of the material. The results show that both the frictional strength and friction coefficient of the composite have been improved. Biswas et al. [17] prepared a graphene/liquid crystal polymer nanocomposite. By improving the preparation process and improving the dispersibility of graphene, the modulus is increased by 25% and 55% when the volume fraction of graphene was 1% and 5%, respectively. Gouda et al. [18] studied the effects of the synergistic effect of graphene and carbon nanotubes on the mechanical properties such as tensile strength, toughness and hardness of polymer nanocomposites.

3.2. Creep properties

At the same time, graphene can also be used to improve the inherent defects of creep deformation of polymer materials, and also has an important impact on the service life (creep recovery behavior) of polymer materials. Wang [19] systematically carried out research on creep resistance and recovery behavior of graphene modified polymer nanocomposites, on the basis of not affecting the static mechanical properties of materials and other important functional properties. Targeting two different polymer systems (thermoplastic polystyrene and elastomeric silicone rubber), graphene/polymer nanocomposites were prepared by different processes, and different nanocomposites were systematically studied, in order to improve the mechanical properties such as creep resistance and recovery of polymer materials. The creep recovery behavior under stress and temperature conditions was explored. The mechanism of interaction between graphene and polymer segments was explored. Combined with relevant theoretical models, the structure-activity relationship between viscoelastic behavior and microstructure of the composites was analyzed.

4. Conclusion

Graphene has been widely used in improving the mechanical properties of polymers, and has achieved remarkable results. The literature has shown that graphene can significantly improve the static mechanical properties, dynamic mechanical properties and creep properties of polymer composites. At the same time, the preparation process and the distribution state of graphene have an important influence on the improvement of mechanical properties.

References

[1] Lv Cheng, MD Smilation of Effects of Chemisorption on the Interfacial Bonding and Characteristics of Graphene/ploymer Composites, Master Degree Thesis, China University of Petroleum, 2011.

[2] Wu Chao, Controlled Preparation and Properties of Polymer-Based Graphene Composites, Doctor Degree Thesis, Shanghai Jiao Tong University, 2014.

[3] Zhang Mengmeng, Graphene Oxide Grafted with Hyperbranched Polymers and Its Application in Modification of Cyanate Ester Resin, Doctor Degree Thesis, Northwestern Polytechnical University, 2015.

[4] Wu Peng, Wang Huina, Li Baoyin, Wang Xu, Liu Xiangyang, Characterization of Fluorinated Graphene and Its Reinforced Aromatic Polyamide Film , Polymer Materials Science and Engineering, 2016, 32(9): 59-64.

[5] Li Jiamei, Wang Gang, Chen Lixing, Xu Yuming, Bian Jun, Lu Yun, Preparation and Mechanical Properties of Polystyrene Nanocomposites Modified Collaboratively by Functionalized Graphene Sheets and Carbon Nanotubes, China Plastics Industry, 2015, 43(4): 74-78.

[6] Su Le, Wei Chun, Fu Jun, Wang Wu, Wei Jingliang, Huang Xiaohua, Lv Jian, Study on Mechanical and Tribological Properties of TLCP/GO/PF Hybrid Composites, New Chemical Materials, 2015, 43(2): 108-110.

[7] Cano M, Khan U, Sainsbury T, O'Neill A, Wang Z M, McGovern I T, Master W K, Benito A M, Coleman J N, Improving the Mechanical Properties of Graphene Oxide Based Materials

by Covalent Attachment of Polymer Chains, Carbon, 2013, 52: 363-371.
[8] Kuila T, Khanra P, Mishra A K, Kim N H, Lee J H, Functionalized-graphene/ethylene vinyl acetate co-polymer composites for improved mechanical and thermal properties, Polymer Testing, 2012, 31: 282-289.
[9] Fang M, Wang K G, Lu H B, Yang Y L, Nutt S, Covalent polymer functionalization of graphene nanosheets and mechanical properties of composites, J. Mater. Chem., 2009, 19: 7098-7105.
[10] Bao Chenlu, Preparation, Structure and Mechnism Research of Graphene and Typical Polymer-based Nanocomposites, Doctor Degree Thesis, University of Science and Technology of China, 2012.
[11] Zhang Yong, Study on Mechanical Enhancement Mechanism of Carbon Nanotube-graphene Hydrid Reinforced Composites, Master Degree Thesis, Harbin Institute of Technology, 2016.
[12] Huang Yifu, Ruan Wenhong, Zhang Mingqiu, Rong Minzhi, Studies on the Dynamic Mechanical Properties of Boron Cross-linked Graphene Oxide-Polyvinyl Alcohol Composite , 29th Annual Academic Conference of China Chemical Society, 2014.
[13] Compton O C, Cranford S W, Putz K W, An Z, Brinson L C, Buehler M J, Nguyen S B T, Tuning the Mechanical Properties of Graphene Oxide Paper and Its Associated Polymer Nanocomposites by Controlling Cooperative Intersheet Hydrogen Bonding, ACS Nano, 2012, 6(3): 2008-2019.
[14] Li X G, Warzywoda J, McKenna G B, Mechanical Responses of a Polymer Graphene-sheet Nano-sandwich, Polymer, 2014, 55: 4976-4982.
[15] Shen M Y, Chang T Y, Hsieh T H, Li Y L, Chiang C L, Yang H, Yip M C, Mechanical Properties and Tensile Fatigue of Graphene Nanoplatelets Reinforced Polymer Nanocomposites, Journal of Nanomaterials, 2013, 2013(4): 7758-7765.
[16] Yan C, Kim K S, Lee S K, Bae S H, Hong B H, Kim J H, Lee H J, Ahn J H, Mechanical and Environmental Stability of Polymer Thin-Film-Coated Graphene, ACS Nano, 2012, 6(3): 2096-2103.
[17] Biswas S, Fukushima H, Drzal L T, Mechanical and electrical property enhancement in exfoliated graphene nanoplatelet/liquid crystalline polymer nanocomposites, Composites: Part A, 2011, 42: 371–375.
[18] Gouda P S S, Kulkarni R, Kurbet S N, Jawali D, Effects of multi walled carbon nanotubes and graphene on the mechanical properties of hybrid polymer composites, Adv. Mat. Lett., 2013, 4(4): 261-270.
[19] Wang Xu, The Preparation, Creep and Recovery Behavior of Graphene-filled-polymer Nanocomposites, Master Degree Thesis, Hangzhou Normal University, 2015.

IOP Conf. Series: Materials Science and Engineering **538** (2019) 012002 doi:10.1088/1757-899X/538/1/012002

Research Status and Control Measures of Transverse Cracks for Micro-alloyed Continuous Casting Billet

Qi Li[a,*], Jian Wu[a], Li Bao[a], Yu Gu[a]

[a]School of Materials Science and Engineering, Northeastern University, Shenyang, Liaoning 110000, PR China

*Corresponding author: 474323718@qq.com

Abstract. The transverse cracks of the micro-alloyed steel continuous casting billet seriously damage the quality of the continuous casting billet and cause great losses to the enterprise. The influence of the hot ductility, microstructure, and micro-alloying elements on transverse crack was discussed, which has been researched in-depth at home and abroad. Through a great deal of studies, various control measures of transverse crack are proposed, such as secondary cooling technology, chamfering mold, refining the surface structure of the casting slab, and double phase transformation process. However, the transverse cracks have not been fundamentally solved. Therefore, the formation mechanism and control measures of the transverse crack are summarized for providing with the idea of eliminating transverse crack.

1. Introduction

Micro-alloyed steel is widely used in various engineering structural steels and is the main product in the modern steel industry[1]. The main feature of micro-alloyed steel is the addition of micro-alloying elements such as niobium, vanadium and titanium to the steel in order to control the formation of related carbonitride to regulate the recrystallization behavior of deformed austenite and prevent grain growth, and to produce precipitation strengthening and grain refinement in matrix, thereby increasing the strength and toughness of the steel.

However, due to the addition of micro-alloying elements, the crack sensitivity of the billet increases, and surface transverse crack and corner crack are formed in most cases, which makes the billet have a high defect rate after rolling and prejudices the production schedule and cost. Therefore, how to optimize the continuous casting process and micro-alloying elements addition, and reduce the transverse crack of continuous casting billet is an urgent problem for domestic and foreign enterprises.

2. Factors of the Transverse Cracks Formation

In the continuous casting process, as the molten steel is injected into the crystallizer and solidifies to form a shell, the billet is subjected to various stresses such as phase transformation stress, molten steel static pressure, thermal stress and mechanical stress. When the stress is greater than the critical stress that the slab can withstand, the solidified shell of the slab will crack. Therefore, the factors of the slab crack are numerous, and most of them are the combined effects of multiple factors.

2.1. Influence of the Hot Ductility on the Transverse Cracks
Suzuki et al. proposed and improved the mechanism of three brittle zones[2-4]. B. Mintz et al.[5-9] conducted many detailed and in-depth studies of the relationship between the third brittle zone and the transverse crack. The reason for forming the ductility trough below the Ae3 temperature is that the strain-induced pro-eutectoid ferrite precipitates along the austenite grain boundary, causing stress concentration around the MnS; moreover, the ductility trough is caused by the grain boundary slip above Ae3 temperature. On the high temperature side of the ductility trough, plastic recovery occurs due to dynamic recrystallization, but the precipitation of micro-alloyed carbonitride hinders dynamic recrystallization of steel, especially Nb (C, N), and widens the ductility trough of steel containing Nb. On the low temperature side of the ductility trough, plastic recovery is dependent on the formation of a large amount of ferrite in the grain boundaries and grains. In case the proportion of ferrite in the austenite-ferrite two-phase region is higher than 45% during straightening stage, the ductility of the slab can be effectively improved and the formation of transverse cracks can be reduced.

2.2. Influence of the Microstructure on the Transverse Cracks
As early as 1974, Rain et al. [10] proposed that the formation and development of transverse cracks in continuous casting billets are always accompanied by coarse austenite grains. Tsai et al. [11] found that in the event the austenite grain diameter is larger than 1 mm, the probability of transverse cracks in continuous casting billet is almost 100%. Mainly due to the different grains orientation on both sides of the grain boundary, the dislocation structure is very complicated, and the difficulty of crack passing through the grain boundary is correspondingly larger. Therefore, the finer the grain size, the larger grain boundary area will be, and the hindrances provided by the grain boundary are more during the expansion process of cracks [12-14].
When the surface temperature of the slab is cooled to below the transformation temperature of austenite to ferrite, the pro-eutectoid ferrite has sufficient time to precipitate along the austenite at a slow cooling rate, and continuously connected into pieces, thus forming film-like pro-eutectoid ferrite with a thickness of about 10 to 25 μm[15]. The film-like pro-eutectoid ferrite can cause the continuity of the matrix to be destroyed under high temperature conditions, and the strength of the ferrite is only about one quarter of the austenite strength, therefore the stress of the continuous casting billet during straightening segment is easy to concentrate on relatively soft ferrite and film-like pro-eutectoid ferrite[16]. As the stress acting on the ferrite is greater than the ultimate stress that it can withstand, tiny pores are formed in the ferrite. Under the constant action of various stresses, the tiny pores will gradually grow, leading to grain boundaries slip, and finally transverse cracks extending along the austenite grain boundary form on the surface of the slab[17].

2.3. Influence of the Microalloy elements on the Transverse Cracks
Alloying elements such as Nb, V and Ti in micro-alloyed steels are forming elements of strong carbides, nitrides, and carbonitrides[18-20]. In the continuous casting process, as the temperature of the slab decreasing, the elements such as Nb, V and Ti in the solid solution react easily with the C and N elements to form stable carbides, nitrides or carbonitrides that precipitate in a chain form on the austenite grain boundary. The bonding force between the austenite grains reduces, and the slip between the grain boundaries is hindered, so that the stress concentration of the slab is easily formed at the grain boundary during deformation of the slab, which is also the reason that the crack generates along the grain boundary.

3. Control Measures of the Transverse Cracks
A large number of researchers have studied the control methods of transverse cracks in continuous casting billet. Based on the mechanism of transverse cracking formation, by controlling the parameters of continuous casting process, the microstructure and precipitation phase of the continuous casting billet are adjusted to reduce the transverse cracks of the continuous casting billet.

3.1. Secondary Cooling Technology
The secondary cooling technology is important for reducing transverse cracks. During bending or straightening process in the third brittle zone (600-900℃), transverse cracks form easily. In contrast, bending or straightening at a temperature higher or lower than third brittle zone, the formation of the transverse cracks can be minimized. Zhu Guosen et al.[21] used the cooling method of blocking the edge nozzle or the edge of the reinforcing plate to improve the transverse crack of the slab.
Sun Yanhui et al.[22] believed that the surface of the continuous casting slab unevenly cooled, which produces $\gamma \rightarrow \alpha \rightarrow \gamma$ repeated phase transformation, accompanied by various micro-alloyed carbonitrides precipitated at the grain boundary, eventually leading to transverse cracks of continuous casting billet under the action of straightening stress. By controlling the amount of water with secondary cooling technology, the weak cooling is used to ensure the uniform transverse cooling rate of the slab and avoid the brittle temperature zone to control the surface cracks. He Dongfeng et al.[23] proposed a secondary cooling middle-scope water spray spraying technology , by which the slab surface temperature became more uniform and the third brittle zone is avoided, compared to secondary cooling wide-scope water spraying by which the surface temperature gradient was too large to separate from the third brittle zone.
A large number of studies[22-26] have shown that through the secondary cooling technology, rapid cooling or slow cooling, avoiding the brittle zone of the slab, can effectively reduce the generation of transverse cracks in the continuous casting billet, which has been widely used.

3.2. Chamfering Mold
Some studies have shown that[27,28], the uneven heat transfer in the corner of the slab, exacerbates the internal thermal cracking caused by the thermal shrinkage of the slab shell in the mold. Liu Yang[29] and Wang Wenjun [30] compared the right angle and chamfer mold by two-dimensional mathematical simulation and actual production, and pointed out that the chamfer mold can increase the corner temperature of the slab by 70℃ during bending and straightening, thereby reducing the thermal stress in this area. Yang Chunzheng[31] mainly studied the temperature, thermal stress and mechanical stress distribution of the billet during straightening under different chamfer lengths and angles. Much progress has been made in the study of the geometrical characteristics of the mold corners. The practice shows that the chamfering mold can reduce the corner crack of the micro-alloyed billet to 1.6%[30], but this defect cannot be eliminated.

3.3. Refining the Surface Structure of the Casting Slab
In 2003, Sumitomo Metal[32] published the SSC (surface structure control) cooling mode to eliminate the corner transverse cracks by controlling the structure and precipitation at the surface and corner. After cooling by SSC and testing, the value of reduction in area has exceeded 60% under various temperature conditions. The hot ductility of the slab has been significantly improved, and the third brittle zone has basically disappeared. The reason why the SSC cooling can improve the hot ductility of the steel is as follows. On the one hand, the size and distribution of the precipitation can be improved by SSC. The precipitations uniformly distribute in the grain boundaries and the grains, and the size of the precipitations are larger. On the other hand, the microstructure of the cast slab can be improved. The intergranular ferrite almost disappears, thereby cracks and expansion along the grain boundary can be avoided, and the corner transverse cracks are fundamentally eliminated.

3.4. Double Phase Transformation Process
The refinement of the original austenite grains in the second cooling stage is another way to solve the transverse cracks. For the hypoeutectoid steel, the double phase transformation method controlled by thermal cycles (ie, transformation of austenite to ferrite can be realized by rapid cooling, and ferrite transforms into austenite by reheating ferrite) can achieve austenite grain refinement. The researches of Lee et al.[33] show that after the samples are heated to 1300℃ and rapidly cooled to 600℃ and then back to 1100℃ to make the sample cyclic transformation, austenite grain size decreases from 88

μm to 33 μm. Du Chenwei et al.[34] calculated the temperature field and cooling rate especially for surface center and corner along the casting direction based on the two-dimensional heat transfer and solidification model of slab continuous casting, and studied the influence of temperature distribution and cooling rate distribution on the surface transverse cracks. The results show that, before straightening segment the temperatures of the surface and corner are always in the range of 700-900℃, resulting in there is enough time for Ni (C, N) and AlN precipitating. In addition, as the temperature reduces to 800℃, the film-like pro-eutectoid ferrite becomes the mainly reason of the deterioration of the hot ductility at the cooling rate of 0.1-0.5℃/s. However, Li Yunfeng[35] believed that the existing continuous casting machines are difficult to achieve the double phase transformation process, and the control process is cumbersome.

4. Conclusions

The formation mechanism and control methods of the surface transverse cracks of the continuous casting billet have reviewed based on many studies of researchers. The transverse cracks of continuous casting billet is mainly caused by microstructure and precipitation phase. The design of secondary cooling technology and chamfering model structure is widely used, which is relatively the most effective and lowest cost. However, it cannot fundamentally eliminate the surface transverse cracks. Therefore, the mechanism and control methods of the generation of transverse cracks requires further study.

References

[1] HOU Jing, WANG Fei, ZHAO Guoying, et al. The resarech situation and development tendency of mirco-alloyed steel[J]. Materials Review, 2007, 21(6): 91.
[2] M.Wolf. Fine intergranular surface cracks in bloom casting[J]. Transactions of the Iron and Steel Institute of Japan, 1984,24: 351-358.
[3] H G Suzuki, S Nishimura and S Yamaguchi. Characteristics of hot ductility in steels subjected to the melting and solidification[J]. Transactions ISIJ. 982,22:48-56.
[4] B G Thomas, J K Brimacombe.The formation of panel cracks in steel ingots[J]. ISS Transaction, 1986, 7: 7-20.
[5] B Mintz. Importance of Ar3 temperature in controlling ductility and width of hot ductility trough in steels and its relationship to transverse cracking[J]. Materials Science and Technology, 1996, 12(2): 132-138.
[6] B. Mintz. Understanding the low temperature end of the hot ductility trough in steels [J]. Materials Science and Technology.2008, 24(1):112-120.
[7] B. Mintz. Influence of nitrogen on hot ductility of steels and its relationship to problem of transverse cracking [J].Ironmaking and Steelmaking,2000, 27(5)： 343-347.
[8] B. Mintz and D. N. Crowther. Hot ductility of steels and its relationship to the problem of transverse cracking in continuous casting [J]. International Materials Reviews, 2010, 55(3): 168-196.
[9] B. Mintz, R. Abushosha and M. Shaker. Influence of deformation induced ferrite, grain boundary sliding and dynamic recrystallisation on hot ductility of 0.1–0.75%C steels [J]. Mater. Sci. Technol., 1993, 9: 907–914.
[10] Rain Dippenaar. Strand surface cracks-the role of abnormally large Prior-austenite grains[J]. Iron&Steel Technology, 2007, 4(7): 105-115.
[11] H.T Tsai, H Yin, M Lowry. Analysis of transverse corner cracks on slabs and countermeasures[J]. Iron & steel technology, 2006,3,(7):23-31
[12] J. Reiter, C. Bernhard , H. Presslinger. Austenite grain size in the continuous casting process: Metallographic methods and evaluation[J]. Materials Characterization. 2008, 59(6): 737-746.
[13] I. Hrivank. Grain growth and embrittlement of steel welds[A]. Mathematical Modelling of Welding Phenomena. London: The Institate of Materials. 1995-2001.
[14] Li Juyan. Study on controlling of transverse corner crack occurrence for CC slabs at Pansteel

[D]. Chongqing University, 2010.
[15] J.M.Gregg, K.D.H.Bhadeshiah. Solid-state nucleation of acicular ferrite on minerals added tomolten steel[J]. Acta Materialia, 1997, 45(2): 739-748.
[16] Ma Fanjun, Wen Guanghua, Tang Ping, et al. Research Present of Corner Transversal Crack on the Nb,V and Ti Micro-alloyed Steel Slab [J]. Materials Review, 2010, 24(3): 89-95.
[17] Zhao Su, Zhang Li. Research progress of transverse corner cracks in continuous cast slabs of microalloyed steel[J]. World Steels, 2011, (1): 42-49.
[18] B.Mintz, R.Abushosha. Effectiveness of hot tensile test in simulating straightening in continuous casting[J]. Materials Science and Technology, 1992, 8(2): 171-178.
[19] Yang Liu, Li Yang, Xue Zhengliang et al. Corner transverse crack sensitivity of Ti-Nb micro-alloyed steel slabs during continuous casting[J]. Journal of Iron and Steel Research, 2010. 30(10): 807-815.
[20] B.H.Chen, H.Yu. Hot ductility of VN and V-Nb microalloyed steels[J]. International Journal of Minerals, Metallurgy, and Materials, 2012.19(6):525-529.
[21] Zhu Guosen, Zhu Zhiyuan, Wang Yanfeng, et al. Control of Transverse Corner Cracking of Nb Steel Slab [J]. Iron & Steel, 2006, 41(12): 30-32.
[22] SUN Yan-hui, ZHAO Chang-liang, CAI Kai-ke, et al. Research of Causes of Surface Network Cracks on Continuous Casting Slabs [J]. China Metallurgy. 2008. 18(4).
[23] Dong-feng HE, Sheng CHANG1, Hong-bing WANG. Controlling Transverse Cracks of Slab Based on Edge Control Technology [J]. Journal of Iron and Steel Research, International, 2015, 22 (Supplement 1): 42-47.
[24] Wang Xinhua, Liu Xinyu, Lv Wenjing. Carbide and Nitride Precipitation and Hot Ductility of Continuous Cast Steel Slabs Containing Nb, V, Ti[J]. Journal of Iron and Steel Research. 1998. 10(6).
[25] R.Hammer, G.Bochher, A.Diener. Technological Measure in Slab Casting to Improve Surface Condition[J]. Stahl u.Eisen, 1989, 109: 277-284.
[26] B Patrick,V.Ludlow.Development of Casting Practices to Minimize Transverse Craking in Microalloyed steels[J].La Revue de Metallurgie-CIT,1994.91:1081-1089.
[27] Sugitani Y. Studies of surface cracking formation by spray cooling in continuous casting [C]. Concast Seminar. Zurich:1973:168.
[28] Diener A. Continuous casting of microalloyed steels at Hoesch AG [C]. Concast Seminar. NewYork:1980:57
[29] LIU Yang, WANG Wen-jun, ZHU Zhi-yuan, et al. Removal of Transverse Corner Crack of Continuous Casting Slab by Chamfering Mold [J]. Iron & Steel, 2012, 47(4): 47-50.
[30] Wang Wen-jun, Li Ben-hai, Zhu Zhi-yuan, et al. Development and Application of Mould With Chamfered Corners for Continuous Casting of Slabs [J]. Journal of Iron and Steel Research, 2012, 24(9).
[31] Yang Chunzheng. Formation Mechanism of Slab Transversecorner Crack and Technological Development of Thechamfered Mold in a Slab Continuous Caster for Typical Micro-alloyed Steels[D]. Shanghai: Shanghai University, 2011.
[32] T.Kato, Y.Ito, M.Kawamoto, A.Yamanaka, et al. Prevention of slab surface transverse cracking by microstructure control [J]. ISIJ International, 2003, 43(11): 1742- 1750.
[33] Lee U H, Park T E.Assessment of hot ductility with various thermal histories as an alternative method of in situ solidification[J].ISIJ International,2010,20(4):540.
[34] DU Chen-wei, WANG Chun-lei, LI Bo et al. Effect of temperature field and cooling rate along casting direction on surface transverse cracks of microalloyed stee [J]. Journal of Iron and Steel Research, 2018,30(7): 523-528.
[35] Li Yunfeng. Study on Refinement of Prior Austenite Grain of Slab Subsurface for Microalloyed Steel Casting [D]. Chongqing: Chongqing University, 2014.

IOP Conf. Series: Materials Science and Engineering **538** (2019) 012003 doi:10.1088/1757-899X/538/1/012003

Design and Finite Element Analysis of Ultrasonic Composite Abrasive Vibration Polishing Equipment

Hongqiang Wan, Peiying Han, Quanli Han, Shuai Ge, Fancong Li

School of Mechanical Engineering, Xi'an Technology University, Xi'an, China

413162931@qq.com; 1003575444 @qq.com; hanqljx@163.com; 260686289@qq.com; sufeiling@qq.com

Abstract. In order to develop a workpiece surface polishing equipment with high performance and high precision, the overall structural design and parameter selection of ultrasonic composite abrasive vibration polishing equipment were mainly completed. Secondly, static analysis of the crankshaft connecting rod piston mechanism is performed by ANSYS Workbench software. We can know that the crankshaft connecting rod piston mechanism has the most concentrated stress at the contact between the small end of the connecting rod and the connecting rod shaft. And the stress value is 121.04MPa but less than the allowable stress value of the connecting rod. Therefore, the strength of the crankshaft connecting rod piston mechanism meets the design requirements, and the modal analysis of the crankshaft is carried out, and the crankshaft right crank displacement deformation amount is the most. It provides a theoretical basis for the optimization design of the crankshaft later and provides a reference for the research of ultrasonic composite abrasive vibration polishing equipment.

1. Introduction

With the development of modern science and technology, the demands on surface quality of the workpiece is much higher in areas such as aerospace, medical, and optical. In order to improve the quality and efficiency of machined surface polishing, it is necessary to design a workpiece surface polishing equipment.

According to experimental study, the workpiece surface quality after compound ultrasonic vibration polishing is better than the workpiece surface after a single compound ultrasonic vibration polishing. Therefore, scholars at home and abroad is dedicated to developing ultrasonic composite vibration polishing equipment[1-4]. For example, Suzuki H, and othershave[5] come up with a second - order longitudinal vibration and four - order bending vibration of composite ultrasonic polishing method allows the polished surface roughness of workpiece reaches 8nm. Han Lei[6] proposed ultrasonic vibration assisted polishing equipment, which can improve the workpiece surface while polishing workpiece strength.

This article proposes a ultrasonic composite abrasive vibration polishing equipment, which is a processing technology of compound surface which combines three kinds of methods together: ultrasonic vibration cutting, machining of mechanical vibration and free abrasive machining. For Static and modal analysis simulation of mechanical vibration unit is carried out by ANSYS Workbench to verify the security and stability of mechanical vibration unit when work.

2. Overall Structure and Main Design Parameters of Equipment

Ultrasonic composite abrasive vibration polishing equipment is developed combing ultrasonic vibration cutting, machining of mechanical vibration and free abrasive machining. Its main mode of operation is the High speed vertical reciprocating motion of fixture and For machining workpieces drove by longitudinal vibration Unit. At the same time, stirring shaft in polishing slot allows abrasive suspensions rotating to achieve workpiece processing. In the bottom of the polishing slot is equipped with ultrasonic generator to make polishing slurry portrait vibrate with High frequency, which realize the polishing on the surface of the workpiece combined with mechanical vibration and abrasive rotation.

Soildworks software is a subsidiary of Dassault Systemes S. A. We can use it to Create part models more quickly and assemble virtually[7-8]. As shown in figure 1 is the three-dimensional models modle of the Overall structure of ultrasonic composite abrasive vibration polishing equipment through Soildworks software.

Ultrasonic composite abrasive vibration polishing equipment is mainly constituded by mechanical vibration unit, Z-axis motion unit and abrasive box unit. The three - dimensional models are respectively shown in figure 2, figure 3 and figure 4.

(1) The operating principle of Z-axis motion unit is that the screw rotate through belt drive driven by motors to lead screw nut and screw nut along the screw pump.

(2) The operating principle of abrasive box unit is that wear particle mixing device rotates driven by motors to realize the polishing effect of free abrasive on workpiece surface and which can prevent free abrasive precipitating, ultrasonic oscillator posted on the department of abrasive bottom polish the workpiece surface via ultrasonic vibration.

(3) The operating principle of mechanical vibration unit is that the crankshaft rotate driven by motors and turn the turning of the crankshaft into the vertical movement of a fixture through the piston mechanism to achieve the vertical movement of the workpiece as shown in figure 4.

The main advantages of the device:

(1) It improves the quality and efficiency of machined surface polishing combing ultrasonic vibration cutting, machining of mechanical vibration and free abrasive machining

(2) It can realize the polishing effect of free abrasive on workpiece surface and it can prevent free abrasive precipitating in the meanwhile under the effect of abrasive mixing structure.

3. Finite-element Analysis of Mechanical Vibration Unit

3.1. Static Analysis of Mechanical Vibration Unit

ANSYS Workbench is of a wide used application of finite element analysis software, which can carry on static analysis, dynamic analysis and collision Analysis, and so on[9-10]. So it is very convenient to carry out static analysis on the crankshaft connecting rod piston mechanism in order to verify whether the strength of crankshaft connecting rod piston mechanism in ultrasonic composite abrasive vibration polishing equipment meets the design requirements. The following is the implementation process.

(1) Import model. Save the crankshaft connecting rod piston mechanism model drawn by Soildworks in the Parasoild format and import it into ANSYS Workbench for static analysis.

(2) Set the parameters of crankshaft. Select 45 steel as the material of crankshaft and its elastic modulus is 206GPa, its poisson ratio is 0.3, and its density is 7.85g/cm^3. Select straight carbon steel as the material of bolt and piston pin and its elastic modulus is 206GPa, its poisson ratio is 0.24, and its density is 7.3g/cm^3. Select cast iron as the material of piston and its elastic modulus is 126GPa, its poisson ratio is 0.24, and its density is 7.0g/cm^3.

(3) Meshing. Partite in the method of using hexahedral mesh. Set unit mesh size of the crankshaft, connecting rod and piston is 4mm, set unit mesh size of the plain bearings, bolts and piston pins is 4mm. The number of nodes after meshing is 182403, the number of units is 61643. Mesh diagram as shown in figure 5.

IOP Conf. Series: Materials Science and Engineering **538** (2019) 012003 doi:10.1088/1757-899X/538/1/012003

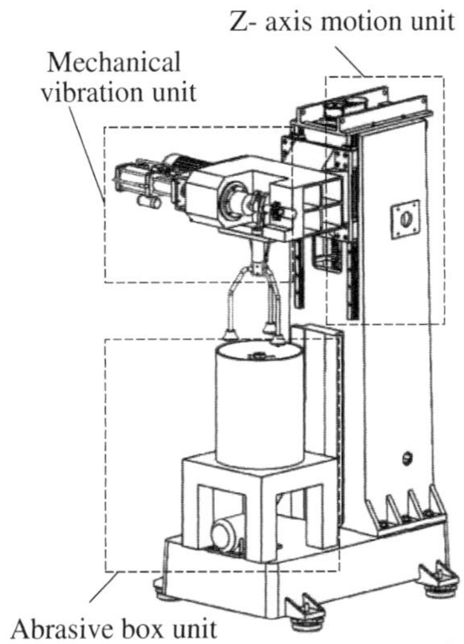

Figure 1. Overall structure of ultrasonic composite abrasive vibration polishing equipment.

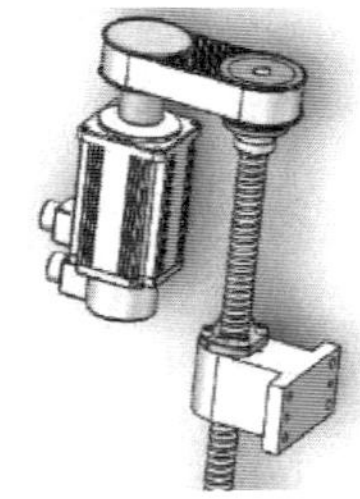

Figure 2. Abrasive box unit

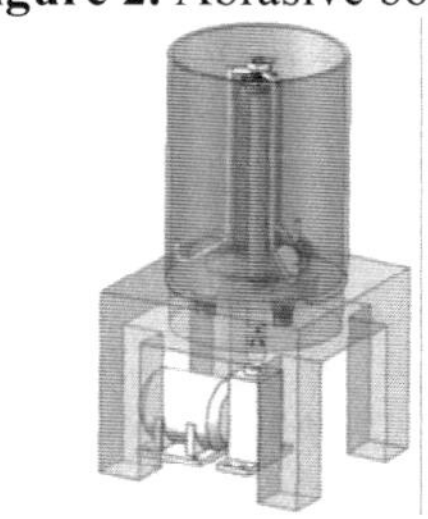

Figure 3. Z - axis motion unit.

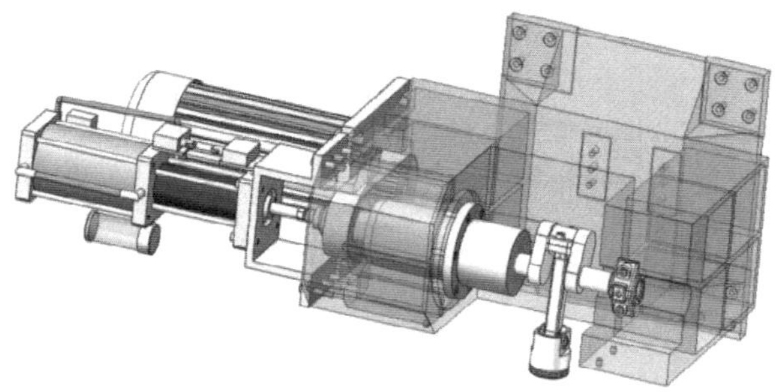

Figure 4. Mechanical vibration unit

Figure 5. Crankshaft connecting rod piston mechanism mesh map.

(4) Imposed loads and constraints. Set the pistons and bearing for all constraints and impose 15529N·mm by the right side of the crankshaft.

(5) View solution. Figure 6 is overall equivalent stress imagery of the crankshaft connecting rod piston mechanism. Figure 7 is the overall displacement image of crankshaft connecting rod. Figure 8 and figure 9 are respectively equivalent stress imagery of the crankshaft and connecting rod.

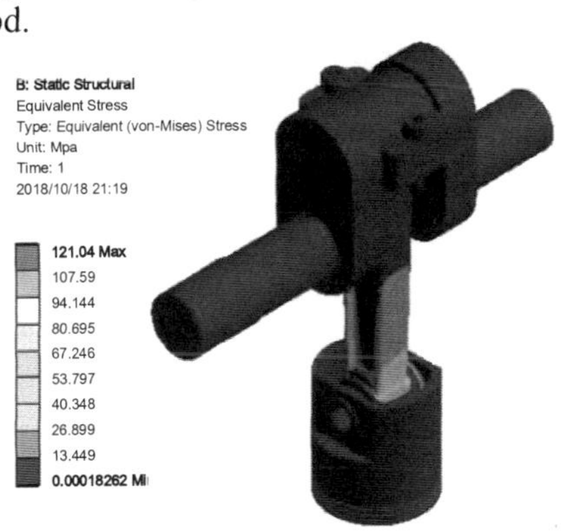

Figure 6. Overall equivalent stress imagery.

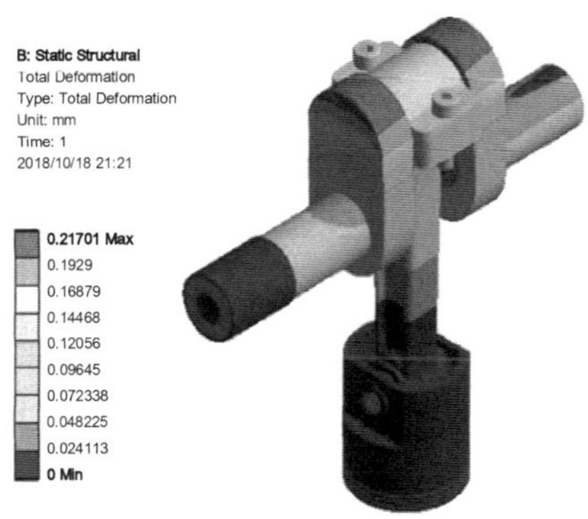

Figure 7. Overall displacement image.

Figure 8. Equivalent stress imagery of the crankshaft.

Figure 9. Equivalent stress imagery of the connecting rod.

According to figure 3, the equivalent stress variation range in the crankshaft connecting rod piston mechanism is the most obvious, and according to figure 4, overall displacement is not obvious. So we study the overall equivalent stress conditions mainly. To clearly see the stress concentration in which part, we will separate analysis the stress of the crankshaft and connecting rods. According to figure 5, the stress on crank shaft is the most concentrated, and the largest stress is 15.631MPa which is less than the maximum allowable stress (120MPa). So the strength of crankshaft meet design requirements. According to figure 6, the stress on the contact between connecting rod and its small end is the most concentrated, and the largest stress is 121.04MPa which is equal to the maximum equivalent stress values of connecting rod and crankshaft of a piston mechanism. So the maximum equivalent stress of the crankshaft connecting rod piston mechanism is on its crankshaft. Because the maximum equivalent stress of the connecting rods is less than the connecting rod (585.92MPa), the strength of connecting rod meets design requirements, and the entire design of mechanical vibration unit is rational.

3.2. Modal Analysis of the Crankshaft

Modal analysis of crankshaft aims to study the characteristics of crankshaft vibration to provide a reference for optimal design of crankshaft in the future. Import the crankshaft in ANSYS Workbench, set the crankshaft material properties, divide the grid and imposing constraints and loads. According to the actual working situation of the crankshaft, left end of the crankshaft is fixed, and runout of the right end of the crankshaft is restricted by sliding bearing. Because the crankshaft vibration analysis and lower order modes of the crankshaft are closely connected, this article solves the top 6 order mode of the crankshaft through ANSYS Workbench. The natural frequency of the order for the modal as shown in table 1.

Table 1. Natural frequency of the first 6 order parameters

Mode	Frequency[Hz]	Mode	Frequency[Hz]	Mode	Frequency[Hz]
1	0.	**3**	2794.7	**5**	3711.2
2	1485.8	**4**	3121.4	**6**	5792.1

According to table 1, we can get the crankshaft vibration frequency, and vibration modal analysis of the plansfrom of each order modes figure 8. Natural frequency of the first order is 0, at which point the crankshaft rigid body vibrate, as shown in figure 8. The second order natural frequency is 1485.8Hz, and the crankshaft overall bending vibrate. The third and the fifth order is torsional vibration of crankshaft. The 4th order is longitudinal vibration of crankshaft. The 6th order is the transverse vibration of crankshaft. According to figure 10, the vibration and deformation of the third order mode is the largest, and it is at the bottom of the crank handle. According to modal vibration graph of 1-6 order modes, the vibration and deformation of crankshaft is primary at the right end of the crank.

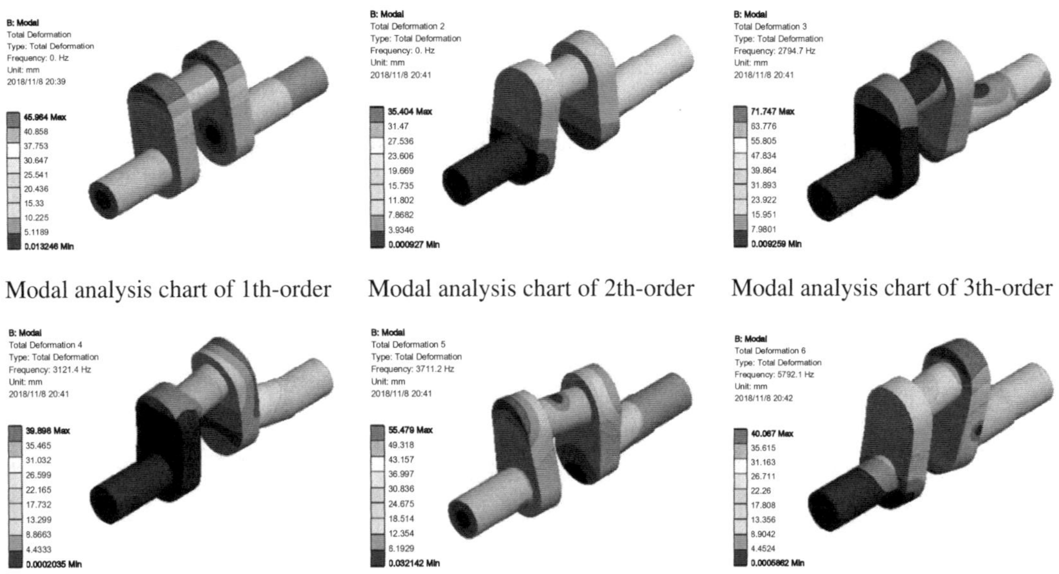

Modal analysis chart of 1th-order Modal analysis chart of 2th-order Modal analysis chart of 3th-order

Modal analysis chart of 4th-order Modal analysis chart of 5th-order Modal analysis chart of 6th-order

Figure 10. Natural frequency of the first 6 order parameters.

4. Conclusion

(1) Through reasonable parameter selection and structure design, the development of ultrasonic composite abrasive vibration polishing equipment is completed, the parameters of the device is got, and the three - dimensional model of ultrasonic composite abrasive equipment is drawn using Soildworks. In order to study whether the strength of the mechanical vibration unit meets the design requirements

(2) Use ANSYS Workbench software to static analysis the crankshaft connecting rod piston mechanism, and the maximum equivalent stress of the crankshaft and connecting rods are less than the allowable stress the value, so the strength of crankshaft connecting rod piston mechanism meets the design requirements.

(3) Use ANSYS Workbench to modal analysis the crankshaft and we find that the largest vibration and deformation of crankshaft occurred in the 3rd - order, and the deformation of right crank of crankshaft is the largest. In order to reduce the amount of deformation, we can shorten the out of the right side of the crankshaft axis length and increase the constraint on the right side of the crankshaft runout, or we can thicken the crank.

This study provide a new design concept and theoretical basis for similar ultrasonic composite vibration polishing equipment. The next research work is to explore the principle of material removal on workpiece surface and deduce the expression formula of displacement and stress on workpiece surface under the action of concentrated force and average distributed force perpendicular to workpiece surface.

Acknowledgement

This study is funded by Scientific research project of Key Laboratory of Shaanxi Provincial Department of Education (16JS047). And by Shaanxi key laboratory research project (17JS060).

References

[1] WAN Hongqiang, HAN Peiying, GE Shuai, LI Fancong. Development of Ultrasonic Vibration Polishing Method on Workpiece Surface. Diamond & Abrasives Engineering, 2018, 38(02): 94-100.

IOP Conf. Series: Materials Science and Engineering **538** (2019) 012003 doi:10.1088/1757-899X/538/1/012003

[2] Yang Xiaopeng. Design and Experimental Research on Spindle of Ultrasonic-EDM Combined Machining for Metal Matrix Ceramic Coating Materials. Shanxi: Taiyuan University of Technology, 2018.
[3] Zhang Yu. Research and Application of Multi-energy Mechanism and Optimal Control in Ultrasonic Composite Electrical Processing. Jiangsu: Yangzhou University, 2018.
[4] Sun Jixin. Design and Application of Precision Ultrasound and Compound Electrolytic Machining Technology for Hard-to-Machine Parts. Jiangsu: Yangzhou University, 2018.
[5] SUZUKI H, HAMADA S, OKINO T, et al. Ultraprecision Finishing of Micro-aspheric Surface by Ultrasonic Two-axis Vibration Assisted Polishing. CIRP Annals-Manufacturing Technology, 2010, 59(1): 347-350.
[6] Han Lei. FEM Analysis and Experimental Research on the Technique of Ultrasonic Flexural Vibration Assisted Lapping/Polishing of Hard and Brittle Materials. Changchun: Jilin University, 2015.
[7] Gao Jianqiang, Yu Ping, Gao Shilong, Zhang Yinpigeon. The Design and Simulation of Centrifugal Glass-Wool Board Drilling Mechanism. Mechanical Design & Manufacture, 2014 (01): 7-10.
[8] Chen Long, Fu Yanzeng, Yang Shuo, Jiang Xue. Research on DF Deep Hole Machining System Based on the Soildworks Animation Simulation. Journal of Hebei Institute of Architectural and Civil Engineering, 2017, 35 (03): 119-122.
[9] Gao Yaodong, Yan Penghe. Motion Simulation and Design Optimization of Hydraulic Support by ANSYS. Mechanical Design & Manufacture, 2018 (08): 202-205.
[10] Wu Min, Xie Longhan. Ansys Workbench Finite Element Analysis and Simulation . Beijing: Electronic Industry Press, 2014.

Effect of Crystal Orientation on Cyclic Plastic Behaviors of Partially Recrystallized Single Crystal Nickel-based Superalloy

D H Wei[1], W J Zhang[1], H Y Huang[1], L P Zhang[1], R Y Chen[1], Y W Wu[2], H J Shi[3], X F Ma[1,*]

[1]Sino-French Institute of Nuclear Engineering and Technology, Sun Yat-sen University, Zhuhai 519082, China
[2]Huadian electric power research institute co., Ltd, Hangzhou 310030, China
[3]AML, School of Aerospace, Tsinghua University, Beijing 100084, China

[*]Corresponding author: maxf6@mail.sysu.edu.cn

Abstract. Surface recrystallization (RX) is one of the main causes of fatigue failure of heavy gas turbine single crystal (SX) blade and thus a threat to safe long-time operation. Using crystal plasticity model, the effects of primary and secondary crystal orientation on the cyclic plastic deformation behaviors of recrystallized SX nickel-based superalloy was investigated via crystal plasticity finite element simulation. The results show that primary orientation has significant effect on the accumulated plastic shear stain γ of the recrystallized SX superalloy: the maximum accumulated plastic shear strain γ_{max} occurs in SX matrix near RX transverse grain boundary; the maximum accumulated plastic shear strain γ_{max} of RX-SX system increases with increasing primary orientation(β); the maximum accumulated plastic shear strain γ_{max} at β=15° is 31.6% larger than that at β=0°; the maximum accumulated plastic shear strain γ_{max} decreases first and then increases with the increasing of the secondary orientation(θ) from 0° to 90°, with the lowest value of γ_{max} at θ=43°, which is 1.4% smaller than that at θ=0° and θ=90°. The study shows that the crystal orientation can be optimized to reduce the effect of recrystallized grain on the fatigue property of SX nickel-based superalloy.

1. Introduction

Single crystal (SX) nickel-based superalloys have been widely used in aircraft turbines due to excellent high temperature fatigue and creep properties [1]. However, the absence of grain boundaries leads to anisotropic and crystallographic orientation dependent mechanical properties [2,3]. In most case, turbine blades are produced with the primary crystallographic orientation along airfoil stacking line. While the other two orientations are not controlled, named randomness orientation, and considered random variables. Although the primary crystallographic orientation is controlled in the manufacturing process of turbine blades, it is difficult to guarantee the airfoil stacking line align with the primary crystallographic orientation. Arakere et al. [3] studied the influence of variation in crystal orientation on blade stress and fatigue life. The result demonstrates that the control of primary crystallographic orientation has significant effect on a component's resistance to fatigue crack growth. Wen et al. [4] studied the influence of loading direction on the three points bending creep and uni-axial tensile creep via three-dimensional finite element analysis. The results show that both of the bending depth and deformation velocity associated with crystallographic orientation [001] are the largest among the three different orientations. Early study [5] assumes that secondary crystallographic orientation has no significant effect on the mechanical properties of SX nickel-based superalloys. However, subsequent studies show that secondary crystallographic orientation has significant effect on the fatigue life of SX nickel-based superalloys [6,7]. Kiyak et al. [6] reported that fatigue cracks grow about two times faster in the [100] direction than in the [110] direction in the Paris regime. He et al. [7] investigated the effects of secondary orientation on plastic deformation and fatigue crack initiation in a SX nickel-based superalloy. The result shows that specimens with different secondary orientation exhibit different crack initiation modes.

During manufacturing process or high temperature application, recrystallization (RX) layers may be formed on the surfaces of turbine components, e.g. leading and trailing edges due to susceptibility to environmental influence. Because the orientations of RX grains are inconsistent with that of single crystal matrix, the RX region is prone to be the initiation source of fatigue crack. The influences of RX grains on the properties of SX superalloys have been widely studied [8-9]. Ma et al. [10] investigated the influence of surface recrystallization on the low cycle fatigue behaviour of a SX superalloy, showing that fatigue life is significantly reduced by surface recrystallization. Zhang et al.[11] studied the fractured turbine blades and indicated that surface recrystallization can lead to the failure of superalloy blades. Therefore, recrystallization should be taken into consider in order to quantify the effect of crystallographic orientation on the strength and fatigue properties of SX superalloy. Up to now, the research about effect of primary orientation and secondary orientation on the cyclic plastic behavior and fatigue properties of recrystallized SX superalloy is rarely reported.

It is reported that fatigue crack is prone to initiate on the slip planes with maximum shear strain via slip band mechanism, or defects near surface or from grain boundaries due to local stress/strain mismatch [12-14]. In this paper, the maximum accumulated plastic shear strain γ_{max} is used to quantify fatigue crack incubation under cyclic loading condition. The influence of primary orientation and secondary orientation on the maximum accumulated plastic shear strain γ_{max} of recrystallized SX superalloy is investigated based on finite element analysis.

2. Crystal Plasticity Theory

The kinematics of crystal plasticity theory presented here is developed by Hill [15], Asaro and Rice [16,17]. The deformation of a crystalline material under external load consists of elastic deformation and inelastic deformation. The total deformation gradient F can be separated into elastic and plastic part and given by

$$F = F^* \cdot F^P \tag{1}$$

where F^* denotes the part of elastic deformation, F^P denotes the part of plastic deformation. The rate of change of F^P is related to shearing rate $\dot{\gamma}^{(\alpha)}$ of α slip system as follow:

$$\dot{F}^P \cdot F^{P-1} = \sum_{\alpha} \dot{\gamma}^{(\alpha)} s^{(\alpha)} m^{(\alpha)} \tag{2}$$

The shearing rate $\dot{\gamma}^{(\alpha)}$ of α slip system can be expressed by the resolved shear stress $\tau^{(\alpha)}$, in a rate-dependent way as

$$\dot{\gamma}^{(\alpha)} = \dot{a}^{(\alpha)} f^{(\alpha)}(\tau^{(\alpha)} / g^{(\alpha)}) \tag{3}$$

$$f^{(\alpha)}(x) = x|x|^{n-1} \tag{4}$$

where $\dot{a}^{(\alpha)}$ is the reference shearing rate on α slip system, $g^{(\alpha)}$ is the current strength of α slip system, and n is the rate sensitivity exponent. $g^{(\alpha)}$ can be given by

$$\dot{g}^{(\alpha)} = h_{\alpha\beta}\dot{\gamma}^{(\beta)} \tag{5}$$

where

$$h_{\alpha\alpha} = h(\gamma) = h_0 \operatorname{sech}^2 \left| \frac{h_0 \gamma}{\tau_s - \tau_0} \right| \tag{6}$$

$$\gamma = \sum_{\alpha} \int_0^t \left| \dot{\gamma}^{(\alpha)} \right| dt \tag{7}$$

$$h_{\alpha\beta} = q h(\gamma) \quad (\alpha \neq \beta) \tag{8}$$

where $h_{\alpha\beta}$ is latent hardening moduli, q is latent hardening constant, For detailed in-depth review of the crystal plasticity theory, see the work of Huang [18].

3. Finite Element Simulation

3.1. Finite Element Model

The finite element geometry model of SX superalloy was constructed with carbide and two RX grains using the finite element software ABAQUS, as shown in Fig.1. A ABAQUS user-defined crystal plasticity material subroutine(UMAT) was used. The geometry of single crystal matrix is 400 μm high, 300 μm wide and 5 μm thick. The length (L) and width of two RX grains are both 100 μm. The carbide is ellipsoidal and the lengths of ellipsoidal semi-major and semi-minor axis are 30μm and 12μm, respectively. The typical finite element meshes are shown in Fig.2. 3D hexagonal linear incompatible element (C3D8I in ABAQUS) was used. The boundary conditions applied to the mesh on the top surface nodes of the mesh include a cyclically varying uniform displacement with a minimum value(-1.0μm) and a maximum value (1.0μm). load ratio is R = -1. In order to prevent rigid body movement, all nodes on the bottom surface of the mesh were fixed in the vertical (Y-axis) direction. The nodes at the bottom-left corner were fixed in the horizontal direction (X-axis).

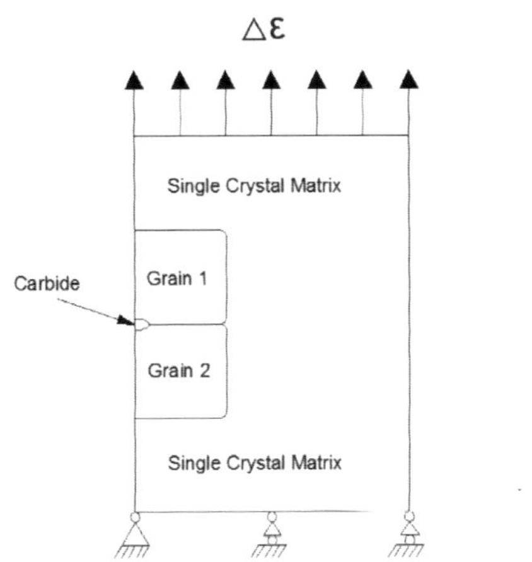

Figure 1. Finite element geometry model of single crystal superalloy with recrystallized grains

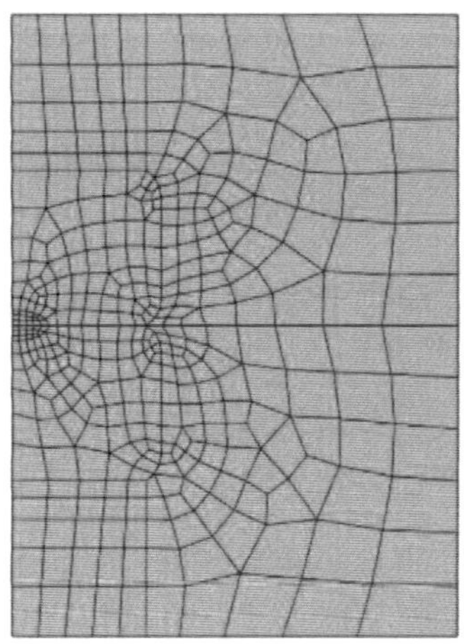

Figure 2. Finite element mesh of the single crystal superalloy with recrystallized grains

3.2. *Material Constants*

The material parameters used in this model are derived from SX nickel-based superalloys PWA1480 and SRR99 [19], as shown in Table 1. C_{11}, C_{12} and C_{44} represent the elastic constants of the material in three different directions, respectively; $\dot{a}$ is the reference shear strain rate in the slip system flow rule; n is the rate sensitivity exponent; h_0, τ_0 and τ_s are initial hardening modulus, initial yield stress and stage I stress, respectively. Details regarding these constants have been documented by Huang[18] and Feng et al. [19].

Table 1. Crystal plasticity model parameters

C_{11}/MPa	C_{12}/MPa	C_{44}/MPa	$\dot{a}$	n	h_0/MPa	τ_s/MPa	τ_0/MPa	q	c/MPa	d
169536.0	107540.0	88105.7	0.001	3.6	548	100.64	60.64	1.0	413300	1960

4. Definition of Primary Orientation and Secondary Orientation

In order to simplify the following discussion, the primary orientation is defined as β. The secondary orientation is defined as θ, as shown in Fig.3. The orientations of the single crystal matrix and two RX grains are different in the local coordinate system. The orientations of two RX grains are constant. The orientations of the single crystal matrix and two RX grains in the local coordinate system at β=0° and θ=0° are shown in Table 2.

Table 2. Orientations of single crystal matrix material and RX grains

Local coordinate	Single crystal matrix	Grain1	Grain2
X' axis	[100]	[100]	[01-1]
Y' axis	[010]	[011]	[111]

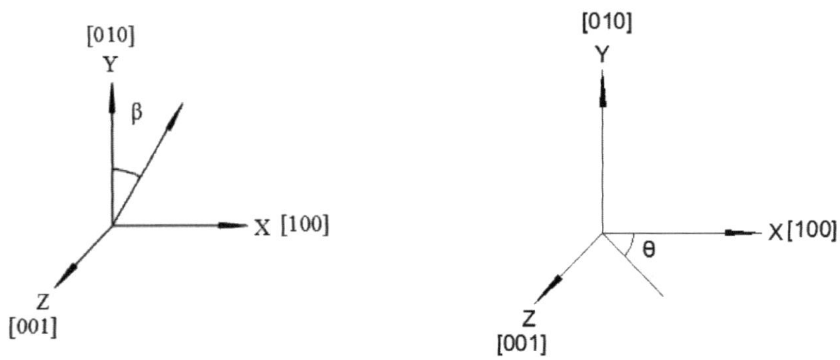

Figure 3. Definition of (a) primary orientation β (b) secondary orientation θ

5. Results and Discussion

Figure 4 (a) and (b) show the distribution of Von Mises stress and accumulated plastic shear strain after five cycles at β=0° and θ=0°. It can be seen from Fig. 4(a) that the maximum Mises stress occurs in the RX grain and near the transverse boundary of RX grain. The reason is that dislocation slip is hindered and dislocation piles up at the transverse grain boundary of the RX grain, resulting in stress concentration. It can be seen from 4(b) that the maximum accumulated plastic shear strain γ_{max} is in the matrix. The reason may be that the region adjacent to the RX grains has larger hydrostatic pressure and plastic deformation is therefore reduced [20].

Fig.5 shows the dependence of the maximum accumulated plastic shear strain γ_{max} in recrystallization-single crystal system on primary orientation (β). The maximum accumulated plastic shear strain γ_{max} increases with increasing of primary orientation (β). The influence of β is remarkable and the maximum accumulated plastic shear strain γ_{max} at $\beta=15^o$ is 31.6% larger than that at $\beta=0^o$. In theory, this finite element model has better fatigue performance at $\beta=0^o$ than that at $\beta=15^o$. Hou et al. [21] studied the dependence of Mises stress, the maximum resolved shear stress and fatigue life of cooled blade on primary orientation based on finite element analysis method. The results show that the Mises stress and the maximum resolved shear stress increase with increasing of primary orientation(β), similar to our result. However, the fatigue life decreases with increasing primary orientation(β).

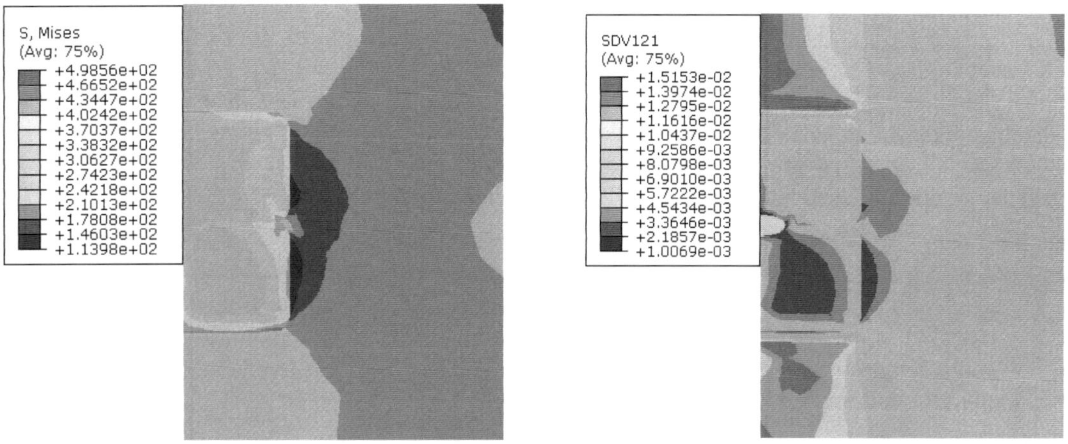

(a) (b)

Figure 4. Distribution of the Mises stress and accumulated plastic shear strain after 5 cycles (a) Mises stress (b) accumulated plastic shear strain

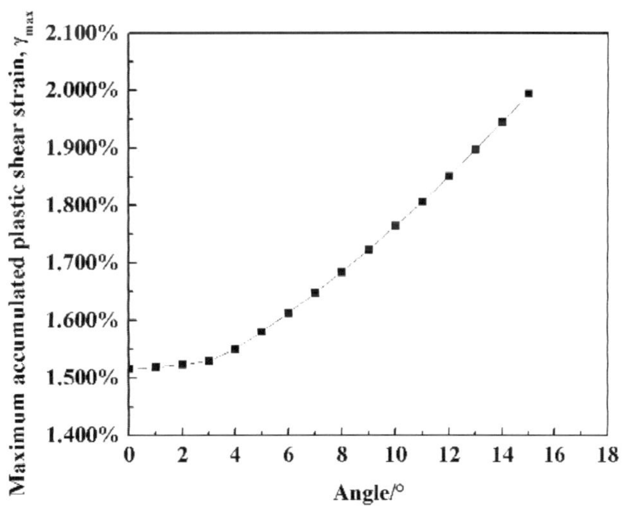

Figure 5. The dependence of the maximum accumulated plastic shear strain in recrystallization-single crystal system on primary orientation(β)

The dependence of the maximum accumulated plastic shear strain γ_{max} in recrystallization-single crystal system on secondary orientation (θ) is shown in Fig. 6. The value at $\theta=0^o$ is the same as that of $\theta=90^o$ due to symmetry. The maximum accumulated plastic strain γ_{max} at 43° is the smallest, and it is 1.4% smaller than that at 0° and 90°, even if the influence of secondary orientation (θ) is not remarkable. It implies that the RX-SX system has better fatigue performance with $\theta=43^o$ than that at 0°

and 90°. Hou et al. [21] also reported the influence of secondary orientation on Mises stress and the maximum resolved shear stress τ_{s_max} in single crystal blade, with the relative variation shown in Fig.7. They noted that the influence of secondary orientation on fatigue life is much more obvious. The relative variations of creep displacement of blades studied by Yue Z F [22] are also shown in Fig. 7 for the sake of comparison. It can be seen that the mechanical response/deformation of single crystal reach an extreme value with secondary orientation of 40°-45°. Arakere et al. [3] analyzed the SX nickel base turbine blade models with different primary and secondary orientations by finite element method. They found that for θ=45±15 degree cracks were arrested after some growth or did not initiate at all, indicating that these are preferential secondary orientations for optimal fatigue performance. It is consistent with the result we obtained here. Zhang Y. et al. [23] investigated the effect of secondary orientations on fatigue crack growth (FCG) behaviors of a SX nickel-based superalloy and proposed an energy release rate (G) coefficient model. The results show that secondary orientation of ~45° benefits the crack resistance and the proposed G coefficient model can effectively describe the difference of crack growth rates caused by different secondary orientations.

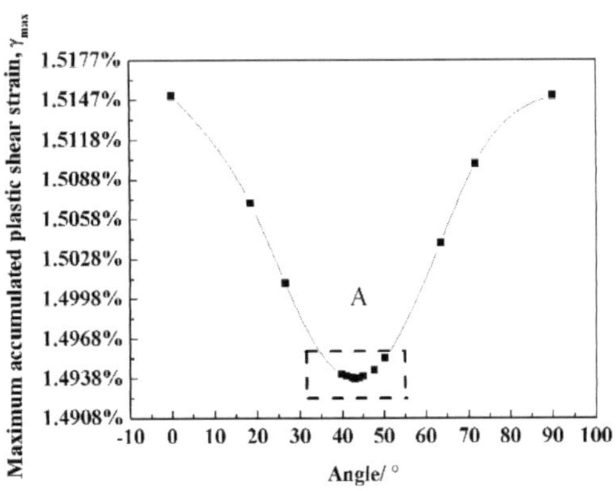

(a)

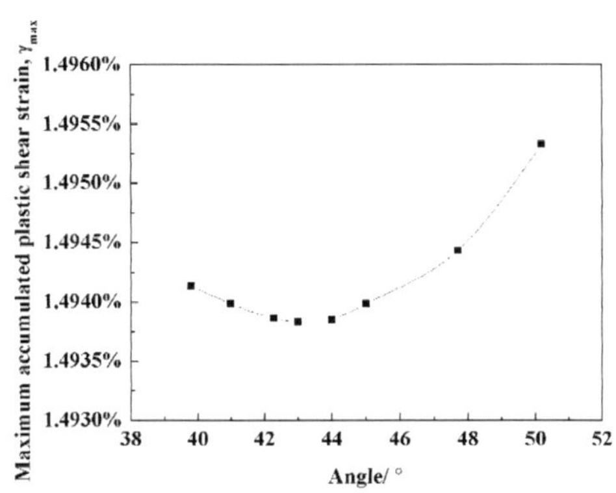

(b)

Figure 6. The dependence of the maximum accumulated plastic shear strain in recrystallization-single crystal system on the secondary orientation (θ) (a) range 0°-90° (b) range A

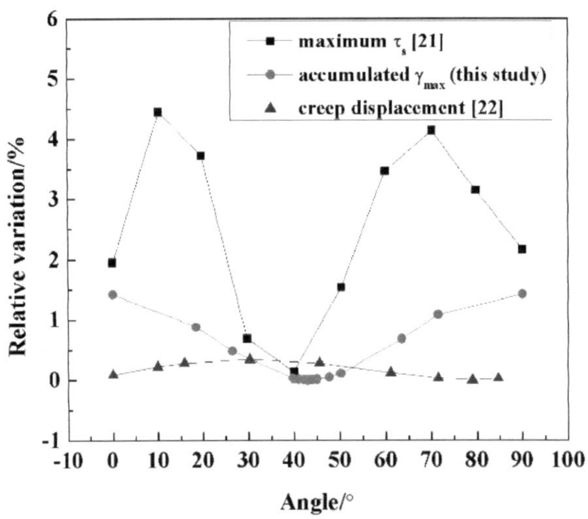

Figure 7. Relative variation of maximum resolved shear stress τ_s [21], creep displacement [22], and maximum accumulated plastic shear strain γ_{max} at different secondary orientations

6. Conclusions

By crystal plasticity finite element method, cyclic plastic deformation behaviors of recrystallized single crystal nickel-based superalloy were studied. The influences of primary orientation and secondary orientation on cyclic plasticity were examined. The following conclusions have been drawn:

1) The maximum accumulated plastic shear strain occurs in the matrix near the RX transverse grain boundary in recrystallization-single crystal system.
2) The influence of primary orientation (β) on the maximum accumulated plastic shear strain in the recrystallized single crystal superalloy is remarkable. The maximum accumulated plastic shear strain increases with increase of primary orientation (β).The maximum accumulated plastic shear strain γ_{max} at $\beta=15^o$ is 31.6% larger than that at $\beta=0^o$.
3) The maximum accumulated plastic shear strain decreases first and then increases with the increase of secondary orientation from 0° to 90° in recrystallized single crystal superalloy. The maximum accumulated plastic strain γ_{max} at secondary orientation $\theta=43^o$ is 1.4% smaller than that at 0° and 90°.
4) The maximum accumulated plastic shear strain γ_{max} reaches minimum value at $\beta=0^o$ and $\theta=43^o$. In other words, the corresponding primary orientation is [010], and the secondary orientation is [1.05 01], which corresponds to optimal fatigue performance due to low γ_{max}.

References

[1] Suzuki S, Sakaguchi M , Inoue H. Temperature dependent fatigue crack propagation in a single crystal Ni-base superalloy affected by primary and secondary orientations[J]. Materials Science & Engineering A, 2018, 724.

[2] ZHOU Z J, WANG L, WANG D, et al. Effect of secondary orientation on room temperature tensile behaviors of Ni-base single crystal superalloys [J]. Materials Science & Engineering A, 2016, 659: 130-142.

[3] ARAKERE N K, SWANSON G. Effect of Crystal Orientation on Fatigue Failure of Single Crystal Nickel Base Turbine Blade Superalloys [J]. 2002, (78576): V004T001A004.

[4] Wen S, Zeng X, And Z K, et al. Investigation the effect of crystal orientation of nickel-based single crystal superalloys on bending creep tests [J]. Material wissenschaft Und Werkstofftechnik, 2014, 45(3).

[5] GEL M, DUHL D N, GIAMEI A F. The development of single crystal superalloy turbine blades; proceedings of the Superalloys, F, 1980 [C].

[6] FEDELICH B, KIYAK Y, MAY T, et al. Simulation of Crack Growth Under Low Cycle Fatigue at High Temperature in a Single Crystal Superalloy [J]. Engineering Fracture Mechanics, 2008, 75(8): 2418-2443.

[7] He Z, Qiu W, Fan Y N, et al. Effects of secondary orientation on fatigue crack initiation in a single crystal superalloy[J]. Fatigue & Fracture of Engineering Materials & Structures, 2017.

[8] Ma X, Shi H J. In situ SEM studies of the low cycle fatigue behavior of DZ4 superalloy at elevated temperature: Effect of partial recrystallization [J]. International Journal of Fatigue, 2014, 61: 255-263.

[9] Bing Z, Delin L, Chunhu T, et al. Influence of Surface Recrystallization on Intermediate-Temperature Stress Rupture Property and Fracture Behavior of Single Crystal Superalloy SRR99[J]. Journal of Aeronautical Materials, 2012, 551(6): 149–153.

[10] MA X, SHI H J, GU J, et al. Influence of surface recrystallization on the low cycle fatigue behaviour of a single crystal superalloy [J]. Fatigue & Fracture of Engineering Materials & Structures, 2015, 38(3): 340-351.

[11] ZHANG W F, GAO W, ZHAO A G, et al. Recrystallization and Fatigue Failure of Blades Made of a Directionally Solidified Alloy[J]. Acta Aeronautica et Astronautica Sinica, 2003, 24(4): 377-381.

[12] ABIKCHI M, BILLOT T, CR PIN J, et al. Fatigue life and initiation mechanisms in wrought Inconel 718 DA for different microstructures; proceedings of the 13th international

conference on fracture, Beijing, China, F 2013-06-16, 2013 [C].
[13] ZHU L, WU Z R, HU X T, et al. Investigation of small fatigue crack initiation and growth behaviour of nickel base superalloy GH4169 [J]. Fatigue & Fracture of Engineering Materials & Structures, 2016, 39(9): 1150-1160.
[14] JANG Y, JIN S, JEONG Y, et al. Fatigue Crack Initiation Mechanism for Cast 319-T7 Aluminum Alloy [J]. Metallurgical & Materials Transactions A, 2009, 40(7): 1579-1587.
[15] HILL R. On Constitutive Macro-Variables for Heterogeneous Solids at Finite Strain [J]. Proceedings of the Royal Society of London, 1972, 326(1565): 131-147.
[16] ASARO R J. Micromechanics of Crystals and Polycrystals [J]. Advances in Appl Mech, 1983, 23(08): 1-115.
[17] ASARO R J, RICE J R. Strain localization in ductile single crystals ☆ [J]. Journal of the Mechanics & Physics of Solids, 1977, 25(5): 309-338.
[18] HUANG Y. A User-Material Subroutine Incorporating Single Crystal Plasticity in the ABAQUS Finite Element Program [J]. 1991:
[19] LU F, ZHANG G, ZHANG K S. Discussion of cyclic plasticity and viscoplasticity of single crystal nickel-based superalloy in large strain analysis: comparison of anisotropic macroscopic model and crystallographic model [J]. International Journal of Mechanical Sciences, 2004, 46(8): 1157-1171.
[20] WANG, DANIEWICZ, S. R, et al. Three-dimensional finite element analysis using crystal plasticity for a parameter study of fatigue crack incubation in a 7075 aluminum alloy [J]. International Journal of Fatigue, 2009, 31(4): 659-667.
[21] HOU N X, GOU W X, WEN Z X, et al. The influence of crystal orientations on fatigue life of single crystal cooled turbine blade [J]. Materials Science & Engineering A, 2008, 492(1–2): 413-418.
[22] YUE Z F, LV Z Z, YANG Z G. Influence of deviation and randomness of crystallographic orientations on the strength and life of nickel-based single crystal superalloy turbine baldes[J].Aerospace Power, 2003, 18(4): 477-480.
[23] ZHANG Y, QIU W, SHI H J, et al. Effects of secondary orientations on long fatigue crack growth in a single crystal superalloy [J]. Engineering Fracture Mechanics, 2015, 136: 172-184.

IOP Conf. Series: Materials Science and Engineering **538** (2019) 012005 doi:10.1088/1757-899X/538/1/012005

Preparation of RGO /Conductive Polymer Nanostructures and Application of Their Supercapacitors

Yu Gu, Qi Li

School of Materials, Northeastern University, Shenyang, Liaoning 110000, China

Corresponding author: 285668529@qq.com

Abstract. Reduction of the nanocomposite of the conductive polymer composite in the GO/conductive polymer and the RGO imparts excellent mechanical, thermal, electrical and electrochemical properties to the composite. Therefore, the RGO/conductive polymer has become the main research direction of new functional materials. The polypyrrole/GO composite was prepared by in-situ polymerization using graphene oxide as the matrix and polypyrrole as the guest of the difference layer. The additive type, the mass fraction of GO emulsion, the molar ratio of NaOH to pyrrole, the molar ratio of oxidant to $FeCl_3$ to pyrrole and the reaction temperature were investigated systematically for polypyrrole/GO. The influence of the structure and properties of GO composites. The supercapacitors properties of nanocomposites as electrodes were investigated.

1. Introduction

The conductive polymer/reduced graphene oxide (RGO) composite material refers to a composite system obtained by self-assembly of a polymer as an organic phase and an inorganic substance having a layered structure on a nanometer scale. The sheet space of the graphene oxide pair can be expanded to some extent after being properly treated, and the structure of the surface of the sheet is also greatly changed, so it can be called a natural micro reactor for preparing a polymer/graphene oxide (GO)[1]. Compared with traditional composite materials, the conductive polymer/graphene oxide (GO) nanocomposites have superior to the same components due to the nano-effects of the nanoparticles themselves and the strong interfacial interaction between the nanoparticles and the matrix. The mechanical, thermal, electrical and other properties of polymer/inorganic composites open up new avenues for high-performance, multi-functional composites[2]. The research of polymer/graphene oxide (GO) nanocomposites has become the frontier and hotspot of current material science research, with important scientific significance and broad application prospects.

A conductive polymer is a synthetic polymer that is electrically conductive between a semiconductor and a metal or that is electrically conductive and even comparable to a metal. In 1976, Japanese scientist Hideki Shirakawa and EJ.Louis and AG MacDiarmid cooperated for the first time to prepare polyacetylene with a conductivity of 10^3 S/cm, which led to great enthusiasm for competing research and development of conductive polymers[3]. In the past decade, scientists have synthesized hundreds of conductive polymers. Conductive polymers have become an indispensable new material with their unique electrical properties, and their research on the preparation, structure, properties and applications has also developed into an emerging discipline.

Depending on the conductivity mechanism, conductive polymers can be classified into three major categories: ion-conducting polymers, electron-conducting polymers, and redox-type conductive polymers[4]. The ionic conductive polymer is mainly a polymer electrolyte, and its carriers are positive and negative ions that can migrate between polymer molecules. Due to the interaction between the polymer chain and the ions, the ions are immobilized on the polymer chain (solvation), and the migration in the polymer is not as free as in the low molecular solvent, but can follow the chain of the polymer chain. The segment motion continuously changes the interacting object (segment) to form ion migration. Therefore, the segment movement of the polymer chain promotes the diffusion of ions in the polymer electrolyte, which is the essential cause of the high conductivity of the ion conductive polymer. Electron-conducting polymers, also known as conjugated conductive polymers, charge transfer is achieved by the migration of delocalized electrons on large conjugated chains in macromolecules. A conjugated two-track overlapping structure formed by alternating double bonds and single bonds in a polymer molecule constitutes a channel for delocalized electron migration. At present, many electron conductive polymers are mainly studied, such as polyacetylene, polyaromatic heterocyclic compounds (such as polyaniline, polypyrrole, etc.) and derivatives thereof, polyaryl rings and polyaryl fused ring compounds and derivatives thereof. A redox type conductive polymer such as polyethylene ferrocene or the like has an active center capable of undergoing a reversible redox reaction on the skeleton, and thus can transfer a charge by a redox reaction.

Polyaniline/graphene oxide (PANI/GO) intercalated nanocomposites prepared by in-situ polymerization methods have been used by researchers[5], and their XRD patterns have been shown to reflect the (001) series diffraction peak with layered ordered structure. However, people also obtained PANI/GO intercalated nanocomposites by ion exchange of polyaniline with hexadecylamine ions in hexadecylamine/graphene oxide[6] complex, and found that XRD was a wide diffraction peak appears in 5 ~ 70 degree, and when the polyaniline content increases, the peak moves to a small angle. In the polyaniline/graphite oxide intercalation nanocomposite, the polyaniline molecular chain exists in the extended chain conformation between the graphite oxide layers, or the benzene ring plane is parallel to the graphite oxide sheet layer, or is sandwiched in the form of a zigzag chain in the graphite oxide. Between layers, the infrared spectrum indicates that polyaniline is present in the form of a semi-oxidized semi-reduced structure in polyaniline/graphite nanocomposites. The powder conductivity of graphite oxide is very low, while the electrical conductivity of PANI/graphene oxide intercalated nanocomposites is 1.45×10^{-3} S/cm, which is much higher than that of reduced graphene oxide.

2. Experimental Scheme and Analysis of Its Principle

Pretreatment of graphene oxide suspension with 0.1 M NaOH solution, mass fraction of pyrrole monomer (2ml), mass fraction of GO (1%), concentration and amount of $FeCl_3$ solution (2.5M, 30ml), reaction temperature PPY / GO nanocomposites were prepared at (0 ° C). PPY/GO was then reduced in hydrazine hydrate solution to make PPY / RGO. The GO layer spacing of PPY/GO was 2.98 nm, and PPY was embedded in the GO layer. At the same time, the GO layer is partially deformed. PPY/GO has a well-layered ordered structure of graphene oxide, and finally forms an intercalation type composite. The nanostructure of the composite remains unchanged after being reduced. The intercalated PPY/RGO has higher conductivity. The conductivity of PPY depends on the length of its molecular chain. The molecular chain length increases the delocalization of the two electrons, the conductivity is enhanced, and the molecular chain is short. Reduced domain and reduced conductivity [7] The barrier effect of two-dimensional constrained space between GO slices can effectively inhibit the cross-linking of PPY molecular chains, improve the degree of symmetry of PPY molecular chains, and delocalize two electrons. Increased in conductivity and improved conductivity, so intercalated PPY/GO High conductivity.

PPY not only has good electrical conductivity and high stability, but its electrical activity has also attracted widespread attention. Conductive polymers have been the research hotspot of new battery materials, and most conductive polymer/layered inorganic nanocomposites are also considered as potential applications for secondary battery electrodes, but related to conductive polymers/layered

inorganic nanocomposites. The electrochemical properties of materials are relatively rare. Therefore, it is one of the tasks of this paper to study the electrochemical properties of PPY/RGO nanocomposites by electrochemical analysis.
GO was investigated under the same conditions of the amount of pyrrole monomer (2ml), molar ratio of NaOH to pyrrole monomer (4:3), reaction temperature (0°C), concentration and amount of FeC13 (2.5M, 30ml). The effect of mass fraction on the structure and conductivity of the product (PPY/GO) [8]. With the increase of RGO mass fraction, the electrical conductivity of the product has a large increase, but when the mass fraction of RGO exceeds 1%, the electrical conductivity of the product decreases rapidly. Although RGO is a non-conducting substance, the conductivity is on the order of $10^{-3} \sim 10^{-4}$S / cm[9], the conductivity is very poor, but the use of the RGO layer structure, through the conductive polymer PPY and RGO nano Composite, a composite material with a certain conductivity can be obtained. The conductivity of PPY/RGO nanocomposites is increased by three orders of magnitude relative to RGO.
Under the conditions of the amount of pyrrole monomer (2ml), the molar ratio of NaOH to pyrrole monomer (4:3), the mass fraction of graphene oxide (1%), the concentration and amount of $FeCl_3$ (2.5M, 30ml), The effects of polymerization temperature on the structure and electrical conductivity of PPY/GO nanocomposites were investigated. It can be seen that as the polymerization temperature increases, the yield of PPY/GO nanocomposites increases, but the electrical conductivity decreases. The higher the reaction temperature, the faster the polymerization rate of pyrrole, and the more PPY is produced in the same time, so the yield is increased. When the polymerization temperature rises, the conductivity decreases because the average degree of polymerization of PPY obtained at low temperature is high, and the aggregation state between the molecular chains is more favorable for electron transport, thereby exhibiting higher conductivity. When the reaction temperature is 0 °C, the product is a typical intercalated PPY/GO nanocomposite. As the polymerization temperature increases, the product is gradually converted into exfoliated PPY/GO nanocomposites [10]. The crystallinity of PPY and the regularity of molecular chain in PPY/GO nanocomposites decrease with the increase of polymerization temperature, which also causes the decrease of the conductivity of PPY/RGO with the increase of reaction temperature.
Pyrrole is polymerized by $FeCl_3$. In the resulting conductive polypyrrole, every three pyrrole rings are one repeating unit with one negative ion (Cl^-), and the positive charge on the macromolecular chain is distributed. $FeCl_3$ on the nitrogen atom is both an oxidant and a dopant in the reaction system, so the amount of $FeCl_3$ is the molar ratio of the pyrrole monomer will simultaneously affect the amount of PPY produced during the polymerization reaction and the conductivity of the PPY. Obviously, in the preparation process of PPY/GO[11], the molar ratio of $FeCl_3$ to pyrrole monomer also has a great influence on the structure and properties of the product. In order to study the extent of this effect, we investigated the molar ratio of $FeCl_3$ to PY under the same conditions of the amount of pyrrole monomer (2ml), the mass fraction of graphite oxide (1%) and the reaction temperature (0℃). Effect of pyrrole monomer on the structure and properties of PPY/RGO.
The resistivity of the PPY / RGO nanocomposite obtained when the molar ratio of FeC13 to PY is different is also different. It can be seen that the electrical conductivity of PPY/RGO nanocomposites increases first and then decreases with the increase of $FeCl_3$: pyrrole monomer. When $FeCl_3$: pyrrole monomer = 2.4, the conductivity of the obtained PPY/GO nanocomposites reaches the maximum value.
When the conditions such as the amount of potassium permanganate, the amount of graphite, the volume of concentrated sulfuric acid, and the reaction time of low temperature remain unchanged, the method of adding water in the high-temperature reaction process is changed, and the product obtained by adding water in both continuous and discontinuous manners is produced. Rate and resistivity. It can be seen that the yield and resistivity of the product obtained when continuously adding water are much larger than the yield and resistivity of the product obtained when the amount of water added is discontinuous and less. It can also be seen from the spectra of the products obtained by the two water addition methods that the degree of oxidation of the product when continuously adding water is much

higher than that of the product when the water is continuously added. The product obtained by continuous water addition has a typical crystal structure of graphite oxide, and there are two distinct diffraction peaks in the spectrum. The product obtained when the water is not continuously added, although the characteristic diffraction peak of graphite oxide also appears, but still exists. The characteristic diffraction peak of the original graphite surface with high strength. When the temperature is high, the continuous addition of water can control the temperature of the mixture below °C, which is beneficial to the hydrolysis of the graphite layer compound which is deeply oxidized to form the graphite oxide with higher oxidation degree. When the method of non-continuous water addition is used and the amount of water added is small at a time, the temperature of the mixed liquid is too high to reach about °C, and the graphite intercalation compound is decomposed by heat, and the water molecules, sulfuric acid molecules and hydrogen sulfate ions which have been inserted between the graphite layers are separated. Partial intercalation occurs and escapes quickly from the graphite layer. Some of the carbon network plane spontaneously recovers to the original graphite layer structure, and some carbon network planes are delaminated. Therefore, there are still strong natural flake graphite surface features in the spectrum of the product. The diffraction peak is around. In summary, it can be considered that there is a hydrolysis reaction of the deeply oxidized graphite intercalation compound in the high temperature reaction process. A thermal decomposition reaction, when the temperature of the mixture is appropriate, hydrolysis dominant, and the mixed solution temperature. When the degree is too high, the thermal decomposition and delamination expansion are dominant. When the discontinuous and one-time addition of water is small, the depth of the oxidized graphite intercalation compound is restored to the original graphite layer structure and delamination expansion occurs, resulting in a decrease in interlayer water and functional groups, and thus the yield of the obtained product is the resistivity has a large drop. Under the same conditions of potassium permanganate dosage, sodium nitrate dosage, concentrated sulfuric acid volume and low temperature reaction time, the effects of graphite dosage on product yield, electrical resistivity and structure were investigated. It was found that when graphite oxide was prepared from graphite and graphite, respectively, the mixture after low temperature reaction was viscous dark green, and the amount of graphite was and the mixture was black. When the amount of graphite is increased, the severity of the high temperature reaction is lowered. After adding the solution, the amount of graphite is golden yellow characteristic of graphite oxide. When the amount of graphite is yellow, the product is dark brown, while the amount of graphite is dark brown, but a small amount of golden yellow precipitate also appears. During the experiment, it was found that the flake graphite was added to concentrated sulfuric acid cooled to ° C, and the reaction liquid was black. As the amount of potassium permanganate added increases, the temperature of the reaction liquid rises, the viscosity increases, and a very viscous dark green mixture can be obtained at the end of the low temperature reaction. In the middle temperature reaction, the color of the solution changes from dark green to brown. When the distilled water is added for high temperature reaction, the mixture boils. After adding the play solution, a golden yellow precipitate appears in the mixed solution, accompanied by a pungent odor. The dried sample has high hardness and absorbs water. Strong, hard to grind.

3. Conclusions

The conductivity, electrochemical stability and thermal properties of the intercalated PPY/RGO nanocomposites are superior, which is attributed to the PPY/RGO PPY molecular chain with good layered structure and high regularity. With the increase of RGO mass fraction, the regularity of PPY molecular chain in PPY/RGO is improved, and the electrical conductivity is increased. When the mass fraction of GO exceeds 1%, the layered order of PPY/RGO nanocomposites decreases. The conductivity is reduced. As the reaction temperature increases and the molar ratio of NaOH and pyrrole monomer decreases, the PPY/RGO nanocomposite gradually becomes intercalated and the electrical conductivity decreases. When $FeCl_3$: the pyrrole monomer < 2.4, the electrical conductivity of the PPY/RGO nanocomposite increases with the increase of $FeCl_3$: pyrrole monomer, and $FeCl_3$ mainly acts as a dopant. $FeCl_3$: pyrrole monomer > 2.4, the content of $FeCl_3$ in the reaction system is too high, which will cause excessive oxidation of pyrrole and reduce the electrical conductivity of PPY/GO nanocomposites. When $FeCl_3$: pyrrole monomer = 2.4, the conductivity of PPY/GO nanocomposites is the largest. The mass fraction of GO, the molar ratio of NaOH to pyrrole monomer, the molar ratio of $FeCl_3$ to PY, and the reaction temperature are the main factors affecting the structure and properties of PPY/GO nanocomposites. With NaOH as additive, the mass fraction of graphite oxide suspension is 1%, NaOH: pyrrole monomer = 4:3 (molar ratio), $FeC1_3$: pyrrole monomer = 2.4: 1 (molar ratio), reaction temperature is 0 °C Under the process conditions, an intercalated PPY/GO nanocomposite with a layered ordered structure can be prepared with an electrical conductivity of 0.5 S/cm.

Reference

[1] Lei Wang. Intercalated Polymer-Layered Inorganic Nanocomposites: [Michigan State University Doctor Thesis].Michigan State: Michigan State University, 1999:1-7.
[2] Shan S G, Gao D S, Chen W, et al. The Nanocomposite Film of Polymer Intercalation in V_2O_5 Xerogel [J]. J Wuhan Univ Technol Mater Sci Ed, 2000, 15(3): 1.
[3] Lei Wang, Jon Schindler, Joyce Albritton Thomas, et al. Entrapment of Polypyrrole Chains between MoS2 Layers via an In Situ Oxidative Polymerization Encapsulation Reaction [J]. Chem Mater, 1999, 7:1753-1755.
[4] T Nakajima, A Mabuchi. A New Structure Model of Graphite Oxide [J]. Carbon, 1988, 26(3): 357-161.
[5] Pinggui Liu, Kecheng Gong, Xiao Peng. Synthesis of Polyaniline-Intercalated Graphite Oxide by an in Situ Oxidative Polymerization reaction [J]. Carbon, 1999, 37: 706-707.
[6] Y Matsuo, K Tahara, Y Sugie. Structure and Thermal Properties of Poly (ethylene oxide)-Intercalated Graphite Oxide [J]. Carbon, 1997, 35(1): 113-120.
[7] Y Matsuo. K Tatase, Y Sugie. Preparation and Characterization of Poly (vinyl alcohol) and $Cu(OH)_2$-Poly(vinyl alcohol)-intercalated Graphite oxide[J]. Chew Mater, 1998, 10:2266-2269.
[8] Chen W, Xu Q, Yuan R Z. The Influence of Polymer State on the Electrical Properties of Polymer/Layered-Silicate Nanocomposites [J]. Comp Sci&Tech, 2001, 61(7): 935-993.
[9] V Mehrotra, E F Giannelis. Nanometer Scale Multilayers of Electroactive Polymers: Intercalation of Polypyrrole in Mica-type Silicates [J]. Solid State Ionics, 1992, 51:115-122.
[10] V Mehrotra, E F Giannelis. Metal-insulator Molecular Multilayers of Electroactive Polymer: Intercalation of Polyaniline in Mica-type Layered Silicates [J]. Solid State Common, 1991, 77(2): 155-158.
[11] Kim JW, Liu F, Choi H J, et al. Intercalated Polypyrrole/Na^+-montmorillonite Nanocomposite Via An Inverted Emulsion Pathway Method[J]. Polymer, 2003, 44(1): 289-293.

IOP Conf. Series: Materials Science and Engineering **538** (2019) 012006 doi:10.1088/1757-899X/538/1/012006

Crystal Plasticity Simulation of Hot Deformation Texture of Titanium Alloy Considering Alpha-beta Transformation

D H Wei[1], W J Zhang[1], L P Zhang[1], H Y Huang[1], R Y Chen[1], Y W Wu[2], X F Ma[1,*], Z H Xiong[1,*]

[1]Sino-French Institute of Nuclear Engineering and Technology, Sun Yat-sen University, Zhuhai 519082, China
[2]Huadian electric power research institute co., Ltd, Hangzhou 310030, China

[*]Corresponding author: maxf6@mail.sysu.edu.cn; xiongzhh5@mail.sysu.edu.cn

Abstract. Titanium alloys are usually subjected to hot deformation to optimize its microstructure and mechanical property. In the hot forming process, such as extrusion, texture is of great importance to both manufacturing and performance. In this study, an alpha-beta titanium was used to study the texture evolution during hot extrusion. Large area electronic backscattering scan was used to examine grain orientation distribution. Experimentally measured alpha and beta textures were compared with a rate-dependent crystal plasticity based simulation. The model is capable of predicting the deformation textures of both alpha and beta phases at extrusion temperature. The model also takes into account the transformation of metastable beta phase to alpha phase in the cooling process, considering variant selection. The transformation texture components mostly have [0001] along extrusion direction, forming the primary component of alpha extruded texture. The model works well in predicting hot deformation texture of titanium alloy.

1. Introduction

Titanium alloys, with lightweight and excellent mechanical properties, have been widely used in aeronautic industry, marine ships, nuclear components, medical instruments, and so on [1]. Alpha-beta titanium alloy has excellent elevated strength, formability, and corrosive resistance, is becoming an important material in future nuclear industry. For titanium alloy with alpha and beta phases, it is sensitive to hot deformation process, which affects dramatically the microstructure, such as grain/colony size, lamellar spacing, and texture. Apparently, it further determines the mechanical property of alloy. Hence, both industry and academia realize that it is crucial to have computational tools for predicting the interrelationship between processing, microstructure, and property. In the United States, both the Materials Genome Initiative and Integrated Computational Materials Engineering (ICME) community have been making efforts to develop methodology integrating the resources of theory, simulation and experiments.

For titanium alloys with low crystal symmetry of hcp structure, hot extrusion are used as a typical processing, typically in the $\alpha+\beta$ temperature field where there is higher fraction of β phase than that at room temperature. During the post-cooling process after hot deformation, the metastable β phase will partially decomposes into α phase, following the well-known Burgers orientation relationship (BOR), i.e. (0001)||(110) and [-2110]||[-111]. Hence, to predict the microstructure evolution during hot

IOP Conf. Series: Materials Science and Engineering **538** (2019) 012006 doi:10.1088/1757-899X/538/1/012006

processing, the β→α transformation must be taken into account. For hot extrusion of titanium alloy, the final texture should consider three important aspects: (i) α deformation texture at the forming temperature, (ii) β deformation texture at the forming temperature, and (iii) α transformation texture formed during the β→α transformation, in which there are 12 possible variants according to the crystal symmetry of α and β phase. Usually, bias selection of certain α variants would occur, which will affect the transformation α texture.

For a computational toolset for predicting the hot texture of titanium alloy, crystal plasticity based models have been widely used in previous studies, as they can capture the plastic deformation mechanisms of slip. For modeling the texture evolution during large deformation of polycrystalline materials, approaches in literature include classical Taylor models [2,3], relaxed-constraint (RC) Taylor model [4], multiple grain models [5], viscoplastic self-consistent (VPSC) model [6,7], and crystal plasticity finite element (CPFE) methods [8-11]. The CPFE approach is quite promising in capturing microstructural effects, however it requires significant computational resources to model an industrial extrusion process. There have been some efforts to reduce computational cost, it is increasily accepted nowadays to use uncoupled continuum FEM and crystal-plasticity model for simulating extrusion or forging process [12]. For instance, Glavicic et al. [13] modeled the forging process of Ti6Al4V and location-specific texture distribution in the forged pancake, by using commercial finite element software to analyze crystal orientation due to metal flow and Taylor-type polycrystalline plasticity codes to calculate the rotation due to crystal plasticity.

This paper reports a crystal plasticity model based toolset for hot texture of titanium alloy. The toolset can be used to predict the "processing-microstructure", i.e. simulating the texture evolution during and subsequent to the hot extrusion of α+β titanium alloy, taking into account β→α transformation process [14]. The validity of this computationally efficient, sequential simulation approach has been validated by successfully reproducing the experimental α extrusion texture.

2. Materials and Experimental

In this study, an α+β titanium alloy was used. It has a high volume fraction of primary α grains, i.e. 70-90% (grain size of ~5 μm). Fig. 1 is the (0001) pole figure obtained from EBSD, indicating slight texture (with an intensity of ~4x random on average) in the as-received state. The extrusion was conducted at 900 °C at an extrusion ratio of 14. The extruded bar was sectioned along the longitudinal section, grinded and polished. Large area EBSD scan was conducted to obtain the grain orientation information. Fig. 1 shows the pole figure from EBSD after extrusion. It is seen that the 0001 poles are preferentially aligned along the extrusion direction.

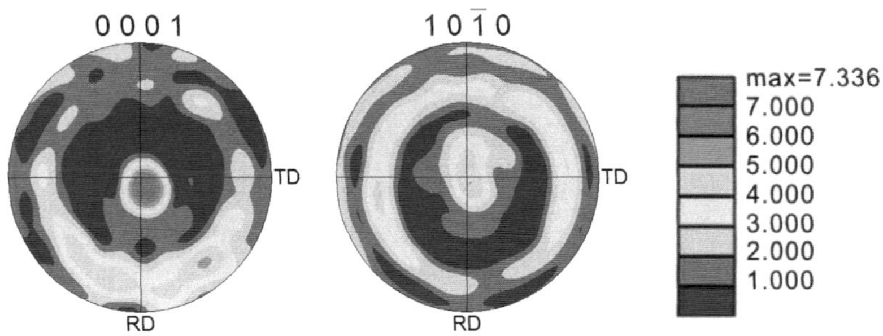

Figure 1. Alpha texture of extruded titanium alloy indicated by EBSD IPF ("ED" is outwards)

3. Crystal Plasticity Theory and Finite Element Simulation

3.1. Crystal Plasticity Theory

For the texture prediction, a rate dependent crystal plasticity model is used, based on the classical crystal plasticity theory developed by Hill [15], Asaro and Rice [16]. The deformation of a crystalline material under external load consists of elastic deformation and inelastic deformation. The total deformation gradient F can be separated into elastic and plastic part and given by

$$\underset{\sim}{F} = \underset{\sim}{F}^* \cdot \underset{\sim}{F}^p, \qquad \det\left(\underset{\sim}{F}^p\right)=1,\ \det\left(\underset{\sim}{F}^*\right)>1 \tag{1}$$

Where F^* denotes the part of elastic deformation, F^P denotes the part of plastic deformation. The rate of change of F^P is related to shearing rate $\dot{\gamma}^{(\alpha)}$ of α slip system as follow:

$$\dot{F}^P \cdot F^{P-1} = \sum_{\alpha} \dot{\gamma}^{(\alpha)} s^{(\alpha)} m^{(\alpha)} \tag{2}$$

where $\dot{\gamma}^{(\alpha)}$ is the resolved shear strain rate on α slip system, $\underset{\sim}{m}^{(\alpha)}$ is the slip normal direction on α slip system, $\underset{\sim}{s}^{(\alpha)}$ is the slip direction on α slip system.

The resolved shear stress $\tau^{(\alpha)}$ on each slip system is obtained via the Schmid's law. The shearing rate $\dot{\gamma}^{(\alpha)}$ of α slip system is expressed by the resolved shear stress $\tau^{(\alpha)}$, given by:

$$\dot{\gamma}^{(\alpha)}=\dot{\gamma}_0\left[\frac{\tau^{(\alpha)}}{g^{(\alpha)}}\right]\left[\left|\frac{\tau^{(\alpha)}}{g^{(\alpha)}}\right|\right]^{1/m-1} \qquad (\text{when } \tau^{(\alpha)} > g^{(\alpha)}) \tag{3}$$

where $\dot{\gamma}^{(\alpha)}$ and $\tau^{(\alpha)}$ are the shear strain rate and resolved shear stress on slip system α respectively, γ_0 is the reference shearing rate, $g^{(\alpha)}$ is the slip resistance on slip system α, m is the rate sensitivity exponent [17-18].

3.2. *Finite Element Model*

For computing the hot extrusion texture, finite element simulation of the extrusion process was firstly conducted. The actual dimensions of extrusion die and billet, extrusion rate etc was used in the simulation. The stress and strain values along the central axis of billet were obtained from the stress and strain contour. At the billet central axis, the deformation of alpha and beta phase can be regarded as bi-axial compression along x and y axis, perpendicular to the extrusion direction (z axis). The strain values were used as boundary condition for crystal plasticity finite element model in Fig. 2, in which 1024 C3D8R elements were used. Each element represents a grain. The deformation texture of alpha and beta phase can be obtained from crystal plasticity simulation of finite element model in Fig. 2.

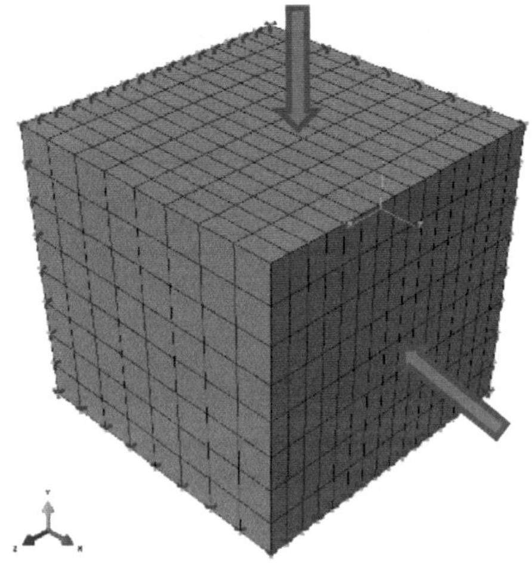

Figure 2. Finite element model for predicting the extrusion texture near extrusion billet central axis

4. Results and Discussion

Crystal plasticity finite element simulation of titanium hot extrusion with ratio of R=14, indicates the alpha deformation texture as shown in Fig. 3. It is seen that alpha texture has 10-10 along the extrusion direction [17-18], which is distinct from the experiment texture in Fig. 1. The experimental texture shows that, however, 0001 along the extrusion direction, meaning deformation texture is not dominant in the final alpha texture.

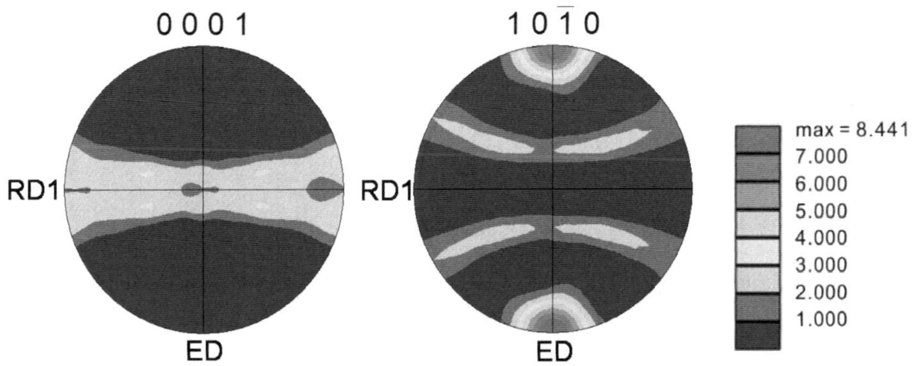

Figure 3. Predicted alpha deformation texture of titanium alloy

Hence, in the following section, we take into account the beta-alpha transformation during the post-extrusion cooling process. First of all, we calculated the deformation texture of beta phase using the finite element model in Fig. 2, then calculate the alpha transformation texture according to the beta-alpha Burgers relationship. In this transformation process, we firstly assume no preference of alpha variant, i.e. 12 alpha variants have the same probability of occurrence, from which the alpha transformation can be obtained. Subsequently, the alpha transformation texture is summed up with the deformation texture by weight, to get the final alpha texture, as shown in Fig. 4. It is seen that predicted alpha texture has 0001 pole along the extrusion direction, which is consistent with Fig. 1. However, the texture intensity is only 3.5, lower than the experimental intensity of 7.3. Nevertheless, it shows that transformation alpha texture in the final alpha texture of extruded titanium.

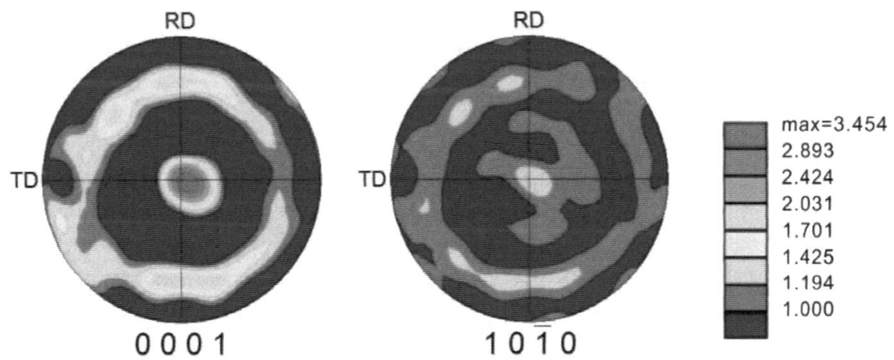

Figure 4. Predicted alpha extrusion texture assuming 12 alpha variant without bias selection

The deviation of texture intensity in Fig. 4 against experimental indicates that alpha variant selection occurred during the beta-alpha transformation. It must be taken into account while computing the final extrusion texture. This is how we calculate the alpha extrusion texture. It is no doubt that beta-alpha transformation texture will greatly affect the final texture. The above simulation has indicated that alpha transformation texture is dominated by 0001texture, while alpha deformation texture is dominated by 10-10 texture. The weight of the abovementioned two components depends on forming temperature, or phase fraction. Different alpha variant selection rule will alter the final alpha extrusion texture 0001 intensity.

Previous studies on the variant selection phenomena have reported a number of potential mechanisms, e.g. (i) the common {110} rule [14], i.e. the adjacent transformed β grains with common {110} pole will lead to preferential selection of α variants. Based on this variant selection rule, alpha transformation texture was calculated by numerically decomposing beta deformation texture according to α variant selection rule. The transformation texture was sum up with the deformation texture in Fig. 3 to obtain the final alpha texture shown in Fig. 5. It shows evident 0001 texture along the extrusion direction, with quite comparable intensity with the experimental result in Fig. 1. It also implies that the common {110} rule has high promise in describing the texture evolution associated with beta-alpha transformation [14].

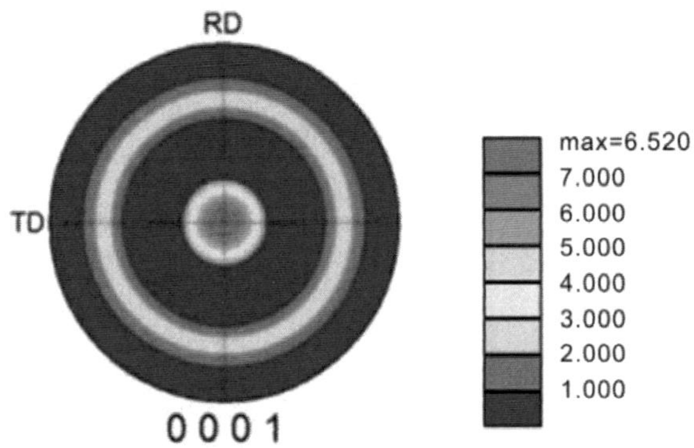

Figure 5. Alpha extrusion texture composed of deformation texture and transformation texture taking into account {011} variant selection rule

5. Conclusions

By crystal plasticity finite element method, the hot extrusion texture of alpha-beta titanium alloy were studied. The following conclusions have been drawn:

1) The crystal plasticity modelling based toolbox is capable of predicting the deformation and transformation texture evolution of alpha and beta phase during extrusion.
2) Alpha deformation texture has 10-10 along the extrusion direction, which is different from the 0001 alpha extrusion texture.
3) Beta-alpha transformation during cooling must be considered, due to its effect on 0001 alpha transformation texture, which is dominant in the final alpha extrusion texture.
4) Variant selection is important in the beta-alpha transformation process. By taking into account the transformation texture via proper alpha variant selection rule, the model can provide satisfactory prediction of extrusion texture. Future study will be centred on microscopic study of alpha variants experimentally in comparison with prediction.

Acknowledgments

The authors would like to acknowledge the support from Guangdong Education Department Fund (2016KQNCX004 and 2016KQNCX005) and Guangdong Provincial key S&T Special Project (2017B020235001). We also acknowledge the support from University of Michigan and Sun Yat-Sen University.

References

[1] Lütjering G, Williams J C, Titanium, Berlin; New York: Springer, 2003, 15-22
[2] Taylor G I. Plastic strain in metals. J Inst Metals, 1938, 62: 307-324

[3] Vanhoutte P. Simulation of rolling and shear texture of brass by taylor theory adapted for mechanical twinning. Acta Metall, 1978, 26(4): 591-604
[4] Honeff H, Mecking H, in: Nagashima S (Ed.) Proc. Int. Conf. on Texture Mater., 6th, The Iron and Steel Institute of Japan, 1981, 347-355
[5] Crumbach M, Pomana G, Wagner P, et al., in: G. G, D. M (Eds.) Recrystallization and Grain Growth Springer, Berlin, 2001, 1061-1068
[6] Lebensohn R A, Canova G R. A self-consistent approach for modelling texture development of two-phase polycrystals: Application to titanium alloys. Acta Mater, 1997, 45(9): 3687-3694
[7] Tomé C N, Lebensohn R A, Kocks U F. A model for texture development dominated by deformation twinning: Application to zirconium alloys. Acta Metall Mater, 1991, 39(11): 2667-2680
[8] Mayeur J R, McDowell D L. A three-dimensional crystal plasticity model for duplex Ti-6Al-4V. Int J Plasticity, 2007, 23(9): 1457-1485
[9] Kim D K, Kim J, Park W, et al. Three-dimensional crystal plasticity finite element analysis of microstructure and texture evolution during channel die compression of IF steel[J]. Comput Mater Sci, 2015, 100:52-60.
[10] Mayama T, Noda M, Chiba R, et al. Crystal plasticity analysis of texture development in magnesium alloy during extrusion. Int J Plasticity, 2011, 27(12): 1916-1935
[11] Yao J Y, Wang B S, Deng L P, et al. Simulation of texture evolution and deformation mechanism in Mg-3Al-1Zn alloy during uniaxial compression. Sci China: Tech Sci, 2015, 58(12): 2052-2059.
[12] Dunst D, Mecking H. Analysis of experimental and theoretical rolling textures of two-phase titanium alloys. Z Metallkd, 1996, 87(6): 498-507
[13] Glavicic M G, Goetz R L, Barker D R, et al. Modeling of texture evolution during hot forging of alpha/beta titanium alloys. Metall Mater Trans A, 2008, 39A(4): 887-896
[14] Bhattacharyya D, Viswanathan G B, Denkenberger R, et al. The role of crystallographic and geometrical relationships between alpha and beta phases in an alpha/beta titanium alloy. Acta Mater, 2003, 51(16): 4679-4691
[15] Hill R. On constitutive macro-variables for heterogeneous solids at finite strain [J]. Proceedings of the Royal Society of London, 1972, 326(1565): 131-147.
[16] Asaro R J. Micromechanics of Crystals and Polycrystals [J]. Advances in Appl Mech, 1983, 23(08): 1-115.
[17] Wei P T, Lu C, Tieu K, et al. Modelling of Texture Evolution in High Pressure Torsion by Crystal Plasticity Finite Element Method [J]. Applied Mechanics and Materials, 2015, 764-765:56-60.
[18] Sheikh H, Ebrahimi R. Investigation on texture evolution during cyclic expansion-extrusion (CEE) technique using crystal plasticity finite element modeling [J]. J Mater Sci, 2016, 51(22): 10178-10190.

An Effective Methodology to Design Scale Model for Microwave Transmitting Composites

Liming Yuan, Fei Dai, Yonggang Xu, Dandan Gu, Feiming Wei, Yuan Zhang

Science and Technology on Electromagnetic Scattering Laboratory, Shanghai 200438, PR China

Email: lming_y@163.com (Liming Yuan)

Abstract. In order to solve the problem on designing scale model for microwave transmitting composites, an effective method has been proposed in this manuscript based on the transmission and reflection (TR) coefficients. By optimizing parameters including the number of layers, the thickness and the electromagnetic parameters of each layer, the TR coefficients of designed scale material could be identical with that of the theoretical scale composite. An example is given according to the proposed method by optimizing the TR coefficients as well as only the reflection coefficient and two designed scale materials are obtained. A combined model of a transmitting hemisphere shell and a PEC paraboloid is constructed to verify the validation of the designed scale materials. FEKO is employed to simulate the monostatic radar cross section (RCS) and comparison is performed between the theoretical scale model and the designed scale models. Result reveals that the model including the designed scale material by optimizing the TR coefficients is more consistent with the theoretical scale model than that including the designed scale material by only optimizing the reflection coefficient. This work has important theoretical and practical significance in electromagnetic measurement.
Keywords: Scale model; microwave transmitting composites; TR coefficients; RCS

1. Introduction

The scale measurement has become a significant method to acquire the radar cross section (RCS) of an object because of it many merits such as low cost, excellent controllability and so on [1,2]. The classic scale theory was proposed by J. A. Stratton J. A. Sinclair on the basis of the linear theory of Maxwell's equations [3,4]. It is that $\sigma_{scale} = \sigma_{full} + 20\lg p$ when the scale system has the same electrical size with the full scale system. Here, p is the scale ratio, σ_{scale} is the RCS of the scale model at the scale frequency and σ_{full} is the RCS of the full scale model at the full scale frequency. In the scale measurement, one of the keys is to construct an accurate scale model. For metallic object or perfect electric conductor (PEC), there is little difficulty in constructing a satisfactory model. However, more and more composites with good electromagnetic loss properties are applied to tailor the scattering characteristic. The composites always have frequency dependent properties [5], which would bring great difficulty to construct the scale material. Some exist studies have given an approximation method to design the scale composite [6~9]. In these studies, composites are placed on metal substrate and then the transmitting wave can be neglected as investigating the scattering properties of an object. Therefore, there is obvious difference from the existing studies when composites are transmitting and without metal substrate.

In this paper, a method is studied by optimizing the transmission and reflection (TR) coefficients. An example is given by the proposed method. In order to verify the designed scale material, a combined model is constructed and monostatic RCS is simulated by FEKO. Furthermore, Comparison is performed between the designed scale model and the theoretical scale model.

2. Methodology Considerations

2.1. Calculating the TR Coefficients of A Multilayer Composite

At the scale frequency, multilayer structure is always used to make the TR coefficients the same with that of the theoretical scale composite. Figure 1 shows the schematic description for calculating the TR coefficients of an n-layer composite irradiated obliquely by plane microwave. The problem can be solved by the transmission-line method and details on derivation procedure can be found in Ref. [10].

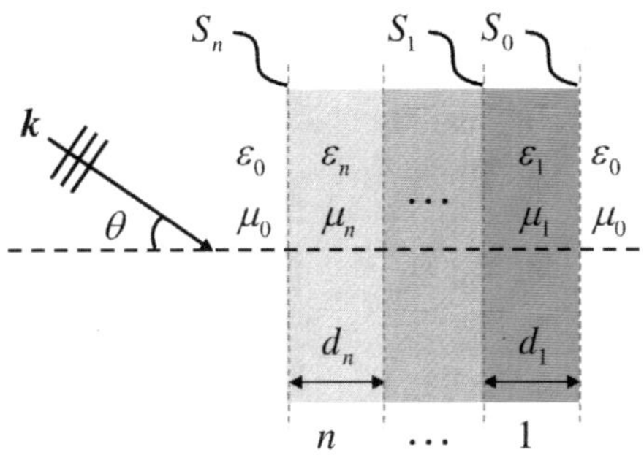

Figure 1. the TR calculation sketch.

The TR coefficients in dB of the multilayer composite can be expressed by following formulae,

$$T_{total} = 20\lg\left|\prod_{k=0}^{n} T_k\right| \tag{1}$$

$$R_{total} = 20\lg\left|\frac{Z_{in,n} - Z_0'}{Z_{in,n} + Z_0'}\right| \tag{2}$$

in which, T_k is the transmission coefficient of S_k , $Z_{in,n}$ the input impedance of the composite, Z_0' is the impedance of free space as the composite is irradiated obliquely by plane microwave.

On the basis of the above formulae, the scale model of a microwave transmitting composite can be designed by the TR coefficients. By optimizing parameters including the number of layers, the thickness and the electromagnetic parameters of each layer, the TR coefficients can be made as identical as possible with the theoretical scale composite in wide angle range.

2.2. Optimizing the TR Coefficients of the Scale Material

The TR coefficients of a multilayer composite are determined by parameters including the number of layers, the thickness and the electromagnetic parameters of each layer. The optimizing procedure mainly includes the following four steps.

(i) Input data. The data includes the TR coefficients of the theoretical scale composite in wide angle band, and for the designed scale composite, the maximum number of layers, the maximum thickness, the minimum thickness meeting engineering requirement and the number of loop calculations.

(ii) Prepare electromagnetic parameters at the scale frequency. Mix selected inclusion with a binder in different proportions and measure the electromagnetic parameters. Furthermore, fit the measured data by the effective medium theory and obtain the formula for calculating the electromagnetic parameters. Here, the GEMT formula based on the dilution process model [11] is used to fit the measured electromagnetic parameters. The expression is written as follows,

$$\frac{\phi_l - \phi_h}{\phi_l + 2\phi_h + V(\phi_l - \phi_h)} = p_{cal} \frac{\phi_m - \phi_h}{\phi_m + 2\phi_h + V(\phi_l - \phi_h)} \quad (3a)$$

$$p_{cal} = 1 - p_l / p_h \quad (3b)$$

where, ϕ_m, ϕ_l and ϕ_h denote the electromagnetic parameter of the matrix medium and the composite of low and high inclusion concentration, respectively. Both ϕ_m and ϕ_h are obtained by experimental measurement. p_l and p_h are the inclusion concentrations of the composites of low and high inclusion concentration, respectively. V is a fitting parameter of a great freedom to let the GEMT formula much more flexible. Here, $V = ap_{cal}^2 + bp_{cal} + c$, in which a, b and c are dimensionless parameters and obtained by fitting experimental data. After ϕ_m, ϕ_h, a, b and c are obtained, ϕ_l can be calculated by Eq.(3).

(iii) Perform loop calculations. In each loop calculation, generate parameters including the number of layers, the thickness and the inclusion concentration of each layer by normal distribution random functions. Record these parameters. The electromagnetic parameters of each layer can be calculated by the formulae introduced in step (ii). Calculate the TR coefficients of the multilayer composite in corresponding angle band. Then calculate and record the deviation by comparing with the TR coefficients of the theoretical scale composite. Here, the deviation is defined as follows,

$$\delta^k = \sqrt{\sum_{n=1}^{N} \left(\left| T_n^k - T_n^{Theo} \right|^2 + \left| R_n^k - R_n^{Theo} \right|^2 \right)} \quad (4)$$

where N are the number of incident angles; T_n^k, R_n^k, T_n^{Theo} and R_n^{Theo} are elements of the TR coefficients of the kth calculation and the theoretical scale composite, respectively.

(iv) Search the optimal result. After the loop calculations, find out the minimum deviation and obtain the corresponding configuration parameters of the optimal scale absorber, which include the number of layers, the thickness and the inclusion concentration of each layer.

3. Results and Discussions

Here an example is given to verify the validation of the proposed method. The full scale frequency is 1.0 GHz, at which, a transmitting composite has a thickness of 10.0 mm and its permittivity and permeability are 14.49-0.12j and 3.56-1.12j, respectively. Then the scale frequency is 10 GHz as the scale ratio is 1:10. The theoretical scale composite has the same electromagnetic parameters with the full scale composite but its thickness is 1.0 mm. Here TE wave is just taken into account.

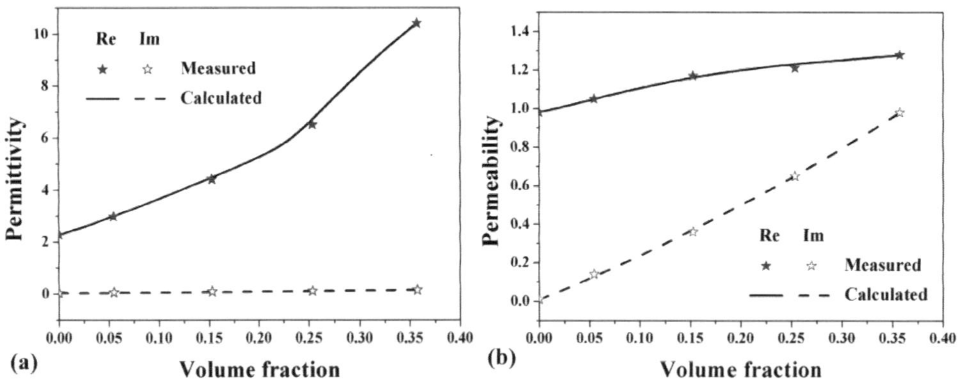

Figure 2. The fitting result of the electromagnetic parameters.

According to the optimizing procedure in section 2.2, it needs to build up a selection of electromagnetic parameters at 10 GHz. Spherical carbonyl iron powders (CIP) are selected as the inclusion and mixed with wax by volume fractions of 0%, 5%, 15%, 25% and 35%. The electromagnetic parameters of the as-prepared mixtures could be measured by transmission/reflection method [12]. Then fit the measured electromagnetic parameters according to step (ii) in section 2.2. The calculated result is plotted in Figure 2. It can be seen that there is an excellent agreement between the calculated results and the measured data. Then electromagnetic parameters of a mixture of any inclusion concentration can be calculated by the fitting formulae.

According to the optimized procedure in section 2.2, the designed scale material is obtained and assigned as M01. Besides, another designed scale material is obtained by only optimizing the reflection coefficient instead of the TR coefficients and assigned as M02. Table 1 gives the related information of M01 and M02.

Table 1. The designed scale materials.

	Permittivity	Permeability	Thickness	Inclusion concentration
M01	5.36 - 0.09j	1.20 - 0.51j	2.3 mm	21vol%
M02	7.36 - 0.12j	1.24 - 0.70j	1.2 mm	27vol%

Figure 3 plots the TR coefficients of M01 and M02. It can been seen that the transmission property of M01 agrees well with the theoretical scale composite at any incident angle, while there is an obvious deviation between M02 and the theoretical scale composite. As the incident angle is small, about <20°, the reflection property of M01 and M02 both agree well with the theoretical scale composite, while the incident angle is in about 20°~80°, there is a little deviation between the both designed scale materials and the theoretical scale composite.

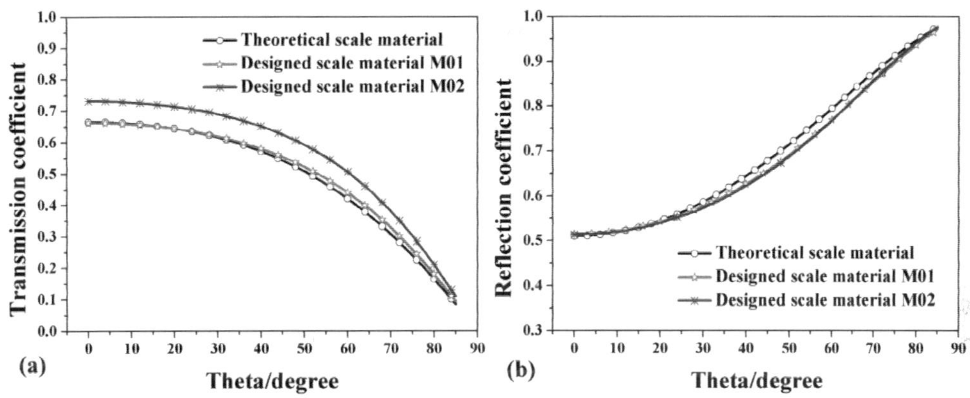

Figure 3. The TR coefficients of the theoretical scale material and the designed scale materials.

In order to verify the designed result, the commercial software FEKO is employed to simulate the monostatic RCS of a combined model of a transmitting hemisphere shell and a PEC paraboloid. The symmetric axis of the PEC paraboloid and the hemisphere shell are the same, and the peak point of the PEC paraboloid is located at the centre of the hemisphere shell. The theoretical scale hemisphere shell has a radius of 50 mm and a thickness of 1.0 mm; the radii of the designed scale shell M01 and M02 are also 50 mm, but the thicknesses are 2.3 mm and 1.2, respectively. The PEC paraboloid reflects the transmitting wave back. The focal depth and radius of the PEC paraboloid are both 30 mm. This combined model includes many electromagnetic scattering mechanisms such as specular-reflection, multiple-reflection, creeping wave scattering and so on. Figure 4 shows the simulated results. It can be seen there are similar change tendency and some deviation between the design scale models and the theoretical scale model. But the deviation of M01 is much smaller than M02. The mean deviation of M01 and M02 are 0.79 dB and 1.13 dB, respectively. The deviations are within the RCS measurement tolerance.

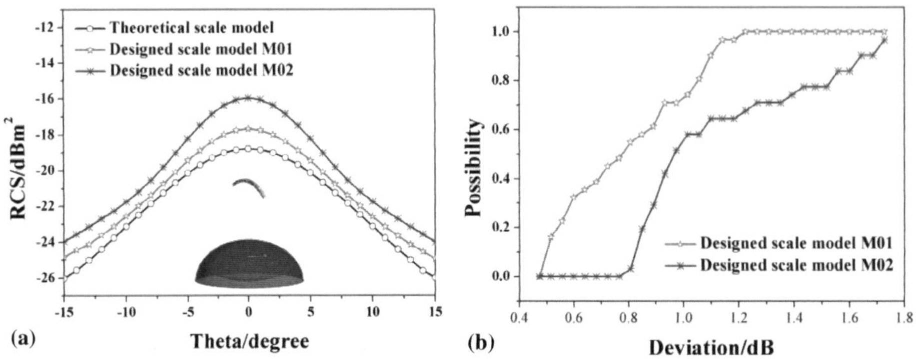

Figure 4. The simulated RCS of the combined models.

4. Conclusions

In this paper, an effective method based on the TR has coefficients been proposed to solve the problem on designing scale model for microwave transmitting composites and the optimizing procedure has been introduced. An example is given according to the proposed method by optimizing TR coefficients as well as only the reflection coefficient and two designed scale materials are obtained. A combined model of a transmitting hemisphere shell and a PEC paraboloid is employed to verify the validation of the designed result. The monostatic RCS of these models are simulated by the commercial software FEKO and comparison is performed between the theoretical scale model and the designed models. Result reveals that the model including the designed material by optimizing the TR coefficients is more consistent with the theoretical scale model than that including the designed material by only optimizing the reflection coefficient. Meanwhile, the method can break the limitation of physical and geometrical parameters by the absolute scale theories. This work will have important theoretical and practical significance in electromagnetic measurement.

Acknowledgements

This work was supported by Shanghai Municipal Science and Technology Commission under Grant Nos.18ZR1437400, 17YF1418400, 17QB1404100, 17210730900 and 16XD1423300, and the National Civil Aircraft Project of China under Grant MJ-2017-F-11.

References

[1] White M O 1998 *Electronics & communication engineering journal* **10** 169
[2] Zhao Y et al 2012 *IEEE Transactions on Antennas and Propagation* **60** 5890
[3] Sinclair G 1948 *Proc. Inst. Radio Eng.* **36** 1364
[4] Suhl H 1957 *Journal of Physics and Chemistry of Solids* **1** 209
[5] Lagarkov AN et al 2009 *Journal of Magnetism & Magnetic Materials* **321** 2082
[6] De Adana F S et al 2003 *IEE Proceedings-Radar, Sonar and Navigation* **150** 375
[7] Susetio A et al 2015 *Antennas and Propagation (APCAP), 2015 IEEE 4th Asia-Pacific Conference*, June 30 - July 3, 2015, Kuta, Indonesia, p. 568
[8] Yuan L M et al 2017 *Results in Physics* **7** 1698
[9] Yuan L M et al 2018 *Chinese Physics B* **27** 44101
[10] Liao Y et al 2016 *Proc. of the 2016 International Symposium on Electromagnetic Compatibility - EMC EUROPE 2016*, September 5-9, 2016, Wroclaw, Poland, p.315
[11] Yuan L M et al 2017 *Advanced Composites Letters* **25** 189
[12] Baker-Jarvis J et al 1990 *IEEE Transactions on microwave theory and techniques* **38** 1096

IOP Conf. Series: Materials Science and Engineering **538** (2019) 012008 doi:10.1088/1757-899X/538/1/012008

Photo-Electric Response of Polyaniline/ZnO Nanocomposites Based on Inorganic/Organic Heterojunctions

Lei-lei Han[1], Chao-ying Meng[1 a], and Shan Fan

[1]Wenhua College, 430074 Wuhan, China

Corresponding author: 253508105@qq.com

Abstract. The samples were analyzed and characterized by scanning electronic microscope (SEM) and X-ray diffraction (XRD).The XRD results showed that the diffraction peaks of PANI/ZnO nanocomposites agreed with the pure ZnO. The as-prepared nanocomposites with different amounts of PANI dopants exhibited different sensing properties. When the weight ratio of PANI is 5%, the photoelectric response of nanocomposites to UV light reaches maximum (1938). Compared with ZnO, the response shows 60 times improvement. Additionally, the probable photoelectric response mechanism was also proposed. Firstly, the existence of p-n heterojunctions formed between p-type PANI and n-type ZnO interfaces was confirmed by the IV characteristics. It is believed that the p-n heterojunctions can promote the charge separation and further increase the amount of absorbed negative oxygen ions. Secondly, the highly dispersed stated of PANI could suppress the formation of interfacial recombination centers and accelerate p-type hole transporting. The photocurrent of composites is improved significantly. And we hope this work can help for designing more photoelectric materials with the higher response.

1. Introduction

Conductive polymers and metal oxide composites are widely used in photocatalysis , solar cells , gas sensors and optoelectronic applications.[1-3] Such inorganic-organic heterojunction structures are capable of complementing each other while overcoming their respective disadvantages, and their schottky structure or p-n structure can promote separation of carriers. [4] Among inorganic metal oxides, zinc oxide is widely used in catalysis, optics and electronics, mainly due to its wide band gap (3.3 ev), low power threshold and exciton binding energy (~60 meV). And low cost. [5] Common conductive polymers are polyaniline, polypyrrole, polythiophene, etc., and polyaniline is one of the most widely used p-type semiconductor materials, with unique electrical, optical, and good stability. [6] Therefore, polyaniline/zinc oxide composites have a good application prospect in photoelectric performance. Ameen [7] has reported that a polyaniline/zinc oxide composite prepared by a simple method has good rectifying properties and can be used for light emitting diodes.

Zhang [8] successfully prepared a mixed photocatalyst of zinc oxide and monolayer molecular polyaniline, which has good catalytic performance for degrading methylene blue under ultraviolet and visible light. Mridha [9] studied the photoelectric properties and carrier transport mechanism of polyaniline/zinc oxide based on inorganic/organic heterojunction structure. Lin [10] research shows that zinc oxide and conductive polymer composite can improve carrier mobility and improve conductivity. The results show that zinc oxide and polyaniline composites have been widely used in UV photodetection. [11]

Although there are many ways to prepare such composite materials, it is important to improve the migration of photogenerated carriers and reduce recombination. [8] However, the interfacial effects based on may be lead to accelerate the recombination.So it is necessary to control the content of zinc oxide or polyaniline. In the paper, we have prepared PANI/ZnO nanocomposites by simple mechanical mixing method and tested the photoelectric response. By testing the photoelectric response of the composites with different composition ratios, the optimum content of the polyaniline in the composite has been determined. In the paper, the percolation value has been determined based on quickly testing the photocurrent. We hope this work can help us to explore more materials with higher photoelectric response.

2. Experiment

2.1. Materials Preparation
Polyaniline and zinc oxide are commercially available and have an average particle size of 1 μm and 80 nm, respectively. The reagents used in the experiments were all analytically pure reagents and were not subjected to any chemical treatment. The two powders were mixed together with a certain amount of organic solvent. The ball was milled in a QM-3SP04 ball mill of Nanjing University Instrument Factory for 3 h (rotation speed: 250 rpm) until a uniformly mixed sample was obtained. The sample was printed on a ceramic substrate printed with gold electrodes. The device has a shape of 8 mm * 6 mm and a thickness of 10 μm. The printing machine was used by the WJ-LCD4050CHL type printing machine of Shenzhen Jinxing Company. These samples were calcined at 200 ° C for 2 h to remove the organic solvent. After the appeal step, a porous film can be obtained. [12] Preparation of six different polyaniline composite materials, the product numbers are listed in Table 1:

Table 1. Numbers of the samples

The Number	The content of Polyaniline (wt %)
ZP000	0%
ZP001	1%
ZP005	5%
ZP010	10%
ZP050	50%
ZP100	100%

2.2. Characterization
In the experiment, the phase of the prepared material was analyzed by X-ray diffractometer (XRD; X'pert PRO; PANalytical B.V., Almelo, the Netherlands), and the experimental conditions were: Cu Kα ray, (λ = 0.1542 nm), 2θ ranges from 10° to 90°. The morphology of the samples was characterized by a field emission scanning electron microscope (FESEM, FEI Sirion 200) produced by the Dutch company.

2.3. Photoconductivity Measurement
The test chamber was made of stainless steel, the test environment temperature was 20 °C, the relative humidity was ≦20%, and the ultraviolet light LED source used as the test source. The parameters were: wavelength 365 nm, and the light density is 36 W/m^2.
The whole test process was detected in real time. The test process was as follows: (1) put the device into the test chamber and let dry air pass for 30 min; (2) add 1v bias voltage after 10 s of the test; (3) 20 seconds to start adding light; 4) Stop adding light for 180 seconds; (5) Remove the bias voltage for 350 s. The test flow for all samples lasted 360 s.
The whole test process was detected in real time. The test process was as follows: (1) put the device into the test chamber and let dry air pass for 30 minutes; (2) add 1v bias voltage after 10 seconds of

the test; (3) 20 seconds to start adding light; 4) Stop adding light for 180 seconds; (5) Remove the bias voltage for 350 seconds. The test flow for all samples lasted 360 seconds.

To obtain the *I–V* characteristics, current was continuously measured as function of voltage. In the light irradiation experiments, we used UV LED array (365nm, $36W/m^2$) and the lamp-to-sample distance was 1cm.

3. Results and Discussion

3.1. Characterization

Figure 1 is an SEM image of different samples. It can be seen from the figure that the particle size of zinc oxide particles is 40~60 nm, and the particle size of polyaniline particles is about 500 nm. It can be seen from the SEM image that polyaniline is surrounded by zinc oxide having a small particle size and the dispersion is relatively uniform.

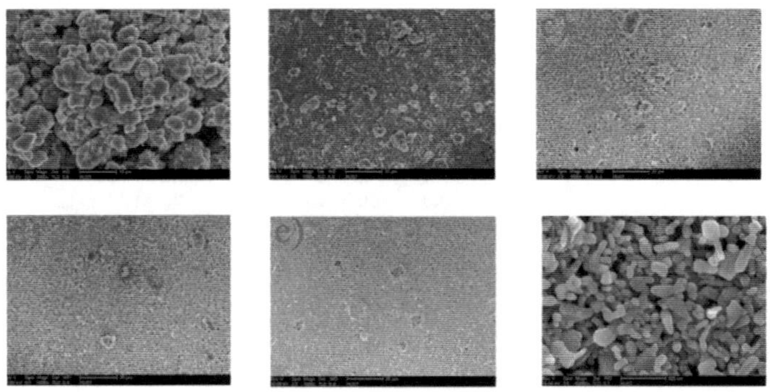

Figure 1. The SEM images of the different samples
(a) PANI (b) ZP050 (c) ZP010 (d) ZP005 (e) ZP001 (f) ZnO

The crystal phases of the different samples were analyzed by X-ray diffraction XRD, and the results are shown in Fig. 2. It can be obtained by analysis that the polyaniline exhibits an amorphous state, and the diffraction peaks of the hexagonal wurtzite structure ZnO appear in other samples, which is consistent with the standard XRD card of ZnO (JCPDS 80-0074). The three diffraction peaks 2θ in the figure are: 31.9°, 34.5°, and 36.4°, respectively corresponding to the (100), (002), and (101) crystal planes of zinc oxide. The intensity of the diffraction peak in the composite increases as the zinc oxide content increases. After XRD analysis of the sample, the crystal structure of zinc oxide was not changed in the composite after doping with polyaniline.

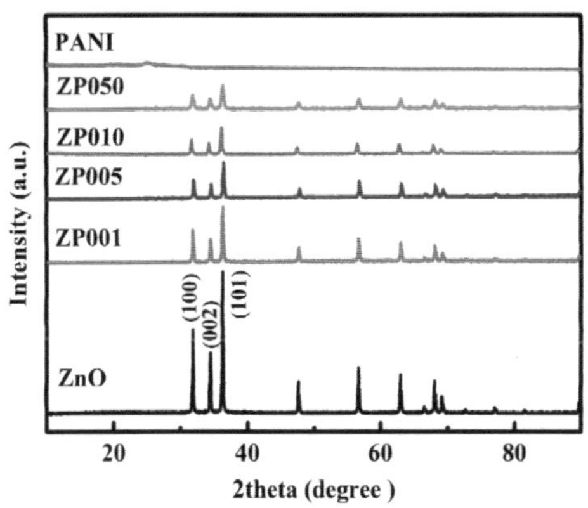

Figure 2. The XRD patterns of the different samples

3.2. Photoconductivity Response of the PANI/ZnO Nanocomposites

Figure 3 is the response of different samples to ultraviolet light at room temperature. It can be seen from the figure that polyaniline has almost no response to ultraviolet light. To better discuss the photoelectric response, the response values are defined as follows:

$$R = (I_{max} - I_0) / I_0 \tag{1}$$

I_0: the initial current before the light
I_{max}: the maximum current after the light

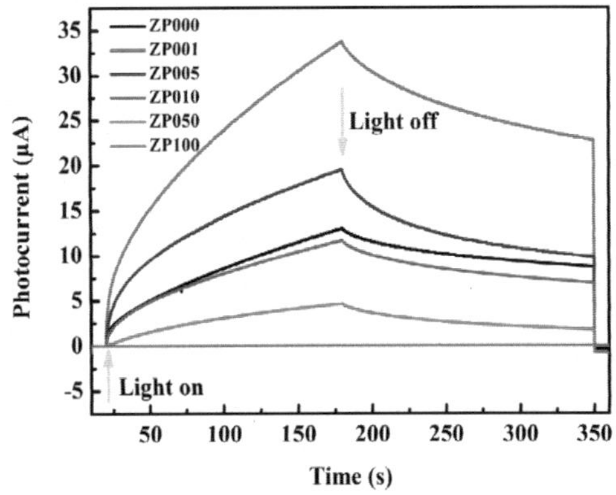

Figure 3. Photoconductivity responses of the PANI/ZnO nanocomposites

Figure 4 is the magnitude of response of different samples to ultraviolet light at room temperature. It can be seen that PANI is not responsive to ultraviolet light at room temperature, while zinc oxide and PANI/ZnO composites are responsive to ultraviolet light at room temperature. When the mass content of PANI was 0%, 1%, 5%, 10% and 50%, the response values of PANI/ZnO composites were 36, 43, 1938, 724, 259 and 1.03, respectively. It can be seen that with the mass of PAN increasing, the response performance has gradually increased. However,continuously increasing the content of PANI , it will result in the decreased response. When the mass content of PANI is 5%, the response reaches the maximum. Therefore, the optimum mass percentage of PANI in the composite is 5%.

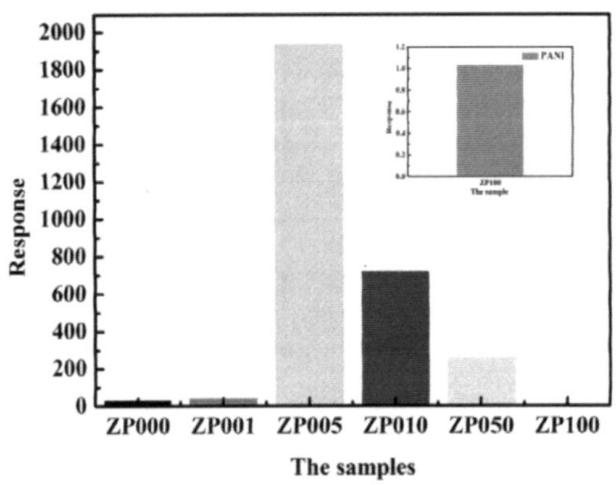

Figure 4. The response of the different samples

Figure 5 illustrates the *I-V* characteristics of the ZP010 nanocomposites in the dark and UV irradiation conditions at room temperature, respectively. The linear curve in the Figure 5 (a) can indicate that the ohmic contact is fromed between the ZnO and PANI (ZP010). The non-linear curves in Figure 5 (b) exhibit almost symmetric characteristic with an extended plateau suggestive of tunneling conduction through the PANI/ZnO nanocomposites. This result confirms that a p-n junction was successfully formed.

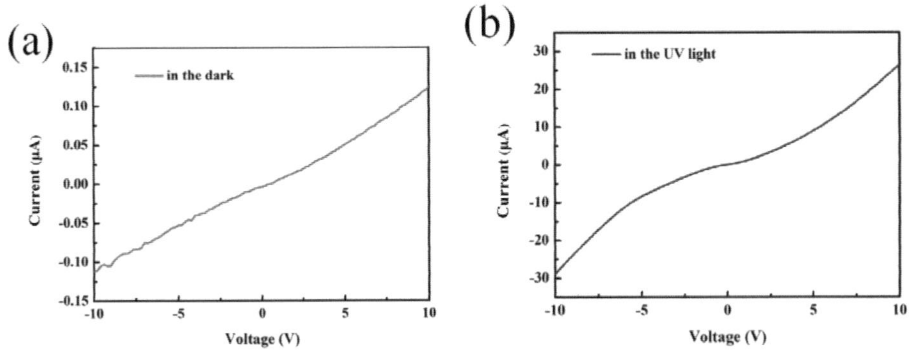

Figure 5. *I-V* characteristics of ZP010
(a) in the dark (b) in the UV light

3.3. Mechanism Analysis

The UV source (365 nm) used in the experiment can excite the electronic transitions in ZnO and PANI to the conduction band (CB) and the lowest occupied orbit (LUMO), while the valence band of ZnO (VB) and PANI are the highest. A positively charged hole appears in the occupied orbit (HOMO). Because of the relative positional relationship between the two, electrons will transition from the lowest occupied orbital of PANI to the conduction band of ZnO, and the hole will jump from the valence band of ZnO to the highest occupied orbit of PANI. [13] PANI as a p-type hole transport material [14] accelerates the transport of holes from the valence band of ZnO. In addition, oxygen molecules adsorbed on the surface of ZnO are very important for photoelectric reaction. Oxygen molecules trap electrons on the ZnO conduction band to form oxygen anions. [15] During the light phase, oxygen anions will react with photogenerated holes. The adsorbed oxygen is released while reducing the barrier of the material. The reduction of the interface barrier facilitates the migration separation of carriers. (Response mechanism diagram is shown in Figure 6.

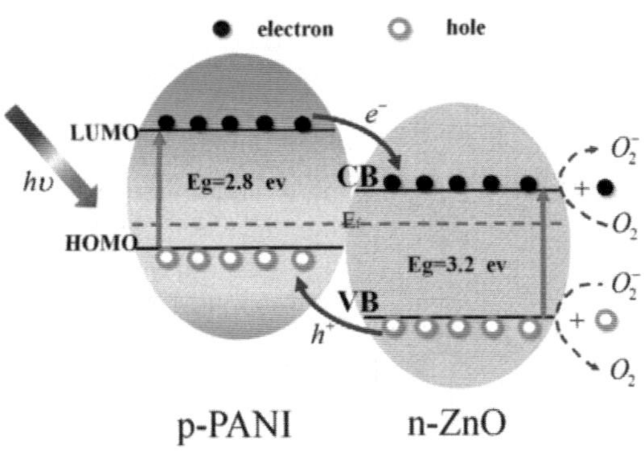

Figure 6. The Band Diagram and Mechanism

When the mass content of PANI is less than 5%, it leads to increase the p-n heterojunction with the increasing the content of PANI. And it is contributed to separate the carriers. When the mass content of PANI is 5%, the response reaches the maximum response, and if the content of PANI continues to increase, the response of the composites to the UV light is reduced. With increasing the content of PANI, it will increase the specific surface area and interface density. And that it is lead to increase the recombination probability of photogenerated electron-holes. Therefore, the excessively high amount of PANI doping will be detrimental to the response of UV light.

4. Conclusion

This paper introduces a simple method fior preparing PANI-doped ZnO composites, and tests the response of PANI-doped ZnO composites with different doping ratios to ultraviolet light at room temperature. When the mass percentage of PANI is 5%, the response of the composites to UV light reaches the maximum. Based on the results, it can be attributed to two factors. One: the interface between p-type PANI and n-type ZnO forms a p-n heterojunction, which contributes to the separation of photogenerated hole-electron pairs. Secondly, the dispersed PANI can inhibit the formation of the interfacial recombination center and accelerate the transport of holes from the ZnO valence band, so that the photocurrent of the composite material is significantly improved. Combining experimental results with mechanism studies, we hope to help develop more p-n heterojunction composites that have higher response to UV light at room temperature.

References

[1] L. Li, C. R. Zhang, Appl. Surf. Sci. 257, 3 (2010)
[2] Adikaari, D. Dissanayake, et al, *IEEE* J. Sel. Top. Quant. Elec. **16**, 6 (2010)
[3] H. Li, J. Gong, et al, Synth. Met. **160**, 5-6 (2010)
[4] O. Game, U. Singh, et al, Nanoscale **6**, 1 (2014)
[5] W. W. Wu, S. Bai, et al, Sci. Adv. Mat. **2**, 3 (2010)
[6] J. H. Jun, K. Cho, et al, Org. Electron. **9**, 4 (2008)
[7] S. Ameen, M. S. Akhtar, et al, Superlattices Microstruct. **46**, 6 (2009)
[8] H. Zhang, R. Zong, et al, J. Phys.Chem. C **113**, 11 (2009)
[9] S. Mridha and D. Basak, Appl. Phys. Lett. **92**, (2008)
[10] Y. J. Lin, C. L. Tsai, et al, Appl. Phys. Lett. **100**, 25 (2012)
[11] J. H. Lin, J. J. Zeng, et al, Appl. Phys. Lett. **100**, 15 (2012)
[12] J. Wu, H. Y. Li, et al, J. Appl. Phys. **110,** (2011)
[13] P. A. C. Quist, W. J. E. Beek, et al, J. Phys. Chem. B 110, (2006)
[14] Y. Xiao, G. Han, et al, J. Power Sources **267**, (2014)
[15] C. S. Lao, M. C. Park, et al, J. American Chem. Soc. **129**, (2007)

Influence of Sintering Temperature on Microhardness and Tribological Properties of Equi-Atomic Ti-Al- Mo-Si-W Multicomponent Alloy

L. Rudolf. Kanyane*, A. Patricia Popoola, and Nicholus Malatji

Department of Chemical, Metallurgical and Materials Engineering, Tshwane University of Technology, P.M.B. X680, Pretoria, South Africa

corresponding author's e-mail: lrkanyane@gmail.com, malatjin@tut.ac.za.

Abstract. This work aim to investigate the effect of sintering temperature on microhardness and tribological properties of novel equi-atomic TiAlMoSiW HEA fabricated via spark plasma sintering. The influence of Spark plasma sintering temperature on morphological evolution and phase formation was also investigated. The microstructure and the phases formed for the developed HEA were examined using scanning electron microscopy (SEM) and X-ray diffractometry (XRD) respectively. The microhardness and tribological properties were studied using a diamond base microhardness tester Rtec tribometer. It was noticed that sintering temperature has an effect on microhardness and tribological properties. Phase analysis of the samples displayed that the alloy exhibited a bcc matrix with secondary phase precipitate of ordered fcc $TiSi_2$ phase. The developed HEA showed improved mechanical properties as the sintering temperature increases.

1. Introduction

Due to industrial development, advanced engineering materials with good strength to weight ratio and wear resistant materials are needed [1]. HEAs are promising multi principal element alloys with good mechanical properties for advanced engineering application [2-5]. Today, HEAs remain the promising multi principal element alloys with combination of outstanding chemical and mechanical properties for advanced engineering application at both room and elevated temperatures [6-8]. Numerous methods are followed by many authors in fabricating these class of alloys. Other authors used techniques such as spark plasma sintering, sputtering, additive manufacturing (LAM) and casting until forming hard and advanced coatings using laser cladding techniques and others [7]. However, in these work spark plasma sintering (SPS) is used as a fabrication technique. SPS has an ability to consolidate a material which possess good mechanical properties. These is achieved by controlling the grains size of the developed alloy as compared to other conventional methods [9, 10].

Chuang et al, [11] investigated the wear resistant of $Co_{1.5}CrFeNi_{1.5}Ti$ and $Al_{0.2}Co_{1.5}CrFeNi_{1.5}Ti$ high entropy alloys. Both alloys showed outstanding wear properties at room temperature. The developed alloys showed better wear properties as compared to conventional wear resistant steels. In many investigated tribological studies, abrasive conditions and adhesive conditions are used to conduct tests for wear behaviour of HEAs. Wu et al. [8] studied the effect of alloying composition ratio on wear behaviour of $Al_xCoCrCuFeNi$ HEAs. The author discovered that higher Al content in the fabricated

IOP Conf. Series: Materials Science and Engineering **538** (2019) 012009 doi:10.1088/1757-899X/538/1/012009

alloy results in worn surface been smooth and it also yields fine debris with a high oxygen content, which also results in a large improvement in wear resistance of the developed alloy.

Therefore, in this study, an attempt was made to fabricate Ti-Al-Mo-Si-W HEAs under different sintering temperature. Therefore, in depth characterization of the synthesized HEAs was undertaken to understand the effect of sintering temperature on microhardness and tribological properties, microstructures and formation of phases of the developed HEAs.

2. Methodology

Ti-Al-Mo-Si-W equi atomic HEA powders were mixed mixer and poured into a graphite die of 40mm diameter, the powder were sintered using an SPS system (FCT Systeme GmbH Rauenstein, Gewerbepark 96528 Frankenblick). Temperatures of 800, 900 and 1000°C at a heating rate of 100°C/min with a holding time of 8min were used as sintering parameters. A constant pressure of 50 MPa was applied during the sintering process. The fabricated HEAs samples were cut transversely, mounted in an epoxy- based resin, polished using metallography procedures and etched with kroll agent solution for investigating the microstructural evolution of the developed alloys. The microstructures of sintered and worn Ti-Al-Mo-Si-W was characterized by means of SEM. The phases present in the synthesized high entropy alloys were identified using an x-ray diffractometer and present phases were revealed using X-Pert High Score Plus software. Vickers microhardness (HVN) measurements of the fabricated samples was obtained using an Emco TEST microhardness tester. The indenting load of 100kgf and a dwell time of 10s was used. The fabricated HEAs samples surfaces were indented randomly five different positions and the average was recorded. Tribological studies were examined using Rtech tribometer with a stainless steel contact ball. The applied force was 150N for a duration of 5min.

3. Results and Discussion

Table 1 present the pattern used to sintered the multicomponent HEAs

Table 1. Summarized pattern of fabricated TiAlMoSiW HEAs.

Alloy	Temperature (°C)	Pressure (kPa)	Holding Time (min)	Heating Rate (°C/min)
Ti-Al-Mo-Si-W	[1A] 800 [1B] 900 [1C] 1000	50	8	100

SEM and XRD Analysis

SEM micrographs of the fabricated Ti-Al-Mo-Si-W high entropy alloy at 800 °C, 900 °C and 1000 °C were taken and recorded in Figure 1. The reported set of samples were sintered at 50MPa pressure and sintering holding time of 8 min each. For each sintering temperature effect, the SEM results of the HEAs surface has no pores or crack for the samples sintered at all temperatures. However, the micrograph revels minimal cavities were 1000 °C has more cavities as compared to samples sintered at 800°C. Due to low melting point of Al, the cavities could be due to the burning off Al. The alloys display morphology of different phases with gray, white and small black precipitated spots contrast phases present. From the three distinctive areas are visible in all micrographs present in Figure 1a-c, which are irregular bulk areas of gray, white phases and irregular black hole areas can be observed. The gray and white phases are possibly the main phase in the alloy due to the fact that they account for a higher volume fraction on the surface.

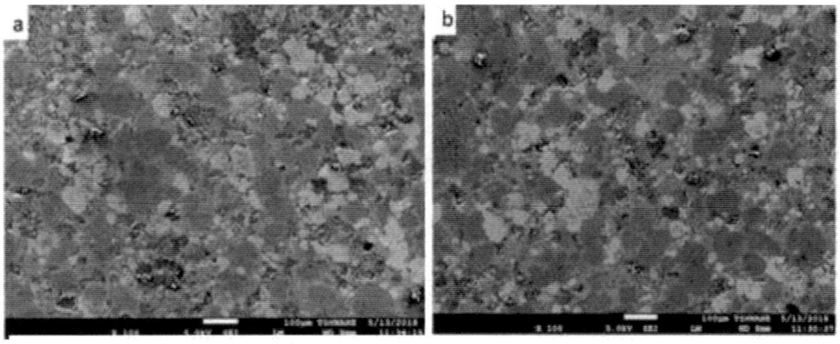

Figure 1. SEM image of the sintered HEA a) 800oC and b) 1000oC

Table 2. The calculated parameters (ΔS_{mix} ($JK^{-1}mol^{-1}$), ΔH_{mix} (KJ/mol), T_m (K) δ (%) and Ω for the investigated alloys.

Alloy	ΔH_{mix}	ΔS_{mix}	T_m	δ & Ω
Ti-Al-Mo-Si-W	-24.32	13.38	2230	7.97 &1.22

Table 2 presents the empirical values of mixing entropy and enthalpy of Ti-Al-Mo-Si-W high entropy alloy. It is anticipated that HEAs forms a solid solution phase (BCC/FCC) when the defined parameters are $\Omega \geq 1.1$ and δ (%) $\leq 6.6\%$ [13-15]. Therefore, from the calculated results, the defined parameters of Ω and δ (%) for the TiAlSiMoW high entropy alloy system was found to be 1.22 and 7.79 % respectively. It is with no doubt that Ω parameter matches the proposed value while δ(%) defined parameter breaks the simple crystalline structure formation rules for the multi-principle HEAs proposed by Zhang et al. [15]. Intermetallics or complex phases are expected to be found in the alloy system as confirmed by the XRD results with intermetallic of Titanium silicide present in the alloy from figure 2 below. Figure 2 present the phases present in the developed Ti-Al-Mo-Si-W sintered at 800, 900 and 1000°C. The XRD shows that the bulk Ti-Al-Mo-Si-W HEAs consists of combination of BCC and FCC phase with the present of $TiSi_2$. The combination of BCC and FCC in the alloy is expected to give better mechanical properties. The developed compound was due to high entropy of mixing of the binary components (Ti, Si). The compound is also known for superior properties such as high strength, high melting point, low density, outstanding oxidation resistance and these properties gives the compound a potential aero-space engineering material [10].The combination of both BCC and FCC in the alloy system is expected to give better mechanical properties.

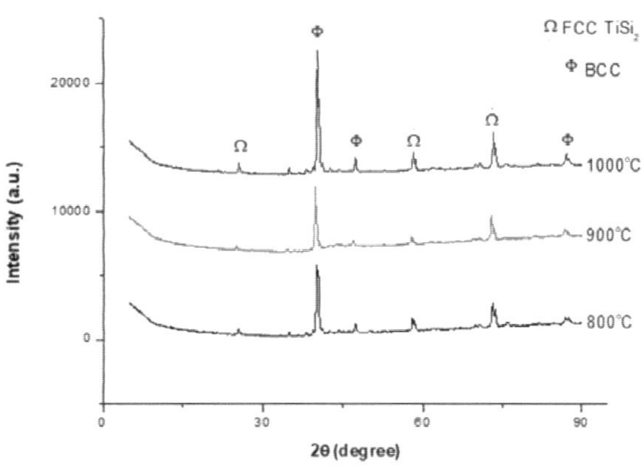

Figure 2. XRD patterns of sintered Ti-Al-Mo-Si-W high entropy alloys

Microhardness and CoF Results

Figure 3 displays the microhardness and CoF properties of the developed high entropy alloy of Ti-Al-Mo-Si-W sintered at different temperatures. The relationship between microhardness and CoF can not be ignored. It is evident that as the sintering temperature increases, the microhardness of the fabricated Ti-Al-Mo-Si-W HEA increases while the CoF decreases. An increase in microhardness values is ascribed to particle to particle rearrangement and uniform distribution of present phases in the alloy [15]. These homogenous distribution of secondary phase precipitates on the grain broundaries which tends to restricts the grains growth and hence it empedes dislocations motion leading to improved microhardness values. The obtained results are inline with Makena et al. [16], Otto et al. [17] and Shongwe et al.[18]. Maximum microhardness of 802.93 $HV_{0.1}$ was evident at sintering temperature of 1000°C with minimum CoF of 0.0916μ.

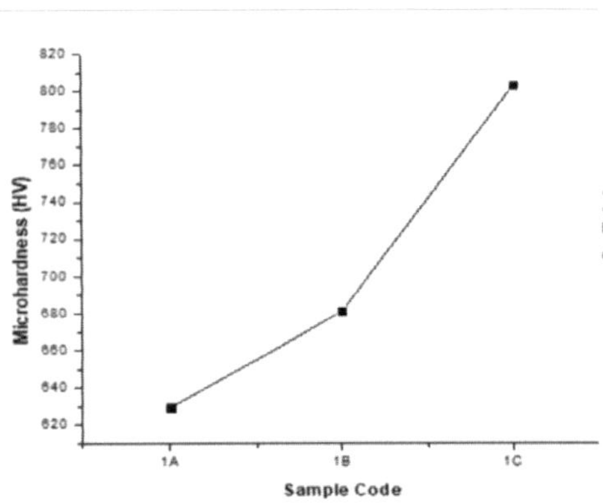

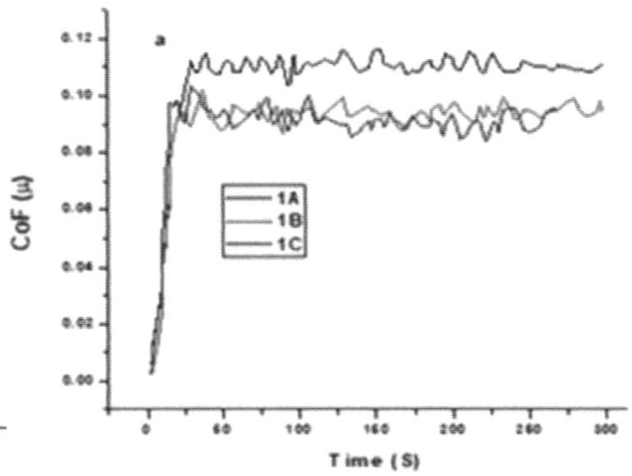

Figure 3. Microhardness and CoF results of the sintered HEA

SEM Analysis of the Worn Surface

The SEM images in figure 4 illustrate the effect of sintering temperature on the wear properties of the developed equi-atomic TiAlMoSiW HEAs. The SEM image shows material carry overs of the HEAs due to minimal plastic deformation occurs in the direction of the sliding movement. For a high load of 150N form the contact ball, the developed HEAs exhibit good wear resistance as there are no defaults on the surface. Good microhardness properties of the synthesized HEAs could be responsible for the outstanding wear properties. Also the combination of both BCC and FCC in the alloy system which was found in the Xrd results proved to have better mechanical properties. All the fabricated HEAs resulted in a smooth surface which displays no plastic deformation by the contact ball The high density, high microhardness and high amount of W could be the reason for minimal wear loss with no evident of furrows tracks as a result of plastic deformation. Only material cracks be detected in specimen 4a and 4c. The SEM image at all sintering temperatures also displays minimal cracks on the surface after wear. The surface shows no degree of deformation which resulted to outstanding wear resistance properties.

Conclusion

TiAlMoSiW HEAs were successfully consolidated via spark plasma sintering and the effect of sintering temperature on the microstructural evolution, microhardness and tribology properties of the synthesized HEAs were investigated. Generally, the synthesized HEAs showed outstanding mechanical properties.

- Maximum microhardness of 802.09HV was achieved at a sintering temperature of 1000°C.
- SEM of the worn samples presented stable surface with minimal deformation

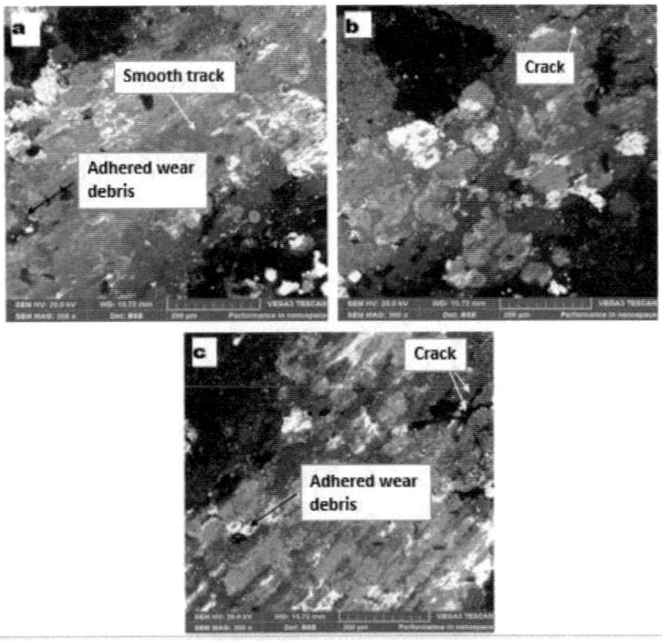

Figure 4. SEM micrographs of wear tested Equi-Atomic TiAlMoSiW high entropy alloy at a) 800 °C b) 900°C c) 1000°C

References

[1] Svensson, D.O., *High Entropy Alloys: Breakthrough Materials for Aero Engine Applications?* 2014, Master's Thesis, Chalmers University of Technology, Sweden.

[2] Haase, C. and L.A. Barrales-Mora, *Influence of deformation and annealing twinning on the microstructure and texture evolution of face-centered cubic high-entropy alloys.* Acta Materialia, 2018. **150**: p. 88-103.

[3] Kumar, D., et al., *Understanding the Effect of Tungsten on Corrosion Behavior of AlCuCrFeMnW x High-Entropy Alloys in 3.5 wt.% NaCl Solution.* Journal of Materials Engineering and Performance, 2018. **27**(9): p. 4481-4488.

[4] Li, M., et al., *Evaluation of microstructure and mechanical property variations in Al x CoCrFeNi high entropy alloys produced by a high-throughput laser deposition method.* Intermetallics, 2018. **95**: p. 110-118.

[5] Waseem, O.A., et al., *A combinatorial approach for the synthesis and analysis of Al x Cr y Mo z NbTiZr high-entropy alloys:* Oxidation *behavior.* Journal of Materials Research, 2018. **33**(19): p. 3226-3234.

[6] Mohanty, S., et al., *Powder metallurgical processing of equiatomic AlCoCrFeNi high entropy alloy: microstructure and* mechanical *properties.* Materials Science and Engineering: A, 2017. **679**: p. 299-313.

[7] Shi, Y., et al., *Corrosion of Al xCoCrFeNi high-entropy alloys: Al-content and potential scan-rate dependent pitting behavior.* Corrosion Science, 2017. **119**: p. 33-45.

[8] Xiao, D., et al., *Microstructure, mechanical and corrosion behaviors of AlCoCuFeNi-(Cr, Ti) high entropy alloys.* Materials & Design, 2017. **116**: p. 438-447.

[9] Sivaprahasam, D., S. Chandrasekar, and R. Sundaresan, *Microstructure and mechanical properties of nanocrystalline WC–12Co consolidated by spark plasma sintering.* International Journal of Refractory Metals and Hard Materials, 2007. **25**(2): p. 144-152.

[10] Borkar, T. and R. Banerjee, *Influence of spark plasma sintering (SPS) processing parameters on microstructure and mechanical properties of nickel.* Materials Science and Engineering: A,

2014. 618: p. 176-181.
[11] Chen, W., et al., *Alloying behavior, microstructure and mechanical properties in a FeNiCrCo0. 3Al0. 7 high entropy alloy.* Materials & Design, 2013. **51**: p. 854-860.
[12] Ige, O., et al., *Erosion-corrosion characteristics of spark plasma sintered pure nickel in simulated mine water.* Tribology International, 2017. **109**: p. 441-446.
[13] Zhang, Y., et al., *Solid - solution phase formation rules for multi - component alloys.* Advanced Engineering Materials, 2008. **10**(6): p. 534-538.
[14] Sheng, G. and C.T. Liu, *Phase stability in high entropy alloys: formation of solid-solution phase or amorphous phase.* Progress in Natural Science: Materials International, 2011. **21**(6): p. 433-446.
[15] Zhang, Y., et al., *Guidelines in predicting phase formation of high-entropy alloys.* Mrs Communications, 2014. **4**(2): p. 57-62.
[16] Makena, M.I., et al., *Effect of sintering parameters on densification, corrosion and wear behaviour of Ni-50Fe alloy prepared by spark plasma sintering.* Journal of Alloys and Compounds, 2017. **699**: p. 1166-1179.
[17] Otto, F., et al., *The influences of temperature and microstructure on the tensile properties of a CoCrFeMnNi high-entropy alloy.* Acta Materialia, 2013. **61**(15): p. 5743-5755.
[18] Shongwe, M.B., et al., *Effect of sintering temperature on the microstructure and mechanical properties of Fe–30% Ni alloys produced by spark plasma sintering.* Journal of Alloys and Compounds, 2015. **649**: p. 824-832.

IOP Conf. Series: Materials Science and Engineering **538** (2019) 012010 doi:10.1088/1757-899X/538/1/012010

Polyvinyl fluoride (PVF); Its Properties, Applications, and Manufacturing Prospects

M. H. Alaaeddin[1,*], S. M. Sapuan[1,*], M.Y.M Zuhri[1], E.S Zainudin[1,2], Faris M. AL- Oqla[3]

[1]Department of Mechanical and Manufacturing Engineering, Universiti Putra Malaysia, 43400 UPM Serdang, Selangor, Malaysia.
[2]Laboratory of Biocomposite Technology, Institute of Tropical Forestry and Forest Products, Universiti Putra Malaysia, 43400 UPM Serdang, Selangor, Malaysia
[3]Department of Mechanical Engineering, Faculty of Engineering, The Hashemite University, 13133 Zarqa, Jordan

E-mails: m.h.alaaeddin@gmail.com; sapuan@upm.edu.my +6 01139395070

Abstract. Polyvinyl fluoride (PVF) is a distinctive polymer with excessively studied properties. The wide utilization of this polymer has dramatically escalated due to its outstanding properties (e.g., mechanical strength, chemical resistance, high stability, low permeability, indoor functionality, outdoor durability, light transmissibility, adherence, stain resistance, sunlight degradation, and high temperature endurance). An overview of the chemical, physical, electrical, and thermal properties of PVF is provided. The current applications of PVF/Tedlar®, its polymerization, and the manufacturing processes are also addressed in this work. A comparative analysis has been conducted to explore common and uncommon properties of PVF, fluoropolymers, and competing polymers such as polyvinylidene fluoride (PVDF), polytetrafluoroethylene (PTFE), and polyvinyl chloride (PVC). Overall, PVF is anticipated to have further promising manufacturing prospects in numerous industries and in various fields such as in medical equipment, renewable energy applications, sensitive devices, advanced industrial platforms, and nanotechnology material composites. Therefore, further in-depth analyses and oriented numerical information on the performance of this polymer are required.

1. Introduction

Polyvinyl fluoride is considered to be an exclusive polymer or special plastic employed in numerous industries and manufacturing processes since 1962. It is therefore an important type of polymer due to its excellent properties and high performance; e.g. chemical resistance, stain resistance, outdoor durability, high stability, and adherence [1]. In recent years, different studies have been conducted to scrutinize the wide range of properties of PVF. It is considered as an ideal material with unique electrical, chemical, and mechanical properties [2]. PVF is mainly designed for surface protection due to its toughness, lightness, and high resistivity to moisture, weathering, and ultraviolet rays. It is also inert to chemicals, staining agents, and solvents [1-3]. The functional properties of PVF were investigated in numerous studies which had asserted unswerving facts on its conductivity odds and ferroelectricity. These attributes continuously reveal vital characterizations for the enthalpies and dielectric constant measurements of this polymer and its composites [4]. PVF is applied in a form of films under multiple names: i.e., Teslar®, Tedlar®, and R films. However, PVF films or Tedlar® are the two most common names for polyvinyl fluoride films [1,5]. The high resistivity and excellent properties of PVF (such as resistance to sunlight degradation and solvents, chemical inertness, high tensile strength, toughness, and dimensional stability) makes it competent for specific functions [3,6]. Initially, PVF films were mainly used for very specific applications such as in PV encapsulation processes, pigmented films for house sidings, and aircraft interiors. This then expanded to printed circuit boards, industrial sidings, roofing films, clear films for skylights with ultraviolet absorber sets, and similar applications [5-7].

2. PVF Structure and Properties

As a fluoropolymer, polyvinyl fluoride (PVF) is a thermoplastic polymer with a chemical formula of (C2H3F)n. The polymer consists of a repeating VF unit and is similar in structure to other compounds (e.g. polyvinylidene fluoride PVDF, polyvinyl chloride PVC, polyethylene PE, and polytetrafluoroethylene PTFE) [8,9]. Fluoropolymers are considered to be homopolymers and copolymers; the fluorinated fluoropolymers contain carbon (C), fluorine (F), hydrogen (H), or other atoms such as Chlorine (CI) and vinyl fluoride (VF). H2C⋍ F undergoes free-radical polymerization leading to C–F and C–H bonds since F is usually the highest electronegative from all elements; it is easy to convert to F–. Likewise, the C–F bond strength is higher than the C–H [10,11]. H is smaller than F and F is smaller than CI [10]. The distinctive structure of PVF justifies the reason behind its good properties, making it an ideal release for fiber reinforced polymers (FRP). It has the ability to meet the expected long range needs for composites, manufacturing operations, and other specific application requirements [10,11]. Fig. 1 provides the ball-and-stick model for the PVF, PVDF, PTFE, PE, and PVC chemical structures.

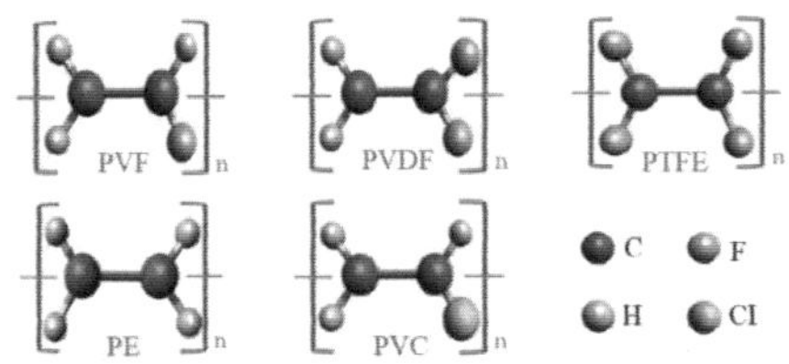

Figure 1. Ball-and-stick model for the structure of referred polymers

PVF properties vary according to manufacturing purposes, but they remain close to original properties; e.g. the commercial one is atactic and the melting point peaks at 190°C. Several transitions are anticipated below the melting temperature in PVF. The polymer is semicrystalline with varied degrees of crystallinity (from 20% to 60%) and a planar zigzag conformation. The solubility in all solvents below 100°C is very low and inert to chemical treatments. PVF is considered to be visible, near-infrared regions, and transparent to radiation in UV [12].

3. Polymerization and Manufacturing of VF

Both the polymerization process and thermal performance in VF significantly contribute to the degree of crystallinity in this polymer. The existing relationship between the polymer structure and the degree of crystallinity and melting point justifies the reason behind the variations found in the melting points, affirming that the head-to-head linkage affects the degree of crystallinity, ranging from 20% to 60%; whereas a 12% head-to-head linkage is anticipated in commercial PVF with an estimated melting point of 190°C [12]. In the polymerization procedure, there are various methods and techniques used to process VF; e.g. suspension polymerization, bulk polymerization, emulsion polymerization, radiation-induced polymerization, and graft polymerization [11,13]. VF is subjected to free-radical polymerization where the thermal initiator is the starter of the polymerization reaction, followed by propagation and termination. With a wide variety of initiators and different polymerization methods, benzoyl peroxide is used to initiate the polymerization process to produce oriented PVF films [13,14]. The initial polymerization includes a saturated solution of VF heated in toluene at 67°C under 600 MPa for 16 hours [12,13]. A combination of oxygen and triisobutylborane can be used to polymerize VF at less pressure and temperature. The temperature in the polymerization process is anticipated to range between 0 to 85°C. The melting point notoriously drops from 230°C at 0°C polymerization to 200°C at 85°C polymerization, considering the degree of crystallinity and melting temperature. This occurs due to existing divergences of monomer repercussions during the polymerization process. VF volatility insists on interfering in moderate high pressure in the polymerization process; whereby, free-radical initiators are incarnated by the photopolymerization process [13]. However, it is worth mentioning that the copolymers and terpolymers of VF can be prepared with a wide range of monomers; in the last few years, the interpolymers of VF have been classified as highly fluorinated monomers (e.g. perfluoroethylvinylether, hexafluoropropylene, and perfluorobutylethylene) [7,13]. As a result, the yielded polyvinyl fluoride demonstrates high softening temperatures and finite solubility, making it a unique polymer with special characteristics. This opened a major market for its application, leading to the development of the polymerization process and the designed commercialized polymerization of VF. The table below provides information on the different types of polymerizations and techniques [12,15].

Table 1. Different types of polymerizations, techniques, adjutants, and initiators [12,13]

Polymerization	Technique	Adjutant	Initiator
Suspension polymerization	VF is suspended in water	Dispersion stabilizer	Organic peroxide
Bulk polymerization	Irradiation with UV	Di-tert-butyl peroxide	
Emulsion polymerization	Emulsion method (highly reduced pressure and lower temperature)	Emulsifiers	Aqueous emulsion
Radiation-Induced polymerization	Radiation (decomposes into free radical)	Gas-phase	Benzyol, lauroyl
Graft polymerization	Radiation polymerization	Grafting with monomers	γ-rays

In the manufacturing process of PVF, converting vinyl fluoride (VF) into polyvinyl fluoride (PVF) is the main process and intended result. At the beginning of the process, VF must be stored in high-pressure storage tanks to be later pumped into a reactor unit. The reacting function should be performed to create PVF slurry in water with unreacted VF. The water then flows into separators to remove the VF. The separators pump the VF using compressors and return it back to the reactor. The procedure is followed by a steam injection process to raise the temperature of PVF and evaporate what remains from VF in the slurry. Subsequently, the heated admixture flows into a small slurry flash tank and there, the residual VF must be released into the atmosphere. The incombustible PVF slurry then flows into a designed insulated tank and out of three tanks for slurry holding. The designed functions of these three tanks form an integrative complementary method; if one reaches a high level, the hot slurry flows into the other. The seal loop is designed in the second slurry tank in order to prevent VF from evaporating as steam when passing from the flash tank into the slurry tank. Once the process is complete, the PVF slurry is impelled from the slurry tanks to a PVF film manufacturing area to produce Tedlar® and undergo further processing [13].

4. PVF Applications and Composite Films

Polyvinyl fluoride offers such tremendous advantages for several applications which can be used in various modern fields and manufacturing platforms. It is an idealistic film for fiber-reinforced polymer (FRP) and carbon-fiber-reinforced polymer (CFRP). It sufficiently works to protect materials under certain conditions and performs well under high pressures and temperatures [16,17]. PVF is used as an encapsulant for the front and back cover of photovoltaic applications and is embedded within the assembly process of solar panels. In 1960s, PVF films were first employed in aircraft interiors, cabin parts, and electronic industries [17,18]. With good capability to withstand various temperatures (up to 204°C), PVF films are useful for numerous aircraft interior parts and are highly recommended for hand lay-up and vacuum bagging, as well as for other related applications. They are available in different colors and forms; whether adhesive materials that function on one side or other bonding materials for laminates. PVF eliminates the production cost and reduces the need for additional processing and other

IOP Conf. Series: Materials Science and Engineering **538** (2019) 012010 doi:10.1088/1757-899X/538/1/012010

finishing materials due to its durability and favorable appearance. It is also available in the forms of oriented and unoriented films; the oriented films are made using an extrusion method of Polyvinyl fluoride. In the extrusion process, PVF must be dispersed in a latent solvent and trailed by both transverse directions and biaxial orientations to enhance the mechanical properties of the anticipated films. As for unoriented films, they are more compliant and formable than the oriented ones. Unoriented films are slightly stretched after being cast onto a carrier web. They have a higher elongation at break with lower tensile strength. Nonetheless, PVF adhesives have a wide range of applications which can be applied for coating and lamination substrates (e.g. vinyl fabric, aluminum, galvanized steel, thermoplastic, vinyl wall covering, and cellulosic substrates). The coated polyvinyl fluoride is also useful for different functioning materials in metals, wood, and plastics. It works as a reliable laminated film for substrates. The excellent properties and high specifications of PVF makes it highly recommended for critical applications; this spurred its use in indoor and outdoor applications, either for decorative purposes, protective, or both. In the PV manufacturing process, PVF composite films represent the main materials for the encapsulation process, which must be applied to ensure the safety and reliability of PV modules for long-term function and lasting efficiency [12-15,17,19].

5. PVF and Competing Polymers

Polymers differ from each other through small or large variances that affect their structure. Some polymers slightly differ in their composition or chemical structure from polyvinyl fluoride (e.g. polyvinylidene fluoride, polytetrafluoroethylene, and polyvinyl chloride (not crystalline), while other polymers from the fluoropolymers family consist of semicrystalline materials or homopolymers. These polymers share special characteristics with polyvinyl fluoride in some chemical, mechanical, thermal, or physical properties. They may have independencies or share some basic assumptions in their piezoelectric and pyroelectric properties [20-22].

5.1. Polyvinylidene Fluoride (PVDF)

PVDF is one of the important pyroelectric polymers that could contribute to the concept of heat-generated electricity with a wide range of implementations; from industrial to biomedical utilizations . The pyroelectricity of PVDF can be enhanced using a certain inclusion that is accurately balanced after determining its properties; however, not all additives or inclusions can enhance the pyroelectricity of PVDF. The enhancement process can be increased by the intension of pyroelectric coefficient due to the addition of nonferroelectric inclusion under certain conditions, with the observation of the electric dipole moment [23]. PVDF shares competing characteristics in weather resistance. It does improve the coating quality and adhesion, and it can also be used for the preparation of superhydrophobic membranes and coatings. Besides the high heat resistance properties, PVDF possesses superior mechanical and chemical features which justify its extensive use in the preparation of micro and ultra-filtration membranes [24,25]. PVDF is further used in the fabricating process of electrospun nanofibrous microfiltration-membranes (ENMs), and is a suitable candidate for downstream bioprocessing, bioseparation and biomedical applications. It is a reliable polymer for pH-responsive membranes, water treatment, and purification systems. Nevertheless, the use of this polymer can vary in different directions and applications [26]. Its favorable characteristics such as its mechanical properties, chemical stabilities, high dielectric permittivity, and weatherability represent a key role in determining the methods that can be employed in various optimization conditions for each process. The crystals of PVDF consist of five different polymorphs which come in the form of α, β, γ, δ, and ε [27,28]. However, the crystallization of altered PVDF polymorphs is essential in order to analyze the anticipated changes in its properties upon each process, despite the degree of crystallinity (DOC) or the molecular chain orientation (MCO) which can be evaluated differently [29].

5.2. Polytetrafluoroethylene (PTFE)

Teflon® or PTFE was first discovered in 1938 [11]. It is a fluoropolymer, the fluorocarbon-based chemical formula of (CF2-CF2)n, as illustrated in Fig. 1. The polymer has low or anti-friction properties, naturally hydrophobic, and non-biodegradable. It is easy to fabricate into a wide range of forms for various applications and industries (e.g. in the healthcare industry, it is used in the fabrication of nerve regeneration tubes, vascular prostheses, maxillofacial surgeries, subcutaneous augmentation materials, and other applications) [30,31]. Furthermore, PTFE is well-known for its good properties and considered as an efficient internal lubricant. The processed and modified features produce a friction film between sliding components to reduce adhesion, whereby the composite is normally treated in a melt mixing process [32]. PTFE thin films with high molecular weight, high melting point of up to 327°C, stability of up to 260°C, elevated chemical and biological inertness, and increased resistance to corrosive solvents and concentrated acids have promising benefits and competitive properties that could be employed in distinct applications [33]. It is one of the most important polymers with a wide range of extensive applications in different sciences and engineering fields. In particular, PTFE with sharp reversible transition at 600 K, changing from the original white substance to a transparent one [34], has a decreased coefficient of friction (CoF). Its low CoF ranges between (0.05-0.08, static) [35]. The C–F bone in PTFE is the strongest bond; it provides the polymer with steady chemical, physical, and thermal properties [11]. These properties lead to the use of PTFE in several important applications for reactive and corrosive chemicals. To alter the surface of PTFE without changing the properties of bulk, the plasma modification technique or the NH3 and H2 plasma treatment is recommended for several reasons: it is considerably easier to control compared to other complex and hazardous techniques, it has a lower processing cost compared to the electron beam and other similar techniques, and more importantly, it is environmentally friendly with a biofunctionalization treatment of the surface. It has much less impact on the environment compared to other techniques [35].

5.3. Polyvinyl Chloride (PVC)

Polyvinyl chloride (PVC) is anticipated to be among the most widely utilized thermoplastic polymer in the world [36]. It displays acceptable mechanical properties, adequate chemical resistance, outstanding process ability, and thermal stability with vast and diversified manufacturing tasks and applications [37]. It was asserted that PVC properties significantly contribute to numerous industrial applications and a wide range of composites. However, there is a need to improve these properties (the mechanical ones in particular). The polymer has shown a positive response to recent improvements accomplished; for example, PVC's mechanical properties as a function of graphene oxide loading increase both the Young's modulus and tensile strength. [38]. In addition, PVC membrane properties improved by the introduction of amino groups, further providing active chemical sites to its surface. In composite materials (CMs), PVC has a strong relationship with different types of materials and natural fiber composites (NFCs). It is a highly recommended polymer for reinforcement and blending processes; it shows adequate portability and high compactness when integrated with other materials. However, due to the massive production of PVC and increasing methods of fabrication, widespread debate has been raised to find alternatives in which a reduction in the use of this non-biodegradable polymer can occur [39]. Currently, nanotechnology studies are focusing on the electric and dielectric characterizations of PVC which have reflected on industrial practices, nanocomposite materials, and polymer nanocomposites such as: PVC/CdSE, PVC/ZnS, and PVC/PVDF/ZnO. This opens further prospects for this polymer in nanocomposite technologies [40].

6. Conclusions

PVF has distinguished properties and exceptional performance under harsh conditions. PVF can last in high temperatures and in volatile weather conditions with excellent resistance to sunlight degradation and high solar transmittance rate. In addition, the chemically inert polymer is highly effective in resisting solvents, chemicals, corrosion, and water absorption. Therefore, these important properties have designated this polymer to be utilized in numerous industrial applications; e.g. indoor and outdoor coatings, encapsulation processes, as well as decorative and functional uses. The polymer is meant to be applied where chemical resistance, mechanical strength, stain resistance, release properties, outdoor durability, thermal stability, low permeability, and adherence are required. Despite the complex

IOP Conf. Series: Materials Science and Engineering **538** (2019) 012010 doi:10.1088/1757-899X/538/1/012010

polymerization process of PVF, the polymer is anticipated to have a promising future in many other industries and in various fields such as in medical equipment, energy devices, advanced educational platforms, and nanotechnology composites.

Acknowledgments
Acknowledgment goes to the Ministry of Higher Education-Malaysia for the research grant HiCoE project vote number 6369107 and University Putra Malaysia for facilitating the necessary means.

References
[1] S. Ebnesajjad, Introduction to fluoropolymers: materials, technology and applications. William Andrew, 2013.
[2] I. Kaur, V. Kumari, B. Sharma, and N. Gupta, "Characterization and applications of PVF film grafted with binary mixture of methacrylic acid and 4-vinyl pyridine by gamma radiations: Effect of swift heavy ions," Appl. Radiat. Isot., vol. 79, pp. 118-130, 2013.
[3] A. Gupta, R. Bajpai, and J. Keller, "Morphological, crystallization and plasticization studies on isomorphic blends of poly (vinyl fluoride)(PVF) and poly (vinylidene difluoride)(PVDF): using microhardness, XRD and SEM techniques," J. Mater. Sci., vol. 41, no. 18, pp. 5857-5861, 2006.
[4] N. Kumar and R. Nath, "Thermal and ferroelectric properties of potassium nitrate: polyvinyl fluoride composite films," IEEE Trans. Dielectr. Electr. Insul., vol. 12, no. 6, pp. 1145-1150, 2005.
[5] G. Oreski and G. Wallner, "Delamination behaviour of multi-layer films for PV encapsulation," Sol. Energy Mater Sol. Cells, vol. 89, no. 2-3, pp. 139-151, 2005.
[6] V. Simril and B. A. Curry, "The properties of polyvinyl fluoride film," J. Appl. Polym. Sci., vol. 4, no. 10, pp. 62-68, 1960.
[7] S. Ebnesajjad, Fluoroplastics, Volume 2: Melt Processible Fluoropolymers-The Definitive User's Guide and Data Book. William Andrew, 2015.
[8] A. Shu et al., "Cratering studies in Polyvinylidene Fluoride (PVDF) thin films," Planet. Space. Sci., vol. 89, pp. 29-35, 2013.
[9] C. L. Yaws, Thermophysical properties of chemicals and hydrocarbons. William Andrew, 2008.
[10] J. E. House and K. A. House, Descriptive inorganic chemistry. Academic Press, 2015.
[11] J. Gardiner, "Fluoropolymers: origin, production, and industrial and commercial applications," Aust. J. Chem., vol. 68, no. 1, pp. 13-22, 2015.
[12] S. Ebnesajjad, "Introduction to fluoropolymers," in Applied Plastics Engineering Handbook: Elsevier, 2017, pp. 55-71.
[13] S. Ebnesajjad, Polyvinyl fluoride: Technology and applications of PVF. William Andrew, 2012.
[14] S. Ebnesajjad and A. H. Landrock, Adhesives technology handbook. William Andrew, 2014.
[15] T. Nguyen, "Degradation of poly [vinyl fluoride] and poly [vinylidene fluoride]," Polym. Rev., vol. 25, no. 2, pp. 227-275, 1985.
[16] V. Karbhari, "Rehabilitation of metallic civil infrastructure using fiber-reinforced polymer (FRP) composites: a materials and systems overview at the adhesive bond level," in Rehabilitation of Metallic Civil Infrastructure Using Fiber Reinforced Polymer (FRP) Composites: Elsevier, 2014, pp. 3-10.
[17] M. Alaaeddin, S. Sapuan, M. Zuhri, E. Zainudin, and F. M. AL-Oqla, "Properties and Common Industrial Applications of Polyvinyl fluoride (PVF) and Polyvinylidene fluoride (PVDF)," in IOP Conf. Ser.: Mater. Sci. Eng., 2018, vol. 409, no. 1, p. 012021: IOP Publishing.
[18] J. G. Drobny, "Fluorine-containing polymers," in Brydson's Plastics Materials: Elsevier, 2017, pp. 389-425.
[19] M. DeBergalis, "Fluoropolymer films in the photovoltaic industry," J. Fluor. Chem., vol. 125, no. 8, pp. 1255-1257, 2004.
[20] M. Biron, Thermoplastics and thermoplastic composites. William Andrew, 2018.

[21] J. G. Drobny, Technology of fluoropolymers. CRC Press, 2008.
[22] D. Ghazanfari, D. Bastani, and S. A. Mousavi, "Preparation and characterization of poly (vinyl chloride)(PVC) based membrane for wastewater treatment," J. Water Process Eng., vol. 16, pp. 98-107, 2017.
[23] K. Al Abdullah et al., "The enhancement of PVDF pyroelectricity (pyroelectric coefficient and dipole moment) by inclusions," Energy Procedia, vol. 119, pp. 545-555, 2017.
[24] J. E. Dohany, "Fluorine-containing polymers, poly (vinylidene fluoride)," Kirk-Othmer Encycl. Chem. Technol., 2000.
[25] J. T. Goldbach et al., "Commercial synthesis and applications of poly (vinylidene fluoride)," Fluorinated Polymers Vol. 2: Applications, 2016.
[26] F. Liu, N. A. Hashim, Y. Liu, M. M. Abed, and K. Li, "Progress in the production and modification of PVDF membranes," J. Membr. Sci., vol. 375, no. 1-2, pp. 1-27, 2011.
[27] W. Ma, J. Zhang, S. Chen, and X. Wang, "β-Phase of poly (vinylidene fluoride) formation in poly (vinylidene fluoride)/poly (methyl methacrylate) blend from solutions," Appl. Surf. Sci., vol. 254, no. 17, pp. 5635-5642, 2008.
[28] J. Wang, J. Wu, W. Xu, Q. Zhang, and Q. Fu, "Preparation of poly (vinylidene fluoride) films with excellent electric property, improved dielectric property and dominant polar crystalline forms by adding a quaternary phosphorus salt functionalized graphene," Compos. Sci. Technol., vol. 91, pp. 1-7, 2014.
[29] Y. J. Park et al., "Molecular and crystalline microstructure of ferroelectric poly (vinylidene fluoride-co-trifluoroethylene) ultrathin films on bare and self-assembled monolayer-modified Au substrates," Macromolecules, vol. 41, no. 1, pp. 109-119, 2008.
[30] I. U. Ahad et al., "Extreme ultraviolet (EUV) surface modification of polytetrafluoroethylene (PTFE) for control of biocompatibility," Nucl. Instrum. Methods Phys. Res., Sect. B., vol. 364, pp. 98-107, 2015.
[31] V. Rudnev, A. Vaganov-Vil'kins, T. Yarovaya, and A. Pavlov, "Polytetrafluoroethylene-oxide coatings on aluminum alloys," Surf. Coat. Technol., vol. 307, pp. 1249-1254, 2016.
[32] H. Hunke et al., "Plasma modified Polytetrafluoroethylene (PTFE) lubrication of α-olefin-copolymer impact-modified Polyamide 66," Wear, vol. 338, pp. 122-132, 2015.
[33] N. Benipal, J. Qi, J. C. Gentile, and W. Li, "Direct glycerol fuel cell with polytetrafluoroethylene (PTFE) thin film separator," Renewable energy, vol. 105, pp. 647-655, 2017.
[34] Y. Zhang, D. Zhang, J. Wu, J. Li, and Z. He, "Non-Fourier heat conduction and phase transition in laser ablation of polytetrafluoroethylene (PTFE)," Acta Astronaut., vol. 140, pp. 338-350, 2017.
[35] H. Hunke, N. Soin, T. Shah, E. Kramer, K. Witan, and E. Siores, "Influence of plasma pre-treatment of polytetrafluoroethylene (PTFE) micropowders on the mechanical and tribological performance of Polyethersulfone (PESU)–PTFE composites," Wear, vol. 328, pp. 480-487, 2015.
[36] L. Ye, C. Qi, J. Hong, and X. Ma, "Life cycle assessment of polyvinyl chloride production and its recyclability in China," J. Cleaner Prod., vol. 142, pp. 2965-2972, 2017.
[37] R. Singh and D. Pant, "Polyvinyl chloride degradation by hybrid (chemical and biological) modification," Polym. Degrad. Stab., vol. 123, pp. 80-87, 2016.
[38] S. Rostam, A. K. Ali, and F. H. AbdalMuhammad, "Experimental Investigation of Mechanical Properties of PVC Polymer under Different Heating and Cooling Conditions," J. Eng., vol. 2016, 2016.
[39] R. A. Khan et al., "Fabrication and characterization of jute fabric-reinforced PVC-based composite," J. Thermoplast. Compos. Mater., vol. 25, no. 1, pp. 45-58, 2012.
[40] M. Pandey, G. M. Joshi, A. Mukherjee, and P. Thomas, "Electrical properties and thermal degradation of poly (vinyl chloride)/polyvinylidene fluoride/ZnO polymer nanocomposites," Polym. Int., vol. 65, no. 9, pp. 1098-1106, 2016.

IOP Conf. Series: Materials Science and Engineering **538** (2019) 012011 doi:10.1088/1757-899X/538/1/012011

EDM-drilling Characteristics with the Rotation Motion of SS304

Quanli Han, Hongqiang Wan, Dongchen Han

School of Mechanical and Electrical Engineering, Xian Technological University, Xian, China

hanqljx@163.com, 413162931@qq.com, hbear202@sina.com

Abstract. This investigation presents an attempt to explore a novel rotary electrical discharge cutting process for high processing quality of work-piece and a hard-to-machine work-piece. An experiment was carried out on SS304 with rational work-piece and new electrode of slot on outer-side surface. Processing quality of work-piece in condition described above is superior to that in the traditional EDM in terms of shape and roughness. The result is attributed to the mixing effect of high probability of normal discharge and speed of electrolyte flow, resulting into good transport efficiency of chip, finally bringing into machining phenomenon with characteristics of high frequency and small removal. The result from compared experiments is promising for manufactures and engineers to improve material removal rate in traditional EDM.

1. Introduction

Hole has been one of the base objects machined in many products, which is conventionally done by drilling. During the processing of drilling, the tool moves in the helical where the work-piece remains stationary. The most obstacles in drilling is the inefficient removal of debris and unsatisfied surface quality needed post process, resulting from the slower transportation speed of chips and the smaller space holding chips attributed to machining principle. Due to the difficult-to-machine materials, the EDM has been used to drill the hole, because the non-contact between the work-piece and tool can make hole-drilling possible. Recently the EDM-drilling has been applied to special requirements of products in industrial fields such as cooling holes and diffusers in turbine blades as well as new alloy materials metallic glasses, and non-conductive material such as S-SiC and SiC[1-3].

In EDM-drilling, the researchers were making their effort to attempt the hole of high aspect ratio. Ultra-high aspect micro holes could be done by tools insulated on the sidewall to prevent secondary sparks by means of a coating [4]. A new strategy relative to the use of helical-shaped electrodes had been proposed and applied resulting in the reduction in machining times and the additional flute-depth [5]. A simplistic analytical model had been introduced to evaluate the effectiveness of low frequency work-piece vibration in WEDM-drilling, because the position of debris particles being above the work-piece can be flushed away from machined zone effectively [6]. The orbital motion of the electrode enlarged the holes diameter resulting in enhanced flushing effect while eliminated exponential reduction in material removal rates in EDM-drilling [7]. The electrode resultant motion of inclined feeding and vibration in the longitudinal direction leaded to an increase in the depth of the hole as compared with horizontal-feeding method [8]. Since electrode materials characteristics and wear in EDM-drilling had great influenced on the tool feed, the relationship between electrode side gap-width and machining feed was investigated; the relationship between electrode wear ratio and machining feed was sought and electrode wear in EDM-drilling was modeled[9-11]. Machining parameters in EDM-drilling were optimized to increase the productivity and accuracy, through adjusting the single discharge energy for given diameters, rotation speed of electrode [12-13]. Energy consumption was used for performance indicator like processing quality, comparing the energy in Micro-EDM and pulse laser drilling [14]. The tapered holes especially reverse tapered holes and array holes could be generated in EDM-drilling on the new machine or new electrode designed [15-16]. To improve the quality, the two steps process was researched to thin the recast layer of work-piece in hole-drilling, while a modified 'volume removal per discharge' approach had been applied to compensate the longitudinal wear of electrode [17-18]. However, up to now most research on EDM-drilling was confined to conventional mechanical model, and the understanding of EDM-drilling did not break through the technology of machine drilling, even through change of driving energy compared with convention drilling.

This paper put forward that the new drilling movement consists in the rotation motion of work-piece and driller motion in a straight line, which creates the helical motion needed in the hole-drilling. This movement above designed maybe led to enrich EDM-drilling knowledge, whose processing efficiency was addressed both qualitatively and quantitatively.

2. Experiment Setup

Experiments were done on die sinking machine EA22M (Mitsubishi, made in Japan) used the brass electrode and the spark machining oils. In machining, the electrode tool moves only in line motion, while the work-piece rotates around its axis. Because the position accuracy in this machine was in the scale of micrometer so that the holes axis of electrode tool was on the same line as the work-piece one designed in practice, even the worse condition was that the difference between the two axes above was strictly controlled in the narrow range accepted. The reserve tool path became helical like the convention drilling, and the motion direction of tool and work-piece confined to the rule of right hand named.

The tool axial feed speed was chosen from the data suggested by machine manufacture, whose magnitude was in the range of 0.25-0.27μm/s. And the work-piece was rotated at the speed of 8000 r/min set by experimenters; this magnitude could take changes because the obstacle force from the electrical fluid. So the helical pitch was defined that the product of diverse of relative rotation speed in ration and line speed of tool, which is computed based the date above, equating to 0.0019μm. The Table 1 showed the processing parameters used in EDM-drilling, and the drillings were carried out in the oil, and the Table 2 listed the characteristics of work-piece.

Table1. Processing parameters in experiments.

Discharge current. (i_e) A	Pulse duration(t_e) μs	Duty factor n(t_e) %	Open circuit voltage(u_e) V	Working fluid
8	15	50	200	Oil

Table2. Characteristics of work-piece in experiments.

Name	Type	Hardness (*HV*)	Yield strong MPa	Tensile strength Mpa	Elongation (δ) %
SS304	ANN	197	210	530	44

In order to investigate the effect of electrode tool rotation on holes quality in EDM-drilling, the brass electrode was specially designed in the shape. The circular surface of columnar copper electrode was created the slot with the cross section of semicircle shape to improve the transportation capability of debris, and the structure parameter of design was seen in Fig.1. To investigate transportation effect of slot, the same size electrode with non-slot was done. Those electrodes were used in the same processing parameters.

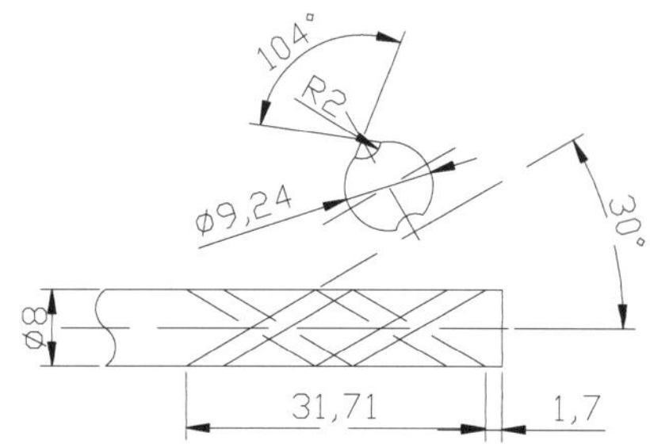

Figure 1. Structure parameters of electrode used in experiments.

The rotary actuator of work-piece was designed according to the mechanical schematic diagram shown in Figure 2. The motor drove the first shaft on which the first dish was fixed, so the first dish rotated could drive the second dish, which the work-piece was lied on. Once the motor rotated, the work-piece followed rotational motion in certain angular velocity. Because the axis of work-piece was on the line of the second shaft, the relative motion between the electrode tool and work-piece was the cutting movement of drilling.
Sheet of SS304 in the width of 3mm was chosen as work-piece in experiment, whose properties listed in Table 2. And the sheet raw was cut into the shape of rectangular section; the length and width of cross section were 40mm and 20mm respectively. The rectangular section blocks were grinded to flatten the up and down surfaces, and those blocks were assembled into the square section whose length was 40mm. The work-piece was clamped on the working dish, that is, the second dish in the Figure 2.

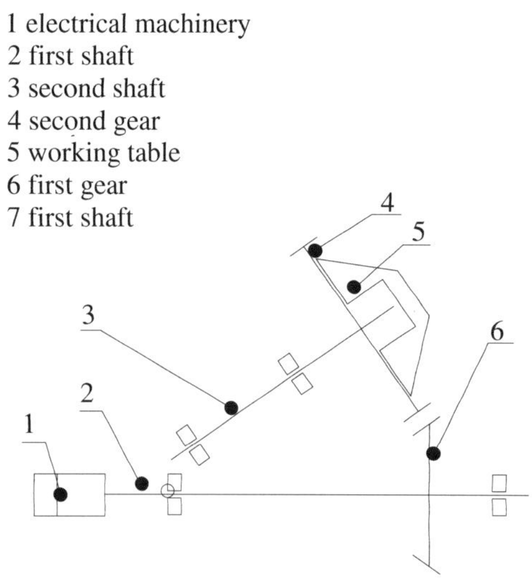

Figure 2. Sketch map of kinematic mechanism of workpiece.

HFS new Rockwell hardness tester was used to get the hardness of work-piece machined, and the load of 150Kgf was applied, five points on one sample were choose randomly when measuring and the average of them was made.

3. Results and Discussions

The experimental results were shown in Figure 3. The geometrical shape measured was listed in Fig. a, which tells that the difference of work-piece used by the new and traditional electrodes. The change of work-pieces shape in the condition of new electrode and rational work-piece is less than that gotten in traditional EDM. The similar trend is found in surface roughness in rational work-piece and new electrode, when comparing with traditional EDM. What is usually overlooked that the difference of surface hardness between new machining and traditional conditions.

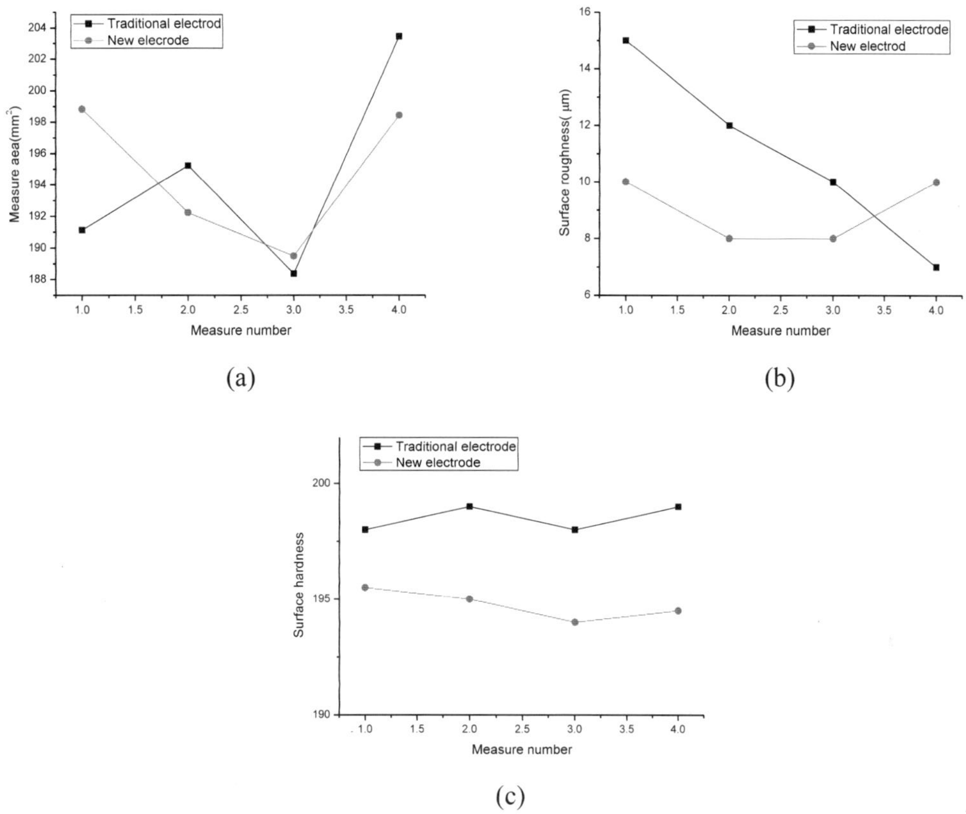

Figure 3. Results of experiment.

It is well-known and accepted understandings in EDM is that the eroded chips in the gap between the workpiece and electrode will hinder the sound discharge even if the electrolyte exists, because the discharge instability, which is induced by debris at the gap, lowered the discharge efficiency. These can be attributed to chip removal rate, which means transport efficiency of chip going out of gap of workpiece and electrode. It is crucial adjective in EDM to have a good chip flow resulting in high chip removal rate.

Rational workpiece in new condition of EDM improves the movement velocity of chips, whose original resource results from the relative movement between the workpiece and electrode. Once the movement described above happens, additional movement on chips helps themselves move away from the gap. Removal chip depends on the flush of electrolyte, whose internal kinetic energy was consumed by removal chip. Corresponding energy is absorbed by chips and then the chips rich in energy can move out of gap where the specific energy of chip is lower, which is said from the law of conservation of energy.

New electrode can improve the space of electrolyte, according to the result of the above analysis, the whole energy of electrolyte rises rapidly, while the chip density falls because of the increase in the volume of electrolyte with the use of new electrode. The two sides said above contribute to the following effect, that is, the kinetic energy of each chip in the new condition improves larger than that in traditional EDM. Transportation of chip becomes easily in condition of new electrode with the help from the high kinetic energy of each chip, when comparing with traditional EDM. One obvious result is the increase in material removal rate, which brings into normal discharge taken place. The higher points on surface machined can be processed in discharge, and then the difference between points becomes less with the machining time, so that the error of shape and the surface roughness can be smaller quickly with the increase in density of normal discharge. This can be attributed to processing quality of work-piece used new electrode in terms of shape and roughness, when making the compare with traditional EDM.

The hardness on surface in EDM depends on the depth of deformed layer resulted from heat during discharge, and the heat created in discharge is indirect proportion to the probability of abnormal discharge and speed of electrolyte flow. The common finding accepted is thought that the effect of electrolyte flow on heat is great more important than that from abnormal discharge, another fact in EDM is the limit of whole depth of deformed layer. Those can be used as the reason for the processing quality of work-piece used new electrode in terms of hardness compared with traditional EDM.

4. Conclusions

Experiments with new electrode and rational work-piece were conducted so that the following findings have been gotten. The processing quality of work-piece in new condition is more promising than that in traditional EDM, especially in terms of shape and roughness. This facts result from the increase in the probability of the sound discharge, which comes from the help of high transportation efficiency of chip. All of the above is believed to have originated from mixing effect of rational work--piece and slotted electrode.

Acknowledgments

We are greatly indebted to Doctor Wang yitiao, for his valuable instructions and suggestions on experiment. And this study was funded by Shaanxi key laboratory research project (17JS060).

References

[1] M. Kliuev, M .Boccadoro, R.Perez, W.Dal Bo, J.Stirnimann, F.Kuster, K . Wegener. Edm drilling and shaping of cooling holes in inconel 718 turbine blades. Procedia CIRP 42(2016)322-327

[2] M. Kliuev , U. Maradia ,M. Boccadoro, R. Perez , J. Stirnimann , K. Wegener . experimental study of EDM-drilling and shapeing of ssic and sic. Precedia cirp 42(2016) 191-196

[3] Sandeen Kuriakose, Promod Kumar Patowari, Jatin Bhatt. Machinability study of Zr-Cu-Ti metallic glass by micro hole drilling using microusm. J. Mat. Pro. Tec. 240(2017)42-51

[4] E. Ferraris, V. Castiglioni, F. Ceyssens, M. Annoni, B. Lauwers, D. Reynaerts. EdM-drilling of ultrahigh aspect ratio micro holes with insulted tools. CIRP Annals Menu. Tech. 62(2013)191-194

[5] Sanchez a, Endik Perez , Ruben Gil , Borja Izquierdo , Naiara Ortega, Inigo, Pombo, Soraya Plaza, Jose. Experimental study on micro EDM-drilling of Ti6Al4V using helical electrode. precision engineering 38(2014) 821-827.

[6] M.P. Jahan. Y.S. Wong, M .Rahman. Evaluation of the effectiveness of low frequency workpiece bibration in deep hole micro edm drilling of tungsten carbide. Manu. Prec. 1492012) 343-359

[7] Eberhard Bamberg, Sumet Heamawatanachai. Orbital electrode actuation to improve efficiency of drilling micro-holes by micro-EDM journal of materials processing technology 209 (2009) 1826–1834

IOP Conf. Series: Materials Science and Engineering **538** (2019) 012011 doi:10.1088/1757-899X/538/1/012011

[8] Y. S. Liao, H. W. Liang. Study of vibration assisted inclined feed micro EDM-drilling. Procedia CIRP 42(2016) 552-556

[9] Yulei Fu,Takumu Miymoato, Wataru Natsu. Wangsheng Zhao Zuyan Yu Study on influence of electrode material on hole drilling in micro-EDM. Procedia CIRP 42(2016)516-520

[10] G. DUrso, C. Merl. Workpiece and electrode influence on micro-EDM drilling performance. Precision Engineering.38 (2014) 903-914

[11] Cheol-soo lee, Eun-young Heo, Jong Min Kim in-hugh Choi Dong Won Kim. Electrode wear estimation model for EDM-drilling. Robotics and computer integrated manufacturing 36(2015) 70-75

[12] Mristor Has, M Munz. Optimization of EDM-drilling process to increase the productivity and geometrical accuracy. Procedia CIRP. 42(2016)537-542

[13] Una Yahagia , Tomohiro Koyanob , Masanori Kuniedab, Xiaodong Yang. Micro Drilling EDM with High Rotation Speed of Tool Electrode Using the Electrostatic Induction Feeding Method. Procedia CIRP.1(2012)162-165

[14] Alessandro Franco, Choudhury Abul Anam Rashed, Luca Romoli. Analysis of energy consumption in micro-drilling processes. Journal of Cleaner Production, Volume 137, 20 November 2016, Pages 1260-1269

[15] Long Zhang, Hao Tong, Yong Li. Precision machining of micro tool electrodes in micro edm for drilling array micro holes. Precision engineering 39(2015) 100-106

[16] C. Diver, J. Atkinson, H.J. Helml, L. Li. Micro-EDM drilling of tapered holes for industrial applications.Journal of Materials Processing Technology, Volume 149, Issues 1 – 3, 10 June 2004, Pages 296-303

[17] Li C. J, Li Y, Tong H, Zhao L. Thinning process of recast layer in hole drilling and trimming by EDM. Procedia CIRP 42(2016) 575-579

[18] C.K. Nirala, P. Saha. Precise EDM-drilling using real-time indirect tool wear compensation Journal of Materials Processing Technology 240 (2017) 176–189

IOP Conf. Series: Materials Science and Engineering **538** (2019) 012012 doi:10.1088/1757-899X/538/1/012012

Fabrication of Micro-Hole Array on Glass Material by Abrasive Jet Machining

Choung-Lii Chao[1,a,*], Wan-Hsuan Wang[1,b], Tian-Ming Chao[1,c], Wen-Chen Chou[1,d] and Kung-Jeng Ma[2,e]

[1]Department of Mechanical and Electro-Mechanical Engineering, Tam-Kang University
[2]Department of Mechanical Engineering, Chun-Hwa University

[a]clchao@mail.tku.edu.tw, [b]802370014@s02.tku.edu.tw, [c]jordan1992728@gmail.com, [d]106096@mail.tku.edu.tw, [e]kjma@chu.edu.tw

Abstract. Brittle materials such as glasses, silicon, and silicon carbide are normally categorized as difficult to machine materials for its high hardness and brittleness. However, they have attracted more and more attentions and been playing critical roles in many scientific/engineering applications for their advanced physic/optical/electronic properties. Micro-patterns such as micro-holes (array) of various sizes and shapes are frequently required to be generated on brittle materials. Many researchers have tried different approaches such as laser ablation, ultrasonic machining, and rotary ultrasonic machining to produce micro-holes in brittle materials. This research applied abrasive jet machining to fabricate micro-holes array on glass. Efforts have been made to investigate the effect of grit-size, stand-off distance, pressure on the material removal rate and the obtained holes accuracy. Micro-hole arrays of various shapes and with characteristic dimension ranging from 0.2 mm to 2 mm are successfully produced in glass plate of 0.4 mm thickness.

1. Introduction

As the demands for machining micro holes/patterns in brittle materials are steadily increasing, many attempts have been conducted by applying various machining techniques such as laser ablation, ultrasonic machining, rotary ultrasonic machining, chemical etching, focus ion beam and abrasive jet machining. Laser ablation is a very versatile technique and has the advantage of fast machining a wide range of materials[1-3]. However, the melt ejection, re-deposited debris, micro-cracks generated by the heat and thermal stress involved in the machining process frequently make the surface quality and holes accuracy unacceptable. Although short wavelength or pulse duration such as femtosecond laser can minimize the photo-thermal effect, it is, in many cases, just too expensive to be economically viable. Ultrasonic machining is considered to be a relatively effective way to generate micro-holes in brittle materials [4,5] but the cost for making the ultrasonic tools can be very expensive. In comparison to laser ablation and ultrasonic machining, abrasive jet machining is a much cheaper and effective way to produce micro holes/patterns in brittle materials[6-10]. Unlike conventional sand blasting which is mainly used for cleaning, abrasive jet machining is frequently used to cut complicated shapes in brittle materials. A summary of the pros and cons of frequently used micro hole-drilling processes are listed in Table 1.

This research applied abrasive jet machining to fabricate micro-holes array on glass. Efforts have been made to investigate the effect of grit-size, stand-off distance, pressure on the material removal rate and the obtained holes accuracy.

Table 1. Summary of the pros and cons of frequently used micro hole-drilling processes

	Wet Chemical Etching	**Dry Etching**	**EDM**	**Laser drilling**	**Ultrasonic machining**	**AJM**
Pros	1. Cheap 2. Almost no mechanical damage	1. Avoid dangerous acids and solvents 2. Use a small amount of chemicals 3. High resolution and cleanliness 4. Less undercuts 5. Easy to automate	1. Can machining hard materials to very tight tolerances. 2. Non-contact process 3. Surface finish is good 4. Small hole down to micrometers scale can be easily drilled	1. Non-contact 2. High aspect ratio possible. 3. Holes at shallow angles. 4. Drilling of difficult to process materials 5. High process speed good dimensional accuracy.	1. Produces very little heat 2. Various hole shape due to the vibrator motion of the tool 3. Relatively high MRR	1. ***Low cost*** 2. ***High productivity*** 3. Very suitable for machining brittle, and heat resistant materials. 4. Can effectively produce complex hole shapes
Cons	1. Undercut due to isotopic etching 2. Process control is critical (temperature sensitivity) 3. High chemical disposal costs 4. ***Environmental pollution***	1. Some gases used are quite toxic and corrosive 2. Re-deposition of non-volatile compounds ***3. Expensive equipment***	1. Small material removal rate 2. ***Conductive materials only*** 3. Reproducing sharp corners on the work piece is difficult due to electrode wear. 4. Over cut	1. Expensive equipment. 2. ***HAZ (Heat affected zone)*** 3. Melt ejection and debris	1. High tool wear 2. Sub-surface damage (micro-cracks) 3. Difficult to machine deep holes 4. ***Limitations in productivity***	1. ***Taper in the obtained holes*** 2. Dust collection 3. Abrasive particles are embedded in the working surface.

2. Experimental Setup

A photo-mask was fabricated based on the designed hole patterns and was used to transfer those patterns to the mask material via photo-lithography process(Figure 1) before abrasive jet machining could get started. Shown in Fig. 2(a) and Fig. 2(b) are the optical-mask used in this study and the glass substrate with patterned mask respectively. Corning glass wafer SGW3 (Alkaline Earth Boro-Aluminosilicate) with Young's modulus of 73.6 (GPa), density of 2.38 (g/cm^3), Vickers hardness of 640 (kgf/mm^2) [11] and thickness of 0.4 mm was used in this study as the workpiece material.

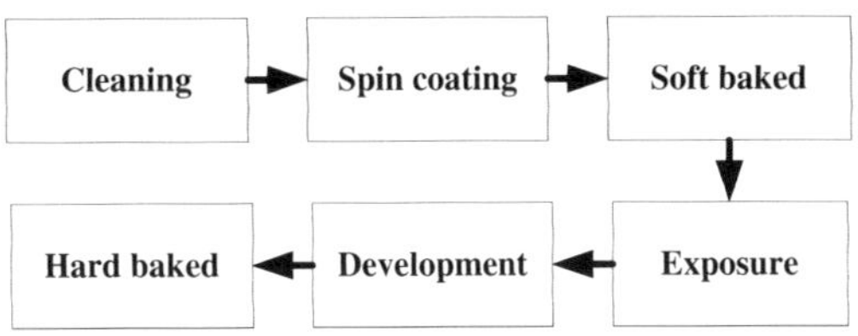

Figure 1. photo-lithography process

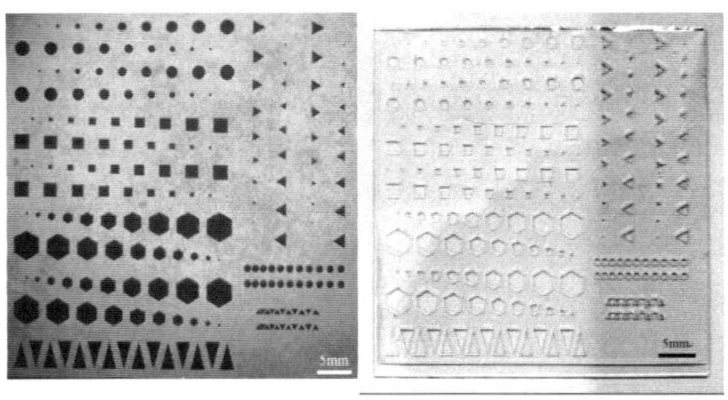

(a) (b)

Figure 2. (a) Optical Mask and (b) glass substrate with patterned mask (SU-8)

A negative photoresist SU-8 was selected as the mask material for its good adhesion properties, easy processing and reasonable erosion durability. Alumina powder with grain size of 31-26μm (WA320) and 22-18μm (WA400) was used as abrasive and the machining parameters were listed in Table 2.
The machined specimens were examined by optical microscope and a laser confocal microscope for holes shapes, erosion depth and sidewall taper angle (Figure 3).

Table 2. Process parameters for abrasive jet machining

Nozzle diameter (mm)	8
Abrasive	Al_2O_3 #320 (26-31μm) and #400 (18-22μm)
Air pressure (MPa)	0.05, 0.1, 0.15….0.5
Stand-off-distance (mm)	2.5, 5, 10, 15, 20, 25, 30, 35, 40, 50, 60, 70
Mask patterns	Round, Square, Triangle (30°, 60°, 75°, 90°, 120°) Spacing (mm): 0.2, 0.25, 0.3…. 0.5
Scanning passes	160, 200, 240
Machining duration (sec)	20, 40, 60, 80

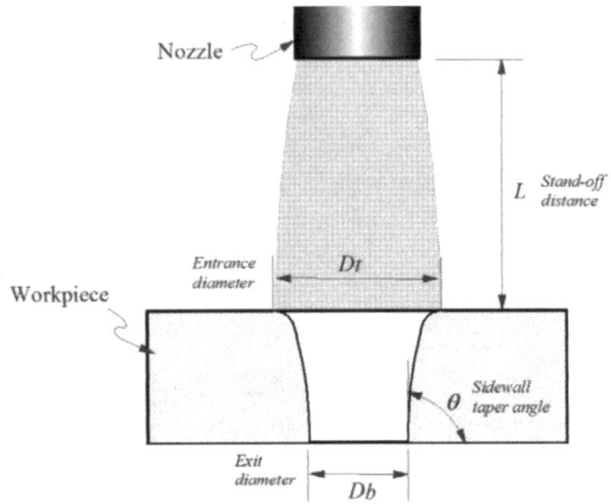

Figure 3. Schematic representation of the abrasive jet machining (without using mask)

3. Results and Discussions

Apart from hole shapes and sizes, the influences of machining parameters such as stand-off distance, air pressure, grit size, scanning speed/passes on the achieved material removal rate, sidewall taper angle and holes accuracy (as defined in Figure 4) were studied and the results were listed below.

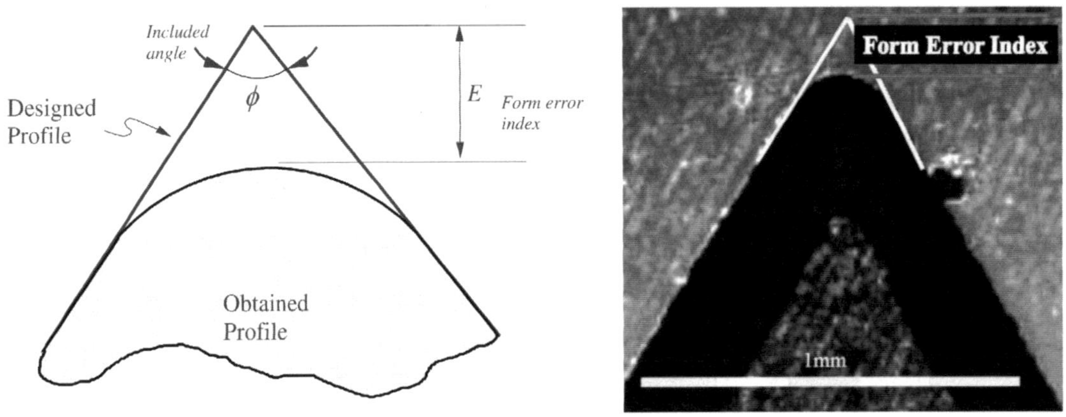

Figure 4. Form Error Index

3.1. Air Pressure

Shown in Figure 5 are the erosion depth under different air pressures. The stand-off distance and machining duration was 35 mm and 40 sec respectively. When pressure is under 0.2 MPa, the erosion depth increases slowly with increasing pressure. The increase rate reaches a stable value when pressure is between 0.2 to 0.35 MPa, and gradually declines when further increasing air pressure. Little change in erosion depth can be made when air pressure gets beyond 0.5 MPa. Air pressure is fixed at 0.5 MPa in most of the tests conducted in this study.

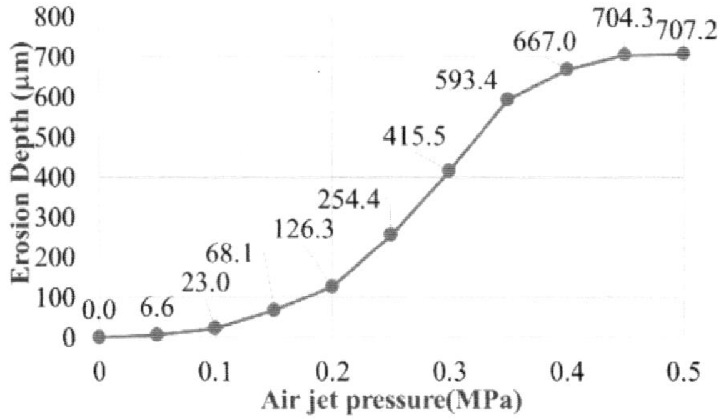

Abrasive	Al_2O_3 #320
Nozzle diameter	8mm
Machining time	40s
Stand-off distance	35mm
Impact angle	90°

Figure 5 Air pressure vs. Penetration depth

3.2. Stand-off Distance

A serial of abrasive jet machining tests with stand-off distance ranged from 2.5 mm to 30 mm were carried out without using mask just to investigate the influence of standoff distance on the obtained erosion depth and shape of the crater. The measured profiles of the machined craters were shown in Figure 6. The air pressure and machining duration was 0.4 MPa and 8 sec respectively. It is worth noting here that the profile of the generated craters gradually change from an "U" shape to a "V" shape as the erosion gets deeper. Thus, the sidewall taper angle gets smaller as the erosion depth gets higher. Owing to the divergent nature of abrasive jet, the entrance diameter increases with stand-off distance. Mask is a necessity in abrasive jet machining if effective control of dimension accuracy is required.

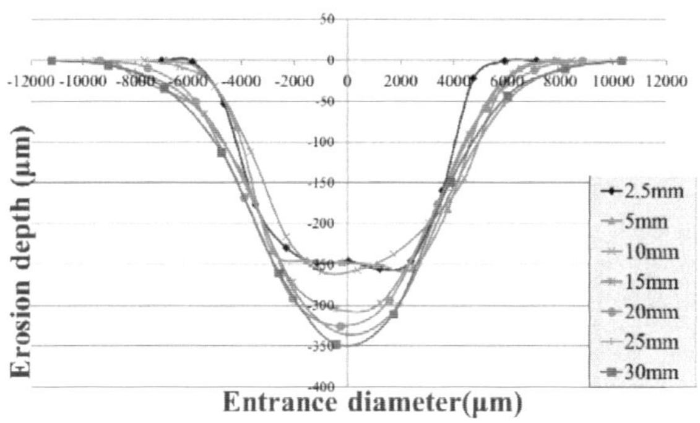

Abrasive	Al_2O_3 #320
Nozzle diameter	8mm
Machining time	40s
Air pressure	0.4MPa
Impact angle	90°

Figure 6 Erosion depth with different Stand-off distances

With air pressure and machining duration fixed at 0.4 MPa and 40 sec respectively, the material removal and erosion depth obtained by setting up various nozzle standoff distances are shown in Fig. 7.

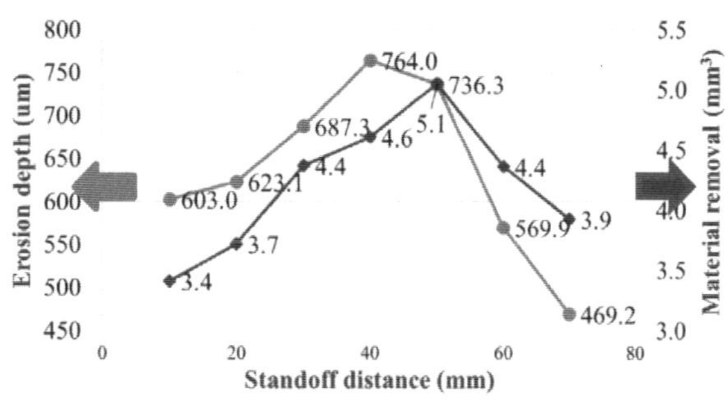

Figure 7 Material removal /Penetration depth vs. Stand-off distance

The erosion depth reaches its peak (764μm) and 2nd highest (736μm) when standoff distance equals to 40 mm and 50 mm respectively. The difference is less than 4%. The material removal reaches its peak (5.1 mm^3) and 2nd highest (4.6 mm^3) when standoff distance equals to 50 mm and 40 mm respectively. The difference is almost 10%. Standoff distance is then fixed at 50 mm in most of the tests conducted in this study.

3.3. Other Machining Parameters and Shape Effect.

Shown in Figure 8 are the optical micrographs of the obtained round holes with WA320 abrasive, 0.5 MPa, 50 mm stand-off distance and 160 passes at a scanning speed of 16.7 mm/sec. The measured sidewall taper angle of specimens machined with different scanning speeds are plotted against various designed hole sizes and shown in Figure 9. In comparison to small holes, big holes have deeper erosion depth and sidewall profile is more bias to a "V" shape than those obtained in small holes. Thus, small holes have higher sidewall taper angle than big holes, as shown clearly in this figure. Although scanning speed does not show too much influence on the sidewall taper angle at the hole with diameters ≥1.2 mm, it does have some effects on the smaller holes.

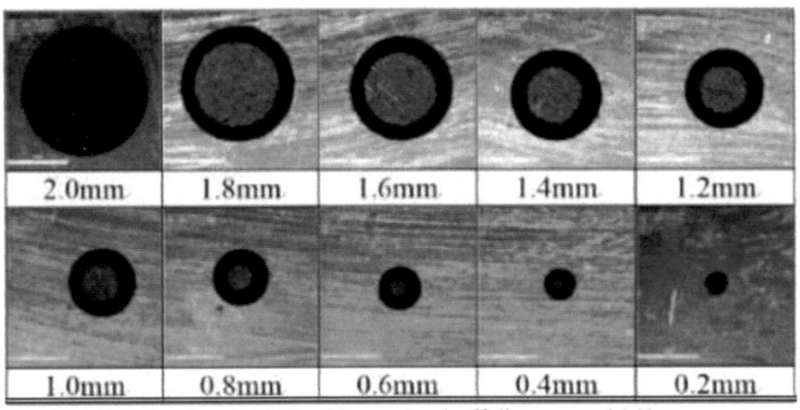

[Process parameter: WA320 abrasive, 0.5 MPa, 50 mm stand-off distance and 160 passes at a scanning speed of 16.7 mm/sec]

Figure 8. Optical micrographs of the obtained round holes.

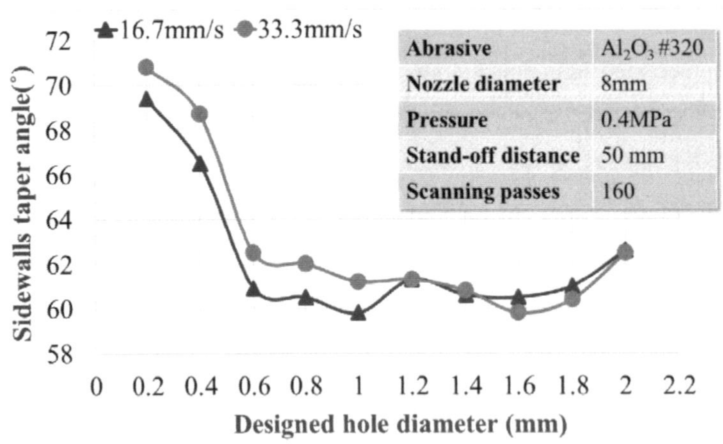

Figure 9. Sidewall taper angle vs. hole diameter at various scanning speeds

To study the influence of abrasive size and included angle on the achieved sidewall taper angle and form accuracy, triangular holes with various included angle are machined by WA320 and WA400 alumina abrasives and the result is shown in Figure 10.

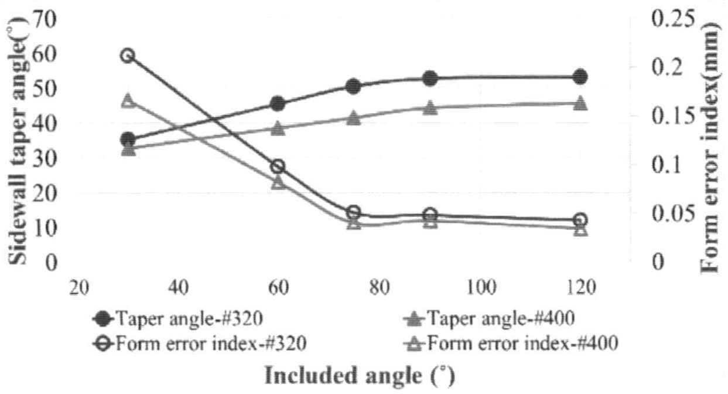

Impact angle	90°
Scanning speed	16.7 mm/s
Scanning passes	240

Figure 10. Influence of abrasive size and included angle on the achieved sidewall taper angle and form accuracy

It gets more and more difficult for abrasive particles to crush into the tip area when the included angle gets smaller. As a result, the form error index gets higher as the included angle gets smaller. Since finer abrasives have better chances to "cut" into the tip area, the smaller the abrasive gets the higher form accuracy (less form error) it can achieve.

In comparison to fine abrasive, coarse abrasive have deeper erosion depth and sidewall profile is more bias to a "V" shape than those generated by fine abrasive. Thus, fine abrasive generates higher sidewall taper angle than those generated by coarse abrasive, as shown clearly in this figure.

Since how narrow a spacing can be achieved in abrasive jet machining is an indicator for deciding how closely a holes array be arranged. Round (1 mm) and triangle (1 mm length) holes of various spacing were machined for 160 passes at 16.7 mm/sec scanning speed. The optical micrographs of the holes arrays and the measured spacing are shown in Fig. 11 and 12. The difference between designed and obtained spacing increases with increasing designed spacing. No matter it's a round hole or triangle holes array. However, if the differences (errors) are represented in error percentage [(designed-obtained)/designed *100%], as shown in Fig. 13, it is obvious that error percentage decreases with increasing designed spacing.

Figure 11. Optical micrographs of the obtained holes arrays

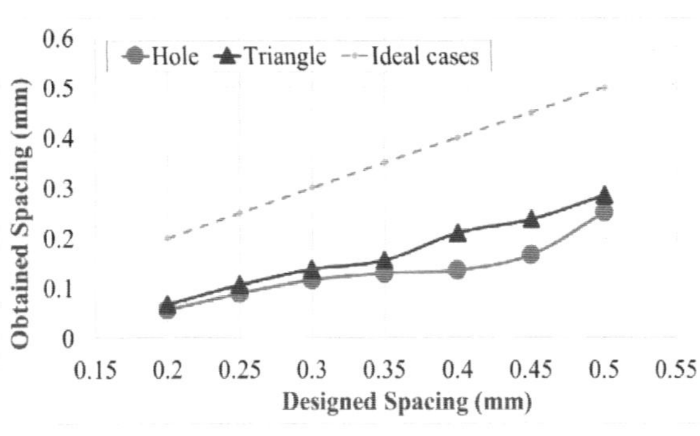

Abrasive	Al_2O_3 #320
Nozzle diameter	8mm
Pressure	0.4MPa
Stand-off distance	50 mm
Scanning speed	160 passes @ 16.7 mm/s

Figure 12. The obtaincd hole spacing plotted against designed hole spacing

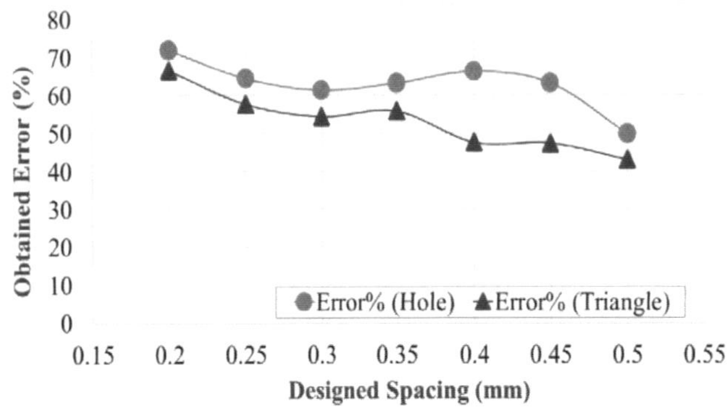

Abrasive	Al_2O_3 #320
Nozzle diameter	8mm
Pressure	0.4MPa
Stand-off distance	50 mm
Scanning speed	160 passes @ 16.7 mm/s

Figure 13. Error percentage decreases with increasing designed spacing

4. Conclusion

To conclude, when generating micro-holes on glass by abrasive jet, parameters such as grit-size, stand-off distance and pressure all have profound effects on the achievable material removal rate and the obtained holes accuracy. Since finer abrasives have better chances to "cut" into the tip area, the smaller the abrasive gets the higher form accuracy it can achieve. In comparison to fine abrasive, coarse abrasive have deeper erosion depth and sidewall profile is more bias to a "V" shape than those generated by fine abrasive. Thus, fine abrasive generates higher sidewall taper angle than those generated by coarse abrasive. Under the same machining conditions, big holes have deeper erosion depth than small hole and sidewall is more bias to form a "V" shape profile than those obtained in small holes. Thus, small holes have higher sidewall taper angle than big holes. As the included angle gets smaller, it's getting more and more difficult for abrasive particles to crush into the tip area. As a result, the form error gets higher as the included angle gets smaller. Based on these results, a "hybrid process" that is applying coarse abrasive first for high material removal and followed by fine abrasive to improve form accuracy should be a very promising way to strike a balance between efficiency and accuracy.

References

[1] Z. K. Wang, H. Y. Zheng, W. L. Seow, and X. C. Wang, "Investigation on material removal efficiency in debris-free laser ablation of brittle substrates," Journal of Materials Processing Technology, vol. 219, pp. 133–142, May 2015..

[2] D. Zhang and L. Guan, "4.06 - Laser Ablation," in Laser Ablation, vol. 4, S. Hashmi, G. F. Batalha, C. J. V. Tyne, and B. Yilbas, Eds. Oxford: Elsevier, 2014, pp. 125–169.

[3] A. K. Nath, "9.06 - Laser Drilling of Metallic and Nonmetallic Substrates," in Comprehensive Materials Processing, S. Hashmi, G. F. Batalha, C. J. V. Tyne, and B. Yilbas, Eds. Oxford: Elsevier, 2014, pp. 115–175.

[4] S. Kuriakose, P. K. Patowari, and J. Bhatt, "Machinability study of Zr-Cu-Ti metallic glass by micro hole drilling using micro-USM," Journal of Materials Processing Technology, vol. 240, pp. 42–51, Feb. 2017.

[5] A. A. Tseng, Y.-T. Chen, C.-L. Chao, K.-J. Ma, and T. P. Chen, "Recent developments on microablation of glass materials using excimer lasers," Optics and Lasers in Engineering, vol. 45, no. 10, pp. 975–992, Oct. 2007.

[6] H. Nouraei, K. Kowsari, J. K. Spelt, and M. Papini, "Surface evolution models for abrasive slurry jet micro-machining of channels and holes in glass," Wear, vol. 309, no. 1, pp. 65–73, Jan. 2014.

[7] D.-S. Park, M.-W. Cho, H. Lee, and W.-S. Cho, "Micro-grooving of glass using micro-abrasive jet machining," Journal of Materials Processing Technology, vol. 146, no. 2, pp. 234–240, Feb. 2004.

[8] A. Kumar and S. S. Hiremath, "Machining of Micro-holes on Sodalime Glass Using Developed Micro-abrasive Jet Machine (μ-AJM)," Procedia Technology, vol. 25, pp. 1234–1241, Jan. 2016.

[9] H. Wensink, "Fabrication of microstructures by powder blasting," Ph. D. Thesis, University of Twente, Enschede, 2002. Retrieved from http://doc.utwente.nl/61698/

[10] M. Wakuda, Y. Yamauchi, and S. Kanzaki, "Effect of workpiece properties on machinability in abrasive jet machining of ceramic materials," Precision Engineering, vol. 26, no. 2, pp. 193–198, Apr. 2002. https://doi.org/10.1016/S0141-6359(01)00114-3

[11] Corning. (2015, January). Semiconductor Glass Products. Retrieved from https://www.corning.com/media/worldwide/csm/documents/Corning%20Semiconductor%20Glass%20Wafers%20PI%20Sheet_rev%2009022014.pdf

IOP Conf. Series: Materials Science and Engineering **538** (2019) 012013 doi:10.1088/1757-899X/538/1/012013

Study of the Hardening in Middle Shaft During Cold Forging

Yuan-Ping Luh[1], Huang-Li Wang[1,a], Jian-Rong Ciou[1] and Jia-Fu Jhang[2]

[1]Graduate Institute of Manufacturing Technology, National Taipei University of Technology, Taipei City 106, Taiwan

[2]Shye Shyang Mechanical Industrial Co., Ltd, Hsin Chu Hsien, Taiwan

[a]Corresponding author: lle5890606@hotmail.com

Abstract. Cold forging is a common processing method. It is fast, precise, and produces high strength. The purpose of this study was to process products by using a cold-forging technique. This study is aimed at achieving hardness of a certain standard without using heat treatment; computer-aided engineering (CAE) analyses were conducted with different diameter settings but the same methods and molds. Tests were conducted to understand the influence of the changes in material diameters on their material hardness values. We expected to produce workpieces of a specific hardness at the lowest cost. An actual cold forging case was conducted. First, Deform-3D (finite element analysis software) was utilized to conduct a computer-aided engineering forging analysis. How the internal stress distribution and equipment output tonnage changed as the diameter of the material changed was observed and compared against the actual hardness. This study discovered that the location where large elastic deformation occurred (i.e., where the stress concentration occurred), the hardness of the cold forging piece at that location increased. The simulation and experiment results revealed that a large area reduction rate of the intermediate shaft formation resulted in dense forging streamlines, high hardness, and substantial hardening effects.

1. Introduction

Metal forming processes using forging and extrusion comprise cold forging, warm forging, and hot forging. Cold forging is processing without heating. In other words, it increases the hardness and toughness in a ductile metal through elastic deformation and dislocation [1-2]. Hot forging is primarily used in processing large mechanical parts and materials with a large work-hardening coefficient. The forging temperature must be maintained at a temperature identical to or higher than the recrystallization temperature of a material. Compared with cold forging, hot forging increases a metal’s plasticity and reduces forging stress. However, hot forging easily produces oxide layers, and relevant sizes cannot be easily controlled [3]. Warm forging is an improved method comprising the strengths of both hot and cold forging. Although materials are heated using this method, they do not reach a metallographic state. Moreover, their load is smaller than that from cold forging, scales do not develop from high-temperature oxidation, and their processing size can be controlled [4-6].

Cold forging enhances the mechanical characteristics of materials, so it is commonly utilized in the auto parts manufacturing industry; and warheads, rockets, weapons, and military equipment parts for aerospace industry, and the defense industry. Although forging increases material hardness, to reach a certain hardness, heat treatment must be adopted to alter the internal structure of a material. To increase surface hardness, the surface structure must be altered through methods such as surface blasting or polishing. Although the aforementioned techniques are commonly used to increase material hardness, these post-forging treatments for hardness result in additional time and production costs. Therefore, this study used a real case to verify methods not involving post-forging treatments to increase material hardness. The forged object in this study was an intermediate shaft found in cars with electric power steering (EPS). Deform-3D was used to analyze the locations of stress concentration [7], the machine output tonnage, and the hardness of the workpiece forged in this study under different diameters to evaluate whether the acquired hardness met the target value.

2. Literature Review

In 2002, Janicek et al. [8] used workpieces of various diameters and bolts of the same size as the workpieces to process materials without heat treatment and observe the hardness range. Shi and Yi [9] stated that when materials are under plastic deformation, under a local plastic deformation condition, the material organization and stress relocation can be utilized to reduce stress concentration, whereas under ununiformed plastic deformation, a balance can be reached through work hardening and mechanical experiments.

Due to plastic deformation caused by the molding process, the hardness of a workpiece formed through cold forging increases, and the hardness influences the mechanical properties of the workpiece. In the present study, a method was developed to optimize the shape of materials and molds to reduce changes in the hardness distribution. The proposed method was suitable for backward extrusion. A finite element analysis (FEA) model that described the material mechanics behavior during the processing was developed to predict the effective strain distribution in the workpiece. The correlation between the effective strain and hardness was analyzed to determine the hardness distribution [10-11].

Tumer and Sonmez [10] proposed two critical design guidelines for improving the hardness distribution in a material based on optimization procedures. First, a small gap between the mold wall and preformed workpiece led to a small variation in the hardness distribution. Second, sufficient lubrication resulted in a homogeneous hardness distribution. In other words, the coefficient of friction at the mold–workpiece interface exhibited a negative correlation with the degree of variation in hardness distribution.

Zottis et al. [12] proposed to evaluate the strain distributions and mechanical performance levels of stretch products by using hardness testing and numerical analysis approaches. These researchers also incorporated the influences of the nonuniform friction between materials and molds into finite element simulations to verify the post-stretching influence as well as the hardness and residual stress distributions. Moreover, with their refined equations and known values of flow stress, they were able to use effective plastic strain to predict material hardness.

This study produced nonaxisymmetric steel elements (crossover U-joints) by using two different processing techniques, namely cold forging and hot forging. This study analyzed and compared the mechanical properties, such as hardness and stretch resistance, of workpieces produced using these two techniques. The material's microstructure was observed in the experiments, and was mutually validated and compared with the material mechanical testing results [13-14].

At least 20 different tribology tests have been used to determine friction coefficients in cold-forging processing. Because such tests have employed different test settings, test methods, and levels of abstraction, comparability among them is questionable. In the present study, six established test principles were compared using the same tribology system. Substantial differences were observed among friction coefficients, and the experimental results suggested that no single tribology type fits all types of cold-forging processing [15].

3. Experiment Design

After cold-forging processing, we tested the material hardness; results were subsequently compared with the output tonnage of devices and a stress–strain diagram generated with a CAE method to analyze the trend of hardness changes (Figure 1).

IOP Conf. Series: Materials Science and Engineering **538** (2019) 012013 doi:10.1088/1757-899X/538/1/012013

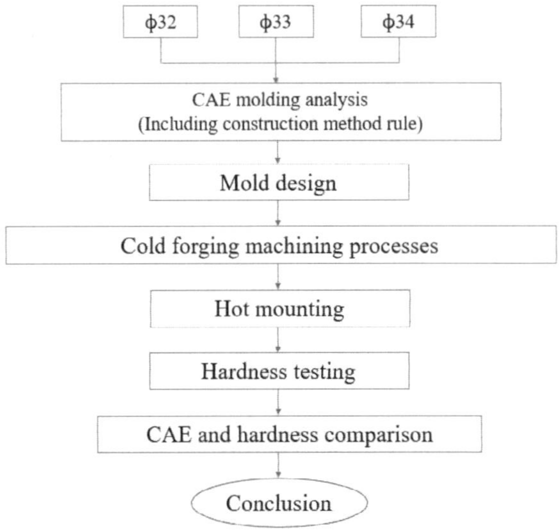

Figure 1. Experiment flowchart.

The experimental material used in the experiment was S45C carbon steel. The raw material has a hardness greater than 190 HB, is a commonly used steel type, has broad applications, and exhibits advantages such as favorable strength, toughness, and performance in turning processing. The material content and mechanical properties are shown in Table 1.

Table 1. Main components and mechanical properties of S45C medium-carbon steel.

Main components				
C	Si	Mn	S	P
0.42~0.48 %	0.15~0.35 %	0.60~0.90 %	≤0.035 %	≤0.030 %
Mechanical property				
Tensile Strength	Yield Strength	Percent Elongation	Percent Reduction in Area	hardness
≥58kgf/mm²	≥35kgf/mm²	≥20%	≥45%	167~229HB

3.1. Material Preprocess

Before the experiment, the materials underwent spheroidizing annealing, phosphate coating, and saponification. After spheroidizing annealing, the hardness of the materials was controlled to 170 ± 10 HV. The spheroidizing annealing treatment reduced the mechanical properties of the materials such as the tensile strength and hardness, and it reduced the impacts from forging, increasing the service life of the mold.

Phosphate coating and saponification enhanced the ductility of the materials and provided the surface with a lubricated coating. During forging, the materials experienced considerable impact and substantial friction against the mold surface, which rapidly increased the temperature. Therefore, a lack of lubrication may result in a failed product. For example, during cold forging, the use of phosphate coating and saponification reduces the resistance of metal flow, reduces deformation loading, controls the surface quality of an object, harmonizes the surface metal flow, reduces the situation in which a workpiece has sticky films, and reduces the surface fiction between the material and mold. Using phosphate coating enables a workpiece to be produced more smoothly, and the service life of the mold to be extended.

3.2. Experiment Equipment

An upright forger 650T was used in this study (Figure 2) (the specifications are shown in Table 2).

IOP Conf. Series: Materials Science and Engineering **538** (2019) 012013 doi:10.1088/1757-899X/538/1/012013

Figure 2. Upright forger 650T and its specifications.

Table 2. Equipment specifications.

SPECIFICATION	
Capacity	650 ton
Rating point	8 mm
Stroke length	200 mm
Die height	750 mm
Bolster area (L.R x F.B)	800 x 800 mm
Bolster thickness	100 mm
Slide area (L.R x F.B)	780 x 780 mm
Working energy	5200 Kg-m@ 35 spm
Max .upper die weight	1.2 ton
Air source	5 Kg/cm^2

3.3. Condition Settings in CAE Analysis

Deform-3D was adopted in a CAE simulation analysis. Deform-3D is FEA software used for processing simulation systems. It specifically analyzes 3D flow during the formation of metals. It has a wide range of applications in plastic processing, such as forging, extrusion, calendering, free forging, bending, and shearing[16].

Because of discrepancies between CAE analysis results and actual situations, and because analyses must be concluded within reasonable timeframes, some simplifying assumptions must be made regarding CAE, but the essential information must remain intact. The parameter assumptions were as follows:

1. Molds were considered as rigid bodies.
2. Blooms were considered to be plastic and to have homogenous and isotropic properties.
3. The weight, inertia effect, and elastic deformation problems of blooms were disregarded.
4. The interface friction between molds and materials were assumed to follow the constant shear friction law. The friction coefficient was simplified to a constant value.
5. The velocity curve of the molding machines' motion was simplified to a curve with a constant speed of 10 mm/sec.

3.4. Size of the Forged Product

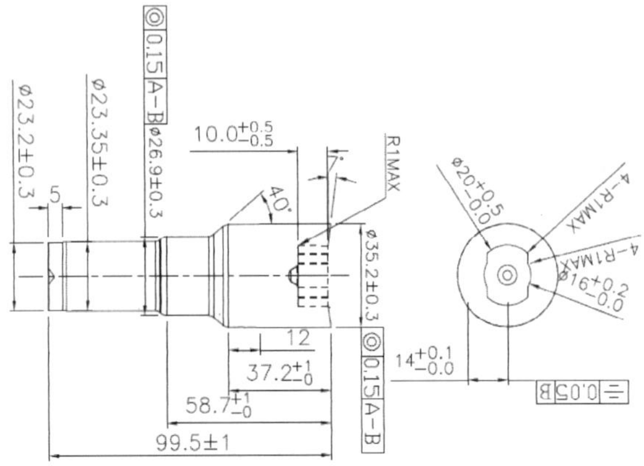

Figure 3. Dimensions of the forged product.

The specified sizes of cold-forged products are shown in Fig. 3. After products had been processed to their required sizes, other mechanical processing methods were applied; the products conformed to the requirements of an electrical power steering system and were later installed into such a system.

4. CAE Simulation Analysis

4.1. Method Planning

After conducting a computerized adaptive testing simulation analysis to determine factors such as mold cost, mold design limitations, and facility ejection specifications, this study used the method shown in Table 3 on φ32, φ33, and φ34. Consequently, the molding stress was lower, the forging tonnage of the machine was within bearable range, and the mold was under a lower loading, extending the service life.

Table 3. Method planning.

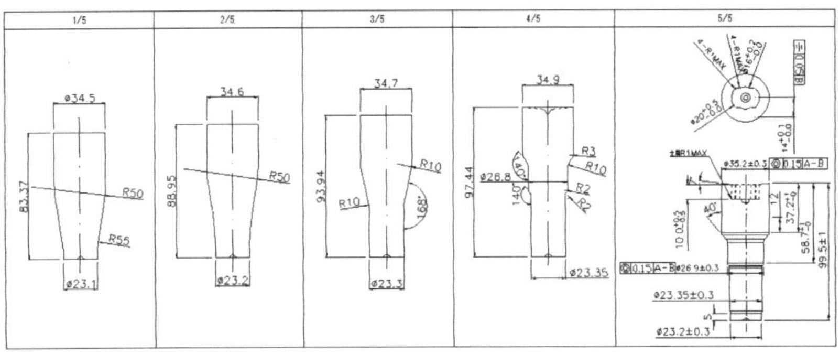

4.2. Parameter Settings in CAE Analysis

The surfaces of molds were finely processed and bright-polished. To simplify the simulation process, the constant shear coefficients were all determined based on the built-in parameters of DEFORMTM. Also, R angles smaller than 0.5 r were disregarded, and the smallest R angle was R5. The minimum edge length criterion was set to be 1/3 of the smallest R angle. In this study, a small edge length indicated fine finished products. The minimum edge length adopted in this study was 0.61.

1. Lattice: 40000 (Min Edge Length = 0.61).
2. Material temperature: 20 °C.
3. Mold temperature: 20 °C.
4. Constant shear friction coefficient μ: 0.12.

IOP Conf. Series: Materials Science and Engineering **538** (2019) 012013 doi:10.1088/1757-899X/538/1/012013

4.3. CAE Simulation Analysis Results

The CAE simulation analysis in this study used materials with the following diameters: φ32, φ33, and φ34. Figures 4–8 present the CAE simulation analysis results.

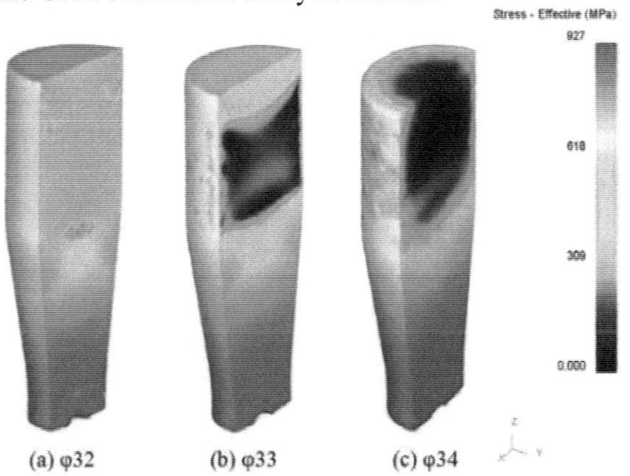

Figure 4. First-pass equivalent stress map.

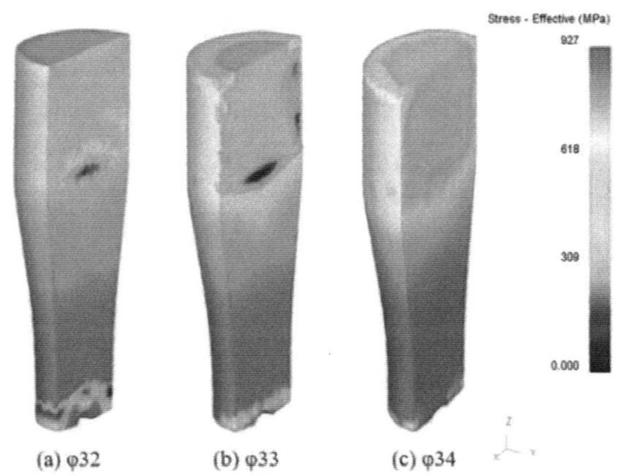

Figure 5. Second-pass equivalent stress map.

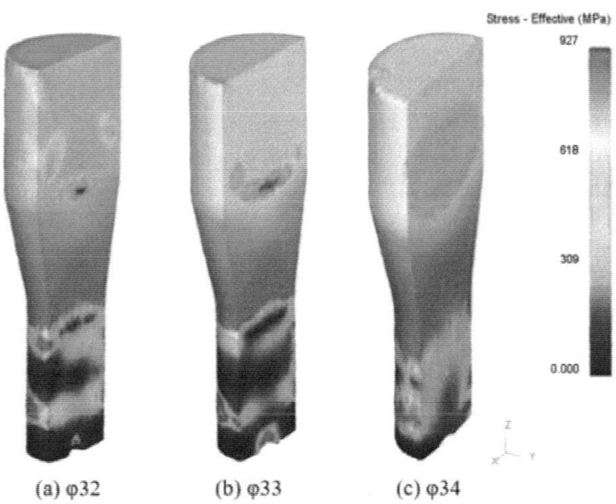

Figure 6. Third-pass equivalent stress map.

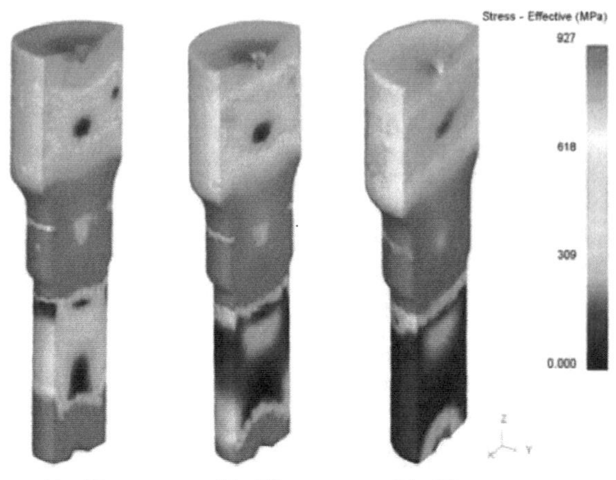

Figure 7. Fourth-pass equivalent stress map.

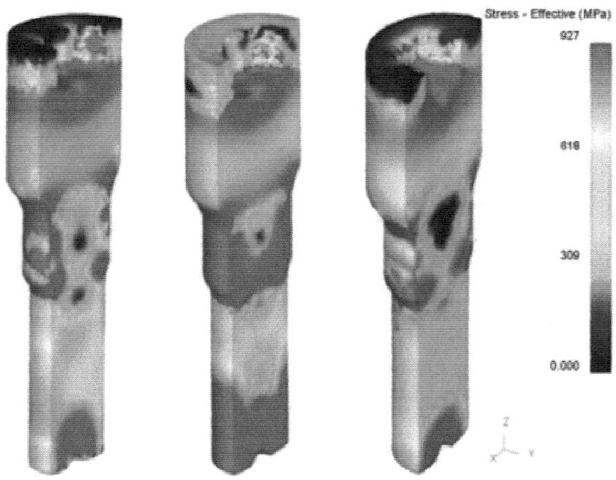

Figure 8. Fifth-pass equivalent stress map.

The impacts of a hammer and the upper mold falling from a certain height were simulated to create plastic deformation in the materials for molding. The area between the blue and red blocks shown in the figures was utilized to evaluate the received stress and determine whether problems such as overly large stress or stress concentration occurred during the processing. The area was also used to understand if a forging piece was broken or if spew occurred. Figure 5 shows that the blue area received less stress than did the red area.

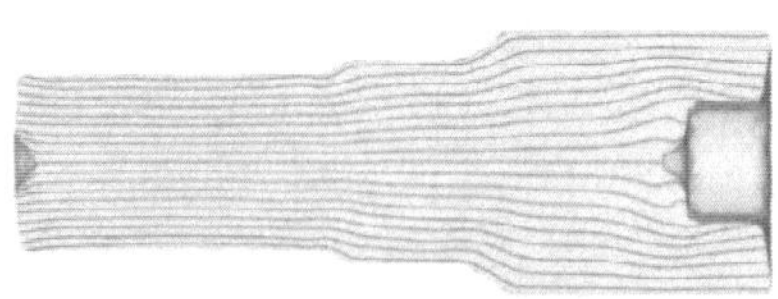

Figure 9. Forging streamline simulation map.

The forging streamlines formed in the CAE simulation were utilized to evaluate whether the method planning was adequate. An uneven streamline distribution reveals uneven metal flow, resulting in reduced mechanical performance, especially the impact toughness and fatigue performance. The actual product may have had such problems as misalignment, folding, and parting surface cracks. The simulation map in Figure 9 shows that the forging streamlines were smoothly distributed, revealing that the mold design and method planning of the present study were reasonable and would not lead to defects during production.

4.4. CAE Analysis of the Machine Output Tonnage

Figures 10–14 present the simulated machine output tonnage. The tonnage of φ32, φ33, and φ34 varied slightly. To verify the simulated data, cold forging was conducted, and the results were compared with the simulated data.

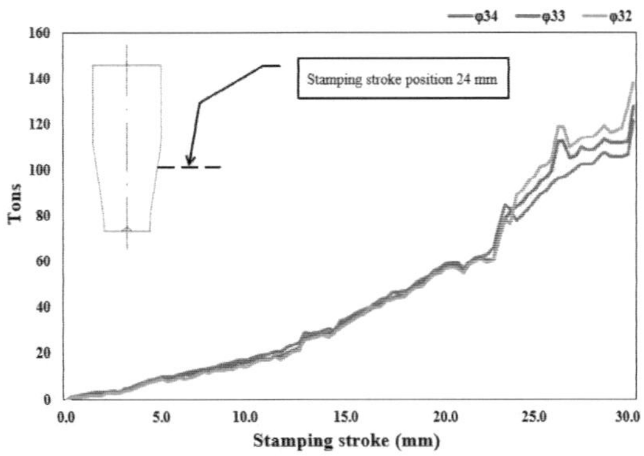

Figure 10. First-pass machine output tonnage.

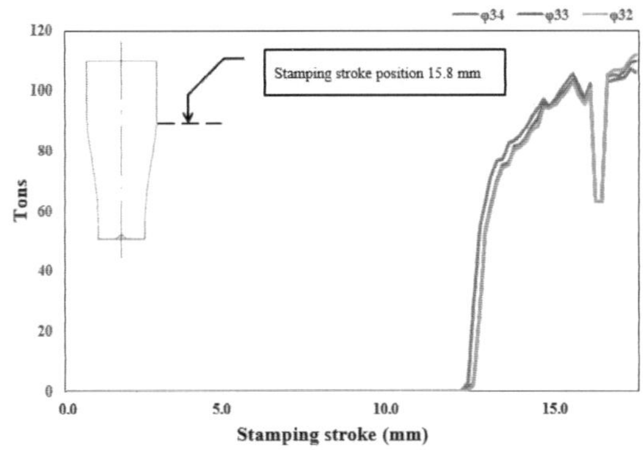

Figure 11. Second-pass machine output tonnage.

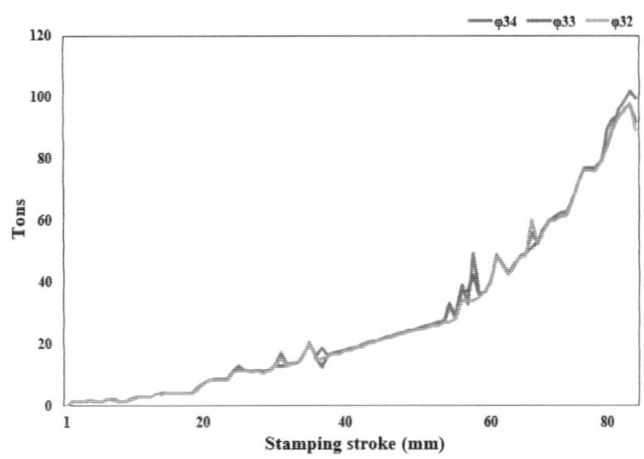

Figure 12. Third-pass machine output tonnage.

Figure 10 shows that when the machine stamping process reached approximately 24 mm, the tonnage of φ34 was the highest. When mapping the simulation position in the CAE simulation, this study saw that before the stamping process reached 24 mm, the molding stress was almost completely concentrated at the nosing. Therefore, this study inferred that the reduction rate correlated to the machine output tonnage of the nosing. After the stamping process reached 24 mm, the tonnage of φ32 was the highest. Therefore, this study inferred that the material deformation of φ32 extruding to φ34.5 was larger than that from φ34 extruding to φ34.5, and the machine output tonnage was also higher.

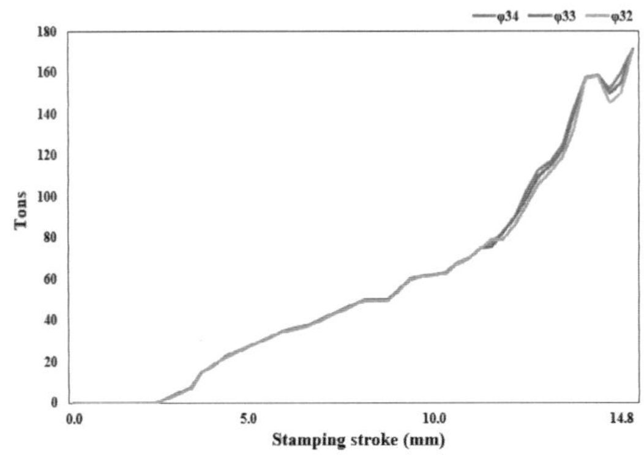

Figure 13. Fourth-pass machine output tonnage.

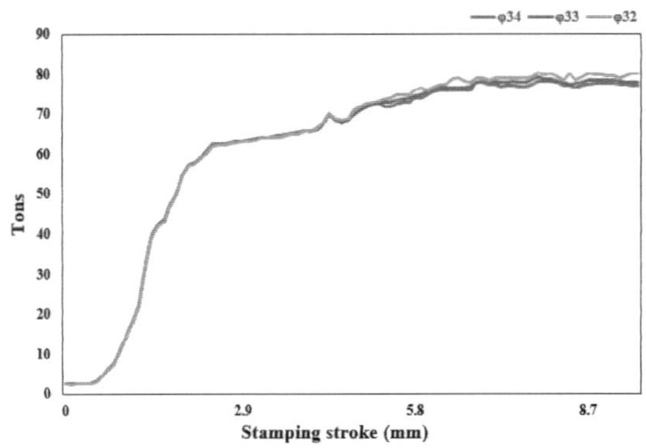

Figure 14. Fifth-pass machine output tonnage.

As shown in Figure 11, when the machine stamping process was approximately 15.8 mm, the tonnage of φ34 was the highest. For the same reasons, a large reduction rate led to high machine output tonnage. When the machine stamping process reached approximately 15.8 mm, φ32 demonstrated the highest output tonnage.

The fifth-pass machine output tonnage simulated using CAE was the highest for φ32; however, the differences were small among the various diameters. The reason may be that during the previous passes, because the deformation from φ32 to φ34.9 was larger than those of the other diameters, the forging streamline should have been denser and its mechanical characteristics enhanced. Therefore, the machine output tonnage was higher than that of the other diameters.

5. Experiment Results

5.1. Comparison of Finished Product and Methods

The following figures show the finished product after cold forging. Their measurements revealed that the sizes of φ32, φ33, and φ34 all satisfied the designed size of the method, and the errors were within the set tolerance. The first-pass of φ34 showed that the phosphate coating at the nosing was completely worn, whereas that of φ32 still had some phosphate coating. A comparison of the CAE simulation and the actual results showed that the largest diameter resulted in the largest stress in the nosing and caused considerable wear in the phosphate coating and mold.

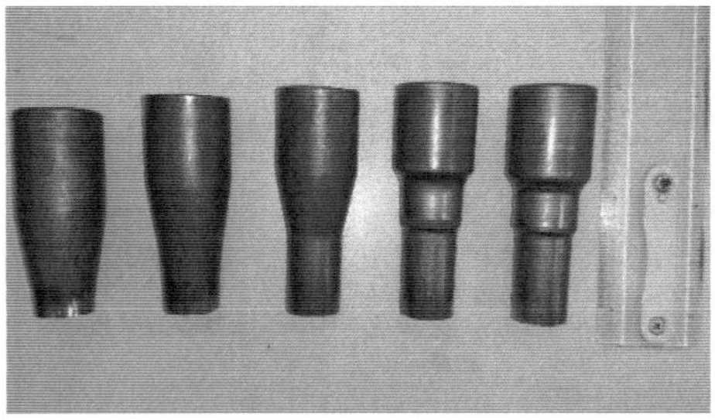

Figure 15. Side view of the ϕ 34 finished product.

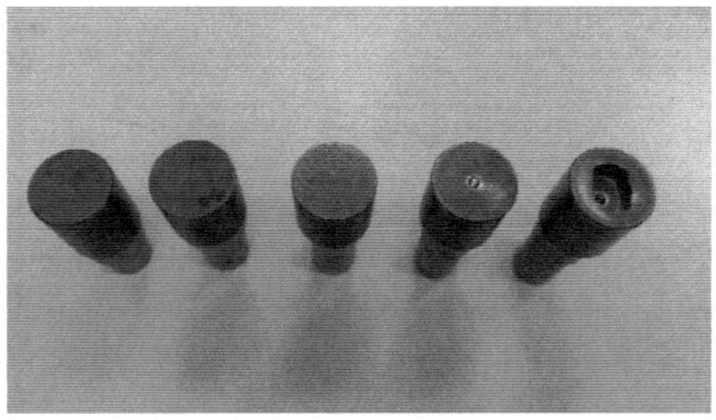

Figure 16. Top view of the ϕ 34 finished product.

5.2. Comparison of Finished Product and Methods

After heat embedment, this study used an optical microscope to observe the metallographic change before and after forging. Figure 17(a) shows the metallographic change after forging. The white tissue resembles ferrite, and the black, austenite. During the formation process, the metallographic tissues were extruded, resulting in their crystal grain shape changing from their original rough and large grain boundaries to smaller and narrower boundaries.

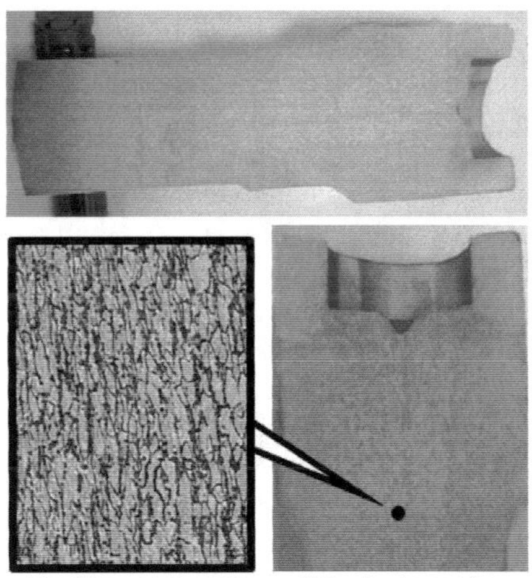

Figure 17. Metallographic observation and forging streamline.

Figure 17(b) illustrates the distribution of the forging streamlines, whereby the flow direction was the same as that of the CAE simulation (Figure 9). Additionally, no misalignment, folding, or parting surface cracks were observed. The metal material flowed homogeneously, and the distribution of the forging streamlines was smooth. Inspecting the metallographic revealed that the tissues of the materials became tighter after extrusion, which subsequently enhanced the mechanical performance.

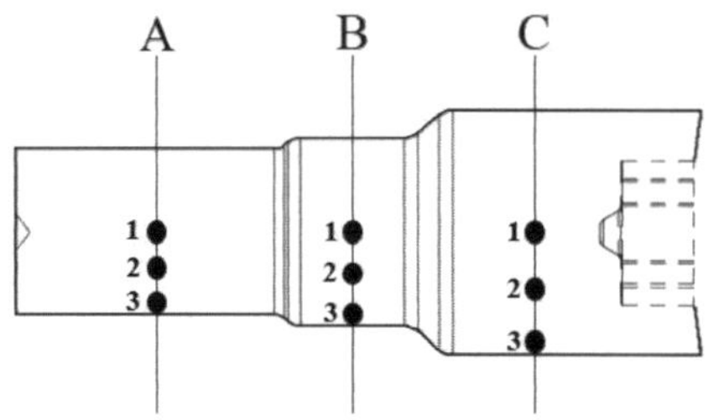

Figure 18. Hardness test locations.

Table 4. Hardness measurements.

Hardness (HV)			
	φ32	φ33	φ34
A1	256	263	278
A2	264	274	290
A3	271	279	296
Average	263.7	272	288
B1	239	249	255
B2	250	262	273
B3	253	267	288
Average	247.3	259.3	272
C1	264	256	249
C2	275	267	256
C3	278	275	267
Average	272.3	266	257.3

Table 4 lists the hardness measurements. As shown, the hardness at locations A1, B1, and C1 was lower than that at A3, B3, and C3, respectively. A comparison with the CAE simulation results revealed that when the internal molding stress was low and the deformation small, the internal hardness was lower than the external hardness. This trend was also observed in φ32, φ33, and φ34.

6. Summary and Conclusions

1. The experiment results showed that using an intermediate shaft of varying diameters could reflect the hardness distribution in a workpiece undergoing cold forging. In other words, an ideal hardness in a forging piece could be achieved without heat treatment.

2. The first-pass CAE simulation revealed that the φ34 nosing into φ23.1 required a molding stress larger than the φ32 nosing into φ23.1. A comparison of the hardness of the simulation and finished product revealed that it had undergone cold forging. Moreover, this study learned that the locations in the CAE simulation that reflected large stress demonstrated high hardness after cold forging.

3. The actual hardness values revealed that after cold forging, the internal hardness of a round bar material was lower than the external hardness. Because the exterior impact was larger than that in the interior, the exterior experienced larger stress, resulting in higher hardness.

4. The hardness data in Figure 18 and Table 3 show that the hardness at the end of location A on φ34 was higher than that at the end of location A on φ32. However, the hardness at the end of location C on φ34 was lower than that at the end of location C on φ32. The CAE simulation results revealed that when the material of the cold forging piece experienced large deformation, its hardness increased (given that the materials and molds were the same; only the diameters of the material were different).

5. The hardness at the end of location A on φ34 was higher than that at end of location C, whereas the situation showed the reverse for φ32. In other words, the hardness at the end of location A on φ32 was lower than that at the end of location C on φ32. The machine output tonnage curve graph and

IOP Conf. Series: Materials Science and Engineering **538** (2019) 012013 doi:10.1088/1757-899X/538/1/012013

equivalent stress graph of the CAE simulation revealed that because the energy required to extrude from φ32 to φ35 was large, it created relatively high hardness.
6. Although an ideal hardness could be achieved through the application of materials of different diameters, the locations of stress concentration should be considered to prevent large deformation that wears the phosphate coating on the surface of materials and subsequently wears the mold or reduces their service life.

References

[1] M.G. Rathi, N.A. Jakhade, Int. J. Sci. Res. Publ. 4, ISSN 2250-3153, 1-7 (2014).
[2] YP Luh, JR Ciou, HL Wang, JF Jhang, 21st Natl. Conf Mech. Mach. Des. 26-27 (2018).
[3] Q-C. Hsu, R.-S. Lee, J. Mater. Process. Technol. 69, 264-272 (1997).
[4] Y.-j. Zhang, W.-j. Hui, H. Dong, J. Iron Steel Res. Int. 14, 189-194 (2007).
[5] W. Shichun, J. Mech. Work. Technol. 6, 333-345 (1982).
[6] F.T. Mahi, D. Schmoeckel, Encyclopedia Mater. 5437-5438 (2001).
[7] J. Walters, W.T. Wu, A. Arvind, G. Li, D. Lambert, J. Tang, J. Mater. Process. Technol. 98, 205-211 (2000).
[8] L. Janicek, J. Petruska, B. Maros, S. Rusz, "Cold forming of bolts without thermal treatment," J. Mater. Process. Technol. 125-126, 341-346 (2002).
[9] D. Shi, X. Yi, Ordnance Mater. Sci. Eng. 1–4 (1990).
[10] H. Tumer, F. O. Sonmez, J. Mater. Process. Technol. 209, 1538-1549 (2009).
[11] M. Rumiński, P. Skubisz, Ł. Lisiecki, Key Eng. Mater. 622-623, 179-185 (2014).
[12] J. Zottis, C.A.T.S. Diehl, A.D.S. Rocha, J. Mater. Research. Technol. 7, 103-202 (2018).
[13] M. Milutinović, S. Baloš, M. Plančak, D. Movrin, J. Mater. Process. Tech. 249, 179-192 (2017).
[14] H. Tumer, FO Sonmez, J. Mater. Process. Tech. 209, 1538-1549 (2009).
[15] P. Groche, P. Kramer, N. Bay, P. Christiansen, L. Dubar, K. Hayakawa, C. Hu, K. Kitamura, P. Moreau, CIRP Ann. 67, 261-264 (2018).
[16] CL HU, Z. Zhao, GR Yin, ZF Yuan, XL Xu, Trans. Nonferrous Metals Soc. China. 19, s552-s557 (2009).

IOP Conf. Series: Materials Science and Engineering **538** (2019) 012014 doi:10.1088/1757-899X/538/1/012014

Manufacturing Processes of Sensorial Materials: Sensors Placement and Experimental Validation

S. Mekid[a] and H. Daraghma

King Fahd University of Petroleum and Minerals, Department of Mechanical Engineering, Dhahran, 31261, Saudi Arabia

[a] Corresponding author: smekid@kfupm.edu.sa

Abstract. Nervous and sensorial materials are gaining more attention with in-material host of both sensors and actuators together to build functional materials for robotics and structural health monitoring. This paper presents a case study of a cantilever beam made with different materials including metals, polymers and powder-based materials using variety of manufacturing procedures with inserted sensors inside. The part becomes sensorial in this case and may be more functional for structural health monitoring and smart materials with dynamic functions. The results show that not only the location of the sensors is extremely important inside the material to capture true material's behavior, but also the integrity of both sensors and material that must be satisfied. Example with experimental validation have shown agreement between theoretical and measured values.

1. Introduction

Most materials can be turned into sensorial materials if the latter can host inserted sensors at the time of fabrication, or, at post fabrication. Actuators can also be added in the neighborhood of the sensors to build nervous materials [1]. A review of past techniques to embed sensors e.g. fiber optics (FBG), fiber PZT in materials has been carried out [2] showing possibilities to embed while manufacturing composite materials but little difficult to insert as a post process in metals e.g. Aluminum.

In particular, embedding FBG sensors into a polymer matrix is not a difficult process. The great potential of optical fiber sensors for the structural health monitoring was highlighted by Minakuchi and Takeda [3]. The authors reviewed research work mainly in Brillouin-based distributed sensing; and presented damage detection, life cycle monitoring and shape reconstruction systems applicable to large-scale composite structures. Luyckx and co-authors [4] provided an overview of some of the technical issues related to embedding sensors in composite structures. The output of an embedded FBG sensor was related to the strain of the structure through a monitoring scheme. In addition, the authors focused on temperature compensation methods when measuring strain with FBGs. In addition, embedding a combination of shape memory alloy wires and FBG sensors in glass fiber reinforced composites was investigated. More recently, the possibility to embed polymer optical fiber gratings [5] in composite materials was explored, and promising results were reported at temperatures and strains. Metallic materials remain widely used in many industries such as automotive [6] and aerospace [7]. Li et al. [8] explored the possibility to embed FBGs in aluminum and copper metal foils using ultrasonic welding processes and evaluated their sensing characteristics. The authors reported that it was not possible to embed the bare fiber in copper. This was attributed to the high hardness of copper. The

IOP Conf. Series: Materials Science and Engineering **538** (2019) 012014 doi:10.1088/1757-899X/538/1/012014

chemical plated fiber was embedded in aluminum; however, it was destroyed, because of the welding pressure, vibration and friction. The interesting method would be to have a sensor that almost naturally embedded at the subsurface or well inside at the time of manufacturing the part e.g. composite material, powder based materials or 3D printing with most materials ranging from metals to plastic [8, 9].

Embedding methods are various and challenging hence need more investigations. The embedding manufacturing processes vary depending on the material e.g. manual setting, gluing, and ultrasonic consolidation in plastic or metallic materials, powder based material with compaction followed by sintering, and rapid prototyping with laser sintering or 3D printing using polymer materials e.g. ABS, PLA. When the fiber optics are embedded inside materials e.g. plastic or metallic, the integrity has to be verified. Conditions of non-violation of integrity are on two levels; the sensors integrity when embedded in the host material, and the material integrity at the interface between the sensors and the host material [10]. Sensors placement in optimal places is important to properly recover data for various objectives e.g. to evaluate deformation and request to repair damage due to an unexpected impacting force on the structure. Consequently, this force needs to be identified in magnitude and location as part of structural health monitoring (SHM). Optimal sensors placement over a structure is one of the most active and interesting research for SHM. We are interested in the sensors placement within the material, but this can be initially at the surface, subsurface or inside at known depth.

The paper discusses corresponding manufacturing techniques for different materials to insert sensors at the subsurface of the material or in-depth in specific locations with a condition of sensors integrity and reliability in reporting material behavior.

2. Sensors location inside materials

While sensors deployment in host material is of importance, their local position becomes more important depending on the specific desired sensing needed in a particular position inside the material part. The offered local positions can be at various levels e.g. the surface, sub-surface or in depth inside the material.

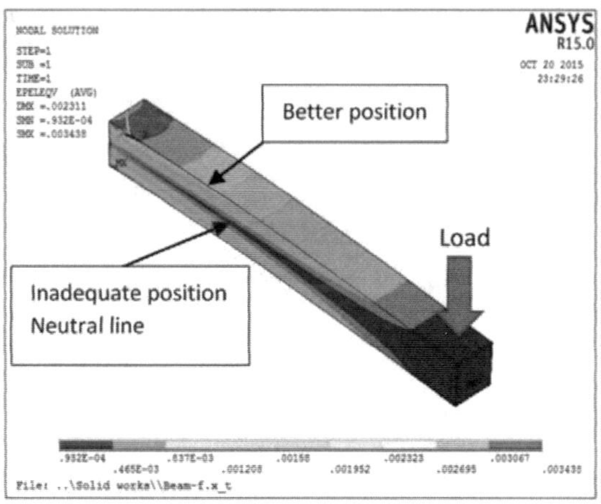

Figure 1. Sensor position suitable positions for strain measurement under bending condition.

This will depend on the type of sensing and where to capture sensing data. Temperature sensors can be placed anywhere in the material to collect temperatures while strain sensors are better placed at the surface or subsurface to collect strain away from neutral line, otherwise, no worth information will be recorded. Fig. 1 shows an example of strain distribution with better and inadequate positions of the sensors in bending cantilever. Embedding sensors inside the material needs careful attention depending on the host material and the type of the sensor. The corresponding embedding process is required. Recent tests have shown that embedding fiber optics can be easier in polymer, nylon but need specific manufacturing process for metals e.g. Aluminum [9].

IOP Conf. Series: Materials Science and Engineering **538** (2019) 012014 doi:10.1088/1757-899X/538/1/012014

3. Experimental tests

The purpose of inserting sensors inside materials is to turn them as sensorial materials capable of capturing true behaviors of any needed aspect depending on the application. This includes structural health monitoring and smart materials with dynamic specified behaviors for robotics and closed loop control. Hence, the integrity of the sensor and the material has to be verified to be able to report true information.

3.1. Embedding fiber optics in polymers

The purpose of the following experimental tests is to build a sample material with embedded fiber optic using 3D printing with the need to check both the material and the sensor integrity so that the material is not affected by the insertion of the sensors and the sensors are within the right operating conditions. The ABS filament is heat up at 250 oC using a standard 3D printing machine as shown in Fig.2. As the printing progresses, the placement of the fiber optics is manual. The fibers are lined up with the trajectory of the filament at the right position. The filling option in the 3D printer has to be high density to secure complete filling of the polymer without empty space.

Figure 2. a) 3D printing machine. b) ABS beam with embedded fiber optic.

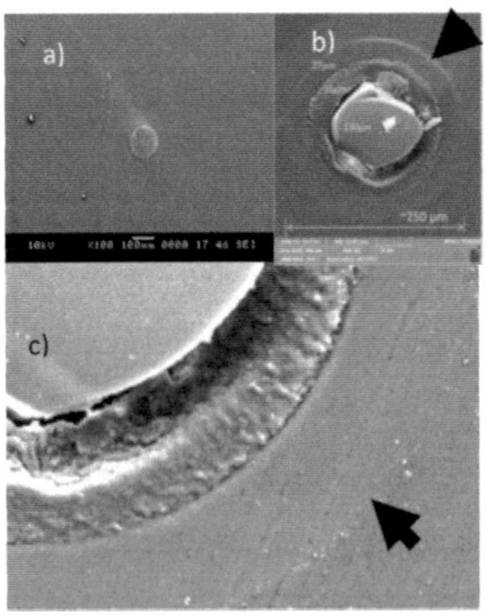

Figure 3. Cross section of the fiber embedded inside the polymer (Courtesy of Dr Khan).

To ensure reliable readings from fiber optics, the embedding of the sensors have to be well encapsulated and adhering to the material. This will allow the sensor to report the real behavior of the material surrounding it. By examining the SEM pictures of the cross section of the embedded fiber inside the printed polymer shown in Fig. 3, it appears that ABS material is firmly and completely adhering the surrounding the fiber outer jacket. The close picture in Fig. 3 d) shows a complete melted material around the outer jacket of the fiber. To validate this observation, an experiment has been prepared with an aluminum beam having fiber optics embedded at the subsurface and having three Bragg sensors. All sensors are connected to data acquisition system run using national instrument interface.

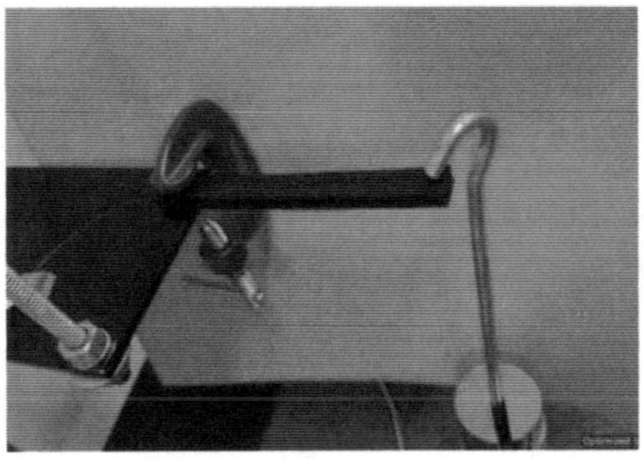

Figure 4. Experimental setup for ABS beam.

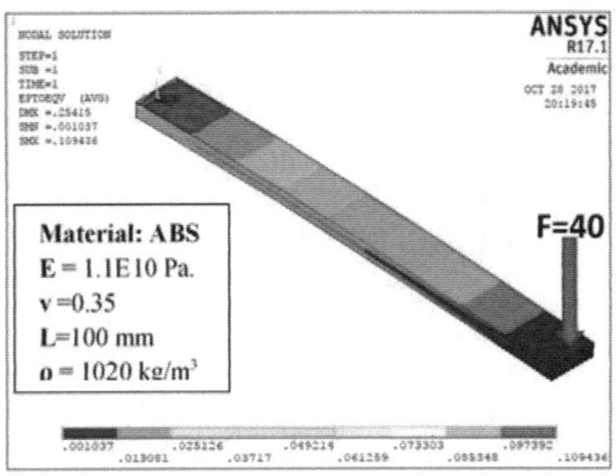

Figure 5. FEM simulation of the ABS cantilever beam.

Similar modeling and simulation of the ABS material has been carried out. The beam has been printed using ABS material and fiber optic with three FBG sensors. The fiber optic is added during the last passes of the printing head to finish the manufacturing. The beam is mounted with an applied load of 40 N at the tip as shown in Fig. 5. Strain measurements have been carried out and compared to FEM simulation as shown in Fig. 5. The comparison shows agreement between simulation and experimental measurements.

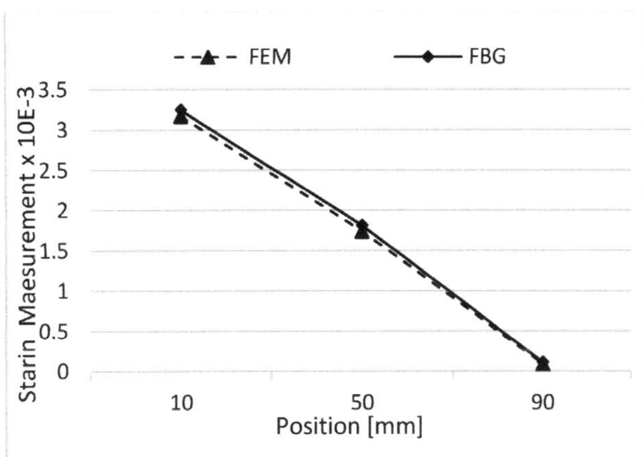

Figure 6. Strain comparison between FEM and fiber optic measurements in ABS.

IOP Conf. Series: Materials Science and Engineering **538** (2019) 012014 doi:10.1088/1757-899X/538/1/012014

3.2. Embedding fiber optics in metals

The fiber can be properly embedded inside a metallic structure at subsurface as discussed previously using two techniques; 1- gluing the fiber inside a slit cut on the metal, or 2- consolidate the fiber optic in the material using ultrasonic vibration followed by a small load on the fiber.

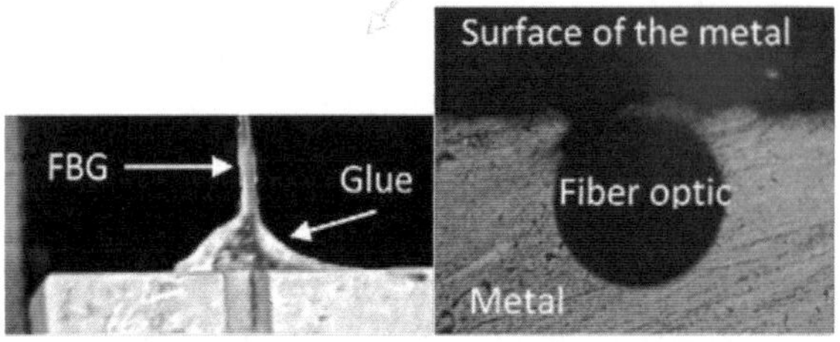

Figure 7. Fiber optic embedding inside Aluminium with glue or by Ultrasonic embedding.

Fig. 7 a) shows an example of the fiber optic glued in a manufactured slit on the surface, while Fig.7 b) shows embedded fiber optic using ultrasonic consolidation. The operating frequency was 20kHz, short wavelength within one second processing time. The host material was aluminum.

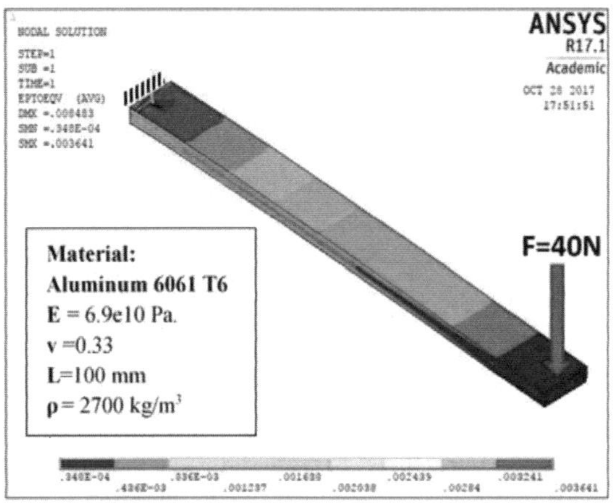

Figure 8. FEM simulation of the Aluminum cantilever beam.

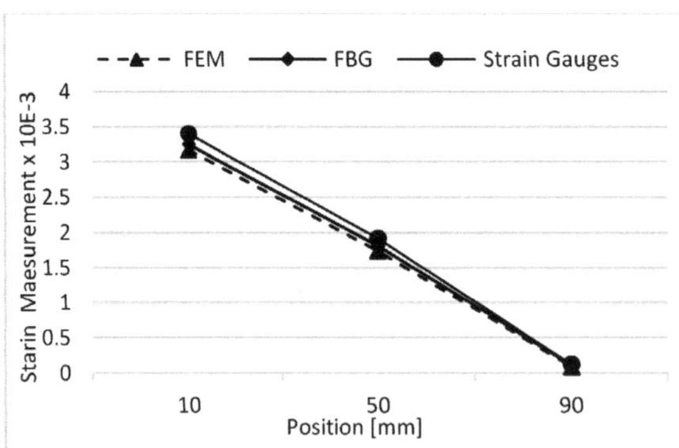

Figure 9. Strain comparison between FEM, fiber optic and strain gauge measurements.

To validate the measurement effectiveness of the embedded fiber optic Bragg sensors, an experiment has been prepared with an aluminum beam having fiber optics embedded at the subsurface and having three Bragg sensors. Three-strain gage sensors have been installed on the top of the FBGs sensors. All sensors are connected to data acquisition system using national instrument interface.

To be able to evaluate the measurements by the FBG sensors, the beam has been modeled and simulated by finite elements to compute the strain distribution at the surface over the length of the beam. A force of 40N is applied at the tip. The results are shown in Fig. 8. The comparison of the measurements from the strain gauges, FBGs and the simulation show good agreement as shown in Fig. 9.

3.3. Embedding carbon fiber sensors in polymers

Carbon fiber based material exhibits excellent mechanical properties and linear piezo resistivity behavior. Hence, it is used as a sensor that can measure strain for example. The fiber optic comes in various densities e.g. 1K, 3k, 5k, ...etc. We have selected 3K type of carbon fiber material that has been embedded inside ABS material, while printing a beam.

An experiment was carried out to compare the reading in deflection for an ABS beam with and without embedded carbon fiber. The beam is cantilever with rectangular cross section i.e. 12 mm width, 9 mm height and 100 mm length (Fig.10,11). The maximum deflection depending on the position over the length can be determined using Eq.1.

$$\gamma_{max} = PL^3/3EI \tag{1}$$

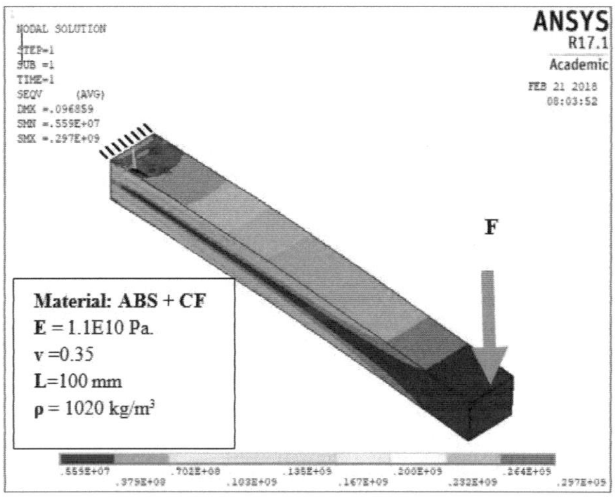

Figure 10. FEM modeling of the ABS beam and sample with embedded carbon fiber.

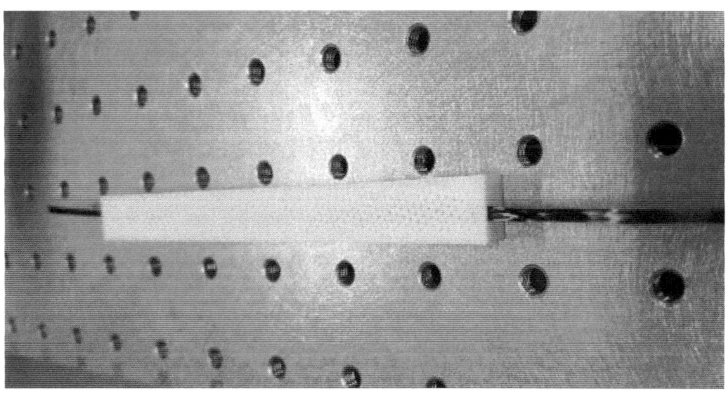

Figure 11. Sample with embedded carbon fiber.

Fig. 12 presents a comparison of deflection at the tip of the cantilever beam with and without carbon fiber inserted. Some discrepancy exists between beams. There is a slight effect of the carbon fiber of type 3K added to the structure in bending. The beam is filled at about 90%. The carbon fiber needs specific installation to make sure the ports can read properly. These measurements have been done several times to reach the expected results.

The carbon fiber has been tested on its own to understand its behavior. It is known that the behavior is little different from the case where it is embedded.

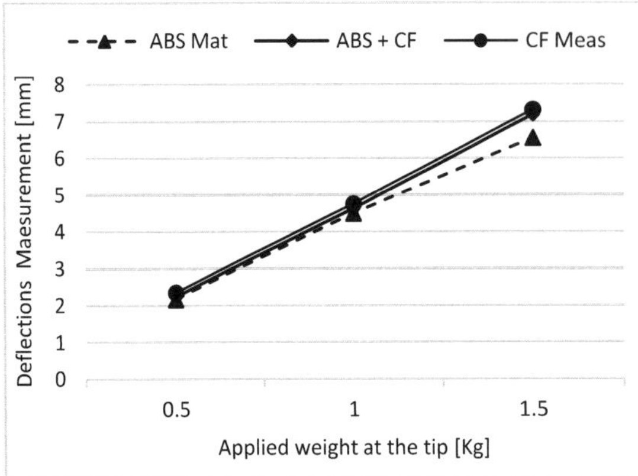

Figure 12. Deflection comparison between various cases.

4. Conclusion

This paper has shown that both fiber optic and carbon fibers seem to be very suitable to be embedded within a wide range of materials. Both sensors can be embedded while keeping their integrity safe. The host material builds a good sticking interface around the sensors which makes the data reading describing the parts behavior reliable. It is important to locate the sensors where useful readings are expected. This constitutes a good application for structural health monitoring and also robotic elements for precision handling or positioning under various shapes.

Acknowledgments
The authors would like to acknowledge the support provided by King Abdulaziz City for Science and Technology (KACST) through King Fahd University of Petroleum & Minerals (KFUPM) for funding this project No.11-ADV2133-04 as part of the National Science, Technology and Innovation Plan.

References

[1] Mekid S and Kwon O, 2009 Nervous materials: a new approach of using advanced materials for better control, reliability and safety of structures, Sci. Adv. Mater.1, pp276–85
[2] S Mekid, N. Saheb, S. M. A. Khan, and K. K. Qureshi, "Towards sensor array materials: can failure be delayed?", Sci. Technol. Adv. Mater., vol. 16, no. 3, p. 34607, 2015.
[3] Minakuchi S and Takeda N 2013 Recent advancement in optical fiber sensing for aerospace composite Photonic Sensors 3(4) pp 345-354.
[4] Luyckx G Voet E Lammens N and Degrieck J 2011 Strain measurements of composite laminates with embedded fibre bragg gratings: Criticism and opportunities for research Sensors 11(1) pp 384-408
[5] Wei Sun, Lifang Jia, Zhenyu Wang, Ziguang Jia, Optical fiber sensor encapsulated by polyurethane, Optik, Volume 165, 2018, pp 124-131,
[6] Miller W S Zhuang L Bottema J Wittebrood A J De Smet P Haszler A and Vieregge, A 2000 Recent development in aluminum alloys for the automotive industry Materials Science and Engineering A 280(1) pp37-49.
[7] Dursun T and Soutis C Recent developments in advanced aircraft aluminium alloys Materials

and Design 56, pp 862-871.
[8] Li Y Liu W Feng Y and Zhang H 2012 Ultrasonic embedding of nickel-coated fiber Bragg grating in aluminum and associated sensing characteristics Optical Fiber Technology 18(1) pp 7-13.
[9] S Mekid, H Daraghma, Experimental ultrasonic sub-surface consolidation of fiber Bragg grating for sensorial materials, Journal of Materials Processing Technology, 2018, 252, pp673-679
[10] Saheb, Nouari, and Mekid Samir. "Fiber-Embedded Metallic Materials: From Sensing towards Nervous Behavior." Materials 8, no. 11 (2015): pp7938-7961.

IOP Conf. Series: Materials Science and Engineering **538** (2019) 012015 doi:10.1088/1757-899X/538/1/012015

Aluminium Lined, Carbon Composite Overwrapped Pressure Vessel

Yunusa Mujaheed[a], Zu Lei, Chen Dan and Tapa Arnauld Robert

School of Materials Sciences and Engineering, Wuhan University of Technology, Wuhan 430070, China

[a] Corresponding author: Ekamujaheed@yahoo.com

Abstract. A numerical analysis of filament-reinforced internally pressurized cylindrical vessel with over-wrapped metallic liner is presented, the study deals with simulation of aluminium liner pressure vessel overwrapping by composite T800/epoxy was selected, ABAQUS Software was use in which 20 layers of composite were analysed. The stress state of the pressure vessel under the internal pressure of 33MPa was analysed, and the bursting strength was predicted to be 92.3MPa. MATLAB software was used to simulate the winding process, the 5 axis numerical control winding machine was used for the actual winding, and the fibres were arranged in a regular and uniform way, and the sliding phenomenon was not occurred. After solidifying, the bursting strength was 85.4MPa, and the difference between the actual value and the predicted value was 6.9%. The experimental results were in good agreement with the simulation results.

1. Introduction

Composites pressure vessels are widely used for many applications. Their light weight is a major benefit for portable uses, such as medical gases for health care applications using vessels volumes in the range 1-5L. On the other hand, their high mechanical strength allows storing very high pressure gas, which make the particular suitable for transportation of large quantity from a production to distribution site, using vessel volumes higher than 150L. The combination of these two advantages lead to become a standard technology for 700 bar hydrogen (CGH2) or 200 bar natural gas (CNG) vehicle using on board storages 50-150 L.

For all these applications, it paramount to ensure the structural integrity of the vessels during their entire lifetime, mechanical damage such as impacts or drops is one of the most feared [1]; as it is well known that composite materials are very sensitive to transverse impacts [2].

Pressure vessels have long been manufactured by filament winding process. In this process, bands of resin impregnated fibre are wound over a cylindrical mandrel using a computer control fibre placement machine. Fibre application to the metallic mandrel is usually accomplished by a transverse feed head over a rotating surface, capable of achieving any desired winding orientation. Traditionally, composite analysis techniques rely on the assumption of plane stress, and therefore they are only applicable to thin laminates. Originally, many

of the classical thin shell theories are based on Kirchhoff-Love hypotheses [3]. Composite vessel has been studied in this present work. The composite vessel is made of inside aluminium liner overwrapping with carbon fibre/Epoxy composite. The analysis demonstrated that for thin-wall and thick-wall single-layered cylinders, the optimum winding angle is in the range of 54^0-57^0. Xu et al. [4] utilized finite element analysis to predict burst pressure of the composite storage vessel. In this present research attempt have been made to:

- Determine the burst pressure and the maximum shell displacement for asymmetric fiber orientations of composite vessel.
- Determine optimum winding angle according to diverse failure criteria.

2. Theoretical Studies

2.1. Optimum Winding Angle

Netting analysis is one of the most popular analytical techniques used for investigating the behaviour of multi-layered composite materials, especially for filament wound pressure vessels. The main assumption in netting analysis is that all loads are carried by the fibres neglecting the stiffness of the matrix and internal pressure subjected to the vessel produces a hoop-to-axial-stress ratio of 2:1 [5] Figure 1 indicates the thin-wall cylindrical vessel with thickness of ''t'' and length of 'L'' which is subjected into the internal pressure ''p''. In order to calculate the axial (longitudinal) and hoop stresses at the cylinder with the radius of ''R'', Eq. **(1)** can be utilized [6]:

$$N_\phi = \frac{PR}{2}, N_\theta = PR \tag{1}$$

Axial and hoop stress can be calculated as in Eq. (2)

$$\sigma_{axial} = \frac{N_\phi}{t} = \frac{PR}{2t}, \sigma_{hoop} = \frac{N_\theta}{t} = \frac{PR}{t} \tag{2}$$

Axial and hoop stresses can be calculated according to equilibrium across the cut section too as Eq. (3) and (4)

$$PL(2R) = 2\sigma_{HOOP}Lt \rightarrow \sigma_{hoop} = \frac{PR}{t} \tag{3}$$

$$P\pi r^2 = \sigma_{axial}(2\pi Rt) \rightarrow \sigma_{axial} \frac{PR}{2t} \tag{4}$$

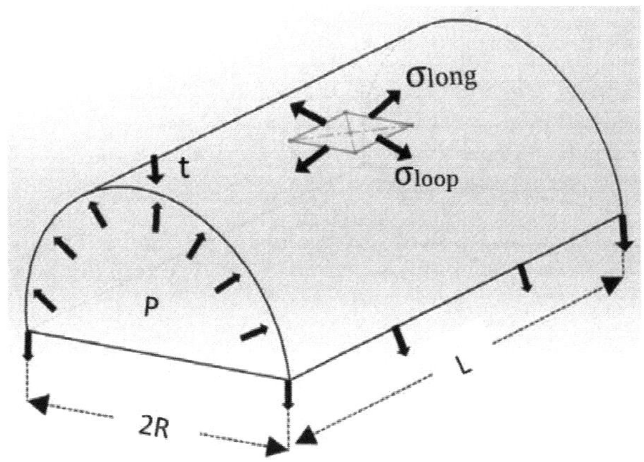

Figure 1. Thin cylindrical shell

Optimum winding angle can be estimated by using ultimate tensile strength along with applying netting analysis respect to 2:1 hoop-to-axial stress ratio as it is shown at Eq.(5) and (6)., as it is shown in figure.

$$N_\theta = \sigma_u t \sin^2\alpha, N_\phi = \sigma_u t \cos^2\alpha \quad (5)$$

$$\frac{N_\theta}{N_\phi} = \tan^2\alpha = 2 \rightarrow \alpha = \arctan(\sqrt{2}) = 54.7^0 \quad (6)$$

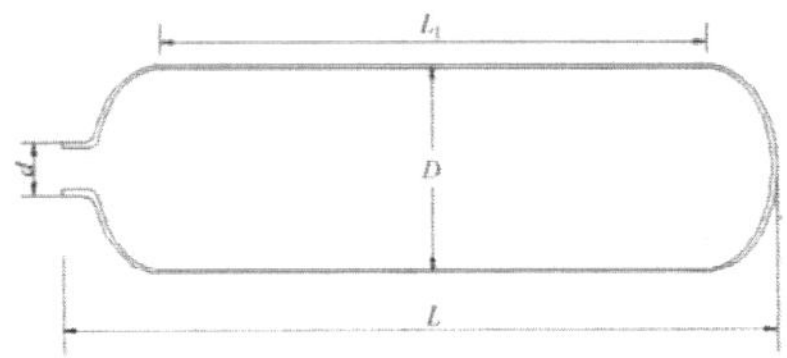

Figure 2. Schematic diagram of the aluminium lining structure

3. Simulation Details

The considered pressure vessel is the vertical one designed for low capacity applications. The vessel is sketched with 505 mm length, 166 mm diameter at the centre point and the wall thickness of (2.4-6) mm base on the geometry. The material of this cylindrical vessel consists of the aluminium alloy inside layer reinforced with composite material T800/epoxy of 20 layers. Table 1 and Table 2 indicates material property of CFRP and aluminum alloy respectively.

Table 1. Basic mechanical properties of selected composite systems

Strength(Mpa)	Modulus(Gpa)	Poisson ratio υ_{12}
$X_t = 2344.96$	$E_{xt} = 144.89$	0.33
$X_c = 1315.18$	$E_{xc} = 138.04$	
$Y_t = 58.07$	$E_{yt} = 8.58$	
$Y_c = 140.84$	$E_{yc} = 8.18$	
$S = 84.46$	$G = 5.09$	

Table 2. Mechanical properties of Al 6061 and CFRP [8]

	Density(kg/m^3)	E_1 (Gpa)	E_2(Gpa)	υ_{12}	G_{12}(Gpa)	τ(Mpa)
AL6061	2750	70	70	0.3	27	600
CFRP	1570	135	8	0.27	3.8	

4. Results and Discussion

4.1. Matlab Results

During the winding process, the yarn starts from one end of the pole hole and passes through a complete loop back to a distance that is offset from the starting point by a yarn width; the lining is then filled through several complete cycles. This article uses the matlab software performs trajectory simulation on the yarn winding first based on mold geometry dimensions with MATLAB's command box input header and geometric parameters of the barrel part are then build by code programming, geometric model of the lining. The equation establishes the geometric model of the winding trajectory.

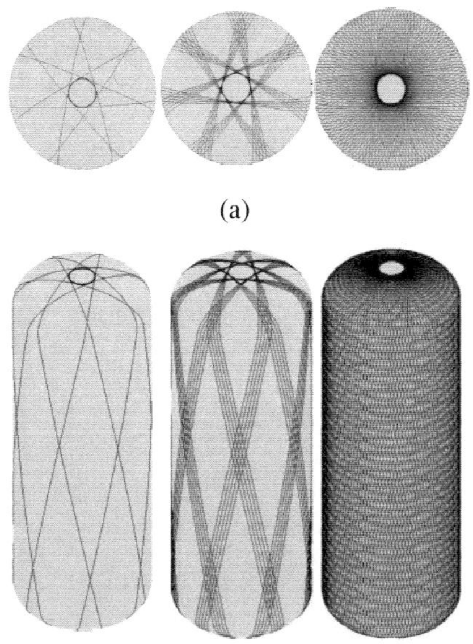

(a)

(b)

Figure 3. Simulation diagram of the winding process considering the bandwidth. (a) Head winding process, (b) Overall winding process

When a complete cycle is over, the yarn is wound into the next stage. Figure 3 (a) intermediate graphics as shown, the webs start at position 1 and are arranged in a counter clockwise direction next to the hysteresis until position 5 is reached. From position 1 to position 5 The number of intermediate yarns, that is, the number of complete loops of the yarn evenly covered with the mold, and at this time the mold has been covered with two layers of overlapping fibres. The arrangement of the yarn after the winding is completed is shown in Figure 3 (a). 'The picture shows. As can be seen from the simulated perspective (b), the yarns are at positions 1, 2, 3, 4, 5 and 1 , because the winding parameters and the number of tangent points are determined on the mold.', and 6′, 5′, 4′, 3′There are overlaps at 2 the width of the aligned individual yarns in the circumferential direction is not necessarily an integer multiple of the circumference of the mold section. Fibre both the bandwidth and the die size are fixed, so this overlap is not effectively avoided.

4.2. Abaqus Result

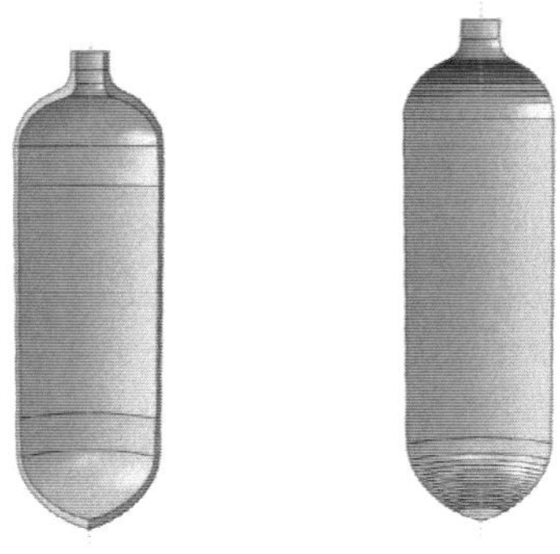

(a) Aluminium lining (b) Composite layer

Figure 4. Geometric model of the pressure vessel in ABAQUS

4.2.1. Finite Element Model. The finite element model of the composite pressure vessel is divided into two parts, the inner liner and the composite layer. First based on Figure 4. Shows the geometric model

of the lining in the ABAQUS software, considering the composite pressure studied in this paper the symmetry of the force vessel in the structure, material and load, and the selection of the pressure vessel to the whole 1/4 of the ring is limited met model.
The winding method of the pressure vessel is spirally and circumferentially wound in order, and the layering method of the barrel section is set for $[11.0^0{}_2/98.9^0{}_2/11.0^0{}_2/98.9^0{}_2/11.0^0{}_2/98.9^0{}_2/11.0^0{}_2/98.9^0{}_2/11.0^0{}_2/98.9^0{}_2/]$, the header section has a total of 10 layers. Composite material layup, its single layer thickness and the barrel section includes spiral and ring direction total 20-layer composite layup, wherein the circumferentially wound single layer has a thickness of 0.15 mm, which is slightly smaller than the spiral wound layer thickness. The paving structure of the pressure vessel barrel section is shown in Figure 5.

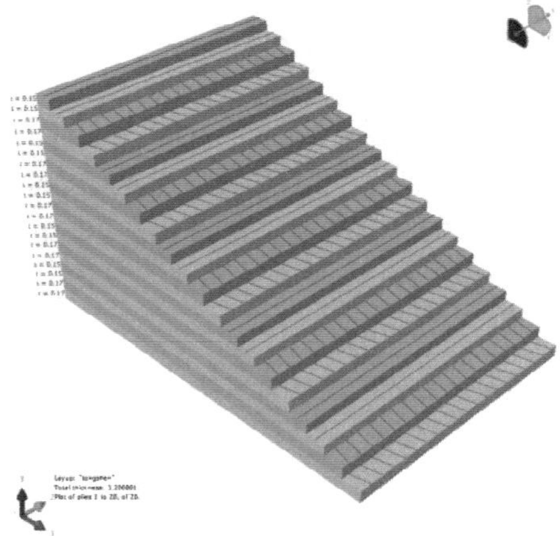

Figure 5. Laminated structure of the composite layer of the barrel section

The S4R shell element is selected as the basic unit of the composite layer, and the lining adopts the C3D8R continuous entity unit. S4R is a linear quadrilateral curved shell unit and it is also a general type of shell element, it is very adaptable, well it can be used for both the simulation of thick shell problems and the simulation of thin shell problems. This unit can be defined, the direction of the local material usually using the finite element method. The anisotropic material parameters of the composite layer are defined according to the direction of the local material. So the unit can be used as a finite element of the composite layer. The basic unit of analysis C3D8R is a 3D solid element and is also a reduced integration unit with 8 Nodes, suitable for computational analysis of large deformations. Applying the unit can guarantee the accuracy of a certain grid. Accurately solve the problem of large distortion of the grid.

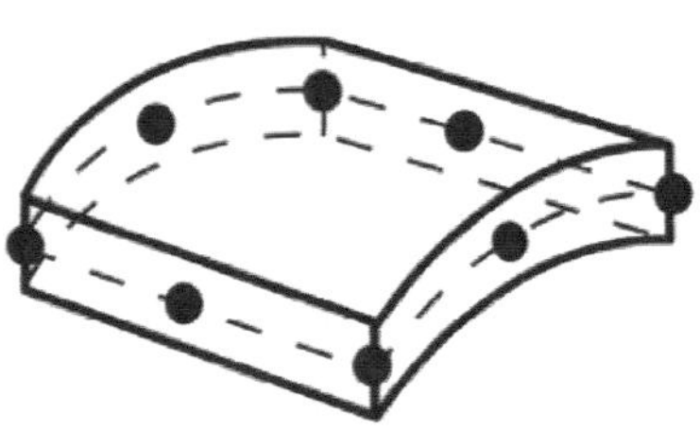

Figure 6. S4R shell geometry

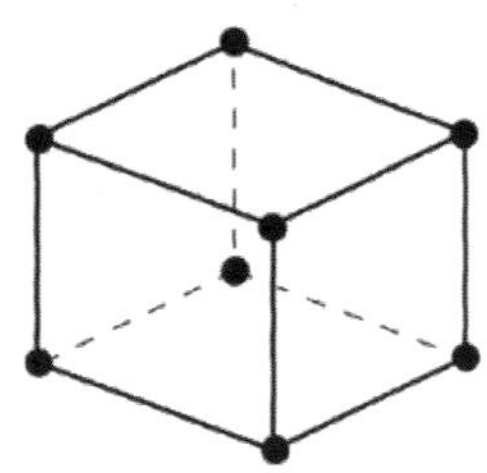

Figure 7. C3D8R continuous solid element geometry

When meshing, you need to consider both the accuracy of the calculation and the efficiency of the calculation in the pressure vessel model. The density of the grid is reasonably set in different parts, appropriate meshing in areas where the calculated data changes greatly. The density is tighter, and the area where the data changes little is calculated to appropriately reduce the mesh density.

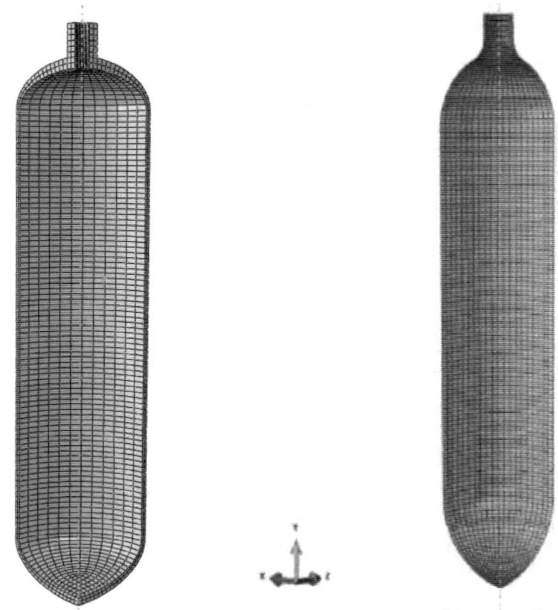

(a) Aluminium liner (b) composite layer

Figure 8. Pressure vessel meshing

4.2.2. Calculation Results and Analysis. A symmetric constraint is applied to the symmetry plane of the finite element model to limit the displacement of the model in the axial direction on the end face of the pole hole, and a binding constraint is imposed between the composite layer and the liner. Taking into account the self-tightening process, the load step is first loaded to the pre-tightening pressure [9] and then unloaded. Reloaded into the container for blasting, analysing the stress state of the pressure vessel when the second load is 33 MPa.

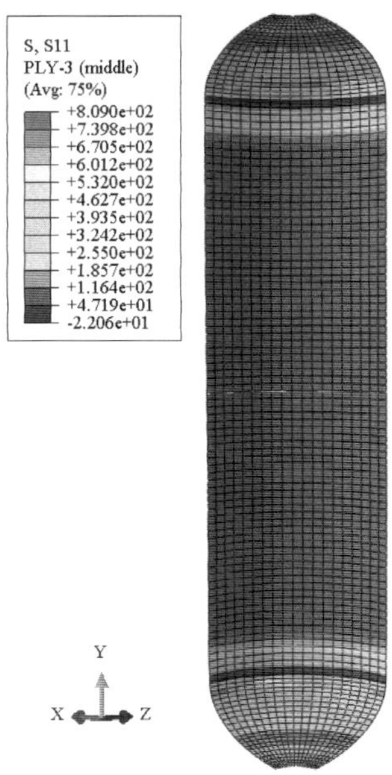

Figure 9. Fibre circumferential layer stress distribution

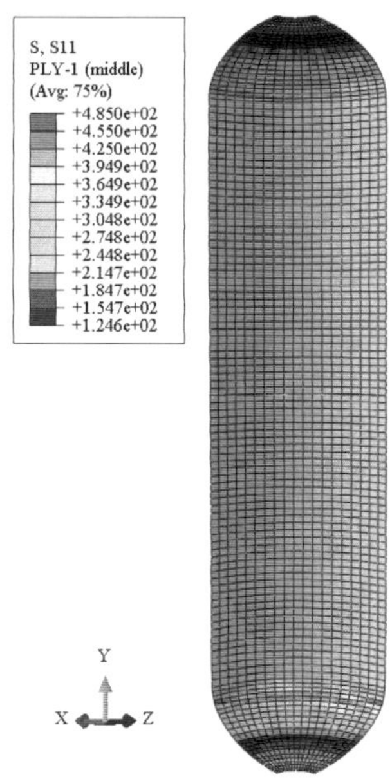

Figure 10. Fibre spiral layer stress distribution

IOP Conf. Series: Materials Science and Engineering **538** (2019) 012015 doi:10.1088/1757-899X/538/1/012015

Analysis of the stress cloud Figure 9 and Figure 10 show that the composite layer has a higher stress level in the barrel section, and the stress level of the circumferential layer of the barrel section is higher than the stress level of the spiral layer, and the maximum stress of the circumferential layer at 809.0MPa, the maximum spiral layer stress is 485.0MPa. This shows the stress of the spiral layer and the circumferential layer. The horizontal difference is large, because the hoop tensile stress is greater than the axial stress during the compression process, and the fibre side the direction of the elastic modulus is the largest, and the ring is subjected to a larger tensile load, so the circumferential stress level is the highest.

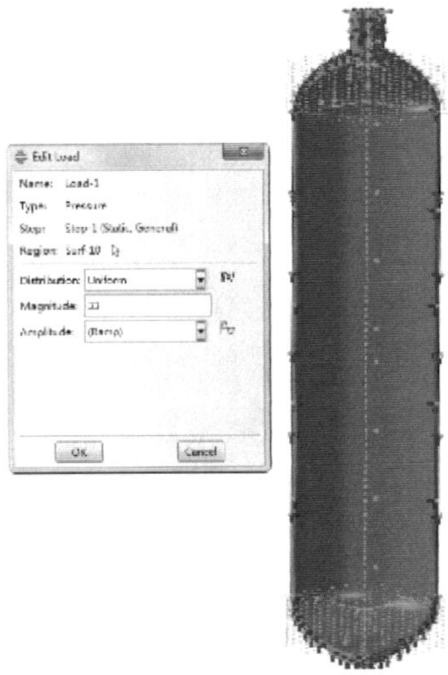

Figure 11. Load and boundary condition of the whole model

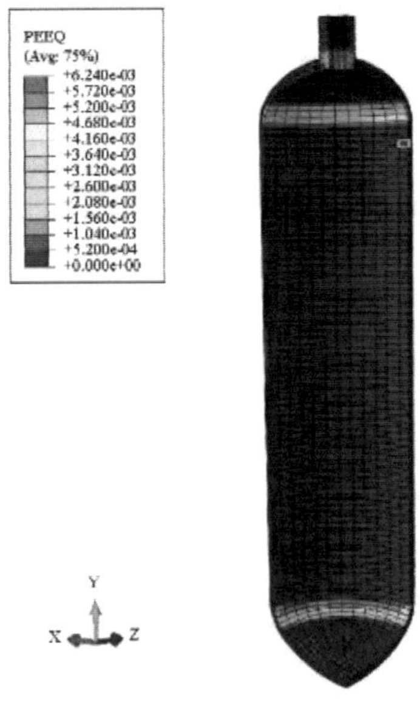

Figure 12. Lined PEEQ distribution

Analyse the stress cloud Figure 11 and Figure 12 the lining also has a high stress level in the barrel section. Stress concentrations are created at the transition between the head and the barrel. This is because the lining has been reinforced in the head section. From the junction of the barrel and the head to the top of the head, the thickness is gradually thicker, and the thickness of the top is 2.5 times that of the barrel due to the particularity of the head structure, is such that the stress level is lower relative to the barrel section when subjected to a load. Simultaneously, since the composite layer is a spiral plus circumferential layup structure, a structure is created in the transition section between the barrel body and the pole hole. The discontinuity causes the inner liner to create a stress concentration in this portion. The equivalent plasticity of the liner body, the maximum value of the change (PEEQ) is 6.040e03, and the head segment PEEQ=0. This shows that the inner lining of the barrel section has been in the yielding state while the head portion is still in an elastic state. At this time, the composite layer in the barrel section bears pressure load. Therefore, under the pressure of 33MPa after the pressure vessel is self-tightening, the circumferential direction of the composite layer is the main bearing. The structure is loaded, and the self-tightening process can give full play to the structural advantages of the composite material.

4.2.3. Prediction of Blasting Pressure. In this paper, the maximum strain criterion is used to predict the burst pressure of the pressure vessel, and the pressure capacity under the maximum strain criterion. The criterion for bursting pressure is [10]

$$\varepsilon_t = \varepsilon_f^h$$

In the middle ε_t -maximum strain in the direction of the fibre in the hoop layer ε_f^h -strain limit. A large number of practices and studies have shown [11, 12], circumferential composites during the blasting of composite pressure vessels the fracture strain of the layer is about 85.4% of the strain at break of the pure fibre, and the strain at break of the known pure fibre is 0.018× 0.854 = 0.01537. Calculate the second applied load×Therefore, the fracture strain of the circumferential composite layer is 0.01537 the hoop strain in the process is plotted and the hoop strain under different loads is plotted.

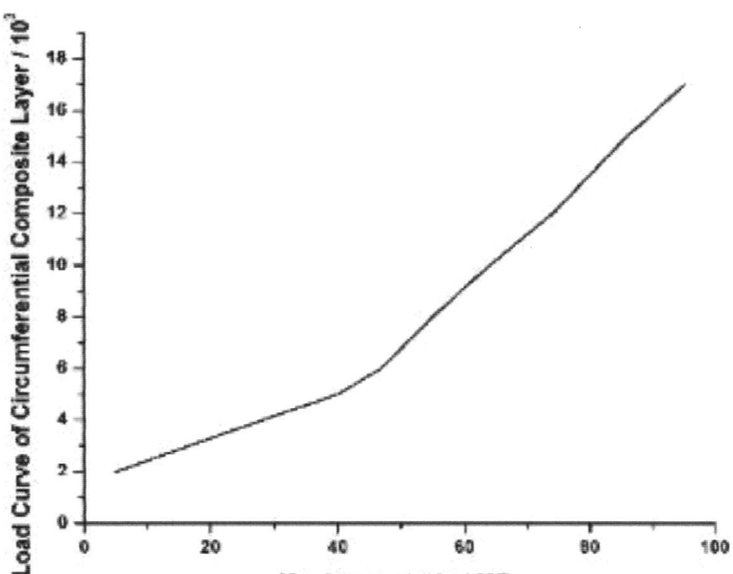

Figure 13. Maximum strain versus load curve of circumferential composite layer

It can be seen from Figure 13 that when the maximum strain of the circumferential composite layer is 0.01537, corresponding to this time the strength is 92MPa. The curve in the figure does not pass the origin, which is due to the self-tightening action of the pressure vessel, resulting in the load. When the charge is 0, there is still strain in the circumferential composite layer. Circumferential composite when the load is 92Mpa the strain of the layer is shown in Figure 14.

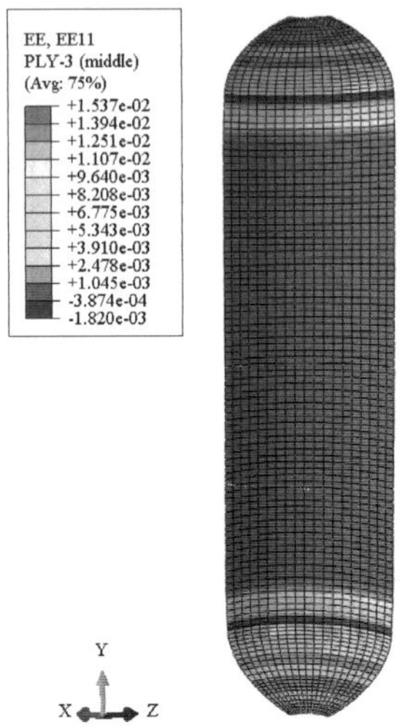

Figure 14. Strain distribution of the circumferential composite layer of 92MPa

Therefore, according to the maximum strain criterion, when the second water pressure is applied to 92MPa, the pressure vessel is blasted in the barrel section.

5. Conclusions

(1) By analysing the stacking rule of the non-geodetic winding of the yarn in the head section, the thickness and angle of the head part are calculated. (2) The geometric model of the composite material is established according to the thickness value and the winding angle of different parallel circles, and the geometric model of the pressure vessel is established by combining the geometrical dimensions of the aluminium liner. Then input the basic mechanical properties of the composite material, the elastoplastic stress-strain parameters of the aluminium lining, and establish the structural model of the composite pressure vessel. (3) Select the appropriate structural units, mesh the structural model, and after applying the constraint, analyse the stress-strain state of the pressure vessel when it is loaded to 33 MPa for the second time. The maximum strain criterion is used to predict the burst pressure of the pressure vessel. The results show that the pressure vessel blasts when the internal pressure is 92 MPa. In the previous paper, the matching experiments of fibre and resin materials were carried out, and the optimal system was selected to test the basic mechanical properties of the selected composite system. Then, the non-geodetic winding linear pattern of composite pressure vessel was studied. According to the stable winding and uniform filling of two process conditions, the suitable winding type was determined. The finite element model was established and the pressure vessel was simulated. The force state predicts the burst strength. In this paper, the pressure vessel is wound and molded by inputting the non-geodetic winding parameters established above, and the accuracy of the simulated line type is verified. The blasting test is performed on the cured pressure vessel to verify the linear design theory based on non-geodes and based on the correctness of the finite element strength analysis method.

References

[1] P. Blanc-Vannet, "<Burst pressure reduction of various thermoset composite pressure vessels after impact on the cylindrical part.pdf>," 2017.
[2] A. K. Agrawal S, " Impact damage on fibre reinforced polymer matrix composite - a review," J Compos Mater, vol. 48, pp. 317-32, 2014.
[3] G. C. Rosato DV, "Filament winding, its development, manufacture, applications and design.," ed: New York: Interscience, 1964.
[4] Z. J. Xu P, Liu P, "Finite element analysis of burst pressure of composite hydrogen storage

IOP Conf. Series: Materials Science and Engineering **538** (2019) 012015 doi:10.1088/1757-899X/538/1/012015

vessels," Materials & Design, vol. 30, pp. 2295-01, 2009.
[5] G. A. G. Evans J T, Evans J & Gibson A Composite angle ply laminates and netting analysis vol. 458. Proceedings of the Royal Society of London: Mathematical, Physical and Engineering Sciences, 2002.
[6] A. Onder, O. Sayman, T. Dogan, and N. Tarakcioglu, "Burst failure load of composite pressure vessels," Composite Structures, vol. 89, pp. 159-166, 2009.
[7] O. A., "First failure pressure of composite pressure vessels " Master of Science Thesis, Graduate School of Natural and Applied Sciences Dokuz Eylül University., 2007
[8] Z. Z. Yingjun W, Minqing S & Sirong Z . "Finite element modeling of carbon fiber reinforced polymer pressure vessel," presented at the International Conference on Educational and Network Technology (ICENT) 2010.
[9] DOT-CFFC Standard, Carbon-Fiber, Aluminum Cylinders[S], 2007.
[10] C. S. Shu JC, Chang JB An enhanced analysis method for composite overwrapped pressure vessels[C]. USA: AIAA publishers,, 1995.
[11] W. X. XU Minglin, AN Mingkang, "Prediction of bursting pressure of UHMWPE filament wound aluminum-lined composite gas cylinder[J]," Fiber Composite Materials, vol. 28, pp. 19-22, 2011.
[12] Z. X. Chang Xinlong, Liu Xinguo, et al. , "Finite Element Analysis and Blasting Pressure Prediction of Composite Gas Cylinders[J]," Journal of Rocket Propulsion, vol. 34, pp. 27-31, 2008.

IOP Conf. Series: Materials Science and Engineering **538** (2019) 012016 doi:10.1088/1757-899X/538/1/012016

Development of Hot Stamping Technology for High Strength Steel Parts with Tailored Properties

Li Bao[a,b,*], Qidong Nie[b], Bo Wang[b], Xiaoping, You[b]

[a]School of Material science and engineering, Northeastern University, Shenyang, Liaoning 110000, China
[b]School of Mechanical and Electrical Engineering, Qiqihar University, Qiqihar, Heilongjiang 161000, China

*Corresponding author: lwjxs2012@163.com

Abstract. In the context of lightweight vehicles, tailored properties high-strength steel parts play a vital role in improving safety under car collisions. Hot stamping technology provides good formability for manufacturing high-strength steel parts with tailored properties, and the hot stamping parts have high strength, small springback and good industrial promotion prospect. This paper summarizes the current state-of-the-art processes for the implementation of hot stamped parts with tailored properties, through intrinsic strategy and tailored welded / rolled blanks.

1. Introduction

With environmental issues and energy issues becoming increasingly prominent, automotive lightweighting has emerged as a result of industrialization based on energy conservation and emission reduction. The data shows that the use of high-strength or ultra-high-strength steel plates, the original thickness of 1.0 ~ 1.2 mm body panels can be reduced to 0.7 ~ 0.8 mm, the overall quality of the body can be reduced by 15% to 20%, fuel economy 8 ~ 15%[1-2]. Although some alloy materials or other carbon fiber materials can be used for lightweight production of automobiles, their cost is too high, which is not conducive to cost control. Therefore, the use of high-strength or ultra-high-strength steel sheets for the manufacture of automotive body-in-white parts is one of the important ways to achieve lightweight vehicles. However, due to the poor mold ability of high-strength steel sheets at room temperature, many of the process problems such as large deformation resistance, cracking, and large rebound are common under the traditional cold stamping process. Therefore, the hot stamping process can solve the above problems and obtain industrial applications. The hot stamping process is a process for forming a martensitic structure by heating a high-strength steel sheet material to a temperature above 900℃, allowing it to be completely austenitized and then rapidly cooled and quenched. Although the traditional hot stamping process can achieve the formation of high-strength steel, only obtaining a uniform martensite phase does not meet the needs of many tailored properties automotive parts, such as automotive B-pillar parts. The upper part of the B-pillar requires high strength to support and ensure that the occupant has enough space to resist external casing intrusion during collision to ensure vehicle integrity and occupant safety; the lower part of the B-pillar requires lower strength and high plasticity to absorb the impact energy of the car, thus maximizing the safety of the occupants. Therefore, the molding process of such parts requiring custom mechanical properties has become the focus of many scholars. According to the analysis of advanced processes in the existing literature, the hot stamping process of tailored properties automotive parts is divided into intrinsic strategies and tailored welded / rolled blanks.

2. Intrinsic Strategy

Intrinsic strategy is composed of differential temperature heating, differential cooing rate method and secondary annealing.

2.1. Differential Temperature Heating

The differential temperature heating method controls the heating temperature of different parts of the sheet. Bernd-Arno Behrens[3] regulates the performance of stamping parts by means of selective heating. In the traditional heating furnace, the appropriate shielding device is added to achieve different heating temperatures at different positions of the blank, and the hot stamping parts with gradient properties are obtained after quenching in the mold. The method of spray cooling is also adopted, and the performance quenching of the hot stamping parts and the performance gradient of the stamping parts are also realized. K. Mori[4-5] uses a grooved gradient quenching die and a shunt resistor heating method to prepare hot stamped parts with a gradient of strength distribution. The high-strength region is heated and in-mold quenched by means of bypass resistance heating, and the copper electrode is in contact with the blank for the low-strength region. And the current is introduced into the bypass by the high conductivity of copper, and the corresponding region of the billet is not heated. The U-shaped part using the bypass resistance heating method was subjected to a bending test, and the strength of the rounded corner reached 1.5 GPa, and the energy consumption and the maximum punching pressure were smaller than those of the uniform heating process. Wilsius J etc.[6] heats different areas of the sheet to different temperatures in the furnace, and the tensile strength of the final sheet temperature is 700 MPa, and the tensile strength of the higher temperature zone is 1600 MPa, which is customized for the performance of the formed parts. H. Hoffmann et al.[7] studied the cooling system of hot stamping dies. They point out that when the steel plate is heated above the austenitizing temperature and rapidly cooled to ensure a martensitic transformation, the cooling system is not set in the original production process. They used quenched alloy steel 22MnB5 as the stamping sheet, integrated the cooling system with the stamping die. And the simulation and experimental study on the distance between the cooling pipe and the mold profile show that the cooling effect is best when the cooling pipe of the cooling system is close to the mold profile. Bardelcik A et al.[8] studied samples of boron steel hot stamping at five different cooling rates for maximum tensile strength analysis at different tensile rates. The difference in volume fraction of martensite and bainite in each sample and the relationship between hardness and martensite and bainite volume fraction were obtained. The experimental data was compared with the constitutive model data with specific performance, and the principle of tensile fracture was also explained.

2.2. Differential Cooling Method

The differential cooling method is a method of realizing the distribution of the tailored properties of the molded part by changing the cooling rate of the mold to the sheet material. There are three methods for achieving differential cooling.

2.2.1. Differential Temperature Mold. The method controls the temperature difference of the sheet material by the local mold temperature. R. George et al. [9-10] of the University of Waterloo, Canada, designed a segmented thermoforming mold with local heating and cooling. Through the control of the cooling rate of the sheet during the sheet forming process, the hot-formed parts with high-strength and low-strength areas were obtained. Finally, the Vickers hardness after forming was predicted by LS-DYNA and compared with the experimental results. The comparison results show that in the fully cooled region, the predictions are similar to the actual measurements, and the prediction results in the transition zone are significantly different from the actual measurements. Feuser et al.[11] developed a full-size B-pillar with low strength and high intermediate strength at both ends. The part of the mold heated independently at both ends has a heating temperature of 550℃ and a dwell time of 15s. The final Vickers hardness is 250-270 HV, and the tensile strength is also reduced to 50% of the martensitic structure strength. Graff S et al.[12] controlled the temperature of the hot zone of the mold at 500℃ and the cold zone to room temperature, and the tensile strength of the final hot zone formed part was reduced by 50% compared with the cold zone. Changing the mold temperature can more accurately control the cooling rate of the sheet, making the final microstructure of the formed part more controllable and the performance transition more gradual.

2.2.2. Differential Thermal Conductivity Mold. The method is to change the mold material to have different thermal conductivity. D. M. Karbasian and A.E. Tekkaya of Germany[13] have changed the thermoforming control strategy and proposed three new methods of different thermal conductivity of different parts of the mold, different surface contact surface roughness, and the use of tailor welded steel sheets to realize the hot-formed parts with variable gradient characteristics. Casas et al.[14] using different thermal conductivity control the heat exchange rate between the sheet and the mold during forming and quenching. The thermal conductivity of the new material area of the mold is reduced by 90% compared with the thermal conductivity of the new material area, and the maximum and minimum tensile strengths of different parts of the final formed parts are 1500 MPa and 600 MPa respectively, indicating that this method can customize the performance of parts. But it is difficult to accurately control the cooling rate, and changing the cooling rate requires changing the mold material, which has high production cost and is not suitable for actual production. Kolleck and Veit[15] use ceramic insulation materials to make the heat transfer coefficient of different parts of the mold different, and the tensile strength of the sheet is changed from 1600 MPa to 650 MPa.

2.2.3. Differential Contact Area Regulation. The method is by controlling the contact area of the mold with the sheet. Maeno et al.[16] proposed a measure to improve the formability of hot stamped parts by increasing the stamping rate and appropriately increasing the local die gap by affecting the stamping rate and die gap on hot stamping. K. Mori and Y. Okuda[17] designed a thermoforming mold based on contact area control. The local contact area between the sheet and the mold is different. During the quenching and cooling process of the sheet, the part in contact with the sheet has a better cooling effect, and the non-contact area has a lower cooling rate, thereby realizing the custom attribute characteristics of the sheet.

3. Tailored Welded / Rolled Blanks

The method of changing the partial thickness of the sheet is to use a differential plate and a tailor welded plate. Choi J W et al.[18] simulated and experimentally analyzed stampings with different custom performances. The parts obtained by laser tailor welding and the hot stamping parts obtained by local quenching were studied and analyzed according to the phase transition characteristics and deformation behavior, and the performance was also analyzed through experimental tests. Bührle and Fritz[19] In order to ensure the mechanical properties of the hot-formed blank welded joints, it is proposed to use the laser ablation method to clean the Al-Si coating at the weld before welding. Tang Bingtao[20] of Shandong University of Architecture established a multi-field coupling numerical model for hot stamping of laser tailor welded blanks, and verified the reliability of the model by using the thickness and microstructure test results of B-pillar stiffeners after thermoforming. Lei Chengxi[21] of Harbin Institute of Technology established the constitutive relationship of TRB at high temperature by using the single tensile test results of different thickness and strain rate of B1500HS steel plates. However, there is a problem with this method. Due to the presence of weld seams in the tailor welded blank, the weld seam is easily cracked during the collision. And differential plates and patch plates increase their own weight when achieving component gradient strength.

4. Conclusions

Although there are many methods to achieve hot stamping parts with tailored properties, the first method- intrinsic strategy is more suitable for industrial production to realize, because this method will completed to obtain the parts in the process of molding and quenching integration in once hot stamping process. However, the second method requires more process steps. Moreover, in terms of achieving differential temperature heating, it can be realized by means of rapid heating such as induction heating, which can not only obtain the parts with tailored properties, but also realize grain refinement, furthermore, the research will be focus on it .

Acknowledgements

This work is supported by the Heilong jiang Provincial Higher Education Fund Basic Research Project (135109310); Heilongjiang Province Higher Education Teaching Reform Research Project (SJGY20180561); Qiqihar University Educational Science Research Project (02472).

Reference

[1] LIN Jianping, WANG Liying, TIAN Haobin, et al. Research and progress of hot stamping of ultrahigh strength steel[J]. Hot Working Technology,2008(21):140-144.
[2] PENG Yan, LIU Caiyi, HAO Luhan, XING Pengda. Review of performance gradient distribution hot forming technology[J]. Journal of Mechanical Engineering,2016,52(08):67-75.
[3] Bernd-Arno Behrens, Anas Bouguecha, Christoph Michael Gaebel. Hot stamping of load adjusted structural parts[J]. Procedia Engineering. 2014, 81:1756- 1761.
[4] Ken-ichiro MORI. Smart hot stamping of ultra-high strength steel parts[J]. Transactions of Nonferrous Metals Society of China,2012,22.
[5] K. Mori, T. Maeno, K. Mongkolkaji. Tailored die quenching of steel parts having strength distribution using bypass resistance heating in hot stamping[J]. Journal of Materials Processing Tech.,2013,213(3).
[6] Wilsius J, Tavernier B Abou-Khalil D, 2011. Experimental and numerical investigation of various hot stamped B-pillar concepts based on Usibor® 1500P. In:3rd International Conference on Hot Sheet Metal Forming of High-Performance Steel, Kassel, Germany, pp. 427-435.
[7] Hoffmann H, So H, Steinbeiss H. Design of Hot Stamping Tools with Cooling System[J]. CIRP Annals - Manufacturing Technology, 2007, 56(1):269-272.
[8] Bardelcik A, Worswick M J, Winkler S, et al. A strain rate sensitive constitutive model for quenched boron steel with tailored properties[J]. International Journal of Impact Engineering, 2012, 50(Complete):49-62.
[9] R. George, A. Bardelcik,M.J. Worswick. Hot forming of boron steels using heated and cooled tooling for tailored properties[J]. Journal of Materials Processing Tech.,2012,212(11).
[10] R. George. Hot forming of boron steels with tailored mechanical properties:experiments and numerical simulations[J]. 2011.
[11] Feuser P, Schweike T. Partially hot-formed parts from 22MnB5-process window material characteristics and component test results[C]. 10th International Conference on Technology of Plasticity, Aachen, Germany, 2011:408-413.
[12] Graff S, Gerber T, Lenze F J, Sikora S. About the simulation of microstructure evolutionin the hot sheet stamping process and the correlation of resulting mechanical properties and crash-performance[C]. Proceedings of the 3rd International Conference on Hot Sheet Metal Forming of High-Performance Steel, Kassel, Germany, 20ll:323-330.
[13] H. Karbasian, A.E. Tekkaya. A review on hot stamping[J]. Journal of Materials Processing Tech.,2010,210(15).
[14] Casas B, Latre D, Rodriguez N, Valls I, 2008. Tailor made tool materials for the present and upcoming tooling solutions in hot sheet metal forming. In:1st International Conference on

Hot Sheet Metal Forming of High Performance Steels, Kassel, Germany, pp. 23-35.
[15] Kolleck R, Veit R. Möglichkeiten zur lokalen Beeinflussung der Bauteilfestigkeit beim Presshärten[C]. Proceedings of the 3th Erlanger Workshop Warmblechumformung, 2008:65-72.
[16] Tomoyoshi Maeno,Ken-ichiro Mori,Takaaki Nagai. Improvement in formability by control of temperature in hot stamping of ultra-high strength steel parts[J]. CIRP Annals - Manufacturing Technology,2014,63(1).
[17] K. Mori, Y. Okuda. Tailor die quenching in hot stamping for producing ultra-high strength steel formed parts having strength distribution[J]. CIRP Annals - Manufacturing Technology,2010,59(1).
[18] Choi J W, Bok H H, Barlat F, et al. Experimental and Numerical Analyses of Hot Stamped Parts with Tailored Properties[J]. Isij International, 2013,53(6):1047-1056.
[19] J. Bührle, M. Fritz. New Trends in Laser within the Process Chain of Hotforming Production[J]. Physics Procedia,2013,41.
[20] TANG Bingtao, LI Chenchen, ZHENG Wei, HUANG Lili, YUAN Zhengjun. Finite element model of hot stamping of tailor welded high strength steel and its experimental verification[J]. Journal of Shanghai Jiaotong University,2016,50(07):994-999.
[21] LEI Chengxi, XING Zhongwen, XU Weili WU, et al. Flow behavior of high strength steel TRB at elevated temperature[J]. Materials Science and Technology,2015,23(03):66-70.

IOP Conf. Series: Materials Science and Engineering **538** (2019) 012017 doi:10.1088/1757-899X/538/1/012017

Research on High Efficiency and Low Destructive Manufacture Process of Complex Structure Honeycomb Core

Cao Xiang[1,*], Wang Yujie[1], Sun Chao[1], Shen Xin[1]

[1]CAC Chengdu Aircraft Industrial (Group) Co., Ltd., Chengdu 610092, China

[*]Corresponding author: 31635649@qq.com

Abstract. With the development of modern aircraft, complex structure honeycomb core parts are widely used. As a kind of difficult machining material, it's easy to destroy the surface quality of the parts during machining the honeycomb core parts because of the special structure. The article studies the technology to increase efficiency and reduce damage during the machining process of complex structure honeycomb core. The problem can be effectively solved by using proper clampings, proper cutting parameters and proper tool path interval.

1. Introduction

Honeycomb core is a kind of material with the properties of high specific strength, good heat resistance, impact resistance and good insulation etc., which is widely used in the aerospace field. For example, F22 fighter' honeycomb core ratio has reached 40%[1-3].

With the development of modern aircraft, the complex honeycomb are widely used. As a kind of difficult machining material, the processing of complex structure honeycomb parts are much more difficult than ruled surfaces honeycomb parts[4-6]. It's easy to damage the surface of the part and cause the part to be out of tolerance during machining the complex structure honeycomb parts, this reduces the performance and quality of the honeycomb core. To a certain extent, the application of honeycomb core in a wider range is limited[7-9]. For this reason, it's important to improve the manufacture process of complex structure honeycomb parts. Therefore, this paper analyzes the structural characteristics and processing difficulties of complex honeycomb parts. Designed the clamping scheme for complex honeycomb parts, solves the problem of reliable fixing. Combining with orthogonal test and analysis of single factor, the rules of influence on surface quality are obtained under different processing parameters. Based on constant scallop-height method, the tool path interval of direction-parallel machining is studied. The correctness of the research is verified by engineering application.

2. Characteristics Analysis of Complex Honeycomb Machining

2.1. Clamping Requirements

The structure of the parts is special, the profile is complex, and the material is made up of multiple honeycombs, which are easily deformed and the holding state is difficult to control. There are sharp corners, thin edges and chamfering edges at the edge of the part, as shown in Figure 1. In the processing, it is easy to pull up the parts, tearing and milling the edges of the parts, causing the parts to be out of tolerance or even scrapped, and it also brings difficulties to the protection of the parts after the processing.

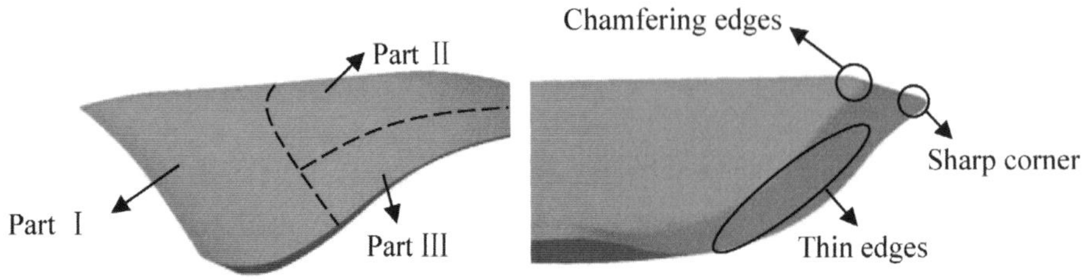

Figure 1. Structure of the typical honeycomb part.

Since the part has a double profile structure, the positioning of the second surface is mainly based on the machined surface of the first surface, the curvature of the two faces of the part is large, and the profile is a curved structure. Therefore, higher requirements are placed on the positioning of the parts.

2.2. Analysis of Cutting Parameters

In the actual machining process, the selection of milling processing parameters for complex honeycomb cores relies on empirical values, and the quality and precision of the machined surface are difficult to control, which is prone to problems such as the parts being entangled, the honeycomb structure being damaged, and the burrs being severe. In order to ensure better surface quality of the parts, it is often used to reduce the speed and small depth of cut, resulting in low processing efficiency. Faced with increasingly heavy production tasks, how to determine efficient processing parameters has become an urgent problem to be solved.

2.3. Tool Path Interval of Direction-Parallel Machining

The processing of complex curved surface of honeycomb core parts through multi-contoured and direction paralleled machining. The tool path interval is an important factor affecting part accuracy, efficiency and surface quality. It is easy to obtain better surface quality with lower tool path interval value, but the processing time increases and efficiency decreases. While the excessive tool path interval will increase the scallop height of the part surface and affect the accuracy of the part. Therefore, the tool path interval of direction-parallel machining must be reasonably determined, and the maximum value can be obtained under the premise of satisfying the processing precision requirement, thereby improving the processing efficiency.

3. Manufacture Process of Complex Structure Honeycomb Core

3.1. Clamping Scheme

The fixing and positioning method of honeycomb parts is different from metal parts beacause of the different properties of honeycomb materials than metallic materials, honeycomb parts through double-faced adhesive tape to achieve fast fixing. The part blank is a V-shaped spliced honeycomb with a V-angle of 120°-160°. Since the blank state is difficult to control, it is easy to appear that the blank surface can not fully fit the tooling surface because of the deformation of the blank or the incorrect geometric size. If the blank is clamped and fixed to the tooling by forced pressing, it is easy to cause the parts to rebound and deform during the machining process, resulting in the scrapping of the parts. In order to solve this problem, the repairing tooling is designed, as shown in Figure 2, and the blank surface with the correct geometrical dimensions is milled by the workpiece, and then the parts are clamped to ensure that the bottom surface of the part and the tooling surface are completely fitted.

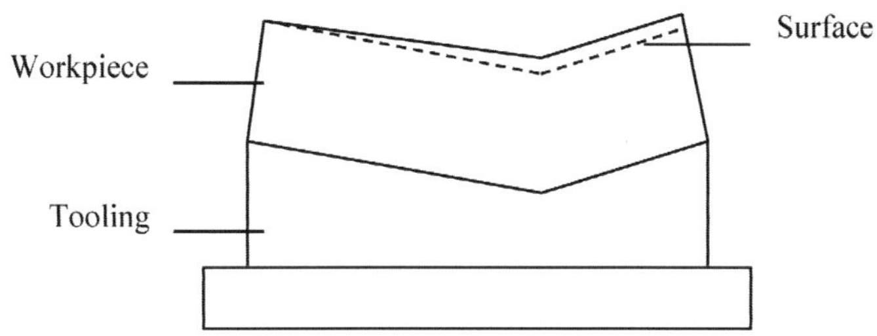

Figure 2. Repairing tooling.

In addition, since the width of the contour surface of the part is extremely small, it can be simplified into a curved structure, and it is difficult to locate by aligning the edge of the part with the contour of the tooling when the clamping is fixed. Therefore, a special tooling with a positioning stop as shown in Figure 3 is designed to realize quick positioning and clamping of the parts.

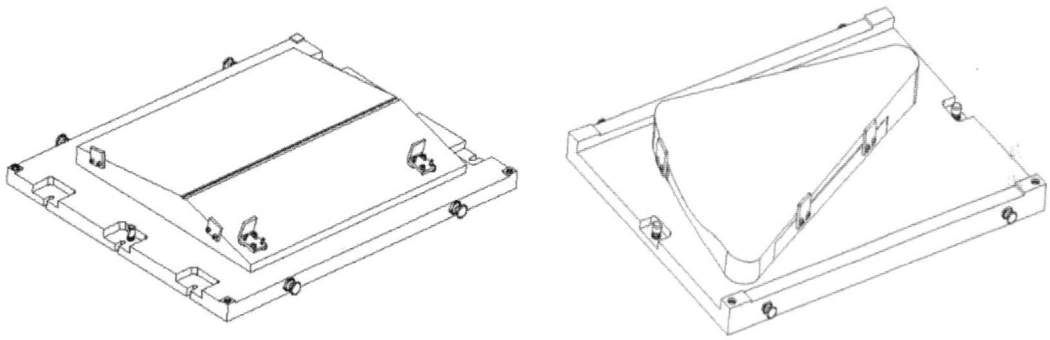

Figure 3. Structure of special tooling.

3.2. Surface Quality Analysis Based on Cutting Parameters

In order to analyze the influence of cutting parameters on the surface quality of honeycomb parts, a set of orthogonal experiments of honeycomb processing are designed. This experiments use climb milling and dry cutting to machining nomex honeycomb core on the five-axis CNC machine tool, the experimental tool is a typical honeycomb milling cutter with tool diameter 45mm. Because the honeycomb is a porous structure, it is difficult to measure surface roughness with a surface roughness meter like the metal part. Therefore, according to the situation that the surface quality of parts in actual production often occurs to dividing the quality grade, as shown in table 1.

Table 1. Honeycomb surface quality grade.

Number	Surface quality	Quality grade
1	Structure damage	0
2	Burrs in surface	5
3	Structural integrity and excellent surface	10

In this paper, four-factor four-level orthogonal test is carried out to obtain the surface quality. The four-factor include cutting depth, cutting width, feed rate and spindle speed. The cutting parameters and experimental results are shown in table 2.

Table 2. Experimental parameters and surface quality results.

Experiment number	Spindle speed (r/min)	Feed rate (mm/min)	Cutting width (mm)	Cutting depth (mm)	Surface quality grade
1	6000	800	1	3	0
2	6000	2000	2	5	0
3	6000	5000	10	15	0
4	6000	8000	20	20	0
5	10000	800	1	3	5
6	10000	2000	2	5	5
7	10000	5000	10	15	0
8	10000	8000	20	20	0
9	14000	800	1	3	10
10	14000	2000	2	5	5
11	14000	5000	10	15	0
12	14000	8000	20	20	0
13	18000	800	1	3	10
14	18000	2000	2	5	10
15	18000	5000	10	15	5
16	18000	8000	20	20	0

In order to obtain the influence of cutting parameters on surface quality, analyze the experimental results by single factor analysis. The surface quality is represented by W, cutting parameters are represented by A_i, B_i, C_i and D_i, i is parameter level of orthogonal test. Add the surface quality grade of the same parameter level, and then calculate the average, as shown in the following equation:

$$W_{A_1} = \frac{\sum_{i=1}^{n} W_{A_1 B_i C_i D_i}}{m}, i = 1, 2, 3, \cdots\cdots, n-1, n \tag{1}$$

Where m is the amount of surface quality result corresponding to A_1, n represents the amount of parameter level of orthogonal test. The W_{A2}, W_{A3} and W_{A4} can be obtained in the same principle. Then use the same calculation for other parameters, the results are shown in table 3.

Table 3. Surface quality results of single factor.

Parameter level	Spindle speed (r/min)	Surface quality results	Parameter level	Feed rate (mm/min)	Surface quality grade
1	6000	0	1	800	3.75
2	10000	2.5	2	2000	3.75
3	14000	3.75	3	5000	1.25
4	18000	6.25	4	8000	0
Parameter level	**Cutting width (mm)**	**Surface quality results**	**Parameter level**	**Cutting depth (mm)**	**Surface quality grade**
1	1	5	1	3	2.5
2	2	6.25	2	5	6.25
3	10	3.75	3	15	3.75
4	20	2.5	4	20	1.25

Through the analysis of the surface quality results of single factor, the influence of various processing parameters on the surface quality is obtained, as shown in Figure 4. The higher surface quality grade, the better surface quality can be obtained, it can be seen that the high spindle speed and low feed rate cutting scheme is adopted in the honeycomb processing, and the great processing effect is more easily obtained. Therefore, in order to improve the processing efficiency under the premise of ensuring the processing quality of parts, it is necessary to consider the comprehensive effect of spindle speed, feed rate, cutting depth and cutting width.

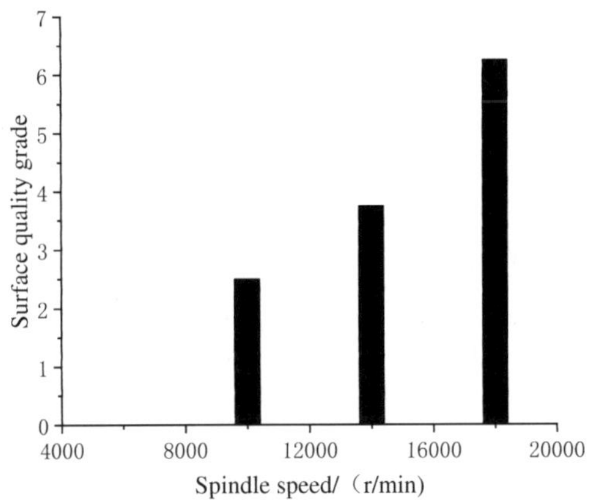

(a) The influence of spindle speed on surface quality

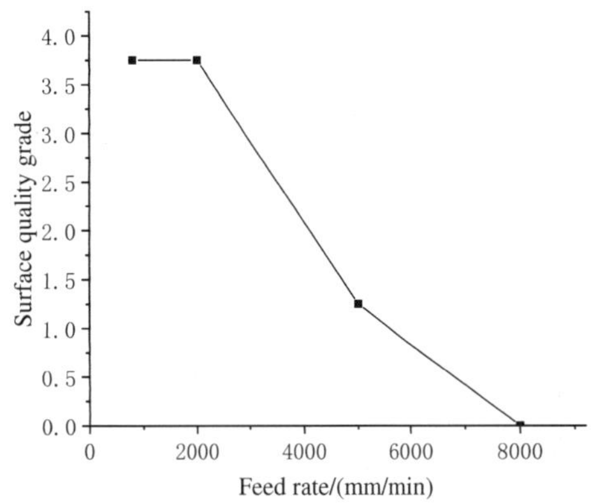

(b) The influence of feed rate on surface quality

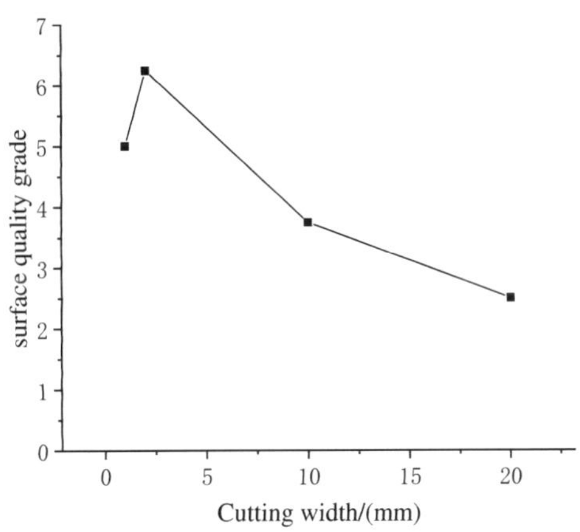

(c) The influence of cutting width on surface quality

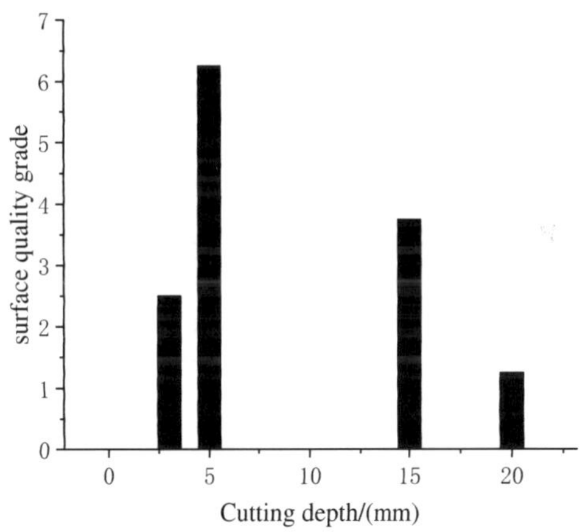

(d) The influence of cutting depth on surface quality

Figure 4. Effect of single factor on surface quality.

3.3. Tool Path Interval Analysis Based on Constant Scallop-Height Method

The processing of complex curved surface of honeycomb core parts through multi-contoured and direction paralleled machining. The tool path interval is an important factor affecting part accuracy and efficiency. The front inclination angle of honeycomb milling cutter is required when direction paralleled machining, and projection from the tool feed direction an ellipse can be obtained. And the direction paralleled machining is as shown in Figure 5.

IOP Conf. Series: Materials Science and Engineering **538** (2019) 012017 doi:10.1088/1757-899X/538/1/012017

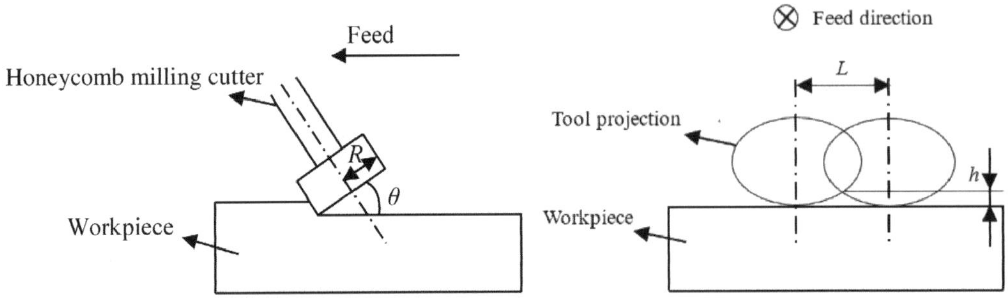

Figure 5. Schematic diagram of direction paralleled machining.

From the geometric relationship on the graph, the tool path interval of direction-parallel machining can be obtained as equation (2):

$$L = \sqrt{4R^2 - \frac{4(R\sin\theta - h)^2}{\sin^2\theta}} \tag{2}$$

Where R is the tool radius and θ is the tool front inclination angle, the scallop height of machining surface is h. Therefore, determining the tool radius, the tool front inclination angle and the scallop height, the optimum tool path interval of direction parallel machining can be obtained.

4. Practical Application Verification

In the production process of the parts, the complex honeycombs machining was carried out by using the key technologies proposed in this paper, the cutting parameters and verification results are shown in table 4.

Table 4. Practical application verification results.

Cutting method	Spindle speed (r/min)	Feed rate (mm/min)	Cutting width (mm)	Cutting depth (mm)	Surface quality grade
Climb milling and dry cutting	18000	2500	5	8	10
Cutting method	**front inclination angle (°)**		**Tool path interval (mm)**		**Efficiency Improved (%)**
Direction paralleled machining	15		8.25		27
	25		6.53		
	30		5.98		
	40		5.29		

The machining results show that the machining time is shortened about 27% while the machining stability and quality are ensured, the precision of the part surface is ±0.1mm and the contour accuracy is ±1mm, which satisfaction the manufacturing requirements. The surface quality of the parts is excellent, and there are no burrs or broken edges.

5. Conclusions

In this paper, the characteristics of complex honeycomb structure and processing difficulties are analyzed in depth, and the corresponding NC processing scheme is formulated. The effects of cutting parameters on the surface quality and processing efficiency of the parts were studied, and the calculation method of the tool path interval was analyzed by the constant scallop-height method. This solution enables efficient, high-quality machining of complex structure honeycomb core.

Acknowledgments
The results presented in this paper were outcome of the research project funded by the National Natural Science Foundation of China (No. 51761145103).

References

[1] Hong Hocheng, Machining Technology for Composite Materials, Cambrige:Woodhead Publishing, 2012.
[2] Gao Tao, Luo Jinwei, and Lin Yong, "Research on CNC processing technology for honeycomb core based on the ultrasonic machine tool," Machinery, 2013(1): 41-43.
[3] Zhang J, Tian W, "Brief analysis of digital manufacturing of honeycomb Core Parts," Aeronautical Manufacturing Technology, 2010 (6): 87-89.
[4] Jin C, Liu G, Ke Y, "Clamping technology optimization of high speed machining of NOMEX honeycomb core based on milling force model," Journal-Zhejiang university engineering science, 2007, 41(4): 547.
[5] Li Y, He Y, "CNC machining of Nomex honeycomb core parts," Aviation and aerospace, 2006 (3): 10-15.
[6] Li Bintai, Xing Liying, and Bao Jianwen, "Research and development progress of national key laboratory of advanced composites on advanced aeronautical resin matrix composites," Journal of Aeronautical Materials, 2016, 36(3).
[7] Hou X, Zhu Y, "NC Machining of honeycomb structure," Aeronautical Manufacturing Technology, 2009 (S1): 60-63.
[8] Wu Shengyou, Hu Xiaoping, and Kong Qian, "The research on ultrasonic cutting of honeycomb composite material with circular cutter," Journal of Hangzhou Dianzi University, 2015, 35(5).
[9] Huang Xiuxiu, Hu Xiaoping, and Yu Baohua, "Research on ultrasonic cutting Mechanism of nomex honeycomb composites based on fracture mechanics," Journal of Mechanical Engineering, 2015, 51(23).

IOP Conf. Series: Materials Science and Engineering **538** (2019) 012018 doi:10.1088/1757-899X/538/1/012018

Corrosion failure analysis of Lightweight Tinplate in Protein Beverage

Fei-fanXU[1,2*], Yun-guangSHI[1], Jan-weiYANG[1], Song-taoYAN[3]
(1.Shougang Research Institute of Technology,69 Yangzhuang Street, Shijingshan District, Beijing100043, China
2.Beijing key Laboratory of Green Recyclable Process for Iron &Steel Production Technology, 69 Yangzhuang Street, Shijingshan District, Beijing100043, China
3.National Center for Materials Service Safety, University of Science and Technology Beijing, 30 Xueyuan Road, Haidian District, Beijing 100083, China)

Correspondingauthor:xuff8908@126.com

Abstract. Electrochemical impedance spectroscopy (EIS), scanning electron microscopy (SEM), energy dispersive X-ray (EDX) and X-ray photoelectron spectroscopy (XPS) have been used to investigate the corrosion mechanism of lightweight tinplate in protein beverage. It is found that proteins in beverages can decomposed into hydrogen sulfide (H2S) and then hydrolyzed to H+ and S2-. Hydrogen evolution reaction between H+ and carbon steel could produce a large amount of H2, the gas assembled at the interface result in the exfoliation of corrosion product layer and accelerate the tinplate's corrosion ultimately. S2- is oxidized in the medium and generated a small fraction of S which lead to the sulfur blackening phenomenon in beverage cans.

Key words: Corrosion mechanism; Tinplate; hydrogen sulfide; Protein beverage

1.Introduction

Tinplate is a kind of non-homogeneous material with a stratified structure, basically consisting of a thin sheet of low carbon steel and two layers of commercial pure tin on both sides.Combining the formability, strength and corrosion resistance ofcarbon steel with great appearance of tin, tinplate is widely used as containers for kinds of products in the food processing industry[1-3]. However, tinplate cans may have some significant problems such as corrosion failure, sulfur blackening which depends on the acidity, food matrix, temperature of storage and duration, those problems can affected the nutritional value of the canned food[4]. Foodstuff can be polluted by this process. Though tin is not considered as a poisonous metal, the pollution problems may also cause the rejection of the consumers[5-6]. Many industries involving in tinplate cans for example tinplate producers, food canners and quality testing institution are concerned with the corrosion mechanism of tinplate greatly [7].

Compared with the pure tin, tinplate's corrosion process is complex due to its heterogeneity and stratified structure. Now, most studies about tinplate corrosion focus on three aspects: the films applied on, the corrosion mechanism of coating/alloy systems and the corrosion behavior in specific solution. Several hybrid sol-gel formulations applied on tinplate and indicated that sol-gel with high inorganic precursor ratios are able to form a continuous film, which is necessary to provides good barriers properties [8]. Through the analysis of the corrosion of chromium based coating for packing steel and found that the Cr(VI) derived coatings were fully resistant to corrosion driven disbondment [9]. The amount of alloy layer increased with the reflowing temperature and time and then the corrosion potential of detinned tinplate shifted positively and the corrosion rate decreased [10]. The corrosion behavior of tinplate in NaCl solution indicated that the tinplate in 0.5mol/L NaCl solution was mainly the dissolution of carbon steel substrate because of the defects in the tin layer, and the corrosion product was mainly γ-FeOOH [11]. In addition, the corrosion resistance was strongly depended on the electrolyte after considered tinplate in different aggressive mediums [6]. But, few studies of tinplate corrosion behaviors in beverage solution have been reported and the corrosion mechanism in protein beverage is still unclear. Protein beverage is a popular packaged beverage whose market share increases quickly by the development of the economy. Therefore, it is significant to explored the tinplate corrosion mechanism in protein beverage.

Recently, the in-situ measurement electrochemical impedance spectroscopy (EIS) has become one of the main methods to study the corrosion mechanism of metal or coating/metal systems, besides, it was used to evaluate the corrosion resistance of tinplate quantitatively[12-15]. X-ray photoelectron spectroscopy (XPS)combined with the low energy Ar+ sputtering technique has been used to investigate the chemical states of elements on the tinplate surface [16].

The aim of this work is to analyze the corrosion process of the lightweight tinplate in protein beverage by a simulation solution(PH=6):sodium sulfide and acetic acid (6.3mol/L). At last, the corrosion mechanism of tinplate in protein beverage was proposed.

2.Experimental

The tinplate samples (15mm×15mm) were provided by the ShougangJingtangIron&Steel Company, China. Tin content on both side of the carbon steel is 1.1 g/m^2. Samples were cleared using ultrasonic bath in acetone. After, they were divided into thirteen parts and immersion for 10min, 30min, 45min, 1h, 2h, 4h, 6h, 1d, 3d, 5d, 6d, 7d, 9d, respectively. Experiment were conducted at 25℃in incubator.

EIS measurements were carried out by PARSTAT 4000+ electrochemical workstation using the tinplate sampleas the working electrode (WE),calomel electrode as the counter electrode (CE) andhigh-purity platinum electrode for the reference electrode (RE). The EIS measurements were performed under the open circuit potential. A 10mV alternating-current signal was imposed with frequencies stepped from 0.01 Hzto 100 kHz. The working area of the tinplate is 1cm^2. ZSimpWin Software has been used to fitting the results. The electrolyte solution used nitrogen to remove oxygen before using and sealed the electrolytic cell during electrochemical measurement. SEM images were obtained by using Zeiss Ultra 55 with an InLens detector to observe the morphology of the samples on surface and cross-section, EDS has been used to analysis the composition of the rust point semi-quantitatively (Acceleration voltage 12KV). XPS measurements were conducted by ESCALAB 250Xi system with monochromatic Al Kα(1486.71eV) radiation. XPS depth profiling was carried out by argon ion bombardment (AIB). Standard sample of C ls (284.8eV) has been used to calibrate the instrumental binding energy scales according to ISO 15472 [17]. An ion accelerating voltage of 2KV was used to analyze the corrosion products on the surface and the sputtering rate was 0.036nm/s.

3. Results and Discussion

3.1 Electrochemical behavior

The impedance spectra measured for tinplate in simulation solution with various immersion time are plotted in Fig.1(a) and (b), electrochemical equivalent circuit in Fig.1(c)and (d) were applied to explain the behavior of the tinplate-electrolyte system, the resistance value obtained from the electrochemical equivalent circuit are shown in Fig.1(e) and (f). Z′ and Z″ were the real partand the imaginary part of the impedance, respectively. R_s is the solution resistance,R_{ct} is change transfer resistance of tin layer,R_c is the micro resistance of tin layer, R'_{ct} is change transfer resistance of carbon steel, $C_{dl}$is the interfacial capacitance of tinplate and $C'_{dl}$is the double layer capacitance of carbon steel.

According to the EIS results, the corrosion behavior of the tinplate insimulationsolutioncanbe divided into two main stages with the immersion time being prolonged: stage one is the corrosion of the tin layer and the stage two is the corrosion of the carbon steel.

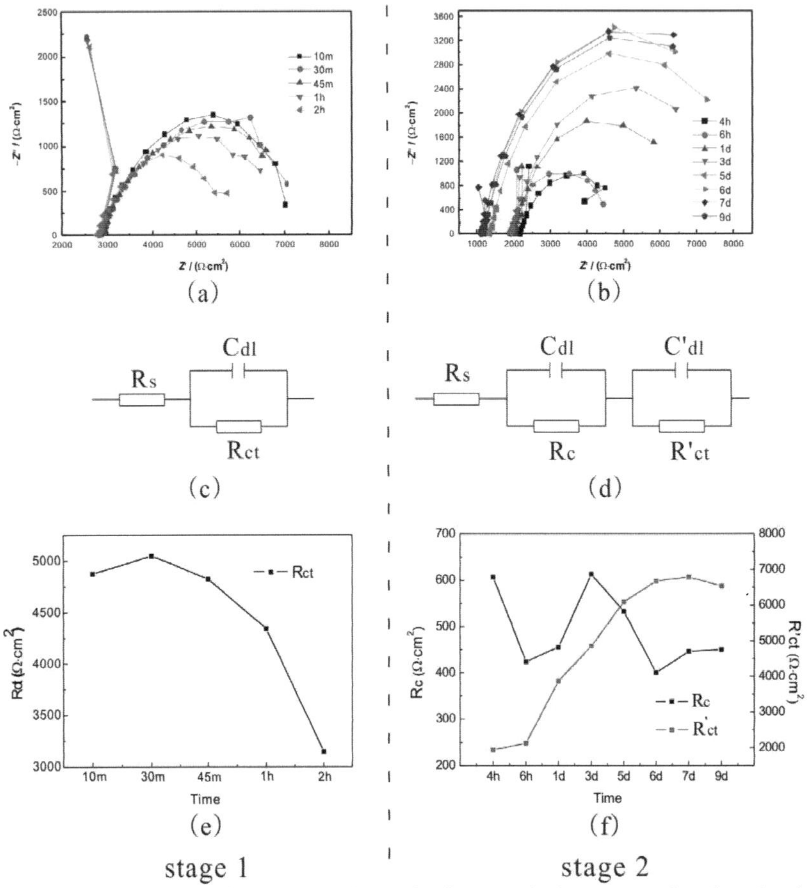

Figure 1. There are two stage of tinplate EIS results in protein beverage simulated solution: (a) (b) Nyquist plots; (c) (d) Electrochemical equivalent circuit; (e) (f) Evolution resistance of electrochemical equivalent circuit

As shown in Fig.1 (a) and (b), a characteristic arc could be seen on each EIS plots at the high frequency (100 kHz ~ 30 kHz). This is due to the poor conductivity of the solution at the beginning and the arc decrease gradually with the immersion time prolonging.

Stage1: Corrosion of the tin layer

At the initial stage of immersion time between 10min and 2hour, the EIS plots containing one time-constant are observed(Fig.2 (a)), indicating that the layer has a good protective performance and there is no electrolyte permeation through the tin layer. Protein beverage could decomposition out gas like H_2S when heated at the sterilization temperature, so the primary type of corrosion is electrochemical corrosion between H_2S and tin layer at this stage. At this period of time, the capacitive arc radius decreases gradually and means the resistance of the tin layerbecome smaller. R_{ct}inFig 1 (e) and $R_c$inFig 1 (f) are the fitting result of tin layer resistance at different immersion time. An order of magnitude gap means the tin layer has been destroied by the beverage.

Stage2: Corrosion of the carbon steel

After 4 hours, the EIS plots with two time-constant are presented on Fig.2 (b). The time-constant at the high frequency is caused by the resistance of the tin layer. Fig.1 (f) showed the tin layer resistance $R_c$maintainedat 400 - 600 Ω·cm^2 and indicated the poor protective effect of tin layer beyond the carbon steel. The time-constant at the low frequency is a result of interfacial capacitance of tin layer and charge transfer resistance of tinplate which reflecting the corrosion of carbon steel. At this stage, the electrolyte has permeated through the tin layer and formed interfacial capacitance at the carbon steel/tin layer interface. Fig.2 (b) shows that the capacitive arc radius increases gradually when the immersion time is more than 4 hours, besides, the resistance of carbon steel R'_{ct} increases gradually with the prolong time. This phenomenon mainly related to the formation and coverage of the corrosion product. The corrosion products formed by the reaction between carbon steel and corrosive medium assembled at the surface of the substrate, which can inhibit the penetration of the corrosive medium effectively and reduce the corrosion rate. Because of the poor adhesion between the corrosion product layer and the carbon steel, the longer the immersion time the greater the risk of peel off.

3.2 Surface and Cross-section characterization

Fig.2 is the SEM results of Lightweight tinplate during the corrosion process at 2 hours and 9 days. Fig.2(a) and (c), Fig.2(b) and (d) are the surface and cross-section morphologies of the samples, respectively.

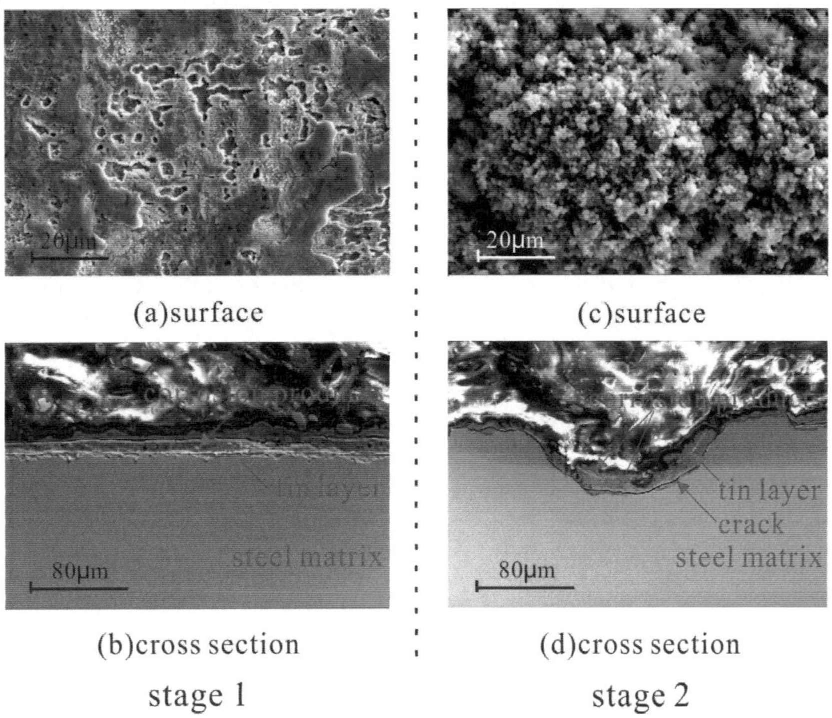

Figure2.SEM results: Stage 1, (a)surface and (b) cross section; Stage 2, (c)surface and (d) cross section.

Spotted corrosion products were observed on the surface after being exposed to protein beverage for 2 hours(Fig.2(a)) and cluster corrosion products formed on the surface for 9 days(Fig.2(c)). The increase of the corrosion products means the deepening of the corrosion process. Cross section could show the process of ions corrosion to the tin layer and carbon steel clearly. In Fig.2(b), a white strip in the red rectangular frame is the tin layer and above it covered by a thick corrosion product layer. There were no corrosion products between the tin layer and carbon steel have been discovered what indicated that the corrosive ions have not permeated through the tin layer. Fig.2(a) and (b) corresponding to the stage one when corrosion between H_2S and tin layer, the chemical reaction occurs as follows:

$$H_2S(g) \rightleftharpoons 2H^+(aq)+S^{2-}(aq) \quad (1)$$

$$Sn(s)-2e^- \rightarrow Sn^{2+}(aq) \quad (2)$$

$$Sn^{2+}(aq)-2e^- \rightarrow Sn^{4+}(aq) \quad (3)$$

$$2H^+(aq)+2e^- \rightarrow H_2(g)\uparrow \quad (4)$$

It is worth noting that there are two-layers of corrosion products clipped the tin layer as shown on Fig.2 (d). Therefore, it is proposed that the presence of two-layer corrosion product results from the corrosive ions permeated through the tin layer. Because the metal activity of iron is better than that of tin, once the tin layer is corroded, the first stage of corrosion stopped with the second stage corrosion beginning. The chemical equation would be changed below:

$$HS^-(aq) \rightleftharpoons H^+(aq)+S^{2-}(aq) \quad (5)$$

$$Fe(s)-2e^- \rightarrow Fe^{2+}(aq) \quad (6)$$

$$Fe^{2+}(aq) -e^- \rightarrow Fe^{3+}(aq) \quad (7)$$

$$2H^+(aq)+2e^- \rightarrow H_2(g)\uparrow \quad (4)$$

Because Fe is more abundant and active than Sn, the hydrogen produced at the second stage is abundant. Amount of hydrogen accumulated at the interface between the carbon steel and corrosion products, which lead to the separation of the corrosion products layer eventually as shown in the Fig.2 (d).

EDS has been used to investigate the component of the corrosion products. S, Sn, O, Fe were detected as the mainly elements in the corrosion products and shown in table 1.

Table 1. Contents of main elements in corrosion products (atomic fraction)

Element / Time	S	Sn	O	Fe
2 hours(stage 1)	5.6	21.8	43	29.6
9 days(stage 2)	8.4	14.5	33	44.1

The corrosion products (stage 1) measured contained 29.6% Fe and 21.8% Sn, while stage 2 contained 44.1% Fe and 14.5% Sn. Due to the chemical equation (6) (7) instead of (2) and (3) took place at the second stage, the Fe and Sn content changes significantly. S element was mainly from the protein beverage and XPS has beenused to analyze the valence state of sulfur in corrosion products for different immersion time, as shown in Fig.3.

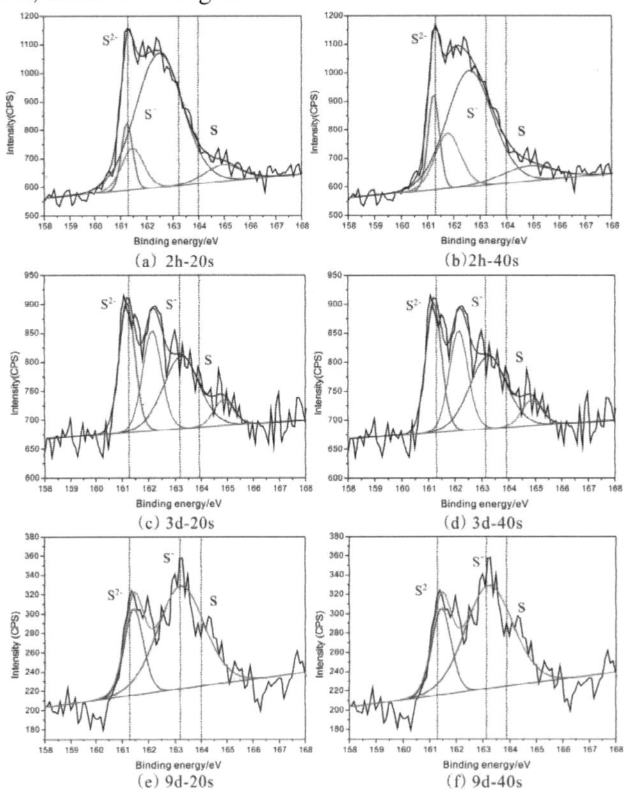

Figure3. Scanning spectra of XPS for S element: (a) (c) (e)sputtering 20s for immersion 2hours, 3days and 9days; (b) (d) (f)sputtering 40s for immersion 2hours, 3days and 9days

Since ions can be freely exchanged and diffused in solution, the valence state of S element at the two corrosion product layers are consistent. Fig.3 shows the XPS spectrum of S element acquired on the corrosion products of the tinplate surface. From S 2p XPS spectra acquired at sputtering time 20s and 40s, it is obvious to see that the content of different S valence state on the corrosion products is various. It is noticed that there are only two main peaks in the Fig.3 (a)(b)(c)(d), and beside the two main peaks has a shoulder peak in the Fig.3(e)(f). Though peak fitting, the component with the binding energy of 164.0eV is attributed to nonmetallic S, while those with the binding energy of 163.2eV and 161.3eV can be attributed to $S^-$and S^{2-}[18]. From Fig.3(a)(b)(c)(d), it is clear that $S^{2-}$and S^-component are the major parts of the sulfur valence state while S component exists on the surface with a small amount when immersion for 3 days. After immersion for 9 days, S^-has become the major part of the S element, while S^{2-}concentration reduces sharply and nonmetallic S concentration increase gradually.
$S^{2-}$is highly reductive, and it's easy to lose electrons in the medium into nonmetal S. The chemical reaction took place as follows:

$$S^{2-}(aq)-e^- \rightarrow S^-(aq) \quad (8)$$

$$S^-(aq)-e^- \rightarrow S(s) \quad (9)$$

The chemicals reaction equation (8) and (9) explain the phenomenon that $S^-$and S contents increases gradually with the prolong immersion time.
Compared with Fig.3(a)(b), Fig.3(c)(d) and Fig.3(e)(f), it is clear to see that with the prolongation of the immersion time, the peak intensity of the sulfur element measured by the XPS is reduced from1150 CPS at 2h to 350 CPS at 9d on the maximum peak strength. This phenomenon is the double result of the decrease of sulfur element concentration in the solution and the diffusion of sulfur element to the matrix. At each etched time, the contents of S element binding state almost have no changes.

4.Conclusion

A comparison between tinplate samples immersion for different times in simulation solution:sodium sulfide and acetic acid (PH=6), with regards to their resistance to protein beverage corrosion, has been completed to show that with the immersion time being prolonged:
●EIS diagram of tinplate in the simulation solution shows the transition from single capacitance arc to double capacitance reactance arc, and the corrosion process can be divided into two stages: first stage is the corrosion of the tin layer and the second stage is the corrosion of carbon steel after the tin layer has been corroded.
●SEM images indicated that ions in the medium can permeated through the tin layer and then corroded the carbon steel. Besides, the hydrogen produced at this stage can separated the corrosion productive layer and carbon steel matrix easily.
●EDS and XPS results revealed that Fe has replaced Sn as the main metal element in the corrosion process and the peak strength of sulfur in corrosion products decreases gradually while the ratio of high valence atoms of sulfur is increasing.
The sulfur corrosion resistance mechanism of lightweight tinplate in protein beverage is proposed that: Decomposition of proteins in beverages to produce hydrogen sulfide (H_2S), etc. H_2S hydrolysis into H^+ and $S^{2-}$and then hydrogen evolution reaction of $H^+$with tin layer under the electrochemical action which product H_2 and leaving S^{2-} in the medium. Once the tin layer is corroded or damaged, the second stage of hydrogen evolution occurs and accompanied by the production of a large amount of H_2. The accumulation of H_2 at the interface between corrosion product and carbon steel matrix would result in the exfoliation of corrosion product layer and accelerate the corrosion.S^{2-} is oxidized to produce a small amount of S which lead to the sulfur blackening phenomenon in beverage cans.

Acknowledgements
The authors would like to thank the colleagues in the Testing Department for the high resolution SEM images and the XPS analysis.

References

[1] S. Blunden, T. Wallace, Tin in canned food: a review and understanding of occurrence and effect [J]. Food Chem. Toxicol. 41(2003):1651-1662.
[2] L. Perring, M.Basic Dvorzak. Determination of total tin in canned food using inductively coupled plasma atomic emission spectroscopy[J]. Analytical and Bioanalysis Chemistry. 374(2002):235-243.
[3] A.M.Arenas, A Conde, J.J.D.Eamborenea. Cerium: A suitable green corrosion inhibitor for tinplate[J]. Corrosion Science. 44(2002):511-520.
[4] A.S.Tselesh. Anodic behaviour of tin in citrate solutions: The IR and XPS study on the composition of the passive layer[J]. Thin Solid Films. 516(2008):6253-6260.
[5] D. Xia, S Song, W. Gong, Y. Jiang, Z. Gao, J. Wang, Detection of corrosion-induced metal release from tinplate cans using a novel electrochemical sensor and inductively coupled plasma mass spectrometer[J]. Food Eng. 113(2012)11-18.
[6] D. Alvarez, A. Collazo, X. R. Novoa, C. Perez. Electrochemical behavior of organic/inorganic films applied on tinplate in different aggressive media [J]. Progress in Organic Coatings. 77(2014)2066-2075.
[7] J.I.Martins. Corrosion problems in tinplates cans for storing contact glues for shoes[J]. Engineering Failure Analysis. 26(2012)258-265.
[8] D. Alvarez, A. Collazo,C. Perez. Structural characterization and barrier properties of hybrid sol-gel films applied on tinplate[J]. Surface & Coatings Technology. 321(2017)108-117.
[9] N.Wint, A.C.A.de Vooys, H.N.McMurray. The corrosion of chromium based coating for packing steel[J]. Electochimica Acta. 203(2016)326-336.
[10] Xian-qiu HUANG, Feng-jun LANG, Ying MA, etal. Effects of reflowing temperature and time on alloy layer of tinplate and its electrochemical behavior in 3.5% NaCl solution[J]. Trans.Nonferrous Met.Soc. China. 24(2014)1978-1988.
[11]Gao Z, Song S, Xu Y, Electrochemical impedance spectroscopy analysis of coating deterioration process with Kohonen neural networks [J]. Journal of Chinese Society for Corrosion and Protection. 25(2005):106-109.
[12]Xia Da-hai, Song Shi-zhe, Wang Ji-hui et a1. Corrosion behavior of tinplate in NaC1 solution [J]. Transactions of Nonferrous Metals Society of China. 22(2012): 717-724.
[13] Bastidas J M, Cabaiiesb J M, Catalib R. Evaluation of prolonged exposure of lacquered tinplate cans to a citrate buffer solution using electrochemical techniques [J]. Progress in Organic Coatings. 30(1997): 9-14.
[14] Zhang X G, Wang J H, Zhou Y W. Analytical modeling for corrosion-induced cover cracking of corrosive reinforced concrete structures [J]. Transactions of Tianjin University. 18(2012): 285-290.
[15] Xia D H, Song S Z, Gong W Q el a1. Detection of corrosion-induced metal release from tinplate cans using a novel electrochemical sensor and inductively coupled plasma mass spectrometer [J]. Journal of Food Engineering.113(2012):11-18.
[16] Sheng Chen, Long Xie, Fei Xue. X-ray photoelectron spectroscopy investigation of commercial passivated tinplate surface layer[J]. Applied Surface Science. 276(2013)454-457.
[17] Committee CH-016, Spectroscopy. ISO 15472:2010, Surface chemical analysis - X-ray photoelectron spectrometers- Calibration of energy scales[S]. Standards Australia, Sydney, 2010.
[18] Wagner C D, Riggs W M, et a1. Handbook of X-ray photoelectron spectroscopy [M]. Perkin-Elmer Corporation Physical Electronics Division, 1979.

IOP Conf. Series: Materials Science and Engineering **538** (2019) 012019 doi:10.1088/1757-899X/538/1/012019

A Study on the Surface of RF Magnetron Sputtered CeO_2 Thin Films at Relatively Low Substrate Temperature

Noormariah Muslim[1,a], Muhammad Nur Syafi'ie Md Idris[2], Ying Woan Soon[1] and Nyuk Yoong Voo[1]

[1]Physical and Geological Sciences, Faculty of Science, Universiti Brunei Darussalam, Jalan Tungku Link, Gadong BE 1410, Brunei Darussalam
[2]Centre for Advanced Material and Energy Sciences, Universiti Brunei Darussalam, Jalan Tungku Link, Gadong BE 1410, Brunei Darussalam

[a]Corresponding author: noormariah.muslim@ubd.edu.bn

Abstract. Influences of substrate temperature on crystal structure, surface topography and surface morphology of cerium dioxide or ceria (CeO_2) thin films deposited using radio frequency (RF) magnetron sputtering were studied. The substrate temperature was varied i.e. room temperature, 100, 200 and 300 °C. The characteristics of the film surface could be controlled by tuning the substrate temperature. As studied in the X-ray diffraction (XRD) analysis, the CeO_2 thin films exhibited a cubic fluorite structure with predominant (200) orientation at substrate temperature of room temperature and 200 °C, while (111) preferred orientation at substrate temperature of 100 and 300 °C. The CeO_2 films featured uniform surface topography and surface morphology, as observed from the atomic force microscopy and scanning electron microscopy analyses, respectively. The surface roughness and the average grain size of the films fluctuated with increasing substrate temperature. This phenomenon was suggested due to the difference in the preferential orientation of the films as in the XRD analysis.

1. Introduction

Cerium dioxide or ceria (CeO_2) thin films have been known to be one of the most promising materials for future advanced devices in optical coatings, optics, microelectronics, catalysis, and biomedical applications [1,2]. This is due to properties of the CeO_2 i.e. high visible and infrared transmission, high refractive index, good adhesion, as well as high mechanical, chemical and thermal stability [3].
Several preparation techniques have been used to obtain CeO_2 thin films such as electron beam evaporation [4,5], chemical vapour deposition [3], magnetron sputtering [6-10], and wet chemical reaction methods [11,12]. Nevertheless, in this work, magnetron sputtering techniques i.e. radio frequency (RF) magnetron sputtering was employed because it provides possibilities to alter film thickness, crystalline structure, porosity, and refractive index of CeO_2 thin films [10]. Many investigations have been reported for the depositions of CeO_2 thin films using RF magnetron sputtering. However, relatively few studies on effects of the substrate temperature on the surface of CeO_2 films were reported. Sets of substrate temperature being applied in those studies were 600 °C [7], 400 and 600 °C [6], and 200, 400, 600 and 800 °C [13]. Hence, depositions at relatively low substrate temperature i.e. from room temperature to 300 °C have not been studied yet.

Thus, in this study, effects of substrate temperature on the crystal structure, surface topography and surface morphology of RF magnetron sputtered CeO_2 thin films were analyzed.

2. Experimental Details

2.1. Film Preparation

CeO_2 thin films were prepared on Corning 2947 glass substrates by means of RF magnetron sputtering. A CeO_2 target (99.99 % purity) with a diameter of 76 mm was used. Prior to depositions of the films, the glass substrates were cleaned with ethanol in an ultrasonic bath for 15 mins and rinsed in distilled water. Then, the glass substrates were dried with nitrogen gas of 99.9 % purity and then clamped on a substrate holder in the deposition chamber. The chamber was evacuated to 10^{-4} Pa using a turbo molecular pump. The CeO_2 target was pre-sputtered in an argon (Ar) atmosphere for 10 mins to remove any impurities or oxide layers on the surface of the target. During the film depositions, the working pressure and RF sputtering power were kept constant at 0.8 – 0.9 Pa and 120 W, respectively. Commercial Ar gas flow rate was supplied as sputtering gas into the deposition chamber during the depositions and maintained at 30 sccm which was controlled by a mass flow controller. The substrate temperature was varied from room temperature to 300 °C. Film thickness was kept constant at 100 nm and monitored with a quartz thickness meter supplied by Inficon.

2.2. Film Characterisation

The structural, topological and morphological properties of the films were measured. The structural properties of the deposited CeO_2 films were studied using Shimadzu X-Ray Diffractometer 7000 with a monochromatic high-intensity CuKα radiation (λ = 1.54056Å) radiation operated at 40 kV and 30 mA. The X-ray diffraction (XRD) patterns of the films were recorded at a scanning rate of 2 °/min in the range of 2θ = 10 – 80 ° with an incident angle of 1 °. The surface topography of the films were investigated using a commercially available AFM system from CSM instruments in contact tapping mode with a constant load force on the cantilever of 2.0 nN. The surface morphology of the films was analysed using a JEOL JSM-7610F field emission scanning electron microscope.

3. Results and Discussion

3.1. Crystal Structure

Figure 1 shows the XRD patterns for CeO_2 thin films deposited on glass substrates at different substrate temperature with 100 nm film thickness. The crystallinity of the CeO_2 thin films was quite obvious as indicated by the presence of well-defined diffraction peaks. The deposited films were polycrystalline with cubic fluorite structure (ICDD card no. 00-043-1002). The films deposited at substrate temperature of room temperature and 200 °C exhibited (200) preferred orientation at 2θ = 33.0 °, while as for films deposited at substrate temperature of 100 °C and 300 °C showed (111) preferred orientation at 2θ = 28.4 °. Both diffraction peaks of (111) and (200) planes revealed obvious changes to the peak intensities. This observation was explained by Netterfield et al. [1] as overall ion bombardment produced increased disorder and caused densification in the films which then promoted (200) growth relative to (111). In addition, the XRD peaks with predominant (200) and (111) orientations could be used to estimate the average crystallite size of the CeO_2 films using the Scherrer formula. The average crystallite sizes of the films deposited at different substrate temperature (S.T.) with respect to peaks at 2θ = 28.4 ° and 2θ = 33.0 ° were tabulated in table 1. As observed, the average crsytallite sizes of the deposited CeO_2 films were almost consistent. In the case of films deposited at substrate temperature of room temperature, the CeO_2 films showed smaller average crystallite size than other prepared films. This implies that there was only a slight increase of about 3 nm in the average crystallite size of the CeO_2 films as the substrate temperature elevated from room temperature to 100, 200 and 300 °C. Meanwhile, all deposited CeO_2 thin films at increasing substrate temperature had common peaks at 2θ = 47.5 °, 56.2 °, 69.0 °, 76.2 °, 78.9 ° which correspond to (220), (311), (400),

(331) and (420) orientation planes, respectively. Similar XRD patterns of the prepared CeO_2 films at substrate temperature of 100 and 300 °C were also reported for CeO_2 thin films deposited at substrate temperature of above 600 °C [13]. Thus, this shows that preparations of CeO_2 films at low substrate temperature were adequate to present high film crystallinity.

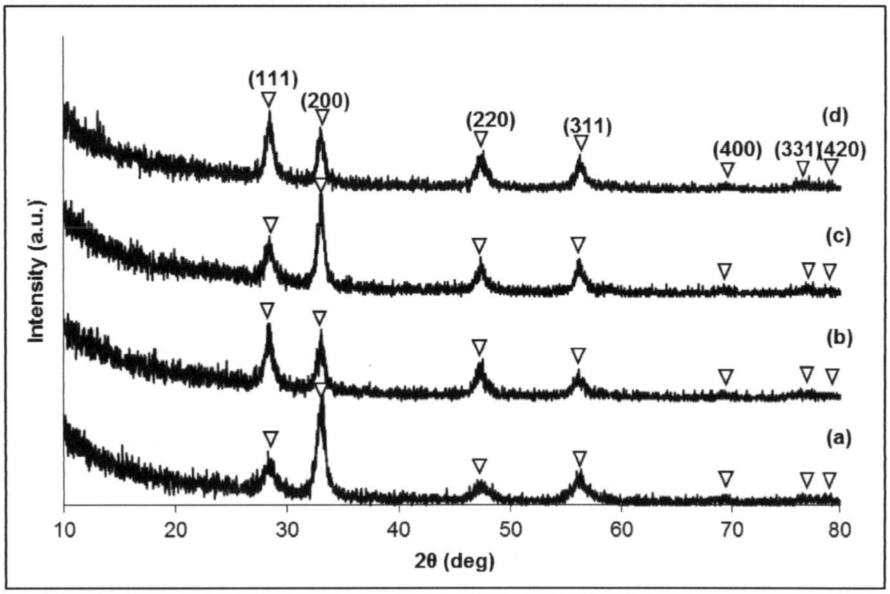

Figure 1. XRD patterns of the cerium dioxide films obtained at different substrate temperature (a) room temperature, (b) 100, (c) 200, and (d) 300 °C.

Table 1. The average crystallite sizes of CeO_2 thin films deposited at different substrate temperature with respect to $2\theta = 28.4$ ° and $2\theta = 33.0$ ° XRD peaks.

CeO_2 films	Average Crystallite Size (nm)	
	$2\theta = 28.4$ °	$2\theta = 33.0$ °
S.T. at room temperature	17.1	18.3
S.T. at 100 °C	21.6	21.2
S.T. at 200 °C	20.0	23.3
S.T. at 300 °C	20.7	22.4

Note: S.T. = Substrate temperature

3.2. Surface Topography

Figure 2 shows AFM images of the deposited CeO_2 thin films. The images were obtained using a scan area of 4 μm^2. As the substrate temperature increased, different grain growth on the film surfaces was observed. The root-mean-square (RMS) surface roughness of the deposited CeO_2 films at substrate temperature of room temperature, 100, 200 and 300 °C were 3.22, 1.98, 2.95 and 3.13 nm, respectively. The RMS surface roughness was found to reduce as the substrate temperature increased to 100 °C but then increase as the substrate temperature elevated to 200 and 300 °C. This might be due to the difference in the preferential orientation of the deposited films as in the XRD analysis. According to Kumar et al. [12], grain growth towards different shape and size could be led by the preferential orientation change in the XRD patterns. In addition, variation of the surface roughness of the films deposited at increasing substrate temperature could also cause the occurrence of grain growth and the rate of surface migration of adatoms [14,15].

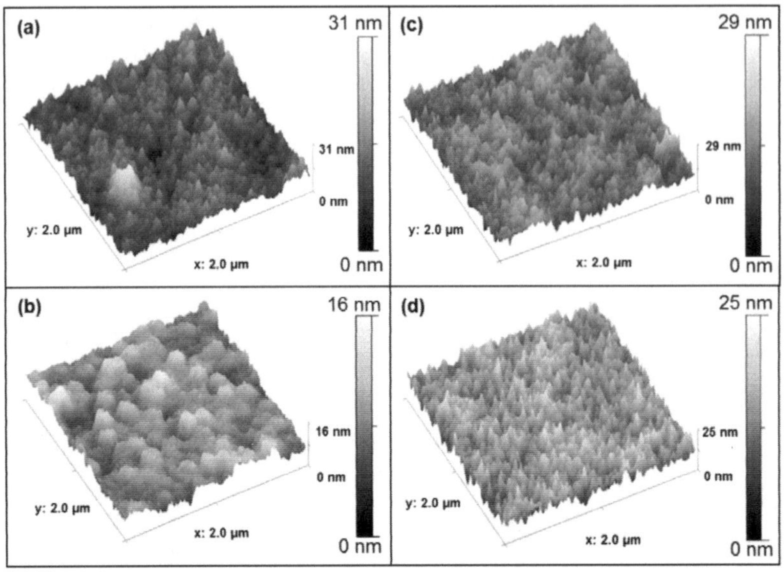

Figure 2. AFM images of the upper surface of the cerium dioxide films sputtered at substrate temperature of (a) room temperature, (b) 100, (c) 200, and (d) 300 °C.

3.3. Surface Morphology

FE-SEM images of the deposited CeO_2 thin films were recorded at an acceleration voltage of 2.00 kV and a magnification of 200,000, as shown in figure 3. Uniform surface morphology of the deposited films with uniform grain-size distribution could be observed. The estimated average grain sizes of the CeO_2 films deposited at room temperature, 100, 200, and 300 °C were 19.2, 15.8, 18.4, and 16.7 nm, respectively. The average grain size decreased as the substrate temperature increased. However, the films deposited at room temperature and 200 °C showed almost the same grain sizes, whereas the films deposited at 100 and 300 °C also showed almost similar grain sizes. This behavior was consistent with the XRD and surface topography analyses.

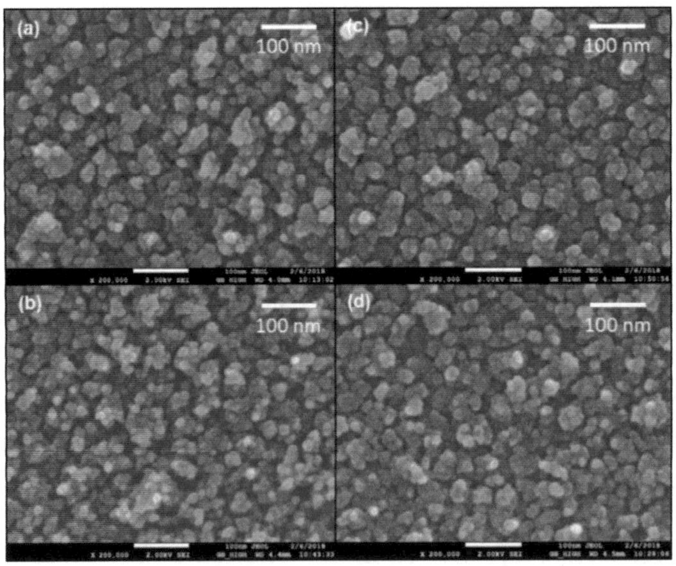

Figure 3. FE-SEM images of the cerium dioxide films sputtered at substrate temperature of (a) room temperature, (b) 100, (c) 200, and (d) 300 °C.

4. Conclusion

CeO_2 thin films were successfully prepared by using RF magnetron sputtering at different substrate temperature. The surface characteristics of the films were studied. As observed, the RF magnetron sputtering is a good alternative deposition technique. This is because CeO_2 thin films could be engineered according to desired applications simply by controlling the substrate temperature. The prepared CeO_2 thin films were polycrystalline with several prominent diffraction peaks. The RMS surface roughness reduced as the substrate temperature increased to 100 °C but then increased as the substrate temperature elevated to 200 and 300 °C. This change was suggested due to the difference in the preferential orientation of the films as in the XRD analysis. The deposited films exhibited uniform surface morphology and the average grain size changed with increasing substrate temperature. Crystallizations of the CeO_2 films were achieved at substrate temperature of room temperature to 300 °C. Thus, low temperature production of CeO_2 thin films with high crystallinity could be obtained.

Acknowledgments
This research was supported in part by the Brunei Research Council under Grant No. S&T 17.

References

[1] R. P. Netterfield, W.G. Sainty, P.J. Martin and S.H. Sie: Applied Optics Vol. 24 (1985), p. 2267-2272.
[2] P. Ratnasamy, D. Srinivas, C.V.V. Satyanarayana, P. Manikandan, R.S. Senthil Kumaran, M. Sachin and V.N. Shetti: Journal of Catalysis Vol. 221 (2004), p. 455-465.
[3] R.G. Toro, G. Malandrino, I.L. Fragalà, R. Lo Nigro, M. Losurdo and G. Bruno: The Journal of Physical Chemistry B Vol. 108 (2004), p. 16357-16364.
[4] S. Debnath, M.R. Islam and M.S.R. Khan: Bulletin of Materials Science Vol. 30 (2007), p. 315-319.
[5] M.S. Anwar, S. Kumar, F. Ahmed, N. Arshi, Y.J. Seo, C.G. Lee and B.H. Koo: Current Applied Physics Vol. 11 (2011), p. S301-S304.
[6] P. Bera and C. Anandan: Surface Review and Letters Vol. 21 (2014), p. 1450054.
[7] T. Toloshniak, Y. Guhel, J. Bernard, A. Besq, S. Marinel and B Boudart: Materials Research Bulletin Vol. 70 (2015), p. 712-718.
[8] D. Barreca, A. Gasparotto, E. Tondello, C. Sada, S. Polizzi and A. Benedetti: Chemical Vapor Deposition Vol. 9 (2003), p. 199-206.
[9] C.Y. Lin, D.Y. Lee, S.Y. Wang, C.C. Lin and T.Y. Tseng: Surface and Coatings Technology Vol. 203 (2008), p. 480-483.
[10] R. Murugan, G. Vijayaprasath, P. Sakthivel, T. Mahalingam and G. Ravi: AIP Conference Proceedings Vol. 1731 (2016), p. 080029.
[11] C.O. Avellaneda, M.A.C. Berton and L.O.S. Bulhões: Solar Energy Materials and Solar Cells Vol. 92 (2008), p. 240-244.
[12] C.S.S. Pavan Kumar, R. Pandeeswari and B.G. Jeyaprakash: Journal of Alloys and Compounds Vol. 602 (2014), p. 180-186.
[13] M.T. Ta, D. Briand, Y. Guhel, J. Bernard, J.C. Pesant and B. Boudart: Thin Solid Films Vol. 517 (2008), p. 450-452.
[14] B. Geetha Priyadarshini, S. Aich and M. Chakraborty: Journal of Materials Science Vol. 46 (2011), p. 2860-2873.
[15] M. Becht, F. Atamny, A. Baiker and K.H. Dahmen: Surface Science Vol. 371 (1997), p. 399-408.

The Residual Stress Distribution of Ti-6Al-4V Thin Wall in the Selective Laser Melting

Changpeng Chen, Haihong Zhu*, Zhongxu Xiao, Shiwen Liu, Jie Yin, Xiaoyan Zeng

Wuhan National Laboratory for Optoelectronics, Huazhong University of Science and Technology, Wuhan, Hubei, 430074, PR China

*Corresponding author: zhuhh@mail.hust.edu.cn; 86-27-87544774

Abstract. High residual stress caused by the high temperature gradient brings undesired effects such as shrinkage and cracking in the SLM process. This study developed a layer-by-layer 3D FE model based on sequentially coupled thermal-structural method to predict the residual stress distribution and the stress variations of Ti-6Al-4V thin wall. The X-component of stress has the greatest impact on the residual stress distribution of the thin wall, followed by the Z-component of stress, and the smallest is the Y-component of stress. The maximum residual stress is observed at the corner of the interface between the part and substrate, the secondary stress is near the top center region of the part. The findings of this study will provide a better understanding of residual stress distribution in the SLM and constructive guidance for process parameter optimization.

1. Introduction

Selective laser melting (SLM), well known as a type of additive manufacturing (AM), is a promising technology for the rapid manufacturing of three-dimensional (3D) parts with complex structures[1, 2]. However, in the SLM process, the local heat input caused by the high-energy laser beam leads to a high temperature gradient, which causes large residual stresses and undesired thermal deformation, and even fails to shape the part [3, 4].

The investigation of the residual stress in the SLM has attracted much attention in the past years. Mercelis and Kruth [4] found that the parts connected to the substrate contained very high levels stress while contained low stress levels when parts removing from the substrate. Hussein et al.[5] predicted high equivalent stress within a layer and alternating compressive and tensile residual stress within each track. Yadroitsev and Yadroitsava[6] found that the maximum stress was at the point where the sample joins with the substrate, the similar results also reported by Gu and He[7]. Ghasri-Khouzani et al.[8] measured a large tensile in-plane stress (up to $\approx$ 400 MPa) near the as-built disk top surfaces and found a marked breakdown of self-similarity in the residual stress distribution using Neutron diffraction method. Anderson et al.[9] used the neutron diffraction to determine the distribution of residual stress in a set of SLMed block. Wang and Chou[10] observed that there was a sharp variation in the average residual stress magnitude at the interface corner between the solid and the overhang. However, the residual stress distribution of Ti-6Al-4V thin wall in the SLM is not clear yet.

In the study, a layer-by-layer 3D FE model based on sequentially coupled thermal-structural method is developed to predict the residual stress distribution and the stress variations of Ti-6Al-4V thin wall.

The findings of this study will provide a better understanding of residual stress distribution in the SLM and constructive guidance for process parameter optimization.

2. The 3D FE Model for the SLM Process

2.1. Basic Equations in the Thermal and Structural Analysis

2.1.1. Governing Equations of Heat Transfer. The governing equations of thermal analysis for three-dimensional heat transfer can be expressed as follows:

$$\rho c\frac{\partial T}{\partial t}=\frac{\partial}{\partial x}\left(k_x\frac{\partial T}{\partial x}\right)+\frac{\partial}{\partial y}\left(k_y\frac{\partial T}{\partial y}\right)+\frac{\partial}{\partial x}\left(k_z\frac{\partial T}{\partial z}\right)+q \quad (1)$$

where, ρ is material density (kg/m^3); c is specific heat capacity (J/kg ^{o}C); T is temperature (^{o}C); t is interaction time of powder and heat source (s); k_x, k_y and k_z are the thermal conductivities (W/m ^{o}C) of x, y and z directions; q is heat generation per unit volume (W/m^3 s);
The initial temperature of finite models is set to ambient temperature, which is 25 ^{o}C. In order to obtain more accurate simulation results, the combined radiative and convective heat transfer coefficient are considered to calculate heat transfer between powder bed and the enviroment, it can be discriped as [11]:

$$h=h_{conv}+\varepsilon\sigma\left(T^2+T_{\mathrm{amb}}^2\right)\left(T+T_{\mathrm{amb}}\right) \quad (2)$$

where h is the combined radiative and convective heat transfer coefficient; h_{conv} is the convection heat transfer coefficient; ε is the emissivity of the powder bed; σ is the Stefan Boltzmann constant.

2.1.2. The Basic Equations of Structural Analysis. During SLM, a large temperature gradient is generated due to high energy input concentrated a small region, which can lead to large thermal stresses and strain. The relationship between stress and strain is defined as:

$$\{\sigma\}=[D]\{\varepsilon^e\} \quad (3)$$

where $\{\sigma\}$ is the stress vector; $[D]$ is the elasticity matrix; $\{\varepsilon^e\}$ is the elastic strain vector. Meanwhile, the simplified ideal elastic-plastic model is used in this model, therefore, $\{\varepsilon^e\}$ can be expressed as follows :

$$\{\varepsilon^e\}=\{\varepsilon\}-\{\varepsilon^p\}-\{\varepsilon^t\} \quad (4)$$

where $\{\varepsilon\}$, $\{\varepsilon^p\}$ and $\{\varepsilon^t\}$ are the total strain vector, the plastic strain vector , and the thermal strain vector, respectivity.
The Von Mises yield criterion is used in this model and equivalent stress, σe is computed as:

$$\sigma_e=\sqrt{\frac{1}{2}\left[\left(\sigma_1-\sigma_2\right)^2+\left(\sigma_2-\sigma_3\right)^2+\left(\sigma_3-\sigma_1\right)^2\right]} \quad (5)$$

where σ_1, σ_2 and σ_3 are the three principal stresses.

2.2. Layer-by-layer Model

2.2.1. Methodology. In the SLM, the three-dimensional size of the components reaches tens or even hundreds of millimeters. However, the laser beam spot diameter is only sub-millimeter, which indicates that tens of thousands or even millions of scanning tracks are needed for fabricating the parts. At the same time, the forming process will take dozens of hours. Therefore, the current computing power is difficult to simulate this process, and develop a model that fully reflect the process is also not practical. A reasonable predicting model must be developed to quickly obtain the residual stress distribution of large-scale parts.

In the study, a layer-by-layer 3D FE model that ignores intra-layer scanning strategies is developeda. In the model, the part is divided into multiple layers along the deposition height direction. Each layer has equal thickness and consists of one or multiple powder layer for improving the computing efficiency. The process that laser scanning the powder layer is simplified into a process that heating the segmentation layer using a volume heat resource (Q). The volume heat source is calculated based on process parameters:

$$Q = m\frac{AP}{DHk} \tag{6}$$

where m is the heat source coefficient, A is the laser absorption coefficient, P is the laser powder, D is the laser beam spot diameter, H is hatch spacing, k is the powder layer thickness.

After the heating phase of the current layer is completed, there is a cooling phase with a powder recoating time (20 s). This process is repeated until the part is completed and the residual stress distribution is obtained after the parts is cooled down to room temperature.

Several assumptions have been applied in the layer-by-layer model for simplifying simulation:

- The whole powder bed is considered to be homogeneous and continuous media.
- The heat transfer at the bottom of the model can be assumed to be negligible due to the fact that the substrate is much larger than the manufacturing zone.
- The effect of scanning strategies in the layers is ignored for improving calculating efficiency.

2.2.2. 3D FEM Model Meshing. Figure 1 depicts the 3D FE model and meshing, the Ti-6Al-4V thin wall has a length of 50 mm, a width 5 mm and a height of 20 mm, the metal substrate of Ti-6Al-4V alloy with dimensions of 100×20×30 mm^3. The the laser scanning region is fine meshed with hexahedral element (0.5×0.5×0.8 mm^3), the coarser mesh is adopted for the surrounding substrate. The process parameters used in the study are listed in Table 1.

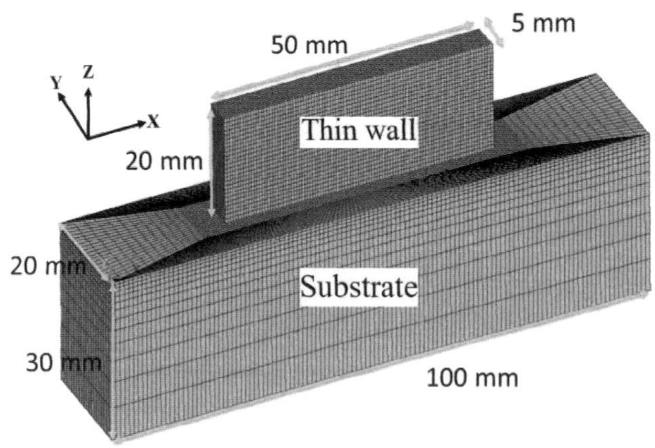

Figure 1. 3D finite element model and meshing

Table 1. Process parameters used in simulation for Ti-6Al-4V

Parameters	Values
Absorptivity	0.42
Ambient temperature (°C)	25
Laser power (W)	300
Scanning velocity (mm/s)	1000
Hatching spacing (μm)	100
Powder layer thickness (μm)	40
Beam diameter (μm)	150

2.2.3. Ti-6Al-4V material properties in residual stress analysis. In order to ensure the prediction accuracy of the residual stress model, the authoritative structual properties of material are obtained, the temperature-dependent material properites in structual analysis are shown in Table 2.

Table 2 The temperature-dependent structual properties for bulk Ti-6Al-4V

Temperature (°C)	**23**	**260**	**316**	**427**	**482**	**538**	**650**	**800**	**825**	**850**	**1650**		
Elastic modulus, E_e (GPa)	125	110	100	100	80	74	55	27	20	5	0.1		
Thermal expansion coefficient, $\alpha\times10^{-6}$(1/°C)	8.78	9.83	10.14	10.71	10.97	11.22	11.68	12.21	12.29	12.37	12.5		
Yield strength, $\sigma_{0.2}$ (MPa)	1000	630	630	525	500	446	300	45	25	5	0.1		
Temperature (°C)	**28**	**107**	**206**	**306**	**402**	**529**	**608**	**706**	**795**	**960**	**1002**	**1034**	**1046**
Poisson's ratio, υ	0.34	0.35	0.35	0.36	0.36	0.37	0.38	0.38	0.39	0.42	0.45	0.44	0.43
Temperature (°C)	**25**	**300**	**500**	**600**	**650**	**700**	**750**	**800**	**850**	**900**	**950**	**975**	**1050**
Elastic modulus, E_p(GPa)	2.87	2.88	1.62	0.41	0.40	0.15	0.01	0.03	0.05	0.05	0.03	0.02	0.02

3. Results and Discussions

3.1. The Residual Stress Distribution of Thin Wall

Figure 2 depicts the residual stress distribution in the three direction after the part is cooled down to room temperature. As shown in Figure 2 (a), it is found that the maximum X-component of stress is observed at the corner of the interface between the part and substrate, and a large tensile X-component of stress is also found near the top region of the thin wall. For compressive stress, the values on both the sides of the thin wall is small while the substrate region under the thin wall is large (≈-500 MPa). Other regions and the bottom region of the substrate are small. The distribution of the Y-component of stress is depicted in Figure 2 (b), the residual stress is small except for the corner of the interface between the part and substrate. For the Z-component of stress, as shown in Figure 2 (c), there is a large compressive stress region inside the thin wall and still a large tensile stress is found the corner of the interface between the part and substrate.

Comparing the residual stress in the X, Y and Z direction, the maximum residual stress is the X-component of stress, followed by the Z-component of stress, the Y-component of stress is the minimum. This is caused by the largest temperature gradient existing in the scanning direction, and Y-component of stress is easily released by deformation due to small size in the thin wall in the Y direction. As shown in Figure 2 (c), a large equivalent stress is found in the top region of the thin wall and it is similar to the X-component of stress as shown in Figure 2 (d). Moreover, the maximum stress is observed at the corner of the interface between the part and substrate due to all three stress components have a maximum at this position. This can be contributed to the high cooling rates and solidification at the bottom of part while the large constrain is imposed by the substrate.

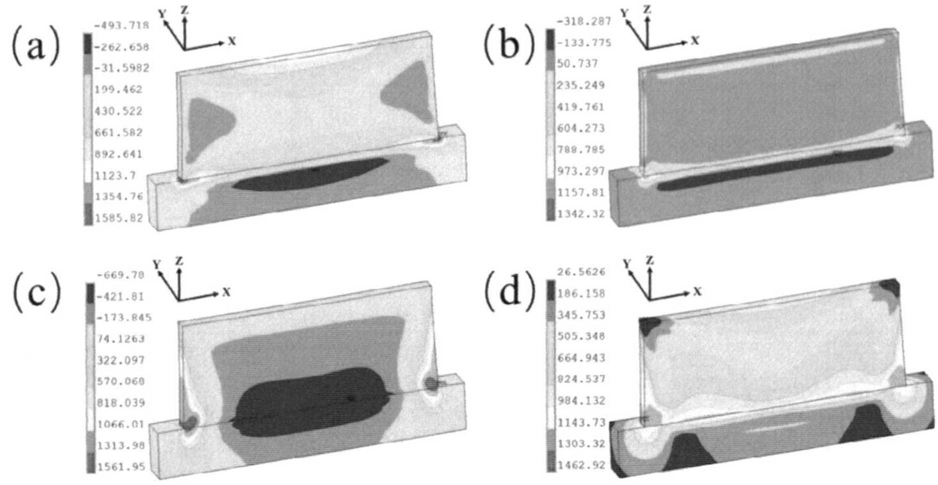

Figure 2. The residual stress distribution of thin wall: (a) X-component of stress, (b) Y-component of stress, (c) Y-component of stress, (d) Equivalent stress

3.2. The Residual Stress Distribution of Thin Wall

The stress variations along the deposition height of the thin wall is shown in Figure 3. It can be seen that both X and Y-component of stress show a transition from compressive stress to tensile stress as the deposition height increases, but the Z-component of stress is always compressive stress. The position of the tensile-compressive transformation is near the junction of the substrate and the thin wall, and a large stress fluctuation is also found at this location.

Furthermore, the maximum tensile stress in the X direction is found at the top surface of the thin wall and the value is 971.60 MPa, the maximum compressive stress is observed at Z = -2.81 mm which is -493.72 MPa. For the values of the Y-component of stress, the maxima of tensile and compressive stress are 310.70 and -307.94 MPa, respectively. The maximum compressive stress in the Z direction is -583.47 MPa and located at the junction of the substrate and the thin wall while the tensile stress can be ignored. As can be seen from the above, the maximum tensile stress is observed in the X direction while maximum compressive stress is found in Z direction. Therefore, the equivalent stresses at the bottom and top of the thin wall are close and the values are 1000.87 and 991.40 MPa respectively. Meanwhile, it is found that the X-component of stress has the greatest impact on equivalent stress in the thin wall, followed by the Z-component of stress, and the smallest is the Y-component of stress.

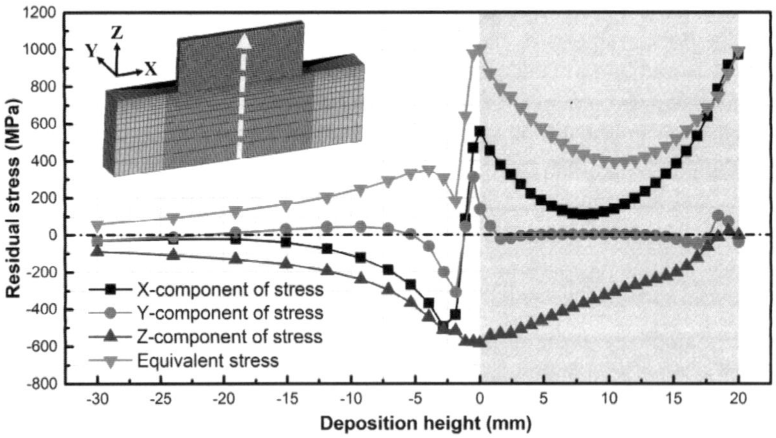

Figure 3. The stress variations along the deposition height of the thin wall

IOP Conf. Series: Materials Science and Engineering **538** (2019) 012020 doi:10.1088/1757-899X/538/1/012020

4. Conclusions

A layer-by-layer 3D FE model based on sequentially coupled thermal-structural method is developed to predict the residual stress distribution and the stress variations of Ti-6Al-4V thin wall. The residual stress in the three direction has different distribution along the deposition height. The X-component of stress has the greatest impact on residual stress distribution of the thin wall, followed by the Z-component of stress, and the smallest is the Y-component of stress. The maximum of residual stress is observed at the corner of the interface between the part and substrate, the secondary stress is near the top center region of the part. The findings of this study will provide a better understanding of residual stress distribution in the SLM and constructive guidance for process parameter optimization.

Acknowledgments

The authors gratefully appreciate the financial support from the National Program on Key Basic Research Project of China (6132801), the National Natural Science Foundation of China (Grant Nos. 61575074), and the Fundamental Research Funds for the Central Universities through Program No.HUST: 2016YXZD005.

References

[1] C.Y. Yap, C.K. Chua, Z.L. Dong, Z.H. Liu, D.Q. Zhang, L.E. Loh, S.L. Sing, Review of selective laser melting: Materials and applications, Applied Physics Reviews 2(4) (2015) 041101.

[2] S.H. Huang, P. Liu, A. Mokasdar, L. Hou, Additive manufacturing and its societal impact: a literature review, The International Journal of Advanced Manufacturing Technology 67(5-8) (2012) 1191-1203.

[3] J.P. Kruth, L. Froyen, J.V. Vaerenbergh, P. Mercelis, M. Rombouts, B. Lauwers, Selective laser melting of iron-based powder, Journal of Materials Processing Technology 149(1-3) (2004) 616-622.

[4] P. Mercelis, J.P. Kruth, Residual stresses in selective laser sintering and selective laser melting, Rapid prototyping journal 12 (2006) 254 - 265.

[5] A. Hussein, L. Hao, C. Yan, R. Everson, Finite element simulation of the temperature and stress fields in single layers built without-support in selective laser melting, Materials and Design 52 (2013) 638-647.

[6] I. Yadroitsev, I. Yadroitsava, Evaluation of residual stress in stainless steel 316L and Ti6Al4V samples produced by selective laser melting, Virtual and Physical Prototyping 10(2) (2015) 67-76.

[7] D. Gu, B. He, Finite element simulation and experimental investigation of residual stresses in selective laser melted Ti–Ni shape memory alloy, Computational Materials Science 117 (2016) 221-232.

[8] M. Ghasri-Khouzani, H. Peng, R. Rogge, R. Attardo, P. Ostiguy, J. Neidig, R. Billo, D. Hoelzle, M.R. Shankar, Experimental measurement of residual stress and distortion in additively manufactured stainless steel components with various dimensions, Materials Science and Engineering: A 707 (2017) 689-700.

[9] L.S. Anderson, A.M. Venter, B. Vrancken, D. Marais, J.V. Humbeeck, T.H. Becker, Investigating the Residual Stress Distribution in Selective Laser Melting Produced Ti-6Al-4V using Neutron Diffraction, Materials Research Proceedings 4 (2018) 73-78.

[10] X. Wang, K. Chou, Effect of support structures on Ti-6Al-4V overhang parts fabricated by powder bed fusion electron beam additive manufacturing, Journal of Materials Processing Technology 257 (2018) 65-78.

[11] J. Yin, H. Zhu, L. Ke, P. Hu, C. He, H. Zhang, Xiaoyan Zeng, A finite element model of thermal evolution in laser micro sintering, The International Journal of Advanced Manufacturing Technology 83(9-12) (2015) 1847-1859.

Effect of Nanosecond Laser Treatment Parameters on Surface Wettability Behaviour of Pure Aluminium

Jian GUO[1], Xiaolei MA[1], Xuekang SI[1], Zeng YANG[1], Jingnan ZHAO[1,*]

[1]Mechanical Engineering College, Tianjin University of Science and Technology, Tianjin, 300222, China

Corresponding author: jingnanzhao@tust.edu.cn

Abstract. In this study, a method of preparing stable superhydrophilic surface on aluminum plate by nanosecond laser is presented. The influences of different laser processing parameters on the wettability of aluminum plate and the effect of preservation temperature on the conversion rate of aluminum plate from superhydrophilic to superhydrophobic after laser ablation were studied. It is demonstrated that the microstructure formed by different laser processing parameters has certain influence on the hydrophilic property of aluminum plate, such as the lower the scanning interval and scanning speed, the easier it is to form a superhydrophilic surface. In addition, the storage temperature has a significant effect on the change of surface contact angle with time. The samples stored in the average temperature of 0°C in the atmospheric environment have an average contact angle of 18.7 °in 60 day after laser treatment, which has always maintained a good hydrophilicity. However, the samples heated 3 hours in the temperature of 150 °C appears the hydrophobicity with an average contact angle of 156.3 °. The mechanism of wettability transition was expounded, and a method to maintain the superhydrophilic durability of metal surface was proposed in this paper.

1. Introduction

In recent years, new properties have been designed on the surface of materials to provide special wettability, such as superhydrophilic and superhydrophobic properties. Wettability can be measured by water droplet contact angle (CA). CA is the angle between the surface of the liquid and the outline of the contact surface (solid surface typically) [1]. When the CA tends to extreme, two kinds of extreme surface are produced super-hydrophilic surface (CA value less 10°) and super hydrophobic surface (CA value over 150°)[1-5]. Hydrophilic materials have potential for use in environmental purification, outdoor protection, medical and health, agriculture, food packaging and other fields [6-8]. Laser as an effective method could fabricate micro/nano structures directly on the surface of many materials and to control the size of micro/nano structure accurately [9-12] Some recent studies on the effect of laser on the wettability of metal surfaces have been reported [9-14]. However, most of these studies focus on the study of surface superhydrophobicity, and there are few independent descriptions of superhydrophilicity.

In this paper, a nanosecond laser method for the preparation of long-term superhydrophilic aluminum plates is presented. The influence of laser parameters on the surface structure and hydrophilicity of aluminum plate was studied. Through the simple adjustment of laser processing parameters, different micro/nano structures can be obtained, so as to produce different hydrophilic effect, and explore the effect of preservation temperature on the long-term retention of aluminum plate superhydrophilic.

IOP Conf. Series: Materials Science and Engineering **538** (2019) 012021 doi:10.1088/1757-899X/538/1/012021

2. Experiment

The material used in this study was Aluminum substrate with thickness of 0.5 mm. The aluminum plate was cut into 10 cm×10 cm square, followed by acetone and ultrasonic cleaning in deionized water for 10 min to remove impurities and oil stains on the surface of the aluminum plate.
An nanosecond laser (HANS laser EP-20-SHG) was employed to conduct the experiments with a laser power of 8 W, wavelength of 532nm, a spot size of 20 μm, a focusing focal length of 224mm, a beam mass of M2<1.4, and a pulse repetition frequency of 10-200KHz. The output of the laser wavelength of 532 nm, repetition frequency of 20 KHz, focusing spot diameter for Ø20 μm. The laser texturing pattern at well-controlled size and laser processing parameters were designed via the graphic interface of nanosecond laser control software. In the process of experiment, (line-by-line in both horizontal and vertical directions with equivalent distances between adjacent scanning lines) were used respectively to investigate the effects of different micro/nano structures on surface hydrophilicity, as shown in Figure 1.

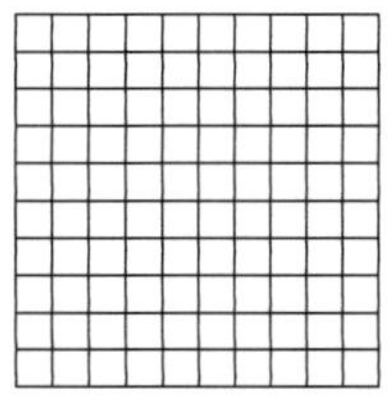

Figure 1. Grid pattern scanning.

In the process of laser scanning, different micro/nano structure were formed on the machining surface by changing the scanning speed V and scanning interval h. The scanning area of each parameter is 10×10 mm2 .The scanning speed V varies from 100 mm/s to 1900mm/s, and the scanning interval from 0.005 mm to 0.02 mm.
After laser fabricated, the surface structure was observed by microscope and field-emission scanning electron microscopy. The surface chemistry was analyzed by X-ray spectrometer. In order to identify the change of sample's wettability, CA tests were taken before and after the laser scanning. The Zhongchen JC2000DM precision automatic CA measuring instrument was used in this study. The CA was measured in an environment with temperature of 22.5℃ and humidity of 47% to 50%. A 25 μl deionized water droplet was dropped to the surface of a sample, then through droplet images to measure the CAs.

3. Results and Discussion

3.1 Surface Analysis

The SEM image of aluminum plate treated by nanosecond laser shows that the change of surface morphology is closely related to laser parameters, such as scanning speed and the scanning interval. The scanning interval is 0.005 mm, and the scanning speed is 100mm/s, forming the micro/nano structure, as shown in Figure 2.(a), typical micro/nano dual structure and columnar convex structure can be observed. The size distribution of these columnar structures is not uniform, because of the low scanning speed and scanning interval laser ablation is serious, the material in the recrystallization process is easy to form irregular columnar structure. With the increase of the scanning speed and scanning interval, the density of this columnar convex structure decreases, and nano-scaled ripples gradually appear, forming periodic square and presenting a uniform distribution as shown in the Figure 2.(b). In the enlarged image of a pair of columnar convex structures, we can observe a large number of micro/nano particles randomly distributed on them, as shown in the Figure.2(c). The size and shape distribution of these columnar bulges depend largely on the scanning speed and spacing of the laser.

IOP Conf. Series: Materials Science and Engineering **538** (2019) 012021 doi:10.1088/1757-899X/538/1/012021

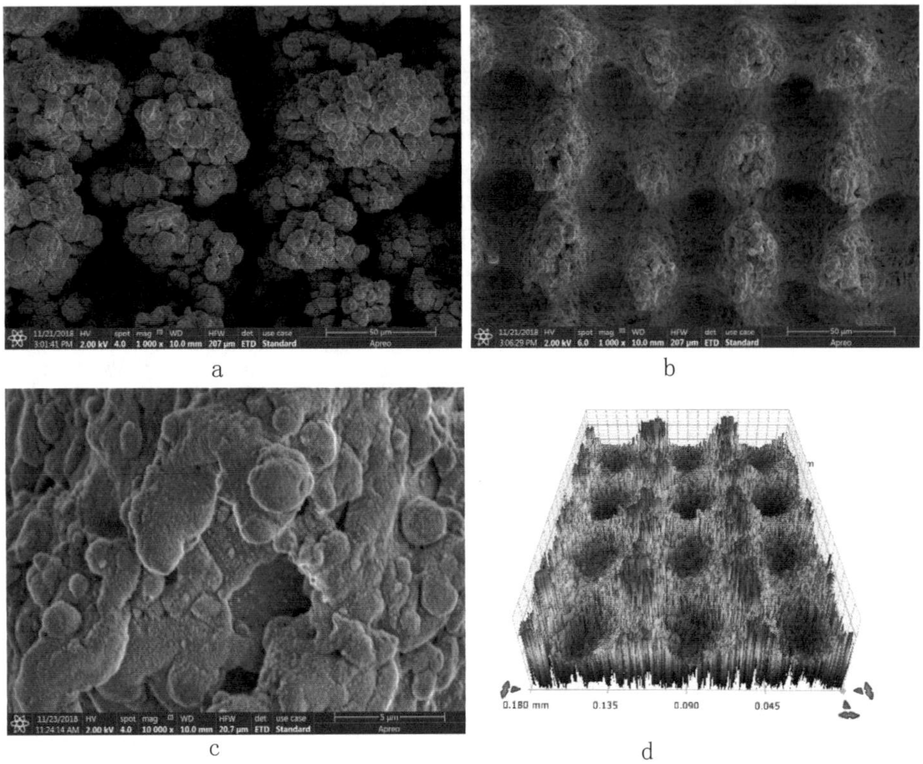

Figure 2. (a-c)SEM images of laser irradiated aluminum surface structures; (a) Scanning interval 0.005mm,scanning speeds 100mm/s;(b,c) Scanning interval 0.005mm,scanning speeds 900mm/s (d) 3D surface morphology obtained by a 3D optical profiometry

In order to largely illustrate the relationship between the laser parameters and the surface microstructure, the morphology of the sample was tested, and the three-dimensional contour map of the surface structure was obtained, as shown in Figure 2.(d). The Figure shows a typical three-dimensional outline of the surface, from which the columnar convex structure and nano-scaled ripples can be intuitively observed, and the period of ripple is around 60–70 um, the surface roughness value is 10.57 um, with the decrease of scanning speed and scanning interval, the surface roughness value increases. Therefore, the morphology of the surface can be tuned by simply adjusting the scanning speed.

3.2. CA Measurement

The effect of laser texture on the wettability of aluminum plate was studied by measuring the static CA of water drop on the surface of aluminum plate after laser ablation. As shown in Figure 3a, the surface texture CA changes formed by different laser ablation parameters, and each data point presents the average value of three measurements, that is, the CA value of water droplets on aluminum plate. The data in the Figure reflect the relationship between CA on aluminum plate surface and scanning interval and scanning speed. From the perspective of scanning speed, when the scanning interval is constant, the CA on the sample surface increases with the increase of scanning speed. Under the condition of constant scanning speed, CA on the sample surface increases with the increase of scanning interval. When the scanning speed is less than 1000 mm/s all the surface CA is less than 10 °, are characterized by super hydrophilicity. When the scanning interval is small, the CA on the surface of the sample is less affected by the change of scanning speed.

The wettability changes caused by laser treatment were analyzed by comparing the raw aluminum v plate with the laser ablation aluminum sample. The hydrophilic surface of the initial polished aluminum plate is 82.7. After laser ablation, CA values of all samples were less than 30.8° , showing hydrophilic behavior. The surface hydrophilicity formed by different laser ablation parameters is also different. In this paper, the wenzel model [15] is considered to explain the behavior of water droplets on rough surfaces. In the Wenzel model $\cos\theta_W = r \cos\theta_e$. The equation shows that the roughness effect amplifies the inherent wettability of the matrix material. Therefore, if the surface of the original aluminum plate is hydrophilic, the surface of the aluminum plate with a certain roughness value after laser ablation will be shown as super hydrophilic, and the microstructure formed on the surface of aluminum plate by grid mode scanning has small spacing and uniform distribution, which is more conducive to the diffusion of water droplets in all directions. In addition, the laser ablation of the aluminum plate surface must have alumina generation. EDS test results O2 35.6% and Al is 64.4%, and also prove that Al2O3 is formed on the aluminum surface by laser ablation, as shown in Figure 3.(b). Al2O3 itself has polar bonds, has large surface energy, is a hydrophilic substance, commonly used in industry as desiccant. The presence of Al2O3 promotes the surface hydrophilic effect of aluminum plates. Therefore, the formation of hydrophilic nanostructures and hydrophilic substances can well explain the mechanism for the change in wettability to a super-hydrophilic state of aluminum plate after laser ablation.

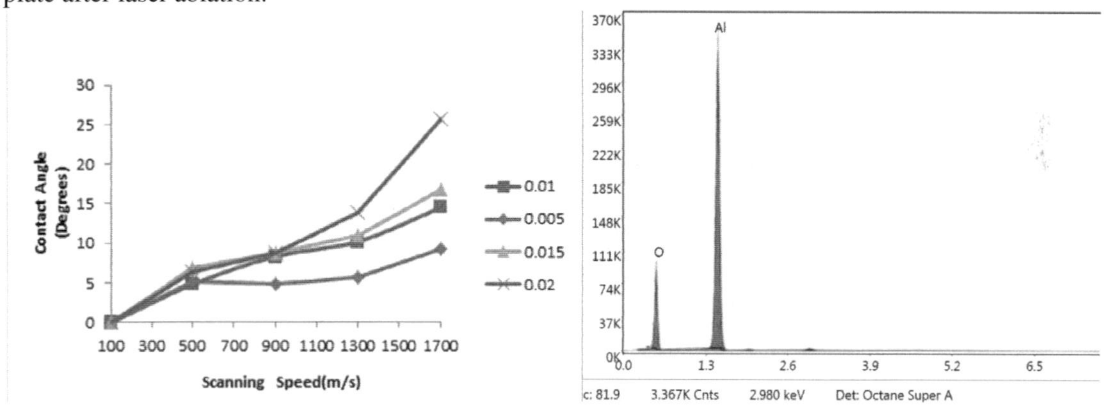

a b

Figure 3. (a) CA evolution of the laser ablation aluminum surfaces about different laser processing parameters. (b) EDS results for processed aluminum surface

Table 1. Change of CA with storage temperature

Storage Temperature (℃)	CA(°)	Storage Time	CA(°)
0	0	60day	18.7
50	0	24h	152
150	0	3h	156.3

In order to explore the influence of temperature on the CA of aluminum plate surface, the samples after laser ablation (Scanning interval 0.005mm,scanning speeds 100mm/s) were storage in different temperature environments, and the changes in the CA were observed after a period of time, as shown in table 1. Stored in the average temperature of 0℃ in the atmospheric environment sample CA was only 18.7 ° in 60 day after, has always maintained a good hydrophilicity. The stared in an oven at 50°C for 24 hour, CA of 152 °, characterized by super hydrophobic. When the storage temperature is 150 degrees Celsius, need only 3 hours CA is 156.3 °, characterized by super hydrophobic. Therefore, the preservation temperature has a great impact on the wettability of aluminum plate. It has been reported that the aluminum plate after laser ablation storage at room temperature in the atmospheric environment in the CA is about 30 days up to 150 °, super hydrophilic surface into a super hydrophobic within 30 days [9]. Chi-vinh Ngo et al [16] proposed that the laser ablated aluminum plate could rapidly change from superhydrophilic to superhydrophobic after simple heat treatment. It is found that the adsorption of organic matter by oxide is accelerated during heat treatment. After heat treatment, the aluminum plate surface appears to strengthen hydrophobic groups, the surface becomes super hydrophobic. Therefore, in a certain temperature range, with the increase of temperature, the adsorption of oxide on organic matter in the air will accelerate, resulting in a rapid transition from super hydrophilic aluminum plate to superhydrophobic. Conversely, the low temperature environment can effectively inhibit the adsorption of organic compounds on the surface of aluminum plate by oxides. In order to maintain the superhydrophilic durability of metal surface, an effective method is proposed.

4. Conclusion

To sum up, the nanosecond laser is used to prepare the super-hydrophilic surface on the aluminum plate, so that the processed aluminum plate surface immediately has the super-hydrophilic surface without any post-treatment. The relationship between sample surface hydrophilicity and laser processing parameters was studied. The optimal laser processing parameters were 0.005 mm scanning interval and 100 mm/s scanning speed. In addition, the storage temperature was explored in this study by observing the wettability of the sample surface with the changing of the preservation temperature. It is found that the transition time from superhydrophilic to superhydrophobic decreases with the increase of storage temperature in a certain range (0-150 ℃), and the transition takes place within a few hours from the original dozens of days to the present. Through this work, we can learn more about the influence of laser processing parameters on the wettability of aluminum plates, as well as the influence of storage temperature in atmospheric environment on the change of surface CA with time.

Acknowledgement

The authors would like to gratefully acknowledge the financial support provided by Tianjin Municipal Education Commission, China under the Grant 2017KDYB20, Tianjin Natural Science Foundation Grant 17JCYBJC42400.

References

[1] Tadanaga K, Morinaga J, Matsuda A, et al. Superhydrophobic−Superhydrophilic Micropatterning on Flowerlike Alumina Coating Film by the Sol−Gel Method[J]. Chemistry of Materials, 2000, 12(3).

[2] Liu H, Feng L, Zhai J, et al. Reversible Wettability of a Chemical Vapor Deposition Prepared ZnO Film between Superhydrophobicity and Superhydrophilicity[J]. Langmuir, 2004, 20(14):5659-5661.

[3] F. Cebeci, Z. Wu, L. Zhai, R. Cohen, M. Rubner, Nanoporosity-driven superhydrophilicity: a means to create multifunctional antifogging coatings, Langmuir .22 (2006) 2856–2862.

[4] V. Rico, P. Romero, J. Hueso, J. Espinós, Wetting angles and photocatalytic activities of illuminated TiO 2 thin films. Catal. Today. 143 (2009) 347–354.

[5] M. Ferrara, L. Pilloni, S. Mazzarelli, Hydrophilic and optical properties of nanostructured titania prepared by sol–gel dip coating, J. Phys. D. Appl. Phys.43 (2010) 95301.

[6] Morehead J, Zou M. Superhydrophilic surface on Cu substrate to enhance lubricant retention[J]. Journal of Adhesion Science and Technology, 2014, 28(8-9):10.

[7] Bennani J, Dillert R, Gesing TM, Bahnemann D. Physical properties, stability, and photocatalytic activity of transparent TiO 2 /SiO 2 films. Separation and Purification Technology. 2009;67:173–9.

[8] Ganjoo S, Azimirad R, Akhavan O, et al. Persistent superhydrophilicity of sol–gel derived nanoporous silica thin films[J]. Journal of Physics D Applied Physics, 2009, 42(2):025302.

[9] Long J, Zhong M, Zhang H, et al. Superhydrophilicity to superhydrophobicity transition of picosecond laser microstructured aluminum in ambient air[J]. J Colloid Interface Sci, 2015, 441(441C):1-9.

[10] Song Y, Wang C, Dong X, et al. Controllable superhydrophobic aluminum surfaces with tunable adhesion fabricated by femtosecond laser[J]. Optics & Laser Technology, 2018, 102:25-31.
[11] Ta D V, Dunn A, Wasley T J, et al. Nanosecond laser textured superhydrophobic metallic surfaces and their chemical sensing applications[J]. Applied Surface Science, 2015, 357:248-254.
[12] F. Chen, D. Zhang, Q. Yang, J. Yong, G. Du, J. Si, F. Yun, X. Hou, Bioinspired wetting surface via laser microfabrication, ACS Appl. Mater. Interfaces 5(2013) 6777–6792.
[13] Qi-Yu X, Lei D, Yun-Long J, et al. Wettability of surfaces with different surface microstructures textured by laser[J]. Acta Physica Sinica, 2015.
[14] Kwon M H, Shin H S , Chu C N . Fabrication of a super-hydrophobic surface on metal using laser ablation and electrodeposition[J]. Applied Surface Science, 2014, 288:222-228.
[15] Wenzel R N. Resistance of solid surfaces to wetting by water[J]. Transactions of the Faraday Society, 1936, 28(8):988-994..
[16] Ngo C V, Chun D M. Control of laser-ablated aluminum surface wettability to superhydrophobic or superhydrophilic through simple heat treatment or water boiling post-processing[J]. Applied Surface Science, 2018, 435:974-982.

IOP Conf. Series: Materials Science and Engineering **538** (2019) 012022 doi:10.1088/1757-899X/538/1/012022

Characteristics of RF Magnetron Sputtered VO_2 Thin Films for Potential Applications of Solar Control Coatings

Noormariah Muslim[1,a], Ying Woan Soon[1] and Nyuk Yoong Voo[1]

[1]Physical and Geological Sciences, Faculty of Science, Universiti Brunei Darussalam, Jalan Tungku Link, Gadong BE 1410, Brunei Darussalam

[a]Corresponding author: noormariah.muslim@ubd.edu.bn

Abstract. Properties of vanadium dioxide or vanadia (VO_2) thin films prepared using radio frequency (RF) magnetron sputtering with a pure VO_2 target were analysed. The properties consisted of crystal structure, surface topography, surface morphology as well as optical and thermochromic properties of VO_2 thin films. The deposition was conducted in an argon atmosphere of flow rate of 30 sccm, substrate temperature of 500 °C and RF sputtering power of 100 W. The film, which was deposited with film thickness of 50 nm, exhibited a single orientation of (110) orientation. The film surface featured quite homogeneous and continuous surface topography, and spherical-like grains that represented grains in VO_2(M) phase. The film had low surface roughness resulting in quite high optical transmittance and relatively low optical reflectance, at room temperature throughout visible and near-infrared regions. Meanwhile, the film also showed a phase transition at ~65.0 °C with hysteresis width of 38 °C.

1. Introduction

Solar control coatings have been widely known for the ability to significantly reduce near-infrared (NIR) transmittance as well as outdoor heat inflow through windows while maintaining or improving visible transmittance when the temperature of the coatings is higher than its transition temperature (Tt). Vanadium dioxide (VO_2) thin films as solar control coatings have been discovered to exhibit Tt of 68 °C for a VO_2 single crystal and are still under investigations to efficiently reduce the Tt in order to produce intelligent coatings for architectural glazing. Nonetheless, accessibilities of deposition techniques to grow stoichiometric VO_2 thin films are still crucial.

Studies on the effects of deposition parameters on crystal structure, surface morphology, surface topography, optical and thermochromic properties of VO_2 films using DC magnetron sputtering have been intensively conducted [1-2]. Similarly, depositions and characterisations of thermochromic VO_2 thin films by using RF magnetron sputtering with pure vanadium (V) [3-5] and vanadium pentoxide (V_2O_5) [6,7] target materials have also attracted much attention.

However, only a few studies reported on RF magnetron sputtering with pure VO_2 target. For instance, Ko et al. [8] reported growth of VO_2 films with (002) preferential orientation on aluminium oxide (Al_2O_3) substrates by RF magnetron sputtering from a VO_2 target with Ar and O_2 gas mixture. The films were prepared at substrate temperature of 550 °C and RF sputtering power of 150 W. The Tt of the prepared VO_2 thin films was ~62.0 °C.

Hence, this work investigates the structural, morphological, optical and thermochromic properties of VO_2 thin films deposited on glass substrates using a RF magnetron sputtering with a VO_2 target as solar control coatings.

2. Experimental Details

2.1. Film Preparation

VO_2 thin films were prepared by means of RF magnetron sputtering using a VO_2 target (99.9 % purity) with a diameter of 76.2 mm and thickness of 3.0 mm. Prior to depositions of the films, Corning 2947 glass substrates were cleaned with ethanol in an ultrasonic bath for 15 mins and rinsed in distilled water. Then, the glass substrates were dried with nitrogen gas of 99.9 % purity and then clamped on a substrate holder in the deposition chamber. The chamber was evacuated to 10^{-4} Pa by a turbomolecular pump. The VO_2 target was pre-sputtered in an argon (Ar) atmosphere for 10 mins to remove any impurities or oxide layers on the surface of the target. During the film depositions, the working pressure and RF sputtering power were kept constant at 1.2 Pa and 100 W, respectively. Commercial Ar gas flow rate was supplied as sputtering gas into the deposition chamber during the depositions and maintained at 30 sccm which was controlled by a mass flow controller. The substrate temperature was varied from room temperature to 500 °C. Film thickness was kept constant at 50 nm and monitored with a quartz thickness meter supplied by Inficon.

2.2. Film Characterisation

Film thicknesses were measured by a surface profilometer (Alpha-Step IQ) with a stylus force of 0.03 N (3.0 mg) and measurement accuracy of ± 1.5 nm. The structural, topological and morphological properties of the films were also measured. The structural properties of the deposited VO_2 films were measured using Shimadzu X-Ray Diffractometer 7000 with a monochromatic high-intensity CuKα radiation (λ = 1.54056Å) radiation operated at 40 kV and 30 mA. The grazing incidence X-ray diffraction (GIXRD) patterns of the films were recorded at a scanning rate of 2 °/min in the range of $2\theta = 20 - 60°$ and an incident angle of 1.0 °. The surface topography of the films were investigated using a commercially available AFM system from CSM instruments in contact tapping mode with a constant load force on the cantilever of 2.0 nN. The surface morphology of the films was analysed using a JEOL JSM-7610F field emission scanning electron microscope. As for studying optical properties for the thermochromic coatings applications, characterisation on transmittance-temperature spectra is the preferred measurement to observe changes in optical properties with respect to temperature. The optical and thermochromic properties of the films were obtained using a CARY 5000 UV-Vis-NIR spectrophotometer (Agilent Technologies) equipped with a SPECAC 4000 Series™ high stability temperature controller to measure optical transmittance against temperature in the wavelength range of 350 – 2500 nm and scanning rate of 250 nm/min. The heating and cooling rate for the measurements was ~5 °C/min. The optical reflectance was also taken in the wavelength range of 300 – 800 nm at room temperature.

3. Results and Discussion

3.1. Crystal Structure of the RF Sputtered VO_2 Thin Films

Figure 1 shows the diffraction pattern of the VO_2 thin films deposited on glass substrates using RF magnetron sputtering and VO_2 target. Only one diffraction peak could be seen which corresponded to VO_2 (110) orientation plane. The (110) crystallite orientation was used to estimate crystallite size of the deposited films by using the Scherrer formula. The estimated crystallite size of the deposited VO_2 thin films was 58.3 nm. Similar GIXRD pattern has been reported in which the VO_2 films were deposited using DC magnetron sputtering and VO_2 target [2].

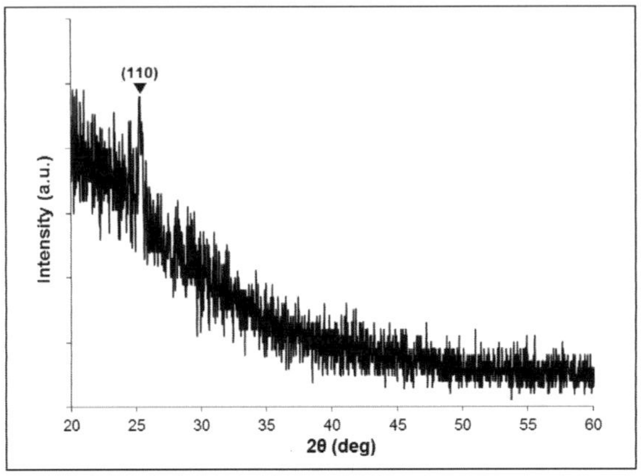

Figure 1. GIXRD patterns of the RF sputtered VO_2 thin films.

3.2. Surface Topography of RF Sputtered VO_2 Thin Films

Surface topography and surface roughness of the films were analyzed by using the representative AFM image of the VO_2 thin films shown in Figure 2, with a scan area of 4 μm^2. To study the AFM image, the dark regions were assumed to represent areas with zero or near zero height value in positive direction, whereas the bright regions represented higher areas such as the top of bulging grains. The film surface was quite homogeneous and continuous. The RMS surface roughness of the films was detected to be ~1.35 nm.

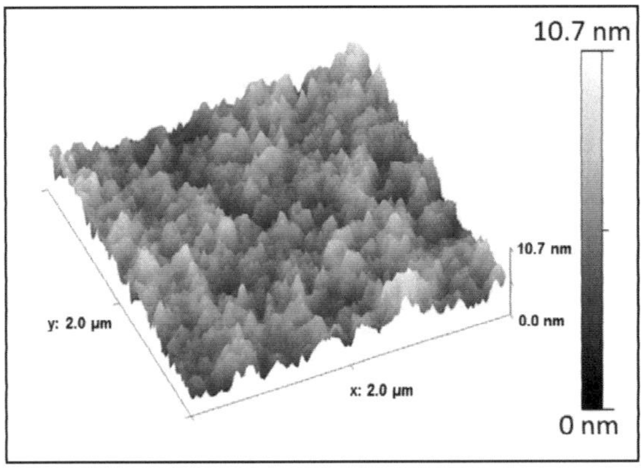

Figure 2. AFM 3D-image of the upper surface of the RF sputtered VO_2 thin films.

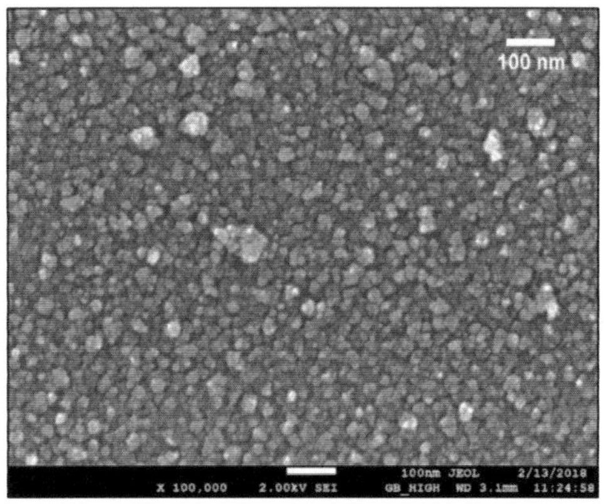

Figure 3. FE-SEM image of the RF sputtered VO_2 thin films.

3.3. Surface Morphology of RF Sputtered VO_2 Thin Films

Figure 3 represents the FE-SEM image of the RF sputtered VO_2 thin films. The film surface composed of uniform grain distributions with the estimated average grain size of 22.1 nm. Spherical-like grains could be seen which represented grains in VO_2(M) phase. As for VO_2(B) and V_2O_5 phases, very flat and no particular features and elongated grains would grow, respectively [9]. The application of VO_2 target for depositions of VO_2 thin films, in this study, using RF magnetron sputtering appeared to have similar surface morphology and spherical grains to the study of DC magnetron sputtered VO_2 thin films [2].

3.4. Optical and Thermochromic Properties of VO_2 Thin Films

Figure 4(a) displays the optical transmittance spectra (in the wavelength range of 350 – 2500 nm) of the RF sputtered VO_2 thin films at room temperature and 100 °C, i.e. in the semiconducting and metallic states, respectively. The deposited VO_2 films had quite high transmittance throughout the wavelength regions. Low surface roughness detected in the AFM analysis might cause low light-scattering loss which then results in high optical transmittance [10]. Nevertheless, as the temperature increased from room temperature to 100 °C, only a slight decrease in the transmittance of the films was measured.

The optical transmittance at a fixed wavelength of 1500 nm during heating and cooling for the RF sputtered VO_2 thin films are illustrated in Figure 4(b). The red and blue lines represent the heating and cooling processes, respectively. Only a slight variation in the hysteresis loop was found. This was also observed for DC magnetron sputtered VO_2 thin films [2]. The films showed Tt of ~65.0 °C with hysteresis width, ω of 38 °C. The Tt was a bit lower by ~3 °C relative to a VO_2 single crystal with Tt of 68 °C.

Moreover, Figure 5 shows the optical reflectance spectrum of the RF sputtered VO_2 thin films, which was only measured at ambient temperature in air. The reflectance of the films was relatively low whereby about 13 % at 400 nm in the visible region and about 10 % at 800 nm in the near-infrared region could be detected. Thus, this could be correlated to surface characteristics of the films, for instance low surface roughness.

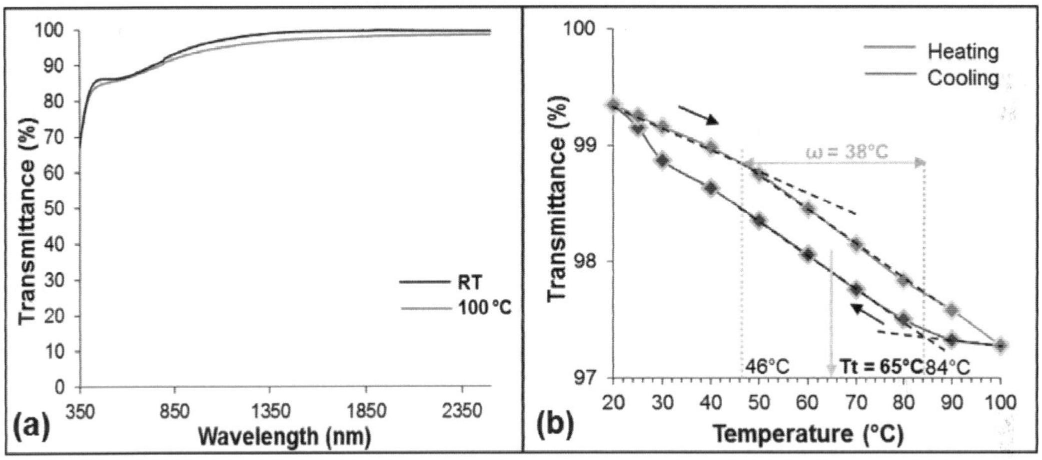

Figure 4(a). Optical transmittance spectra of the VO_2 films at room temperature (RT) and 100 °C, and **(b).** Optical transmittance at a fixed wavelength of 1500 nm as a function of temperature for the films.

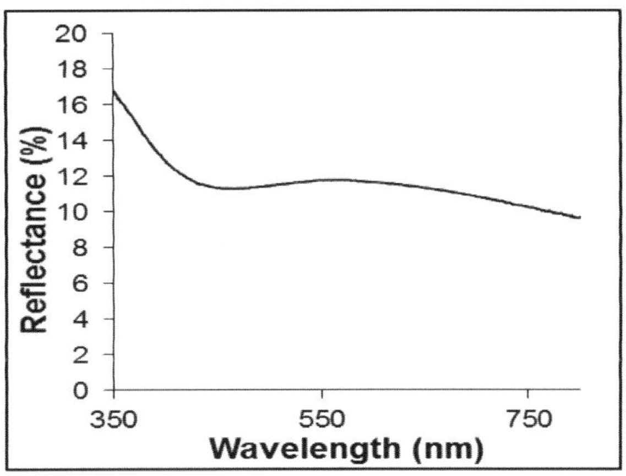

Figure 5. Optical reflectance spectrum of the RF sputtered VO_2 thin films measured at room temperature.

4. Conclusion

VO_2 thin films with a single orientation of (110) orientation were prepared using RF magnetron sputtering with a high purity VO_2 target at substrate temperature of 500 °C. The estimated crystallite size of the films was 58.3 nm. The film surface was quite homogeneous and continuous with RMS surface roughness of ~1.35 nm. The film also composed of uniform grain distributions with the estimated average grain size of 22.1 nm. The films showed quite high transmittance and relatively low reflectance throughout visible and near-infrared regions, at room temperature. Only a slight decrease in the transmittance and a slight variation in the hysteresis loop were measured between room temperature and 100 °C. The films had Tt at ~65.0 °C with hysteresis width of 38 °C. The Tt was a bit lesser than Tt of a VO_2 single crystal. Hence, as for solar control coatings, this study has shown a potential alternative technique to fabricate VO_2 thin films with a lower Tt using RF magnetron sputtering and a VO_2 target in an Ar atmosphere without the presence of any reactive gases such as oxygen.

Acknowledgments
This research was supported in part by the Brunei Research Council under Grant No. S&T 17.

References

[1] N. Muslim, Y.W. Soon, C.M. Lim and N.Y. Voo: IOP Conference Series: Materials Science and Engineering Vol. 409 (2018), p. 012016.
[2] N. Muslim, M.N.S.M. Idris, Y.W. Soon, Y.F.C. Chau, C.M. Lim and N.Y. Voo: IOP Conference Series: Materials Science and Engineering Vol. 409 (2018), p. 012025.
[3] R.E. Marvel, R.R. Harl, V. Craciun, B.R. Rogers and R.F. Haglund Jr: Acta Materialia Vol. 91 (2015), p. 217-226.
[4] Y.Y. Luo, S.S. Pan, S.C. Xu, L. Zhong, H. Wang and G.H. Li: Journal of Alloys and Compounds Vol. 664 (2016), p. 626-631.
[5] E. Gagaoudakis, E. Aperathitis, G. Michail, M. Panagopoulou, D. Katerinopoulou, V. Binas, Y.S. Raptis and G. Kiriakidis: Solar Energy Vol. 165 (2018), p. 115-121.
[6] F. Guinneton, L. Sauques, J.C. Valmalette, F. Cros and J.R. Gavarri: Journal of Physics and Chemistry of Solids Vol. 66 (2005), p. 63-73.
[7] S. Saitzek, F. Guinneton, G. Guirleo, L. Sauques, K. Aguir and J.R. Gavarri: Thin Solid Films Vol. 516 (2008), p. 891-897.
[8] C. Ko, Z. Yang and S. Ramanathan: ACS Applied Materials & Interfaces Vol. 3 (2011), p. 3396-3401.
[9] C. Batista, V. Teixeira and J. Carneiro: Journal of Nano Research Vol. 2 (2008), p. 21-30.
[10] H.N. Cui, V. Teixeira, L.J. Meng, R. Wang, J.Y. Gao and E. Fortunato: Thin Solid Films Vol. 516 (2008), p. 1484-1488.

The Effect of Annealing on Microstructure and Mechanical Properties of Selective Laser Melting AlSi10Mg

Changchun Zhang[1], Haihong Zhu[2,*], Yang Qi[3], Xiaoyan Zeng[4]

Wuhan National Laboratory for Optoelectronics, Huazhong University of Science and Technology, Wuhan, Hubei, P.R. of China

*zhuhh@hust.edu.cn

Abstract. The effect of annealing treatment on SLM AlSi10Mg on microstructure and mechanical properties has been investigated. It is observed that, with the increase of the temperature, the melt pools gradually become blurred and completely disappear when the temperature rises to 320℃. The microhardness and strengths of as-built samples are the highest due to the solid solution strengthening together with grain refinement strengthening. After heat treatment, the microhardness and strengths all decrease due to the weakening of solid solution strengthening but the elongation increases.

1. Introduction

Selective laser melting (SLM) is an additive manufacturing (AM) technique which can directly fabricate 3D parts from metal powders [1]. Compared with traditional manufacturing methods, it has the advantages of the high utilization rate of raw materials, short manufacturing cycle, near net forming, free from the limitation of the complex shape of parts. AlSi10Mg is a widely used aluminum alloy in aerospace and automobile industry for its high specific strength and excellent corrosion resistance. Furthermore, SLM Al-Si alloy has much better corrosion resistance than the cast counterpart, and different planes exhibit various corrosion properties[2,3]. The microstructure of the as-built aluminum parts is reported to be a very fine cellular-dendrite structure, the tensile strength is high but the elongation is poor. In order to balance the strength-elongation and obtain better comprehensive mechanical properties. It is necessary to control the microstructure and properties through heat treatment.

Li et al. [4] investigated the influence of solution and artificial aging heat treatments on the microstructure and mechanical properties of SLM AlSi10Mg parts. After heat treatment, the tensile strength decreased significantly. Takata et al. [5] examined changes in microstructure and mechanical properties of AlSi10Mg alloy at 300°C or 500°C. Heat treatment made fine Si phase precipitates within the columnar α-Al matrix and coarsening of the eutectic Si particles. Heat treatment eliminated the direction dependence of tensile ductility. Rosenthal et al. [6] provided a correlation between the mechanical properties of vertically built AlSi10Mg specimens subjected to T5 stress relief treatment, modified T5 treatment and Hot Isostatic Pressing treatment. It was found that the modified T5 treatment resulted in an increase in strength values beyond those of as-built condition. It can be seen that the current research on the heat treatment of AlSi10Mg mainly focus on the solid solution and aging heat treatment. Very little research has been done on low temperature annealing of AlSi10Mg. Fiocchi et al. [7] investigated changes in microstructure of AlSi10Mg after annealing. However, the

IOP Conf. Series: Materials Science and Engineering **538** (2019) 012023 doi:10.1088/1757-899X/538/1/012023

study of mechanical properties has only hardness but not tensile strength. It is necessary to study the relationship between the microstructure and mechanical properties especially tensile properties.
This work investigated the effect of low temperature annealing on SLM AlSi10Mg. The phase, microstructure, microhardness and tensile properties were discussed.

2. Materials and Experiment Procedures

2.1. Materials
The gas atomized spherical AlSi10Mg powders with an average particle size of 33.7μm were used in the experiments. The chemical composition (wt. %) was 9.15 Si, 0.18 Fe, 0.015 Cu, 0.225 Mn, 0.3 Mg, 0.01 Zn, 0.011 Ti and Al (balance).

2.2. SLM Process
The SLM experiments were conducted on a self-developed machine (LSNF-2), whose details have been given elsewhere [8]. All samples were deposited in an argon environment with the concentrations of O_2 controlled below 200 ppm. The processing parameters of cubic and tensile samples were chosen as follows: laser power (P) 490 W, scanning speed (V) 2000mm/s, layer thickness (δ) 40 μm, hatch spacing (S) 0.10 mm and a zigzag scan pattern with 90° rotation between adjacent layers. The samples were heat treated at 260°C, 280°C, 300°C and 320°C for 2 hours, furnace cooling, respectively.

2.3. Characterization
For microstructure analysis, samples were etched by Keller reagent which consists of 2.5 mL HNO3, 1.5 mL HCl, 1.0 mL HF and 95 mL deionized water. The microstructure of samples was observed by an EPIPHOT 300 optical microscopy (OM) and scanning electron microscopy (FEI NOVA NanoSEM450). Phase analysis was conducted by X-ray diffraction (PANalytical X'pert PRO) with Cu K_α radiation. Si particle size and density in the microstructure were estimated by business software[9]. Microhardness was tested using an HVS-1000 microhardness tester at a load of 200g and holding time of 15s. Tensile test samples were designed according to GB/T228.1-2010 standard (nearly equivalent with ISO 6892-1:2009), which was shown in Fig.1. Tensile tests were carried out using Zwick/Roll tester at the speed of 2mm/min.

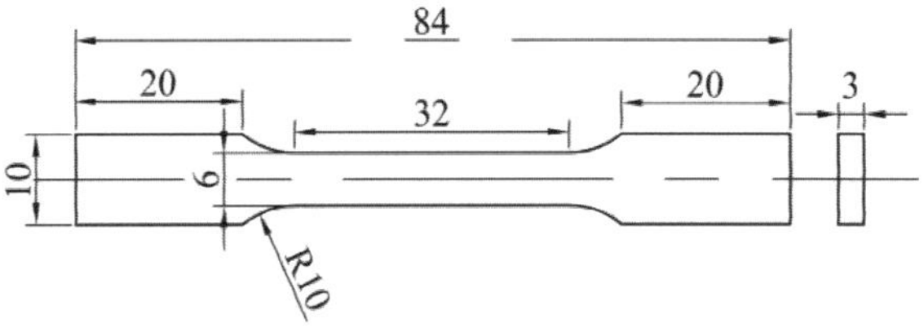

Figure 1. Configuration of tensile samples in this work.

3. Results and Discussions

3.1. Microstructure Characterization
Fig.2 shows the optical micrographs of AlSi10Mg samples of as-built condition as well as different annealing temperatures. The layer-wise feature intrinsic to SLM technology is obvious. The layer to layer and melt track to melt track are closely stacked to form a good metallurgical bonding. With the increase of annealing temperature, the boundaries of melt pool gradually become blurred. When the temperature rises to 320 °C, the boundaries of melt pool disappear completely, as shown in Fig.2 (e). Furthermore, SEM images with higher magnification were shown to investigate the details of the microstructure. The Si of as-built sample presents dendritic morphology and extends to the center of the melt pools, as shown in Fig.3 (a). After annealing treatment, Si changes to particles. Additionally, as the temperature increases, Si particles tend to grow up. It can be seen from Fig.3 (b)-(e).

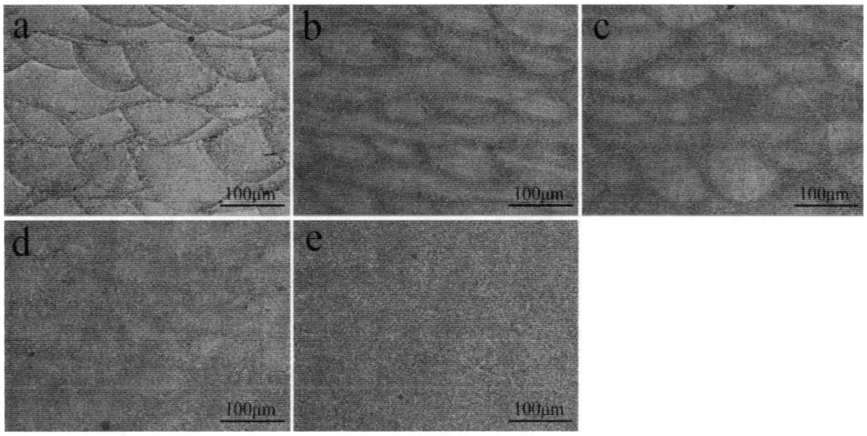

Figure 2. Optical microscopy images of etched AlSi10Mg samples with different annealing temperatures: (a) as-built, (b) 260°C, (c) 280°C, (d) 300°C, (e) 320°C.

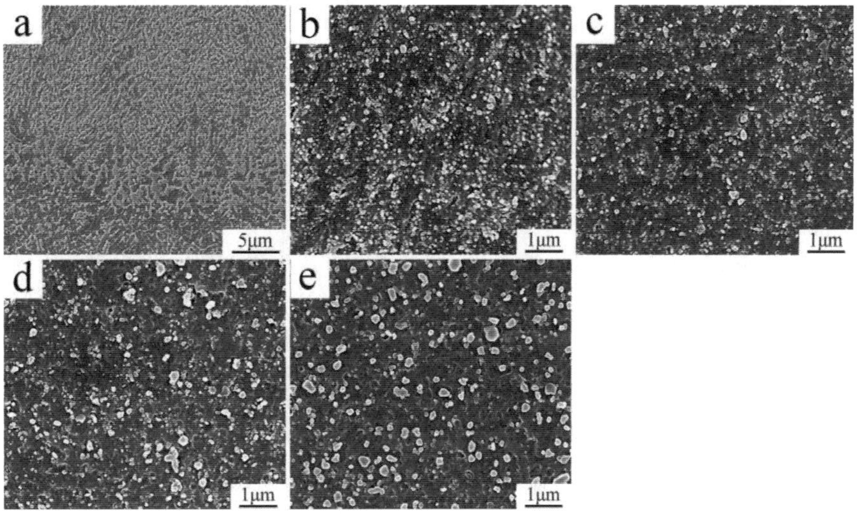

Figure 3. SEM images of etched AlSi10Mg samples with different annealing temperatures: (a) as-built, (b) 260°C, (c) 280°C, (d) 300°C, (e) 320°C.

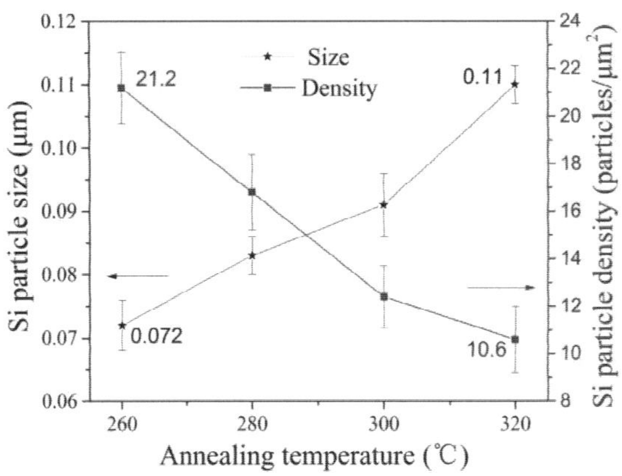

Figure 4. Size and density of Si particles of samples with different annealing temperatures.

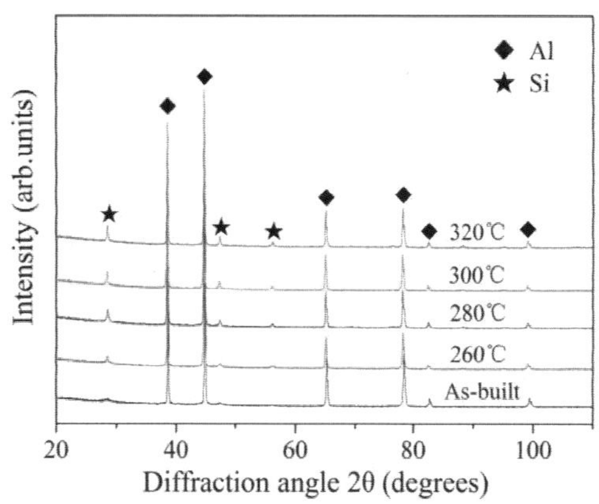

Figure 5. X-ray diffraction patterns of samples with different annealing temperatures.

The Si particle size and density at different annealing temperatures were calculated. The statistical results are shown in Fig.4. The results show that the Si particle size increases and the density decrease gradually with the increase of annealing temperature. This is because the heat treatment promotes atomic diffusion. Dendritic Si breaks into a discontinuous state, agglomerates and grows up to particles. Additionally, solid solution Si precipitates from the supersaturated α-Al matrix. This can be confirmed from the XRD test results shown in Fig.5.

3.2. Phase Analysis

Fig. 5 shows the phase analysis results on the samples after different annealing temperatures. Only Al peak and Si peak were detected in all the samples. With the increase of the annealing temperature, the diffraction peaks of Al phase slightly shift to lower angle, which implies the precipitation of Si phase from the supersaturated α-Al matrix. The higher the temperature, the more Si precipitation.

3.3. Microhardness

As for microhardness, 15 points were tested for each sample, and the average value was taken as the microhardness value of the sample. The test results were shown in Fig. 6. The results show that the microhardness of as-built samples is the highest. Annealing treatment reduces microhardness. The higher the temperature, the lower the microhardness. By calculating, the annealing temperature increases by 20°C, the microhardness decreases by 10-20 HV.

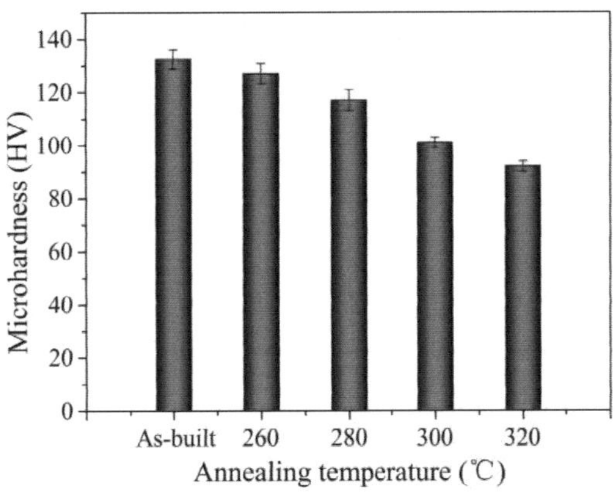

Figure 6. Microhardness of samples with different annealing temperatures.

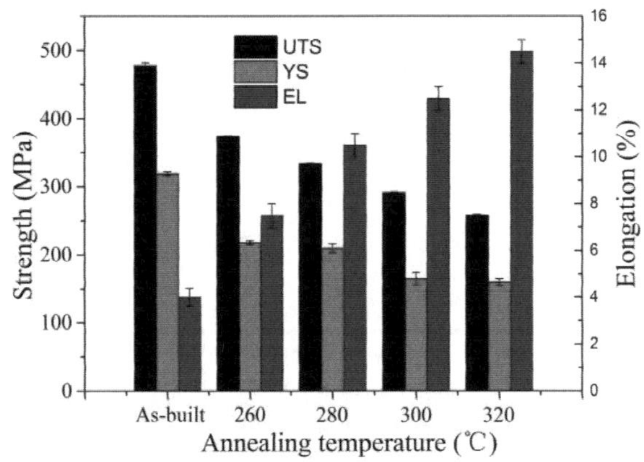

Figure 7. Tensile properties of samples with different annealing temperatures.

3.4. Tensile Properties

Fig.7 shows the tensile results of AlSi10Mg samples with different annealing temperatures. With the increase of annealing temperature, The ultimate tensile strength (UTS) and yield strength (YS) decrease and the elongation (EL) increase, the annealing temperature increases by 20°C, the UTS decreases by 30-40MPa and the EL increases by 2-3%. Due to the high solidification rate, a supersaturated solid solution of aluminum is formed during the SLM process. Furthermore, the grains of the as-built sample are very fine [10]. Therefore, the strength of the as-built samples is the highest due to the solid solution strengthening together with grain refinement strengthening. After annealing treatment, Si precipitates from supersaturated α-Al matrix and grow up to particles. The resistance of lattice distortion to dislocation motion reduced. So the effect of solid solution strengthening is weakened. Similar phenomenon has been reported by Rabadia et al. [11]. This leads to a decrease in strength and increase in ductility.

4. Conclusions

The effect of annealing temperature on microstructure, phase and mechanical properties of SLM AlSi10Mg were investigated in this work. The main conclusions are as follows:

(1) With the increase of the annealing temperature, the melt pools gradually become blurred. When the temperature rises to 320°C, the melt pools disappear completely. The morphology of Si changes from dendritic into particles, agglomerates and grow up.
(2) Annealing treatment makes the Si phase to precipitate from the supersaturated α-Al matrix.
(3) The microhardness of as-built samples is the highest. Annealing treatment reduces microhardness.
(4) The strength of the as-built samples is the highest due to the solid solution strengthening together with grain refinement strengthening. After annealing, Si precipitates from the supersaturated α-Al matrix. The effect of solid solution strengthening is weakened. This leads to a decrease in strength and increase in ductility.

Acknowledgements

This work was supported by the National Natural Science Foundation of China (61575074). The Authors also thank the Analytical and Testing Center of HUST for the XRD and SEM analysis.

References

[1] B.L. Zhang, H. Attar, Selective laser melting of titanium alloys and titanium matrix composites for biomedical applications: a review, Adv. Eng. Mater 18 (2016) 463–475.
[2] Y. Yang, Y. Chen, J. Zhang, X. Gu, P. Qin, N. Dai, X. Li, J.-P. Kruth, L.-C. Zhang, Improved corrosion behavior of ultrafine-grained eutectic Al-12Si alloy produced by selective laser melting, Mater. Des. 146 (2018) 239-248.
[3] Y. Chen, J. Zhang, X. Gu, N. Dai, P. Qin, L.C. Zhang, Distinction of corrosion resistance of selective laser melted Al-12Si alloy on different planes, J. Alloys Compd. 747 (2018) 648–658.
[4] W. Li, S. Li, J. Liu, A. Zhang, Y. Zhou, Q. Wei, C. Yan, Y. Shi, Effect of heat treatment on AlSi10Mg alloy fabricated by selective laser melting: Microstructure evolution, mechanical properties and fracture mechanism, Mater. Sci. Eng. A. 663 (2016) 116–125.
[5] N. Takata, H. Kodaira, K. Sekizawa, A. Suzuki, M. Kobashi, Change in microstructure of selectively laser melted AlSi10Mg alloy with heat treatments, Mater. Sci. Eng. A. 704 (2017) 218–228.
[6] I. Rosenthal, R. Shneck, A. Stern, Heat treatment effect on the mechanical properties and fracture mechanism in AlSi10Mg fabricated by additive manufacturing selective laser melting process, Mater. Sci. Eng. A. 729 (2018) 310–322.
[7] J. Fiocchi, A. Tuissi, P. Bassani, C.A. Biffi, Low temperature annealing dedicated to AlSi10Mg selective laser melting products, J. Alloys Compd. 695 (2017) 3402–3409.
[8] J. Yang, J. Han, H. Yu, J. Yin, M. Gao, Z. Wang, X. Zeng, Role of molten pool mode on formability, microstructure and mechanical properties of selective laser melted Ti-6Al-4V alloy, Mater. Des. 110 (2016) 558–570.
[9] L.C. Zhang, J. Das, H.B. Lu, C. Duhamel, M. Calin, J. Eckert, High strength Ti – Fe – Sn ultrafine composites with large plasticity, Scripta Mater.57 (2007) 101–104.
[10] C. Zhang, H. Zhu, H. Liao, Y. Cheng, Z. Hu, X. Zeng, Effect of heat treatments on fatigue property of selective laser melting AlSi10Mg, Int. J. Fatigue. 116 (2018) 513–522.
[11] C.D. Rabadia, Y.J. Liu, S.F. Jawed, L. Wang, Y.H. Li, X.H. Zhang, T.B. Sercombe, H. Sun, L.C. Zhang, Improved deformation behavior in Ti-Zr-Fe-Mn alloys comprising the C14 type laves and β phases, Mater. Des. 160 (2018) 1059-1070.

IOP Conf. Series: Materials Science and Engineering **538** (2019) 012024 doi:10.1088/1757-899X/538/1/012024

The Influence of Laser Parameters on the Coloring Effect and Microstructure of Stainless Steel Surface

Xiaolei MA[1], Jian GUO[1], Xuekang SI[1], Jingnan ZHAO[1,*]

[1]Mechanical Engineering College, Tianjin University of Science and Technology, Tianjin, 300222, China

Corresponding author: jingnanzhao@tust.edu.cn

Abstract. Nanosecond pulsed, green (532 nm) laser was used to irradiate SUS 304 stainless steel. Black, brown, gray, pink and other colors were obtained on the surface of polished stainless steel by adjusting the three process parameters: scanning interval (*Ls*), scanning speed (*v*) and defocusing distance (*Dd*). Scanning electron microscope (SEM) showed that there were three kinds of structures on the surface of stainless steel: columnar protrusions, droplets and cracks. The results indicated that the three process parameters affected the surface coloring effect and the microstructure formation of stainless steel by changing the total accumulated laser fluence in the scanning area.

1. Introduction

The rapid development of laser technology makes it widely used in metal processing [1-5]. Laser coloring is a kind of surface irradiation technology that makes use of pyrolysis reaction to make the metal matrix react with surrounding gas at high temperature and produces metal oxide[6]. A large number of metal and alloy materials form a variety of structures of oxidation growth layer after laser irradiation, many of which can be distinguished by the color layer can be used as a logo label and unique authenticity markings[6].

Lasers irradiated on metal surfaces are divided into nanosecond, picosecond and femtosecond lasers according to pulse width. There are many reports that three kinds of lasers can be used to fabricate color and surface structure on metal materials. In the past decade, femtosecond lasers have been widely used in the preparation of micro/nano structure on metal surface. The color and optical properties of metal surface have been changed by periodic micro-grating on metal surface and laser induced periodic surface structure (LIPSS) covered by nanostructure[7].

In this paper, we reported a study on coloring stainless steel surface using a nanosecond pulse laser. Black, brown and other color patterns were obtained on the surface of stainless steel by single irradiation with pulsed laser and different surface micro/nano structures are prepared. The influence of three process parameters include defocusing distance, scanning speed and the scanning interval on coloring and surface structure were analyzed. The results showed that their changes would lead to changes in the total accumulated laser fluence in the scanning area, thus affecting the color and morphology of the stainless steel surface.

2. Experiment

SUS304 stainless steel, an austenitic stainless was used for the study and the chemical composition

was shown in Table 1. The sample dimensions were 100 x 100 x 1 mm and the surface has been polished. Prior to the laser irradiation, the specimens were cleaned in deionized water. Ultrasound in ethanol was used to degrease the specimen surface and laser shortly irradiated after preparation.
The experiments were carried out by a Solid-state Laser (Draco-EP15A-SHG-S), which emits pulses at the central wavelength of 532 nm. Laser repetition rate (1-100 kHz) and pulse width (10-200 ns) can be adjusted according to different experimental requirements. The beam was focused onto a specimen using a f160 f-theta lens and the focal spot diameter is about 20 um. The microstructure and topography of the sample surface irradiated by laser were observed by metallographic microscope (CX40M) and the microscopic morphology of the specimens was characterized by SEM (SU1510).

Table 1. The chemical compositions of the SUS 304 stainless steel.

Element (Wt. %)							
C	Cr	Ni	Si	Mn	P	S	Fe
0.04	18.35	8.11	1	1.15	0.044	0.03	Balance

3. Results and Discussion

3.1. Surface Colors and Metallographic Micrograph

The surface colors of the sample are shown in Figure 1. Many colors were successfully prepared on the surface of stainless steel by adjusting the process parameters. This is the same as many previous studies[8-10], which all obtained different colors by adjusting the laser parameters. Figure 2 shows the microscopic images of the above colors under a metallographic microscope. The black sample presented severe ablation under the microscope, and the observation area was in the form of focal mass. The observation area of the gray sample showed a large number of worm-like structures, and the entire substrate was in the grayish white color. The brown sample was composed of very dense colored dots. The path of laser irradiation can be clearly observed in bright brown, and the ablation area was the crescent. The dark red, light blue and pink samples showed a large number of cracks under the microscope.

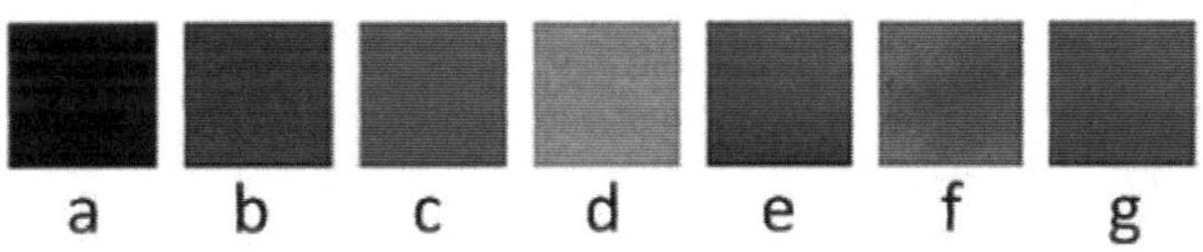

Figure 1. Different colors obtained on the surface of stainless steel by laser irradiation, (a) black (b) brown (c) silver gray (d) light brown (e) dark red (f) light blue and (g) pink.

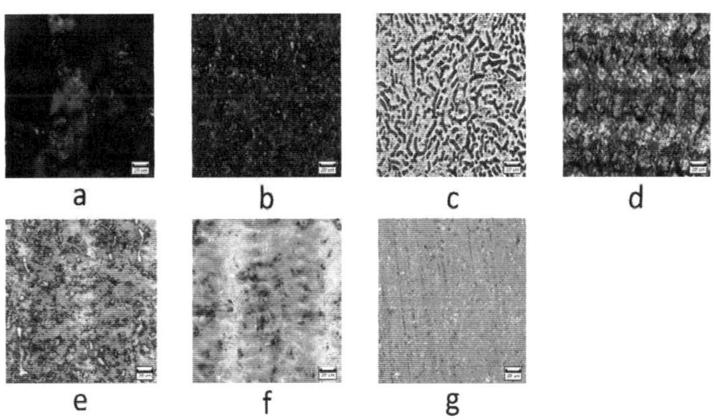

Figure 2. Microscope images of different colors, which corresponds to the seven colors shown in Figure 1 (The magnification is X500).

Different colors showed different morphologies under the microscope mainly because of their different process parameters, which leads to different degrees of reaction of the laser in the process of radiation samples, accompanied by molten, cooling, vaporization and other phenomena of the material[11].

3.2. Microstructure of Sample Surface

In order to further examine the differences of samples, the treated samples by laser and the untreated samples were observed in SEM. A mark of surface polish with untreated sample was shown in Figure 3(a). The corresponding color of Figure 3(b) is black. The SEM image shows that the dense columnar protrusion structures are formed on the sample surface. The illustration at the upper right is the magnification picture, which clearly indicates that these structures are composed of many small groups of particles. The SEM image of brown is shown in Figure 3(c). After laser irradiation, the structure of the sample surface is similar to that of water droplets spreading out. In the center of the irregular structure is the dense sand grain. Figure 3(d) presents a SEM micrograph of a pink sample, which is representative of the typical surface topography. Pervasive, interconnecting, through-thickness cracks were observed[6]. However, these cracks do not affect the appearance of color on the sample surface. In Guo's study[12], various surface microstructures also appeared on the surface of samples with different laser parameters.

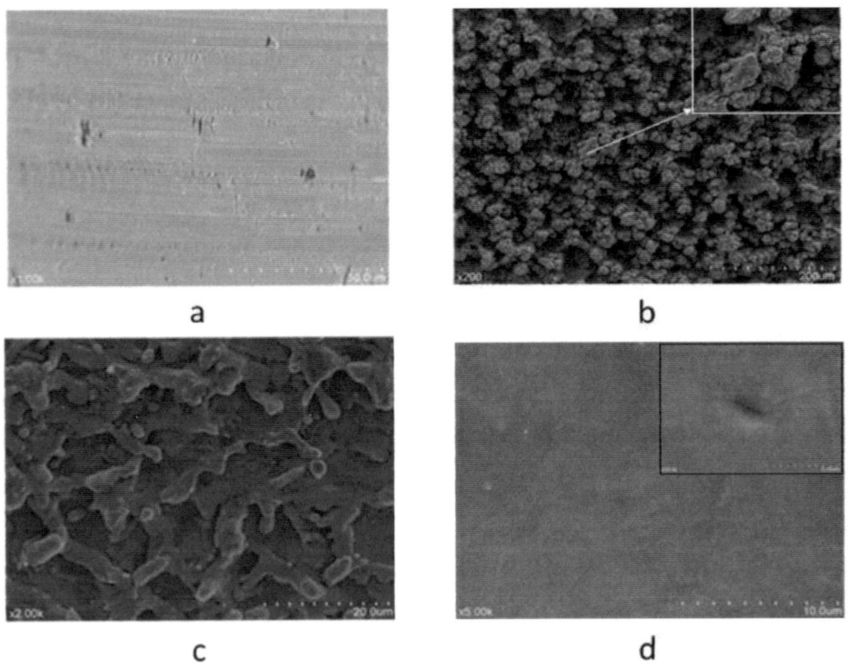

Figure 3. SEM images showing surface microstructure of different samples, (a) untreated stainless steel (b) black sample (c) brown sample (d) pink sample.

The main reason for the above different microstructures is that the total accumulated laser fluence in the scanning region is different and the total accumulated fluence Φ is defined as [13]:

$$\Phi = \phi N = \frac{a^2 E f}{v L_S} \tag{1}$$

Where the laser fluence is given by

$$\phi = \frac{2E}{\pi \omega_0^2} \tag{2}$$

The number of effective laser shots is

$$N = \underbrace{\sqrt{\frac{\pi}{2}}\frac{a\omega_0 f}{v}}_{\text{intra-line}} \underbrace{\sqrt{\frac{\pi}{2}}\frac{a\omega_0}{L_S}}_{\text{inter-line}} \quad (3)$$

where E is the pulse energy, f is the laser repetition rate, v is the scanning speed, Ls is the scanning interval and a is a correction factor due to the larger damaged area and therefore larger regions of laser overlap[14]. This correction factor was determined from semi-logarithmic plots used to determine the laser spot size[15]. The intra-line component of N encompasses the shot overlap within a local region, where f / v is the distance traveled between successive laser pulses in a single laser line. The inter-line component, in comparison, considers the geometrical overlap, $a\omega_0$ / Ls, between successive laser lines. The process parameters of several colors obtained in the experiment are shown in Table 2. It can be seen that Ls and v in accordance with each color are different. According to the formula (1), ϕ is directly related to a, E, f, L, s and v, where the laser spot size (laser spot size is related to the distance from the focus) and Ls directly affect the overlap rate. Therefore, three parameters lead to different total accumulated laser fluence in the scanning area, thus affecting the microstructure of the coloring surface.

Table 2. The process parameters corresponding to different colors.

Colors	Scanning interval	Scanning Speed	Repetition Rate	Defocusing distance
	Ls (mm)	*v (mm/s)*	*f (kHz)*	*Dd (mm)*
Black	0.005	50	25	0
Brown	0.02	400	25	0
Gray	0.005	100	25	4
Light brown	0.005	500	25	0
Dark red	0.005	1650	25	4
Light blue	0.005	2400	25	4
Pink	0.01	350	25	6

4. Conclusion

The work presented in this paper attests to the power of nanosecond pulsed laser surface treatment for stainless steel coloring. Three process parameters are studied in this paper, which are the defocusing distance, the scanning speed, and the scanning interval. The colors obtained on the sample surface have black, brown, gray，light brown，dark red，light blue and pink. At the same time, three kinds of microstructures include columnar protrusions, droplets and cracks, were observed in SEM. Experimental results show that the surface microstructure and coloring effect are closely related to three laser parameters, which can change the total accumulated fluence in the scanning area.

The results provide a strong basis for further development and applications related to laser-assisted surface coloring. Unlike other chemical techniques used for metals surface coloring, the advancement of this process will allow for the nanosecond laser to rapidly process into specific microstructures with different optical colors on the steel substrate in a very simple fashion. Since the relationship among part of laser treatment parameters and stainless steel colors were discovered in this paper, it could be said that more colors can be found on the stainless steel surface as long as these parameters are properly adjusted in the following experiments. The applications of the proposed technique are far-reaching and could potentially impact the advancement of metal coloring industry.

Acknowledgments
The authors would like to gratefully acknowledge the financial support provided by Tianjin Municipal Education Commission, China under the Grant 2017KDYB20, Tianjin Natural Science Foundation Grant 17JCYBJC42400.

References
[1] GORUNOV A I. Fabrication of coatings and bulk products made of a nickel-based material by additive technology laser metal deposition [J]. Russian Metallurgy, 2016, 2016(1): 91-4.
[2] NOVAKOV T, JACKSON M J, ROBINSON G M, et al. Laser sintering of metallic medical materials — a review [J]. International Journal of Advanced Manufacturing Technology, 2017, 93(5-8): 2723-52.
[3] OCA A J L, GONZ LEZCARRASCO J L, LIEBLICH M, et al. Compressive Residual Stresses and Associated Surface Modifications Induced in Ti6Al4V by Laser Shock Processing [J]. Materials Science Forum, 2016, 879(14):08-13.
[4] SOBOTOVA L, BADIDA M. Laser marking as environment technology [J]. Open Engineering, 2017, 7(1).
[5] ZHENG Y, AN Z, SMYREK P, et al. Laser interference patterning and laser-induced periodic surface structure formation on metallic substrates; proceedings of the IEEE International Conference on Manipulation, Manufacturing and Measurement on the Nanoscale, F, 2017 [C].
[6] LAWRENCE S K, ADAMS D P, BAHR D F, et al. Mechanical and electromechanical behavior of oxide coatings grown on stainless steel 304L by nanosecond pulsed laser irradiation [J]. Surface & Coatings Technology, 2013, 235(12): 860-6.
[7] AHSAN M S, AHMED F, KIM Y G, et al. Colorizing stainless steel surface by femtosecond laser induced micro/nano-structures [J]. Applied Surface Science, 2011, 257(17): 7771-7.
[8] GUO-QIANG L, XIAO-HONG L, HONG-DAO Y, et al. Fabrication of colorful metals with femtosecond laser pulses [J]. Chinese Journal of Optics, 2011, 4(1): 000072-76.
[9] WU Y, YANG B, KANG X. Effects of Processing Parameters of Ultra-Fast Laser on Surface Coloring of Stainless Steel [J]. Chinese Journal of Lasers, 2017, 44(03): 90-7.
[10] HUANG T Q, LI J M, H. L C. Laser-Induced Color Marking on 304 Stainless Steel by Nd: YAG Nanosecond Laser [J]. Applied Laser, 2013, 33(6): 586-90.
[11] LIU Y H, HU J D, ZHAO L, et al. Accumulation morphology on the surface of stainless steel irradiated by a nanosecond Nd:YAG pulsed laser [J]. Optics & Laser Technology, 2010, 42(4): 647-52.
[12] GUO L, LIN Y T, ZHANG Z H. Mechanism of laser coloration of stainless steel and color prediction based on neural network [J]. Chinese Journal of Lasers, 2016, 43(11): 118-26.
[13] GUAY J M, LESINA A C, C T G, et al. Topographical coloured plasmonic coins [J]. 2016.
[14] GUAY J M, CALALESINA A, GORDON P G, et al. Plasmonic coloring of noble metals rendered by picosecond laser exposure [J]. Proceedings of the Spie, 2017, 94.
[15] JANDELEIT J, URBASCH G, HOFFMANN H D, et al. Picosecond laser ablation of thin copper films [J]. Applied Physics A, 1996, 63(2): 117-21.

IOP Conf. Series: Materials Science and Engineering **538** (2019) 012025 doi:10.1088/1757-899X/538/1/012025

Very High Cycle Fatigue Behaviour of the 316L Weldment Fabricated By Laser Butt-Welding

Z H Xiong[1], X F Ma[1,*], X Y Qi[2]

[1]Sino-French Institute of Nuclear Engineering and Technology, Sun Yat-sen University, Zhuhai 519082, China
[2]Wuhan Huagong Laser Engineering Co., Ltd., Wuhan 430074, China

[*]Corresponding author: maxf6@mail.sysu.edu.cn

Abstract. In nuclear industry, 316L weldments are widely used in components fabricating, which usually suffer from very-high-cycle fatigue. Therefore, the very high cycle fatigue (VHCF) behaviour of 316L weldment fabricated by laser butt-welding joint were studied using an ultrasonic fatigue testing system. The microscopic examination indicated that an evident pore is observed in the weld seam and the fine columnar-dendrite structure is formed in the central part of the weld seam. Near the weld interface in the weld seam, the weld microstructure shows larger columnar-dendrite. The hardness distribution of the weldment is to coincide with the characteristic of the microstructure. In the central part of the weld seam, the hardness is the highest, and is the lowest in the base metal. Fatigue failure still occurred in the VHCF regime. The fatigue crack initiated from the internal pores caused by laser butt-welding. Although the fatigue strength of weld seam might be decreased due to the pore, the fatigue strength of the weldment was the same as the base metal in this research. That is, the defect with a certain size in the weld seam does not affect the strength of the weldment.

1. Introduction

316L is widely used as the structural material of nuclear components owing to its excellent corrosive resistance and mechanical properties [1,2]. These components usually suffer from VHCF loading during its service life [3]. Since some researchers [4] found that fatigue failure occurred in the very high cycle regime, more and more researchers started to investigate the very high-cycle fatigue behaviour of steels and alloys [5-7]. It was found that the fatigue crack usually initiated from the internal defect in the VHCF regime and a so-called fish-eye can be observed on the fracture surface, which is different from that of in the low and high-cycle regime. However, the VHCF data of the austenitic stainless steel used as the structural materials of nuclear components is still insufficient [8]. For austenitic stainless steel, the fish-eye is difficult to be observed on the fracture surface [9]. Most of the 316NG specimens failed due to the surface crack initiation, only a specimen failed due to internal crack initial caused by the internal inclusion [8]. The previously research results of the author [3,10] also showed that the fatigue crack initiated from the 316L specimen surface in the VHCF regime. Weld fatigue is a very important design consideration for the application of laser welded component in nuclear industry. However, the fatigue behaviour of 316L weldments in very-high-cycle region has rarely been investigated. The VHCF behaviour of some steel and alloy welds [11,12] indicated that the fatigue crack initiated from subsurface or internal detects caused by welding, and fatigue failure still occurred in the VHCF regime. The fatigue strength reduced owing to the existence of welding defects and soft zone along the welds. As reported by Hong et al. [7,13,14], fatigue crack usually initiates from internal or subsurface inclusions or defects in the VHCF regime, and the characteristic of the *S-N* curves differs from those of the low- and high-cycle fatigue.

Therefore, in this research, the VHCF behavior of 316L weldments fabricated by laser butt-welding were investigated by an ultrasonic fatigue testing machine. In addition, the microstructure and hardness distribution of the welding seam were investigated. To clarify fatigue crack initiation mechanism, the fatigue fracture surface was investigated through scanning electron microscopy (SEM).

2. Experimental Procedures

2.1. Materials and Specimen Fabrication

Type 316L austenitic stainless steel (or ASTM A240) was used in this research. The yield strength, ultimate tensile strength, and Vickers hardness were 319 MPa, 614 MPa, and 183, respectively. As shown in Figure 1 (a), the 316L welded joint was fabricated by laser butt-welding. The fiber laser head (IPG, YLS-10000) was held by a Robotic arm (KUKA, KR60HA). The laser beam has a circular shape with the diameter of 0.5 mm. The welding parameter used in this study were presented in Table 1. During the welding proceeding, ultra-high purity argon was used as a protecting gas at a flow rate of 25 L/min. The specimens were machined following the geometry shown in Figure 1 (b). Before fatigue testing, each specimen surface was polished using a 1200 grit abrasive paper.

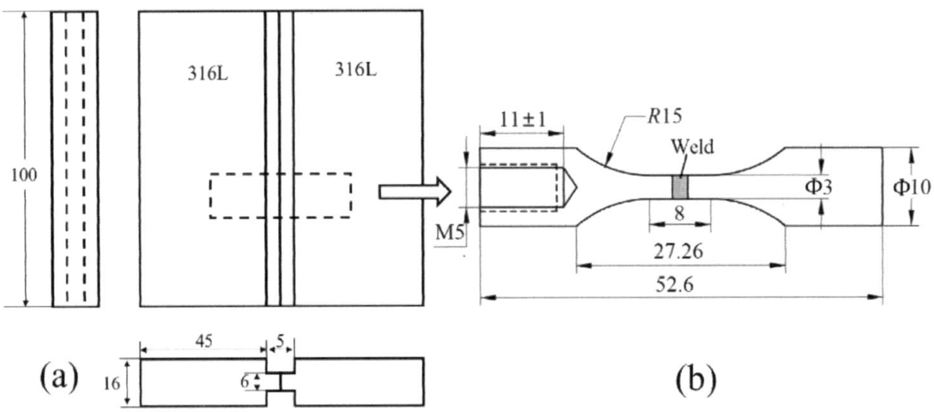

Figure 1. Schematic diagram of specimen fabrication (dimensions: mm) (a) the geometry of the fiber laser welded joint, (b) the size and shape of the ultrasonic fatigue specimen.

Table 1. Laser welding parameter

Laser power (kW)	Welding speed (mm/s)	Beam diameter (mm)	Defocus distance (mm)
6	30	0.5	0

2.2. Measurement of Microstructure and Hardness

The central part of the fatigue specimen with a uniform cross section was cut along the longitudinal direction and mounted in epoxy to study the microstructure and hardness. The samples were ground and polished until a mirror finish was achieved, and were then etched by immersing in an aqueous solution of 15-mL HCl, 5-mL HNO_3, and 100-mL H_2O for approximately 40 min. The specimen microstructures were observed through an optical microscope (VHX-6000). The hardness of the as-received specimens was measured using a Buehler VH1202 indenter under 0.98N before conducting the fatigue test.

2.3. Ultrasonic Fatigue Test

The ultrasonic fatigue tests were performed using an ultrasonic fatigue testing system at 20 kHz with a stress ration of -1 and manually stopped when the fatigue cycles reached up to 10^9. During the fatigue test, loading/arresting interval was applied in addition to the air-cooling of specimen surface to prevent specimen temperature increase. More details information for the ultrasonic fatigue test was reported in literature [15]. After the fatigue test, the fracture surfaces were examined to reveal the fatigue crack initiation and failure mechanism through SEM.

3. Results and Discussions

3.1. Microstructure and Hardness Distribution

The microstructure of the as-received 316L steel is shown in Figure 2(a), where typical austenitic grains are observed. Figure 2 (b) shows the overview image of the weld seam. The weld interface can be observed obviously. An evident pore is observed in the weld seam, which might be caused by the metal evaporation induced by high energy density and protection gas captured by the melted metal. The microstructures of area #1 in the central part of the welding bead is presented in Figure 2(c). The result shows that the fine columnar-dendrite structure is formed in the central part of the weld beam. And the growth directions of such dendrite structure are different, which are distributed symmetrically along the center line. The growth direction is dependent on the heat dissipation path, which is shortest to the base metal. As shown in Figure 2(d), near the weld interface in the weld seam, the weld microstructure shows more columnar-dendrite. This may be due to the fastest heat dissipation in the direction perpendicular to the weld interface. In general, the material properties reduce owing to the grain coarsening in the heat-affected zone. However, in this study, the heat-affected zone was not pronounced, as shown in Figure 2(d), and thus the grains in the base-material area near the welding interface did not grow significantly, which is consistent with the reported results, i.e. the heat-affected zone of 316L weld is not evident [16].

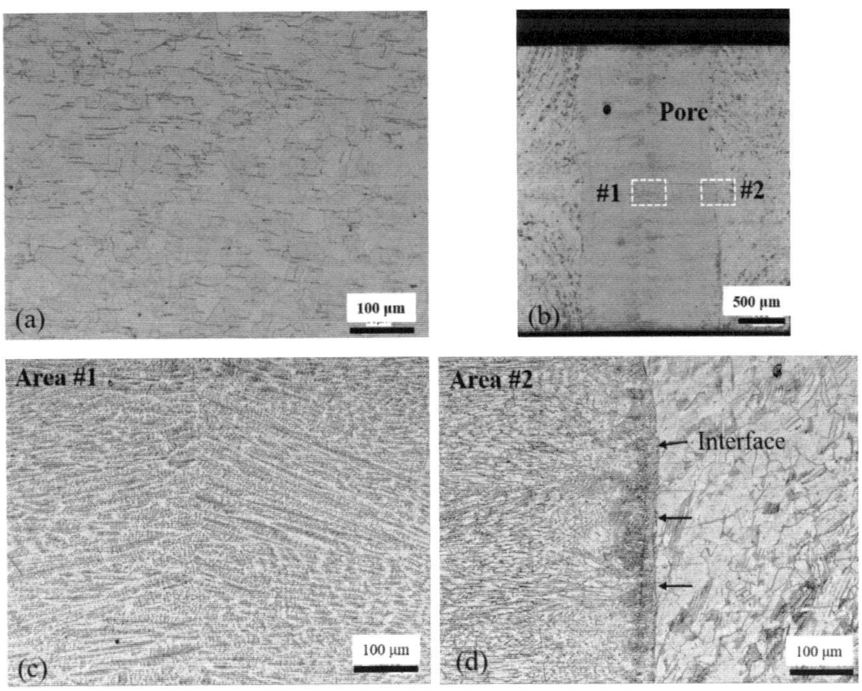

Figure 2. Microstructure of 316L weldment (a) Base metal, (b) Overview of the weld seam, (c) Enlarge view of the area #1 in (b), (d) Enlarge view of the area #2 in (b).

Figure 3 shows the hardness distribution of the weldment. In the central part of the weld seam, the hardness is highest (Hv223 in average) and resulting from the fine columnar-dendrite structure. The hardness of the area near the weld interface in the weld seam is lower than the central part of weld seam but higher than the base metal, which is attributed to the larger columnar-dendrite structure compared to the centre part of the weld seam.

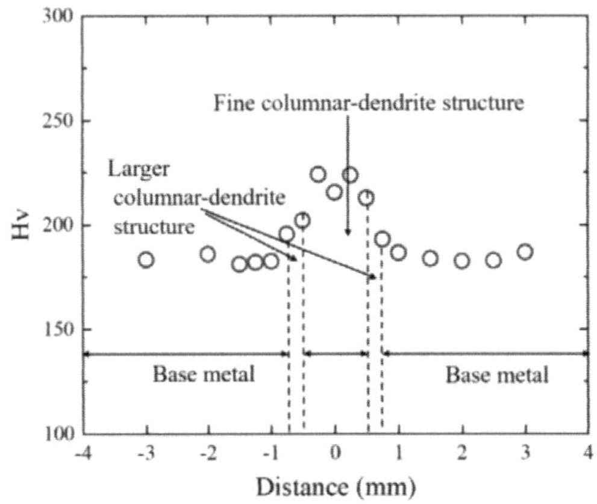

Figure 3. Hardness distribution of the weldment.

3.2. S-N Curve of the Weldment

The relationship between stress amplitude and number of cycles to failure for the 316L weldment is shown in Figure 4. Obviously, the fatigue failure occurred in the very high cycle regime, the plateau region between 10^6-10^7 in convention fatigue was not observed in this research. In the high cycle fatigue regime, most of the specimens failed due to the surface crack initiation, three specimens failed due to subsurface crack initiation. However, in the VHCF regime, most of the specimens failed due to the interior crack initiation except one specimen (surface crack initiation). In other words, the crack tends to initiation from interior of the specimen when the fatigue life beyond 3×10^6 cycles. The similar interior crack initiation behaviour have been observed in other steels and alloys [5,6,13,17]. However, it is different from the results of the investigation on 316L base metal, the fatigue crack initiated from specimen surface in the VHCF regime [3,10].

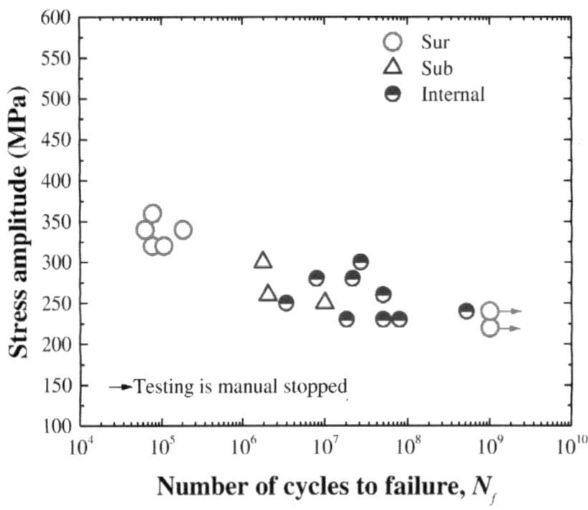

Figure 4. *S-N* curve of the 316L weldment tested at room temperature. Sur: crack initiation from surface, Int: crack initiation from interior, Sub: crack initiation from sub-surface.

Figure 5 shows the comparison between the fatigue strength of 316L weldment and SUS 316L base metal [3]. It should be noted that the fatigue strengths were normalized by its tensile strength in order to eliminate the effect of tensile strength on the fatigue strength. The normalized results showed that the fatigue strength of 316L weldment is almost the same as that of SUS 316L base metal, indicating that the fatigue strength of 316L is not affected by the laser butt-welding process applied in this research, although the pores existed in the weld seam.

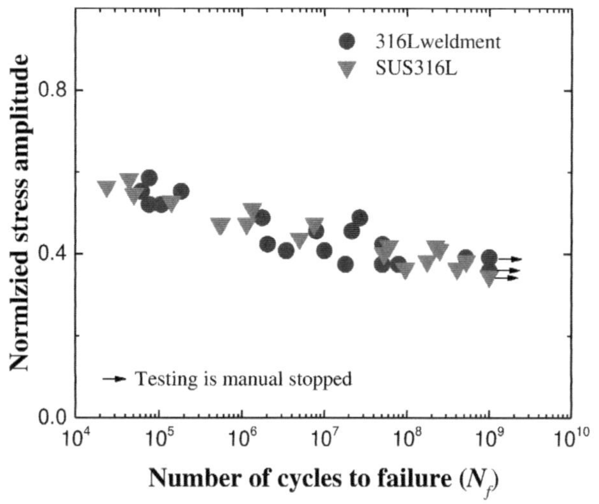

Figure 5. Relationship between normalized stress amplitude and number of cycles to failure

3.3. Fractography

After fatigue tests, the fracture surface of each failed specimen under different stress amplitudes was analyzed through SEM. Most of the specimens were fractured at the weld part. In the low and high cycle fatigue regime, most of the specimens failed due to surface crack initiation. Figure 6 shows the fracture morphology of the specimen tested at 320 MPa and failed at 7.67×10^4 cycles. Obviously, the fatigue crack initiated from the specimen surface and multi-crack initiation sites were observed.

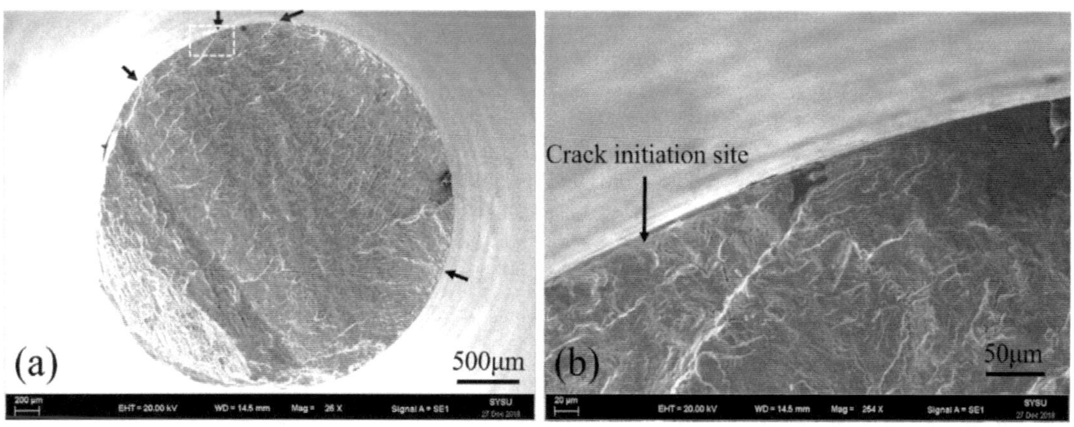

Figure 6. Fracture surface of the specimen tested at 320MPa failed at 7.67×10^4 cycles. (a) Overview of the fracture surface, and (b) Enlargement of the marked rectangle in (a).

Figure 7 shows the fracture surface of the specimen tested at 300MPa failed at 1.75×10^6 cycles. The fatigue crack initiated from the pore caused by the welding at the subsurface. The main crack initiated at the edge of the large pore, where is adjacent to a smaller pore. In the VHCF regime, all of the specimens failed due to the crack initiation from the pores in the specimen. Figure 8 shows the fracture surface of the specimen tested at 230MPa failed at 7.93×10^7 cycles, i.e. in the VHCF regime. A-so called fish-eye was observed in the crack initiation and propagation area, as shown in the dash cycle in Figure 8 (a). A rough area was located in the fish-eye as shown in Figure 8(b), which was different from the reported results [18], i.e. the rough area was located in the central part of the fish-eye. No inclusion, but a pore was observed in the rough area, which was different from the situation of an inclusion as crack initiation site for steels and alloys [14,15,17]. It was also different from the previously results [3,10], no interior crack initiation was observed for 316L in the very high cycle regime. In general, the crack initiation site of the material, i.e. from surface or interior, is contributed to the competition between the surface defect and interior defect. When the size of the interior defect is large enough, the fatigue crack would initiate from interior. Otherwise, it initiates from surface [19]. Hence, in the case of 316L base metal, due to the good control of inclusion, the fatigue crack initiated from specimen surface. On the contrary, in the case of 316L weldment, due to the existing internal pores, the crack initiated from interior.

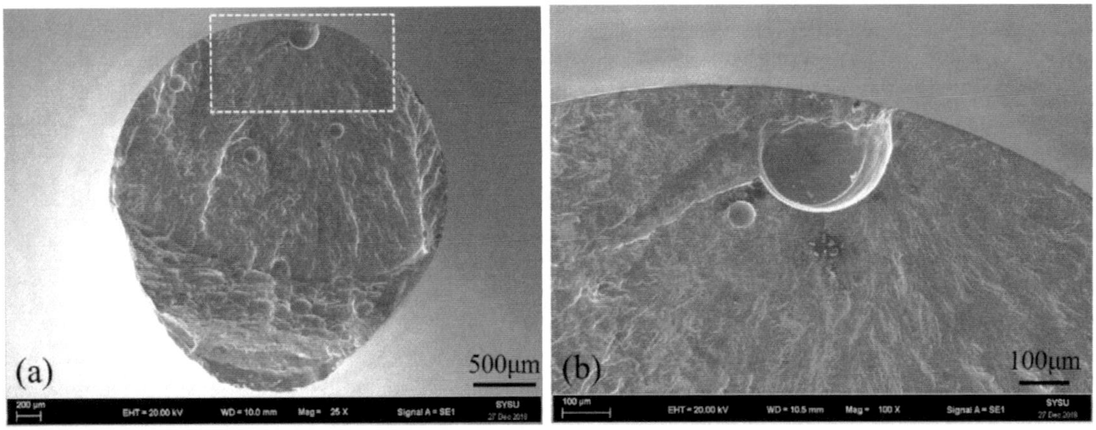

Figure 7. Fracture surface of the specimen tested at 300MPa failed at 1.75×10^6 cycles. (a) Overview of the fracture surface, and (b) Enlargement of the marked rectangle in (a).

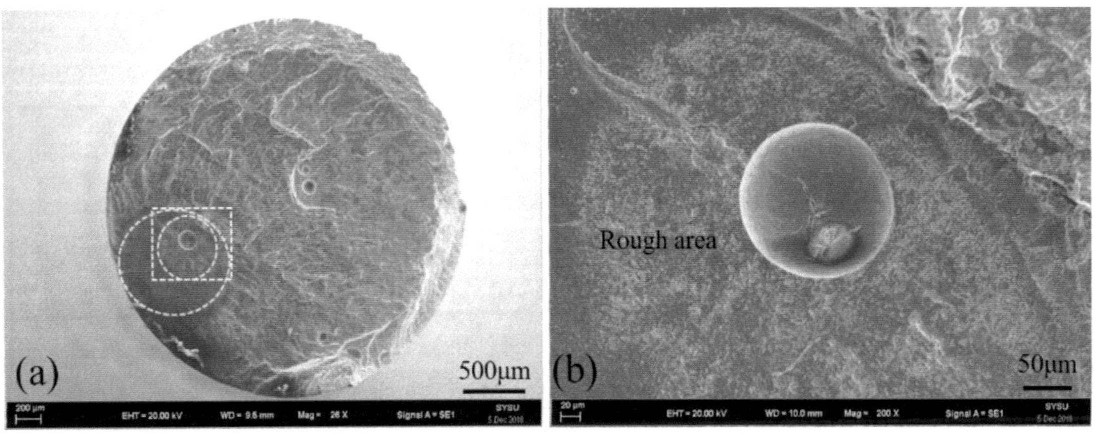

Figure 8. Fracture surface of the specimen tested at 230MPa failed at 7.93×10^7 cycles. (a) Overview of the fracture surface, and (b) Enlargement of the marked rectangle in (a).

4. Conclusions

In this research, the VHCF behaviour of the 316L weldment fabricated by laser butt-welding were investigated by the ultrasonic fatigue testing system. The following conclusions have been drawn:

1) An evident pore is observed in the weld seam, the fine columnar-dendrite structure is formed in the central part of the weld seam and larger columnar-dendrite is observed near the weld interface
2) The hardness distribution is to coincide with the characteristic of the microstructure., the hardness is the highest in the weld seam central and the lowest in the base metal.
3) Fatigue failure still occurred in the VHCF regime. The VHCF crack initiated from the internal pores caused by laser butt-welding. The typical fish-eye was observed at the fracture surface.
4) The fatigue strength of the weldment with pores was the same as the base metal, i.e. the defect with a certain size in the weld seam does not affect the strength of the weldment.

Acknowledgements

This work was financially supported by the Fundamental Research Funds for the Central Universities (Grant No. 17lgpy34), the Natural science foundation of the Guangdong province (Grant 2018A030310102), and the Young Innovative Talent Project of GuangDong Education Department Fund (Gant No. 2016KQNCX004 and 2016KQNCX005). And thanks be to Professor H.Q. Xue and Doctor T. Gao, the ultrasonic fatigue tests were conducted in their lab in Northwestern Polytechnical University.

References

[1] Y. Han, J. Mei, Q. Peng, E. Han, W. Ke, Effect of electropolishing on corrosion of nuclear grade 316L stainless steel in deaerated high temperature water, Corros Sci, 112 (2016) 625-634.
[2] Z. Xiong, T. Naoe, T. Wan, M. Futakawa, K. Maekawa, Mechanical Property Change in the Region of Very High-cycle Fatigue, Procedia Engineering, 101 (2015) 552-560.
[3] T. Naoe, Z. Xiong, M. Futakawa, Gigacycle fatigue behaviour of austenitic stainless steels used for mercury target vessels, J Nucl Mater, 468 (2015) 331-338.
[4] T. Naito, H. Ueda, M. Kikuchi, Observation of fatigue fracture surface of carburized steel, J. Soc. Mater. Sci., Jpn., 32 (1983) 1162-1166.
[5] Z. Ping, G. Gao, R.D.K. Misra, B. Bai, Effect of microstructure on the very high cycle fatigue behavior of a bainite/martensite multiphase steel, Mater Sci Eng A, 630 (2015) 1-7.
[6] S. Siddique, M. Imran, F. Walther, Very high cycle fatigue and fatigue crack propagation behavior of selective laser melted AlSi12 alloy, Int J Fatigue, 94 (2017) 246-254.
[7] Y. Hong, C. Sun, The nature and the mechanism of crack initiation and early growth for very-high-cycle fatigue of metallic materials – An overview, Theoretical & Applied Fracture Mechanics, 92 (2017).
[8] K. Takahashi, T. Ogawa, Evaluation of giga-cycle fatigue properties of austenitic stainless steels using ultrasonic fatigue test, Journal of Solid Mechanics and Materials Engineering, 2 (2008) 366-373.
[9] J. Carstensen, H. Mayer, P. Brøndsted, Very high cycle regime fatigue of thin walled tubes made from austenitic stainless steel, Fatigue Fract Eng M, 25 (2002) 837-844.
[10] T. Naoe, Z. Xiong, M. Futakawa, Temperature measurement for in-situ crack monitoring under high-frequency loading, J Nucl Mater, 506 (2018) 12-18.
[11] W.C. Zhang, M.L. Zhu, K. Wang, F.Z. Xuan, Failure mechanisms and design of dissimilar welds of 9%Cr and CrMoV steels up to very high cycle fatigue regime, Int J Fatigue, (2018).
[12] C. Deng, H. Wang, B. Gong, X. Li, Z. Lei, Effects of microstructural heterogeneity on very high cycle fatigue properties of 7050-T7451 aluminum alloy friction stir butt welds, Int J Fatigue, 83 (2016) 100-108.
[13] Y. Hong, X. Liu, Z. Lei, C. Sun, The formation mechanism of characteristic region at crack

IOP Conf. Series: Materials Science and Engineering **538** (2019) 012025 doi:10.1088/1757-899X/538/1/012025

initiation for very-high-cycle fatigue of high-strength steels, Int J Fatigue, 89 (2016) 108-118.
[14] W. Hui, Y. Zhang, X. Zhao, Z. Chao, K. Wang, S. Wei, D. Han, Very high cycle fatigue properties of Cr–Mo low alloy steel containing V-rich MC type carbides, Mater Sci Eng A, 651 (2016) 311-320.
[15] S. Stanzl-Tschegg, Very high cycle fatigue measuring techniques, Int J Fatigue, 60 (2014) 2-17.
[16] S.X. Li, F.Z. Xuan, S.D. Tu, S.S. Yu, Interfacial fatigue crack growth behavior of diffusion bonded joints of 316L stainless steel, Mater Sci Technol, 18 (2010) 141-144.
[17] X. Liu, C. Sun, Y. Hong, Effects of stress ratio on high-cycle and very-high-cycle fatigue behavior of a Ti–6Al–4V alloy, Acta Metallurgica Sinica, 622 (2015) 228-235.
[18] H. Su, X. Liu, C. Sun, Y. Hong, Nanograin layer formation at crack initiation region for very-high-cycle fatigue of a Ti-6Al-4V alloy: Nanograin Formation for VHCF of TI Alloy, Fatigue Fract Eng M, (2017).
[19] L.S. Xin, Very high cycle fatigue properties of high strength steels: effects of nonmetallic inclusions, Metallurgical Industry Press, BeiJing, 2010.

IOP Conf. Series: Materials Science and Engineering **538** (2019) 012026 doi:10.1088/1757-899X/538/1/012026

Different methods of introduction gold NPs into Perovskite Solar Cells with their efficiencies

Mouna Mohamed Abdoul-latif[1,2], JiaXu[1], Jianxi Yao[1], Songyuan Dai[1]and Mohamed Osman Awaleh[2]

1 State Key Laboratory of Alternate Electrical Power System with Renewable Energy Sources, North China Electric Power University, Beijing 102206, China
2 Center of Studies and Researches of Djibouti, Djibouti
E-mail: latifmouna1@yahoo.com

Abstract. The introduction of gold nanoparticles (Au NPs) into Perovskite Solar Cells doesn't bring a higher enhancement in efficiency than fullerene like C60 or PCBM (phenyl C61-butyric acid methylester) or graphene introduced into the PSCs which have improved the efficiency more than 15%. In this work, the Au NPs prepared by sol-gel method using Turkevish method have been introduce into PSCs by three different ways and their efficiencies have been investigated for looking for the best method. This adding of the Au NPs is done firstly by doping the TiO_2, secondly by depositing Au NPs layer into the TiO_2 layer and thirdly by depositing the Au NPs doped TiO_2 layer into TiO_2 layer. All of them have given some results but still significantly low compared to other materials used. Even though the Au NPs could improve the short-circuit photocurrent density J_{sc}, the contact between Au NPs or TiO_2 and spiro-OMeTAD may also have decreased the PCE.

1. Intoduction

The perovskite deposition can vary from sensitising the TiO_2 electron transporting (called mesoporous perovskite solar cells sensitised), to filling the pores (mesoporous perovskite solar cells with capping layer) where it acts as both light absorber and hole transporting material in one[1]. There is also planar-structured without TiO_2 mesoporous used. Solar cell architectures which have given the best power conversion efficiency are TiO_2mesoporousperovskite solar cell (mesostructured device) and planar heterojunction perovskite solar cell (planar-structured device).In this paper, only mesoporous perovskite solar cells were discussed. Perovskite solar cells (PSCs) using Perovskite as light absorbers in photovoltaic has been developed more during these decade years starting with one deposition of Perovskite, the lead iodide and methylammonium mixed before deposition with different concentration mixture. The crystallography, the absorption, the diffusion length have investigated minutely[2, 3]. The big growth has been bringing in a sequential deposition with its ease synthesis allows bringing up forward this research[4]. During these last years, the improving of PCE has also reached high efficiency by the knowledge of the proprieties of these materials[5, 6]. But also by introducing different materials [7, 8, 9,10, 11] as P3HT (poly(3-hexylthiophene)), PCBM, fullerene, graphite, or metal nanoparticles[12, 13, 14] for enhancing the PCE, FF, J_{sc} or V_{oc}.

IOP Conf. Series: Materials Science and Engineering **538** (2019) 012026 doi:10.1088/1757-899X/538/1/012026

2. Experimental

2.1 Fabrication and Synthesis

The synthesis of Au NPs has gotten with a citrate reduction method[15].

A 16 ml aqueous solution of 4.5 mg $HAuCl_4 \cdot 3H_2O$ (hydrogen tetrachloroaurate (III) trihydrate) was brought to boil on a hot plate, and about 0.5ml of 1% w/w trisodium citrate aqueous solution was added. The solution was stirred for 30 min, and finally, a wine red- coloured solution was obtained. The same procedure has done, with different concentration of $HAuCl_4.3H_2O$ and different concentration of trisodium citrate solution. Immediately after Au NPs have been synthesized, the TiO_2 paste (TiO_2 (18NR-T) or (30NR-D) with ethanol w/w 1:3.5) added to the solution of Au NPs. Also the TiO_2 made with Titanium isopropoxide is used instead of TiO_2 paste (18NR-T) added to the solution of Au NPs.

The synthesis of TiO_2 paste

The synthesis procedure for nano-TiO_2 is as follows: titanium (IV) isopropoxide, nitric acid, ethanol and distilled water mixed under vigorous stirring between half or one hour and at 70 °C. After that, the solution was kept for one day for the ageing process.

Fabrication of Perovskite Solar Cells

Solar cells fabricated on fluorine-doped tin oxide glass substrates with a sheet resistance of 15Ω/square (Pilkington). Chemical etching patterned FTO substrate with Zinc powder and HCl aqueous solution. The conductive substrates cleaned with a detergent solution, de-ionised water and ethanol. A thin dense, compact TiO_2 layer (c-TiO_2) was deposited on the etched FTO substrate by spin coating (3000 rpm, 30s) using a titanium diisopropoxidebis (acetylacetonate) solution in ethanol and HCl at a volumetric ratio of 0.35:5:1 followed by annealing at 500°C. Porous TiO_2 films got by commercially available TiO_2 paste (18-NR-T, Dyesol, average particle size ~ 20 nm) diluted in ethanol (1:3.5 weight ratio) were directly deposited on top of the c-TiO_2 by spin coating at 5000 rpm for 30s and annealed at 500°C. PbI_2 solution is diluted in DMF (dimethylformamide) (462mg/ml) and was kept at 70°C.Thus, the prepared mesoporousTiO_2 films were loaded with PbI_2 for 30s and then spin-coated at 6000 rpm for 30s after that dried 2min and finally annealed at 70°C during 30min.After drying, the films were spin coated at 6000 rpm at 30s after were loaded a solution of CH_3NH_3I in 2-propanol (13mg/ml) for 30s. The HTM (hole-transporting-material) was deposited by spin-coating a solution of spiro-MeOTAD, tert-butylpyridine, lithium bis(trifluoromethylsulphonyl)imide and tris(2-(1H-pyrazol-1-yl)-4-tert-butylpyridine)cobalt(III) bis(tri-fluoromethylsulphonyl)imide in chlorobenzene. Silver (Ag) (66nm) was thermally evaporated on top of the device to form the back contact.

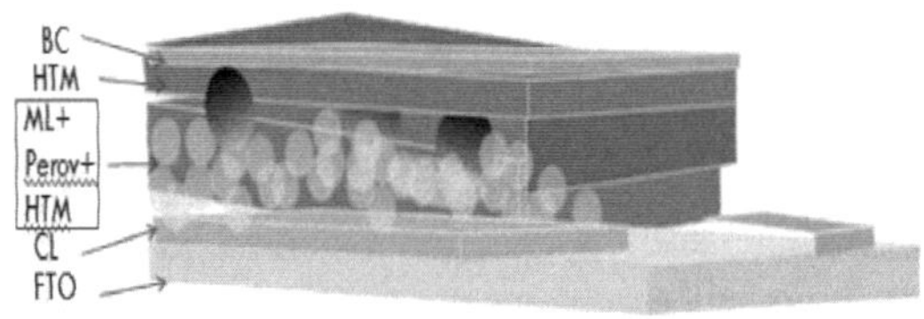

Figure 1. Schematic view of perovskite solar cell with introducing Au NPs

The same fabrication described above used, only the anode has modified by introducing the Au NPs. Firstly, TiO_2 paste (TiO_2 (18NR-T) with ethanol w/w 1:3.5) was added to solution of Au NPs allowing to get Au NPs doped TiO_2. Secondly the Au NPs layer has been deposited into TiO_2 paste (18NR-T) or (30NR-D). Thirdly, the Au NPs doped TiO_2 layer has been deposited into TiO_2 paste (18NR-T) or (30NR-D).

When the deposition of Au-TiO_2 done at 6000 rpm into the TiO_2; the efficiency became insignificant which could be explained by the increasing of thickness. The J_{sc} varies between 8 to 10 mA/cm^2 and 10 to 13mA/cm^2 for Au NPs doped TiO_2 at 2mM $HAuCl_4$.$3H_2O$ added 1.2mL citrate and at 1mM $HAuCl_4$. $3H_2O$ added 2ml citrate respectively.

2.2. Characterization

The photocurrent density –voltage characteristics of the perovskite solar cells were evaluated by a source meter (Keithley 2400) under a solar simulator (Oriel 92250A-1000, Newport, 100 mW cm^{-2}). The absorption of different films was measured using a UV-visible Spectrophotometer (UV-2450, Shimadzu).

3. RESULTS AND DISCUSSION

3.1. First method

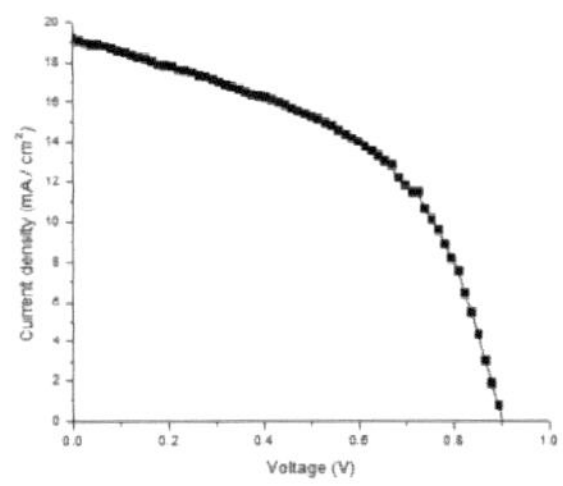

Figure 2. J-V curve for a photovoltaic device (FTO/c-TiO_2/Au-TiO_2/perov/HTM/Ag)

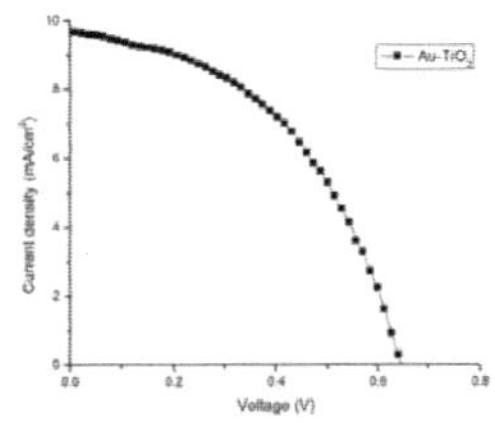

Figure 3. J-V curve for a photovoltaic device (FTO/c-TiO_2/Au-TiO_2/perov/HTM/Ag) with Au nanoparticle (Au NPs made with an 18ml aqueous solution and 20mg of $HAuCl_4 \cdot 3H_2O$)

The short-circuit photocurrent (J_{sc}) decreases with increasing the concentration of $HAuCl_4.3H_2O$ with different citrate volume adding due to increasing the thickness of the Au doped TiO_2ordue to increasing the carrier concentration.

3.2. Second method

Absorption

Figure 4 shows the absorption of perovskite; the speed coating used is same for TiO_2 deposition and TiO_2/Au NPs bilayers, but two ways could explain the increasing of perovskite absorption, the first one is due by increasing the infiltration of PbI_2 by adding the Au NPs layer, Plasmon effect could give the second explanation.

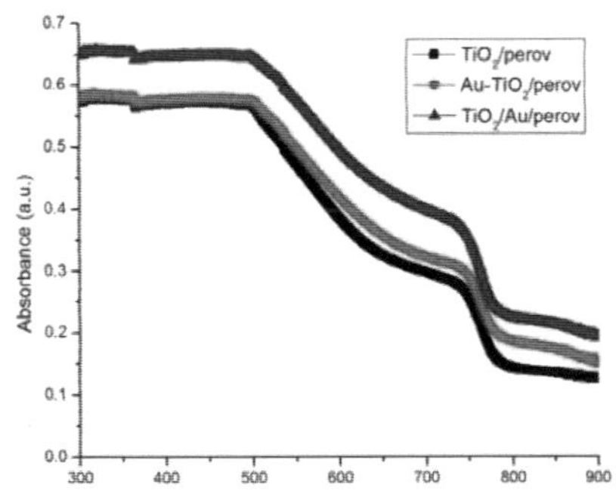

Figure4. Absorbance spectra of perovskite deposited into TiO_2, Au-TiO_2 and TiO_2/Au

IOP Conf. Series: Materials Science and Engineering **538** (2019) 012026 doi:10.1088/1757-899X/538/1/012026

3.2.1. Difference between different adding of citrate

Table 1. Data of J-V characteristic at different volume

Devices	J_{sc} (mA/cm^2)	V_{oc} (V)	FF (%)	PCE
1mM+2ml citrate	14.35	0.79	60	6.8
1mM+3ml citrate	14.79	0.8	68.8	8.2

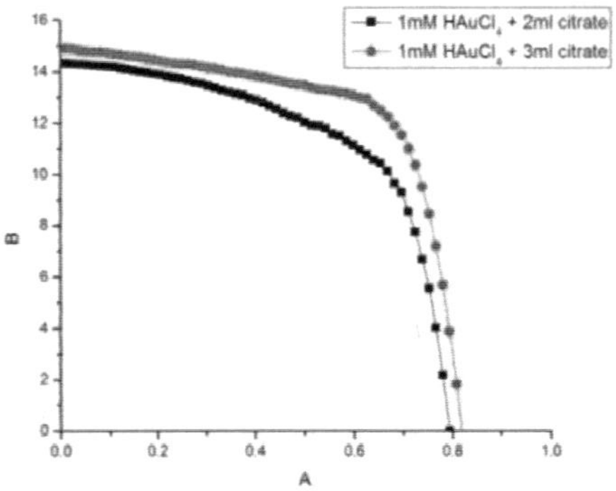

Figure 5. Reverse curve with different volume of citrate

3.2.2. Au NPs synthesised with 45 ml solvent with 11 mg of HAuCl4 3H2O added 3ml of citrate

Table 2. Data of J-V characteristic

Device	J_{sc} (mA/cm^2)	V_{oc} (V)	FF (%)	PCE (%)
11mg/45ml $HAuCl_4$+3ml citrate	16.7	0.73	64.5	7.8

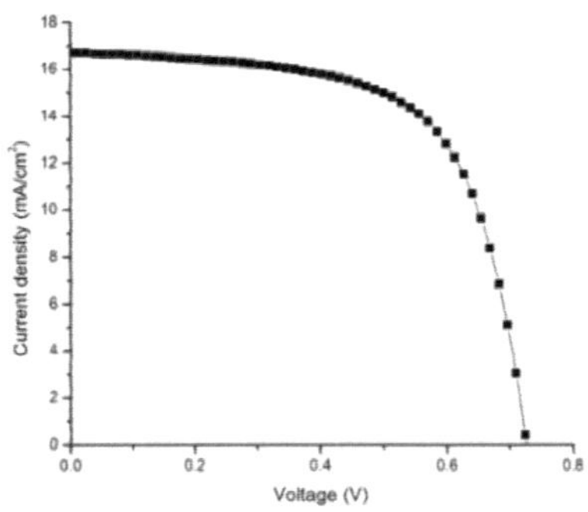

Figure 6. Reverse curve of J-V characteristic

The majority of devices have short circuit density (J_{sc}) between 10 to 15 mA/cm^2 at different citrate solution volume add to solution aqueous of $HAuCl_4.3H_2O$ at 10ml of3mM $HAuCl_4.3H_2O$ for improving the distribution of Au NPs on the surface and approaching the Au NPs between them. However increasing the concentration of $HAuCl_4.3H_2O$ and decreasing the citrate solution volume for getting the big size of Au NPs, the solution became less bright and unstable. Increasing the precursor solvent has allowed improving the short-circuit photocurrent (J_{sc}). Therefore this study could be improved by finding the best rate volume between $HAuCl_4.3H_2O$ and citrate for increasing the size for getting a Plasmon effect allowing the enhancing the PCE.

3.3. Third method

FTO/c-TiO_2/m-TiO_2/AuNPs-TiO_2synthesised /$CH_3NH_3PbI_3$/spiro-OMeTAD/Ag.

Au NPs used to dope TiO_2 is made with 11mg/45ml $HAuCl_4$+3ml citrate.

Au NPs-TiO_2 layer is deposited into mesoporous TiO_2 by spin coating at 3000 rpm during 30s.

Table 2. Data of J-V characteristic

Device	J_{sc} (mA/cm^2)	V_{oc}(V)	FF(%)	PCE(%)
Forward curve	14.55	0.63	52.5	4.8

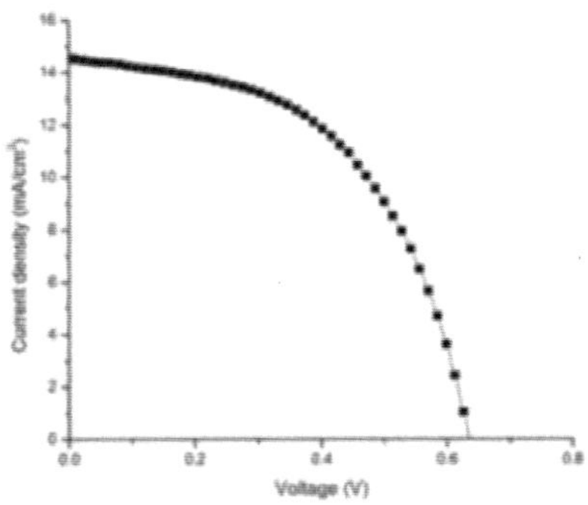

Figure 7. J-V characteristic

The short-circuit photocurrent (J_{sc}) has improved with the deposition a layer of Au-TiO_2 into TiO_2 at 6000 rpm with TiO_2 synthesized with Titanium isopropoxide. This layer should be thin for improving the J_{sc}.

4. Conclusion

This study about introducing Au Nanoparticles by different ways into PSCs is allowed to improve the J_{sc} that is following the literature. However the enhancement of PCE still low compared to improvement excepted in all of the methods used. At the third method the thickness of the second layer of Au NPs doped TiO_2 should be thin for improving the efficiency; therefore the synthesis of Au NPs doped TiO_2 should be improved. Perhaps this study could be developed by using different techniques for more enhancements. In all cases the Plasmon effect need more investigations for looking for a best device of PSC using Au NPs.

5. References

[1] Leijtens T, Lauber B, Eperon G E, Stranks S D and Snaith H J 2014 *Physical Chemistry Letters* 5 1096-1102. From Periodicals

[2] Im, J. H.; Lee, C. R.; Lee, J. W.; Park, S. W.; Park, N. G. Nanoscale 2011, 3, 4088. DOI: 10.1039/c1nr10867k.

[3] Kim, H.S; Lee, C.R; Im, J.H; Lee, K.B; Moehl, T; Marchioro, A; Moon, S.J; Humphry-Baker, R; Yum, J.H; Moser, J.E;Gratzel, M; Park, N.G. Scientific Reports 2012, 2, 591. DOI: 10/1038/srep00591.

[4] Burschka, J.; Pellet, N.; Moon, S.-J.; Humphrey-Baker, R; Gao, P.; Nazeeruddin, M.K; Gratzel, M. Nature 2013, 499, 316-319.

[5] Lee, M.M.; Teuscher, J.; Miyasaka, T.; Murakami, T. N.; Snaith, H.J. Science 2012, 338, 643. DOI: 10.1126/science.1228604.

[6] Etgar, L.; Gao, P.; Xue, Z.; Peng, Q. ; Chandiran, A. K. ; Liu, B. ; Nazeerudin, Md. K. ; Gratzel, M. J. Am. Chem. Soc. 2012, DOI: 10.1021/ja307789s.

[7] Junfeng Yan and Brian R. Saunders**,** Third-generation solar cells: a review and comparison of polymer: fullerene, hybrid polymer and perovskite solar cells, Royal Society of Chemistry, 26th August2014, DOI: 10.1039/c4ra07064j

[8] Abate, A.; Saliba, M.; Hollman, D.J.; Stranks, S.D.; Wojciechowski, K.; Avalio, R.; Grancini, G.; Petrozza, A.; Snaith, H.J. NANO Letters 2014, 14,3247-3254.

[9] Zhu, Z; Ma, J; Wang, Z; Mu, C; Fan, Z; Du, L; Bai, Y; Fan, L; Yan, H; Phillips, D.L ; Yang, S. American Chemical Society 2014, 136, 3760-3763.

[10] Giordano, F; Abate, A; Juan, Baena, J.P.C ; Saliba, M; Matsui, T; Im, S.H ; Zakeerudin, S.M; Nazeeruddin, M..K; Hagfeldt, A; Graetzel, M. nature communications 2016, 7:10379, DOI: 10.1038.

[11] Manseki, K; Ikeya, T; Ban, T ; Sugiura, T ; Yoshida, T. Communication, RSC Advances 2014, 4, 9652.

[12] Zhang, W; Saliba, M; Stranks, S.D; Sun, Y; Shi, X; Wiesner, U; Snaith, H.J. NANO LETTERS 2013, 13, 4505-4510.

[13] Pathak, N.K; Chander, Nikhil; Komarala, V; Sharma, R.P. Springer Science 2016, DOI 10.1007.

[14] Omelyanovich, M; Makarov, S; Milichko, V; Simovski, C. Materials Sciences and Applications, 2016, 7, 836-847.

[15] Nguyen, D.T ; Kim, D-J ; So, M.G ; Kim, K-S. Advanced Powder Technology 2010, 21,111,118.

Numerical Simulation Analysis of Surfacing Temperature Field of Pulling-straightening Roller

Jie Luyang, Chen Hua[1,*], Wang Liyan[2,*]

Changchun University of Technology, Key laboratory of advanced structural materials, ministry of education, Changchun 130000, China

chenhua@ccut.edu.cn: 10521773@qq.com

Abstract. In this paper, the ANSYS software thermal analysis is used to simulate the 0Cr13Ni4MoN surfacing process of the pulling-straightening roller. The Gaussian model was selected as the heat source model, and the direct solution method was used to calculate the temperature field of the surfacing process. The effects of preheating temperatures of 250°C, 300°C and 350°C on the temperature field of the pulling-straightening roller were discussed. The results show that the preheating temperature range of 250~350°C gives a uniform temperature field.

1. Introduction

The pulling-straightening rollers are the main components of the continuous casting and rolling equipment, the role of the pulling-straightening roller is to correct, guide and drive the plastic deformation of the casting blank[1]. The general method used for the repair of the pulling-straightening roller is the overlay welding repair method, that is, the alloy material is deposited on the base material substrate. 0Cr13Ni4MoN has good plasticity, wear resistance, heat resistance, corrosion resistance, etc[2]. It is suitable as a surface coating material for pulling-straightening roller, which can improve the service life of the roller body and the quality of the steel slab.The finite element analysis software ANSYS gradually includes various engineering field solving capabilities in the development process. After having stable and reliable structural strength analysis and thermal analysis functions, it has been developed and expanded to be used more widely. In this paper, the software ANSYS version 14.0 was used to simulate the surfacing temperature field.

2. The Simulation Calculation of Temperature Field

2.1. Material Properties of 0Cr13Ni4MoN

0Cr13Ni4Mo is a nickel-containing, molybdenum ultra-low carbon martensitic stainless steel. It adds nickel and molybdenum to the chromium martensitic stainless steel and reduces the carbon content[3]. Material properties below 500°C including 500°C can be found by data, while properties above 500°C need to be calculated by software. In the numerical simulation of the temperature field of the pulling-straightening roller surfacing process, it is necessary to add some material properties such as density DENS (Kg/m^3), specific heat capacity C (J/Kg·K), heat transfer coefficient KXX [W/(m·K)] and so on. These material parameters are all required for thermal analysis. The thermal conductivity of

0Cr13Ni4Mo in the range from room temperature 20°C to melting temperature 1500°C is 34 W/(m·K), and the specific heat capacity is 683 J/Kg·K.

2.2. *Establishment of Finite Element Model*

Generally, the length of the pulling-straightening roller is long, so a section of the roller is used as a model for simulation. The three-dimensional model for simulation uses a hollow roller, and the hollow roller is cooled by internal circulating water.

Solid model size: the size of the pulling-straightening roller is 200 mm×100 mm×20 mm.The digital model is built using the CAD software CATIA, and then the entity model is imported through the ANSYS and CATIA software data interface. The solid model is shown in Figure1.

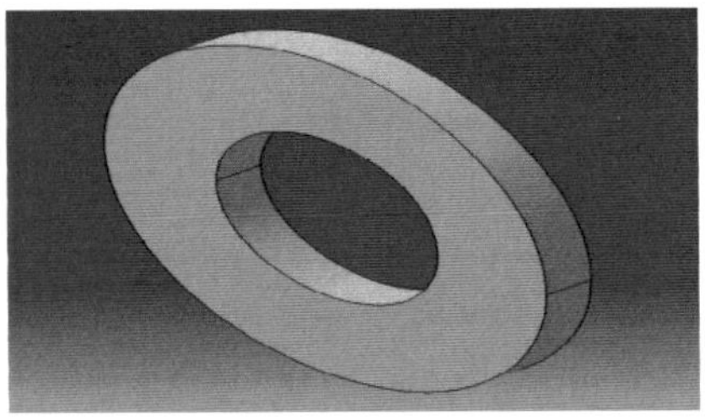

Figure 1. CATIA model

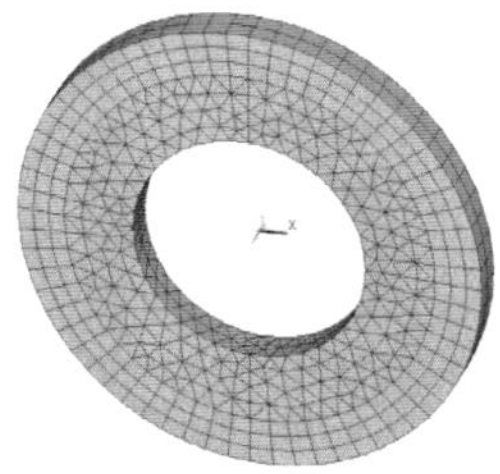

Figure 2. Finite element model

2.3. *Meshing*

The pulling-straightening roller is a hollow cylinder structure, which can be meshed by a mapping meshing method, and the main body of the pulling-straightening roller and the welding layer are divided into regular hexahedrons. The mesh size of the main body of the pulling-straightening roller is about 3 times larger than the mesh size of the overlay layer. The mesh of the main body and the mesh of the overlay layer are connected by a tetrahedron, which can improve the speed of the computer operation and reduce the calculation time , can also guarantee the degree of compliance with the actual calculation. Since the pulling-straightening roller is an axisymmetric structure, a 1/4 model is used for the simulation calculation. The finite element model meshing is shown in the figure2.

2.4. *Selection of Heat Source*

It has been found that Gaussian distribution function can effectively improve the calculation accuracy when penetration ability of manual arc welding and tungsten argon arc welding is small.

Eagar and Tsai's gaussian mathematical model was used to calculate the heat flux distribution on the welding spot [4]. Its function was as follows:

$$q(r)=\frac{3\eta UI}{\pi R^2}*\exp\left(-\frac{3r^2}{R^2}\right) \tag{1}$$

Where: R—the effective heating radius of the arc, the unit is mm; r—the distance from any spot on the weldment to the center of arc heating spot, the unit is mm; η —welding thermal efficiency; U—welding voltage, the unit is V;I—welding current, the unit is A.

Arc welding is a continuous dynamic process. Since the Gaussian heat source is a transient command, if the heat source is to be surfacing in a certain direction during the surfacing process, it needs to be programmed in the APDL language. APDL is the ANSYS parametric design language, it is an interpretive language that can be used to build models based on parameters [5]. In transient thermal analysis, after the initial temperature is specified, the temperature will change with time due to the thermal load [6].

This topic to be discussed under different preheating temperature for the pulling-straightening roller overlaying welding temperature field, the influence of the temperature 250°C, 300°Cand 350°C respectively. The initial conditions for: the pulling-straightening roller matrix correction temperature is

250°C, 300°C, 350°C. The initial temperature of the surfacing layer is 1659°C, and the convective boundary conditions: convective heat transfer coefficient is 45.42 W/(m^2·°C), and air temperature is 20°C.

2.5. Analysis of Temperature Field Results

As shown in Figure.3~5, when the welding current I=340A, the voltage U=32V, the welding speed v=15cm/min, the temperature field simulation results at different times during the welding process are as follows.

In the initial stage of surfacing welding, the heating rate of the matrix is slow, so it is seen that the base metal temperature is around 1400°C in the initial stage, as the heat source is loaded and moved in the circumferential direction, the surface of the roller body in the forward direction of the heat source is affected by the preheating of the welding heat source, and the temperature in the vicinity of the heat source rises rapidly, and the area of the high temperature region is also expanded. In Figure.4, in the welding process, based on the gradual preheating phenomenon of welding heat source on roll body, the temperature at the next moment of the welding heat source will be higher than that at the previous moment, the welding temperature can be well controlled during the surfacing process without major fluctuations. The maximum temperature of the welding temperature field is not stable, and the temperature is gradually increasing as the heat source moves. According to the observation in Figure 5, when the temperature of the temperature field changes greatly, the expansion of the molten pool is very small, and the calorific value of the surface swept by the heat source changes little compared with that before, but the overall temperature of the roller body increases continuously after welding.

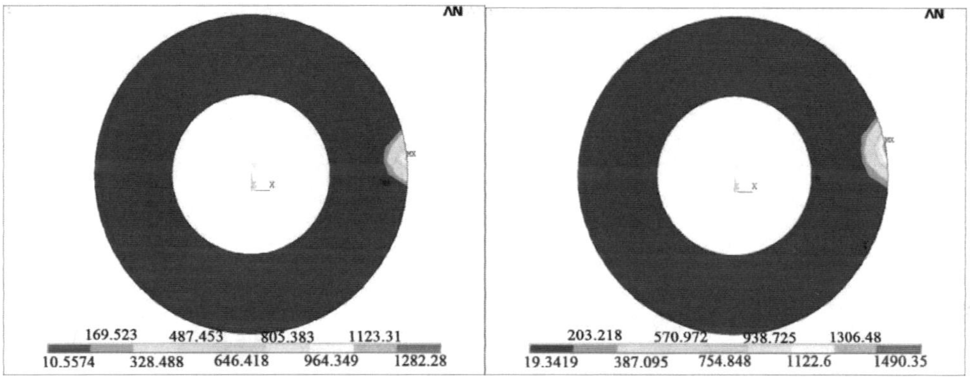

Figure 3. Simulation results of the initial stage of surfacing
(a) Temperature field distribution at 3.9s (b) Temperature field distribution at 7.8s

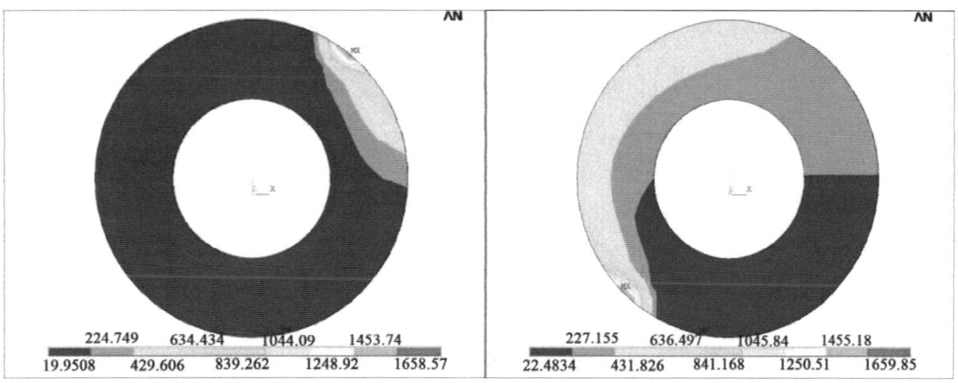

Figure 4. Simulation results of the intermediate stage of the overlay welding
(a)Temperature field distribution at 39.3s (b) Temperature field distribution at 161.0s

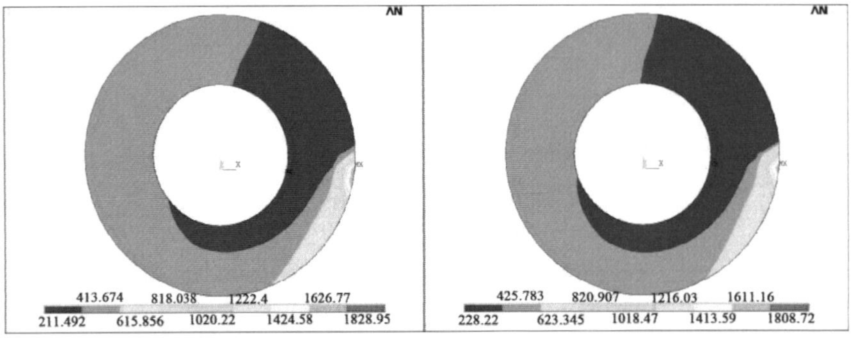

Figure 5. Simulation results at the end of the surfacing process
(a) Temperature field distribution at 247.4s (b) Temperature field distribution at 251.3s

Normally, for large roller bodies, preheat treatment is required before welding, so the influence of temperature on the temperature field is simulated by different preheating conditions. The four preheating temperature parameters selected below are 20°C (no preheating) 250°C, 300°C and 400°C.

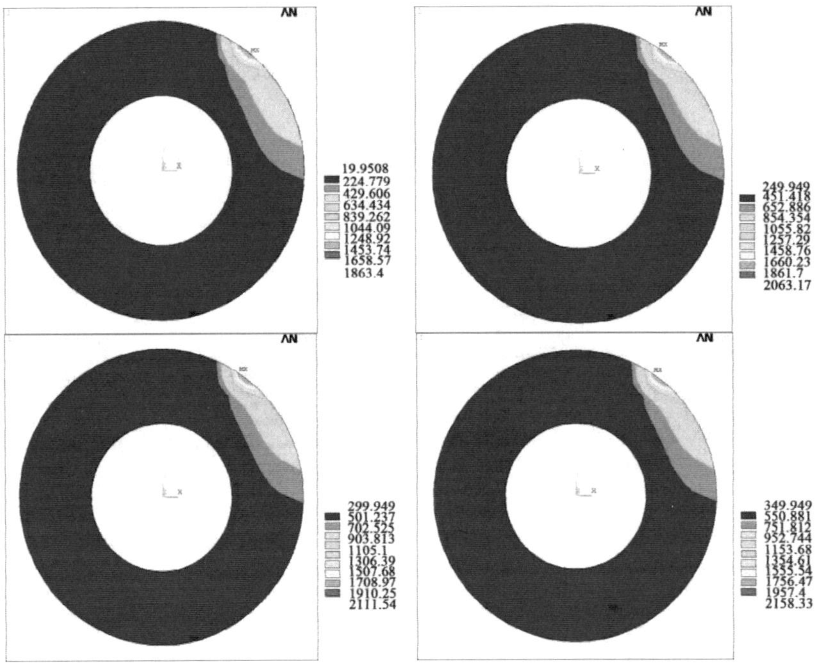

Figure 6. Effect of different preheating temperatures on the welding temperature field
(a)No preheating temperature (b) 250°C (c) 300°C (d) 350°C

Figure 6 shows the effect of different preheating temperatures on the welding temperature field. In Figure 6(b), the preheating temperature before welding is 250°C, and the maximum temperature of the temperature field is about 2060°C at around 40s. In Figure 6(c), (d), the preheating temperatures respectively are 300°C and 350°C, the peak temperature of the temperature field respectively are 2115 °C and 2159°C,it can be inferred that the preheating temperature value directly affects the temperature field after cooling, and they are in a positive proportional relationship.

Figure 7 shows the temperature field-time plot of node 1755, which contains three different preheat temperatures.When the heat source does not move to the node, the temperature of the node will always maintain the temperature of the preheating field. After the heat source moves to the node, the temperature of the node will rise rapidly, the duration of high temperature peak is short, and then it will drop rapidly, and finally the node reaches the equilibrium temperature. In the figure, with the increase of preheating temperature value, the node temperature value increases correspondingly, and the final equilibrium stable temperature value is higher.The peak width of calorific value did not change with the change of preheating temperature value, but the peak value changed greatly with the change of preheating value. The three sets of graphs show that the curve heating rate and the cooling rate are basically equal, and the change is small.

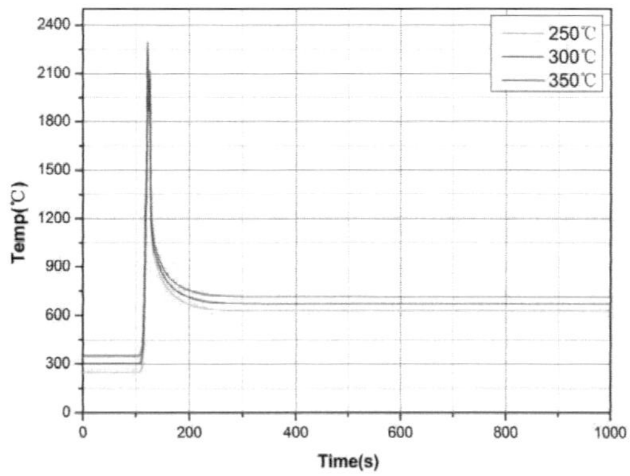

Figure 7. Effect of preheating temperature on temperature field

3. Conclusion

1. When the welding current is I=340A and the welding speed is v=15cm/min, the center temperature of the molten pool is about 1600°C in the initial stage of surfacing. Under the condition of heat source continues to move forward, the internal temperature of the molten pool slowly rise from 1600°C to 1900°C, indicating that each weld to be affected by excessive temperature zone in the former period, makes the molten pool temperature increased slowly.
2. If the surfacing is done without preheating, temperature of overall roll is low and The temperature of roller body is quite different from that of surfacing layer so that the temperature is difficult to rise rapidly. At 250°C, 300°C and 350°C preheat temperature conditions, the difference of heat affected zone temperature decreases. Even though it is in the early stages, the roll body temperature field distribution is more homogeneous.

Acknowledgements
This work is financially supported by the Natural Science Foundation of China (No.51604034) and the Science and Technology Development Plan Project Specification of Jilin province (No.20190302003GX).

References

[1] Zeng Zhiling, Ren Lijun.Thermal-structure coupling analysis of pulling-straightening roller using ANSYS software[J]. Heavy Machinery.
[2] Chen Hua, Wang Liyan, Liu Xiaochun. Study on Microstructure and Properties of Welding of Pulling-straightening Roller[J]. Thermal Process Technology.
[3] Zhang Wenqian. Welding heat transfer [M]. Beijing: Mechanical Industry Press.
[4] Liu Zhe et al. Three-dimensional Dynamic FEM Simulation of Temperature Distribution with Thermo-metallurgical Interaction During Welding [J]. Mechanical science and technology
[5] Zheng Yongyang,Gao Peng,Ansys APDL-use of Finite Element Simulation Structure[J]. Architecture and structural design.
[6] U shio, M Proc. Metallo-thermo-mechanics application to phase transformation incorporated processes [J].

IOP Conf. Series: Materials Science and Engineering **538** (2019) 012028 doi:10.1088/1757-899X/538/1/012028

Study on Aerodynamic Characteristics of Super High Speed Elevator

Zhe Yang, Qing Zhang[a] and Ruijun Zhang

School of Mechanical and Electrical Engineering, Shandong Jianzhu University, Jinan, 250101, Shandong, P. R. China

[a]Corresponding author: zhangqing@sdjzu.edu.cn

Abstract. With the increase of elevator running speed, the influence of aerodynamic force on elevator becomes more and more obvious, and the impact is also increasing gradually. In this paper, the Computational Fluid Dynamic numerical simulation method is used to simulate and analyse the aerodynamic characteristics of elevator car under the influence of different operating speed and distance from side of car to hoistway wall by establishing full-scale model of elevator and hoistway. The results show that when the car is in a symmetrical position, the aerodynamic characteristics on both sides always show obvious symmetry. With the decrease of distance from side of car to hoistway wall and the operating speed of the car, the flow field around the car changes intensified, and the aerodynamic forces on the side of the car will also increase significantly.

1. Introduction

With the continuous emergence of high-rise and super-high-rise buildings, the city's vertical transportation capacity is also constantly improving. As the main vertical transport vehicle, elevators are also developing towards high speed and long journey. Undoubtedly, the application of super high-speed elevator greatly improves people's work efficiency, but its super high operating speed not only brings convenience to people, but also brings more obvious aerodynamic problems to elevators. The unsteady pressure field on the surface of the car will be formed by the high-speed air flow. This not only aggravates the transverse vibration of the elevator, increases the energy consumption of the elevator, but also produces greater aerodynamic noise and reduces the ride comfort of the elevator. Therefore, it is particularly important to study the aerodynamic characteristics of super high-speed elevators.

Many scholars have done a lot of useful explorations in the research of aerodynamic characteristics of elevator super-high-speed elevator. Duan et al. [1] used Digital Particle Image Velocimetry System (DPIV) to measure the instantaneous velocity field around boxes with different end shapes, found that complex unsteady eddy flow does exist in the flow field around the box, but the influence of different shapes on the aerodynamics of the head of the elevator is not obvious, relatively, there are obvious differences at the tail of the elevator; Li et al [2], by comparing and analyzing the aerodynamic characteristics of different types and heights of domes in 6 cases, found that adding 1.0m and 1.4m elliptical domes at the top and bottom of the car can effectively optimize the shape of the car, reduce the resistance, vibration and noise of the elevator when it is running. Considering that the operation conditions of the high-speed railway is similar to that of the car, and it is also a problem of flow around the box, thus, this paper also refers to some relevant literature in the field of high-speed railway. Niu et al. [3] utilized the Sheer Stress Transport (SST) turbulent model and Improved Delayed Detached Eddy Simulation (IDDES) method, simulated the unsteady aerodynamic performance of trains with respect to different lengths of the tapered nose of the train, by comparing and analyzing the obtained results, elucidated the effects of the length of the tapered nose of the train on the drag force, velocity distribution, pressure distribution, and flow structure around the train; Sakuma Y. et al [4,5] chose train as research objects, considering the inviscid fluid dynamic forces, viscous frictional forces, and form drag, theoretically studied the dynamic, stability and boundary layer development of a flexibly interconnected rigid cylinders subjected to fluid dynamic forces.

In this paper, the aerodynamic characteristics of the car under the influence of different operating speeds and different distance from the side of the car to the hoistway wall are numerically simulated and analyzed by using Computational Fluid Dynamics (CFD) method. Based on the aerodynamic data of the car under the influence of various variables obtained by simulation, the effects of elevator operating speed and the distance from the side wall of the car to the hoistway wall on the flow field around the car and the aerodynamic characteristics of the car are compared and analyzed. It provides a theoretical basis for optimizing the aerodynamic characteristics of elevators.

2. Numerical simulation

2.1. Governing equation of flow field

When the elevator runs in the hoistway, it is surrounded by viscous, incompressible and unsteady turbulent flow field with high Reynolds number. According to previous studies [6-8], the standard $k-\varepsilon$ has sufficient accuracy to solve similar problems, and the wall function is used to deal with the near wall area. The governing equation describing the air flow around the car is,

$$\frac{d}{dt}\int_V \rho\Phi dV + \int_S\left[\rho\Phi\left(U-U_g\right)-\Gamma_\Phi grad\Phi\right]dS = \int_V S_\Phi dV \tag{1}$$

where, t is the time; ρ is the density of air; V is the volume of control body; S is the source term; grad is the gradient of Φ; Φ is the flow field fluxes, when $\Phi = 1, U, T, k, \varepsilon$, and represent the continuity equation, momentum equation, energy equation, turbulent kinetic energy k equation and turbulent energy dissipation rate ε equation, respectively; U is the velocity of fluid flow; U_g is the motion velocity of cross section of control body; Γ_Φ and S_Φ is the generalized discrete coefficient and generalized source term, respectively.

2.2. Numerical domain and boundary conditions

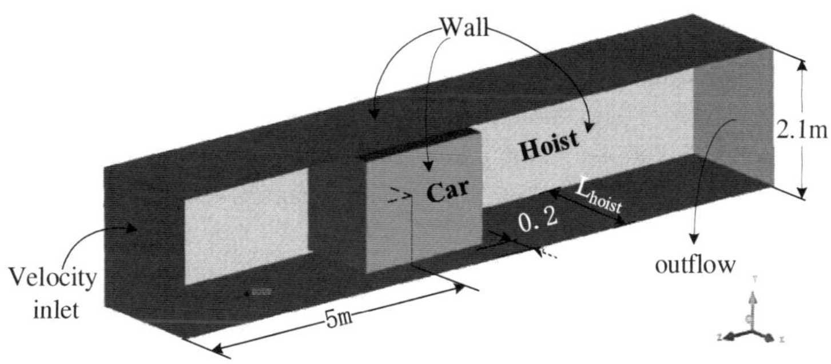

Figure 1. Numerical domain and boundary conditions of flow field

In order to make the simulation results more representative, in the actual simulation calculation process, we simplified the calculation model of the elevator, removed those small and complex structures in the elevator system, such as guide shoes, safety pliers, etc. and only retain the car. In addition, considering the accuracy of simulation results, and in order to highlight the influence of the analyzed factors on the aerodynamic characteristics of elevators, a full-scale elevator car model is adopted in this paper: fixed car size ($1600 \times 1500 \times 2350$mm), and fixed length (12m), fixed depth (2.1m) of hoistway, the width are 1.8m, 2.0m, 2.2m, 2.4m, 2.6m, and 2.8m, respectively; the distance between the car door and the hoistway wall is fixed at 0.2m; the distance from the center of the car to the inlet face is 5m; the operating speed of elevator are 3m/s, 6 m/s,9 m/s,12 m/s,15 m/s and 18 m/s, its numerical domain and boundary conditions is shown in Figure 1.

2.3. Mesh generation of internal flow field of hoistway

Considering the correctness and accuracy of the calculation of the flow field around the car, the mesh refinement is adopted in the area near the car during the meshing process, at the same time, to accurately capture the aerodynamic characteristics of the car surface, a boundary layer is added, as shown in Figure 2. Because of the change of the width of the hoistway, the number of meshes of the flow field will change, so the total number of meshes used in the model is about 0.9-1.2 million.

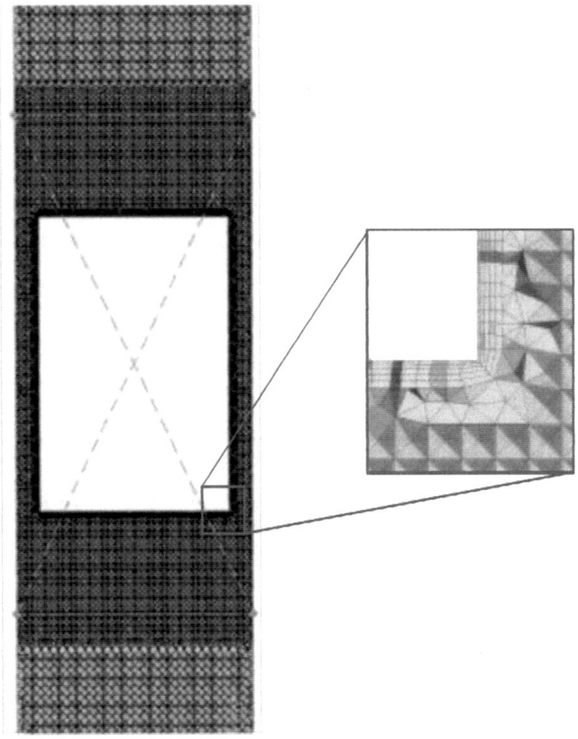

Figure 2. Grid encryption zone and boundary layer

3. Results and discussion

3.1. Effect of operating speed on elevator aerodynamics

Figure 3-5 show the pressure, turbulent kinetic energy (TKE) and velocity contours of the internal flow field of the hositway at different operating speeds. As can be seen from Figures 3-5 the aerodynamic characteristics on both sides of the car show obvious symmetry. However, with the increase of operating speed, both the pressure and the TKE and the air flow rate show an increasing trend. The airflow forms a separate shear layer at the entrance of the annular space, and as the operating speed increases, the velocity gradient, pressure gradient and local TKE all increases, which aggravates the instability of the elevator. It can also be seen from Table 1 that the aerodynamic force of the car increases with the increase of the speed of the car, and is basically proportional to the quadratic of the speed.

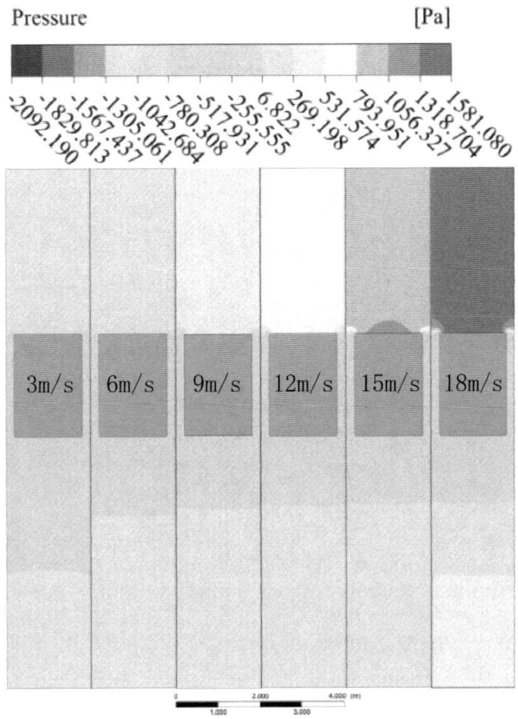

Figure 3. Pressure contours at different operating speeds

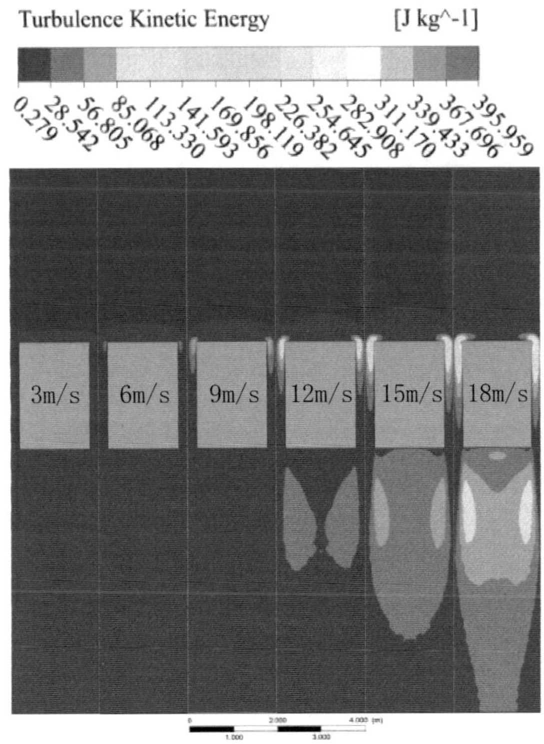

Figure 4. TKE contours at different operating speeds

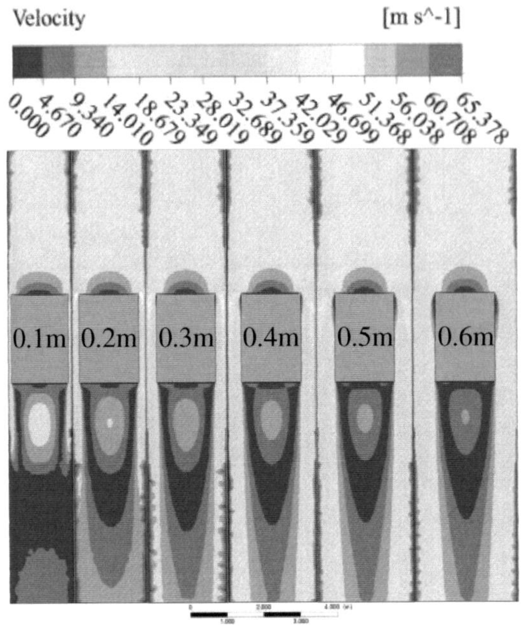

Figure 5. Velocity contours at different operating speeds

Table1. Aerodynamic force on one side of car at different speeds.

Velocity/(m/s)	3	6	9	12	15	18
Lateral force/(N)	13.994	57.006	128.079	227.788	355.931	512.536
Yawing moment/(N·m)	-8.775	-35.976	-82.022	-147.039	-231.132	-334.426

3.2. Effect of distance on elevator aerodynamics

Figure 6-8 show the pressure, turbulent kinetic energy (TKE) and velocity contours of the internal flow field of the hositway at different distances (from the side of car to hoistway wall).

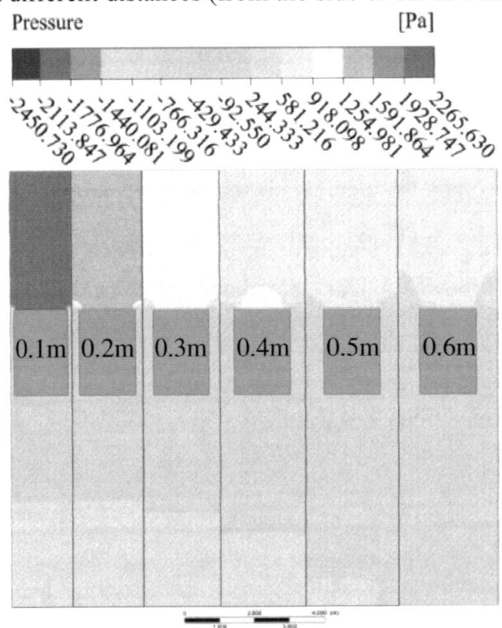

Figure 6. Pressure contours at different distances

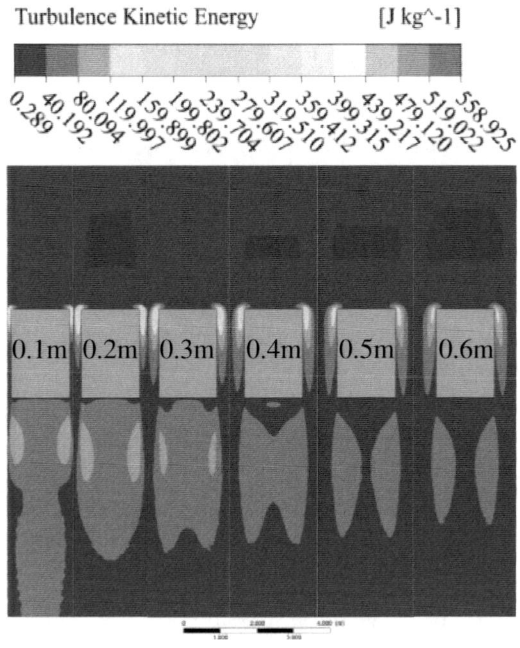

Figure 7. TKE contours at different distances

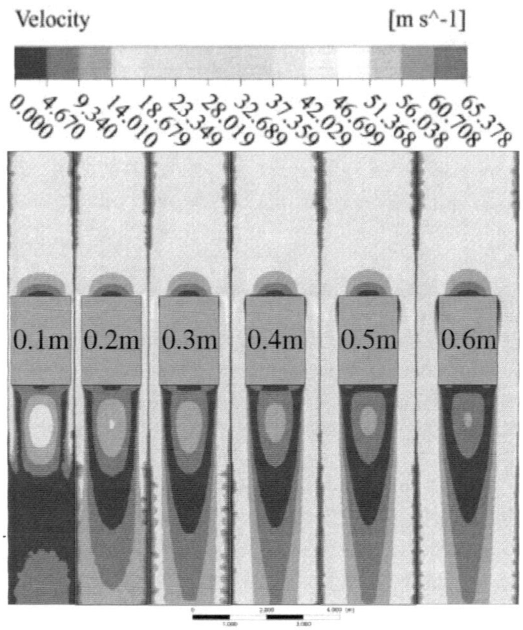

Figure 8. Velocity contours at different distances

From Figure 6-8, it can be seen that the smaller the distance between the side of the car and the wall of the hoistway, the more violent the change of the flow field around the car. However, due to the same distance between the two sides, it can also show obvious symmetry. As the distance increases, the pressure gradient of the flow field decreases, and the local TKE of the car decreases, the velocity of annular space change slightly. Meanwhile, it can be seen from Table 2 that the aerodynamic force is proportional to the quadratic of the distance.

Table2. Aerodynamic force on one side of car at different speeds.

Distance/(m)	0.1	0.2	0.3	0.4	0.5	0.6
Lateral force/(N)	736.778	512.424	363.233	292.263	246.730	227.788
Yawing moment/(N·m)	-413.607	-334.427	-311.234	-288.186	-280.037	-271.572

4. Conclusions

In this paper, the aerodynamic characteristics of elevators under different operating speeds and distances are compared and analysed by using computational fluid dynamics (CFD) method, and the influence rules are briefly analysed. It is found that under different conditions, the two sides of the elevator car still show obvious symmetry, and the aerodynamic force acting on the car is basically proportional to the quadratic power of running speed and distance. Small distance and high velocity will make the flow field change more dramatically, which will make the annular entrance have greater velocity gradient and turbulent kinetic energy, and aggravate the instability of the elevator.

Acknowledgments

This study was funded by the Shandong Province Nature Science Foundation, China (GRANT NO. ZR2017MEE049), the Key Research Development Project of Shandong Province (GRANT NO. 2018GSF122004).

References

[1] Y. Duan, G.J. Shen, Y.G. Zhang, W.B. Su, J Beijing Univ Aeron Astron, 30(5), 444-447, (2004).
[2] X.D. Li, K. Wang, J Harbin Inst Technol, 6, 82-86, (2009).
[3] J.Q. Niu, Y.M. Wang, L. Zhang, Y.P. Yuan, Int. J. Heat Mass Transf., 127, 188-199, (2018).
[4] Y. Sakuma, M.P. Païdoussis, S.J. Price, J. Fluids Struct., 24(7), 932-953, (2008).
[5] Y. Sakuma, M.P. Païdoussis, S.J. Price, J. Wind Eng. Ind. Aerodyn., 126, 38-47, (2014).
[6] J. Muñoz-Paniagua, J. García, A. Crespo, J. Wind Eng. Ind. Aerodyn., 130, 48-61, (2014).
[7] J.K. Choi, K.H. Kim, Tunn. Undergr. Space Technol., 41, 62-73, (2014).
[8] F. Liu, S. Yao, J. Zhang, Y.B. Zhang, Tunn. Undergr. Space Technol., 52, 62-70, (2016).

IOP Conf. Series: Materials Science and Engineering **538** (2019) 012029 doi:10.1088/1757-899X/538/1/012029

The Development of Mode for Solubility of Oxygen in High Pressure Condition

Yaqin Guo[1,*], Duoqiang Liang[2], Yong Deng[3]

[1]School of Mining and Environmental Engineering, Chongqing Vocational Institute of Engineering, Chongqing 402260, China/College of Materials Science and Engineering, Chongqing University, Chongqing 400044, China;
[2]Innovation Center for Metal Resources Utilization and Environment Protection, School of Resources, Environment and Materials, Guangxi Key Laboratory of Processing for Non-ferrous Metal and Featured Materials, Guangxi University, Nanning 530004, China;
[3]State Key Laboratory of Complex Nonferrous Metal Resources Clean Utilization, National Engineering Laboratory for Vacuum Metallurgy, Faculty of Metallurgical and Energy Engineering, Key Laboratory of Vacuum Metallurgy for Nonferrous Metal of Yunnan Province, Kunming University of Science and Technology, Kunming 650093, China

*Email: yaqin.guo@cqu.edu.cn

Abstract. Previous investigation of solubility of oxygen in high pressure is summarized in the paper. Summing-up shows that compared with various methods for such measurement, solubility theory is developed slowly. Models for prediction of oxygen solubility in aqueous acid and/or salt solution have been developed. However, no model can reproduce the existing experimental data, and extrapolate beyond the data range from simple aqueous salt system to complicated brine systems including seawater, simultaneously. A model to cover much wider temperature and pressure space in variable composition brine systems needs much more efforts. At last, the main consensuses and differences as well as the further work is reviewed.

1. Introduction

Oxygen leaching is an important hydrometallurgical process that is frequently considered for the pretreatment and upgrading of ores at elevated temperatures and pressures. The solubility of oxygen in the aqueous phase determines the availability of oxidant during leaching and, in the case of high operating temperatures, the volume of oxygen that is evolved during cooling of the leached solution. A detailed knowledge of temperature and pressure effects on oxygen solubility enables the leaching processes to be modeled and controlled more effectively[1-2].

2. Mode for Theories

Some authors suggested some methods or models for predicting the effect of ions (or salts) on the gas solubility. Two models (the setschenow salt effect parameter and the ionic strength effect) are introduced for this purpose.

2.1. Mode 1: The Setschenow Salt Effect Parameter

Studies concerning the effect of dissolved salts on gas solubility are well presented by Ting Gong et al[3]. The magnitude of the effect of salts on the activity coefficient of a dissolved gas in aqueous solution depends strongly on the properties of all solute species, which include the nonelectrolyte (or gas) and the salt. At a given temperature and pressure and when there is a negligible chemical interaction between solute species, the logarithm of the dissolved gas activity coefficient can be represented as:

$$\log \gamma_B = k_s C_s + k_B C_B \tag{1}$$

In Expression (1), is the salt effect parameter, is the solute-solute gas interaction parameter. It is considered that dissolved gas activity in the pure solvent is same as that in the solution.

$$\alpha_B = \gamma_B S_B = \gamma_B^0 S_B^0 \tag{2}$$

In Eq. (2), r_{B} and S_{B} are in the solution, r_{B}^0 and S_{B}^0 in the pure solvent. Also, and are their corresponding activity coefficients. Therefore, the following equation results:

$$\log \frac{\gamma_B^0}{\gamma_B} = \log \frac{S_B^0}{S_B} = k_s C_s + k_B (C_B - C_B^0) \tag{3}$$

If the gas solubility is low, the term with can be neglected, thus:

$$\log \frac{\gamma_B^0}{\gamma_B} = \log \frac{S_B^0}{S_B} = k_s C_s \tag{4}$$

The Setschenow salt effect parameter, k_{SCA}, is defined as follows:

$$k_{\mathrm{SCA}} = \frac{1}{Cr} \log(\frac{S_B^0}{S_B}) \tag{5}$$

In Eq. (5), is the total electrolyte concentration.

The salt that increase the activity coefficient of the dissolved gas is said to “salt out” the gas and the salt that decrease the activity coefficient of the dissolved gas is said to “salt in” the gas. Salting out means decreasing gas solubility, whereas salting in implies increasing its solubility.

The solubility of gases in water is usually decreased by the addition of other solutes, particularly electrolytes. The extent of this salting out varies with different salts, gases, solvents and temperature.

2.2. Mode 2: The Effect of Ionic Strength

Based on the Setschenow salt effect parameter, Braibanti et al [4] proposed a method to estimate the solubility in electrolyte solutions. In this method, the Henry’s law constant in the solution in the solution is related to that in water at the same temperature as shown in the following expression:

$$\log \frac{K_B}{K_B^0} = \log \frac{S_B^0}{S_B} = h_1 I_1 + h_2 I_2 \tag{6}$$

In Eq. (6), 1 is the ionic strength of the solution defined by:

$$I_i = \frac{1}{2} C_i (Z_+ + Z) \tag{7}$$

The effect of the different species on gas solubility in an electrolyte solution are expressed by

$$h = h_+ + h_- + h_G \tag{8}$$

In Expression (8), $h_{+,}$ h_{-}, and h_G are the contribution or effects on gas solubility by the positive ions, of negative ions, as well as the dissolved gas in solution. Thus, the total effect of the gas solubility is simply the sum of the effects of the dissolved molecular or ionic species.
Until now, the Setschenow Salt Effect Parameter has been most useful in the prediction of gas solubility in salt solutions. However, it is inconvenient, because the parameter ($h_{+,}$ h_{-}, and h_G) in the model are independent of the composition of aqueous salt solution.

3. Mode for Experiments

3.1. Mode 1: In Pure Water

The solubility of oxygen in water is known precisely. The work is considered to the most precise measurement of gas solubility to date. They have proposed an equation to predict the amount of oxygen dissolved in water, in the temperature range from 273.15 k to 373.15 k, with a standard deviation of 0.017%. For that reason, the solubility of oxygen of oxygen in water can serve as a reference to ensure the reliability of the solubility apparatus, and more importantly, the experimental technique of the investigator. After examination of the water reference system, gas solubility in sulfuric acid and copper sulfate solutions were measured[5].

3.2. Mode 2: In Sulphuric Acid/Or Salt Solution

Qin Li et al[6] investigated the solubility of oxygen in sulfuric acid solutions containing cupric and/or nickel sulfate at atmospheric and high pressures, and found that the logarithm of oxygen solubility in aqueous acid-salt solution is lineally related to molar concentration of bivalent cations and densities of acid-salt solutions. A simple model is then suggested to predict oxygen solubilities in the acid-salt solutions. This model has more advantages, compared with the models proposed by Tan et al[7]. The effects of temperature and pressure on oxygen solubility are also discussed. In the solutions studied, the oxygen solubility as a function of temperature follows a similar pattern to that in water. For different pressures, Hery's law is obeyed in the range of this research (<5 atm)
Very few workers have measures the solubility of oxygen in aqueous sulfuric acid solutions. Those who did, did not cover sufficient common ranges of concentration and temperature for direct comparison. The data available are thus, considered tentative only.
In the review of oxygen solubility data done by IUPAC, it was shown that by 1980 there were only six recorded sets of measurements for the solubility of oxygen in aqueous sulfuric acid solutions. Half of the measurements were carried out just after 1900. The rest were reported in 1965 and 1967. The data are too scattered for a direct comparison by interpolation or extrapolation. However, all the results indicate the following trends:
At a given temperature and partial pressure of oxygen, the solubility decreases with an increase in the sulfuric acid concentration. The solubility increases with the increase in the partial pressure of oxygen at a constant temperature and concentration. As expected, the gas solubility increased with the increase in partial pressure of oxygen at a constant temperature as expressed by Henry's law.
A complication of oxygen solubility in various aqueous solutions was produced in 1981 by Battino et al. surprisingly, only six reports concerning the solubility of oxygen in aqueous sulfuric acid solution were found. Most of his initial work was performed near the turn of the century. Additional data were made by Frank et al.[8], at five temperatures between 323 and 523 k and for concentrations of 0.5, 1.0, and 1.5 mol/l sulfuric acid solutions, show a decrease in the solubility from 323 to 373 k. this is followed by an increase from 373 to 723, 473 and to 523 k. thus, a minimum solubility was indicated at 373 k.
Gubblins and Walker measured the solubility of oxygen in sulfuric acid solutions at higher concentrations of the acid. The results indicated that oxygen solubility decreases to a minimum at about 80% by mass H_2SO_4 is approached.
Shimizu et al[9] is the only person who measured the solubility of oxygen in aqueous sulfuric acid solutions containing salts. He measured the solubility at oxygen partial pressure of 0.25 and 1.0 MPa (2.5 and 10 atm) at six temperatures between 323 and 473 k. the concentration of the solution was

0.0125 mol/l sulfuric acid and saturated with either nickel sulfide (NiS_2), copper sulfide (Cu_2S) or cobalt sulfide (Co_4S_3). The actual sulfide concentrations were not given. The sulfuric acid-metal sulfide decreased the oxygen solubility to less than 50 percent its solubility in water. The variation of oxygen solubility with temperature and partial pressure of oxygen followed the same trend as those plain sulfuric acid solutions.

Sinha et al[10] carried out a thermodynamic analysis of the temperature dependent equilibrium between oxygen solubility activity in pure water and oxygen partial pressure. He used experimental oxygen solubility sources to determine a set of self consistent values for the chemical potential, entropy and partial molar heat capacity of dissolved oxygen, then combine with established thermodynamic data for gaseous oxygen to develop a single thermodynamic equation describing the relationship between gaseous oxygen fugacity, dissolved oxygen activity, Temperature and the equilibrium constant *k*. This equation was consistent with *k* values calculated from published experimental solubility data for temperatures from 273 to 616 K and pressures to 60 atm, encompassing the conditions encountered during oxygen leaching operations. Within these ranges, it was shown that the gas fugacity and dissolved oxygen activity coefficients were sufficiently close to unity that the *k*-equation may be used to relate the moral concentration of dissolved O_2 for any combination of temperature T(K) and (atm).

$$C_{aq}=P_{0_2}\exp\left\{\frac{0.046T^2+203.357\ln(\frac{T}{298})-(299.378+0.092T)(T-298)-20.591\times10^3}{(8.3144)T}\right\} \quad (9)$$

4. Main Conclusions

Great numbers of data on solubility of oxygen in high pressure conditions have been measured, and some models have been built, the trend of change can be sure to predict. But a more accurate data and model need further investigation.

(1) The main consensuses and differences are as follows:

(a) The solubility of oxygen in water approaches a minimum at about 373.15 k. Oxygen solubility in aqueous sulfuric acid-salt sulfate solutions has a minimum solubility at a temperature of about 368 k.

(b) The aqueous solution did not closely obey Raoult's law, hence the activity coefficient for the electrolytes were calculated from the vapor pressure data.

(c) The solubility of oxygen in the aqueous sulfuric acid-salt leach solutions closely obeys Henry's law.

(d) As the temperature is increased from 368 k to 458 k, there is generally a large increase in the solubility of oxygen in the aqueous sulfuric acid-salt leach solution.

(e) In a given temperature and pressure ranges, the densities of the solutions decrease as the temperature is increased following the same pattern as that for pure water, and increase with an increase in concentration of iron, zinc and/ or sulfuric acid.

The main difference is that some workers state that the salt out of oxygen from the aqueous solutions was found to be function of the type of electrolyte added. However, the Setschenow salt effect parameter, for a given electrolyte, was found to decrease slightly as the concentration of the electrolyte was increased., but some insist that marked reduction in the oxygen solubility is observed at all temperatures as the quantity of dissolved zinc sulfate is increased in the solution.

(2) The further work

(a) A more universal model is to be developed for more accurate predict solubility of oxygen in different media and (or) condition.

(b) A detail method and program is to be made to fully automate measurement of solubility of oxygen for the chemical industry and metallurgical industry.

(c) Parameters about solubility of oxygen such as activity should be measure accurate to better model building.

Acknowledge
This work was supported by the Scientific and Technological Research Program of Chongqing Education Commission (KJ1503305), State Key Laboratory of Complex Nonferrous Metal Resources Clean Utilization / Kunming University of Science and Technology (CNMRCUKF1603), Chongqing Vocational Institute of Engineering (JG182022).

References

[1] Mazuelos, Alfonso. Jesus Garcia-Tinajero, Carlos. Romero, Rafael. et al. Oxygen solubility in copper bioleaching solutions [J], HYDROMETALLURGY, 2017(1):1-7
[2] Astrom, E. Bjorkvall, J. Bjorkman, B. etc. Vanadium oxide solubility in CaO-SiO_2-VOX system [J], IRONMAKING & STEELMAKING, 2017(1):42-51
[3] Ting Gong. Preparation and oxygen sensing properties of water-soluble silicon nanoparticles assembled SBA-15 [J]. Materials Research Bulletin, 2019(3):1-6
[4] A. Braibanti, E. Fisicaro, C. Compari. Solubility of oxygen and inert substances in water [J]. Polyhedron, 2000(11):2457-2461
[5] Desmond Tromans. Temperature and pressure dependent solubility of oxygen in water: a thermodynamic analysis [J]. Hydrometallurgy 1998 (48):327-342
[6] Qian Li, Christopher Batchelor-McAuley, Nathan S. Lawrence, etc. Anomalous solubility of oxygen in acetonitrile/water mixture containing tetra-n-butylammonium perchlorate supporting electrolyte; the solubility and diffusion coefficient of oxygen in anhydrous acetonitrile and aqueous mixtures [J]. Journal of Electroanalytical Chemistry, 2013(1):328-335
[7] Zhi-Qiang Tan, Guang-Hua Gao. Solubility of oxygen in aqueous sodium carbonate solution at pressures up to 10 MPa [J]. Fluid Phase Equilibria, 2001(4):375-382
[8] Frank J. Millero, Fen Huang, Arthur L. Laferiere. Solubility of oxygen in the major sea salts as a function of concentration and temperature [J]. Marine Chemistry,2002(6):217-230
[9] Ryohei Shimizu, Mikio Yagi, Azusa Kikuchi. Suppression of riboflavin-sensitized singlet oxygen generation by l-ascorbic acid, 3-O-ethyl-l-ascorbic acid and Trolox [J]. Journal of Photochemistry and Photobiology B: Biology, 2019(2):116-122
[10] Shivendra Sinha, Devabrata Mishra. Aqueous process intensification through enhanced oxygen mass transfer using oxygen vector: An application to cleaner leaching [J]. Journal of Cleaner Production,2018(3):452-462

IOP Conf. Series: Materials Science and Engineering **538** (2019) 012030 doi:10.1088/1757-899X/538/1/012030

Sensitivity Analysis of Piston Wind in Hoistway of Super High-Speed Elevator

Shuai Qiao[1], Ruijun Zhang[1,a] and Luzhong Zhang[2]

[1]School of Mechanical and Electrical Engineering, Shandong Jianzhu University, Jinan, Shandong, 250101, P.R. China

[2]Shandong Electronic Information Products Inspection Institute, Jinan, Shandong, 250014, P.R. China

[a]Corresponding author: zhangruijun@sdjzu.edu.cn

Abstract. The piston wind will be produced when the car runs at high speed in the narrow and long hoistway, which will produce aerodynamic noise with the car enclosure and the hoistway walls, thus affecting ride comfort. In order to put forward better noise reduction measures, a piston wind model of super high-speed elevator with ventilation holes during the whole operation process was established firstly in this paper. Then, based on this theoretical model, the influence of the variation of ventilation-hole area, hoistway blockage ratio and car length on the piston wind was analysed separately. The analysis proved that the larger the area of ventilation holes, the more obvious the weakening of piston wind was when the building code permits. With the increase of blockage ratio, the intensity of piston wind increases obviously, but the duration of maximum piston wind does not change obviously. As the length of the car increases, there is no significant change in the movement of the piston wind, and this small change can be ignored in engineering applications.

1. Introduction

Because of the great aspect ratio of elevator hoistway, the airflow in the hoistway can be approximately regarded as one-dimensional flow. When the elevator goes up and down, the air in the windward zone of the car is squeezed forward by the car due to the limitation of the hoistway wall, which is called the piston effect, and this forward flow is called the piston wind. When the elevator is running, the forced gas in the hoistway can't be discharged in time, and part of the airflow flows to the end of the car through the annular space between the car enclosures and the hoistway walls. During this process, the piston wind rubs violently against the car enclosures and the hoistway walls, which produces aerodynamic noise.

Shi et al. considered that when the elevator reaches a certain speed, the aerodynamic noise generated by the hoistway airflow will become the main part of the interior environmental noise of the elevator car [1]. Based on the theoretical analysis of piston effect, Zhu et al. established a typical physical model of vertical evacuation passage in high-rise buildings, and used large eddy simulation (LES) method to simulate the airflow characteristics in hoistway [2]. Cross et al. studied the effect of changing the blockage rate of metro tunnels on the ventilation airflow under the piston effect [3]. Strege et al. based on field data measurement of high-rise buildings in different cities, explored the size of chimney effect in stairwell and elevator hoistway and vertical movement of airflow in well [4].

Therefore, through the study of the airflow characteristics in tunnel (hoistway), the idea of putting forward effective noise reduction measures has been fully affirmed and valued by academia and engineering circles [5]-[8].

Aiming at the influence of piston wind generated by car running in narrow hoistway on ride comfort of super high-speed elevator, this paper first establishes a piston wind model of super high-speed elevator with ventilation holes in the whole process of car running. Then, based on this theoretical model, the influence of the change of ventilation-hole area, blockage ratio and car length on piston wind in the hoistway is analyzed separately.

2. The Theoretical Model of Piston Wind for Super High-Speed Elevator

2.1. Model Simplification

In order to reduce the influence of aerodynamic noise on elevator ride comfort, the design of ventilation holes should be taken into account in the civil construction of buildings with elevators above 3 m/s. Here, the ventilation holes of super high-speed elevator hoistway in a project are simplified as the ventilation holes on both sides of the hoistway (equal horizontal height and equal area). The running speed of the car is a 6 m/s curve as shown in Fig. 1, the acceleration of the elevator at this speed is 1.5 m/s^2 and the jerk of the elevator is 1.2 m/s^3.

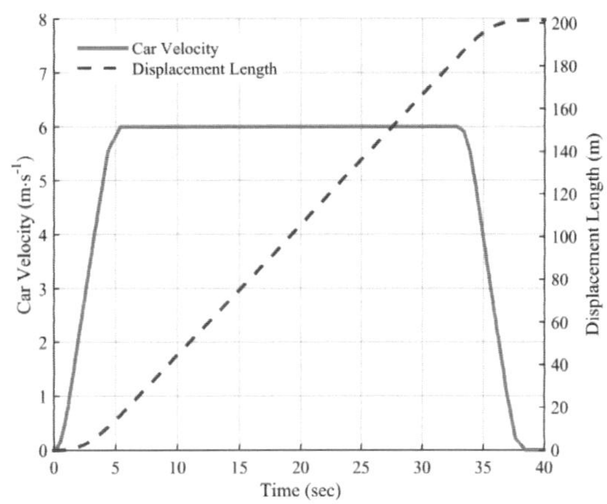

Figure 1. The operation speed curve of the elevator car

In order to better study the effect of ventilation holes on reducing the piston wind, it is assumed that there is no natural wind outside the hoistway.

2.2. Establishment of Theoretical Model

For the model shown in Fig. 2, the continuity equation is used to obtain:

$$v_B \mathrm{A_{hoist}} = v_F \mathrm{A_F} + v_T \mathrm{A_{hoist}} \tag{1}$$

where, $\mathrm{A_F}$ is the sum of the cross-sectional area of two ventilation holes.

Based on Bernoulli's principle for unsteady flows and the law of conservation of energy in stationary fluid, the following conclusions are obtained:

$$\mathrm{k_F}\frac{v_F^2}{2} + \xi_{45}\frac{v_B^2}{2} + \mathrm{l_F}\frac{dv_F}{dt} = \xi_{46}\frac{v_B^2}{2} + \mathrm{k_T}\frac{v_T^2}{2} + \mathrm{L_2}\frac{dv_T}{dt} \tag{2}$$

In which, $\mathrm{k}_F = \lambda_F \frac{\mathrm{l_F}}{\mathrm{d_F}} + \xi_F$, $\mathrm{k}_T = \lambda \frac{L_2}{\mathrm{D}} + \xi_6$, ξ_{45} is the local drag coefficient of airflow diverting into ventilation holes, λ_{F} is the drag coefficient along the ventilation hole, ξ_{F} is the local drag coefficient at cross-section 5-5, l_F is the length of the ventilation hole, d_F is the hydraulic diameter of the ventilation hole, ξ_{46} is the local drag coefficient of airflow from cross-section 4-4 to cross-section 6-6.

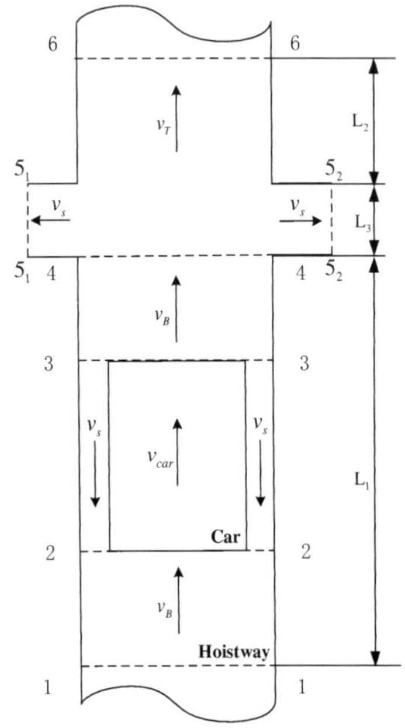

Figure 2. Before the car passes through the ventilation holes

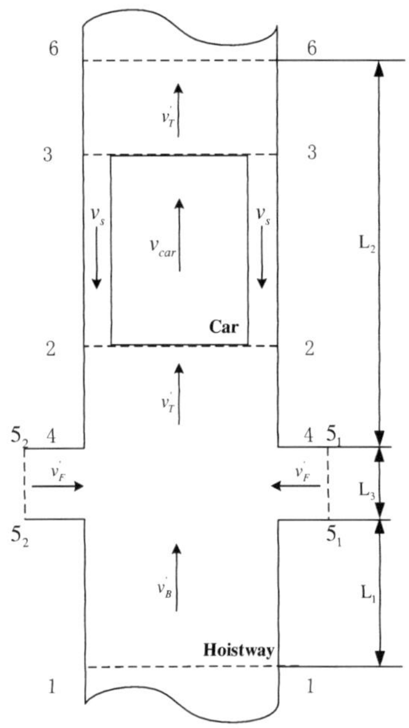

Figure 3. After the car passes through the ventilation holes

When the car starts to run, the displacement of the car is as follows,

$$S = \int_0^{t_1} v_{car} dt \tag{3}$$

It can be obtained from Bernoulli's principle for unsteady flows and the continuity equation,

$$\mathrm{A}\frac{dv_B}{dt}+\mathrm{L}_2\frac{dv_T}{dt}+\mathrm{k}_\mathrm{B}\frac{v_B^2}{2}+\mathrm{k}_\mathrm{T}\frac{v_T^2}{2}=\frac{\mathrm{l}_{23}}{1-\alpha}\frac{dv_{car}}{dt}+\mathrm{k}_{23}\frac{\left(v_{car}-v_B\right)^2}{2\left(1-\alpha\right)^2}+\frac{\lambda \mathrm{l}_{23}\left(\alpha v_{car}-v_B\right)^2}{2\mathrm{D}\left(1-\alpha\right)^2} \tag{4}$$

where, $\mathrm{k}_B=\xi_1+\lambda\frac{\mathrm{L}_1-\mathrm{l}_{23}}{\mathrm{D}}$, $\mathrm{k}_{23}=\xi_2+\lambda_1\frac{\mathrm{l}_{23}}{\mathrm{d}}+\xi_3$, $\mathrm{A}=\frac{\mathrm{L}_1-\alpha\mathrm{L}_1+\alpha\mathrm{l}_{23}}{1-\alpha}$.

Therefore, the variation of piston wind with time during the operation of the car in section L_1 of the hoistway is shown as follows:

$$\begin{cases} v_B\mathrm{A}_{\mathrm{hoist}}=v_F\mathrm{A}_\mathrm{F}+v_T\mathrm{A}_{\mathrm{hoist}} \\ \mathrm{k}_\mathrm{F}\frac{v_F^2}{2}+\xi_{45}\frac{v_B^2}{2}+\mathrm{l}_\mathrm{F}\frac{dv_F}{dt}=\xi_{46}\frac{v_B^2}{2}+\mathrm{k}_\mathrm{T}\frac{v_T^2}{2}+\mathrm{L}_2\frac{dv_T}{dt} \\ \mathrm{A}\frac{dv_B}{dt}+\mathrm{L}_2\frac{dv_T}{dt}+\mathrm{k}_\mathrm{B}\frac{v_B^2}{2}+\mathrm{k}_\mathrm{T}\frac{v_T^2}{2}=\frac{\mathrm{l}_{23}}{1-\alpha}\frac{dv_{car}}{dt}+\mathrm{k}_{23}\frac{\left(v_{car}-v_B\right)^2}{2\left(1-\alpha\right)^2}+\frac{\lambda \mathrm{l}_{23}\left(\alpha v_{car}-v_B\right)^2}{2\mathrm{D}\left(1-\alpha\right)^2} \end{cases} \tag{5}$$

The computational domain of the above mathematical model is $[0,t_1],\left(S=\int_0^{t_1} v_{car}dt < L_1\right)$, and its initial value is $v_B|_{t=0}=0$, $v_T|_{t=0}=0$, $v_F|_{t=0}=0$.

Similarly, the variation of piston wind with time during the operation of the car in section L_2 of the hoistway in the model shown in Fig. 3 is shown as follows:

$$\begin{cases} v_T^{'}\mathrm{A}_{\mathrm{hoist}}=v_F^{'}\mathrm{A}_\mathrm{F}+v_B^{'}\mathrm{A}_{\mathrm{hoist}} \\ \mathrm{k}_\mathrm{B}^{'}\frac{v_B^{'2}}{2}+\mathrm{L}_1\frac{dv_B^{'}}{dt}=\mathrm{k}_\mathrm{F}^{'}\frac{v_F^{'2}}{2}+\mathrm{l}_\mathrm{F}\frac{dv_F^{'}}{dt} \\ \mathrm{A}\frac{dv_T^{'}}{dt}+\mathrm{L}_1\frac{dv_B^{'}}{dt}+\mathrm{k}_\mathrm{B}^{'}\frac{v_B^{'2}}{2}+\mathrm{k}_\mathrm{T}^{'}\frac{v_T^{'2}}{2}=\frac{\mathrm{l}_{23}}{1-\alpha}\frac{dv_{car}}{dt}+\mathrm{k}_{23}\frac{\left(v_{car}-v_T^{'}\right)^2}{2\left(1-\alpha\right)^2}+\frac{\lambda \mathrm{l}_{23}\left(\alpha v_{car}-v_T^{'}\right)^2}{2\mathrm{D}\left(1-\alpha\right)^2} \end{cases} \tag{6}$$

where, $\mathrm{k}_\mathrm{F}^{'}=\lambda_\mathrm{F}\frac{\mathrm{l}_\mathrm{F}}{\mathrm{d}_\mathrm{F}}+\xi_\mathrm{F}^{'}+\xi_{54}$, $\mathrm{k}_\mathrm{B}^{'}=\xi_1+\lambda\frac{\mathrm{L}_1}{\mathrm{D}}+\xi_{14}$, $\mathrm{A}=\frac{\mathrm{L}_2-\alpha\mathrm{L}_2+\alpha\mathrm{l}_{23}}{1-\alpha}$, $\mathrm{k}_{23}=\xi_2+\xi_3+\lambda_1\frac{\mathrm{l}_{23}}{\mathrm{d}}$, $\mathrm{k}_\mathrm{T}^{'}=\xi_{14}+\xi_6+\lambda\frac{\mathrm{L}_2-\mathrm{l}_{23}}{\mathrm{D}}$, A_F is the sum of the cross-sectional area of two ventilation holes, ξ_{54} is the local drag coefficient when the air enters the ventilation hole, λ_F is the drag coefficient along the ventilation hole, $\xi_\mathrm{F}^{'}$ is the local drag coefficient at cross-section 5-5, l_F is the length of the ventilation hole, d_F is the hydraulic diameter of the ventilation hole, ξ_{14} is the local drag coefficient of airflow from cross-section 1-1 to cross-section 4-4.

The computational domain of the above mathematical model is $[t_1,t_2],\left(\int_0^{t_2} v_{\mathrm{car}}dt = L-l_{23}\right)$, and its initial value is $v_\mathrm{B}^{'}=v_B|_{t=t_1}$, $v_T^{'}=v_T|_{t=t_1}$, $v_F^{'}=v_F|_{t=t_1}$.

3. Parametric Impact Analysis

3.1. *The Effect of Ventilation-Hole Area on Piston Wind*

In order to better observe the influence of ventilation-hole area on piston wind, the motion of piston wind in the range of ventilation holes area from 1.0 m^2 to 2.6 m^2 is observed, and the piston wind operation state without ventilation hole is compared to check the effect of ventilation hole on reducing piston wind intensity.

In Fig. 4, before the car passes through the ventilation holes, the drag of the parallel connection between the ventilation holes and the L_2 section of the hoistway (the car is in the L_1 section) is generally smaller than that of the L_2 section of the hoistway itself, so the piston wind in the hoistway section where the car is located is higher than that in the hoistway without the ventilation holes. When the car passes through the ventilation holes, that is to say, after reaching L_2 section, the drag of the parallel connection between the ventilation holes and L_1 section of the hoistway is smaller than that of

IOP Conf. Series: Materials Science and Engineering **538** (2019) 012030 doi:10.1088/1757-899X/538/1/012030

L_1 itself. Therefore, when the car decelerates in L_2 section of the hoistway, the piston wind speed decreases rapidly, which is much faster than that without the ventilation holes.

Fig. 4 shows that the drag of parallel connection between L_2 section and ventilation holes decreases and the piston wind speed increases with the increase of ventilation-hole area before the car passes through the ventilation holes. When the car passes through the ventilation holes, the drag of parallel connection between L_1 section and the ventilation holes decreases with the increase of the ventilation holes area, so the decline speed of the piston wind is obviously higher than that of the piston wind without the ventilation holes.

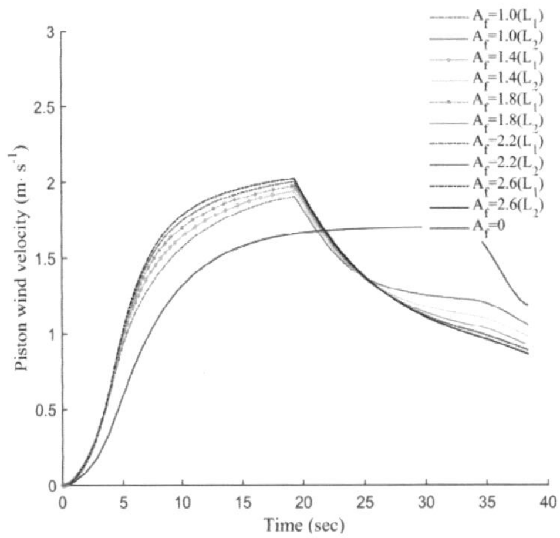

Figure 4. The effects of different ventilation holes areas on piston wind

3.2. The Effect of Different Blockage Ratios on Piston Wind

In order to study the influence of blockage ratio on piston wind in elevator hoistway with ventilation holes, the change of blockage ratio from 0.1 to 0.9 is selected.

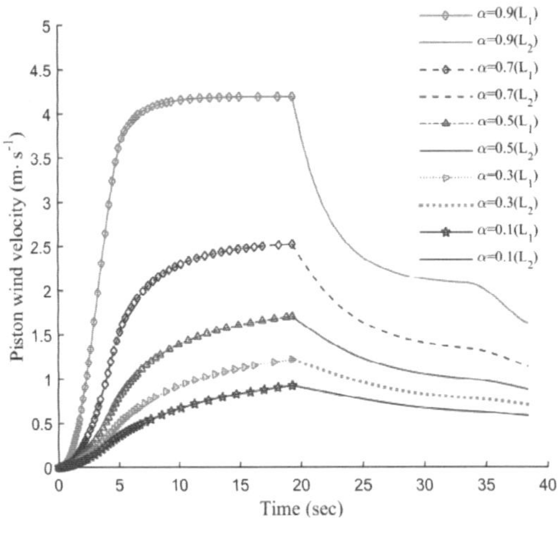

Figure 5. The effects of different blockage ratios on piston wind

Fig. 5 shows that with the increase of blockage ratio, the piston effect is strengthened and the maximum velocity of piston wind is increased obviously, but the variation of the maximum velocity duration of piston wind is not obvious. And in the process of elevator acceleration to uniform speed, the piston wind speed increases rapidly until it reaches the maximum speed and tends to be flat; and when the car passes through the ventilation holes, the piston wind speed begins to decline rapidly.

3.3. The Effect of Different Car Lengths on Piston Wind

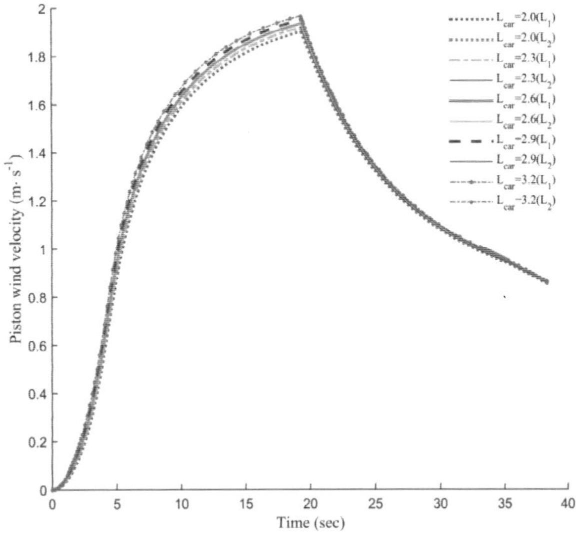

Figure 6. The effects of different car lengths on piston wind

In order to better study the effect of car length on the piston wind, the range of car height from 2 m to 3.2 m is analysed here, which covers the height dimensions of most elevators in the current elevator market.

It can be seen from Fig. 6 that as the height of the car increases continuously, the maximum speed of the piston wind increases, but the growth rate is small. After the car passes through the ventilation holes, the piston wind speed drops rapidly, but when the car length changes, the downward trend of the piston wind is almost invisible, that is, the variation range of the piston wind is negligible in engineering applications.

4. Conclusion

The main conclusions of this study follow.

(1) Based on Bernoulli's principle for unsteady flows, a theoretical model of super high-speed elevator with ventilation holes was established.

(2) With the increase of ventilation-hole area, the piston wind speed increased rapidly due to the decrease of drag of the parallel connection until it was higher than the piston wind speed without ventilation holes; after passing through the ventilation holes, the piston wind speed decreased rapidly, and the decreasing trend is significantly higher than that without ventilation holes.

(3) With the increase of blockage ratio, the maximum velocity of piston wind increased obviously, but the duration of maximum piston wind did not change obviously.

(4) As the length of the car increased, the maximum speed and duration of the piston wind changed very little. Considering the running speed and operation environment of the elevator, this small change could be ignored.

Acknowledgments

This study was funded by the Shandong Province Natural Science Foundation, China (GRANT NO. ZR2017MEE049) and the Key Research Development Project of Shandong Province (GRANT NO. 2018GSF 122004).

References

[1] L. Shi, Y. Liu, S. Jin, et al, J. Hydrodyn. Ser. B. 19(6), 720–725(2007).
[2] Y.J. Zhu, Y.J. Zhao, Building Science. 27(1),20-23(2011).

[3] D. Cross, B. Hughes, D. Ingham, et al, J. Wind Eng. Ind. Aerodyn. 146, 195–206 (2015).
[4] S. Strege, M. Ferreira, Fire Technol. 53(1), 211–226 (2016).
[5] CY. Yuan, M.Q. Li, J. Vibroeng. 19(6), 4759-4782(2017).
[6] J. Zhu, Z. Hu, D. Thompson, Proc. Inst. Mech. Eng. Part F-J. Rail Rapid Transit. 232(3), 759–773(2017).
[7] H. Hu, B. Tang, Y. Zhang, *Complexity*. 1–14 (2018).
[8] X. Xiang, L. Xue, B. Wang, W. Zou, Build. Environ. 132, 245–254(2018).

Cutting Simulation Analysis of the Cutter Head of the Shield Machine

Jingnan ZHAO[1], Xuekang SI[1], Jian GUO[1], Xiaolei MA[1] ,Yuzhao CHEN[1], Zichao LI[1], Yinghuai Dong[1,*]

[1]Mechanical Engineering College, Tianjin University of Science and Technology, Tianjin, 300222, China

Corresponding author: jingnanzhao@tust.edu.cn

Abstract. In the process of shield machine tunneling, the interaction between cutter head and soil is very complex. In order to meet the requirements of tunneling, the cutter head was designed according to the construction conditions of Tianjin metro and the Archimedes spiral layout method. The granite model and cutter head model were respectively established by the ABAQUS. The granite model was based on the classic Drucker - Pragel model, and the failure criterion including element deletion function was employ to simulate the yield and failure of rock. The force of the cutter head under cutting condition was analyzed. The results show that the maximum equivalent stress of the cutter head is 8.216950 MPa, and the allowable stress of the cutter head material is 230 MPa, which was less than the allowable stress of the cutter head. The results indicated that the cutter head can meet the actual work requirements and has high structural reliability. The research process of cutter head arrangement provides a method for solving practical engineering problems, and the numerical simulation results provide a reference for the design of cutter head.

1. Introduction

Shield machine is widely used in municipal, mining, transportation and hydraulic tunnel construction due to its advantages of environmental protection, rapidity and small impact on the surrounding environment. The cutter head plays a role in cutting the soil layer and supporting the cutting surface of the shield machine, and is an important part of the shield machine[1].

At present, there are many researches on the simulated cutting of the cutters head of shield machine in China. B Cai et al[2] studied the key parameters of cutter arrangement and simulated the cutting process of cutter groups under three geological conditions. Y M Xia et al[3] analyzed the influence of penetration degree and cutter head rotation speed on total torque of cutter head through single factor analysis, and verified it by comparing actual engineering data. S J Li et al[4] used the finite element numerical simulation analysis method to study the dynamic resistance characteristics of the cutter at the beginning and stationary stages of soil cutting. K Chen et al[5] used ANSYS to analyze the stress distribution of the cutter head, and obtained the distribution law of the axial force of the blade blade.

In order to study the deformation of soil and the stress of shield cutters in the process of cutting. Firstly, cutter head were arranged by Archimedes spiral method, and then dynamic simulation models of cutter head driving in the rock stratum are established by ABAQUS. Based on the simulation results, the deformation and failure process of the soil and the stress on the cutter head were studied.

2. Cutter Layout Curve of Shield Machine

At present, the most commonly used cutter layout method is Archimedes spiral layout method.

2.1. Spiral of Archimedes

The Archimedes spiral line arranged by the cutter head is shown in Figure 1.

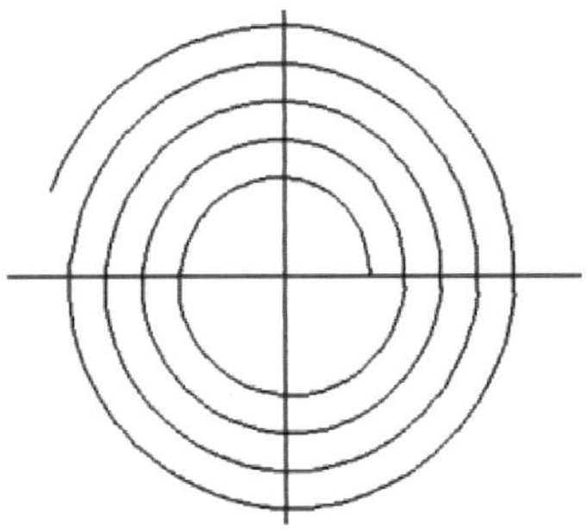

Figure 1. Archimedes spiral of cutter head arrangement

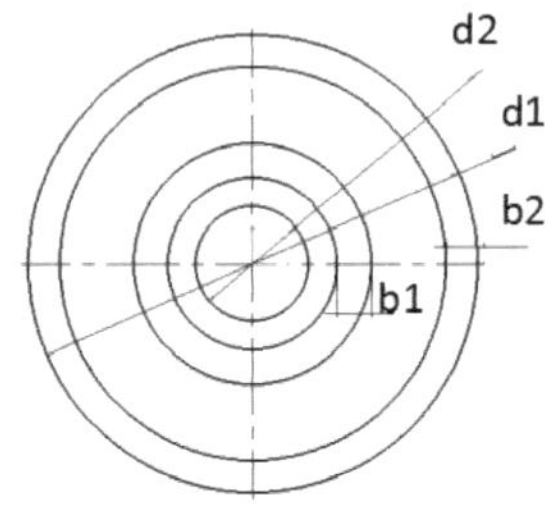

Figure 2. Schematic diagram of cutter layout

The polar coordinate expression of Archimedes spiral is shown in equation (1) [6].

$$\rho = \rho_0 + \alpha\theta \tag{1}$$

where, ρ is polar axis, ρ_0 is initial value, α is constant coefficient; θ is polar angle.

2.2. Determination of Shield Cutter Head

From the rationality of cutter head layout, structure and load, the cutter is arranged on several spokes (usually even number). The layout diagram of the cutter is shown in Figure 2.

The calculation method for the number of main cutters of shield machine[7].

$$N = \frac{{}^{d_1}/_2 - {}^{d_3}/_2 - b_2}{b_1} \tag{2}$$

Where, d_1 is shield machine outer diameter, d_2 is outside diameter of cutter head cutting, d_3 is center cutting length, b_1 is width of the main cutter, b_2 is width of peripheral cutter.
The clearance amount between the cutter of main and the center cutter is expressed as:

$$c = b_1 \times (N - N^*) \tag{3}$$

where, N^* is integer of cutter number.
The initial diameter of the spiral of Archimedes is:

$$\rho_0 = \frac{d_3}{2} + \frac{b_1}{2} \pm c \tag{4}$$

In the formula, the value is - when c is taken, and the value is - when c is rounded off.
For Archimedes' spiral constant coefficient:

$$\alpha = \frac{b_1}{\Delta\theta} \tag{5}$$

In the formula, $\Delta\theta$ is the angle between adjacent spokes.
Therefore, the spiral equation of Archimedes in the cutter arrangement of the cutter head is:

$$\rho = \rho_0 + \alpha\theta = \left(\frac{d_3}{2} + \frac{b_1}{2} \pm c\right) + \frac{b_1}{\Delta\theta} \times \theta \tag{6}$$

2.3. Based on Archimedes Spiral Cutter Head Layout Example

The above theory is illustrated with an example of Tianjin shield machine. The outside diameter of shield head is 6400 mm, the length of the fishtail cutter is 1500 mm, the width of the main's cutter is 100 mm, the width of the peripheral cutter is 150 mm. the use of four spokes, according to formula (2) the number of the layout of the winner cutter is:

$$N = \frac{{d_1}/_2 - {d_3}/_2 - b_2}{b_1} = \frac{{6400}/_2 - {1500}/_2 - 150}{100} = 23 \text{ control}$$

According to formula (3) the overlap between the main cutter and the center cutter is:

$$c = 100 \times (23 - 23) = 0 \text{ mm}$$

According to formula (4) the archimedes' spiral starts value.

$$\rho_0 = \frac{d_3}{2} + \frac{b_1}{2} \pm c = \frac{1500}{2} + \frac{100}{2} - 0 = 800 \text{ mm}$$

According to formula (5) and (6) the expression of the spiral of the cutter head is:

$$\rho = \rho_0 + \alpha\theta = 800 + \frac{200}{\pi} \times \theta$$

The polar coordinate values of cutter layout are shown in Table 1.

Table 1. Polar coordinate values and polar angle of cutter layout

Serial number	Polar coordinates ρ	Polar angle θ	Serial number	Polar coordinates ρ	Polar angle θ	Serial number	Polar coordinates ρ	Polar angle θ
1	800	0	9	1600	4π	17	2400	8π
2	900	0.5π	10	1700	4.5π	18	2500	8.5π
3	1000	1π	11	1800	5π	19	2600	9π
4	1100	1.5π	12	1900	5.5π	20	2700	9.5π
5	1200	2π	13	2000	6π	21	2800	10π
6	1300	2.5π	14	2100	6.5π	22	2900	10.5π
7	1400	3π	15	2200	7π	23	3000	11π
8	1500	3.5π	16	2300	7.5π			

3. Cutting Analysis of Shield Cutter Heads

3.1. Model Parameters and Assembly

For soil materials, Drucker-Prager model and linear elastic model are employ to simulate the model parameters of cutter head cutting granite, as shown in Table 2[8].

Table 2. Model parameters

Material	Q345	Cutting material
Modulus of elasticity/Gpa	210	26
Poisson'ratio	0.3	0.3
Density/(kg/m^3)	7850	2580

IOP Conf. Series: Materials Science and Engineering **538** (2019) 012031 doi:10.1088/1757-899X/538/1/012031

The three-dimensional size of the granite in this model is 7 m x 7 m x 2 m, and the cutter head diameter is 6.4 m. The three-dimensional model of cutter head is shown in Figure 3.

3.2. Damage Failure Criteria
Damage failure of element refers to that after the strength limit of element material is reached, the stiffness of material gradually decays to zero in accordance with a certain rule, and the element gradually loses its bearing capacity, and finally it exits the calculation of finite element model.

3.3. Cutting Principles and Processes
The contact algorithm of symmetric penalty function was used to simulate the interaction between cutter head and soil, the shear failure criterion including element deletion function is applied to avoid soil element grid distortion, the ABAQUS/Explicit display algorithm for complex nonlinear problems was used to simulate the cutting process of the cutter on the cutter head. (1)In the initial state, the cutter head was about to contact with the rock mass; (2)The rotation speed of the cutter head was 2 r/min; (3)The advancing speed of the cutter head was 40 mm/min; (4)The simulation time was 15 s.

3.4. Force on Cutter Head
Figure.4 shows the distribution of Mises stress on the front of the cutter head, and the stress on the cutter head when t = 15s. The stress distribution cloud diagram shows that the stress is relatively concentrated at the cutter, the stress at the center of the cutter head is relatively large, and the stress value at other parts is relatively small.

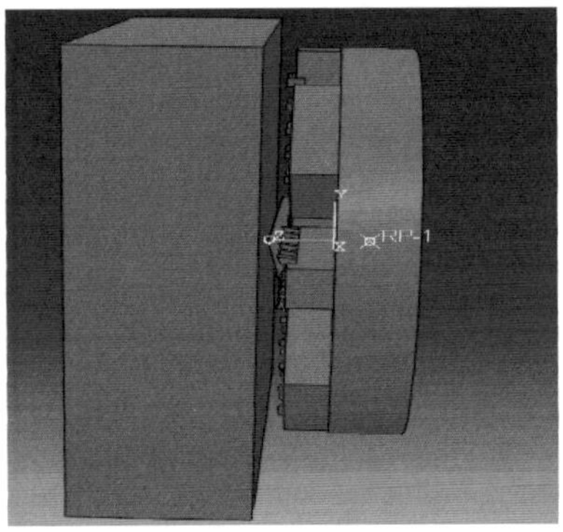

Figure 3. Three-dimensional model of cutter head head cutting

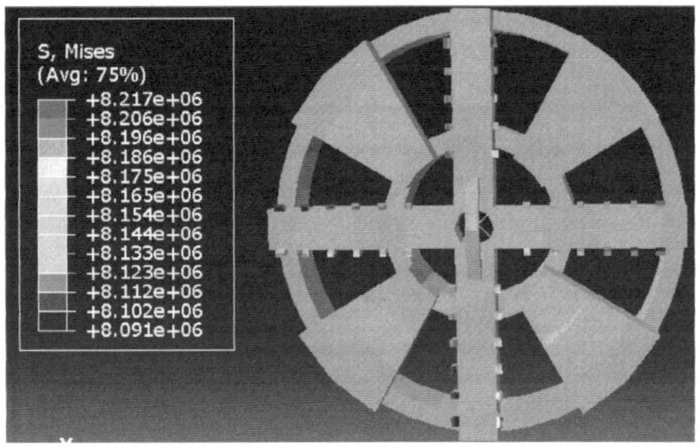

Figure 4. Stress condition of cutter head plate when t = 15s

The maximum equivalent stress of the cutter head is 8.216950 MPa, located at the cutter, as shown in figure 4. The allowable stress of cutter head material is 230 MPa, which meet the intensity requirement[9].

4. Conclusion
In this paper, the cutter head of shield machine is designed by Archimedes spiral layout method. Using The three-dimensional simulation model of shield cutter cutting soil was established by ABAQUS, simulated the process of the cutter head cutting soil. The PEEQ condition of soil and the stress on cutter surface were studied in the process of cutting simulation, the main draw the following conclusions.

(1) By the Archimedes spiral arrangement method, the designed cutter head of the shield machine can completely cut the soil.

(2) The three-dimensional rock breaking simulation model of cutter head was established, and the damage and failure criterion including element deletion function is applied to simulate the

formation and separation of cutting, and the direct numerical simulation of the driving process of cutter head was realized.

(3) Through the analysis of the cutting rock of the cutter head, the maximum stress of the cutter head is 8.216950 MPa, far less than the allowable stress of 230 MPa. The result meets the intensity requirement and the structure was reliable.

(4) The simulate results can provide some basis for the design and research of the cutter head, so as to improve and optimize the design of the cutter head.

Reference

[1] Q J Zhou,Y X Zheng, B J LI, et al. Finite element analysis of shield tunneling cutter head plate modification [J]. Chinese journal of construction machinery, 2008, 6(2): 65-70. (in Chinese)
[2] B CAI, K J Shi, W H Zhu. Research on simulated cutting of shield tunneling cutter head group based on ABAQUS [J]. Tunnel construction, 2017, 37(10): 1334-40.
[3] Y M Xia, X H Zhu, Z B Lin, et al. Research on influencing factors of tunneling load of TBM cutter head head [J]. Modern manufacturing engineering, 2015, 9): 1-6.
[4] S J Li, S Yu, J Chui, et al. Finite element simulation analysis of soil cutting process with shield machine cutter head [J]. Journal of underground space and engineering, 2013, 9(6): 1346-9.
[5] K Chen, C X Su, Y Q Wang. Finite element parametric modeling and analysis of shield cutter head [J]. Tunnel constructio Yn, 2010, 31(12): 57-60. (in Chinese)
[6] G S Chen. Application of Archimedes spiral in shield technology [J]. Heavy industry and lifting technology, 2006, 2): 18-20.
[7] R Y Pei. Analysis and research on cutter wear and layout of shield tunneling machine [D]. Tianjin university, 2009.
[8] C Zheng,W Zhao,H F Zhang, et al. Dynamic numerical simulation of excavation with earth pressure balance shield cutter head [J]. Tunnel construction, 2015, 35(8): 855-60.
[9] Q Fang, Y Li,L Guo. Stress analysis and optimization design of pressure packer for long-distance pipeline based on ANSYS Workbench [J]. Chemical machinery, 2013, 40(2): 197-202.

IOP Conf. Series: Materials Science and Engineering **538** (2019) 012032 doi:10.1088/1757-899X/538/1/012032

Robust Control of High-Speed Elevator Transverse Vibration Based on LMI Optimization

Shunxin Cao[1], Qin He[1,a] and Ruijun Zhang[1]

[1] School of Mechanical and Electrical Engineering, Shandong Jianzhu University, Jinan, Shandong, 250101, P.R. China

[a] Corresponding author: heqin67271@163.com

Abstract. Aiming at the problem of transverse vibration of high-speed elevator caused by surface roughness excitation of guide rail and relative motion between car and car frame, firstly, this paper establishes an eight-degree-of-freedom active control model for the transverse vibration of the high-speed elevator separated of car from the car frame. Secondly, the transverse vibration state equation of the high-speed elevator is derived. The H_2 norm is used to minimize the vibration acceleration of the car system. Based on Linear matrix inequality (LMI) optimization technology, a robust controller for the high-speed elevator is designed. Finally, the robust controller is solved by MATLAB and its simulation analysis is carried out. The results show that the vibration of the car and the car frame is inconsistent in actual operation. At the same time, the simulation results also show that the root mean square value, mean value and maximum value of the acceleration of the transverse vibration of the car and the car frame are reduced by more than 19% after using the robust controller. Therefore, the robust controller designed in this paper can effectively suppress the transverse vibration of the vibration car system.

1. Introduction

Nowadays, with the aggravation of social and economic development, population growth, land resource shortage and other problems, the number of high-rise and super-high-rise buildings is increasing, and the speed of elevator as a vertical transportation tool is also increasing. The results show that the transverse vibration acceleration of the car of a high-speed elevator running at a speed of more than 3m/s is significantly greater than that of a low-speed elevator [1]. The traditional method of vibration reduction is to install a spring between the car frame and the guide shoes [2]. This passive vibration reduction method is simple in structure, easy to realize, economical and reliable, and has satisfactory effect when the speed of the elevator is low, because it does not need external energy. However, with the increase of elevator speed, the limitation of passive vibration reduction is shown. For example, it is only suitable for external disturbance with constant frequency or small change, and it can only provide small vibration reduction force, so it is difficult to meet people's requirements. Active vibration control technology has become a new way to solve elevator vibration because of its good effect and adaptability. Many scholars have made a lot of research on it and achieved a lot of useful results [3-4]. In the research process, these scholars often regard the car and the frame as a whole of rigid connection, and then control the whole, but in the actual operation process of the elevator, the car and the frame are connected by the damping rubber block. With the increase of the elevator speed, the transverse relative displacement between the car and the car frame increases, at this time, the car and the car frame cannot be treated as a rigid body.

In view of the above shortcomings, this paper establishes the dynamic model of active transverse vibration control of high-speed elevator with car and car frame as two rigid bodies, and derives its state equation. The H_2 norm is used to minimize the vibration acceleration of the output car system. Based on LMI optimization technology, a robust controller for transverse vibration of high-speed elevator is designed, and the simulation analysis is carried out by using MATLAB.

2. Active Control Model for Transverse Vibration of High-Speed Elevator

This paper only considers the transverse vibration between the elevator car and the guide rail, that is, the vibration parallel to the direction of the car door. In order to simplify the model and conform to the actual operation conditions of the elevator as far as possible, the following reasonable assumptions are made according to the structure and movement law of the elevator.

(1) In order to simplify the calculation, it is assumed that the centroid of the car and the frame does not shift in the vertical direction during the discussion of the relative motion of the car and the frame, and the car and the frame are referred to as the car system.
(2) Because the roller is close to the guide rail and the relative displacement between them is small in the course of vibration, the guide shoe can be simplified as a mass spring damping system.
(3) The car is connected with the car frame by the damping rubber block. The damping rubber block is simplified as a spring damping system, and the structure and parameters of each damping rubber block are identical.
(4) The structure and parameters of each roller are identical.
(5) Consider only the incentive of guide rail roughness.

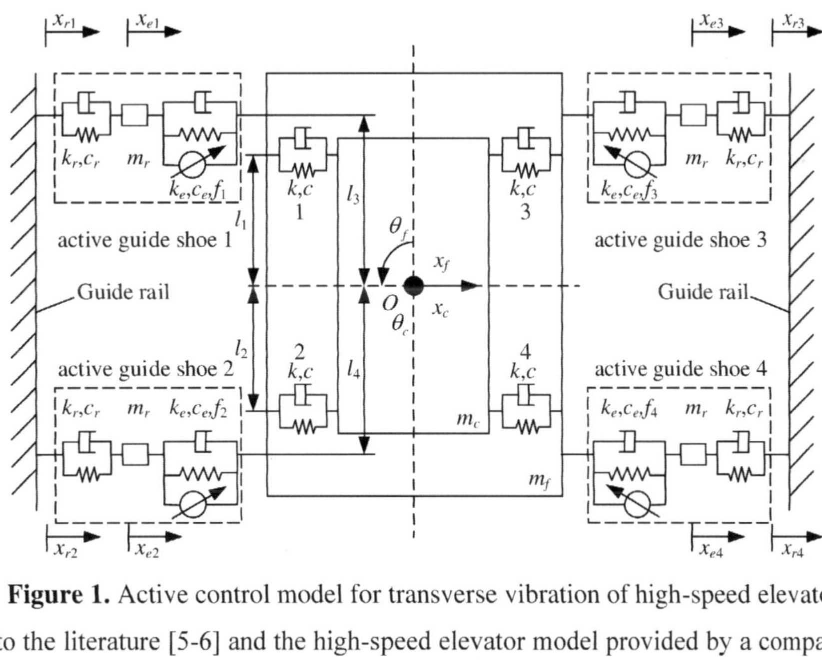

Figure 1. Active control model for transverse vibration of high-speed elevator

According to the literature [5-6] and the high-speed elevator model provided by a company, this paper establishes an active control model for the transverse vibration of the high-speed elevator car system. As shown in Fig. 1, with the center of mass O of the car and frame as the origin, and the x axis transversely, The rotation of the car around its centroid O is θ_c, and the rotation of the car frame around its centroid O is θ_f. In this paper, the vibrations of the car and the frame in the direction of x axis and the vibrations in the direction of rotation about its centroid are called transverse vibrations. At the same time, each guide shoe of the car system has its own degrees of freedom in the horizontal direction, that is, the elevator model has eight degrees of freedom. The mass of the elevator car is m_c, the mass of the elevator car is m_f. The moment of inertia of the car is J_c, the moment of inertia of the car frame is J_f. The horizontal displacement of the centroid of the car is x_c, the horizontal displacement of the centroid of the car frame is x_f, the horizontal displacement of the roller is x_{ei} (i=1,2,3,4). The stiffness coefficient and damping coefficient between roller and guide rail are k_r and c_r, respectively. The stiffness and damping coefficients between roller and car frame are k_e and c_e, respectively. The stiffness coefficient and damping coefficient of the rubber block connected between the car and the car frame are k and c, respectively. The vertical distances of the centroids of the active guide shoes 1, 3 and 2, 4 to the car system are l_3 and l_4, respectively. The vertical distances of the damping rubber blocks 1, 3 and 2, 4 to the car system center of mass are l_1 and l_2, respectively. The guide rail roughness excitation is x_{ri} (i=1,2,3,4). The active control force is f_i (i=1,2,3,4).

According to Lagrange equation, the dynamic equation of active control for transverse vibration of high-speed elevator is established,

$$\begin{cases} M_1\ddot{q} + C\dot{q} + C_1\dot{x}_e + Kq + K_1x_e = GF \\ M_2\ddot{x}_e + C_1^T\dot{q} + (C_e + C_r)\dot{x}_e + G_1C_r\dot{x}_r + K_1^Tq + (K_e + K_r)x_e + G_1K_rx_r = G_1F \end{cases} \quad (1)$$

where,

$q = \begin{bmatrix} x_c & x_f & \theta_c & \theta_f \end{bmatrix}^{\mathrm{T}}$, $x_e = \begin{bmatrix} x_{e1} & x_{e2} & x_{e3} & x_{e4} \end{bmatrix}^{\mathrm{T}}$, $x_r = \begin{bmatrix} x_{r1} & x_{r2} & x_{r3} & x_{r4} \end{bmatrix}^{\mathrm{T}}$, $\Delta l_1 = l_1 - l_2$,

$\Delta l_2 = l_3 - l_4$, $h_1 = l_1^2 + l_2^2$, $h_2 = l_3^2 + l_4^2$, $M_2 = diag(m_r, m_r, m_r, m_r)$, $C_e = diag(c_e, c_e, c_e, c_e)$,

$C_r = diag(c_r, c_r, c_r, c_r)$, $G_1 = diag(-1,-1,-1,-1)$, $K_e = diag(k_e, k_e, k_e, k_e)$, $K_r = diag(k_r, k_r, k_r, k_r)$

$$C_1 = \begin{bmatrix} 0 & 0 & 0 & 0 \\ -c_e & -c_e & -c_e & -c_e \\ 0 & 0 & 0 & 0 \\ c_el_3 & -c_el_4 & c_el_3 & -c_el_4 \end{bmatrix}, K_1 = \begin{bmatrix} 0 & 0 & 0 & 0 \\ -k_e & -k_e & -k_e & -k_e \\ 0 & 0 & 0 & 0 \\ k_el_3 & -k_el_4 & k_el_3 & -k_el_4 \end{bmatrix}, G = \begin{bmatrix} 0 & 0 & 0 & 0 \\ 1 & 1 & 1 & 1 \\ 0 & 0 & 0 & 0 \\ -l_3 & l_4 & l_3 & -l_4 \end{bmatrix},$$

$$M_1 = \begin{bmatrix} m_c & 0 & 0 & 0 \\ 0 & m_f & 0 & 0 \\ 0 & 0 & J_c & 0 \\ 0 & 0 & 0 & J_f \end{bmatrix}, C = \begin{bmatrix} 4c & -4c & -2c\Delta l_1 & 2c\Delta l_1 \\ -4c & 4(c_e + c) & 2c\Delta l_1 & -2c_e\Delta l_2 - 2c\Delta l_1 \\ -2c\Delta l_1 & 2c\Delta l_1 & 2ch_1 & -2ch_1 \\ 2c\Delta l_1 & -2c_e\Delta l_2 - 2c\Delta l_1 & -2ch_1 & 2c_eh_2 + 2ch_1 \end{bmatrix},$$

$$K = \begin{bmatrix} 4k & -4k & -2k\Delta_1 & 2k\Delta_1 \\ -4k & 4(k_e + k) & 2k\Delta_1 & -2k_e\Delta_2 - 2k\Delta_1 \\ -2k\Delta_1 & 2k\Delta_1 & 2kh_1 & -2kh_1 \\ 2k\Delta_1 & 2k_e\Delta_2 - 2k\Delta_1 & -2kh_1 & 2k_eh_2 + 2kh_1 \end{bmatrix}.$$

Select the state vector as $X = \begin{bmatrix} q & \dot{q} & x_e & \dot{x}_e \end{bmatrix}^T \in R^{16\times1}$ .Select the control input vector as $U = \begin{bmatrix} f_1 & f_2 & f_3 & f_4 \end{bmatrix}^T \in R^{4\times1}$. Select the interference input vector as $W = \begin{bmatrix} x_r & \dot{x}_r \end{bmatrix}^T \in R^{8\times1}$ .Therefore, the state space equation of the transverse vibration model of an eight-degree-of-freedom active control elevator car system is expressed as follows,

$$\dot{X} = AX + B_1U + B_2W \quad (2)$$

where,

$$A=\begin{bmatrix} 0_4 & I_4 & 0_4 & 0_4 \\ -M_1^{-1}K & -M_1^{-1}C & -M_1^{-1}K_1 & -M_1^{-1}C_1 \\ 0_4 & 0_4 & 0_4 & I_4 \\ -M_2^{-1}K_1^T & -M_2^{-1}C_1^T & -M_2^{-1}(K_e+K_r) & -M_2^{-1}(C_e+C_r) \end{bmatrix}, B_2=\begin{bmatrix} 0_4 & 0_4 \\ 0_4 & 0_4 \\ 0_4 & 0_4 \\ -M_2^{-1}G_1K_r & -M_2^{-1}G_1C_r \end{bmatrix},$$

$$B_1=\begin{bmatrix} 0_4 & M_1^{-1}G & 0_4 & M_2^{-1}G_1 \end{bmatrix}^{\mathrm{T}}.$$

The working principle of the active guide shoe in the current active control technology: the vibration of the car system can be suppressed by the force opposite to the roller displacement generated by the actuator on the active guide boot. The evaluation index of elevator ride comfort is the vibration acceleration of car system. Therefore, the transverse vibration acceleration of car system is chosen as the evaluation index of active control effect. So, select the output vector as $y=\begin{bmatrix} \ddot{x}_c & \ddot{x}_f & \ddot{\theta}_c & \ddot{\theta}_f \end{bmatrix}^{\mathrm{T}}$, that is,

$$y=C_aX+DU \tag{3}$$

where,

$$C_a=\begin{bmatrix} -M_1^{-1}K & -M_1^{-1}C & -M_1^{-1}K_1 & -M_1^{-1}C_1 \end{bmatrix}, D=-M_1^{-1}G.$$

3. Robust Controller Design

The root mean square value of vibration acceleration is often used in engineering as an index to evaluate the vibration acceleration of car system. According to the physical meaning of H_2 norm, when the input signal of the system is independent of each other, the H_2 norm is the root mean square (RMS) value of the system output [7]. In this paper, the excitation disturbance input is random excitation and independent of each other. For the performance output y_1, the H_2 norm is used to describe, and the output is the root mean square value of the output y_1. When G_1 is used to represent the transfer function from $W(t)$ to $y_1(t)$, the active control problem in this paper can be described as follows,

$$\min\|G_1\|_2 \tag{4}$$

If the state feedback gain of the system is K_2, then the control input is $U=K_2X$. Therefore, the elevator system in this paper can be expressed as follows,

$$\begin{cases} \dot{X}(t)=AX(t)+B_1U(t)+B_2W(t)=(A+B_1K_2)X(t)+B_2W(t) \\ y_1(t)=C_aX(t)+DU(t)=(C_a+DK_2)X(t) \end{cases} \tag{5}$$

that is,

$$\begin{pmatrix} A_{cl} & B_{cl} \\ C_{cl1} & D_{cl1} \end{pmatrix}=\begin{pmatrix} A+B_1K_2 & B_2 \\ C_a+DK_2 & 0 \end{pmatrix} \tag{6}$$

In the case of multi-objective control, the linear time invariant system has a conclusion [8]: If the system state matrix A_{cl} is stable, and $\|G_1\|_2^2<v$, if and only if there exists $P=PT>0$, $S>0$, such that,

$$\begin{pmatrix} A_{cl}^TP+PA_{cl} & PB_2 \\ B_2^TP & -I \end{pmatrix}<0, Trace(S)<v, D_{cl1}=D_{cl2}=0, \begin{pmatrix} P & C_{cl1}^T \\ C_{cl1} & S \end{pmatrix}>0 \tag{7}$$

Let $Q=P$-1 and $Y=KQ$, using S-Procedure and Schur complement formula to get LMI of v, Q, Y, S:

$$\begin{pmatrix} QA^T + AQ + Y^T B_1^T + B_1 Y & B_2 \\ B_2^T & -I \end{pmatrix} < 0 \,, Trace(S) < v \,, D_{cl1} = D_{cl2} = 0 \,, \begin{pmatrix} Q & QC_a^T + Y^T D^T \\ C_a Q + DY & S \end{pmatrix} > 0 \quad (8)$$

Therefore, the robust controller of the high-speed elevator described in this paper can be described as,

$$\text{Solving } \min_{v,Q,Y,S} v \text{ satisfies the above LMI.} \quad (9)$$

4. Solution and Simulation of Robust Controller for High-Speed Elevator

4.1. Solution of State Feedback Matrix

The parameters of 4m/s high-speed elevator are shown in Table 1. The LMI solution toolbox (feasp) in MATLAB is used to solve the system state feedback matrix K_2.

$$K_2 = \begin{bmatrix} K_{21} & K_{22} & K_{23} & K_{24} \end{bmatrix} \quad (10)$$

where,

$$K_{21} = \begin{bmatrix} 4.7382e5 & -5.1007e5 & 1.2773e5 & -1.3013e5 \\ 4.7382e5 & -5.1007e5 & -1.2773e5 & 1.3013e5 \\ 4.7378e5 & -5.1003e5 & 1.0379e6 & -1.0943e6 \\ 4.7378Ee5 & -5.1003e5 & -1.0379e6 & 1.0943e6 \end{bmatrix},$$

$$K_{22} = \begin{bmatrix} 2.8781e2 & -4.3483e2 & -6.9734e2 & 7.5525e2 \\ 2.8781e2 & -4.3483e2 & 6.9734e2 & -7.5525e2 \\ 2.8780e2 & -4.3480e2 & 1.4360e3 & -1.7854e3 \\ 2.8780e2 & -4.3480e2 & -1.4360e3 & 1.7854e3 \end{bmatrix},$$

$$K_{23} = \begin{bmatrix} 1.4181e4 & 8.2959e3 & 2.1706e4 & 7.9235e3 \\ 8.2959e3 & 1.4181e4 & 7.9235e3 & 2.1706e4 \\ 1.2867e4 & 1.6761e4 & 2.1485e4 & 9.8989e2 \\ 1.6761e4 & 1.2867e4 & 9.8989e2 & 2.1485e4 \end{bmatrix}, K_{24} = \begin{bmatrix} 1.0595 & -9.4490e-1 & -2.2619e-1 & 6.2148e-1 \\ -9.4490e-1 & 1.0595 & 6.2148e-1 & -2.2619e-1 \\ 2.4508 & -2.0556 & 1.6155 & -1.5008 \\ -2.0556 & 2.4508 & -1.5008 & 1.6155 \end{bmatrix}.$$

Table 1. Simulation parameters of elevator car system.

Parameters	Numerical value	Parameters	Numerical value
Car mass m_c	**1.2e3 kg**	Damping coefficient c_e	1.2e2 N·s/m
Car moment of inertia J_c	**1.3e3 kg·m^2**	Stiffness coefficient k	5.0e5 N·m^{-1}
Car frame mass m_f	**7.5e2 kg**	Damping coefficient c	3.2e2 N·s/m
Car frame moment of inertia J_f	**3.0e3 kg·m^2**	Distance l_1	1.2 m
Roller mass m_r	**3.1 kg**	Distance l_2	1.2 m
Stiffness coefficient k_r	**7.0e5 N·m^{-1}**	Distance l_3	1.6 m
Damping coefficient c_r	**1.34e2 N·s/m**	Distance l_4	1.6 m
Stiffness coefficient k_e	**5.0e4 N·m^{-1}**	Lifting speed V	4 m·s^{-1}

4.2. Simulation Experiment

The disturbance excitation involved in this paper is the guide rail unevenness excitation. The research shows that the surface unevenness of the guide rail obeys the normal distribution with zero expectation

[9]. The simulation experiment is carried out by using MATLAB. The vibration acceleration response of the car system under the action of uncontrolled and robust controller designed in this section is compared and analyzed, and the validity of the robust controller designed in this paper is verified.
State equation of car system:

$$\begin{cases} \dot{X} = AX + B_1U + B_2W \\ y_1 = C_N X \end{cases} \tag{11}$$

where, $U=K_2X$, $C_N=C_a+DK_2$.
The equation (11) can be discretized by MATLAB:

$$\begin{cases} X(k+1) = A_d X(k) + B_{d1}U(k) + B_{d2}W(k) \\ y_1 = C_d X(k) \end{cases} \tag{12}$$

where, A_d, B_{d1}, B_{d2}, C_d are matrices with the same dimension as A, B_1, B_2, C_N.

4.2.1. Analysis of Simulation Results of Transverse Vibration of Car. The transverse vibration acceleration of the car's centroid position under the same random excitation is controlled by No Control and robust controller designed in this paper. The transverse vibration acceleration image of the car is shown in Figure 2.
The calculation results of root mean square, mean and maximum transverse vibration acceleration of elevator car are shown in Table 2. The contrast analysis shows that after adopting the robust control, the root-mean-square value, the mean value and the maximum value of the transverse vibration acceleration of the car are 25.8%, 26.9% and 19.0% lower than the root-mean-square value, the mean value and the maximum value of the transverse vibration acceleration of the car by no control, respectively. The results show that the robust controller designed in this paper can effectively control the transverse vibration of the car.

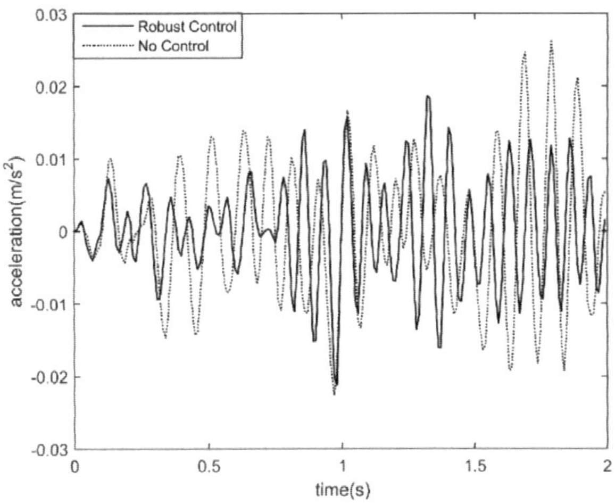

Figure 2. Comparative diagram of transverse vibration acceleration of elevator car

Table 2. Acceleration parameters of car transverse vibration.

Elevator Car	No Control	Robust Control
RMS	0.0097	0.0072
Mean value	0.0078	0.0057
Maximum value	0.0263	0.0213

4.2.2. Analysis of Simulation Results of Transverse Vibration of Car Frame. No control and robust controllers are used to control the transverse vibration acceleration of the centroid position of the car frame under the same random excitation respectively. The transverse vibration acceleration image of the car frame is shown in Figure 3.

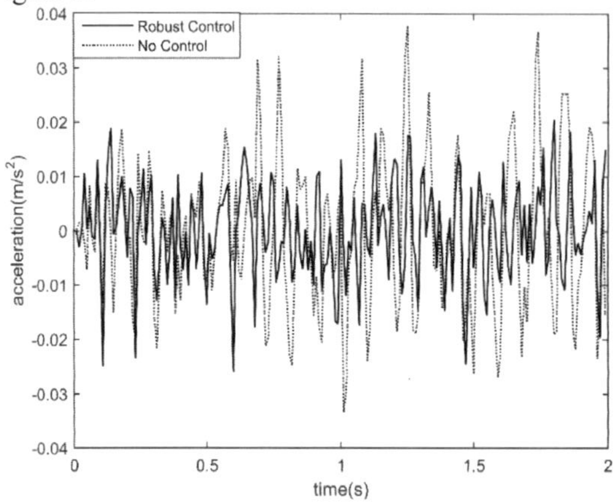

Figure 3. Comparative diagram of transverse vibration acceleration of elevator car frame

The calculation results of root mean square, mean and maximum transverse vibration acceleration of elevator car frame are shown in Table 3. The contrast analysis shows that after adopting the robust control, the root-mean-square value, the mean value and the maximum value of the transverse vibration acceleration of the car frame are 34.5%, 34.8% and 31.5% lower than the root-mean-square value, the mean value and the maximum value of the transverse vibration acceleration of the car frame by no control, respectively. The results show that the robust controller designed in this paper can effectively control the transverse vibration of the car frame.

Table 3. Acceleration parameters of car transverse vibration.

Elevator Car	No Control	Robust Control
RMS	0.0139	0.0091
Mean value	0.0112	0.0073
Maximum value	0.0378	0.0259

5. Conclusion

(1) In view of the relative displacement between the elevator car and the frame during the actual operation of the elevator, this paper establishes the active control model of the transverse vibration of the high speed elevator, which is in accordance with the actual operation of the elevator. The H_2 norm is used to minimize the output acceleration of the car vibration. Based on the LMI optimization technology, the robust controller of the high-speed elevator is designed and solved. The validity of the designed robust controller is verified by the MATLAB simulation.

(2) Through the simulation results, it can be seen that the root-mean-square value, the mean value and the maximum value of the transverse vibration acceleration of the car are smaller than the root mean value, the mean value and the maximum value of the transverse vibration acceleration of the car frame. The conclusion that the vibration of the car and the frame is inconsistent in the actual process is drawn, which further proves the correctness of the model established in this paper.

Acknowledgments
This study was funded by the Shandong Province Nature Science Foundation, China (GRANT NO. ZR2017MEE049), the Key Research Development Project of Shandong Province (GRANT NO. 2018GSF122004).

IOP Conf. Series: Materials Science and Engineering **538** (2019) 012032 doi:10.1088/1757-899X/538/1/012032

References

[1] Q. Zhang, Z. Yang, C. Wang, Y.H. Yang, R.J. Zhang, P I MECH ENG C-J MEC. (to be published)
[2] Y.H. Feng, J.W. Zhang, Chinese Journal of Mechanical Engineering. 20,68~73 (2007)
[3] Y.H. Feng, Journal of Mechanical Strength. 15,1375-1396 (2008)
[4] L. Wang, Mechanical Design and Manufacturing,178-180 (2011)
[5] Q. Zhang, Y.H. Yang, Periodica Polytechnica Mechanical Engineering, 61 (2017).
[6] S.H. Zhang, R.J. Zhang, Q. He, et al., ARCH APPL MECH. (to be published).
[7] J.H. Kim, T. Hagiwara, INT J CONTROL. 90, 1-23 (2016).
[8] H. Chen, M.M. Ma, P.Y. Sun, Control Theory and applications. 24, 790-794 (2007)
[9] H.L Wang, Shanghai Metrology. 43, 18-22 (2016).

IOP Conf. Series: Materials Science and Engineering **538** (2019) 012033 doi:10.1088/1757-899X/538/1/012033

Modeling Automation of Space Payload Experimental Cabinet Based on Secondary Development of Hypermesh

Sinian Huang[1,2,a] and Anping Wang[2]

[1] University of Chinese Academy of Science, Beijing 100049, China
[2] Technology and Engineering Center for Space Utilization, Chinese Academy of Science, Beijing 100094, China

[a] Corresponding author: huangsinian16@csu.ac.cn

Abstract. For the repeatability of the finite element analysis process of the electronic cabinets, thermal control units and other equipment inside the space payload experimental rack, this paper developed a modeling automation system for the cabinets based on the secondary development function of HyperMesh. The system uses Nastran as a solver, and comprehensively use Tcl/Tk language, HyperMesh API function, Process Studio process editor for development. Details the technology and processes in the development process of HyperMesh, providing a new method for the finite element modeling analysis of the space payload experimental cabinets and improving the simulation efficiency.

1. Introduction

The space scientific experimental rack is an important structure for installing payloads in the capsules of the spacecraft and an important part of the space application system [1]. The experiment cabinet contains modules such as electronic chassis, thermal control drawer, and electronic control unit, these modules are part of the space scientific experimental rack, a large number of finite element simulation analysis is needed in the design and manufacturing process to provide a basis for the test of the laboratory cabinet, shorten the development cycle and save costs [2]. In the finite element analysis, the main types of analysis includes static analysis, modal analysis, harmonic response analysis, transient response analysis, etc. The analysis process is basically the same. The pre-processing of finite element modeling analysis of space payload cabinets is usually carried out in HyperMesh. The main processes include geometric model import, geometric cleaning, meshing, material attribute assignment, unit attribute assignment and boundary load setting, etc. The model is established and the finite element solution file is output and imported into the corresponding solver for calculation.

The finite element analysis of the space payload cabinets has many repetitive tasks, large workload, the content is tedious and error prone. Based on the secondary development function of HyperMesh, the corresponding modeling automation system is developed. The system uses Nastran as the solver, which can greatly improve the finite element modeling efficiency of the cabinets model, reduce a lot of repetitive work, and avoid mistakes in manual modeling. HyperMesh is a finite element pre-processing software developed by Altair Corporation of the United States. It is widely used in the field of finite element analysis and has strong meshing ability [3], which can provide users with high quality and efficient meshing technology. HyperMesh allows users to use the Tcl (Tool Command Language) and API functions for secondary development, users can write scripts to perform specific functions. Nastran software is a series of products of MSC in the United States, it is widely used in the aerospace field for simulation analysis due to its high precision and rapid convergence in finite element analysis.

Based on the secondary development function of HyperMesh, and aiming at the finite element analysis process of the space payload cabinets, taking the Process Studio module in HyperWorks as the development environment and using Tcl/Tk and HyperMesh API function[4], a set of space payload cabinets modeling automation system using Nastran as the solver was developed. The key technologies and theories in the secondary development process of HyperMesh are introduced in detail, and the finite element analysis process of the cabinet is automated and streamlined.

2. Development of Space Payload Cabinet Modeling Automation System

2.1. The Secondary Development Technology of HyperMesh

HyperMesh secondary development uses Tcl/Tk as its standard development language, similar to C++ and Java, Tcl is a scripting language [5], with fixed syntax, good readability and no need to compile. It is executed at run time through the interpreter for sequential interpretation of each statement. Tk is Tcl's graphical interface toolkit, it is a set of tools for developing graphical interface applications, where users can create interface components that trigger corresponding command functions and implement corresponding functions [6]. The TcL/Tk can be used as a stand-alone application, and the corresponding functions can be easily implemented and called by writing scripts[7].

HyperMesh provides users with a large number of functions, through which the corresponding functions in HyperMesh can be achieved. In the secondary development process of HyperMesh, two commom functional functions are mainly used, one is Tcl Modify Commands, such as *defaultremeshsurf command function can be used for 2D mesh division; the other is Tcl Query Commands, such as hm_entityinfo can perform attribute query. With the help of the HyperMesh API function, the corresponding secondary development function can be completed.

2.2. Modeling Automation System Development Method

When performing finite element analysis of equipment such as space payload cabinets, it is necessary to clearly analyze the requirements and arrange the analysis process reasonably so that the analysis process is as standardized and efficient as possible. First, select a simple model for finite element modeling analysis, and fully consider the various problems encountered in the modeling process, and ensure the correctness and standardization of manual modeling. During the HyperMesh operation, each step of the operation commands is recorded in the Command.cmf file. The Command file is a standard ASCII file that records a series of operational commands during operation, the command stream strictly follows the HyperMesh function format. Through the finite element modeling analysis of the simple model, for each operation, the valid commands are extracted, and the command function is modified to conform to the Tcl/Tk syntax.

The process automation template is established through the Process Studio, the finite element modeling analysis process and the system operation interface are designed, and the corresponding Tcl code and function are written for the function corresponding to each node in the process tree. Process Studio is a process development tool in HyperWorks that provides users with a visual process editing environment, by writing a process tree to create a corresponding finite element modeling analysis process, the secondary development process is organized. Each node in the process tree has a specific process corresponding to it, by writing a corresponding operation interface, it can form an interaction with the user, which greatly facilitates the user's operation. The well-written modeling automation system is called by Process Manager in HyperMesh, the user can input the corresponding data according to the process prompt to complete the finite element analysis process, which greatly improves the efficiency of finite element analysis. As shown in Figure 1, it is the secondary development method flow of the modeling automation system.

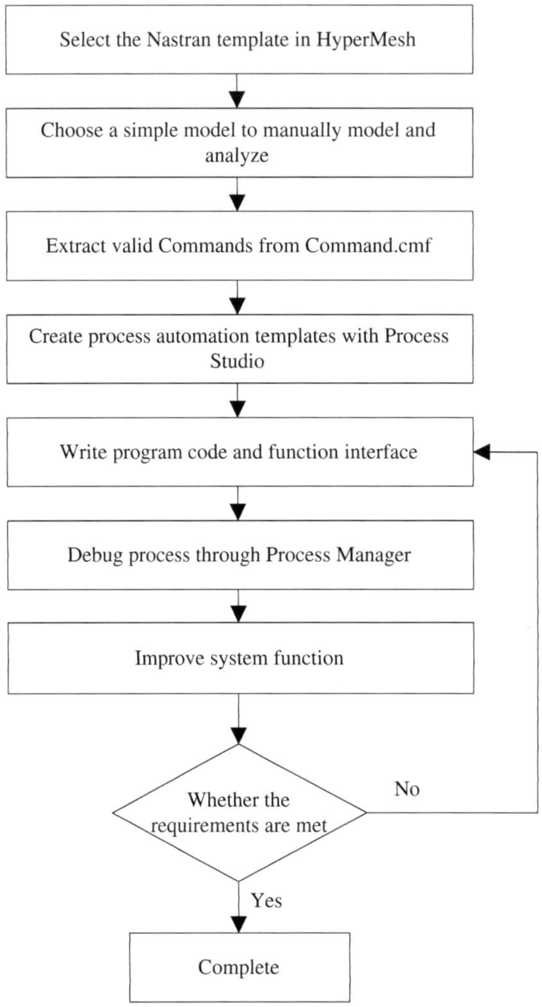

Figure 1. Modeling automation system secondary development method

2.3. Create Process Automation Templates with Process Studio

Manually complete the establishment of the finite element model and extract the valid command functions in the Command.cmf file. Create a process automation template through the Process Studio and add a Task to each function node in the modeling analysis in Process Studio, each node has a corresponding GUI interface. Each of the individual Tasks is written separately to performs a specific function.

Process Studio provides users with different utils control types, users can choose different types of controls for GUI interface design. There are two ways to implement data connections between utils and Tcl scripts. One way is to implement specific function directly by Tcl Scripting, for example, to get a data named "Text field" text box, you can use program statements: set strValue [::hwpmgr::PmgrGetData 0 "Text field.value"]; Another way is to directly call the Tcl program file that has been written through the source command.

3. Application of Modeling Automation System

For the most common electronic cabinets in the space scientific experimental rack, different modeling automation systems were developed according to the different types of finite element analysis requirements of the cabinets. For the common electronic cabinets modal analysis process, based on

IOP Conf. Series: Materials Science and Engineering **538** (2019) 012033 doi:10.1088/1757-899X/538/1/012033

HyperMesh's secondary development function, the modeling automation system using Nastran as the solver was developed by Process Studio to conduct modal analysis.

3.1. Get Command Stream

The command stream in the Command.cmf file is obtained by manual modeling and modified to conform to the Tcl/Tk syntax. The command stream includes automatic import of geometric models, automatic creation of connections, automatic addition of washers to holes and automatic creation of materials, etc. Some of the command streams are shown below.

```
# Import .stp geometry model file
*feinputwithdata2 "#ct/step_reader"
# Add a washer to the selected hole
*trim_by_offset_edges 0 1 $r 0 0 0
# Meshing and the grid order is 1, the size is $ELE_SIZE
*defaultremeshsurf 1 $ELE_SIZE
# Set the material property to MAT1
*dictionaryload materials 1 "D:/Program Files/ templates/feoutput/nastran/general" "MAT1;
# Define unit property
*dictionaryload properties 2 "D:/Program Files/ templates/feoutput/nastran/general"
# Set the modal analysis card and output the first 10th order frequency
*dictionaryload loadcols 2 " D:/Program Files/ templates/feoutput/nastran/general" "EIGRL";
*createmark loadcols 1 " EIGRL"
*attributeupdateint loadcols 2 804 1 1 0 10
……
# Output bdf and hm files
*feoutputwithdata
*writefile
```

3.2. Create Analysis Process and GUI Interface

According to the modal analysis process, Process Studio is used to establish the process automation template of the cabinets modal analysis to develop the modeling automation system. Figure 2 shows the electronic cabinets analysis system process by Process Studio.

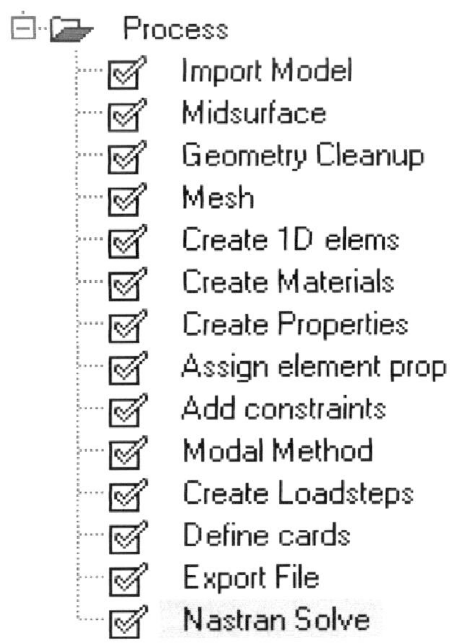

Figure 2. Electronic cabinets analysis system process

In the process structure, each node corresponds to a specific function, after the user completes the operation of a function node, the system automatically jumps to the next function node to guide the user to complete the analysis process. Reducing the number of user switching interfaces and improving the analysis efficiency. Through this modal analysis analysis process, users can perform operations such as geometric model input, automatic creation of connections and meshing.

In the process structure, each function node has a corresponding function operation interface, the user can complete the data input and other operations through the designed function interface. As shown in Figure 3, it is the meshing interface of the modeling automation system. After selecting the specified area and inputting the grid size, two-dimensional grid can be automated, and the grid quality can be checked. It avoids the repeated switch of interface and improves the efficiency of meshing.

After the meshing is completed, material creation and attribute assignment are performed, as shown in Figure 4, which is a material creation interface in the automatic modeling system. In the material creation process, the user only needs to input the material, name, elastic modulus, Poisson's ratio and density, the materials can be created automatically, which greatly simplifies the process.

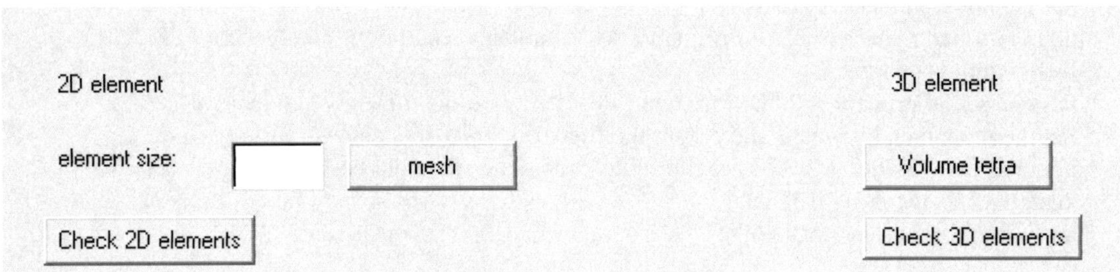

Figure 3. Meshing interface

Figure 4. Material creation interface

In the development process of the modeling automation system, for some functional interfaces that need to be repeatedly switched, it is inconvenient to write the function of the program, and the original function panel can be directly called. For example, if call the command to add washer in HyperMesh, it can use the hm_setpanelproc {hm_callpanel "quick edit"} to call the original function panel directly.

3.3. Application of Modeling Automation System

The modal analysis of the space payload cabinet is carried out by modeling automation system, setting the User Profiles to Nastran in HyperMesh, and calling the system through the Process Manager, the payload modal analysis process can be performed.

Taking the payload cabinet shown in Figure 5 as an example, the 3D geometric model of the cabinet is imported into HyperMesh. The user can be complete to operation of each function node according to the guidance of the modeling automation system, and the entire element analysis process can be successfully completed. According to the system's prompt, the finite element model file and the bdf file with Nastran as the solver can be exported. As shown in Figure 6, the finite element model is established by the cabinet modal analysis system.

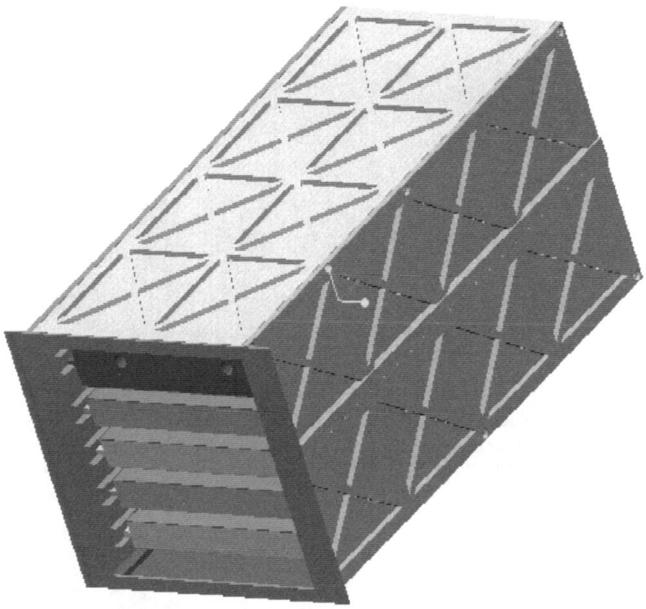

Figure 5. Geometric model of the space payload cabinet

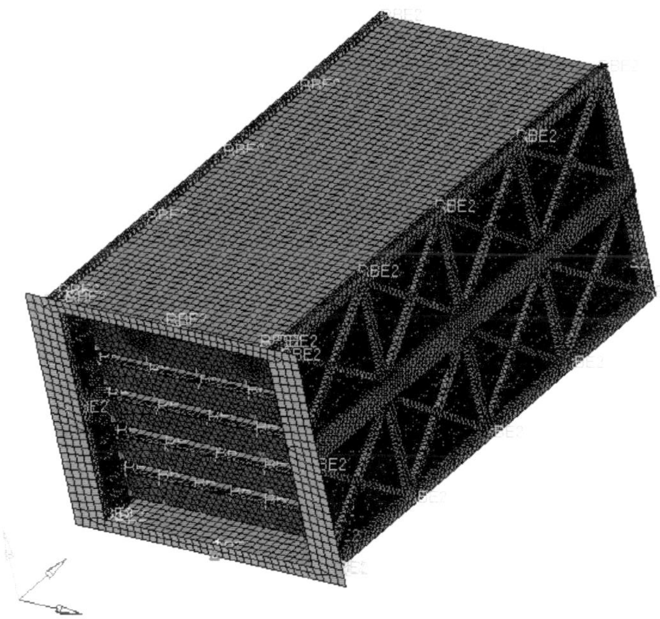

Figure 6. Finite element model of the space payload cabinet

By directly calling solving software Nastran, the finite element model can be automatically solved, and the solution results are saved in the corresponding folder. Next the payload cabinet is post-processed after the solution is completed, the frequency of modal analysis under fixed constraints can be obtained. As shown in Figure 7, it is the first four order modal shapes of the payload cabinet.

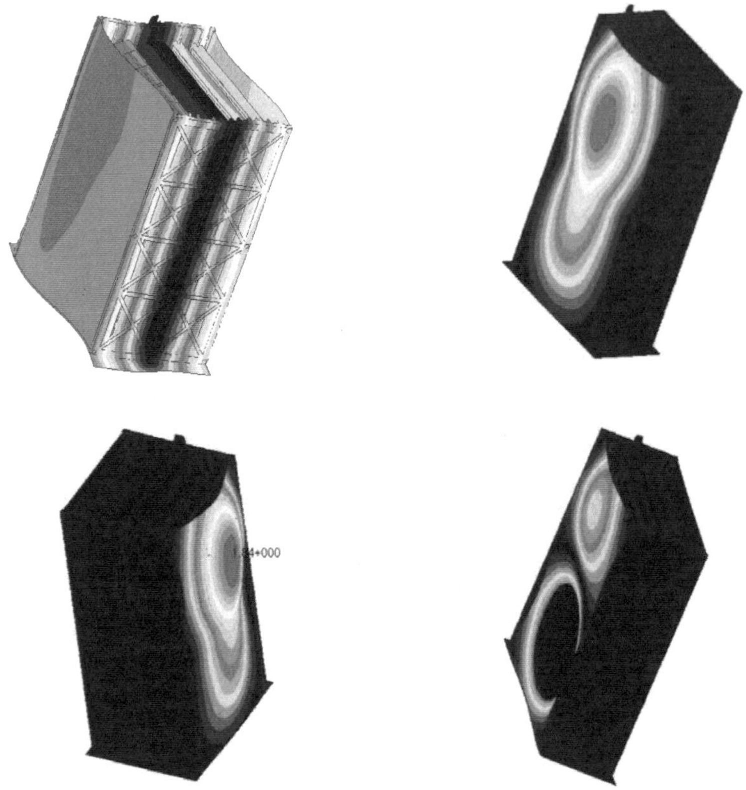

Figure 7. The first four order mode shapes

4. Concluding Remarks

This paper introduces the development method and key technology of space payload cabinets modeling automation system based on secondary development of HyperMesh. For the modules with similar geometry and consistent analysis of the space payload cabinets, comprehensively use Tcl/Tk language, HyperMesh API function, Process Studio to develop the modeling automation system, which provide a new method for finite element analysis of cabinets. The specific application examples are given to verify the feasibility and effectiveness of the method. The flow chart and program of the finite element analysis of the cabinets model are realized, and the modeling efficiency is improved.

References

[1] Wang Dapeng. Optimal Design of Payload Cabinet Structure[J]. Spacecraft Engineering, 2010, 19(2):34-39.

[2] Ding Tao, Peng Wang. Application of HyperMesh Secondary Development in Pre-processing of Bus Finite Element Analysis[J]. Bus Technology and Research, 2015,6:47-49.

[3] Zhu Feng, Li Shuxiao. Modal Analysis of Vehicle Exhaust System by Hypermesh and Workbench[J]. MECHANICAL ENGINEERING & AUTOMATION, 2014,1:62-64.

[4] Sun Jing, Huang Xuefei, et al. Develpment and application of Process Automation Systems based on HyperWorks[J]. RAILWAY COMPUTER APPLICATION, 2012,21(12):30-33.

[5] Yang Jindian. Automated Modeling Procedure of Semi-steel Tire in Finite Element Simulation[J]. Jinan : SHANDONG UNIVERSITY, 2015.

[6] Ousterhout J K, Jones K. Tcl/Tk Introduction to Classic[M]. 2nd Edition. Zhang Yuanzhang, Trans. Beijing : Tsinghua University Press, 2010 : 3-6.

[7] Lu Shanbin, Lv Jie, Chen Wei, et al. Finite Element Fast Modeling of Clinching Joints Based on HyperMesh Secondary Development[J]. JOURNAL OF GRAPHICS, 2014, 35(5):804-808.

[8] Liu Jianxiao, Li Bin, Du Chong. Modeling and optimization analysis on aircraft empennage composite based on secondary development of HyperMesh[J]. Computer Aided Engineering, 2016, 25(3):29-33.

[9] Narasimaha Rao, Vamsi Krishna. Automation for Material and Property Values in HyperMesh Using TCL[J]. Simulation Driven Innovation, 2012.

[10] JERZY KRUKOWSKI, Application of the Rigid Finite Element Method for Modeling an Offshore Pedestal Crane[J]. THE ARCHIVE OF MECHANICAL ENGINEERING, 2013.

[11] Faycal Benyahia, Abdulmohsen Albedha. Elliptical and Circular Bonded Composite Repair under Mechanical and Thermal Loading in Aircraft Structures[J]. Materials Research, 2014, 17(5): 1219-1225.

[12] Marc Alexander Schweitzer. Generalizations of the Finite Element Method. Central European Journal of Mathematics, 10(1), 2012, 3-24.

IOP Conf. Series: Materials Science and Engineering **538** (2019) 012034 doi:10.1088/1757-899X/538/1/012034

Study on Transverse Vibration Characteristics of High-Speed Elevator under Gas-Solid Coupling

Jie Liu, Ruijun Zhang[a] and Qing Zhang

School of Mechanical and Electronic Engineering, Shandong Jianzhu University, Ji'nan, Shandong 250101, China

Corresponding author: zhangruijun@sdjzu.edu.cn

Abstract. The purpose of this paper is to study the transverse vibration characteristics of high-speed elevator under the action of gas-solid coupling. A three-dimensional physical model of the high-speed elevator is established, the aerodynamic loads on the car under different transverse vibration states (translation and rotation) are numerically solved, and then the relationship between the different transverse vibration states of the car and the aerodynamic loads is obtained. The transverse vibration model of the high-speed elevator based on gas-solid coupling is established by transferring the attitude parameters and aerodynamic parameters of the car in real time combined with the guide excitation. Finally, the transverse vibration characteristics of high-speed elevators with or without gas-solid coupling are compared and analysed by solving the model with an example. The results show that the effect of gas-solid coupling on the transverse vibration acceleration peak value of high-speed elevator is greater when the running speed is greater than 4m/s, and the greater the speed, the more obvious the effect is, which provide some theoretical guidance for the analysis and control of the transverse vibration of high-speed elevator.

1. Introduction

With the increasing number of high-rise buildings and the rapid development of science and technology, the elevator is developing in the direction of high-efficiency, safety, comfort and high-speed. Compared with the low-speed and medium-speed elevator, the drastic fluctuation of ambient airflow caused by high-speed elevator in the process of high-speed operation in narrow and long shaft has become the main factor affecting its vibration, especially the transverse vibration, which is also an aspect that needs to be improved urgently to meet the high-speed elevator's superior performance [1]. Good system dynamics and aerodynamics performance are the preconditions to ensure the safe operation and ride comfort of high-speed elevator, and a large number of scholars have carried out relevant research. Considering the deviations of the guide rail and guide shoe, Mei [2] fitted the mathematical model of their irregular excitation, and analysed the horizontal vibration response of high-speed elevator with the numerical method; Zhang [3] studied the transverse compound random vibration of the high-speed elevator by considering the nonlinearity of guide shoe and the random unevenness excitation of the guide rail; Santo[4] studied the transverse non-linear response of high-speed elevator under the deformation of guide rail, and proposed a control strategy for it. In the research on aerodynamics, Li [5] studied the aerodynamic characteristics of high-speed elevator by using CFD numerical calculation method, and put forward the specific optimization schemes accordingly; Aiming at different hoistway models, Chen [6] studied the aerodynamic characteristics of the flow field of super high-speed elevator by using finite element analysis method. It can be seen from the above literatures that the current studies of the dynamics (transverse vibration characteristics) and aerodynamic characteristics of high-speed elevator system are mostly carried out in isolation. With the continuous improvement of elevator speed, the factors neglected in the analysis of transverse vibration characteristics of low-speed and medium-speed elevator need to be reconsidered again. The high-speed elevator runs in a relatively closed hoistway depending on its unique wheel-rail restraint relationship, its motion state (including vibration) changes the airflow field around the car at all times, and at the same time, the airflow affects the response of the car system. That is, the car system and the airflow in the hoistway are an organic coupling system with mutual connection, interaction and energy transfer, and the faster the elevator runs, the more prominent the coupling effect between them will be. Nowadays, with the increasing speed of the elevator, the safety and comfortable operation of existing high-speed elevator and the development of new high-speed elevator need to fully understand their vibration characteristics in real service environment. Therefore, it becomes more and more important to study the transverse vibration characteristics of high-speed elevator under the action of gas-solid coupling.

In this paper, a three-dimensional physical model of the high-speed elevator is established, the aerodynamic loads on the car under different transverse vibration states (translation and rotation) are numerically solved, and the relationship between the aerodynamic loads and the vibration states is analysed. Combined with the guidance system excitation, the transverse vibration model of high-speed elevator is established by real-time transmission of the attitude parameters and aerodynamic parameters of the car, and the coupling effect between the car system and the airflow in the hoistway is realized. Finally, the model is solved by an example, and the transverse vibration characteristics of the high-speed elevator with or without the consideration of gas-solid coupling are compared and analysed.

2. Aerodynamic Loads Analysis

2.1. Analysis of Airflow Field in Hoistway

During the operation of the high-speed elevator, the airflow in the hoistway presents a state of low Mach number and high Reynolds number, which can be regarded as the turbulent motion of incompressible gas. Regardless of energy exchange, the finite volume method of computational fluid dynamics (CFD) is used for analysis. According to the continuity equation, momentum equation and standard k-ε turbulence model, the integral control equation for each control volume of the gas flow in the hoistway is established as follows:

$$\frac{\mathrm{d}}{\mathrm{d}t}\int_{CV}\rho\phi \mathrm{d}V + \int_{A}(\rho\phi\mathbf{u} - \Gamma_{\phi}\mathrm{grad}\phi)\mathrm{d}A = \int_{CV} S_{\phi}\mathrm{d}V \tag{1}$$

where, t is time, ρ is air density, ϕ is generic variable, $\mathbf{u}$ is airflow velocity, Γ_{ϕ} is generalized diffusion coefficient, S_{ϕ} is generalized source term, grad is gradient, CV is control volume, and A is area surrounding the control volume.

2.2. Solution and analysis of aerodynamic loads

The transverse vibration state of the high-speed elevator includes translation and rotation around the car centroid, and different vibration state produces different aerodynamic loads in the hoistway [7]. Aiming at the kind of high-speed elevator whose car size is 1.7×1.7×2.4m and cross section is 2.1×2.1m, this paper establishes its three-dimensional physical model, ignoring the influence of the weight and the overall flow of airflow in the hoistway, defines the hoistway and the lower surface as the boundary of pressure inlet and outlet respectively, and uses tetrahedral mesh to mesh the calculation area.

Because the influence of the car translation on the aerodynamic load is smaller than that of the rotation [7], this paper only considers the influence of the car rotation on the aerodynamic load considering the calculation efficiency. In order to ensure the accuracy of the simulation results, the car rotation angles are taken as 0.5π/180 rad, π/180 rad, 1.5π/180 rad, 2π/180 rad and 2.5π/180 rad, which are combined with the running speed of 4m/s, 6m/s, 8m/s and 10m/s respectively. The transverse forces and torques acting on the car centroid under different rotation angles and running speeds are calculated, and the transverse force and torques of the car in a state of uniform speed are certain values. The calculation results are shown in Figures 1 and 2.

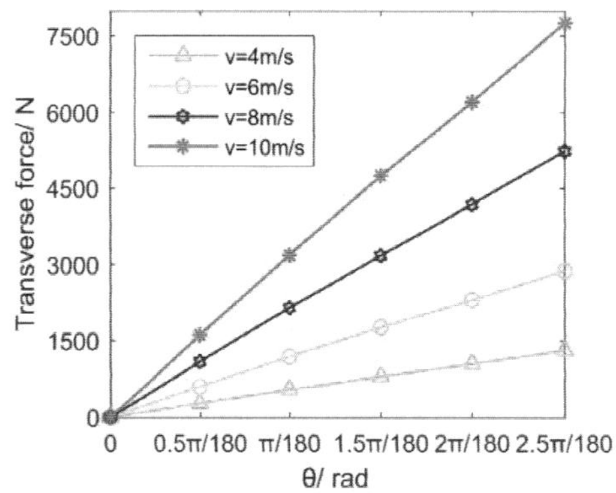

Figure 1. Transverse force generated by car rotation.

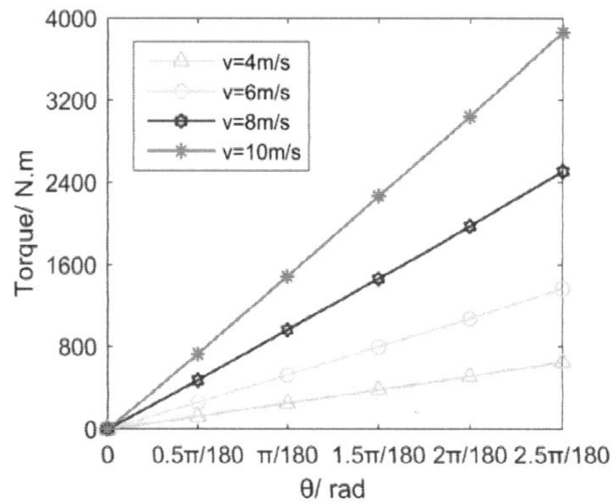

Figure 2. Torque generated by car rotation.

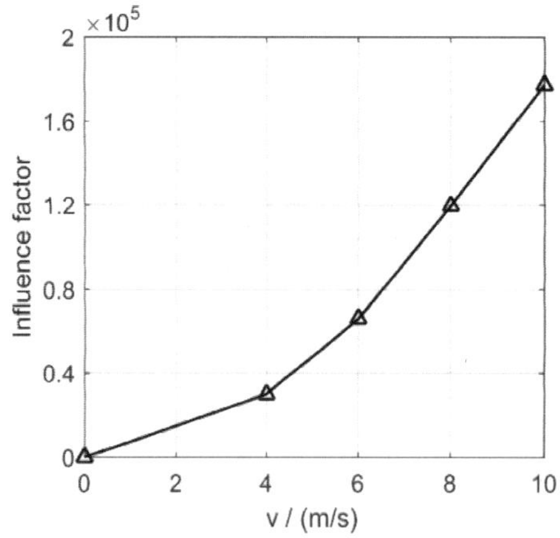

Figure 3. Influence factor of car rotation on transverse force.

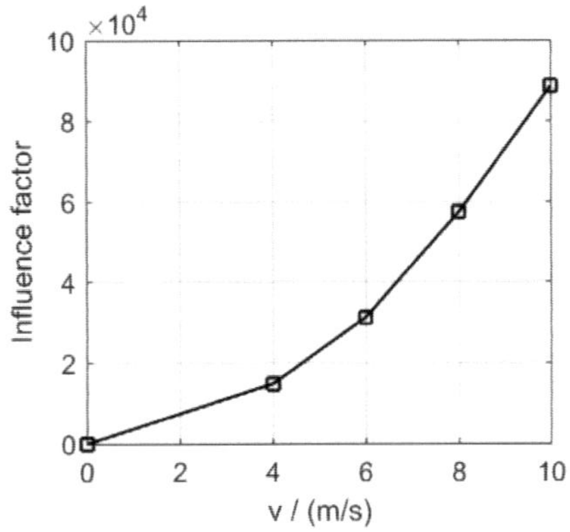

Figure 4. Influence factor of car rotation on Torque.

From Figures 1 and 2, it can be seen that with the increase of elevator speed, the transverse force and torque of the car increase gradually at the same vibration deflection angle. In addition, at the same speed, the bigger the deflection angle, the greater the transverse force and torque of the car, and they all show a near linear relationship (straight line). The linear ratios of each line in Figures 1 and 2 are used to obtain the influence factors of car deflection angle on transverse force and torque at different running speeds, as shown in Figure 3 and Figure 4, respectively. Based on aerodynamic theory, the curves in Figures 3 and 4 are fitted quadratically, and the approximate values of transverse force and torque generated by car rotation of high-speed elevator at different running speeds are obtained respectively as follows:

$$F_{\theta} = -1804v^2\theta \tag{2}$$

$$M_{\theta} = 886.9v^2\theta \tag{3}$$

3. Establishment of Transverse Vibration Model of High-Speed Elevator Under Gas-Solid Coupling

3.1. System Dynamic Model

In the process of coupling action between high-speed elevator and airflow in the hoistway, the car is the main spoiler, and the elastic deformation of the car is very small relative to its own geometric size. Therefore, the elastic deformation of the car can be neglected in the analysis process, and the car system can be regarded as a rigid body. In this paper, the guide wheel-guide shoe system is simplified to a parallel spring-damper system [8]. Under the influence of guide rail excitation, the transverse vibration dynamic model of high-speed elevator system is established as shown in Figure 5, where the car mass is m, the moment of inertia is J, the equivalent stiffness and damping of the guide wheel-guide shoe system are k_1 and c_1, the displacement excitation of the guide rail at the contact point between the guide wheel and the guide rail is y_{di} (i = 1,2,3,4), the vertical distances between the upper and lower guide shoes to the car centroid is l_1 and l_2. According to Lagrange principle, the dynamic equation of transverse vibration of high-speed elevator system is established as follows:

$$\boldsymbol{M}\ddot{\boldsymbol{Y}} + \boldsymbol{C}\dot{\boldsymbol{Y}} + \boldsymbol{K}\boldsymbol{Y} = \boldsymbol{F}_r \tag{4}$$

where, $\boldsymbol{Y} = \begin{bmatrix} y & \theta \end{bmatrix}^{\mathrm{T}}$, $\boldsymbol{M} = diag(m, J)$,$\boldsymbol{C} = \begin{bmatrix} 4c_1 & 2c_1(l_2 - l_1) \\ 2c_1(l_2 - l_1) & 2c_1(l_1^2 + l_2^2) \end{bmatrix}$, $\boldsymbol{K} = \begin{bmatrix} 4k_1 & 2k_1(l_2 - l_1) \\ 2k_1(l_2 - l_1) & 2k_1(l_1^2 + l_2^2) \end{bmatrix}$,

$\boldsymbol{F}_r = \begin{bmatrix} F_f & M_f \end{bmatrix}^{\mathrm{T}}$, $F_f = k_1(y_{d1} + y_{d2} + y_{d3} + y_{d4}) + c(\dot{y}_{d1} + \dot{y}_{d2} + \dot{y}_{d3} + \dot{y}_{d4})$,

$$M_f = -l_1k_1(y_{d1} + y_{d2}) - l_1c_1(\dot{y}_{d1} + \dot{y}_{d2}) + l_2k_1(y_{d3} + y_{d4}) + l_2c_1(\dot{y}_{d3} + \dot{y}_{d4}).$$

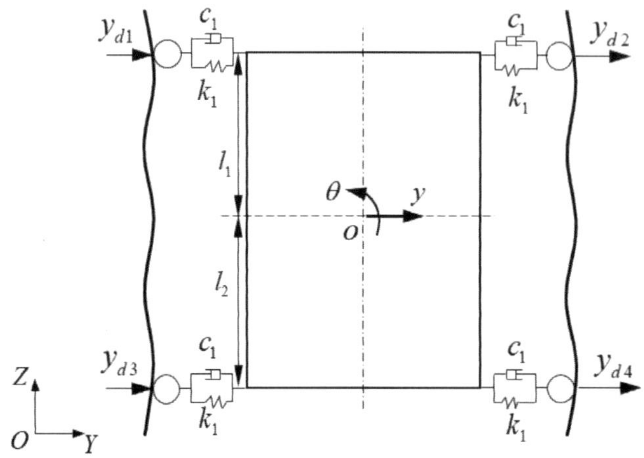

Figure 5. Transverse vibration dynamic model of high-speed elevator.

3.2. *Transverse Vibration Dynamic Model under Gas-Solid Coupling*

Equations (2) and (3) are written in the form of a matrix product and expressed as $\boldsymbol{F}_a$:

$$\boldsymbol{F}_a = \begin{bmatrix} F_\theta \\ M_\theta \end{bmatrix} = \begin{bmatrix} 0 & -1804v^2 \\ 0 & 886.9v^2 \end{bmatrix} \begin{bmatrix} y \\ \theta \end{bmatrix} \tag{5}$$

From the above analysis, it can be seen that different vibration states correspond to different aerodynamic loads during the operation of high-speed elevator, and different loads will produce different vibration states. Therefore, this paper adopts the interactive coupling method to iterate within each coupling time step. According to the above method, the real-time transmission between the car attitude parameters and aerodynamic parameters is carried out, and then the transverse vibration states of the high-speed elevator under the gas-solid coupling are solved. Combined Equations (4) and (5), the transverse vibration dynamic model of high-speed elevator under gas-solid coupling can be obtained as follows:

$$\boldsymbol{M}\ddot{\boldsymbol{Y}}(t+1) + \boldsymbol{C}\dot{\boldsymbol{Y}}(t+1) + \boldsymbol{K}\boldsymbol{Y}(t+1) = \boldsymbol{F}_r(t) + \boldsymbol{F}_a(t) \tag{6}$$

4. Case Study

This paper takes a high-speed elevator of Shandong Fuji Elevator Co., Ltd. as the reference, and the parameters as shown in Table 1.

Table 1. Parameter values of high-speed elevator.

Parameter	Value	Parameter	Value
m/(kg)	1100	l_1/(m)	1.2
J/(kg·m^2)	1850	l_2/(m)	1.2
Δm/(kg)	800	Hoistway cross-section /(m)	2.1×2.1
k_1/(N/m)	1.6e5	Car height /(m)	2.4
c_1(N·s/m)	700	Car cross-section /(m)	1.7×1.7

For the displacement excitation of the guide rail, its unevenness is the main expression form, and is close to the sinusoidal excitation mode [9]. Therefore, this paper uses the unevenness of the sinusoidal form with an amplitude of 0.7mm as the displacement excitation of the guide rail. On the premise of high calculation accuracy, the Newmark-β method in numerical integration form is used to solve the dynamic model iteratively in this paper. When the elevator speed is 4m/s, 7m/s, 10m/s and 13m/s, the transverse vibration acceleration of high-speed elevator without air flow and gas-solid coupling under the medium-load condition is obtained by solving the Equations (4) and (6) respectively, the results are shown in Figures 6-9.

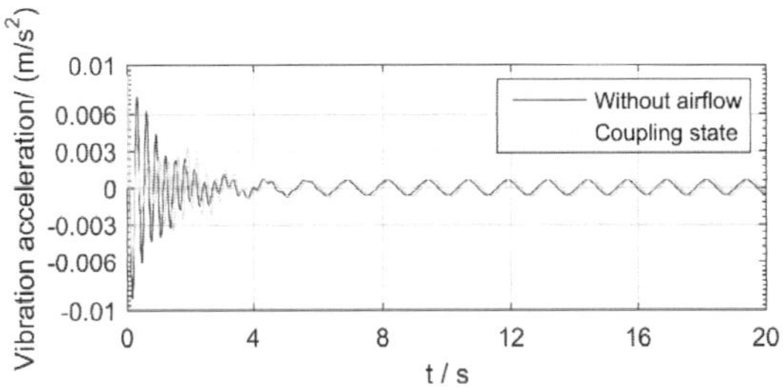

Figure 6. Vibration acceleration at 4m/s.

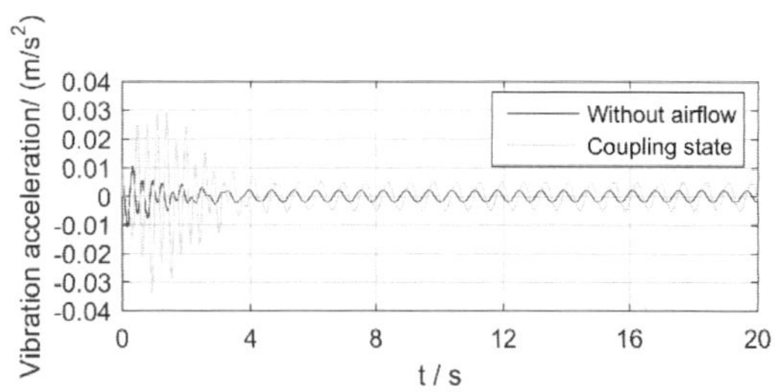

Figure 7. Vibration acceleration at 7m/s.

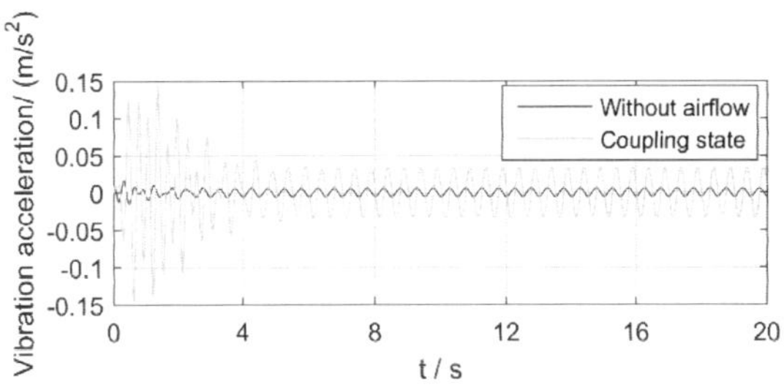

Figure 8. Vibration acceleration at 10m/s.

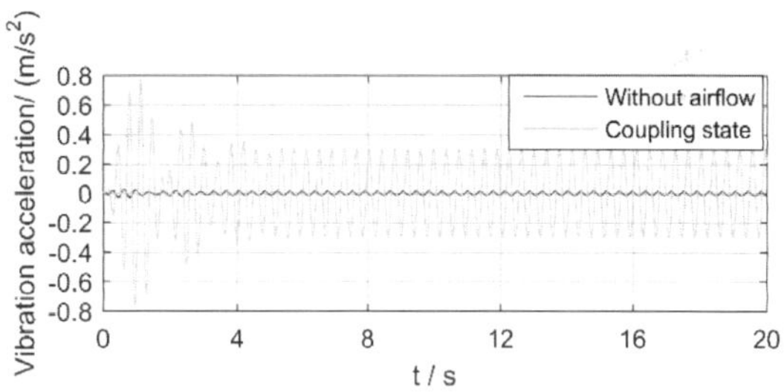

Figure 9. Vibration acceleration at 13m/s.

Figures 6-9 show that when the elevator speed is less than 4m/s, the effect of gas-solid coupling on the transverse vibration acceleration peak value of high-speed elevator is small, while when the speed is greater than 4m/s, the effect of gas-solid coupling on the transverse vibration acceleration peak value of high-speed elevator is increasing with the increase of speed. And when the running speed reaches 13m/s, the transverse vibration acceleration peak value is much higher than the allowable value of 0.1m/s^2, which seriously affects the operation safety and ride comfort of the high-speed elevator.

5. Conclusion

For the high-speed elevator with running speed less than 4m/s, the action of gas-solid coupling hardly affects the transverse vibration acceleration peak value, and guide rail excitation is the main influencing factor, so the emphasis should be placed on the guide rail excitation in the analysis and control of transverse vibration. For the high-speed elevator with speed greater than 4m/s, with the speed increases gradually, the effect of gas-solid coupling on the transverse vibration acceleration peak value of high-speed elevator is increasing. When the speed is greater than 13m/s, the transverse vibration acceleration peak value is much high than 0.1m/s^2, which exceeds the maximum allowed. Therefore, gas-solid coupling is the main factor affecting the transverse vibration of high-speed elevator, and the effect should be fully considered in design, dynamic analysis and vibration control of the high-speed elevator. The above conclusions have some academic guidance and applied merit for the high-speed elevator in its vibration characteristics analysis and vibration damping and noise reduction.

IOP Conf. Series: Materials Science and Engineering **538** (2019) 012034 doi:10.1088/1757-899X/538/1/012034

Acknowledgments
The authors are grateful for the financial support by Shandong Provincial Natural Science Foundation, China (ZR2017MEE049) and the Key Research Development Project of Shandong Province (2018GSF122004).

References
[1] N.V. Gaiko, W.T. van Horssen, J SOUND VIB 424, 272-292 (2018)
[2] D.Q. Mei, X.Q. Du, Z.C. C, Journal of Mechanical Engineering, **45**, 264-270 (2009)
[3] R.J. Zhang, C. Wang, Q. Zhang, J BRAZ SOC MECH SCI, **40**, 190 (2018)
[4] D. R. Santo, J.M. Balthazar, A.M. Tusset, et al., J VIB CONTROL, **24**, 825-838 (2018)
[5] X.D. Li, K. Wang, Journal of Harbin Institute of Technology, **41**, 82-86 (2009)
[6] D.D. Chen, Journal of Mechanical & Electrical Engineering, **33**, 1359-1363 (2016)
[7] H.J. Du, Soochow University (2009)
[8] C. Wang, R.J. Zhang, Q. Zhang, INT J ACOUST VIB, **22,** 218-223 (2017)
[9] Q. Zhang, Y.H. Yang, S.H. Zhang, et al., Journal of Shandong Jianzhu University, **33**, 18-24 (2018)

IOP Conf. Series: Materials Science and Engineering **538** (2019) 012035 doi:10.1088/1757-899X/538/1/012035

Simulation and Optimization of Temperature Field in Large-Sized MOCVD Reactor by Resistance Heating

Lili Zhao[1], Zhiming Li[1,*], Runqiu Guo[2], Lansheng Feng[2], Ligen Lu[1] and Xiaoqian Fu[1]

[1]Shandong Provincial Key Laboratory of Network based Intelligent Computing, School of Information Science and Engineering, University of Jinan, Jinan, Shandong, 250022, China
[2]State Key Discipline Laboratory of Wide Band-Gap Semiconductor Technology, School of Microelectronics, Xidian University, Xi'an, Shanxi, 710071, China

[*]Corresponding author: ise_lizm@ujn.edu.cn

Abstract. The effect of the number of layers and the length of the heat shields on the heating efficiency and temperature distributions in the substrate have been studied by establishing the model of a single-piece 18-inch MOCVD reactor. The results show that the number of layers of the heat shields is directly proportional to the heating efficiency, decreasing the length of the thermal shield can reduce the standard deviation(STD) of the substrate temperature. When the length of the heat shields is 56mm, the coefficient of substrate temperature STD is 21.41℃and the STD is about 45% lower than the traditional susceptor. An area within substrate radius of 200mm, the coefficient of substrate temperature STD is 2.64℃and the STD is about 93% lower than the traditional susceptor. The results obtained will provide theoretical basis for developing the heating structure of large-sized MOCVD reactor.

1. Introduction

MOCVD is an abbreviation for Metal-organic chemical vapor deposition and is an important technology for preparing nitride semiconductor materials[1-2]. With the maturity of the third generation semiconductor technology, the market demand for semiconductor devices is increasing, and the quality of semiconductor thin film materials is becoming more and more demanding[3]. Due to the complexity of MOCVD technology, the film growth process is affected by many factors [4-8], among which the uniformity of substrate temperature distribution is an important factor affecting the growth of the film[9-14].

However, in the conventional resistive heating system, the heating efficiency and temperature uniformity of large-sized substrate are more prominent. In this work, the influence of the number of layers and the length of the heat shields on the heating efficiency and the uniformity of substrate temperature distribution have been studied by establishing an 18-inch MOCVD reactor model of resistance heating, aimed at improving the heating efficiency of the heating system and the uniformity of substrate temperature distribution.

2. Model of MOCVD Reactor

The growth mechanism of MOCVD is complex and affected by many parameters. Usually, the finite element method is used to establish a mathematical model, and the thermal field and the fluid field are coupled and calculated. The mathematical model uses the fluid equation and heat conduction equation of STR's VR-Nitride simulation software.
Heat conduction equation:

$$\rho C_p \frac{\partial T}{\partial t} + \nabla \cdot (-k \nabla T) = Q - \rho C_p \nabla T \tag{1}$$

Fluid equation:

$$\rho \frac{\partial \vec{u}}{\partial t} - \eta \left(\nabla \vec{u} + (\nabla \vec{u})^T \right) + \rho(\vec{u} \cdot \nabla)\vec{u} + \nabla p = 0 \tag{2}$$

Where ρ is density, C_p is specific heat, T is temperature, k is material thermal conductivity, Q is resistance heating, $\vec{u}$ is gas velocity, η is gas viscosity coefficient, and p is pressure. In order to improve the calculation speed in the simulation experiment, the following assumptions are made for the model without affecting the main simulation results:

(1) Neglect of the chemical reaction between gases;
(2) Assume that the gas is ideal gas;
(3) The wall temperature of the reaction chamber is constant;
(4) The thermal expansion of the materials is not considered.

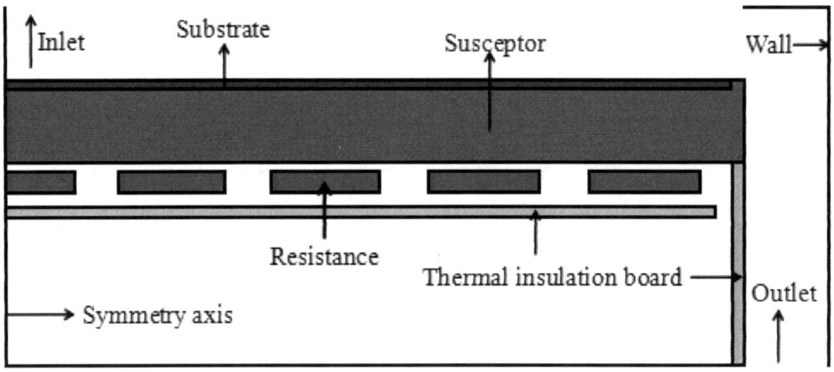

Figure 1. Geometrical model of MOCVD reactor

Figure1. shows a schematic diagram of a two-dimensional axisymmetric structure of the MOCVD reactor, including the wall, an inlet and an outlet, heating resistors, a substrate, and heat shields. The susceptor material is graphite, and the heating resistors under the susceptor and the components of the heat shields are tungsten and molybdenum, and the number of layers of the heat shields can be set to 0 or more. The simulation model susceptor height H is 19.05mm, the substrate radius R is 232.6mm, and the number of the heating resistors is 5, and the resistance power density is 7.5E+007W/m^3. The distance between the adjacent heating resistors is 4mm. The length of the heat shield is 148.7mm and the width is 2mm, and the interval between the heat shields is 2mm.
In the simulation process, H_2 and N_2 are used as carrier gases, NH_3 and TMAl are reactants, and mixed into the reaction chamber from the inlet. The flow rates of H_2, N_2, NH_3 and TMAl are 50slm, 50slm, 20slm and 1slm, respectively. The temperature boundary condition of the wall is set to 2℃, the reactor pressure is 50Torr, and the susceptor rotation speed is 1000 rpm. Unless otherwise specified, these parameters remain unchanged during the simulation.

3. Results and Analysis

3.1. Influence of The Number of Heat Shields on Substrate Temperature

In order to study the influence of the heat shields on the heating efficiency and temperature, the number of heat shields is changed. The length of the heat shields is 148.7mm and the width is 2mm, and the interval between the heat shield is 2mm.

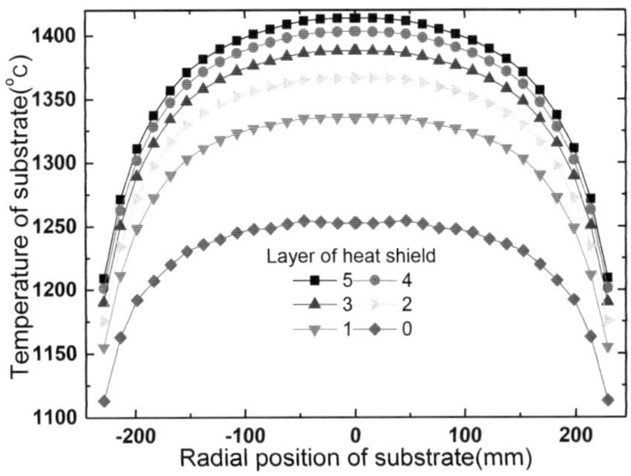

Figure 2. Substrate temperature distribution under different numbers of heat shields

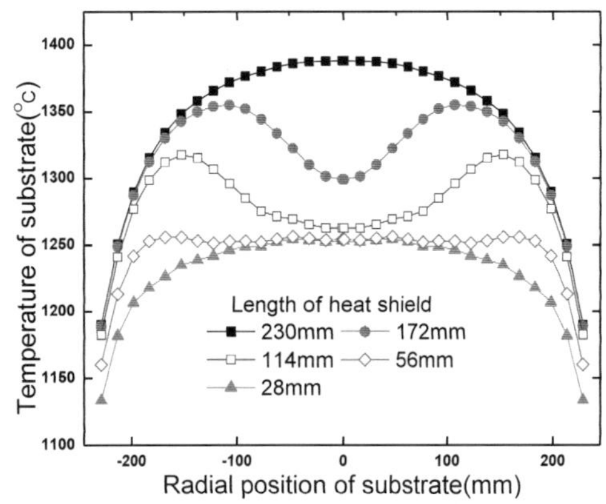

Figure 3. Substrate temperature distributions of different lengths heat shields

Figure 2. shows the temperature distribution of the substrate under different numbers of heat shields.When the number of heat shields is increased from 0 to 4, the corresponding substrate temperature increases from 1227℃to 1380℃, and the standard deviation of temperature distribution is changed from 35.8℃rose to 54.4℃. It can be seen that the heat shields can effectively increase the surface temperature of the substrate, and the heating efficiency is proportional to the number of layers of the heat shields. It was found that when the number of layers of the heat shields exceeds 3, the increase of the substrate temperature slows down with the increase of the number of the heat shields. Taking the heating efficiency needed into account, it is reasonable to take three layers of heat shields.

3.2. Influence of the Length of Heat Shields on Substrate Temperature

Based on the above simulation results, the number of layers of the heat shields is 3, the length of the heat shields is 148.7mm, the width is 2mm, and the interval between the heat shields is 2mm. In order to investigate the effect of the length for the heat shields on the temperature, the length of the heat shields is reduced from the center of the substrate, that is, from left to right.

Figure 3. shows the temperature distributions of the substrate under different lengths of the heat shields. It can be seen that the longer the length of the insulation board, the higher the temperature of the substrate.As the lengths of the heat shields are changed from 230mm, 172mm, 114mm and 56mm to 28mm, the temperature STD in the substrate are 52.1, 40.2℃, 30.3℃, 21.4℃ and 29.4℃. It is not difficult to find that the coefficient of substrate temperature STD is the best when the length of the heat shield is 56 mm.

Figure 4. describes the distribution of flow field and temperature field in two different reactor, the convention reactor(a) and the optimized reactor. Figure 5. illustrates the distribution curves before and after substrate temperature optimization. The results show that the optimized flow field and temperature field are more suitable for the growth of AlN films. Through data comparison, the coefficient of substrate temperature STD was reduced by 45%, especially in the -200-200mm region, the coefficient of substrate temperature STD was reduced by 93%.

IOP Conf. Series: Materials Science and Engineering **538** (2019) 012035 doi:10.1088/1757-899X/538/1/012035

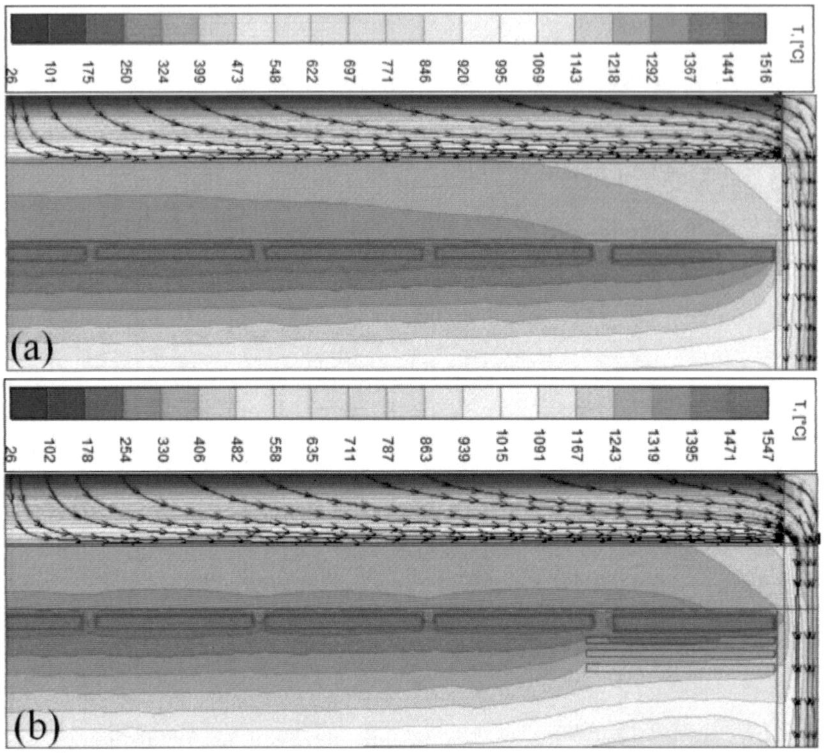

Figure 4. Distributions of flow field and temperature field in conventional reactor (a) and in optimized reactor(b)

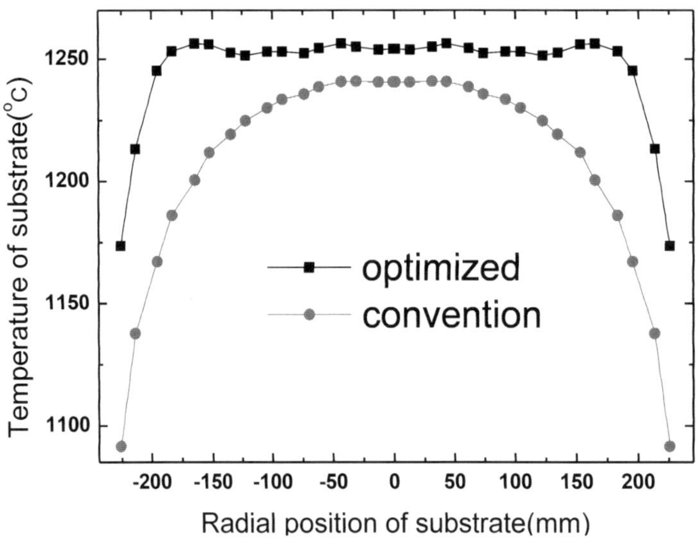

Figure 5. Substrate temperature distributions in conventional reactor and in reactor with optimized heating structure

4. Conclusions

The effect of the number of layers and the length of the heat shields on the heating efficiency and temperature distributions in the substrate have been studied by establishing the model of a single-piece 18-inch MOCVD reactor. The results show that the number of layers of the heat shields is directly proportional to the heating efficiency. And decreasing the length of the thermal shield can reduce the coefficient of substrate temperature STD. When the length of the heat shields is 56mm, the coefficient of substrate temperature STD is 21.41℃and the STD is about 45% lower than the traditional susceptor. An area within substrate radius of 200mm, the coefficient of substrate temperature STD is 2.64℃and the STD is about 93% lower than the traditional susceptor. The results obtained will provide theoretical basis for developing the heating structure of large-sized MOCVD reactor.

Acknowledgments

This work is supported by the National Natural Science Foundation of China under Grant No 61601198, supported by Shandong Provincial Natural Science Foundation of China under Grant No ZR2017MF046 and ZR2018BF025, and supported by National Key R&D Program of China under Grant No 2017YFB0404202.

References

[1] Dobrzanski L, Pakula D, Staszuk M. (2014) Chemical Vapor Deposition in Manufacturing. Springer London. London.
[2] Capper P, Irvine S, Joyce T. (2017) Epitaxial Crystal Growth: Methods and Materials. Springer US. US.
[3] Zuo R, Zhang H. (2017) III-nitride MOCVD growth in the study of chemical reactions and progress. China Light & Lighting. 10:4-9(in Chinese).
[4] Wang B L, Zuo R, Meng S C, et al. (2015) Quantum chemistry study on reaction paths of GaN/AlN grown by MOCVD. Journal of Synthetic Crystals. 44:2237-2244.
[5] Liu B T, Ma P, Li X L, et al. Influence of Al Preflow Time on Surface Morphology and Quality of AlN and GaN on Si (111) Grown by MOCVD[J]. Chinese Physics Letters, 2017, 34:058101.
[6] Loganathan R, Baskar K. (2019) Effect of TMAl flow rate on AlInGaN/GaN heterostructures grown by MOCVD. 178:66-73.
[7] Feng L S, Guo R Q, Zhang J C. (2016) Effect of growth parameters on GaN in a vertical MOCVD reactor. Journal of Xidian University. 43: 178-182.
[8] Feng L S, Guo R Q, Zhang J C. (2017) Effect of reactor geometry on GaN in a vertical MOCVD reactor. Journal of Xidian University. 44: 171-175.
[9] Zhang Z, Fang H S, Yan H, et al. (2015) Influencing factors of GaN growth uniformity through orthogonal test analysis. Applied Thermal Engineering. 91:53-61.
[10] Li K H, Hamad S.A, Sun H D, et al. (2018) Induction-heating MOCVD reactor with significantly improved heating efficiency and reduced harmful magnetic coupling. Journal of Crystal Growth. 488:16-12.
[11] Li Z, Li J, Jiang H, et al. (2013) Effect of thermocouple position on temperature field in nitride MOCVD reactor. Journal of Crystal Growth. 368: 29-34.
[12] Xia H, Xiang D, Mou P. (2015) The array-control heater and non-uniform resistance module design for regulating the temperature profile in a reactor chamber. Journal of Mechanical Science and Technology. 29:593-604.
[13] S. S, K. P, R. L, et al. (2017) Effect of growth temperature on InGaN/GaN heterostructures grown by MOCVD[J]. Journal of Crystal Growth. 468:249-251.
[14] K K R, Deepak A, Chin J H, et al. (2018) Temperature effect on the growth of Au-free InAs and InAs/GaSb heterostructure nanowires on Si substrate by MOCVD. Journal of Crystal Growth. 490:19-24

Modeling and Analysis of Transverse Vibration of Traction Rope of High Speed Traction Elevator

Xiujuan Qi, Ruijun Zhang[a] and Qin He

School of Mechanical and Electrical Engineering, Shandong Jianzhu University, Jinan, 250101, Shandong, P.R. China

Corresponding author: zhangruijun@sdjzu.edu.cn

Abstract. Based on continuum mechanics theory and considering the influence of external excitation on high-speed traction elevator wire rope, Hamlition principle is applied to establish the transverse vibration control equation of variable-length lifting wire rope, and Galerkin method is used to discretize the solution. Finally, the model is simulated by Matlab software, and the transverse vibration law of the wire rope and the influence of damping on the transverse vibration of the wire rope are analyzed. The results show that the transverse vibration strength of the wire rope is basically the same as that of the wire rope when there is no damp effect. Increasing the damp can reduce the transverse vibration strength of the wire rope.

1. Introduction

With the rapid development of super high-rise buildings, the use of high-speed elevators has been promoted. Elevator is the transportation equipment of the vertical high-rise building nowadays. In order to improve the transportation efficiency, the elevator speed is improved. However, the safety of elevators is difficult to guarantee because of the high speed. The most important problem is the vibration of elevator lifting system. The vibration of elevator hoisting system is mainly caused by the vibration of elevator wire rope. So it is important to study the vibration of elevator wire rope for improving the vibration of lifting system.

Wire rope vibration accounts for a large proportion of the factors causing elevator vibration. Nowadays, more and more researchers are studying the vibration of wire rope. Jihu Bao [1] Considering the influence of geometric nonlinearity caused by deformation on variable length wire rope, the control equation of transverse vibration of wire rope is established by energy method, and the model is simulated by MATLAB software. Zhu [2] aiming at the problem of vibration of axially moving chords, neglects the influence of geometric deformation factors and establishes a linear differential equation for transverse vibration of horizontally moving chords with variable length. Considering the influence of balanced wire rope, Juan Wu [3] established the control equation of transverse vibration of hoisting wire rope by Hamilation principle, and analyzed the influence of guide rail unevenness and external horizontal excitation on the transverse vibration of hoisting wire rope. Hiroyuki [4] assumes that the tension and velocity of the rope are constant. The governing equation of the forced vibration of a variable length rope with linear damping is established and solved by using a virtual wave source. Yang [5] considers the coupling effect between building and elevator rope with suspension cage motion, applies Hamilton principle to establish the dynamic model of elevator rope,

IOP Conf. Series: Materials Science and Engineering **538** (2019) 012036 doi:10.1088/1757-899X/538/1/012036

and uses this model to predict the vibration of actual building and elevator rope. Jihu Bao [6] considers the coupling of the cable's axial motion and bending deformation, establishes the governing equation, and studies the nonlinear vibration of flexible hoisting wire rope with time-varying length and axial velocity. Jiang [7] takes into account the time-varying characteristics of rope mass, stiffness, damping and power, establishes the dynamic model of high-speed traction elevator, and simulates the acceleration response under given speed curve.

2. Transverse Vibration Model of Wire Rope

The model establishment and solution in this paper are based on the following four assumptions:
1. Ignore the influence of airflow and friction in wellbore;
2. Ignore the influence of balanced wire rope vibration;
3. Ignore the influence of torsion and longitudinal vibration of wire rope, and the elastic deformation caused by transverse vibration is much smaller than the length of the entire string.
4. In the operation of high-speed elevator, the linear density of wirp rope is ρ , the cross-sectional area is A ,and the elastic modulus is E remain unchanged.

2.1. Vibration Control Equation

Figure 1 is a transverse vibration model of high-speed traction elevator lifting system. In this paper, the wire rope is regarded as a variable length string with axial motion, the linear density is ρ, the cross-sectional area is A, the elastic modulus is E, and the length $l(t)$ varies with time t. The lifting weight is simplified as a rigid weight with mass m suspended at the bottom of the chord line, which is free longitudinally, Constrained by a spring with a lateral stiffness of k and a damper with a damping coefficient of c, the existence of a rigid tank passage in a real high-speed elevator hoisting system is simulated. Assume that there is a horizontal external excitation e_1 at the upper end of the wire rope to simulate the eccentric rotation of the motor or other factors. The horizontal excitation e_2 exists at the lower end to simulate the unevenness of the rigid tank. Choose the tangent point between the traction wheel and the wire rope as the origin, the vertical downward direction of the x-axis is positive, and the horizontal left direction of the y-axis is positive.

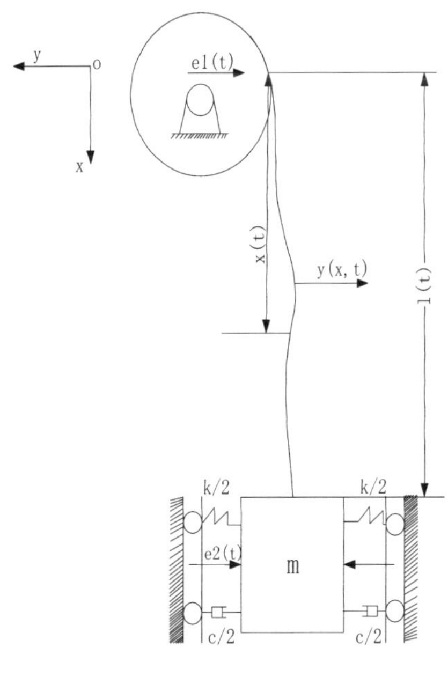

Figure 1. Transverse vibration model of wire rope.

IOP Conf. Series: Materials Science and Engineering **538** (2019) 012036 doi:10.1088/1757-899X/538/1/012036

The kinetic energy of the system can be expressed as:

$$T = \frac{1}{2}\rho\int_0^{l(t)} V^2 dx + \frac{1}{2}mv_1^2\Big|_{x=l(t)} \tag{1}$$

The potential energy of the system can be expressed as:

$$V = \int_0^{l(t)} (\frac{1}{2}Py_x^2 + \frac{1}{2}EA\varepsilon^2)dx + \frac{1}{2}k\left[y(l(t),t)\right]^2 \tag{2}$$

where: P is the tension applied by the wire rope x at the time t, and ε is the wire rope strain.

$$P = \left[m+\rho(l(t)-x)\right](g-a) \tag{3}$$

where: $a = \ddot{l}(t)$ is the overall motion acceleration of the lifting system; g is the acceleration of gravity.
The virtual work of the damping force in the system can be expressed as:

$$\delta W = -c\frac{Dy(l(t),t)}{Dt}\delta y(l(t),t) \tag{4}$$

According to the Hamiltonian principle:

$$\int_{t_1}^{t_2} (\delta T - \delta V + \delta W)dt = 0 \tag{5}$$

After a series of transformations of Formula 5, the transverse vibration control equation of high-speed traction elevator hoisting wire rope can be derived.

$$\rho(y_{tt} + 2vy_{xt} + \dot{v}y_x + v^2 y_{xx}) - (Py_{xx} + P_x y_x) - \frac{3}{2}EAy_{xx}y_x^2 + cy_t = 0,(0 < x < l(t)) \tag{6}$$

The geometric and temporal boundary conditions are:

$$y(0,t) = e_1(t),\ y(l(t),t) = e_2(t)$$

The boundary condition at this time is a non-homogeneous boundary condition, and the function $y(x,t)$ having a non-homogeneous boundary condition is transformed into a function $y_1(x,t)$ having a homogeneous boundary condition, that is, the lateral displacement $y(x,t)$ can be expressed as

$$y(x,t) = y_1(x,t) + y_2(x,t) \tag{7}$$

where: $y_1(x,t)$ is the part that satisfies the corresponding homogeneous boundary condition; $y_2(x,t)$ is the part that does not satisfy the homogeneous boundary condition.
Substituting equation (7) into equation (6) can obtain the transverse vibration control equation of the lifting system with excitation:

$$\begin{aligned}&\rho(y_{1,tt} + 2vy_{1,xt} + \dot{v}y_{1,x} + v^2 y_{1,xx}) - (Py_{1,xx} + P_x y_{1,x}) - \frac{3}{2}EA(y_{1,xx}y_{1,x}^2 + y_{2,xx}y_{2,x}^2 + 2y_{1,x}y_{1,xx}y_{2,x} + 2y_{1,x}y_{2,xx}y_{2,x} + y_{1,xx}y_{2,x}^2)\\&+cy_{1,t} = -\rho(y_{2,tt} + 2vy_{2,xt} + \dot{v}y_{2,x} + v^2 y_{2,xx}) + (Py_{2,xx} + P_x y_{2,x}) + \frac{3}{2}EAy_{2,xx}y_{2,x}^2 - cy_{2,t}\end{aligned} \tag{8}$$

IOP Conf. Series: Materials Science and Engineering **538** (2019) 012036 doi:10.1088/1757-899X/538/1/012036

2.2. Galerkin Method Discretization

The transverse vibration control equations of steel wire ropes are complex partial differential equations with infinite degrees of freedom, and many parameters are time-varying, so it is impossible to obtain accurate analytical solutions. Therefore, the article uses Galerkin method for discretization. An approximate numerical solution is obtained. Let the solution of equation (8) be:

$$y_1(x,t)=\sum_{i=1}^{n}\varphi_i(\xi)q_i(t) \tag{9}$$

where: $q_i(t)$ is the generalized coordinate, $i=1,2,3.....,n,n$ is the modulus; $\varphi_i(\xi)$ is the type function.

$$\varphi_i(\xi)=\sqrt{2}\sin(i\pi\xi)$$

Using the modified Galerkin method to transform the infinite dimensional transverse vibration partial differential equation of the high speed traction elevator lifting system wire rope into a finite dimensional ordinary differential equation .

$$M\ddot{Q}+C\dot{Q}+K_1Q+K_2Q^2+K_3Q^3=F \tag{10}$$

where: $Q=\left[q_1(t),q_2(t),....,q_n(t)\right]^T$ is the generalized coordinate vector; the values of the elements in the matrices M、C、K_1、K_2、K_3 and F are:

$$m_{ij}=\rho\delta_{ij}$$

$$c_{ij}=\frac{2v\rho}{l(t)}\int_0^1(1-\xi)\varphi_i'(\xi)\varphi_j(\xi)d\xi+c\delta_{ij}$$

$$\begin{aligned}k_{1ij}(t)=&\frac{\rho(g-a(t))}{l(t)}\int_0^1\varphi_i'(\xi)\varphi_j(\xi)d\xi+\frac{2v^2(t)\rho}{l^2(t)}\int_0^1(1-\xi)\varphi_i'(\xi)\varphi_j(\xi)d\xi-\frac{v^2(t)\rho}{l^2(t)}\int_0^1(1-\xi)^2\varphi_i'(\xi)\varphi_j'(\xi)d\xi\\&-\frac{3EA(e_2-e_1)^2}{2l^4(t)}\int_0^1\varphi_i''(\xi)\varphi_j(\xi)d\xi-\frac{cv(t)}{l(t)}\int_0^1\xi\varphi_i'(\xi)\varphi_j(\xi)d\xi-\frac{m(g-a(t))}{l^2(t)}\int_0^1\varphi_i''(\xi)\varphi_j(\xi)d\xi\\&+\frac{\rho(g-a(t))}{l(t)}\int_0^1(1-\xi)\varphi_i'(\xi)\varphi_j{}'(\xi)d\xi+\frac{\rho a(t)}{l(t)}\int_0^1(1-\xi)\varphi_i'(\xi)\varphi_j(\xi)d\xi\end{aligned}$$

$$k_{2ij}(t)=-\frac{3EA}{l^4(t)}(e_2-e_1)\int_0^1\varphi_i''(\xi)\varphi_i'(\xi)\varphi_j d\xi$$

$$k_{3ij}(t)=\frac{3EA}{2l^4(t)}(e_2-e_1)\int_0^1\varphi_i''(\xi)\left[\varphi_i'(\xi)\right]^2\varphi_j(\xi)d\xi$$

$$\begin{aligned}f_j(t)=&-\rho e_1''\int_0^1\varphi_j(\xi)d\xi-\rho(e_2''-e_1'')\int_0^1\xi\varphi_j(\xi)d\xi-\frac{2v(t)\rho}{l(t)}(e_2'-e_1')\int_0^1(1-\xi)\varphi_j(\xi)d\xi\\&-\frac{a(t)\rho}{l(t)}(e_2-e_1)\int_0^1(1-\xi)\varphi_j(\xi)d\xi-\frac{2v(t)\rho}{l^2(t)}(e_2-e_1)\int_0^1\xi\varphi_j(\xi)d\xi\\&-\frac{2v^2(t)\rho}{l^2(t)}(e_2-e_1)\int_0^1\varphi_j(\xi)d\xi-\frac{(\rho a(t)-\rho g+c)}{l}(e_2-e_1)\int_0^1\varphi_j(\xi)d\xi\end{aligned}$$

where δ_{ij} is a delta function, and the ordinary differential equation (10) is solved, and the value $q_{j(t)}$ of the generalized coordinate can be obtained. Substituting equations (10) and (8) can obtain the real-time value of transverse vibration.

3. Examples and Results Analysis

Taking high-speed traction elevator as an example, the model is simulated and analyzed. The maximum lifting height of the elevator parameters is $l_{\max} = 150m$, the maximum speed $v_{\max} = 5m/s$, the maximum acceleration $a_{\max} = 1m/s^2$, and the maximum acceleration $J_{\max} = 0.5m/s^3$. The car system mass $m = 400kg$, wire rope elastic modulus $E = 8\times10^{10} N/m^2$, wire rope cross-sectional area A=89.344cm^2, wire rope wire density ρ=0.87kg/m, traction wheel eccentric rotation on the wire rope upper end horizontal excitation is $e_1 = 0.01\sin(3.14t)$, rail irregularity on the wire rope lower end level The incentive is e_2=0.005sin(6.28t). Taking the actual running curve of the elevator as input, the numerical calculation method is used, and the simulation is carried out by using Matlab software.

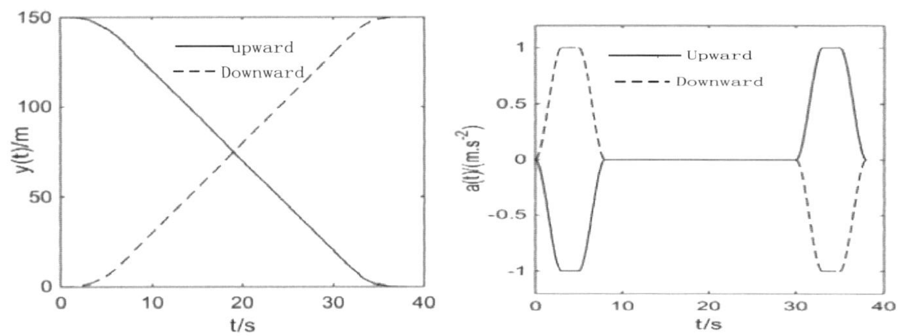

Figure 2. Running curve of the traction elevator.

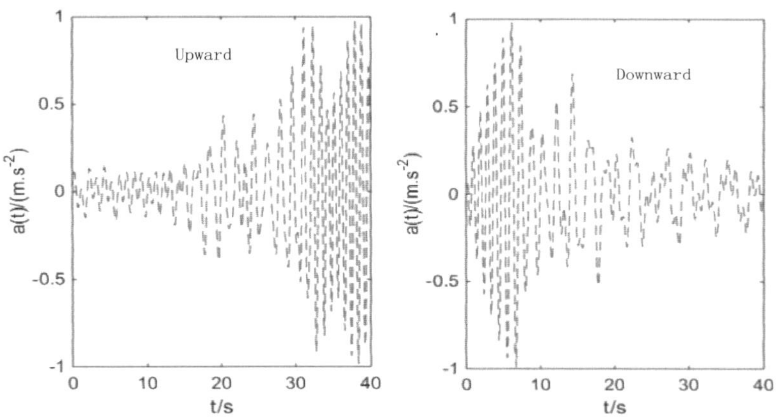

Figure 3. Undamped lower acceleration vibration amplitude.

From Figure 3, it can be seen that the vibration frequency and amplitude of transverse acceleration increase with the shortening of the length of the wire rope when the wire rope goes up without damp, and decrease with the length of the wire rope when the wire rope goes down, and the amplitude and peak value of transverse acceleration vibration are basically the same.

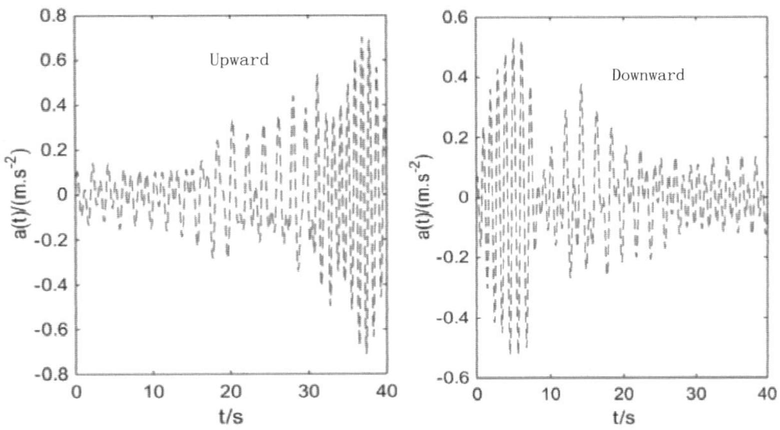

Figure 4. Transverse acceleration vibration response with a damping of 0.5.

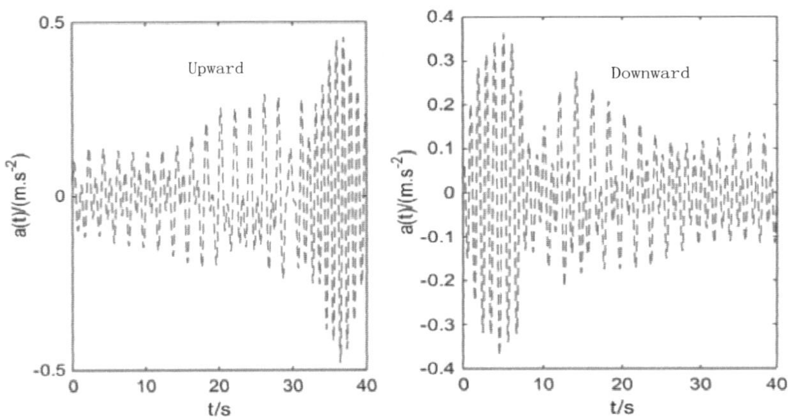

Figure 5. Transverse acceleration vibration response with damping 1.

From Figure 4, it can be concluded that the vibration frequency and amplitude of transverse acceleration increase with the decrease of the length of the wire rope when the wire rope goes up under the action of damping; while the vibration frequency and amplitude of transverse acceleration decrease with the increase of the length of the wire rope when the wire rope goes down, but the vibration amplitude of the upward is larger than that of the downward, which has a greater impact on the transverse acceleration vibration of the downward wire rope. From Fig. 4 and Fig. 5, it can be seen that the vibration amplitude of lateral acceleration varies with different dampers. The average amplitude of lateral acceleration of wire rope decreases with the increase of damping.

4. Conclusion

(1)Aiming at the time-varying characteristics of high-speed traction elevator wire rope, considering the influence of damp on the lateral vibration of wire rope, a partial differential equation for the lateral vibration of high-speed traction wire rope is established by using Hamilton principle.
(2)Taking the actual operation curve of traction elevator as input, the vibration response of wire rope lateral acceleration under different dampers is simulated and analyzed by MATLAB software. The results show that when there is no damper, the vibration amplitude of the upward and downward lateral acceleration is basically the same. When the cable is damped, the transverse acceleration vibration intensity on the wire rope is greater than that on the wire rope. Increasing the damping can

reduce the vibration intensity of the lateral acceleration of the wire rope, which verifies the validity of this method.

(3)The establishment and analysis of the transverse vibration model of the wire rope in this paper have certain reference value for the further study and analysis of the vibration characteristics of the wire rope of the traction elevator and the control of the vibration.

Acknowledgement
This study was funded by the Shandong Province Nature Science Foundation, China (GRANT NO. ZR2017MEE049), the Key Research Development Project of Shandong Province (GRANT NO. 2018GSF122004).

References

[1] J.H. Bao, P. Zhang, Journal of Shanghai Jiaotong University.46(3),341-345(2012).
[2] W.D. Zhu, J. Ni, Journal of Vibration and Acoustics.122(3),295-304(2000).
[3] J. Wu, Z.M. kui, et al, Journal of Vibration and Shock.35(2),184-188(2016).
[4] ITO. Hiroaki, Journal of System Design and Dynamics.2(2),540-549(2008).
[5] D.H. Yang, k.y. Kim,et al , J. Sound Vibr.390,164-191(2017).
[6] J.H. Bao, P. Zhang, et al, International Journal of Acoustics and Vibration.20(3),160-170(2015).
[7] X.M. Jiang, L.Z. Guo,et al, 6th International Workshop of Advanced Manufacturing and Automation (IWAMA).24,221-224(2016).

IOP Conf. Series: Materials Science and Engineering **538** (2019) 012037 doi:10.1088/1757-899X/538/1/012037

The Finite Element Analysis of the Boom of 20-ton Backhoe Hydraulic Excavator Based on ANSYS

Chao-ying Meng[1], Shan Fan[1a], and Lei-lei Han[1]

[1]Wenhua College, 430074 Wuhan, China

Corresponding author: 271732369@qq.com

Abstract. Taking the boom of 20-ton backhoe hydraulic excavator as the research object, this paper analyzed the strength of the boom by ANSYS software, simulated the load condition of the boom under the bucket excavation condition, and obtained the stress distribution cloud diagram of the boom. The analysis results show that the stress of the boom of the excavator is mainly concentrated on the hinge of the boom and the stick cylinder, which is the weakest link.

1. Introduction

Excavators are widely used in many industries of national economic construction. The working device is an important part of the excavator. It is mainly composed of the stick, the boom, the boom cylinder, the bucket cylinder and the bucket [1]. The boom is the main component of the excavator directly subjected to the working load, and its structural strength directly affects the reliability and working performance of the excavator. The analysis and research of the boom is the basis for the development of the whole machine. The stress on the boom is very complicated and constantly changing, and the working environment is also very bad. Therefore, it is difficult to get a real working situation analysis.
At present, a large number of scholars have done a series of research on the force of the working device, including kinematics and dynamics analysis calculation, optimization design, reliability analysis and so on. [2-4]. There is very little analysis of the boom of the backhoe excavator. This article took the boom of a 20-ton backhoe excavator as the research object. The author established the parametric model of the excavator boom by Pro/E and imported it into the ANSYS software to analyze the force of the boom under the bucket excavation condition.

2. The force Analysis of Excavator Boom

The moving speed of each component in the excavator is small and the force is large. Therefore, the force of the boom is mainly analyzed from the static point of view. When the excavator is in the deepest excavation state, the boom is lowered to the lowest point, and the bucket arm has the largest force arm. The stick and the arm cylinder hinge point F, the stick and bucket hinge point Q and the bucket tip point V are three points on a lead line. The position of its working device is shown in Figure 1.From the analysis of various failures of the working device, the maximum stress of the dangerous section of the boom occurs in the bucket excavation condition. Therefore, this paper selected this working condition for force analysis.
During work, the tooth resistance is affected by the cutting resistance and the soil resistance, and the sum of two is called the total resistance of the excavation [5]. However, accurate calculation of the total resistance is more difficult. Therefore, the total drag W is generally decomposed into two

component forces in the calculation: the excavation resistance F_{wt} in the tangential direction and the excavation resistance F_{wn} in the normal direction. At the same time, the working device is also subjected to the force of the oil cylinder. There are three main types: the boom cylinder thrust F_1, the arm cylinder thrust F_2, and the bucket hydraulic cylinder force F_3. The force is shown in Figure 1.

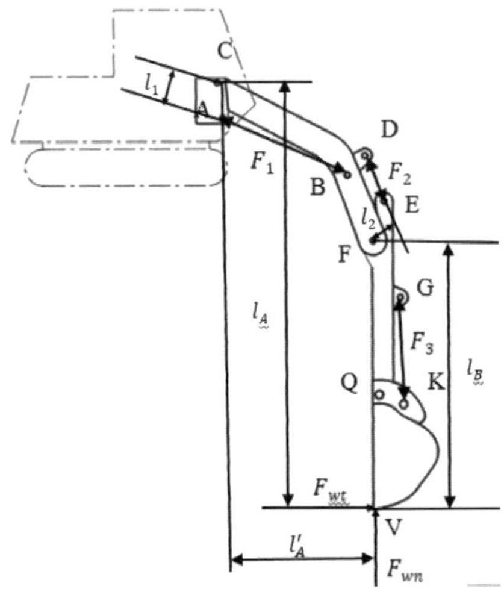

Figure 1. Excavator working device position and force analysis

3. Load Calculation

3.1. The Bucket Excavation Resistance

F_{wt} is the component of the total digging force in the tangential direction, which is the excavation resistance; and F_{wn} is the resistance generated when the bucket is inserted into the soil. In fact, the excavation resistance F_{wt} consists of three parts, which are the additional resistance when loading soil, the frictional resistance of the soil to the bucket, and the resistance to soil cutting [6]. The empirical formula of F_{wt} is as follows:

$$F_{wt} = C\left\{r_1\left[1 - \frac{\cos\emptyset max}{(\cos\emptyset max - \emptyset)}\right]\right\}^{1.35} BAZX + D \quad (1)$$

Where C is the hardness coefficient of the soil, different values for different soil conditions, for grade II soil, C=*50～80*, for grade III soil, C=*90～150*, for grade VI soil, C=*160～320*.This paper assumes that the excavator is used for excavation of class III soil, its value can be *100*; r_1 is the distance between the bucket and the stick hinge to the tip of the bucket, r_1=*1500mm*; $\emptyset_{max}$ is the half of the total corner of the bucket during the excavation; $\emptyset$ is the instantaneous corner of the bucket, take the empirical value $\emptyset_{max} = \emptyset = 55$; B is the cutting edge width influence coefficient, $B = 1 + 2.6b = 1 + 2.6 \times 1.04 = 3.7$; A is the cutting angle variation influence coefficient, $A = 1.3$; Z is the bucket coefficient, $Z = 0.75$; X is the bucket sidewall thickness influence coefficient, $X = 1.15$; D is the force of the cutting edge to squeeze the soil, according to the empirical statistics and the size of the bucket capacity, $D = 1.35 \times 104N$.

Bring the parameters into equation (1), we get:

$$Fwt = 168560N$$

The digging resistance of the bucket is generally less than that of the tangential excavation. The empirical formula is:

$$Fwn = 0.2\ Fwt = 33712N$$

3.2. The Cylinder Thrust

There are three kinds of cylinders on the hydraulic excavator working device—the bucket cylinder, the stick cylinder and the boom cylinder. The determination of the force of these cylinders depends on the type of the excavator and the specific working conditions. Since only the force of the boom is analyzed here, it is only necessary to calculate the boom cylinder thrust F_1 and the arm cylinder thrust F_2.Calculate its size according to the torque balance formula.

Ignore the quality of the bucket and soil, the quality of each component, and the factors affecting the efficiency of the linkage mechanism [7]. According to the moment balance, the formula for calculating F_1 is as follows:

$$F_1 = \frac{F_{wt}l_A + F_{wn}l_A^{'}}{l_1} \tag{2}$$

Where l_1 is the force arm of the boom cylinder thrust F_1 to the hinge point C on the boom; l_A is the force arm of F_{wt}on the hinge point C of the boom; l_A' is the force arm of F_{wn}on the hinge point C of the boom. The specific location is shown in Figure 1.

At this time, the arm and the boom cylinder are both locked, and we have:

$$F_2 = \frac{F_{wt}}{l_2} \tag{3}$$

Where l_2 is the arm of the arm cylinder thrust F_2 to the hinge point F; l_B is the force arm of F_{wt} to the hinge point F. The specific location is shown in Figure 1.

By calculation, we get:

$$F_1 = 1062N\ ;\ F_2 = 256N$$

3.3 The calculation of hinge point reaction force

The force of the stick and the boom is shown in Figure 2. We decomposed the force at F into F_{Fx} in the x direction and F_{Fy} in the y direction. The stick is in equilibrium under the action of these forces, so we have:

$$\begin{cases} F_{Fx} = -F_{wt} - F_2\cos\alpha \\ F_{Fy} = -F_{wn} + F_2\sin\alpha + G_2 + G_3 + G_4 + G_5 \end{cases} \tag{4}$$

Where α is angle between F_2 and x direction; G_2, G_3, G_4, G_5 is the gravity of the stick, bucket, bucket cylinder and connecting rod, its value can be *3.1* tons.

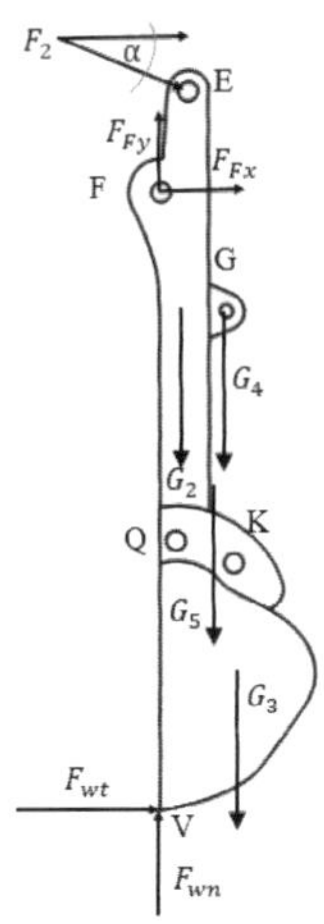

Figure 2. Stick reaming force diagram

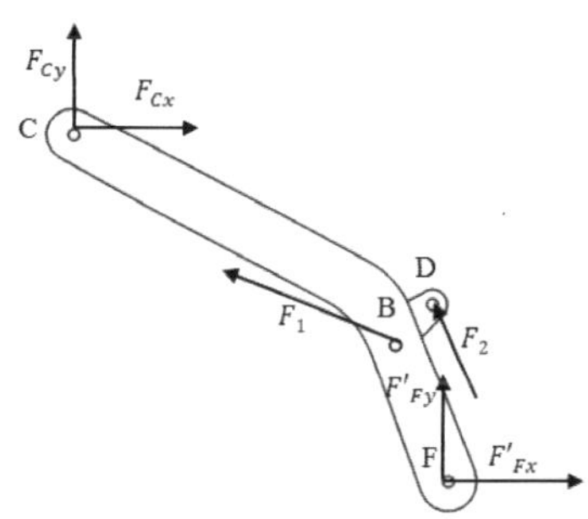

Figure 3. Boom reaming force diagram

The boom was extracted for force analysis, and the force was shown in Figure 3. According to the relationship between the force and the reaction force, the magnitude and direction of the F-force of the boom hinge point could be obtained.
According to the principle of static balance and torque balance, we could calculate the magnitude and direction of the load at the hinge points B, D and F of the boom under dangerous conditions. The specific values are shown in Table 1.

Table 1. Force values of each hinge point of the boom

F_1/KN	F_2/KN	F_{Fx}/KN	F_{Fy}/KN	α/ (°)
1062	256	252	-301	47

4. Finite Element Analysis

4.1. Finite Element Model
This paper used Pro/E software to parametrically model the boom. The boom is a welded part, and its structure is complicated. It has a lot of rounded corners and steps, which makes modeling difficult, so the necessary simplification of the boom was done before modeling. Small components such as rounded corners, steps, and ribs that do not affect the boom were omitted. Then we imported the *3D* model into ANSYS software. The finite element model is shown in Figure 4.

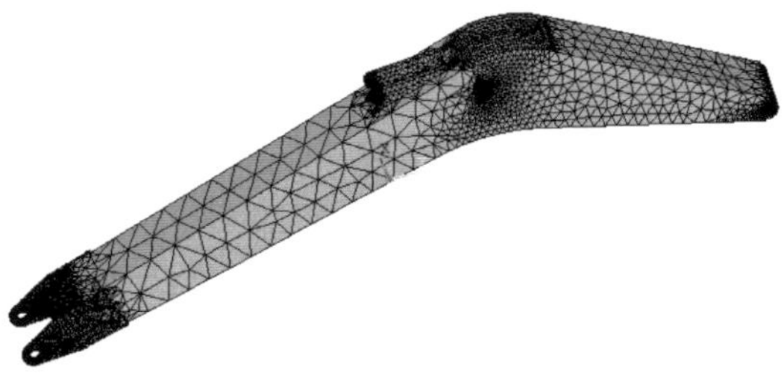

Figure 4. The finite element model of the boom

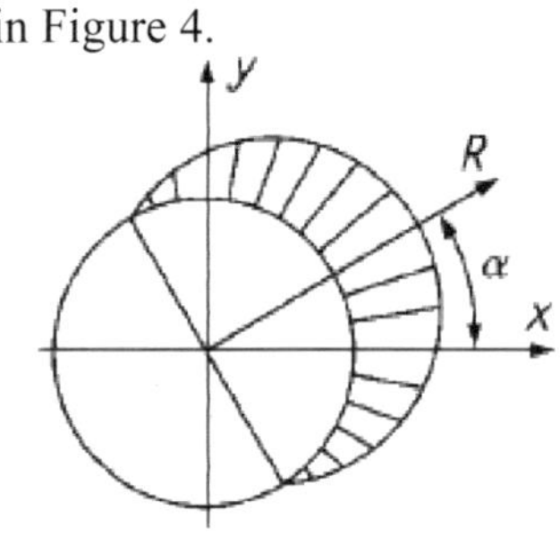

Figure 5. Force distribution hypothesis diagram

4.2 Apply Constraints and Loads

4.2.1 The constraint
When the excavator was in the bucket excavation condition, the boom was hinged to the frame. So we applied full constraint to the inner wall of the boom and frame hinge C. Since the stick and the bucket hinge point and the bucket tip point were on a vertical line, the inner hole at F was restrained in the y direction.

4.2.2 The load condition
The components of the working device were hinged by the pin shaft. If the influence of the eccentric load was ignored in the calculation of the working condition, the boom could be considered to bear only the forces in the *X* and *Y* directions, and the entire boom is a statically fixed structure. In fact, the force at the hinge point is not a concentrated force or a uniformly distributed load. To simulate the actual situation, we assumed that the load was distributed by cosine in the range of *180°* in the X-Y plane. The direction of the distributed force was the normal along the surface of the pin hole, and the load is evenly distributed in the z direction. The distribution of forces is shown in Figure 5.
According to the above assumption, the distribution function *F(θ)* of the load is:

$$F(\theta) = P\cos(\theta - \alpha) \tag{5}$$

Where θ is the angle between the line connecting the point of the pin hole and the center of the circle and the *X*-axis direction, $\theta=0\sim180°$; *P* is the undetermined coefficient; α is the angle between the resultant force R and the *X* axis.
The resultant R size is:

$$R = lr\int_{\alpha-\frac{\pi}{2}}^{\alpha+\frac{\pi}{2}} \cos(\theta-\alpha)\,d\theta = lr\int_{\alpha-\frac{\pi}{2}}^{\alpha+\frac{\pi}{2}} p\cos^2(\theta-\alpha)\,d\theta \tag{6}$$

The resultant force R is known, and the sum of the components of the distribution force *F* in the direction of the resultant force should be equal to R, and the undetermined coefficient can be derived according to the formula (6):

$$P = \frac{2R}{\pi lr} \tag{7}$$

Where *l* is the length of the pin hole; *r* is the radius of the pin hole; R is the resultant force of the cosine distribution load at the pin hole.
According to the formula (6), the distribution function of the load at the pin hole can be obtained, and the undetermined coefficient *P* can be obtained according to the formula (7). For any hinge point, it is only necessary to know the magnitude and direction of the force it is subjected to apply the cosine-changing surface load.
Apply boundary constraints to the boom based on the force of the boom (see Figure 4) and the calculated load. Since the structure and force of each reaming are different, different loads must be applied to the three reaming holes of the boom. At the hinge hole B of the excavator boom and boom cylinder, the boom and the arm cylinder reaming D are at the pin joint, and the surface load was distributed as a cosine function with angle. Therefore, we was applied a surface load that varies by cosine function to the cylindrical surface. At the same time, we applied a concentrated force load in the *x* and *y* directions at the hinge point F.

5. Result Analysis

After pre-processing, the software can complete different preset commands in the module to view different results. The results of the analysis can be presented in a variety of forms in the software, enabling observation of the results from different angles [8-9]. After the solution, the overall deformation map and the stress cloud map were obtained. After solving, the deformation displacement cloud diagram is shown in Figure 6. The stress cloud diagram of the boom and the arm cylinder reaming D is shown in Figure 7. The stress diagram of the boom and the arm reaming F is shown in Figure 8.

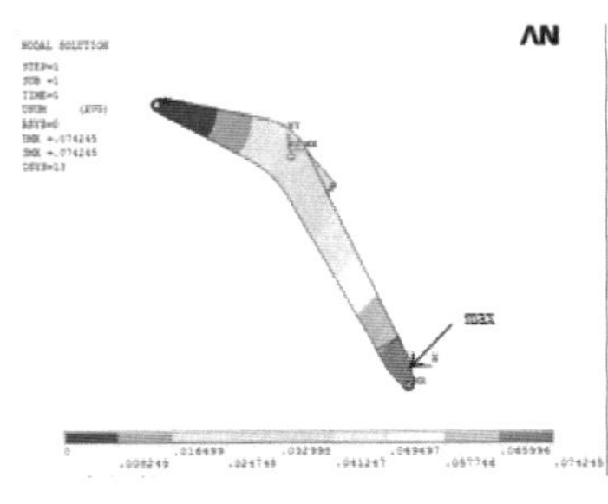

Figure 6. The deformation Displacement

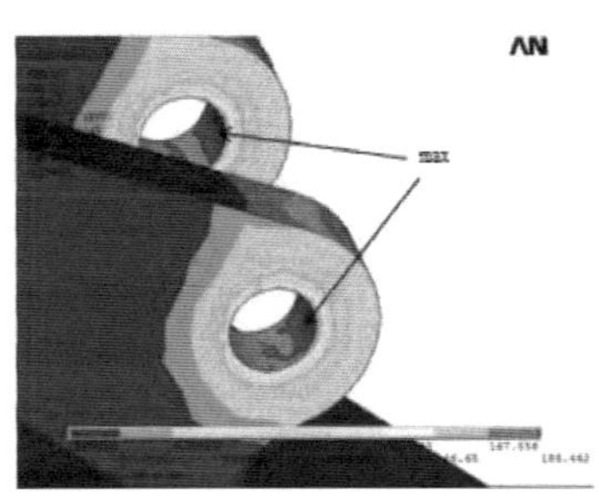

Figure 7. The stress cloud diagram of boom and stick cylinder reaming D

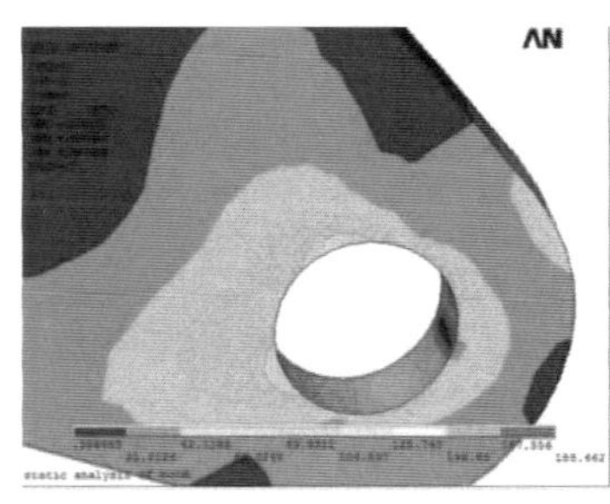

Figure 8. The stress diagram of the boom and the arm reaming F

As can be seen from Figure 6, the boom undergoes compression deformation. The amount of displacement of the boom from top to bottom and from left to right gradually increases. The maximum displacement deformation is 0.074245mm, and the maximum deformation occurs at the reaming hole where the boom and the arm are connected, that is, the lowest point of the boom when the boom is in the bucket excavation condition. The amount of deformation of the boom is in the elastic range, so the design meets the requirements.

As can be seen from Figure 7-8, the maximum stress of the boom occurs at the joint of the boom and the arm of the stick. The maximum stress is *188.462MP*. The material of the boom is *Q235* steel, and the allowable stress is *235MP*. The maximum stress of the boom is close to the allowable stress, but less than the allowable stress, within the safe range,the analysis results meet the requirements.

6. Conclusion

This paper analysed the force of the excavator boom under excavation conditions and performed finite element analysis in *ANSYS* software. The analysis results show that the stress of the boom of the excavator is mainly concentrated in the hinge of the boom and the arm of the stick, which is the weakest link. Therefore, in the design of the boom, it can be improved by adding reinforcing ribs inside the boom, increasing the thickness of the guard plate, and adding a reinforcing plate at the arm cylinder support. It also shows that the finite element analysis of the excavator boom with the useful finite element software *ANSYS* is a reliable method, which provides a valuable reference for product design.

References

[1] D. Kim, J. Kim, K. Lee, C. Park, J. Song, D. Kang, Excavator tele-operation system using a human arm , Automat Constr.**18**, 173–182(2009)

[2] K. Awuah-Offei, S. Frimpong. Cable shovel digging optimization for energy efficiency, Mech Mach Theory.**42**,995-1006.(2007)

[3] S. Frimpong, Cable shovel health and longevity and operat or efficiency in oil sands excavation, Mining and Mineral Engineering.**1**,15-33 (2008).

[4] S. Frimpong, H. Yafei, K. Awuah-Offei, Mechanics of cable shovel-formation interactions in surface mining excavations, J Terramechanics. **42** (2005)

[5] A.R.Dexter, E.A.Czyz, O.P.Gat, A methord for prediction of soil penetration resistance, Soil Till Res. **93**,412-419 (2007)

[6] D. Dopico, A. Luaces, M. González, A soil model for a hydraulic simulator excavator based on real-time multibody dynamics, ACMD. 23-26 (2010)

[7] H. Takahashi, M. Hasegawa, E. Nakano, Analysis on the resistive forces acting on the bucket of a Load-Haul-Dump machine and a wheel loader in the scooping task, Adv Robotics.**13**,97-114(1998)

[8] S.Kim, J Ha, S Jeong, Effect of joint conditions on the dynamic behavior of a grinding wheel Spindle,Int J Mach Tool Manu.**41**,1749-1761(2001)

[9] I. Cotton, Bradken Building, 5 Mining Monthly.**7**, 122(2006)

IOP Conf. Series: Materials Science and Engineering **538** (2019) 012038 doi:10.1088/1757-899X/538/1/012038

Study on Atomization Characteristics of Nozzles in Wet Electrostatic Precipitator

Lichun XIAO[1,*], Xin WANG[1], Yongpeng WEI[2], MengZHANG[1], Yuchuan WANG[1]

[1]College of Environmental and Chemical Engineering, Yanshan University, Qinhuangdao, China
[2]Solid Waste Management Center, Qinhuangdao, China

xlc@ysu.edu.cn

Abstract. Wet electrostatic precipitator (ESP) has been paid more and more attention because of its good collection effect on fine particles in coal-fired power plants. Spraying system is an important part of the wet ESP and its atomization characteristics has an important influence on the collection efficiency. The atomization performance of solid and hollow nozzles was tested under varying pressure. The dust collection efficiency of wet ESP affected by atomization parameters and corona discharge characteristics were studied. Dust removal experiments were carried out by using fly ash from coal-fired power plants. Results show that the best working water pressure of ESP is 0.6 MPa and the atomization effect is best at the same time. Under the same conditions, the dust collection efficiency increased with the decrease of droplets' diameter and the increase of spray water flow rate. When the droplets' size is 70 μm, the dust collection efficiency can reach 99.57%, which can meet the high efficiency of fine particles removal in coal-fired power plants.

1. Introduction

China is currently the largest coal producer and consumer country in the world[1]. The fine particulate matter emitted from coal-fired power plants is the main cause of haze weather[2,3]. Effective control of emission of fine particles from coal-fired power plants will contribute to the improvement of air quality in China[4-6].

Dry electrostatic precipitator is widely used in coal-fired power plants, but the removal effect of fine particles is not satisfactory. Wet ESP can remove fine particles effectively after wet desulfurization, and it has the advantages of small pressure loss, large temperature range and good economic feasibility. In order to improve the collection efficiency for removing fine particles in coal-fired power plants, it is necessary to study the spraying system of wet ESP.

Spraying system plays an important role in wet ESP. On the one hand, it is used to flush the dust collector plate; on the other hand, Droplets can coagulate fine particles and be conducive to their removal[7-9]. Meanwhile, the droplets' diameter has important influence on corona discharge and the spray water flow rate which affects the thickness of ash removal water film[10,11]. In order to make the ESP work in the best working conditions, i.e., low corona voltage, high spark voltage, large corona current and small water consumption, it is necessary to explore the best atomization characteristics[12,13]. The atomization parameters of solid and hollow nozzles were measured and their influence on Volt-Ampere characteristics and dust removal efficiency were studied. The results can provide guidance for the design and operation of wet ESP.

2. Experimental Principle and Device

2.1. Experimental Principle

The working process of wet ESP is as follows: the discharge system generates corona discharge under DC high voltage. The ionization of gas in the electrode space makes the dust particle charge. The charged particles move toward the dust collected plate under the action of electric field force. The water from the spraying system forms a layer of uniform water film on the collected plate. It can flush the dust into the ash hopper.

2.2. Experimental Device

Spray experiment and dust removal experiment were carried out in a wet ESP. As shown in Fig.1, the experimental device is mainly made up as follows: feeder (1), pipeline (2), high voltage power supply system (3), spraying system (4), polar matching system (7), induced draft fan (8), water tank (9), pump (11) and other units.

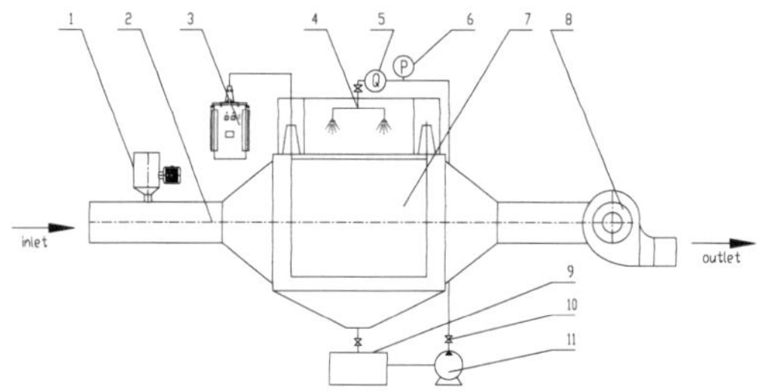

1-feeder 2-pipeline 3- high voltage power supply system 4-spraying system 5-flow meter 6- pressure meter 7-polar matching system 8- induced draft fan 9-water tank 10-stop valve 11- pump

Figure 1. The experimental device of wet ESP.

As shown in Fig.2, Water in the tank 1 is pressurized by pump 2 and ejected from the nozzle 7. The water pressure and flow rate can be adjusted by the regulating valve 3. And they are displayed from flowmeter 4 and pressure gauge 5. Droplets' diameter of the nozzle under different water pressure were measured by a laser particle size analyzer what type is DP-02. The laser particle size analyzer consists of laser launcher 6 and laser receiver 8.

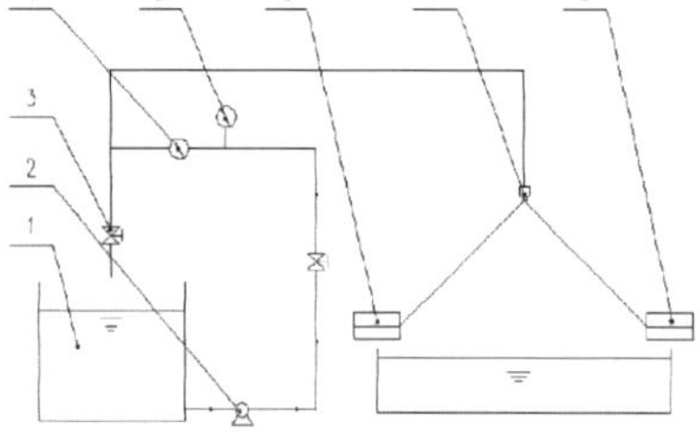

1.water tank 2. pump 3. regulating valve 4. flowmeter 5. pressure gauge 6. laser launcher of laser particle size analyzer 7. nozzle 8. laser receiver of laser particle size analyzer

Figure 2. Measurement system for droplets' diameter.

The stainless-steel atomizing nozzle is chosen as the test nozzle. It is divided into two types what name are solid cone nozzle (1#~3#) and hollow cone nozzle (4#~5#). Their parameters are shown in Table 1. where Q_V is the nozzle flow rate and α is the atomizing angle.

Table1. Nozzle parameters for atomization characteristics test

Nozzle number	solid cone nozzle			Hollow cone nozzle	
	1#	2#	3#	4#	5#
Flow rate Q_V/(L/min)	2.35	3.7	10	3.9	3.0
α (°)	90	65	65	70	90

3. Results and Analysis

3.1. *Determination of Atomization Parameters of Nozzles*

The atomization quality of the nozzle is usually measured by droplet size distribution. Droplet size distribution is usually expressed by mean diameter. For example, d_{50} is the mass median diameter, d_{sv} is the surface area volume mean diameter and d_v is the volume mean diameter. For laser particle size analyzer, the more approximate spherical particles, the more accurate measurement results. Then, the closer values of d_{sv} and d_v are. What's more, the more regular shape of the sample particles is, the more concentrated particle size distribution is.

Fig. 3 shows the relationship between droplets' size and water pressure. It can be seen from the figure, the median diameter d_{50} of droplets spraying from the five kinds of nozzles decreases with the water pressure. This indicates that the atomization effect of the nozzles gets better with the increase of water pressure. Droplets' diameter changes little when the pressure is more than 0.6 MPa. The reason is that the breakage process of droplets is mainly described by Weber number which is shown in equation (1). Droplet collisions will cause the coalescence, rebound and tensile separation. When the Weber number is high, droplets will have great kinetic energy. At this time, the droplets will be easily separated and broken.

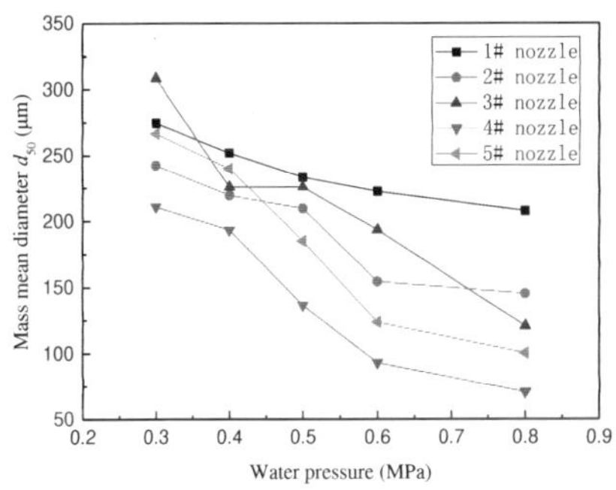

Figure 3. Relationship between droplets diameter and water pressure.

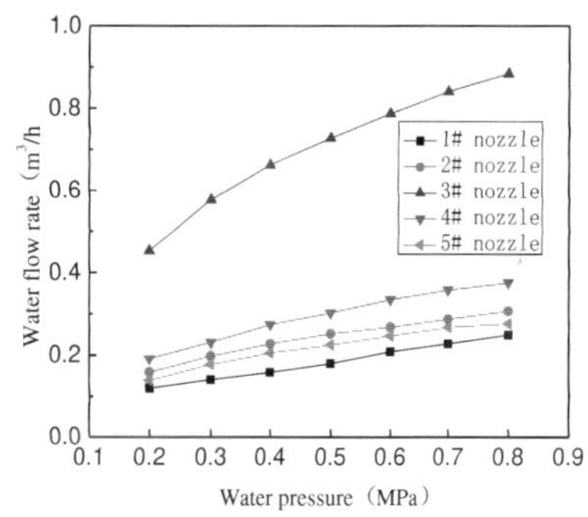

Figure 4. Relationship between spray water flow rate and pressure.

$$\mathrm{We} = \rho v^2 D / \sigma \tag{1}$$

Where ρ is the liquid density, σ is the surface tension of the liquid, D is the droplet diameter, v is the characteristic velocity of the droplets.

When $\rho = 1000\ \mathrm{kg/m^3}$, $\sigma = 7.28\times10^{-2}\ \mathrm{N/m}$, the average velocity of droplets is the characteristic velocity. When the droplets collide each other, whether the droplets occur separation can be expressed by a threshold value *x*. When the value exceeds the value *x*, the droplets will occur tensile separation. The value x is shown as the equation (2).

$$x = \frac{1}{\Delta^3}\left(\frac{6(1+\Delta^3)}{\mathrm{We}}\right)^{1/2} \tag{2}$$

Where Δ is a dimensionless constant. It is the ratio of the maximum droplets' diameter to the minimum droplets' diameter. The larger water pressure is, the closer droplets' diameter is, and the smaller dimensionless constant Δ is, the larger value x is. The droplets will be easy to separate. Therefore, considering the factors of atomization effect and energy consumption, the optimum atomization pressure is 0.6 MPa.

3.2. Spray Water Flow Rate Affected by Water Pressure

As can be seen from Fig.4, the spray water flow rate of the nozzle increases with the increase of water pressure. This trend of change gradually becomes gentle when the pressure increases to a certain range. For the same fluid, the relationship between the spray water flow rate and pressure is shown in Equation (3). From the Equation, it can be seen that for a nozzle, there is a quadratic linear relationship between the flow rate and pressure, i.e., with the increase of pressure, change of the spray water flow rate gradually slows down.

$$Q_1/Q_2 = \sqrt{P_1/P_2} \tag{3}$$

3.3. V-I characteristic Affected by Gas Speed

Secondary voltage and current of the wet ESP can reflect its working state affected by various factors. The relationship between corona voltage and current of the ESP is called volt-ampere(*V-I*) characteristic curve. Fig.5 shows the *V-I* characteristic curve under different gas speed. The experimental conditions were as follows: Fishbone needling line was selected as cathodic line. The water pressure was 0.6 MPa and the gas speed was selected from 0.7 m/s to 1.5 m/s. As can be seen from the figure, the curve is not obviously dispersed. It indicates that the secondary current under the same secondary voltage has little difference with the increase of gas speed. The reason is that: with the increase of gas speed, more large diameter droplets are carried away with the gas, which will lead to a downward trend of secondary current. Nevertheless, the momentum of space ions increases with the gas speed, which will lead to an increasing trend of collision probability of ions and an upward trend of secondary current.

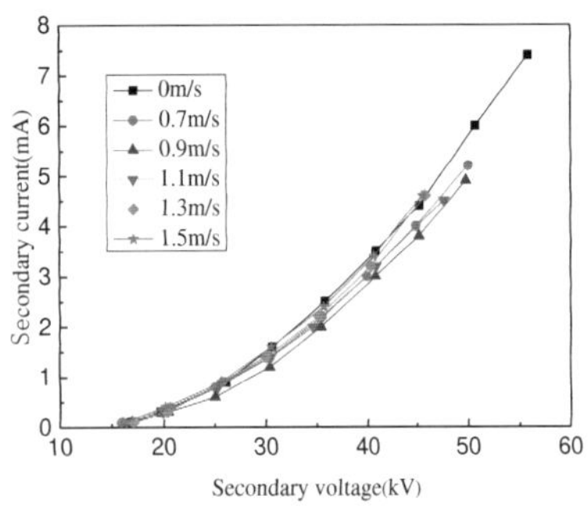

Figure 5. *V-I* characteristic curve affected by gas speed.

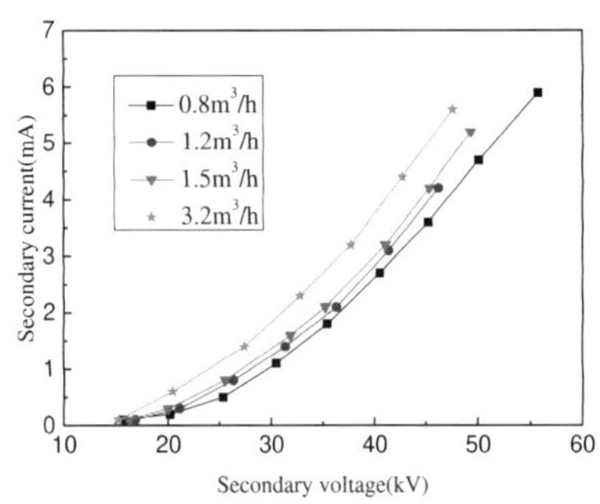

Figure 6. *V-I* characteristic curve affected by water flow.

3.4. *V-I* characteristic Affected by Water Flow

Fig. 6 shows the volt ampere characteristic curve at different spray water flow when the pressure is 0.6 MPa. As can be seen from the figure, with the increase of spray water flow, the secondary current increases gradually on the same secondary voltage. The reason is that the number of droplets in the electric field increases with the water flow rate. The local distortion of the electric field will be enhanced by a great number of droplets. The corona discharge will be accelerated because of the streamer propagation under the guidance of droplets. The collision between ions and droplets in electric field is strengthened, which will produce more electrons and make the corona current increase..

Fig. 7 shows the discharge inception voltage and spark voltage affected by the water flow. The difference between spark voltage and inception voltage is named effective operation voltage. The larger operation voltage, the wider working range of wet ESP. As can be seen from the figure, the discharge inception voltage of the five nozzles is basically the same, but the discharge spark voltage is distinctly different. On the same water pressure, the spark voltage of solid cone nozzle is higher than that of the hollow one. For the same type of nozzle, the larger flow rate is, the lower spark voltage is. This is because the number of small droplets in the electric field space increases with the spray water flow rate. These droplets will distort the electric field and cause the mechanism of electron collision effect become worse. The irregular collisions of some electrons with strong energies will lead to spark discharge.

3.5. Collection Efficiency Affected by Droplets' Diameter and Spray Flow

The atomization quality of the nozzle is usually measured by droplet size distribution. Droplet size distribution is usually expressed by mean diameter. For example, d_{50} is the mass median diameter, d_{sv} is the surface area volume mean diameter and d_v is the volume mean diameter. For laser particle size analyzer, the more approximate spherical particles, the more accurate measurement results. Then, the closer values of d_{sv} and d_v are. What's more, the more regular shape of the sample particles is, the more concentrated particle size distribution is.

Table 2 shows the dust removal efficiency affected by different droplets' diameter. It can be seen from the table that the dust collection efficiency gradually increases with the droplets' diameter. When the droplets' mean diameter d_{50} is 70 μm, the dust collection efficiency can reach 99.57%. The reason is that the smaller droplets' mean diameter is, the more droplets in the electric field are. The probability of collision between droplets and dust will increase.

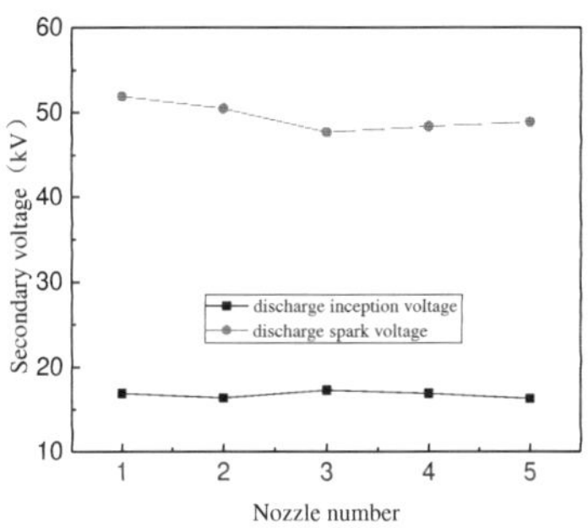

Figure 7. Discharge inception and spark voltage affected by water flow.

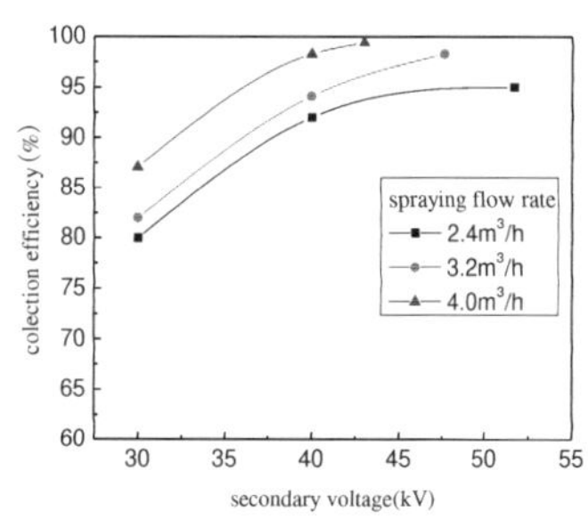

Figure 8. Collection efficiency at different spray flow rate.

Table2. Collection efficiency at different droplets' diameter

Test number	1#	2#	3#	4#	5#
d_{50} (μm)	205	155	115	95	70
Collection efficiency(%)	97.22	98.01	98.87	99.16	99.57

The water consumption of the wet ESP is determined by the spray water flow of the nozzle. Therefore, it is important for the wet ESP to save water by choosing right nozzle.

Fig. 8 show the collection efficiency at different spray flow. The same type of nozzle was selected to change the spray flow in order to eliminate the influence on the efficiency affected by droplets' diameter. It can be seen that the dust removal efficiency increases with the increase of spray flow rate. It is more than 97% when the spraying flow rate is 4.0 m^3/h. The reason is that the secondary current increases with the spray flow rate, it will have advantages to improve the dust removal efficiency. What is more, the thickness of water film increases with the spray flow and dust removal becomes more thorough.

4. Conclusion

The atomization characteristics and corona discharge characteristics of different spraying nozzles were studied in this experiment. Dust removal experiments were carried out with fly ash from the coal-fired boiler in a power plant. The conclusions were obtained as follows: The droplets' diameter decreases with the increase of water pressure. When the water pressure is more than 0.6 MPa, the change of droplets' size turns smooth. The optimum water pressure is 0.6 MPa. The secondary current under the same secondary voltage has little difference with the increase of gas speed. When the spray water flow rate is 3.2 m^3/h, the wet ESP has larger secondary current under the same secondary voltage. With the decrease of droplets' diameter, the dust removal efficiency of wet ESP increases. When droplets' diameter is 70 μm, the dust removal efficiency reaches the highest value which is 99.57%. The dust removal efficiency increases with the spraying flow rate and it is more than 97% when the spraying flow rate is 4.0 m^3/h.

Acknowledgments
This research was supported by the Natural Science Foundation of Hebei pvovince, China(E2015203236).

References

[1] J. Shen, C. Zheng, L. Xu, Atmospheric emission inventory of SO3 from coal-fired power plants in China in the period 2009–2014, ATMOS ENVIRON, vol.197,2019, pp. 14-21.
[2] K. Cong, Y. Zhang, F. Han, Influence of particle sizes on combustion characteristics of coal particles in oxygen-deficient atmosphere, ENERGY, 2018 ,pp. 1-32.
[3] H. Bin, Y. Yang, L. Cai, Y. Zhulin, S. Roszak, Y. Linjun, Experimental study on particles agglomeration by chemical and turbulent agglomeration before electrostatic precipitators, POWDER TECHNOL, vol.335,2018,pp.186-194.
[4] Z. Sun, L. Yang, A. Shen, B. Hu, Improving the removal of fine particles from coal combustion in the effect of turbulent agglomeration enhanced by chemical spray, FUEL, vol.234,2018,pp. 558-566.
[5] Z. Yang, C. Zheng, S. Liu, A combined wet electrostatic precipitator for efficiently eliminating fine particle penetration, FUEL PROCESS TECHNOL, vol.180,2018,pp. 122-129.
[6] X. Li, J. Li, D. Wu, Removal effect of the low-low temperature electrostatic precipitator on polycyclic aromatic hydrocarbons, CHEMOSPHERE, vol.211,2018,pp. 44-49.
[7] J. Mertens, P. Khakharia, P. Rogiers, Prevention of mist formation in amine based carbon capture: field testing using a Wet ElectroStatic Precipitator (WESP) and a Gas-Gas Heater (GGH), in: T. Dixon, L. Laloui, S. Twinning(Eds.), 2017, pp.987-999.
[8] P. Wang, X. Wang, Y. Liu, Particle Removal Efficiency Analysis of WESP in a Typical 660MW Coal-fired Power Plant, 2016, pp.682-686.
[9] Y. Liu, J. Ren, H. Liu, Experimental Study on 1000MW Ultra-supercritical Coal-fired Units with Wet Electrostatic Precipitator Efficiency, 2015, pp.53-57.
[10] Y. Liu, B. Hu, L. Zhou, Improving the Removal of Fine Particles with an Electrostatic Precipitator by Chemical Agglomeration, ENERG FUEL, vol.30,2016,pp. 8441-8447.
[11] B. Hu, Y. Yi, C. Yang, Improving the electrostatic precipitation removal efficiency by desulfurized wastewater evaporation, RSC ADV, vol.6,2016,pp.113703-113711.
[12] T. Chen, C. Tsai, S. Yan, An efficient wet electrostatic precipitator for removing nanoparticles, submicron and micron-sized particles, SEP PURIF TECHNOL, vol.136,2014,pp. 27-35.
[13] A. Bologa, H.R. Paur, H. Seifert, Novel wet electrostatic precipitator for collection of fine aerosol, Journal of Electrostatics 11th International Conference on Electrostatics, vol,67,2009,pp. 150-153.

Measurement of Gas Temperature and Concentration Based on TDLAS

Zijun Chang[1], Qingming Liu[1,a], and Yang Shen[1]

[1]Beijing Institute of Technology, School of Mechatronical Engineering, 100081 Beijing, China

Corresponding author: qmliu@bit.edu.cn

Abstract. The measurement of gas temperature and concentration in combustion process has always been an important part of research. Based on the application of tunable diode laser technology in gas detection, the molecular absorption spectroscopy system was established and the concentration and temperature of water vapor were measured. Based on the spectral parameters in the latest HITRAN 2016 database, a group of programs have been compiled, which adopt the form of VOIGT function and consider frequency shift caused by pressure and temperature and . The spectral absorbance of H_2O was obtained by the computer program. Based on the scanning range of the laser, the spectral lines of water vapor molecule near 1392 nm were screened to obtain the suitable pairs of spectral lines for measuring temperature and concentration. The pairs of spectral lines were validated as gas temperature measurements at high temperature by numerical simulation. TDLAS system is built independently, and TDLAS technology is used to measure the gas concentration field and temperature field in an experimental way. It is of great significance for TDLAS technology in gas monitoring.

1. Introduction

The measurement of gas temperature and concentration is of great significance in aerospace, energy and chemical industry, environmental monitoring and other industries. With the development of molecular spectroscopy and semiconductor laser technology, tunable Diode Laser Absorption Spectroscopy (TDLAS) [1] technology is an optical diagnostic technology and can measure the concentration and temperature of gas components. It has the advantages of high sensitivity and is not easily affected by other gases [2]. It has become a research hotspot at home and abroad. At present, TDLAS technology has been applied in various aspects of industrial production and life, such as production of exhaust gas and flue gas, monitoring of production of toxic and combustible gases, diagnosis of combustion flow field such as engines, and monitoring of air pollutants [3-5].As a spectral detection method with high sensitivity and resolution, TDLAS plays an important role in the fields of analysis chemistry and combustion diagnose [6-7].

Through the analysis of molecular spectra, molecular spectra can be obtained, which can be used as the application basis of TDLAS technology. At present, the spectral line data of TDLAS research at home and abroad are basically from HITRAN database. The spectral line parameters of 49 substances in HITRAN 2016, the seven main atmospheric absorbers in the initial infrared band, include spectral line position, linear intensity, several broadening coefficients and molecular cutting function, pressure frequency shift, Einstein coefficient and other molecular spectral parameters [8]. Existing tools for molecular spectroscopy mapping, such as HITRAN 2012 and HITEMP 2010 databases used by Spectraplot, have not been updated in time, and many molecular spectroscopic maps, such as H_2, SF_6 and C_4H_2, are lacking. When calculating the absorbance, less attention is paid to the pressure frequency shift.

Water vapor, as a common hydrocarbon fuel combustion product, has a wide spectrum distribution and complete data parameters. It is suitable for measuring gas temperature and other aspects. It has important significance in gas diagnosis such as combustion field [9-10]. Because of many factors affecting the measurement results, the selection of water molecular spectral lines is diverse, and there is no standard for selecting spectral pairs. Generally, the selection depends on experiments according to the actual situation. In this paper, based on the HITRAN2016 spectral parameters of water, considering the influence of temperature, pressure and other factors, the absorbance of H_2O was simulated by using MATLAB. The absorption spectra of water at 1392 nm were taken as the object of study, and suitable spectral lines were selected to measure gas temperature and concentration. A TDLAS system platform based on THORLABS ITC4001 semiconductor laser with temperature control mode was built to measure gas temperature and water molecule concentration.

2. Measuring method of temperature and concentration based on TDLAS

Using TDLAS technology to measure gas concentration can be measured by direct absorption method. Direct absorption spectroscopy is to record incident and transmitted light intensity directly according to Lambert's law, and to retrieve gas concentration according to absorbance. According to Lambert Bill's formula:

$$PXS(T)L = \int_{-\infty}^{+\infty} -\ln(\frac{I}{I_0})dv = A \tag{1}$$

P [atm] is the partial pressure of the absorbing species; X is the concentration of gas; of the absorbing species; S [$cm^{-2} \cdot atm^{-1}$] is the strength of the transition , which indicates the absorption intensity of the spectral line; L [cm]is the absorption path length; Io is incident laser intensity or power, And I is transmitted intensity or power.

Therefore, a formula for calculating gas concentration can be obtained:

$$X = \frac{\int_{-\infty}^{+\infty} -\ln(\frac{I}{I_0})dv}{PS(T)L} \tag{2}$$

TDLAS-based gas temperature measurement scheme is to obtain two spectral absorbances under different transition states at the same pressure, molar fraction and optical path in the wavelength scanning scheme. This method needs to scan two different characteristic absorption spectra of gases, and calculate the temperature by two spectra varying with temperature. It is also called bispectrum temperature measurement method. The ratio of spectral absorbance area measured can be expressed as the ratio of spectral line intensity, and the function related to temperature can be obtained.

$$R = \frac{A_1}{A_2} = \frac{S_1(T)}{S_2(T)} = f(T) \tag{3}$$

Therefore, the gas temperature can be calculated from the ratio of the spectral absorbance obtained. If the central wavelength of the two characteristic lines is close enough, and the intensity ratio of the two absorption lines can be expressed as follows:

$$R = \frac{A_1}{A_2} \approx \frac{S_1(T_0)}{S_2(T_0)} \exp[-\frac{hc}{k}(E_1^{"} - E_2^{"})\left(\frac{1}{T} - \frac{1}{T_0}\right)] \tag{4}$$

From equation (4), the formula for calculating gas temperature can be obtained by measuring the ratio of absorption intensity of gas absorption line pairs. The sensitivity of temperature measurement is expressed by the differential of spectral intensity ratio to temperature.

IOP Conf. Series: Materials Science and Engineering **538** (2019) 012039 doi:10.1088/1757-899X/538/1/012039

$$\frac{\partial R/R}{\partial T/T}=(\frac{hc}{k})\frac{\left|E_1^{"}-E_2^{"}\right|}{T} \tag{5}$$

At present, the output wavelength of commercial semiconductor lasers can be tuned by adjusting the magnitude of the current and the temperature of the laser. By changing the modulation current of the input laser, the single wavelength laser is driven to scan the gas absorption spectrum in frequency domain. The spectral absorption signal is determined by comparing the attenuated laser intensity with the reference laser intensity (baseline). Therefore, when the parameters such as pressure, linear intensity and laser traversing distance in gas medium are known, the integral value of spectral absorptivity in frequency domain can be brought into equation (2), and the gas concentration can be finally obtained.

For TDLAS bispectrum temperature measurement, the key is to choose the right pair of lines. For direct absorption measurement, how to determine the reference signal of laser intensity is very important for the final calculation results. Especially for the measurement of weak absorption signal, the rerror caused by the baseline intensity would produce tremendous impact on concentration evaluation.

3. Selection of spectral pairs

According to the scanning center wavelength of the laser at 1392 nm and the HITRAN 2016 database, the absorption spectra of water molecules in the range of 7174-7192 cm^{-1} are shown below.

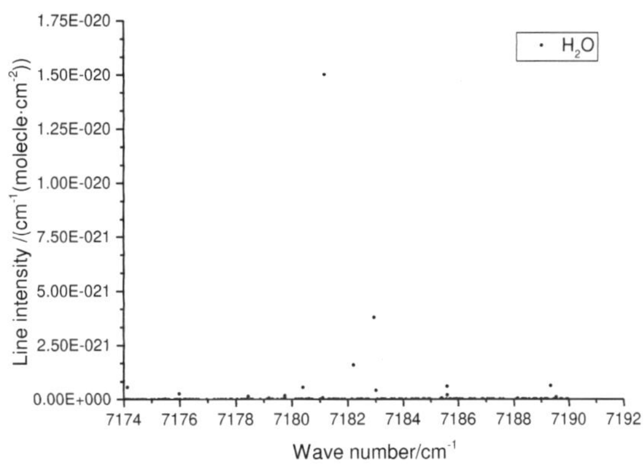

Figure 1. Absorption spectra of water molecules in the range of 7174-7192 cm^{-1}.

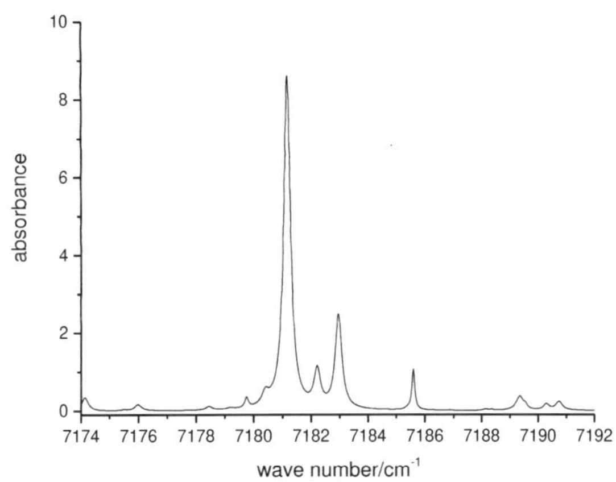

Figure 2. Absorbance of H_2O in the range of 7174-7192 cm^{-1}.

IOP Conf. Series: Materials Science and Engineering **538** (2019) 012039 doi:10.1088/1757-899X/538/1/012039

The spectral absorption map of this section can be calculated by a MATLAB program. Figure 2 shows the spectral absorbance at 296K and 1% concentration at 1atm and 100cm pathlength. There are 13 peaks in the absorption spectrum. The spectral data of this spectral position are derived from HITRAN 2016. Considering that temperature measurement requires a larger spectral intensity, the spectral lines with a spectral intensity more than 4E-23 ($cm^{-1}/(mol \cdot cm^{-2})$) or 0.001 ($cm^{-2} \cdot atm^{-1}$) are selected as the optional spectral lines.Table 1 shows the spectral parameters corresponding to the absorbance peaks of water molecules in 7174-7192 cm^{-1}.

Among the 13 absorption peaks in Table 1, peak5, 9, 10, 11 have overlapping spectral lines, which are not considered. Peak10 has been selected as available in many studies, but it is no longer applicable after the update of HITRAN 2016. It is necessary to re-select the more suitable spectral pairs.Although peak 1 laser can be scanned, it is not easy to control the temperature too high and cause damage to the instrument. It is not considered when there is a better choice. In the remaining spectral pairs, peak12(7190.2890 cm^{-1}) and peak13(7190.7388 cm^{-1}) were selected as the measurement line pairs considering various factors. The main reasons are as follows: the intensity of the two spectral lines is relatively high, and better spectral line measurement data can be obtained; the distance between the two spectral lines is 0.4498 cm^{-1}, which ensures that the measurement results can be scanned by a laser beam, so that the measurement results can be separated under the same experimental conditions and at the same time have a certain wavenumber difference; within a certain range, the intensity of the two spectral lines varies with temperature, and the intensity ratio is different at different temperature. The ratio of absorbance area varies with temperature as a single valued function.

Table 1. Spectral parameters corresponding to absorbance peaks of water molecules in 7174-7192 cm^{-1}.

peak	ν cm^{-1}	ν cm^{-1}	S $cm^{1}/(molcm^{-2})$	E'' cm^{-1}
1	7174.14	7174.13778	5.77E-22	95.1759
2	7175.987	7175.98676	2.68E-22	206.3014
3	7178.446	7178.44584	1.44E-22	602.7735
4	7179.19	7179.18718	5.76E-23	326.6255
5	7179.75	7179.7524	1.698E-22	1216.1944
		7179.7534	5.667E-23	1216.1897
6	7180.4	7180.39972	5.56E-22	224.8384
7	7181.154	7181.15578	1.50E-20	136.7617
8	7182.207	7182.209	1.58E-21	42.3717
9	7182.953	7182.94962	3.77E-21	142.2785
		7183.01579	4.07E-22	134.9016
10	7185.596	7185.596571	1.98E-22	1045.0583
		7185.5973	5.93E-22	1045.0578
11	7189.347	7189.34444	6.21E-22	142.2785
		7189.5420	1.07E-22	1255.1666
12	7190.29	7190.2890	2.41E-22	842.3565
13	7190.74	7190.7388	3.94E-22	275.4971

Figure 3 (a) (b) is linestrength of Peak 12 and Peak 13, and the ratio of line strength to peak 12 and peak 13. According to the sensitivity calculation formula, the energy difference between the low-state energy levels of the two lines is 566.8594, and the temperature sensitivity is 2.76>1.

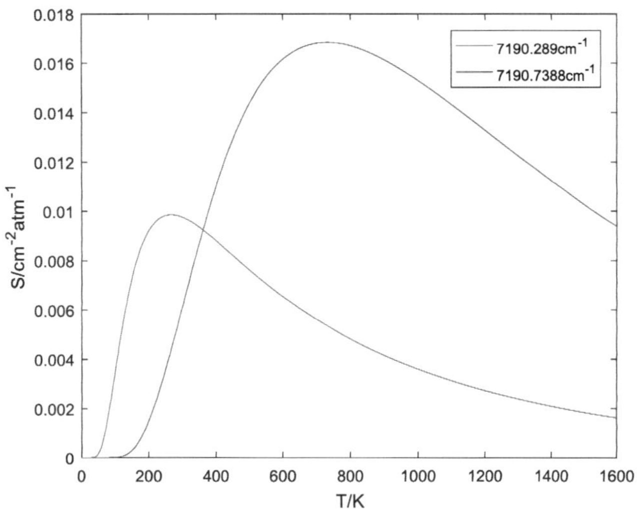

(a)

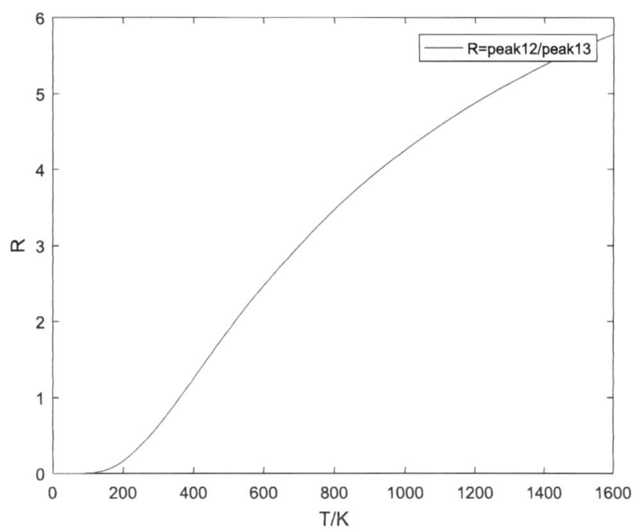

(b)

Figure 3. (a) Linestrength with temperature at v_{12} and v_{13}. (b) linestrength ratio with temperature at v_{12} and v_{13}.

4. Numerical simulation

According to HITRAN 2016, peak12 and peak13 can be selected as spectral lines for temperature measurement. The absorbance data of 10% concentration, 100 cm optical path, 1 atm pressure and 300 K temperature were obtained by numerical simulation. The temperature and concentration were deduced from the data. According to the spectrum method, the absorbance area A1 (7190.2890 cm^{-1}) and A2 (7190.7388 cm-1) need to be measured. In order to increase the accuracy of the experiment, this paper does not integrate the data directly, but fits the Voigt curve by bimodal fitting, and then carries out numerical integration to get the area ratio. The advantage of bimodal fitting is that the absorbance area of two spectral lines can be calculated separately by using the fitting Voigt curve of the two spectral lines. The following figure simulates the theoretical data as well as the spectrogram of the fitting curve based on the data. The numerical integration results are A1= 0.06067 and A2=

0.09514. The calculated temperature is T = 300.55K, the measurement error is 0.55K, the calculated water molecule concentration is X1 = 9.89%, X2 = 9.89%, and the calculation error is 0.11%.

Figure 5 is the data and fitting curve when the temperature is changed to 500 K.The numerical integration results are A1=0.09389 and A2=0.05106.The calculated temperature is T = 492.88K, the measurement error is7.12K, the calculated water concentration is X1 = 10.9%, X2 = 10.9%, and the calculation error is 0.9%.

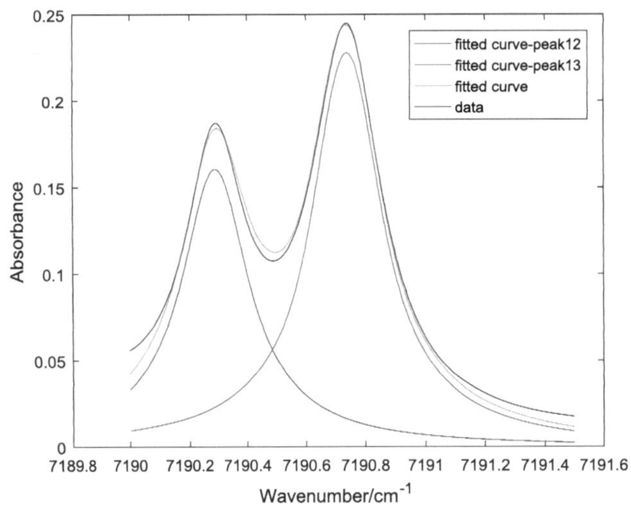

Figure 4. Simulated data and spectrograms of fitting curves at 300K.

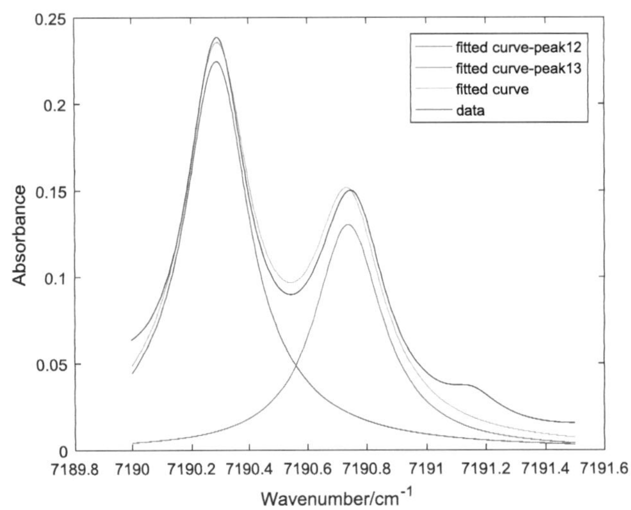

Figure 5. Simulated data and spectrograms of fitting curves at 500K.

Based on the absorbance data of the reconstructed spectral pairs under known conditions, the temperature error is less than 1.5% and the concentration error of water gas is less than 10%. It can be used as a spectral pair for gas temperature and concentration measurement.

5. Experimental verification

The experimental scheme is shown in figure 6. The laser beam is passed through a etalon and a gas pipe containing water vapor, respectively. The etalon is used to obtain the variation of laser intensity with wavelength. The laser emits a beam through a sample cell of 88 cm in length, and a certain concentration of water vapor is introduced into the tube. The gas temperature in the tube reaches 300 K by heating the tube. The laser intensity is collected after the laser beam is absorbed by the gas. The laser modulation condition is that the modulation frequency is 20 Hz, the central current is 65 mA, and the modulation depth is 8%. The heating pipe is 30.4°C and the concentration of water vapor is 2.5%. The transmission intensity under this condition is measured and Figure 7 is obtained.

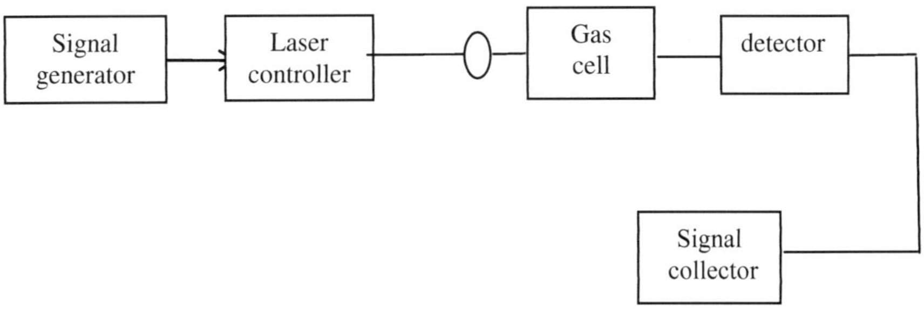

Figure 6. Schematic of measuring device.

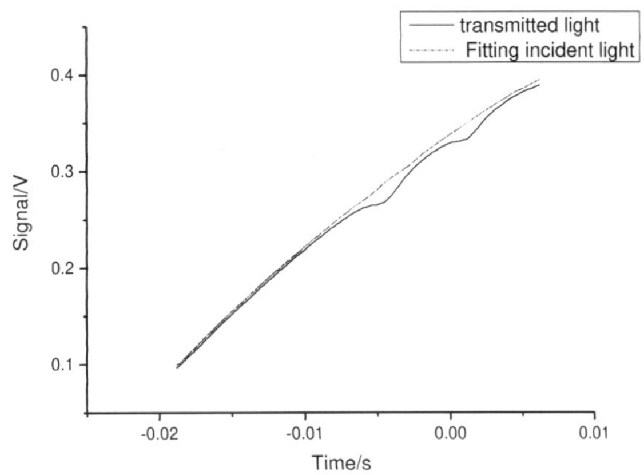

Figure 7. Transmission signals and fitting incident light.

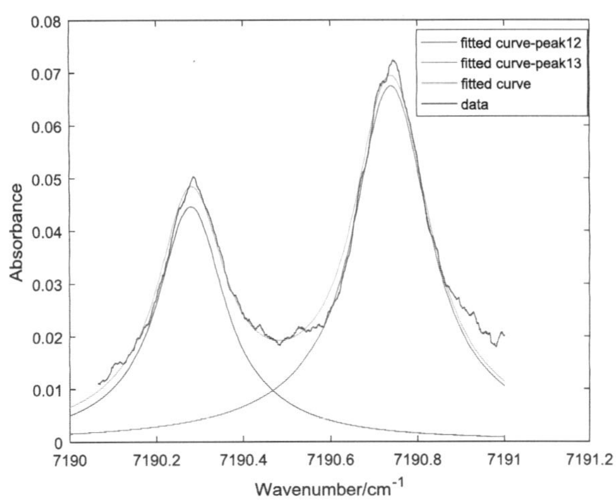

Figure 8. Schematic of fitting curve.

IOP Conf. Series: Materials Science and Engineering **538** (2019) 012039 doi:10.1088/1757-899X/538/1/012039

6. Conclusion

In summary, TDLAS system can realize simultaneous measurement of temperature and gas concentration. It has important application prospects in energy and chemical industry, aerospace, environmental monitoring and other industries. Aiming at the application of TDLAS technology in gas temperature field and concentration field, based on the HITRAN 2016 database spectral line parameters, the spectrum lines at 1392 nm are selected through detailed theoretical and experimental analysis. The absorption lines of 7190.2890cm^{-1} and 7190.7388cm^{-1} spectral lines pairs are used as the absorption lines of spectrum temperature measurement method. The ratio of spectral line intensity to area is compared by numerical simulation, and it is proved that the spectral line meets the requirements of gas temperature and concentration measurement by back-stepping method. TDLAS experimental platform was built to verify that the spectral pair is suitable for gas temperature and concentration measurement.

Acknowledgments

This work is Supported by the National Key Research and Development Program of China (Grant No. 2017 YFC0804705) and the National Natural Science Foundation of China(Grant No.11572044, 11732003). All support is gratefully acknowledged.

References

[1] J. Reid, D. Labrie. Second harmonic detection with tunable diode lasers —comparison of experiment and theory. Applied Physics B. 1981, 26: 203-210.
[2] P. H. Beckwith, C. E. Brown, D. J. Dannagher, et al. High sensitivity detection of transient infrared absorption using tunable diode lasers. Applied Optics. 1987, 26(13): 2643-2649.
[3] F Zh Dong, W Q Liu , J G Liu, et al, on-line real-time monitoring of motor vehicle exhaust (I), Journal of Testing Technology. 19 (2005): 119-127
[4] Z R Zhang, H Xia, F Z Dong, et a1. Simultaneous and on-line detection of multiple gas concentration with tunable diode laser absorption spectroscopy [J]. Optical and Precision Engineering, 2013, (11): 2771-2777.
[5] F D Chen, J G Liu, F R Kan, et a1. Software design based on TDLAS for two-dimensional temperature field distribution inversion [J]. Journal of atmospheric and Environmental Optics, 2017, (05): 393-400.
[6] Sappey A D, Howell J, Masterson P, Hofvander H. Determination of O_2, CO, H_2O concentrations and gas temperature in a coal-fired utility boiler using a wavelength-multiplexed tunable diode laser sensor. Zolo tecknologies inc. Technical papers. 2004.
[7] Bernath P, Rothman L. The HITRAN2016 molecular spectroscopic database [J]. Journal of Quantitative Spectroscopy and Radiative Transfer.
[8] Liu X, Jeffries J B, Hanson R K. Measurement of spectral parameters of water-vapor transition near 1388 and 1345 nm for accurate simulation of high-pressure absorption spectra. Meas. Sci. Technol. 2007 (18): 1185-1194.
[9] Sappey A D, Howell J, Masterson P, Hofvander H. Determination of O_2, CO, H_2O concentrations and gas temperature in a coal-fired utility boiler using a wavelength-multiplexed tunable diode laser sensor. Zolo tecknologies inc. Technical papers. 2004.

IOP Conf. Series: Materials Science and Engineering **538** (2019) 012040 doi:10.1088/1757-899X/538/1/012040

Optimization Techniques and Multi-Objective Analysis in Hybrid Solar-Wind Power Systems for Grid-Connected Supply

M. H. Alaaeddin*, A. Zakaria, J.M. Jani, N. Seyajah

Malaysia Italy Design Institute-University Kuala Lumpur (MIDI UniKL), No119, Jalan 7/91, Taman Shamelin Perkasa, 56100 Kuala Lumpur, Wilayah Persekutuan Kuala Lumpur.
MIDI-UniKL, No119, Jalan 7/91, Taman Shamelin Perkasa, 56100 Kuala Lumpur, Wilayah Persekutuan Kuala Lumpur.

m.h.alaaeddin@gmail.com; dzakaria@unikl.edu.my; Phone +6-03-8946-6262;

Abstract. Renewable energy has become a vital source for generating electricity, mitigating greenhouse gas emissions, and eliminating the reliance on non-renewable forms of energy. This work highlights on the urgent need for clean, efficient, and resilient energy sources through the development of advanced optimization methods. The hybrid solar-wind systems are meant to maximize power generation. Complexity arises when determining the optimum design, size, functionality, materials, distribution, and mechanism of each system as they are combined to ensure system reliability and optimal outcomes. The development of the new optimization methods (such as genetic algorithms, heuristic approaches, and particle swarm techniques) was generated to overcome the encountered barriers in traditional or classical approaches; such as in iterative techniques and linear programming methods. Additionally, the multi-objective analysis has been involved in various studies where vectors, criteria, and alternatives are applied with respect to the main objectives. However, further research efforts are encouraged to innovatively optimize hybrid solar-wind systems; including the optimization of hybrid solar-wind sizing (HSWS), design, capacity, battery storage, energy distribution systems and techno-economical considerations.

1. Introduction

The omnipresence, availability and sustainability of the solar-wind combination have become increasingly attractive with promising technological and economic considerations [1]. Both solar and wind technologies have been explosively urbanized and applied as alternative sources for green energy, with a sustainable supply of electricity for a widespread range of applications [2]. They are employed in separate unit supplies and hybrid systems [3,4] displaying various rates of efficiency [5], the hybrid systems based on solar–wind energy harnessing are known for their ability to produce collective electrical supply. Both systems are known for their different functionalities and various operative methods. The existence of this combination requires different parameters which can be determined according to the modelling approach [6]. Therefore, the solar module with wind turbines will ensure a better supply of energy and more resilient rates of efficiency as both possess the merit of dispatch ability [7]. Thus, hybrid solar–wind power systems have been extensively studied in recent decades due to the increasing demand of energy supply. Different methods of optimization were applied to enhance solar–wind hybrid systems so as to attain more efficient hybrid systems. Certain factors were considered such as: sizing model optimization or hybrid solar–wind system optimization (HSWSO), optimum sizing, design optimization, battery storage management optimization, energy distribution optimization, techno-economical optimization and multi-objective optimization [8–10]. Additionally, the net present values (NPV) as well as the levelized cost of energy/electricity (LCOE) were also examined and improved in numerous research methods [11,12]. On the other hand, the existence of artificial intelligence (AI) techniques have become a realistic alternative for integrating systems and handling complicated issues such as system identification, modelling, prediction, forecasting of meteorological data, optimization and complex system controlling [13].

2. Optimization Techniques of Hybrid Solar-Wind Systems

The optimal design of a hybrid solar–wind system can significantly enhance the technical and economic performance for efficient energy harnessing [5,14]. The multi-optimization techniques are also known for their accuracy and simplicity when encountering complicated optimization problems. The prioritization process analyzes criteria with respect to objectives, and it is based on realized importance and relativity [15]. The optimization approaches in hybrid solar-wind systems can be categorized into two groups: the first would be traditional or classical approaches which apply rigorous procedures such as iterative techniques, linear programming, and similar techniques. The second approaches are known as new generation methods such as particle swarm techniques and genetic algorithm approaches [8]. However, there is a common flow employed when optimizing a hybrid solar-wind system. Figure (1) presents a flowchart on the size optimization of hybrid solar-wind systems.

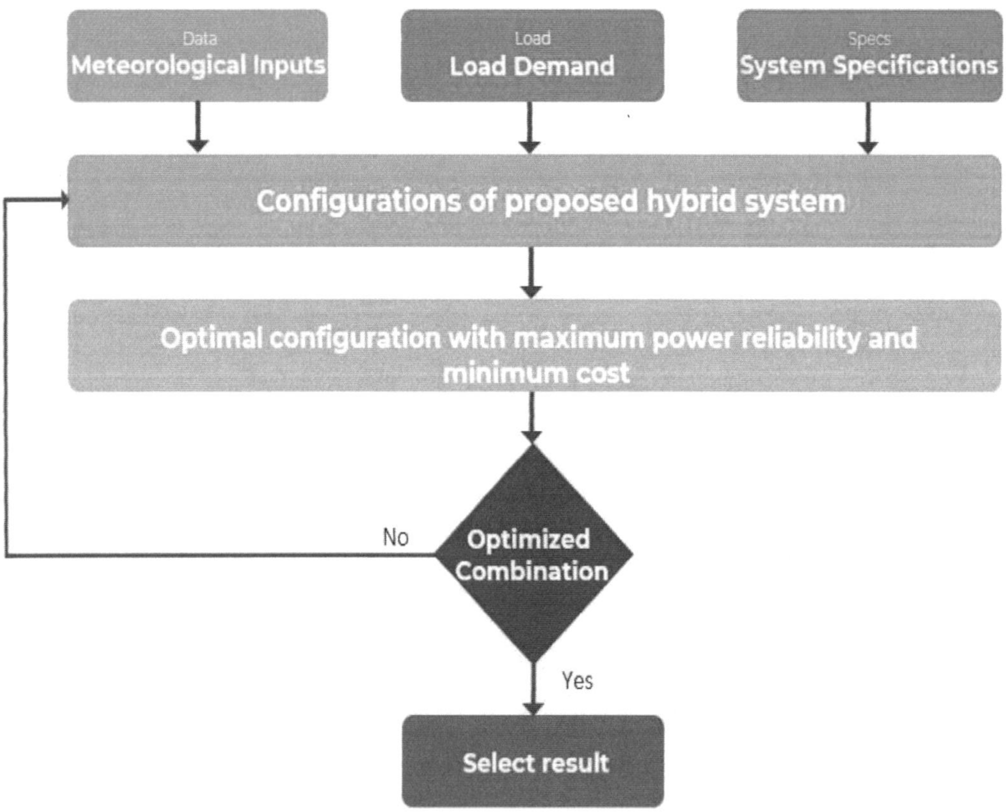

Figure 1. Common implementation process applied in optimal sizing of hybrid systems-Redrawn. [8]

In solar-wind hybrid systems, the meteorological analysis is considered as very essential for evaluating system performance expectancy. The meteorological parameters for a hybrid solar-wind system can be solar radiation, wind speed, wind direction, relative humidity, and air temperature [16]. On the other hand, the appropriate evaluation of the load demand is a crucial requirement in which electrical supply can be fulfilled and the annual cost can be minimized [17]. System specifications will contribute to the determination of the technical components within a system [14]. The demonstration of reliable meteorological inputs, load demand analysis, and the appropriate determination of system specification will lead to an enhanced configuration [9,18] and therefore attaining an optimum arrangement when simulating a system. It can be maintained by comparing the energy production cost and the performance of various systems. This can lead to a superior optimization analysis. There are various software tools to enhance a solar-wind hybrid system by considering certain objectives. The optimization techniques for solar-wind systems can be software-based such as HOMER, HYBRIDS, HYBRID2, and HOGA [9,15]. The merits and demerits of these software methods vary from one type to another, for example, in HOMER, users are not enabled to intuitively select the components of the selected system, in HOGA HYBRIDS, users need to simulate one configuration at a time, the system functions by genetic algorithms, can apply either mono or multi objective analysis, the comprehensive optimization variables can be applied, the system requires higher level of system-configuration. The graphic construction methods, probabilistic approaches and iterative techniques can eliminate the need of time-series data. On the other hand, two parameters are allowed in the process of optimization, the dynamic changing performance of the system cannot be displayed, causes increased computational determinations and suboptimal solutions. The artificial intelligence methods support the global optimum system configuration and consider computational simplicity [9]. The multi-objective evaluation/optimization methods support the simultaneous optimization, these methods are of multiple criteria analysis involving numerous parameters and alternatives to be optimized or evaluated simultaneously [9,15].

3. Modelling and Designing of Hybrid Solar-wind Systems

The modelling process of hybrid solar-wind systems is essential for determining the main elements of the system design while considering specific parameters and various constraints. The modelling process usually targets important factors such as the sizing criteria, ecological statistical aspects, battery autonomy function, and annual cost analysis. However, several researches have highlighted different modelling techniques for hybrid system components. The modelling of a solar system usually concentrates on outputs that generally rely on solar radiation and solar cell efficiency [19,20], whereas the modelling of wind turbines considers other factors such as; the mechanical power, power coefficient, wind velocity, tip-speed-ratio and stall behavior [11,19]. There are two major concerns when designing any hybrid solar-wind system: the reliability of the power supply under various weather conditions and the corresponding system cost [21]. There are additional important and relevant considerations when designing a hybrid solar-wind power system such as; the outline for optimum sizing and resource allocation based on the load demand, which is necessary to cut off initial and operational costs [9,19]. The lifespan of solar modules and wind turbines is almost the same, so when they are combined as a hybrid system, both can last longer with convenient reliability [4,22]..However, the climatic conditions are very important factors to consider when designing hybrid solar-wind systems; both systems function on different weather conditions. Solar modules require high solar radiation, while wind turbines need optimum wind velocity which can fit into the installed system. Such an equation will escalate the designing procedure and therefore increase the complexity of creating an efficient and well-optimized solar-wind power generator [8,23]. Figure 2 provides a schematic diagram of a hybrid solar-wind system based on a grid-connected configuration.

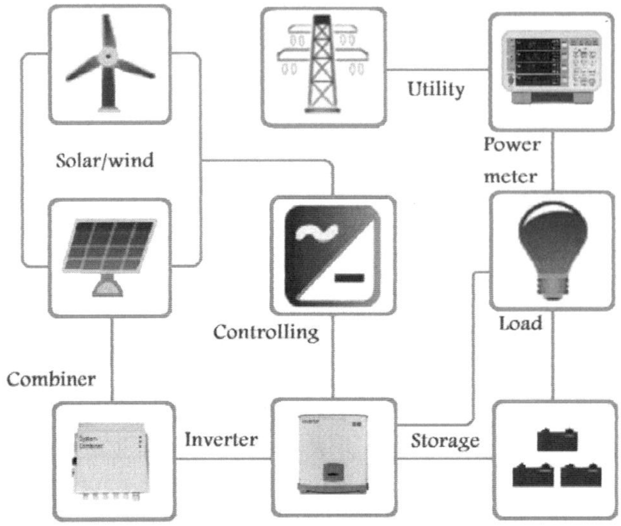

Figure 2. A schematic diagram of a hybrid solar–wind system connected to a utility [24]

In the grid connected system, the hybrid solar-wind system is specifically designed to connect to a local utility. The utility grid is usually based on distributed generation system which has the ability to convert the electrical generated power into the same amplitude, phase with the grid, and frequency as well as afford electrical energy to local loads [25]. The grid connected method is known for its ability to perform better, requiring less maintenance and is cost effective compared to other grids. The simulation of the grid connected PV system improves the voltage profile and evaluates the load sides [26]. The purpose of having a grid connected supply method is to achieve multiple objectives such as: contributing to the overall generation of electricity by offering a clean source of energy to the local utility, offering higher electrical flow to the public consumption by providing a share to the local utility, reducing users' overall cost by dropping electricity expenses as well as increasing their income by paying them when they put energy back into a local grid [27,28]. Therefore, more efforts on the utilization of hybrid solar-wind systems and optimization methods for grid connected supply are highly encouraged and will significantly contribute to the inclusive outcome of clean energy for various industries, utilizations, and local consumption. These efforts will make up the latter implementations of evolutionary development in hybrid solar–wind power systems [29–35].

4. Conclusions

In hybrid solar-wind advancements, dramatic progress has resulted from the optimization researches in the last few years. Numerous optimization techniques and multi-objective analyses were conducted to enhance the efficiency, reliability, and system integrity. The use of novel generations of optimization such as the genetic algorithm, heuristic approaches, and particle swarm techniques are expected to overcome existing weaknesses in the traditional or classical approaches as in iterative techniques and linear programming methods. The multi-objective analysis has also been involved in several studies where vectors, criteria, and alternatives are applied with respect to the main objective. It can be concluded that multiple methods and techniques have been employed to optimize hybrid solar–wind systems, including design methods, size, battery, energy distribution, capacity and techno-economical issues. Despite this fact, more attention on AI techniques is encouraged for hybridizing, modelling, and enhancing hybrid solar-wind power systems and their configurations.

Acknowledgments

Acknowledgment goes to the Universiti Kuala Lumpur Malaysia Italy Design Institute (UniKL MIDI) for facilitating the necessary means to accomplish this research work.

References

[1] Nema P, Nema RK, Rangnekar S. A current and future state of art development of hybrid energy system using wind and PV-solar□: A review. Renew Sustain Energy Rev 2009;13:2096–103. doi:10.1016/j.rser.2008.10.006.

[2] Khare V, Nema S, Baredar P. Status of solar wind renewable energy in India. Renew Sustain Energy Rev 2013;27:1–10. doi:10.1016/j.rser.2013.06.018.

[3] Chakraborty S, Reza SMS, Hasan W. Design and Analysis of Hybrid Solar-Wind Energy System Using CUK & SEPIC Converters for Grid Connected Inverter Application. In Power Electronics and Drive Systems (PEDS), IEEE 11th International Conference on 2015:278–83. doi: 10.1109/PEDS.2015.7203442.

[4] Wagh S, Dhage P, Gavhad A. A Review on Vertical Axis Wind Solar Hybrid Power System. 11th Int. Conf. Sci. Technol. Manag., 2018, p. 85–91.

[5] Diaf S, Notton G, Belhamel M, Haddadi M, Louche A. Design and techno-economical optimization for hybrid PV / wind system under various meteorological conditions 2008;85:968–87. doi:10.1016/j.apenergy.2008.02.012.

[6] Jadallah AA, Mahmood DY, Er Z, Abdulqaedr ZA. Hybridization of solar/wind energy system for power generation in rural aeras. Acta Phys Pol A 2016;130:434–7. doi:10.12693/APhysPolA.130.434.

[7] Maleki A, Hafeznia H, Rosen MA, Pourfayaz F. Optimization of a grid-connected hybrid solar-wind-hydrogen CHP system for residential applications by efficient metaheuristic approaches. Appl Therm Eng 2017;123:1263–77. doi:10.1016/j.applthermaleng.2017.05.100.

[8] Sinha S, Chandel SS. Review of recent trends in optimization techniques for solar photovoltaic-wind based hybrid energy systems. Renew Sustain Energy Rev 2015;50:755–69. doi:10.1016/j.rser.2015.05.040.

[9] Zhou W, Lou C, Li Z, Lu L, Yang H. Current status of research on optimum sizing of stand-alone hybrid solar – wind power generation systems. Appl Energy 2010;87:380–9. doi:10.1016/j.apenergy.2009.08.012.

[10] Ekren BY, Ekren O. Simulation based size optimization of a PV/wind hybrid energy conversion system with battery storage under various load and auxiliary energy conditions. Appl Energy 2009;86:1387–94. doi:10.1016/j.apenergy.2008.12.015.

[11] Bandrova T, Konecny M. Modelling and Optimising the Value of a Hybrid Solar-Wind System. IOP Conf. Ser.: Mater. Sci. Eng. 2017; 197. doi:10.1088/1757-899X/197/1/012035.

[12] Rose A, Stoner R, Pérez-arriaga I. Prospects for grid-connected solar PV in Kenya□: A systems approach. Appl Energy 2017;161:583–90. doi:10.1016/j.apenergy.2015.07.052.

[13] Mellit A, Kalogirou SA. Artificial intelligence techniques for photovoltaic applications□: A review 2008;34:574–632. doi:10.1016/j.pecs.2008.01.001.

[14] Nacer T, Hamidat A, Nadjemi O. Techno-economic impacts analysis of a hybrid grid connected energy system applied for a cattle farm. Energy Procedia 2015;75:963–8. doi:10.1016/j.egypro.2015.07.292.

[15] Fadaee M, Radzi MAM. Multi-objective optimization of a stand-alone hybrid renewable energy system by using evolutionary algorithms: A review. Renew Sustain Energy Rev 2012;16:3364–9. doi:10.1016/j.rser.2012.02.071.

[16] Zaghba L, Borni A, Khennane M, Terki N, Fezzani A, Terki N, et al. Experimental typical meteorological years to study energy performance of a PV grid-connected system. Energy Procedia 2017;119:297–307. doi:10.1016/j.egypro.2017.07.085.

[17] Askarzadeh A. Developing a discrete harmony search algorithm for size optimization of wind-photovoltaic hybrid energy system. Sol Energy 2013;98:190–5. doi:10.1016/j.solener.2013.10.008.

[18] Yang H, Lu L, Zhou W. A novel optimization sizing model for hybrid solar-wind power generation system. Sol energy 2007;81:76–84. doi:10.1016/j.solener.2006.06.010.

[19] Sawle Y, Gupta SC, Bohre AK. PV-wind hybrid system: A review with case study. Cogent Eng

2016;3:1–31. doi:10.1080/23311916.2016.1189305.
[20] Sawle Y, Gupta SC, Bohre AK. Socio-techno-economic design of hybrid renewable energy system using optimization techniques. Renewable Energy. 2018;119:459-72. doi.org/10.1016/j.renene.2017.11.058
[21] Yang H, Zhou W, Lu L, Fang Z. Optimal sizing method for stand-alone hybrid solar – wind system with LPSP technology by using genetic algorithm. Sol energy 2008;82:354–67. doi:10.1016/j.solener.2007.08.005.
[22] Ma T, Yang H, Lu L, Peng J. Optimal design of an autonomous solar-wind-pumped storage power supply system. Appl Energy 2015;160:728–36. doi:10.1016/j.apenergy.2014.11.026.
[23] Chakraborty S, Simoes MG. PV-microgrid operational cost minimization by neural forecasting and heuristic optimization. Conf Rec - IAS Annu Meet (IEEE Ind Appl Soc 2008:1–8. doi:10.1109/08IAS.2008.147.
[24] M H Alaaeddin, A Zakaria, JMJ. Optimizing and evaluating the influencing factors in the design of hybrid solar-wind power systems for grid-connected supply methods. UniKL, 2018.
[25] Renukadevi V, Jayanand B. Harmonic and Reactive Power Compensation of Grid Connected Photovoltaic System. Procedia Technol 2015;21:438–42. doi:10.1016/j.protcy.2015.10.067.
[26] Praiselin WJ, Edward Belwin. Voltage Profile Improvement of Solar PV Grid–Connected Inverter with Micro Grid Operation using PI controller. Energy Procedia 2017;117:104–11. doi:10.1016/j.egypro.2017.05.112.
[27] Li S, Sun F, He H, Chen Y. Optimization for a Grid-Connected Hybrid PV-Wind-Retired HEV Battery Microgrid System. Energy Procedia 2017;105:1634–43. doi:10.1016/j.egypro.2017.03.532.
[28] Coelho RF, Schmitz L, Martins DC. Proposal of a power flow control strategy applied to a hybrid microgrid. Power Electronics Conference and 1st Southern Power Electronics Conference (COBEP/SPEC), IEEE 13th Brazilian. IEEE, 2015. doi: 10.1109/COBEP.2015.7420226.
[29] Ekren O, Ekren BY. Size optimization of a PV/wind hybrid energy conversion system with battery storage using simulated annealing. Appl Energy 2010;87:592–8. doi:10.1016/j.apenergy.2009.05.022.
[30] Yaqub U, Al-nasser A, Sheltami T. Applied Computing and Informatics Implementation of a hybrid wind-solar desalination plant from an Internet of Things (IoT) perspective on a network simulation tool. Appl Comput Informatics 2018:1–5. doi:10.1016/j.aci.2018.03.001.
[31] Bilal BO, Sambou V, Ndiaye PA, Kébé CMF, Ndongo M. Optimal design of a hybrid solar-wind -battery system using the minimization of the annualized cost system and the minimization of the loss of power supply probability (LPSP). Renew energy 2010;35:2388–90. doi:10.1016/j.renene.2010.03.004.
[32] Singh RS, Abbod M, Balachandran W. A design scheme of control/optimization system for hybrid solar—Wind and battery energy storages system. In Power Engineering Conference (UPEC), 2016 51st International Universities, IEEE, 2016;1-6. doi: 10.1109/UPEC.2016.8114093
[33] Şengül Ü, Eren M, Shiraz SE, Gezder V, Şengül AB. Fuzzy TOPSIS method for ranking renewable energy supply systems in Turkey." Renew Energy 2015;75: 617-625. doi.org/10.1016/j.renene.2014.10.045
[34] Mohanraj K, Bharathnarayanan S. Three Level SEPIC Hybrid Wind-Solar Energy Systems. Energy Procedia 2017;117:120–7. doi:10.1016/j.egypro.2017.05.114.
[35] Baños R, Manzano-Agugliaro F, Montoya FG, Gil C, Alcayde A, Gómez J. Optimization methods applied to renewable and sustainable energy: A review. Renew Sustain Energy Rev 2011;15:1753–66. doi:10.1016/j.rser.2010.12.008.

IOP Conf. Series: Materials Science and Engineering **538** (2019) 012041 doi:10.1088/1757-899X/538/1/012041

Correlation Based Parametric Design Method of Lifting Device in Nylon Coated Conveyor Line

Ma Ning, Yang Bo[a] and Shang Junzhi

School of Mechanical Engineering, University of Jinan, 250022 Shandong Jinan, China

Corresponding author: 1603206175@qq.com

Abstract. In order to solve the problem of low efficiency in design variation of the lifting device in nylon coated conveyor line and meet the user's individual and diversified demands, the key technologies and implementation methods of the automatic generation of lifting device were studied. Firstly, the dimension correlation between each part of the lifting device was analyzed and the dimension correlation diagram was established. Secondly, the parts parametric modeling method was given based on the dimension relationship. Then, the structural correlation was analyzed, based on which the assembly modeling was established. Finally, the design process of the lifting device was taken as an example to verify the effectiveness of the method.

1. Introduction

Nylon coated conveying line is an important equipment in extending the operational life and reduce friction losses of the spline shaft by coating nylon on its surface uniformly. The lifting device is one of the indispensable parts of Nylon coated conveying lines. Its deformation is mainly based on lifting weight, lifting height, lifting speed, working level, etc. Its design process is complex. How to realize the automatic design of lifting device is a problem that needs to be solved at present.

Parametric design is the core technology of intelligent CAD system and one of the most widely used hotspot designs. With the development of modern mechanical design, the research cycle of products can be shorten through parametric design [1]. By using certain parametric design method to create parts and assembly templates, modification and optimization can be realized by changing the main parameters [2].

Association design is a parametric design technique based on association geometry, which uses geometric relationships and parameters to spread variation. Because the size and structure related design in the product variant design can improve the design efficiency of the product and ensure the design quality of the product, the correlation-based parametric product design technology has been widely used in product variant design. As Jia proposed a novel modular design method based on the fuzzy correlation analysis [3].Yang et al. put forward a method of building parametrization component library based on automatic assembly technology, which achieved the explicit expression of assembly relations and accelerated product design [4].

2. Size correlation analysis of lifting device

The dimensional information not only determines the size and the position of the part, the coordinates of the set size, but also determines the constraint relationship between the geometric features [5]. It is the basis of constructing the whole assembly. In mechanical products, constraint relationship between parts plays an important role in the fulfill of product function. Therefore, the associated design of the part should be based on constraint-driven association [6]. When the upstream design is changed, the change of design information was transmitted to the downstream and the transmission direction was maintained in time.

The lifting device of the nylon coated conveyor line is a component composed of many parts. As an assembly, assembly relationship exists among parts, that is, the dimensional parameters and the structure of the part have assembly-related characteristics. It is necessary to perform dimensional analysis on the dimensional parameters of the lifting device to determine the correlation between the dimensional elements and the basic design parameters.

2.1. *The basic dimension parameters of lifting device*

Parameters relationship reflects a valid portfolio relationship among the parts of the product, including constraints, characteristics and dimensions of the product. The basic size parameter is the first-order parameter of the parametric drived association, which can be established among different parameters in the same part or among different parts in the assembly. Therefore, it is very important to determine the basic size parameters of the lifting device.

The ball screw is the main part of the lifting device and the change of its design parameters has a great influence on the lifting device. Other parts are designed and selected on the basis of the ball screw. Therefore, the basic parameters of ball screw are the basic parameters of lifting device. In the design process, the ball screw is designed according to the working conditions of the lifting device, such as lifting weight, lifting height, lifting speed, working level. As shown in Fig.1, main design parameters of ball screw are: L0, L3, PB, Da, etc. In the design, the structural size of main parts was determined by Da, so Da can be used as the basic parameter.

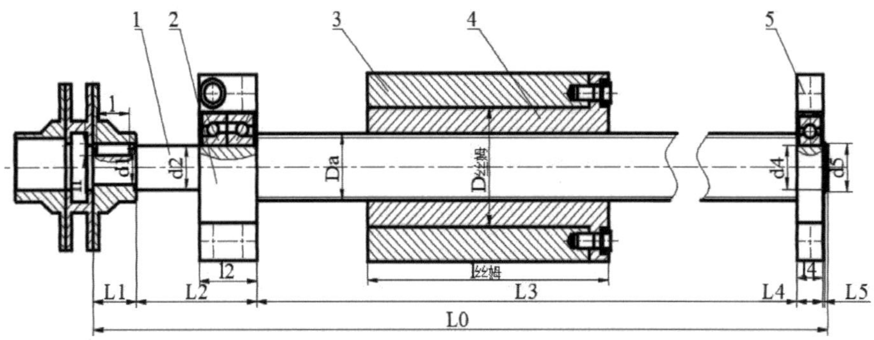

Figure 1. Dimension structure diagram of lifting device (1.ball screw 2.screw support seat I 3.screw mounting seat 4.ball nut 5.screw support seat II).

2.2. *Correlation analysis of dimension parameters*

The key and foundation of assembly parametric design is to establish correlation of dimension variables between parts. In order to simplify the process, we stipulate that: Internal parameters are not related to the part or sub-assembly; External parameters refer to parameters published in the assembly environment; Correlation parameter refers to the parameters related to assembly between the parts that construct the assembly [2].

The association relation of dimension parameter is directional, including vertical relation and horizontal relation.

Vertical relation: It is a one-way relationship that changes in the upstream design will cause changes in the downstream structure. But the change of downstream has no influence on the upstream structure.

Horizontal relation: It is a two-way relationship that the chang of any part will cause changes of the structural dimensions of relating parts.

IOP Conf. Series: Materials Science and Engineering **538** (2019) 012041 doi:10.1088/1757-899X/538/1/012041

In parametric association design, the order and direction of the parameters have a crucial influence, which in turn determines the overall size of the lifting device. Therefore, it is necessary to analyze the correlation of each part in lifting device.

As shown in Fig. 2a, changes in the ball screw (Da) will case L3 to change, which in turn causes change in other dimensional parameters (L1, L2, L4, etc.). Between parts, the size of the ball screw parameters (L1, d1, Da, L5, d5) changes result in changes of other parts size parameters (d1 l2, Da ', L5, d5, etc.) as shown in Fig.2b. This is a vertical relation. The dimensional parameters of the ball screw (L2, d2, L4, d4) vary with the size of the screw support seat I, II parameters (L2, d2, L4, d4), as shown in Fig.2c. This is a horizontal relation.

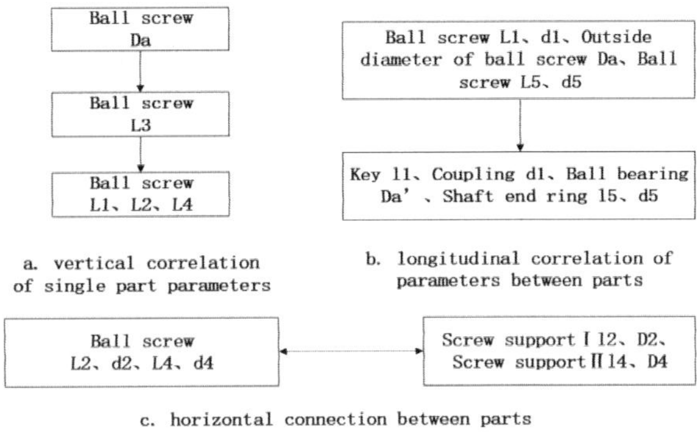

Figure 2. Vertical and horizontal correlation between parameters.

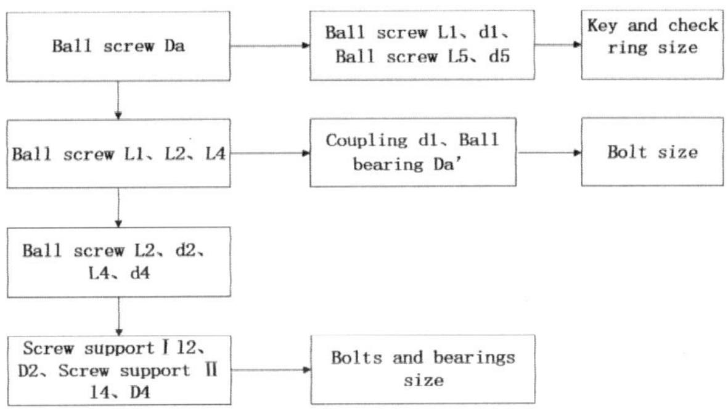

Figure 3. Dimension correlation diagram of lifting device.

2.3. Parts modeling of lifting device based on dimension correlation

The size correlation of the lifting device determines the parameter transfer relation and provides a rule for parametric modeling. For lifting device components, there are standard parts and non-standard parts, both of them should be designed according to the correlation relationship between each part of lifting device. The determination of basic dimension parameters will directly or indirectly drive the design of other parameters.

2.3.1. Establishment of dimension correlation model. Based on basic dimension parameters and correlation between parts in lifting device, the interdependence among the components can be summed up. Thus, a parameter-driven association model can be established, as shown in Fig.3.

2.3.2. Parametric modeling of non-standard parts. Non-standard parts in mechanical products are designed by functional requirements. Its design modeling methods include: dimension parametric modeling, structural parametric modeling [6]. In this paper, the parametric modeling method of parts based on standard template is proposed.

(1) The establishment of the critical parts standard template: According to the functional requirements of the product, a standard model template for each part of the lifting device with universal applicable characteristics is established, and some key parameters are used to describe the relationship between the shape and the size of the model to form a module family. The main model of part module consists of its geometric model and the corresponding characteristic table. The geometric model and its parameters describe the unified characteristics and parameterized dimensions of the module family, and the characteristic table describes the specific dimensions and unique characteristics of each module in the module family.
(2) Based on the dimensional parameterization modeling of the part standard template, only the size of the standard template is changed according to the structural design requirements. In the design process, based on the standard template of the part, the dimensional relationship is used to drive the change of other dimensional parameters, and the part model is updated to complete the design of the part.

3. Structural correlation analysis of lifting device

Functional structure is the core of a product and the analysis of structural correlation relationship lays a foundation for the establishment of skeleton model and assembly model. The skeleton model is used to describe spatial requirements, important installation position and motion of the product. It is the three-dimensional space planning of product assembly, which determines the interface of design object structure. The skeletal model contains transmission mode and direction information for the upstream and downstream transmission of the design unit.

3.1. Basic structural correlation types

In the assembly model, the structural correlation determines the structural characteristics of the whole assembly model. The following four kinds of structure association constraint is the basis of assembly spatial planning.
Performance correlation constraint: It is interrelated constraint among several parts to achieve a certain performance. For example, the performance of ball screw is determined by the constraint of the rotary association between ball screw and ball spring.
Physical correlation constraint: It is physical relation of energy conversion and signal transfer between structures. For example, the torque on the ball screw is transferred to ball screw mounting seat through the ball nut which is transmission constraint of energy signal between them.
Shape constraint: It is a shape position relation between parts, including the link, fit and fixation between structures.
Other correlation constraint: It refers to other constraint relationships that are involved in the product life cycle [7].
The analysis of structural association constraints lays the foundation for product design, provides theoretical support for product design, and can effectively improve design efficiency and save production costs.

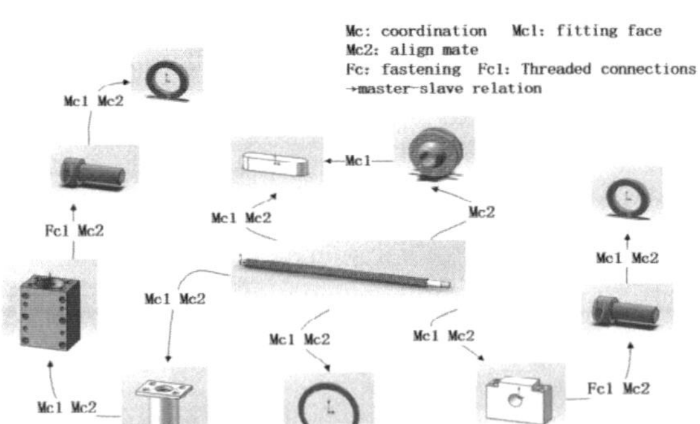

Figure 4. Associated configuration constraint of lifting device structure.

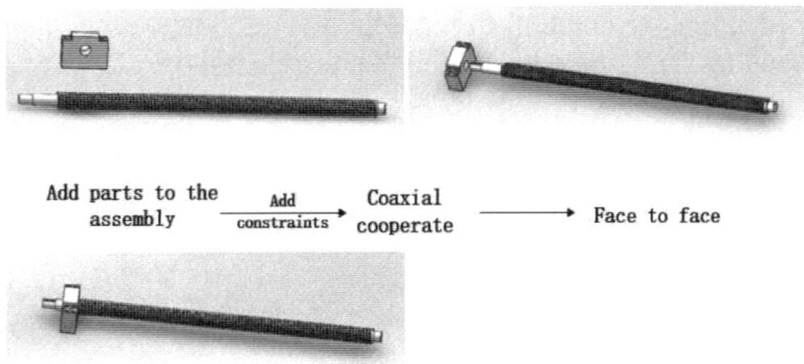

Figure 5. Assembly modeling of screw support stand and ball screw.

3.2. Structural correlation characteristics of lifting device

The assembly modeling was established according to the structural correlation of lifting device. Because the structure of the lifting device is relatively simple, its performance association constraint, physical association constraint and other association constraints are not considered. In this paper, the lifting device is described by the shape and position constraint in the structural correlation constraint.

As shown in Fig.4, the structural shape and position correlation constraints mainly include: surface constraint (Mc1), coaxial constraint (Mc2) and thread fastening (Fc1). It is the key to define the assembly feature and assembly sequence between the parts in assembly [4]. The main and subordinate relations between parts are represented by directed line segments and the parts pointed by the arrow are Sub-part. For example, the shape and position correlation constraint of ball screw and ball nut in the Fig.4 is Mc1 and Mc2 and ball screw is subject to ball nut. In addition to the establishment of the structural correlation model, the assembly correlation relationship is also determined. This relationship does not change with the dimensional element of the part, thus ensuring that the relative position between the parts does not change.

3.3. Product assembly modeling based on structural correlation

The basic elements of the assembly model are determined by the analysis of structural correlation and the geometric elements of the skeleton model: datum plane, datum axis, datum point, etc. The assembly relation of the lifting device is determined by structural relation, on which assembly parameters, parameter constraints and assembly target constraints are determined. The assembly relationship will drive each part to establish the constraint association relationship of the corresponding features, so that the lifting device can automatically complete the assembly.

Fig.5 shows the assembly modeling process of ball screw and screw support seat based on shape and position association constraint. The assembly is realized according to the shape-related constraints associated with the screw support seat and the ball screw, such as the coaxial constraint and the surface constraint. Firstly, the pose of screw support seat should be adjusted and then the radial freedom degree between them should be restricted by the coaxial constraint. According to the surface constraint, the corresponding surface of them are bonded together and the axial freedom between them is limited. These constraint types, as well as the adjustment and positioning of spatial position, are driven by assembly association.

4. Example

In order to realize the parametric design method of lifting device based on correlation, an associated parametric design system in Solidworks environment is developed. The following examples used to illustrate this process.

Initial design parameters of lifting device are: (1) lifting weight is 480 kg; (2) lifting height is 550 mm; (3) lifting speed is 150mm/s; (4) work level is level 5.

The design parameters are first input, and the part model can be generated through background design calculation and automated modeling.

Then in the assembly module environment, the system will automatically generate the lifting device assembly model according to the shape and position constraint relationship between the parts in Fig.4, as shown in Fig.6.

Change propagation design: The lifting height of the device changed from 550 mm to 600 mm. According to the dimension and structural correlation, the corresponding size and structure of ball screw, lifting installation plate and lifting transition plate can be changed automatically, as shown in Fig.7.

Figure 6. Assembly view of lifting device.

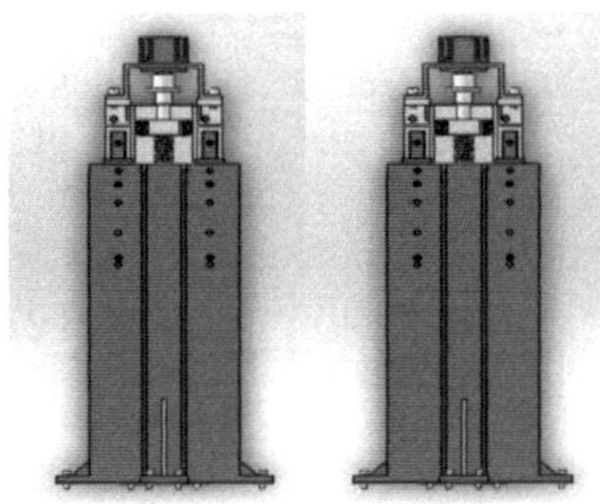

Figure 7. Height change of lifting device.

5. Conclusion

In this paper, a parametric design method based on correlation is proposed for the designing of lifting device in nylon coated conveyor line. The method focus on extracting the dimension parameters associated with the product. By defining the structure relation and dimension constraint relation between parts, the parametric design based on association-driven is formed. Based on which change propagation design can be realized automatically. This method lays a foundation for product design, provides theoretical support for product design, can improve design efficiency and save production costs.

Acknowledgments

The authors would like to thank the National Natural Science Foundation of China (51775239, 50975124) for financial support.

References

[1] Z. Xuechen, Z. Qihui, F. Ruxiang, J. Eng. Graph. 1, (2010)
[2] Z. Ye, C. Zhiying, Mech. Sci. Technol. Aerosp. Eng. 22, (2003)
[3] J. Weiqiang , L. Zhenyu ,L.Daxin,Chin.J.Mech.Eng.51,(2015)
[4] Y. Xiangang, H. Yulin, J. Xin, Chin. Mech. Eng. 24,(2013)
[5] D. Yude, Z. Han, T. Jianrong, J. Chin. Comput. Syst.24,(2003)
[6] X. Yi, D. Gang, Z. Huawei, Comput. Integra. Manuf. Sys. 19,(2013)
[7] S. Mingyu, Performance demand-driven product structure association design and its application in large hydraulic press(Zhejiang:Zhejiang University, 2014).

Development of a Robotic Structural Steel Cutting System

Xijun Zhao[1], Yunchan Chung[2,a]

[1]Department of Design and Engineering, SeoulTech, Seoul, Rep. of Korea
[2]Department of Design and Engineering, SeoulTech, Seoul, Rep. of Korea

ychung@seoultech.ac.kr; zhaoxijun5@sina.com

Abstract. Curved buildings and structures are getting popular and digital fabrication of the structural steels has been successfully tried in several ways. This study presents a development of a robotic system cutting structural steels and marking texts on the steels using digital information of shapes. A configuration of the overall system is explained based on functional decomposition of conventional cutting systems. The cutting system consists of a robotic system and a handling system, and the main functions of the cutting system are classified as translation, orientation, alignment, and cutting. A digital model of the system was created and tested in a robot simulation system. A robot tool-path generation system was also developed. The tool-path generation system uses the DSTV file, an industry standard for the shape of a structural steel, to generate robot job files. This study also provides robot workspaces in which the cutting tool can safely access and cut the workpieces. The proposed system was tested for cutting structural steels clamped in front of the robot. The results show that the generated tool-paths are valid in cutting the structural steel and the possibility of an overall robotic system for cutting structural steels.

1. Introduction

This study presents a development of a robotic structural steel cutting system. The cutting system consists of a robotic system and a handling system. The robotic system cuts structural steel beams and marks text on the steel beams using digital files that include the final shape of the structural steel. The handling system feeds, aligns and clamps structural steel beams allowing the robotic system to cut the beams accurately. This study describes the conceptual design and the development of the structural steel cutting system.

Standard structural steel as shown in Figure 1 should be cut and fabricated for use in the construction of buildings, bridges and other large structures. Curved buildings or structures make it more difficult to fabricate the steel beams because curved structures include angled and curved steel frames made of a variety of shapes as shown in Figure 2. These days with the help of Building Information Modeling (BIM) technology, the curved steel frames are designed as digital models, and the digital models include all the information of the structural steel. Using the digital models, robotic systems can accurately and quickly fabricate the various shapes of structural steel at low cost.

2. Background

The shape, size, mechanical property, etc. of common structural steels are regulated by standards in most countries [1]. Typical shapes of standard structural steel are shown in Figure 1 and they are usually 4 to 12 meters long and 100 to 600 millimeters wide. The most popular one is an H-beam (or

I-beam) with an H-shaped cross-section. Therefore, this study mainly uses the H-beams as examples although it focuses on all types of structural beams.

The structural steels with various shapes as shown in Figure 2 are made by cutting and joining standard structural steels. Cutting structural steels to length has usually done with band saws, and drills and mills have been used to make holes and slots on the steel beams. For cutting curved ends and irregular openings of the steel, Computer Numerical Control (CNC) machines with a cutting torch are typically used [2]. A CNC machine moves a torch head in accordance with programmed cutting instructions. Oxy-fuel, laser and plasma torches are common technologies for cutting the steel beams [3].

As the digital design technology such as BIM is developed, automated structural steel cutting systems are becoming common. PythonX [4] and KraneDonk [5] are the representative robotic solutions for cutting and welding structural steel. They are all good solutions, but no detail technologies have been published. This study describes the functional requirements, and presents an approach of development of the robotic system for cutting structural steel.

Curved buildings and structures are getting popular. The percentage of curved buildings designed by world's 5 largest architecture firms increased to 25% in 2010 [6]. Mun et al. researched the applications of digital fabrication and described a CNC system to fabricate irregular steel workpiece for the curved buildings [7]. Using the CNC system, the welded structural steel could be precisely assembled at construction site.

There have been several studies on robotic cutting system in ship building industry [8, 9]. One of the studies have explained a method to define the working envelop of an articulated robot for H-beam cutting and presented a verification method using a robot simulation system.

Recently most of curved structure designs are performed digitally [10]. For the digitally designed structural steel, robotic cutting system is necessary and possible. This study explains the functions of the robotic structural cutting system and presents a new robotic cutting system.

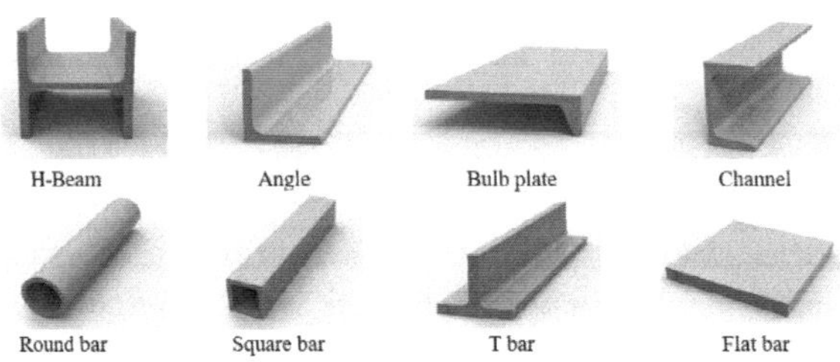

Figure 1. Common standard shapes of structural steel

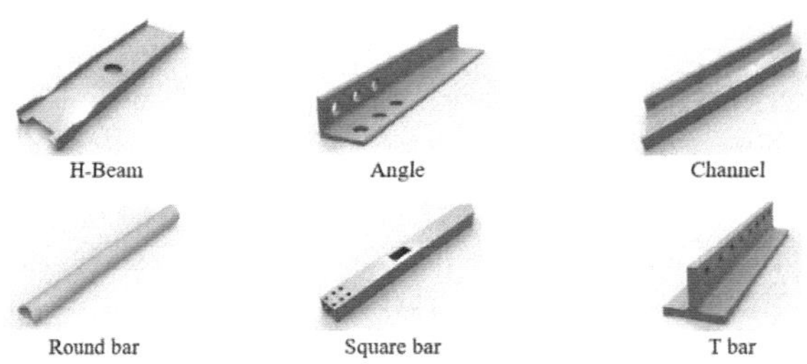

Figure 2. Various shapes of structural steel

3. Overall Function of Structural Steel Cutting System

In order to identify overall function, the authors have analyzed traditional operations and existing systems for the fabrication of the structural steel. Functional decomposition technique [11] have been used for conceptual design of the system. The proposed structural steel cutting system has four sub-functions; cutting, translation, orientation, and alignment.

IOP Conf. Series: Materials Science and Engineering **538** (2019) 012042 doi:10.1088/1757-899X/538/1/012042

The cutting function is the most basic function of the system, which cuts a steel beam into two parts along a contour or makes holes in the steel. The function is accomplished by a cutting tool or torch, which is the end effector of a robot. The cutting tool can be an oxy-fuel, laser, or plasma torch [3, 12, 13]. The presented system employs a plasma torch which makes a clean cut section at a low cost.
The translation function provides linear movement in length direction. As shown in Figure 1 standard structural steels are long. The longest one is 12m in length and the popular one is 10m. The translation function can be achieved by two ways; moving the robot with a cutting tool and moving the structural steel. The proposed system uses a roller conveyor to move the structural steel.
The orientation function makes the cutting tool to access the workpiece surface for cutting in the good posture without any collision, and enables possible kinematics and trajectories of the robot. Figure 3 shows the tool orientations for an H-beam. The bottom and top side flanges of the H-beam are accessed from outside, and the front and back side webs should be accessed from both sides; upside and downside. Rotating the structural steel is an alternative of the orientation function. Otherwise dexterity of robot hands in all direction of the steel is necessary. Due to the heavy weight of the structural steel, large expensive mechanical structures are required to rotate the steel. The proposed system utilizes the dexterity of the robot.
In an automated system the geometry of the workpiece should be same with the geometry of digital model, because the cutting tools move along the programmed paths which are computed based on the digital model. Real structural steels are made with tolerances and have warpages, and the geometry can be different from the digital shape. Therefore the distortion of the steel should be straighten physically. Otherwise, the tool paths, digital information, should be adjusted for the actual shape. The alignment makes the geometry of the digital model and the actual workpiece the same. Typically the alignment is accomplished by two methods. Jig and fixtures are used for straightening overall shape of the steel, and an adaptive tool height control system is used to keep a constant distance between the nozzle of cutting torch and the workpiece surface. The proposed system adopts both methods.

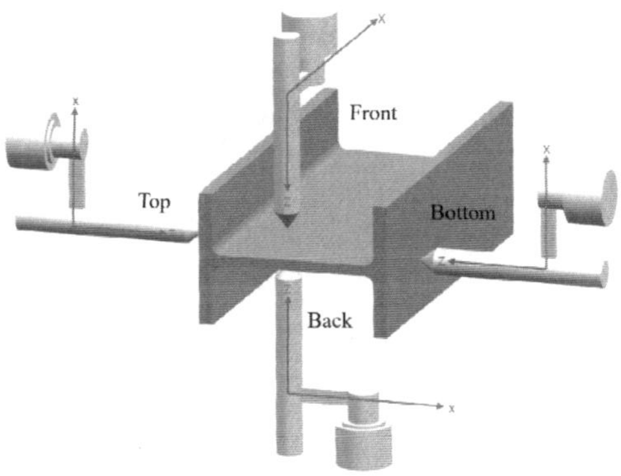

Figure 3. Tool orientations for an H-beam

4. Tool-path Generation System

Robot tool-paths are generated from a digital model of steel beams saved in the DSTV file format. DSTV format is the industrial standard defined by the German Steel Construction Association (Deutsche Stahlbau-Verband) [14]. A DSTV file is a text file in ASCII format, and is designed for the CNC machining. The overall procedure of the tool-path generation system is shown in Figure 4.
At first the shape of an initial steel beam is identified from the given DSTV file, which supports 8 different types of structural steel. The target shape is then extracted. By comparing the two shapes, initial shape and target shape, initial cutting or marking paths can be calculated. Figure 5 shows the initial steel beam as a wire-frame model and the final target shape as a shaded model. The paths that collide with the workpiece must be recomputed, and the presented system uses two strategies. At first it tries to avoid collision by tilting the tool. When the tilt angle exceeds a specified value, it retracts the

tool in surface normal direction. Final tool paths are computed by adding approach and retract paths, and linking separated paths at different locations. Finally the final tool-paths are saved as a job file that operates a robot.

The proposed tool-path generation system has developed and tested through real cutting. The tool-paths were operated by a Yaskawa MA1440 robot [15] with a plasma torch, and collisions were successfully avoided. The results of a test cut are shown in Figure 6.

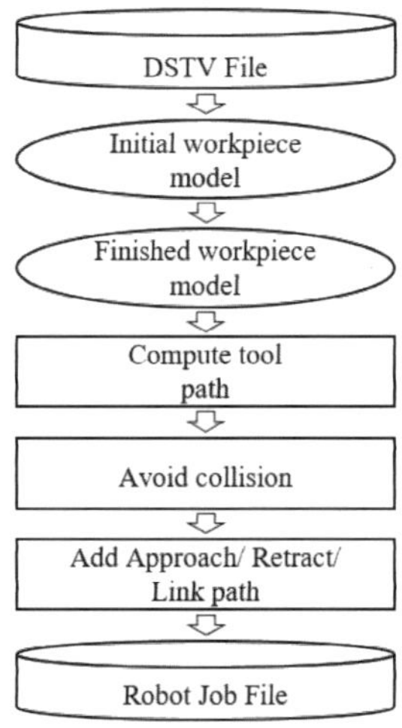

Figure 4. Overall procedure of tool-path generation system

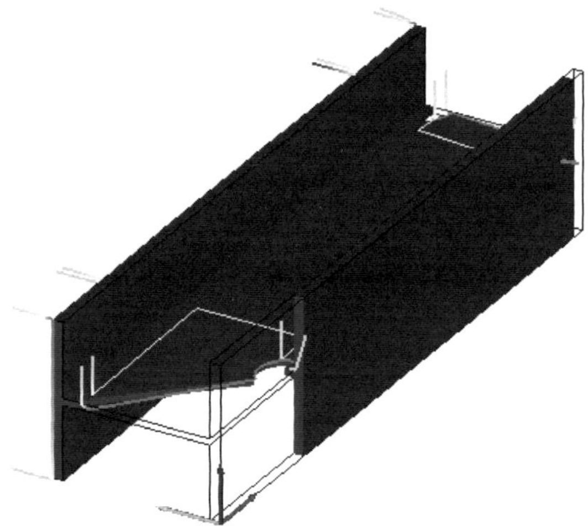

Figure 5. Example of a cut shape and tool-paths

5. Digital model of structural steel cutting system

To validate the proposed overall system a digital model was created as shown in Figure 7 and the overall system was tested in a robot simulation system, RoboDK [16]. The system consists of a Yaskawa MA1440 robot with a plasma torch and a roller conveyor with a jig and fixture system. In the digital simulation system, the overall system runs successfully for all test examples without collision.

In the system the work envelop of a robot is important, because the steel workpiece moves only linear by the roller conveyor and the tool of a robot has to orient all direction. If the work envelop is defined incorrectly, the tool-paths of the robot make collisions with workpiece or robot itself. Otherwise, the tool-paths can be impossible positions for the robot. As shown in Figure 8, the authors have found workspace envelops of four sides; front, back, top, and bottom side of structural steels. In the figure the bottom workspace is two-dimensional because the bottom side of any steel is always aligned at a specific location in the proposed system. The other sides have three-dimensional workspaces because of their different locations and heights depending on the steel. The workspaces allow the robot to cut most standard steel beams. Due to the size of the robot, Yaskawa MA1440, small workspaces are available in the bottom and back sides. Therefore linear translations using the conveyor system are necessary operations for cutting or marking long shapes in the proposed system.

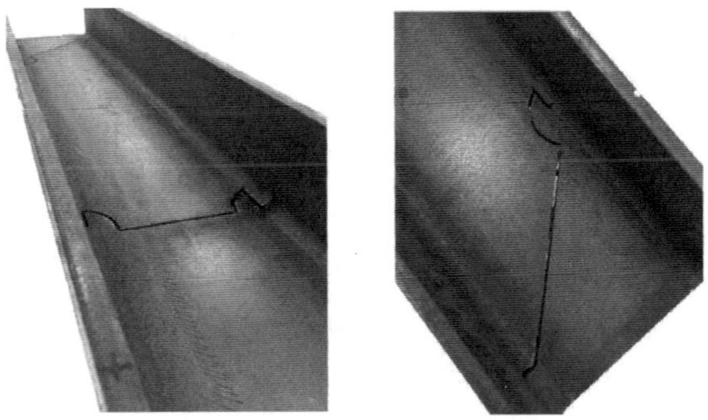

Figure 6. Results of a test cut

Figure 7. Digital model of the proposed cutting system

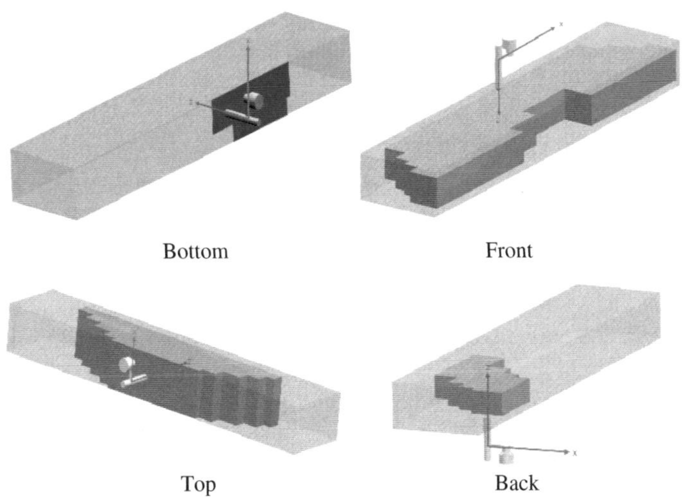

Figure 8. Robot workspaces for cutting four directions

6. Conclusion

In this study the authors have designed a new robotic structural steel cutting system. The overall system has been modeled as a digital model and verified in a robot simulation system. The results show that the possibility of the proposed system.

Tool-path generation system in the cutting system has been developed and tested in which the geometry information of a DSTV file is converted to a robot job file. The actual cutting test shows that the tool-paths generated by the system can be used to cut structural steel.

The authors have tested only cutting the structural steel clamped in front of the robot. It will be difficult to precisely align the coordinates between a robot and a workpiece which is on a conveyor and moves. It will be also difficult to synchronize the control of the conveyor with the robot dynamically.

Acknowledgments

This paper is supported by the R&D project of the "Innovative Design Advanced Manpower Training Program" of the Ministry of Commerce, Industry and Energy.

References

[1] A. Fussel, K. Cowie, S. Hicks, M. Karpenko, Basis for and implications of key changes to 2016 structural steel product standards, SESOC Journal, 30, 1, 38-43 (2017)
[2] K. Simunovic, G. Simunovic, T. Saric, Single and multiple goal optimization of structural steel face milling process considering different methods of cooling/lubricating, Journal of Cleaner Production, 94, 321-329 (2015)
[3] K. Patel, P. Agarwal, Effect of process parameters on structural steel using plasma arc cutting: a review, International Journal of Advanced Engineering and Research Development, 4, 4, 608-611 (2017)
[4] PythonX, https://www.pythonx.com
[5] KranenDonk, https://www.kranendonk.com
[6] G. Lee, Construction Issues of Irregular-Shaped Buildings, Review of Architecture and Building Science, 52, 4, 63-65 (2008)
[7] S. H. Mun, B. H. Shin, G. B. Park, M. H. Nam, S. J. Bae, K. H. Kim, H. J. Lee, D. K. Kim, A Development of Profile Cutting Robot System. Proc. of KSME Conference, Oct. 113-116 (2007)
[8] E. Garcia, M. A. Jimenez, P. G. De Santos and M. Armada, "The evolution of robotics research," in IEEE Robotics & Automation Magazine, 14, 1, 90-103 (2007)
[9] D. Lee, N. Ku, T.W. Kim, J. Kim, K.Y. Lee, Y.S. Son, Development and application of an intelligent welding robot system for shipbuilding, Robotics and Computer-Integrated Manufacturing, 27, 2, 377-388, (2011)
[10] L. Caneparo, Design for Manufacture, Digital Fabrication in Architecture, Engineering and Construction, 77-116, Springer, Dordrecht (2014)
[11] D. Ullman, The Mechanical Design Process, 3rd edition (McGraw Hill, Boston, 2003)
[12] G. Mauer, M. O. Jarligo, D. Marcano, S. Rezanka, D. Zhou, R. Vaßen, Recent developments in plasma spray processes for applications in energy technology, IOP Conference Series: Materials Science and Engineering, 181, Conference 1 (2018)
[13] J. Aldazabal, A. Martin-Meizoso, A. Klimpel, A. Bannister, S. Cicero, Mechanical and Microstructural Features of Plasma Cut Edges in a 15 mm Thick S460M Steel Plate, Metals, 8, 6, 447-460 (2018)
[14] J.P. Gutssch, U. Kammertöns, J. Keil, D. Knierim, Dr. H.G. Liekweg, U. Pfingst, F. Streit, J. Verlies, B. Wiefel, J. Zühlke, Standard Description for Steel Structure Pieces for the Numerical Controls, 7th Edition (1998)
[15] Yaskawa MA 1440 Arc Welding Robot, https://www.motoman.com/industrial-robots/ma1440
[16] RoboDK, https://robodk.com

IOP Conf. Series: Materials Science and Engineering **538** (2019) 012043 doi:10.1088/1757-899X/538/1/012043

Design, Modeling and Analysis of a XY Nanopositioning Stage for High Speed Scanning

Shenglong Lin, Xianmin Zhang and Benliang Zhu

Guangdong Province Key Laboratory of Precision Equipment and Manufacturing Technology, South China University of Technology, Guangdong Guangzhou 510640, China

Corresponding author: zhangxm@scut.edu.cn

Abstract. In order to increase the imaging speed of scanning probe microscopy (SPM), especially atomic force microscopy (AFM) where needs a high-bandwidth moving stage possessing high resonant frequency and low cross-coupling, the paper proposes a kind of XY nanopositioning stage achieving about 10kHz resonant frequency, 15um×15um workspace and well decoupled performance. Considering the design objective, a compliant nanopositioning stage is built with doubly clamped beam and parallelogram hybrid beam for overcoming the problem of low natural frequency and cross-coupling performance. By establishing mathematical model of the proposed stage including stiffness model and resonant frequencies model, the paper solves the highest natural frequency with its optimal dimensions of beams by applying optimization. Finally the designed stage is imported to Workbench for the validation of mathematical model by simulation, where presents the FEA results can nicely match the analytical results.

1. Introduction

Nowadays, micro and nanopositioning stage plays an indispensable role in both commercial and scientific field including SPM [1], biological science [2], nano-metrology [3], nano-machining and nano-fabrication [4], optical calibration [5] and micro-nano manipulation [6] due to its high accuracy and fine repeatability. However, high speed nanopositioning stage is extremely needed in various research fields especially in AFM, one kind of SPM recent years. For instance, in the area of biological research, it's so urgent to acquire biospecimen image in AFM at a video-rate speed because the dynamic process of the sample in nanoscale would be a great benefit for the further study of bioscience [2, 7]. But the dynamic process of sample is always hard to be achieved because of the low resonant frequency of traditional nanopositioning stage in AFM, such as sectored piezoelectric tube scanner [8], However, compliant mechanism with logical design seems to be a solution for the task, which owns advantages of monolithic structure, ease of processing, no friction and no backlash compared with traditional rigid mechanism, utilizing the elastic deformation of flexible hinges providing with accurate movements [9]. In general, the maximum bandwidth of stage is only 1/100th to 1/10th of the resonant frequency in open-loop [10]. Hence, it's necessary to design a nanopositioning stage with high resonant frequency which can undistortedly track to a high frequency signals with proper control strategy. Aiming at high-bandwidth moving stage, many researchers have studied much researches in this field. For example, Brian et al. [11] designed a kind of high-bandwidth

IOP Conf. Series: Materials Science and Engineering **538** (2019) 012043 doi:10.1088/1757-899X/538/1/012043

nanopositioner with natural frequencies of 24.2*kHz* and 6.0 *kHz* along X and Y axes respectively and a stroke of $9\times 9um^2$,which can achieve fine accuracy track under 100 Hz triangular wave. Kam et al. [12] proposed a compliant positioning stage with a $9\times 9um^2$ travel range and resonant frequencies of 4.5k*Hz* and 20 *kHz* in slow and fast axis respectively. Cai et al. [13] designed a compliant scanning platform with parallel structure, whose natural frequencies of X and Y axes are both 5.6*kHz* and the stroke is close to $9\times 9um^2$. Readers are also referred to more concerning stage by paper [14]. Based on the previous studies involved with high speed stage, a high bandwidth, relative large travel range and low cross-coupling nanopositioning stage is designed and optimized in the article, which adopts doubly clamped beam and parallelogram hybrid beam.

The remaining contents of this article are as follow. In Sec. II, the details of the structural design and analysis of stage are shown including the establishment of the initial structure and mathematical models of the developed stage. In Sec. III, this part introduce how to figure out the highest natural frequency of developed stage and obtain the optimal dimension of relevant compliant units by optimization in MATLAB. Then, finite element analysis (FEA) of the optimized mechanism is presented for static and dynamic analysis. In final section, the results of the paper and future works are illustrated.

2. The Design and Analysis Of Stage

In accordance with the kinematic structures, a XY flexure-based stages are categorized into two main configurations: serial-kinematic [2, 15, 16] stage and parallel-kinematic [13, 17-21] stage shown in figure 1. The serial-kinematic stage is created by nesting one flexure-based nanopositioner into another, which is more effective and economic with only one fast axis but generally have disadvantages of cumulative error and parasitic motion that hard to be measured [11]. The parallel-kinematic stage is constructed by symmetrical configurations, usually performing high motion accuracy,high mechanical stiffness and low inertia of the moving sample platform helpfully contributing to high resonance frequency. The latter configuration would be employed in our paper because of its potential in high resonance frequencies along both X and Y axis as well as advantage in temperature compensation. Besides, novel non-raster scan methods such as spiral- [22], cycloid- [23], and Lissajous-scan patterns [24] can be applied on parallel-kinematic stage.

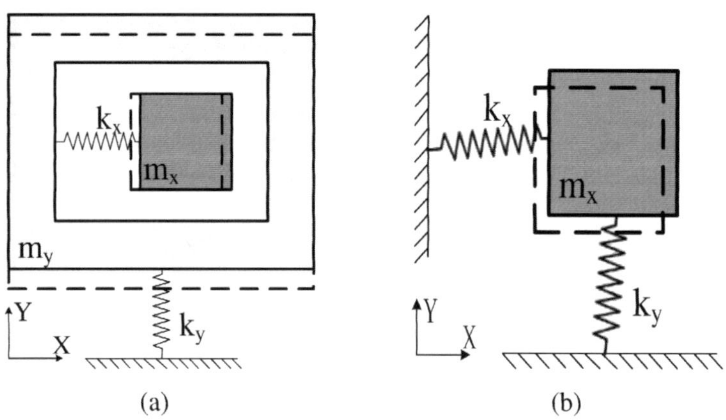

Figure 1. The Serial- and parallel-kinematic stage: (a) Serial-kinematic model; (b) Parallel-kinematic model.

For achieving high resonance frequency while possessing a relative large travel and low coupling errors,the initial structure is presented at figure 2.As shown in figure 2(a),the piezoelectric actuators (PZTs) coloured green are applied to drive the compliant mechanism for nanoscale positioning.The main flexible elements of the stage cover with doubly clamped beam and parallelogram hybrid beam.The former beam is constant-width beam with both ends fixed,providing with a high stiffness compared to parallelogram hybrid beam;the latter beam is made up of a thicken beam in one end in order to strengthen the bending stiffness of the middle part of doubly clamped beam as well as a lumped compliance hinge in another end for increasing the compliance of end-effector motion.The major component of the two kinds of beam is width-constant beam having advantage of large deformation,compact structure and easy processing.In the developed stage,the symmetrical configuration is adopted to reduce the cross-coupling error shown in figure 2(b).

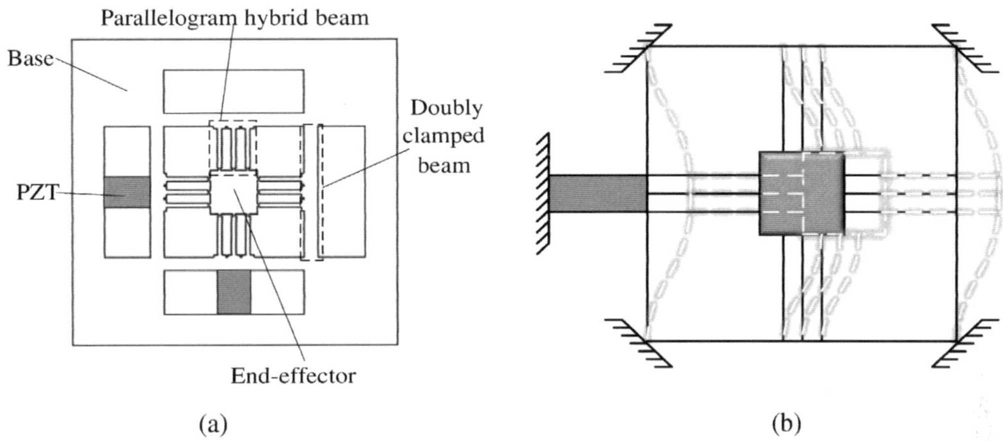

Figure 2. The structure of developed stage: (a)The structure of XY stage; (b)The schematic diagram of XY stage in x direction.

2.1. Stiffness Analysis of Stage

The stiffness of stage is a crucial factor for acquiring high resonance frequency so it's significant to precisely establish the stiffness mathematical model of stage. Due to the symmetrical configuration of stage, only one kinematic chain need to be analyzed in the paper. And the stiffness analysis of each kinematic chain can be divided into two sections, including doubly clamped beam and parallelogram hybrid beam.

As to build the stiffness of cantilever beam, the paper employed Castigliano's second theorem, one of effective method to derive the relationship of external load and deformation when the beams deform elastically and comply with Hooke's law [25]. According to the theorem, the force-displacement relation of the variable cross-section beam in *i* direction can be expressed as

$$u_i = \frac{\partial U}{\partial F_i} \tag{1}$$

where u_i is the displacement in *i* direction, F_i is the external force in *i* direction and U denote the total strain energy of the beam, including bending strain energy, shearing strain energy, axial strain energy and torsional strain energy, which can be shown as

$$U = U_b + U_s + U_a + U_t \tag{2}$$

The beams of proposed stage are mainly influenced by bending moment M_z and force F_y which are presented in figure 3. Therefore, U_a and U_t is equal to zero, U_b and U_s can be presented as

$$U_b = \int_L \frac{M_z^2}{2EI_z} ds \,, \tag{3}$$

$$U_s = \int_L \frac{\alpha F_y^2}{2GA} ds \tag{4}$$

The E, I_z and G denote the Young's modulus of beam material,inertia moment respect to Z axis and the shear modulus of beam material respectively. There would be gain a relationship between force and displacement when we take the partial derivative of U with respect to M_z and F_y.

$$u_y = C_{y-F_y} F_y + C_{y-M_z} M_z \tag{5}$$

$$C_{y-F_y} = \frac{12}{Eh}\int_0^L \frac{x^2}{t(x)^3}dx + \frac{\alpha}{Gh}\int_0^L \frac{1}{t(x)}dx \tag{6}$$

$$C_{y-Mz} = \frac{12}{Eh}\int_0^L \frac{x}{t(x)^3}dx \tag{7}$$

t(x) represent the sectional dimension and h is the height of proposed stage.

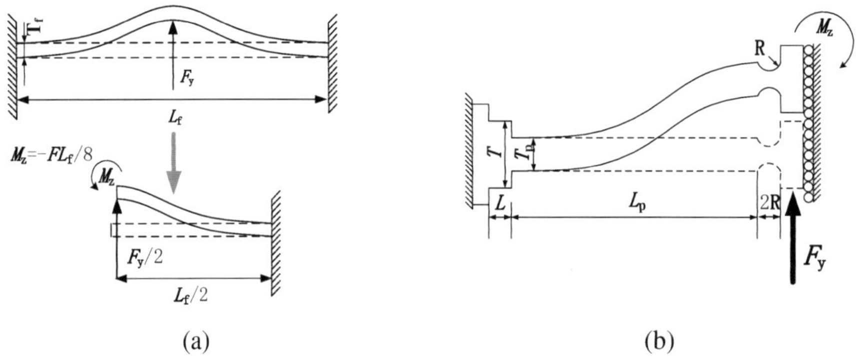

Figure 3. The external loading of beams: (a) Doubly clamped beam and its equivalent form; (b) Parallelogram hybrid beam.

The doubly clamped beam is directly connected to the PZTs and parallelogram hybrid beam, which provide the stage with accurate motion and high stiffness. When the equation (5) comes to the doubly clamped beam, it can be equivalent to the fixed-guided beam shown in figure 3(a), whose form can be directly used in Castigliano's second theorem with invariant thickness. Therefore, the stiffness of single doubly clamped beam can be obtained by

$$C_f = \frac{3L_f^3}{16EhT_f^3} + \frac{\alpha L_f}{4GhT_f} \tag{8}$$

For the parallelogram hybrid beam presented in figure 3(b), the segmented function with respect to its variable thickness would be listed below:

$$t(x) = \begin{cases} T_p - 2\sqrt{x(2R-x)} & x \in [0,2R] \\ T_p & x \in [2R, 2R+L_p] \\ T & x \in [2R+L_p, 2R+L_p+L] \end{cases} \tag{9}$$

Take R, T and L as 0.3(mm), T_p+1.6(mm) and 0.5(mm). Similarly,we can get the compliance of parallelogram hybrid beam by substituting equation (9) into (5).

2.2. The Resonance Frequency Estimation of the Stage

As discussed previously,due to the great significance of improving resonance frequency of stage it's important to correctly build natural frequency model of proposed stage ready for the optimization.The natural frequency of stage contain two aspects including stage itself and stage attached PZTs shown in figure 4. For the stage itself, it's assumed that the stage is a two-degree-of-freedom undamped system, wherefore the Lagrange equation can be obtained as follow [26]:

$$M\ddot{q} + Kq = 0 \tag{10}$$

Here, $M=diag\ (M_x^e, M_y^e)$, $K=diag\ (K_x, K_y)$, $q=[x,y]^T$。 M_x^e and M_y^e represent the equivalent mass of stage in x and y direction respectively while K_x and K_y denote the stiffness along x and y direction respectively. Due to the symmetrical structure, $M_x^e=M_y^e$, $K_x=K_y$. If $q=A\sin(\omega t+\theta)$, $A=[A_x,A_y]^T$, we can get the generalized eigenvalue problem $(K-\omega^2 M)A=0$, then the necessary and sufficient condition for A having non-zero solution is $|K-\omega^2 M|=0$. Finally the angular natural frequency of stage ω_1, ω_2 can be obtained based above equation and the natural frequency of the stage is

$$f_{res} = \frac{\min[\omega_1, \omega_2]}{2\pi} \tag{11}$$

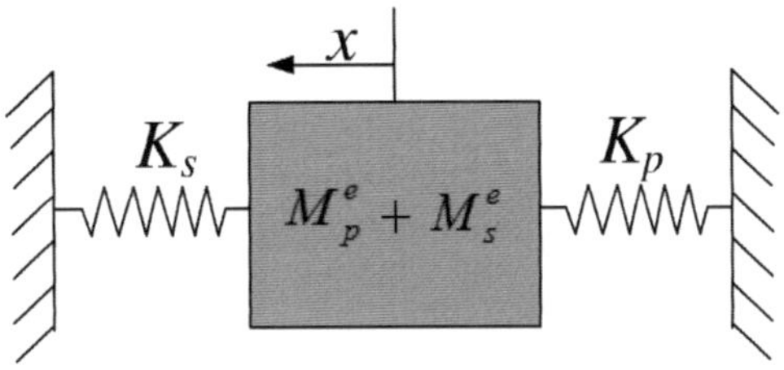

Figure 4. One-degree-of-freedom second order undamped system.

Similarly, when PZTs are considered to the model, the stage can be equivalent to one-degree-of-freedom second order undamped system as shown in figure 4. M_p^e, M_s^e represent the equivalent mass of PZTs and stage respectively while K_p, K_s denote the stiffness of PZTs and stage. Hence, the resonant frequency of stage attached PZTs can be expressed by

$$f_{pzt} = \frac{\sqrt{(K_p + K_s)/(M_p + M_s)}}{2\pi} \tag{12}$$

3. Dimensional Optimization of Stage Model

Generally, a rigid and compact structure would greatly contribute to a high resonant frequency due to its high stiffness but it would sacrifice a part of stage's travel range. In other words, the high stiffness and large stroke of the stage are contradictory. What's more, the accuracy and stiffness also exist conflict [27] but the relationship won't be taken into account in this article. Based on the mentioned above, the paper attempts to optimally design a kind of XY nanopositioning stage whose resonance frequency can achieve 10*kHz* and travel range can be around 15*μm*. Beside,a good decoupled performance between x and y axis are expected.

Table 1. The piezoelectric stack actuators technical parameters.

Model	Length*Width*Height	Driving Voltage	*Nominal Stroke*	Stiffness
NAC2014-H14	7*7*14 mm^3	150V	19.8μm	104N/μm

In order to achieve the objective referred above, optimization theory [28] is employed for obtaining max first resonant frequency by adjusting the dimensions of beam. For the workspace of stage, two

suitable piezoelectric stack actuators would be applied on the article, whose specific parameters are shown in table 1.In the meantime, aluminium alloy 7075 are employed for its Young's modules, density, and yield stress up to 71.7 *GPa*,2810 kg/m^3, and 500 *MPa*, respectively. The target function and optimization variable can be expressed by $minf(x)=-f_{res}$ and $x=[L_f,T_f,L_p,T_p]$, respectively.
The constraints of the optimization model are listed as follow:
1) The constraint of stage's travel range: As a crucial performance index of nanopositioning stage, the stroke of stage is largely influenced by the nominal stroke of piezoelectric stack actuator R_{nom}, stiffness of piezoelectric stack actuator K_p, stiffness of the stage K_s and preload force F_{pre}. So the max displacement of stage in one direction is

$$D=\frac{K_pR_{nom}-F_{pre}}{K_s+K_p}\geq 15\mu m \tag{13}$$

2)Stiffness constraint between doubly clamped beam and parallelogram hybrid beam: Doubly clamped beam provides the stage with displacement and a majority of the stiffness of whole stage while the parallelogram hybrid beam plays an important role in guiding as well as lower cross-coupling errors. Hence, confining the compliance ratio between parallelogram hybrid beam and doubly clamped beam to a proper range would be greatly beneficial for decoupling performance. The stiffness ratio in paper is

$$15\leq\frac{C_p}{C_f}\leq 30 \tag{14}$$

3)The max stress constraint: In the deformation process of beams , the maximum stress should be less than the allowable stress of the material.The maximum stress of doubly clamped beam σ_{max}^f and parallelogram hybrid beam σ_{max}^p can be derived as follow.

$$\sigma_{\max}^f=\frac{3L_fR}{2T_f^2hC_f}\leq[\sigma] \tag{15}$$

$$\sigma_{\max}^p\leq\frac{6K_\theta K_t\arcsin(R/(R+L_p))}{(T_p-2R)^2}\leq[\sigma] \tag{16}$$

R represent the maximum displacement of stage, coefficients K_θ and K_t concerning size and material of flexure hinge[29] while $[\sigma]$ is always set to be a half yield stress of the material.
4)The constraint of valuable range: In order to ensure the sufficient stiffness in the z-axis direction, we set the thickness of the stage to 10*mm*.The lower limit of optimal variables is given to assure that the optimized size is not too small, which will not only affect the assembly of the platform but also increase the difficulty of processing. Moreover, The upper limit also need to be set for a compact stage.

$$\begin{aligned}&20mm<L_f<45mm, 1.5mm<T_f<5mm\\&5mm<L_p<15mm, 0.8mm<T_p<1.5mm\end{aligned} \tag{17}$$

Aiming at solving the optimal value of nonlinear multivariate functions with constraints, the "fmincon" optimization function provided by MATLAB R2016a optimization toolbox is adopted. Taking the initial value as x_0=[30,3,10,1.2]*mm*, the max natural frequency is 10.04kHz without PZTs and 17.00kHz with PZTs when the optimized variable is $\boldsymbol{x}^*$=[26.1 2.5 6.95 0.8]*mm*.

4. Simulation for Validation

After the dimensional optimization of developed stage above, both static and dynamic simulation is indispensable for validating optimization results. In order to verify the accuracy of optimization results, finite element analysis in *ANSYS WORKBENCH 16.0* software is adopted in this paper, by which the stiffness, travel range, dynamic performance, cross-coupling error, and stress analysis can be explicitly presented as follow.

IOP Conf. Series: Materials Science and Engineering **538** (2019) 012043 doi:10.1088/1757-899X/538/1/012043

Table 2. The comparison of theoretical results and FEA results.

	Stiffness(N/um)	Workspace(um^2)	Resonance frequency(Hz)
Theoretical results	22.75	15.07×15.07	10036.78/10036.78
FEA results	22.28	15.13×15.13	9618.00/9626.8
Error(%)	2.07%	0.40%	4.17%/4.08%

Table 2 shows the contrast between theoretical results and FEA results, showing a goodness of fit. The reason accounting for the errors among theoretical results and FEA results mainly results from the assumption of parallelogram hybrid beam whose axis is parallel to the actuation direction being a rigid body. In fact, there still exist tiny deformation in these beam leading to the loss of stiffness and resonance frequency in FEA. The relationship of force and displacement of stage are presented by figure 5(a). As shown in the figure, the X-axis displacement primarily actuated by the X-axis force while the Y-axis force has almost no effect on X-axis displacement exhibiting a good decoupled performance of stage. In figure 5(b), the maximum equivalent stress are far less than he allowable stress of the material, $250MPa$.What's more,by comparing to modal analysis of traditional stage in identical dimensions in figure 6(b), the proposed stage in figure 6(a) performs better decoupling performance in high frequency without sacrificing much natural frequency,the ratio of coupling errors between them is about 1:15.

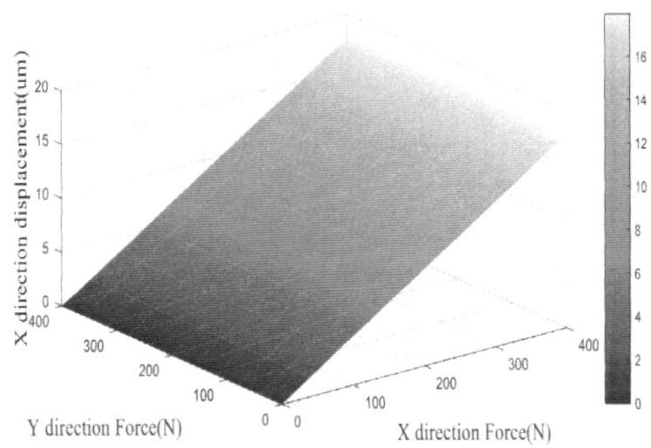

(a)

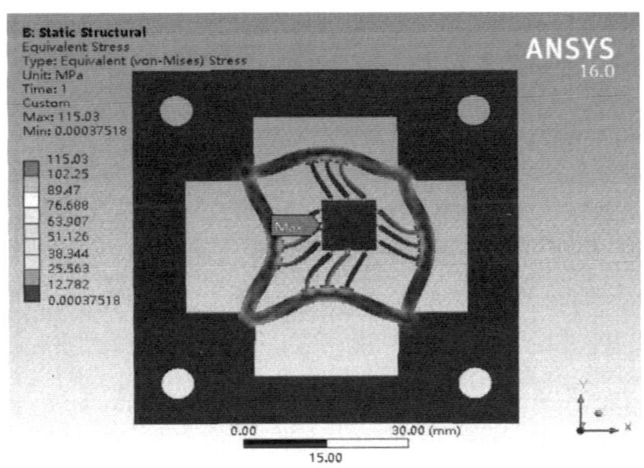

(b)

Figure 5. (a) The force-displacement relationship of stage in max actuation force; (b) The distribution of Equivalent (von-Mises) stress of the stage under the max actuation force.

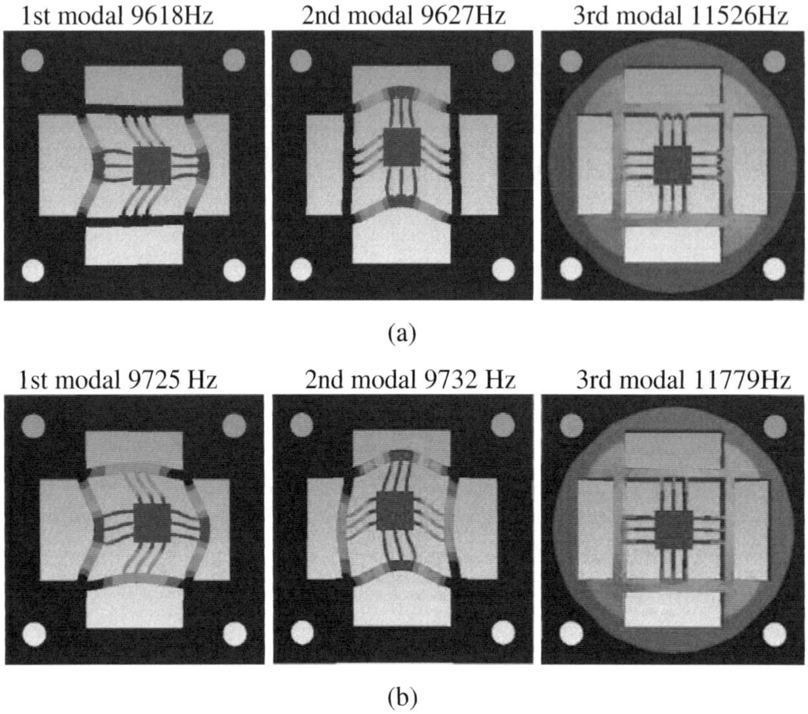

Figure 6. The first three order modal of proposed stage and traditional stage in identical size. (a) The proposed stage using parallelogram hybrid; (b) The traditional stage using parallelogram beam.

5. Result and Future Work

The article proposes a XY nanopositioner by employing optimization method with advantages of high dominant frequency, relative large workspace and low cross-coupling, corresponding to $10kHz$, $15\times15\mu m^2$. In the following work, experiment system on the stage would be built and experiments concerning bandwidth would be conducted with suitable control strategy.

References

[1] R. Wang, X. Zhang, IEEE Trans.Ind Electron **65**, 3 (2017)
[2] T. Ando, N. Kodera, E. Takai, D. Maruyama, K. Saito, A. Toda. PNAS 98, 22 (2001).
[3] A. D. Mazzeoa, A. J. Steina, D. L. Trumpera, R. J. Hockenb, Precision Engineering **33**, 2 (2009).
[4] S. Verma, W. Kim, H. Shakir, IEEE Trans Ind Appl **41**, 5 (2005).
[5] H. Marth, B. Lula, Proc. SPIE **62731K** (2006).
[6] M. L. Culpepper, G. Anderson, Precis. Eng **28**, 4 (2004).
[7] H. Y. Kim, D. H. Ahn, D. G. Gweon, Rev. Sci. Instrum **83**, 5 (2012).
[8] G. Schitter, A. Stemmer, IEEE Trans. Control Syst. Technol **12**, 3 (2004).
[9] M. Liu, X. Zhang, S. Fatikow, Rev. Sci. Instrum **87**, 5 (2016).
[10] G. M. Clayton, S. Tien, K. K. Leang, Q. Zou, S. Devasia, J. Dyn. Sys., Meas., Control **131**, 6 (2009).
[11] B. J. Kenton, K. K. Leang, IEEE/ASME Trans. Mech **17**, 2 (2012).
[12] S. P. Wadikhaye, Y. K. Yong, S. O. R. Moheimani, Rev. Sci. Instrum **85**,10 (2014).
[13] K. Cai, X. He, Y. Tian, X. Liu, D. Zhang, B. Shirinzadeh, Microsystem Technologies **24**, 7 (2018).
[14] Y. K. Yong, S. O. R. Moheimani, B. J. Kenton, K. K. Leang, Rev. Sci. Instrum **83**,12(2012).
[15] K. K. Leang, A. J. Fleming, Asian Journal of Control **11**, 144 (2009).

[16] T. Ando, N. Kodera, T. Uchihashi, A. Miyagi, R. Nakakita, H. Yamashita, e-Journal of Surface Science and Nanotechnology **3, 3** (2006)
[17] T. Lu, D. C. Handley, Y. K. Yong, C. Eales, Industrial Robot: An International Journal 31, 4 (2004).
[18] S. Salapaka, A. Sebastian, J. P. Cleveland, M. V. Salapaka, Rev. Sci. Instrum **73**, 9 (2002).
[19] G. Schitter, K. J. Åström, B. E. Demartini, P. J. Thurner, K. L. Turner, P. K. Hansma, IEEE Trans. Control Syst. Technol **15**, 5 (2007).
[20] Y. K. Yong, S. S. Aphale, S. O. R. Moheimani, IEEE Trans. Nano **8**,1 (2009).
[21] C. X. Li, G. Y. Gu, M. J. Yang, L. M. Zhu, Rev. Sci. Instrum **84**, 12 (2013).
[22] I. A. Mahmood, S. O. R. Moheimani, B. Bhikkaji, IEEE Trans. Nano **10**, 2 (2011).
[23] Y. K. Yong, S. O. R. Moheimani, I. R. Petersen, Nanotechnology **21**, 36 (2010).
[24] A. Bazaei, Y. K. Yong, S. O. R. Moheimani, Rev. Sci. Instrum **83**, 6 (2012).
[25] R. Lin, X. Zhang, S. Fatikow, Rev. Sci. Instrum **84**, 8 (2013).
[26] L. S.Jacobsen, R. S. Ayre, Engineering vibration. McGraw-Hill (1994).
[27] S. Awtar, A. H. Slocum, E. Sevincer, J. Mech. Des **129**, 6 (2007).
[28] J. Nocedal, S. J. Wright, Numerical Optimization. Springer (2006).
[29] J. W. Ryu, D. Gweon, K. S. Moon, Precis. Eng **21**, 1 (1997)

IOP Conf. Series: Materials Science and Engineering **538** (2019) 012044 doi:10.1088/1757-899X/538/1/012044

Design of the Curtain Wall Cleaning Robot

Shan Fan[1], Lei-lei Han[1,a] and Chao-ying Meng[1]

[1]Wenhua college, 430074 Wuhan, China

Corresponding author: 15637664@qq.com

Abstract. The curtain wall cleaning robot suitable for special-shaped wall structure of buildings is designed, and the mechanical structure design and control system design of the curtain wall cleaning robot is completed in the paper.The curtain wall cleaning robot is composed of two parts with the same structure. The hinge mechanism is used to realize the turning action of irregular wall. The suction and movement of the cleaning robot are realized by vacuum adsorption of the sucker. And the motor controlled the rotation of the roller brush. The screw guide rail drives the roller brush up and down to realize cleaning. Based on the mechanical structure design, the control system of curtain wall cleaning robot is designed. And PLC control system is used to control the sequence of action of curtain wall cleaning robot.

1. Introduction

At present, the cleaning of most glass curtain walls in China relies on "spiderman" to complete the cleaning at high altitude. The working environment of workers is very bad, and there are certain safety risks. Cleaning the surface of a building is time-consuming, inefficient and expensive. Therefore, it is necessary to design an efficient and convenient curtain wall cleaning robot to replace manual work. More and more special-shaped buildings have been built, and the shape of such buildings is not a single plane, all with a certain tilt Angle, which requires the cleaning robot to have more functions.

In China, the research on wall mobile robot started very late, but the development speed is very fast. Since 1998, special robot research office of institute of robotics, Harbin Institute of Technology has taken the lead in the research of wall-climbing robots with the support of the state[1]. It has made many positive and meaningful explorations and attempts. The negative pressure adsorption and magnetic adsorption two series of five kinds of wall crawling robot have been successfully designed and manufactured. There are two kinds of cleaning climbing robot: single sucker, double wheel and double crawler, which can clean ceramic tile and glass wall. In foreign countries, the development of curtain wall cleaning robot is slightly earlier, BVE company of Japan has successfully developed a fixed-track automatic window cleaning robot, which mainly relies on the track and cable lifting system installed on the roof of the building to align the window cleaning machine with the window, and then the robot moves vertically up and down the guide groove fixed on the surface of the building for automatic cleaning.BVE company of Japan has successfully developed a fixed-track automatic window cleaning robot, which mainly relies on the track and cable lifting system installed on the roof of the building to align the window cleaning machine with the window, and then the robot moves vertically up and down the guide groove fixed on the surface of the building for automatic cleaning. It's highly automated and efficient, and of course expensive, and it requires Windows cleaning systems to be taken into account at the beginning of a building's design, and tracks to be laid. Otherwise, it cannot adapt to the stepped shape of the wall and Windows, nor is it suitable for the already built

IOP Conf. Series: Materials Science and Engineering **538** (2019) 012044 doi:10.1088/1757-899X/538/1/012044

buildings[2].So not only the scope of use and efficiency of the equipment is severely restricted, but also to the building itself increased a huge cost. Therefore, this paper innovatively designs a curtain wall cleaning robot suitable for special-shaped wall structure of the building, and the hinge mechanism realizes the reversal of special-shaped wall.

2. Overall Design of Curtain Wall Cleaning Robot

2.1. Selection of Various Parts of the Curtain Wall Cleaning Robot

The curtain wall cleaning robot can be divided into five parts: the cleaning part, the adsorption part, the moving part, the turning part and the shell. Morphological matrix was established for the type selection of each mechanism, as shown in Table 1.

Table 1. Morphological matrix

Cleaning device	Disc brush	Roller brush	Strip brush
Adsorption device	Negative pressure adsorption	Magnetic adsorption	Thrust adsorption
Moving device	wheel	caterpillar	Foot type

(1) Cleaning part of the design: such as the disk brush through the high-speed rotation to achieve the purpose of cleaning, which is fast and convenient, but because of the way of cleaning, some corners can not be cleaned. The cleaning method of the brush is to achieve the cleaning purpose by moving up and down the brush, with low efficiency and weak cleaning degree. The rolling brush cleans a range through the rotation of the rolling brush itself, and then drives the rolling brush up and down through the screw rod for cleaning. And the cleaning effect is better[3].
(2) Adsorption part design: the working principle of magnetic adsorption is to make the curtain wall cleaning robot body adsorbed on the wall by magnetic force, and then the curtain wall cleaning robot starts to work. Magnetic adsorption requires that the problem of conducting materials be considered in advance when building a building. The working principle of wind thrust adsorption is to generate an opposite force through the fan, so that the robot is adsorbed on the wall, the greater the thrust required, accompanied by noise. At present, most designs adopt negative pressure adsorption. Sliding seal negative pressure sucker is used here to ensure sealing by controlling the gas pressure in the sealing ring and adopting inflatable rubber sealing skirt edge. Sealing ring can better fill the gap between the sucker and the wall surface, so that the sucker and the wall close together. Therefore, a certain sealing environment is formed in the cavity of the sucker, which is able to adapt to the environment where the wall surface is not completely flat.
(3) Design of moving part: footwork imitates the biological movement method, but the movement speed is slower, the structure is more complex, and the cleaning efficiency is lower. The wheel movement efficiency is higher, but because of the wheel structure question, causes the adsorption to become difficult, has the security hidden trouble. Track moving speed is fast, multiple suckers ensure the safety of sucker adsorption, more appropriate. The movement method used here is for the purpose of moving through the rotation of the triple sucker foot[4].
(4) Flip mechanism design: when facing the non-planar wall, one end of the curtain wall cleaning robot can be raised to climb up the special-shaped wall. For this kind of turnover movement, it can be analogous to the opening and closing of a door, and the two sides are connected by a hinge.
In conclusion, the curtain wall cleaning structure is mainly cleaned by rolling brush, and the screw rod drives the roller to move up and down to achieve the purpose of cleaning. The purpose of movement is achieved by negative pressure adsorption of sliding seal and rotation of sucker foot.

2.2. Working Principle of Curtain Wall Cleaning Robot

Curtain wall cleaning robot is driven by motor, sliding seal adsorption to complete the adsorption movement. Firstly, the pneumatic cylinder is started, and the robot is placed on the glass wall for solid adsorption. The adsorption and release of each sucker foot are controlled by the single chip microcomputer, and the movement of the robot is realized at the same time. When the cleaning robot moves to the position to be cleaned, turn on the drum motor switch through the console, the drum motor runs, and the drum starts cleaning. At the same time, the water storage box starts to spray mist droplets onto the glass wall through the nozzle, and the sliding table motor works, driving the whole clean structure part to complete the up and down reciprocating movement. After cleaning, the sewage is collected by the wiper board and guided to the ground by a small pipe, so as to clean the wall.

In the process of cleaning, the whole cleaning robot also moves forward slowly at a certain speed. When the face can not be directly cleaned, through the console, open the hinge motor. Turn the hinge at an Angle so that one side of the cleaning robot is attached to the wall and the other side is released. The robot then repeats until it crosses over.

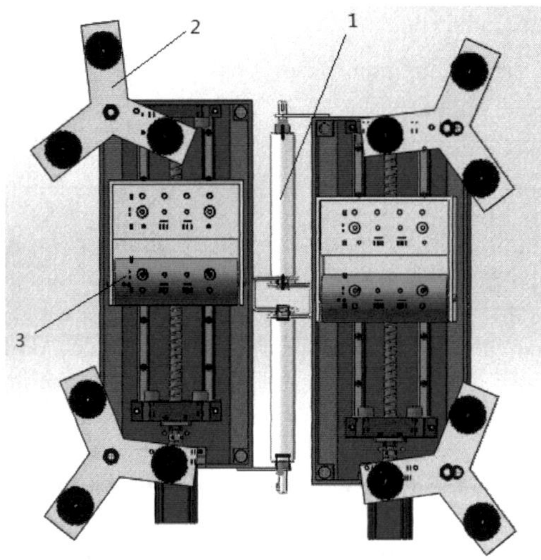

Figure 1. General Assembly Drawing
1- Hinge flip mechanism; 2-Walking mechanism; 3-Cleaning device

3. Mechanical Structure Design of the Curtain Wall Cleaning Robot

3.1. Design of Cleaning Device

Cleaning part of the movement is to allow the roller brush to clean at the same time up and down the movement. The same position for repeated cleaning, simulation of manual cleaning of the glass action, to ensure the thorough cleaning. The cleaning speed of this part is fixed. Here, the transmission speed is 0.3m/s. The total mass of the whole curtain wall cleaning robot does not exceed 10kg, requiring smooth movement, simple structure and continuous movement. Combined with the movement speed and characteristics of the cleaning part of the curtain wall cleaning robot, the sliding table screw mechanism is proposed. The screw rod is connected to the motor through the coupling, and the screw rod rotates to drive the sliding table to complete up and down sliding. When the slider reaches the end buffer, the motor reverses immediately. The screw rod rotates in the opposite direction and the sliding table moves in the opposite direction to realize the reciprocating movement of the clean structure. Sliding table screw structure transmission smooth structure is simple, very consistent with the movement of clean structure[5].

The working parts cleaned by the curtain wall cleaning robot are roller brushes as shown in figure 3-1. Rolling brush through the motor to achieve continuous rotation, cleaning the glass parts. The roller brush is installed on the drum box. A water tank is attached to the drum box to store the water needed for cleaning. Before work, fill the tank with water through the water inlet on the drum box. When working, the water in the water tank is sprayed to the roller working part through the nozzle, so as to complete the cleaning work.

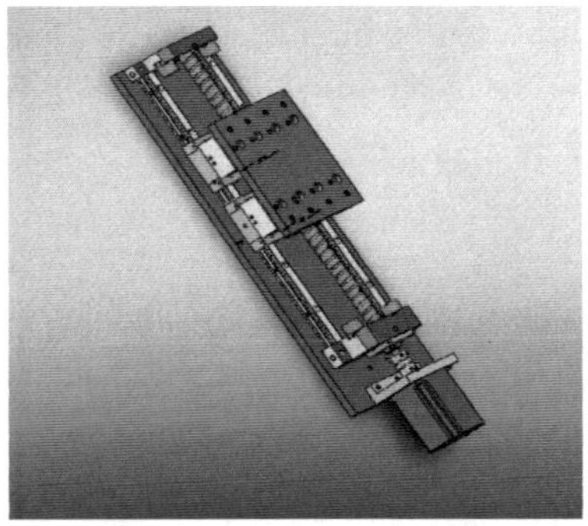

Figure 2. The transmission system of the cleaning device

3.2. Design of Walking Mechanism

A sucker foot consists of three suckers. In the process of movement, one of the three suckers of the sucker foot is fixed. The other two suckers take the fixed sucker as the center of the circle and rotate at a certain Angle to fix the next sucker. Follow the steps above to move. The robot moves forward, backward or left and right by controlling the Angle at which the sucker turns.

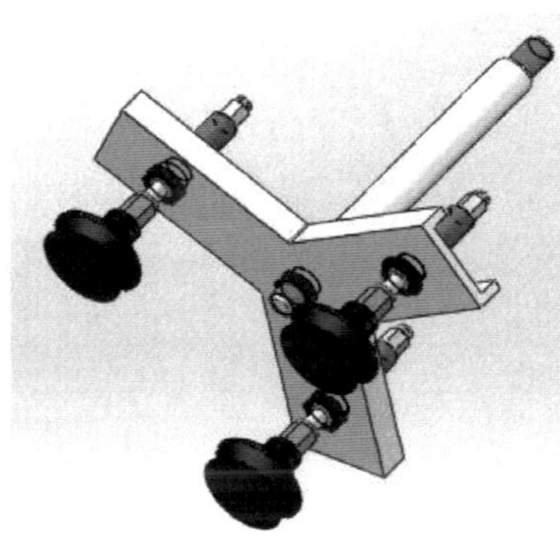

Figure 3. The Sucker Foot

The walking of curtain wall cleaning robot is mainly controlled by several suckers, and the suckers are controlled by the pneumatic system, as shown in figure 3 below[6].

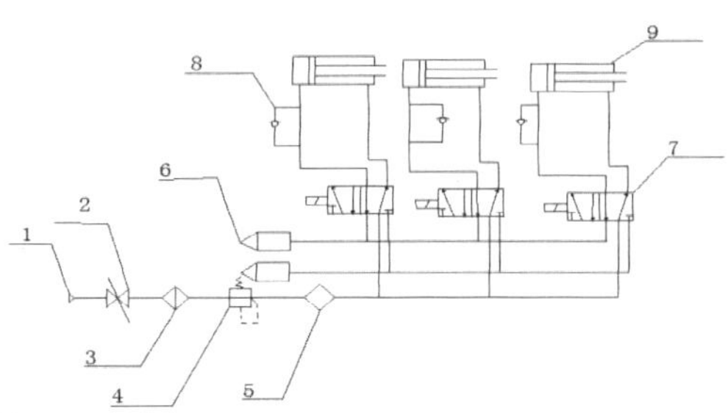

Figure 4. The pneumatic control system
Air source;2- Gate valve;3-Air filter;4-Pressure reducing valve;5- Separator; 6-Muffler; 7-Two five-way solenoid valve; 8-One-way throttle valve; 9-Cylinder

The air source provides power, which passes through gate valve, air filter, pressure reducing valve and separator successively. Through two five - way solenoid valve to control the pneumatic cylinder process and return. Finally, it leaves through the muffler.Process: two five-way solenoid valve is placed in the left position, and the cylinder moves to the right to complete the adsorption action.Return journey: two five-way solenoid valve is placed in the right position, the cylinder moves to the left, and the sucker is released.Air filters are used to filter air to ensure that no debris enters the cylinder and blocks the pipe. Pressure relief valve is used to control the pressure, when the pressure is too high, it will automatically disconnect the channel, protect the pressure device. One-way throttle valves are used to control the flow rate.

3.3. Design of Flip Mechanism

The material of hinge joint shaft is 45 steel. Hinge shaft is mainly used to connect cleaning devices on both sides together, two hinge shaft control half of the mechanism, through the gear drive. It comes into play when it encounters a different surface. The turnover mechanism needs to realize the turnover of the curtain wall cleaning robot and control the Angle of rotation of the shaft. So the motor choice stepper motor, control with a single chip microcomputer. If the pulse equivalent is 0.001, then 10,000 pulses are equal to one cycle, and 27.7 pulses are equal to one degree. By controlling the number of pulses to ensure the rotation Angle of the turnover mechanism, the purpose of climbing the special-shaped wall can be achieved.

4. Design of Curtain Wall Cleaning Robot Control System

4.1. Selection of Curtain Wall Cleaning Robot Controller

The curtain wall cleaning robot has simple structure, fixed action and simple operation. Therefore, the controller also requires simple operation, good reliability and easy maintenance. According to these characteristics, FX2N PLC controller is selected. The specific I/O distribution table is shown in table 2.

Table 2. I/O Distribution Table of PLC

Input point	Code name	Function	Output point	Code name	Function
X0	SA1	manual	Y0	YV1	Motor start
X1	SA2	automatic	Y1	YV2	Roller rotation
X2	SB1	Start button	Y2	KM1	The pump work
X3	SB2	Drum motor switch	Y3	KM2	The slide table moves upward
X4	SB3	The water valve switch	Y4	KM3	Pneumatic pump operation
X5	SB4	Slide table motor switch	Y5	KM4	The slide table moves downward
X6	SB5	Valve switch	Y6	HL1	Power indicator light
X7	SB6	Light switch	Y7	HL2	Water pump indicator light
X8	XB7	Stop	Y8	HL3	Pneumatic pump indicator light
			Y9	HL4	floodlight

4.2. Function Design of Curtain Wall Cleaning Robot Control System

This paper mainly designs the cleaning part control system of curtain wall cleaning robot. Now analyze the cleaning steps in a single cycle. When the curtain wall cleaning robot is at the origin, the water valve opens and begins to discharge water. Then the drum motor opens and the drum starts to turn. The sliding table motor starts to turn forward and the sliding table moves towards the end point. When the end point is reached, the motor reverses and the sliding table moves downward until it

IOP Conf. Series: Materials Science and Engineering **538** (2019) 012044 doi:10.1088/1757-899X/538/1/012044

returns to the origin. Then the sliding table motor is closed, the drum motor and the water valve are closed in turn to complete a cycle.

5. Conclusion

This paper designs the overall scheme of curtain wall cleaning robot and completes the mechanical structure design, including cleaning structure, moving mechanism and turnover structure. Solidworks is used for 3d modeling of curtain wall cleaning robot. Based on PLC, the control system of curtain wall cleaning robot is designed. The innovation of this paper lies in the design of a hinge flip mechanism to achieve the purpose of climbing different surfaces.

References

[1] H.Shao. Research on cleaning wall-climbing robot mechanism and cleaning operation system. Harbin Institute of Technology.15(1999)

[2] S.X.Liu. Research on control technology of wall cleaning and climbing robot. Harbin Institute of Technology,35(2000)

[3] M. M. Cusack and J. G. Thomas. Wall-Climbing Robot for the Inspection and Maintenance of Buildings. Proc. 9th Int. Symp. on Automation and Robotics in Construction. 48,2 (1992)

[4] K. Sato, M. Watanabe, Y. Fukaawa. On-wall Traveling Robot for Nuclear Power Plant. Conference on Robots and Remote System. Charleston S.C. in the USA.13, 3(1989)

[5] C.Almon. A window climbing robot.The 4th Mechanical and Manufacturing Engineering(ICME 2013), 118(2013)

[6] ZH.Y.Qian,Y.Z.Zhao, Zh.Fu ,et al. Design and realization of a non-actuated glass-curtain wall-cleaning robot prototype with dual suction cups. The International Journal of Advanced Manufacturing Technology,301-302,1(2006)

IOP Conf. Series: Materials Science and Engineering **538** (2019) 012045 doi:10.1088/1757-899X/538/1/012045

Design and Analysis of A Multifunctional Wheelchair

Jin Rui and Qiang Gao

School of Mechanical & Electronic Engineering, 178 Ganjiang East Road, Soochow University, Suzhou 215021, P.R. China

Email: gaoqiang@suda.edu.cn

Abstract: Based on existing designs of multifunctional wheelchairs, a new wheelchair designed with body posture adjustment function is proposed to satisfy the needs of the elderly and the disabled. This paper shows the overall design of the wheelchair and the establishment of its 3D model at first; then, the kinematics analysis of the wheelchair's backrest adjustment mechanism, and verifying the smoothness of its reclining process with establishing a virtual wheelchair prototype for motion simulation in ADAMS; finally, finite element analysis of key components of the wheelchair to make sure their requirements of structural strength.

1. Introduction

Recently, the aging in China is characterized by the continuously growth of the elderly population, disability and chronic disease, empty-nesters and lost only-child old [1]. A report, published by the National Office for Aging in 2016, estimated that the number of disabled and semi-disabled elderly in China reached 40.63 million, of which about 10 million are severely disabled and mentally retarded [2]. The demands of life, health and rehabilitation care for elderly, disabled and demented people are significantly increasing [3]. However, the training of professional nurses, the systematic research and development of geriatric nursing projects are a gap in China [4].

Research shows that, for the elderly and the disabled who rely on wheelchairs in daily life, changing body posture regularly can relieve their physical fatigue. As a result, the occurrence of complications, such as pressure sores could be reduced [5]. To develop a wheelchair with function of automatic posture adjustment becomes a solution to the absence of professional nursing system. Compared with the traditional way of nursing, multifunctional wheelchairs do not require a large amount of work and long-term care [6]. In order to meet the requirements above, some medical equipment enterprises and relevant scientific research institutes have developed various kinds of old-age and disabled assistance devices, represented by combined nursing beds and multifunctional intelligent wheelchairs. The nursing bed is a multifunctional device that combines the mobile function of electric wheelchair with the nursing function of nursing bed, e.g. the Panasonic's robotic device [7], and the robotic chair/bed system E-bed [8]. The multifunctional wheelchair is capability of assisting patients in posture adjustment besides the basic mobility function. For example, Wy'East Medical Corporation invented a new multifunctional stretcher chair named Totalift [9], and Roger Bostelman et al. developed a multipurpose robotic wheelchair and rehabilitation device for the home [10]. The multifunctional wheelchair has excellent advantages such as compact structure, high flexibility and strong versatility, thus especially suitable for home care [11].

IOP Conf. Series: Materials Science and Engineering **538** (2019) 012045 doi:10.1088/1757-899X/538/1/012045

In order to provide efficient nursing care for the elderly and the disabled, a multifunctional wheelchair is designed in this research which can automatically adjust human body posture without any auxiliary equipment. The wheelchair can not only serve as a means of transportation for disabled people, but more importantly, it can prevent complications caused by long-term bedridden and sitting, reduce the work intensity of nursing staff, and improve life quality of the disabled.

2. Design Scheme

This designed multifunctional wheelchair not only fulfil the needs of the disabled, but also satisfy the requirements of stability and safety. Wheelchair has sitting and lying states, and usually it keeps sitting state during normal use. Patients can adjust their backrest and legs posture according to their own comfort needs. Standard height which is defined as 1750mm was used to compute size parameters of each part of the multifunctional wheelchair. The overall structure of the multifunctional wheelchair is shown in Figure 1. It is mainly composed of the base frame, the backrest mechanism, and the leg flexion and extension mechanism. These three parts bear most of the load during the transformation of the wheelchair. Backrest adjusting mechanism, leg flexing and stretching mechanism are used to adjust patient's body posture. The wheelchair mechanisms are independent and are not interfered with each other in the process of transformation.

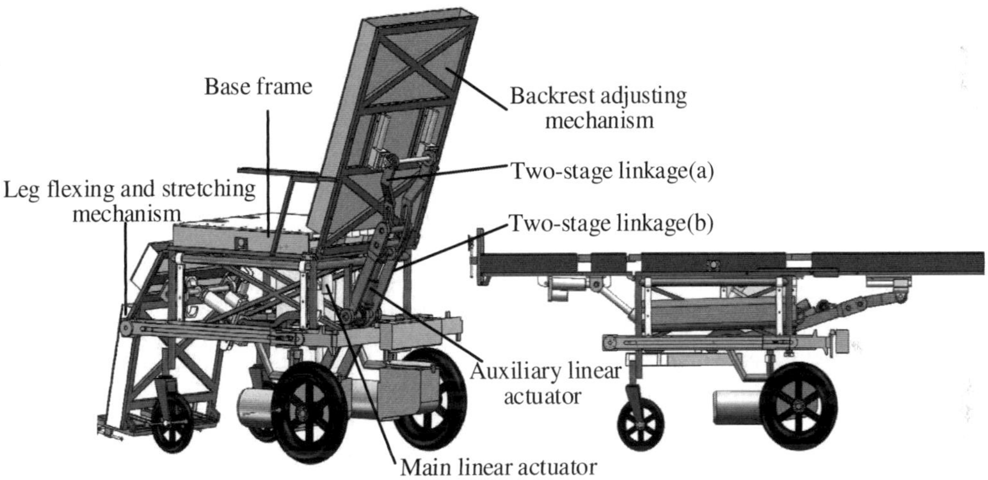

Figure 1. Illustration of the Multifunctional wheelchair. (a) sitting state. (b)lying state.

The adjusting part of the backrest adopts a two-stage linkage mechanism to ensure that the backrest can achieve the required angle. On the other hand, this design can reduce the space occupation of the backrest, making the structure more compact. The linear actuator installed on the frame of wheelchair base drives the two-stage linkage to rotate, and the other installed on the two-stage linkage drives the linkage to expand, realizing the adjustment of back angle. The leg flexion and extension mechanism is divided into the thigh and calf. The thigh support plate is directly driven to a specific position by the linear actuator on the wheelchair base frame. The leg support plate is the deformed portion of the thigh support plate, which is supported and lowered by the linear actuator.

3. Motion Analysis and Simulation of the Mechanism

The wheelchair should maintain stability during deformation to avoid discomfort, thus the speed and acceleration of each action mechanism are required to be smooth. The kinematics analysis and calculation of the backrest adjusting mechanism is taken as an example, and then the kinematics simulation analysis of wheelchair is carried out in ADAMS, with the movement speed of the linear actuator as the parameter, to verify the rationality of the mechanism design.

3.1. Kinematics Analysis and Calculation

The simplified structure of the main linear actuator part of the backrest adjustment mechanism is shown in Figure 2(a). It is correspondent to the part of Figure 1(a), called Two-stage linkage(a). In the model, AB indicates the shortest state of the main linear actuator while AB′ indicates the extended state, and BC stands for the main part of the two-section linkage. A denotes the connecting location of the frame and the linear actuator, B denotes the connecting location between the extended axis of linear actuator and linkage, and C denotes the joint connecting the frame of wheelchair base and linkage. The angular velocity of the linkage ω_1 can be obtained using:

$$\omega_1 = \frac{1}{\sqrt{1-\left(\frac{L_1+L_2-(d_n+v_1t)^2}{2L_1L_2}\right)^2}}\left[\frac{-1}{2L_1L_2}\left(2v_1^2t+2d_nv_1\right)\right] \tag{1}$$

Where, d_n is the original length of the linear actuator, v_1 denotes the movement speed and t denotes the movement time. L_1 denotes the distance between A and C, L_2 denotes the distance between B and C.

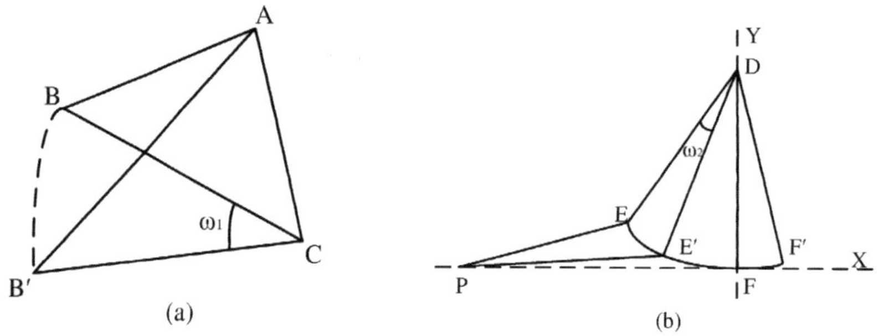

(a) the main linear actuator part. (b) the auxiliary linear actuator part.
Figure 2. Simplified schematic diagram of backrest adjusting mechanism.

The simplified structure of the auxiliary linear actuator part of the backrest adjustment mechanism is shown in Figure 2(b). It is correspondent to the part of Figure 1(a), called Two-stage linkage(b). PE indicates the linear actuator. In the model, the arc chute is simplified to a fixed track arc $\widehat{BE}$ as the initial state, $\widehat{B'E'}$ is the actual position state after the axis of the linear actuator extends. The angular velocity of the linkage ω_2 can be obtained using:

$$\omega_2 = \frac{2d_mv_2+2v_2^2t}{\sqrt{A^2+B^2-\left[(d_m+v_2t)-C\right]^2}} \tag{2}$$

$$A=-2r^2+2y_pr, B=2x_pr, C=2r^2+x_p-2y_pr \tag{3}$$

Where, (x_p, y_p) is the coordinates of point P, the variable r denotes the radius of the arc segment. d_m is the original length of the linear actuator, v_2 denotes the movement speed of the linear actuator and t denotes the movement time.

3.2. Motion Simulation

A virtual prototype is built to carry out kinematics simulation analysis of the wheelchair. The reclining process of the backrest adjusting mechanism is controlled by the back main linear actuator and the back auxiliary linear actuator. The speed of the main linear actuator is 14 mm/s, and that of the auxiliary linear actuator is 4.1 mm/s, the motion time of both is 10 seconds. After the simulation, the angle change of backrest were measured, and the velocity and acceleration curves were output. It can be seen from Figure 3 that the backrest angle adjustment range is between 0° and 70°, and the operation process is stable without mutation. According to the above simulation analysis, the design of the wheelchair backrest adjusting mechanism is reasonable and meets the expectation.

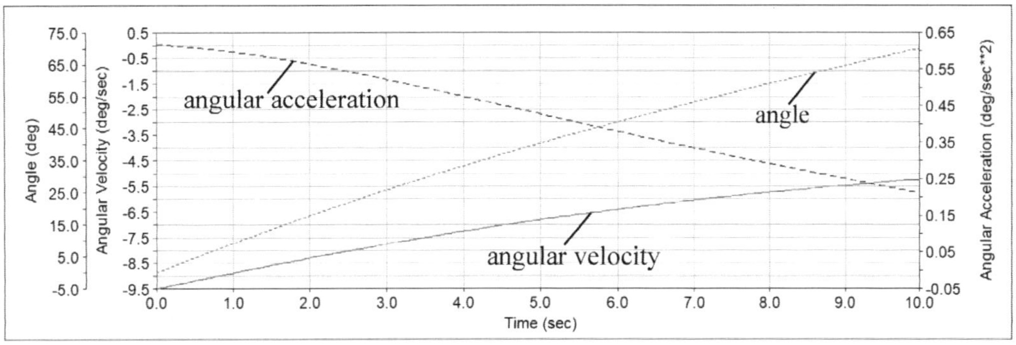

Figure 3. Kinematics characteristic curve of backrest

4. Finite Element Analysis of the Two-Section Linkage

Since the safety of the wheelchair is very important, static analysis of the two-section linkage was conducted, and the analysis results were compared with material strength standards to evaluate their safety. The two-stage linkage of the backrest are made of 6082 aluminum alloy, of which tensile strength is 310 MPa, the yield strength is 260 MPa and the elongation is 10% [12].

As a support and transmission part, the two-section linkage is subjected to the maximum load when the wheelchair is lying flat. Mass including the upper body, the backrest structure and the connecting parts act on the two-section linkage through the support points. The load distribution is shown in Table 1. ADAMS was used to calculate that the forces on the two-stage linkage are about 481N and 920N in the horizontal and vertical direction, respectively.

Table1. Load distribution of two-section linkage.

	Weight	Remarks
Upper body	440N	above the pelvis
Backrest adjusting mechanism	130N	frame
Parts	7N	slide rail
Total	577N	/

The two-section linkage is modeled according to the original size and imported into the ANSYS Workbench for materialization. After setting the material properties, the mesh is divided, and finally the load and constraints are applied for finite element analysis. The number of grids and nodes of the linkage is 1166679 and 1743020 respectively. According to the analysis and calculation results in Figure 4 and Figure 5, the two-section linkage shape variable is 2.2 mm in the acceptable range; the maximum stress is 131.9 MPa, which is less than the ultimate stress of the material. Therefore, the structural strength of the component is qualified.

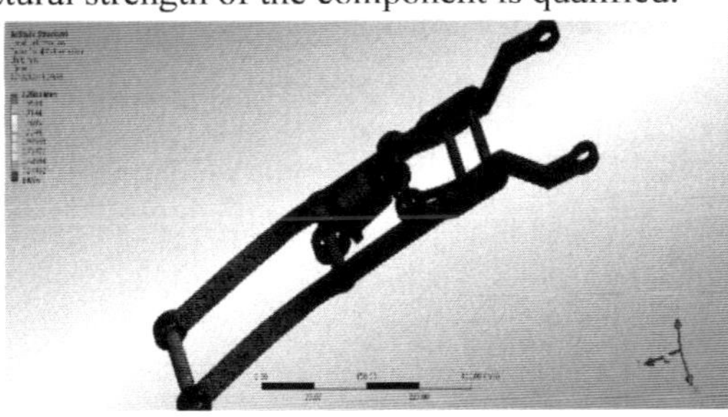

Figure 4. Deformation cloud.

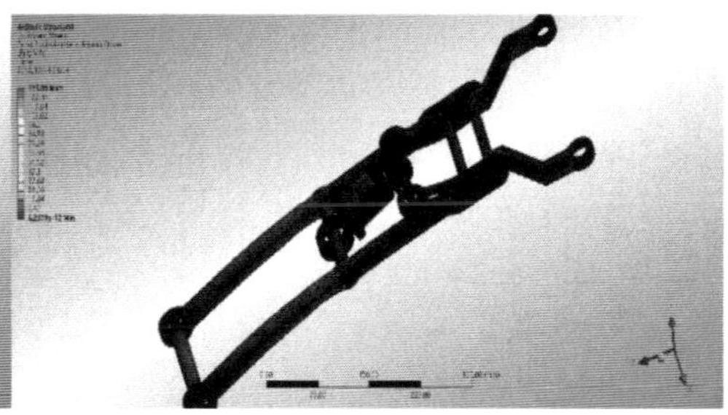

Figure 5. Stress cloud.

IOP Conf. Series: Materials Science and Engineering **538** (2019) 012045 doi:10.1088/1757-899X/538/1/012045

5. Conclusion

In this paper, a new multifunctional wheelchair with automatic postural adjustment is designed to satisfy the user's nursing needs and reduce the workload of nursing staff. The kinetic characteristic curve obtained through kinematics analysis and simulation indicates that the process of wheelchair backrest reclining is smooth and the rationality of mechanism design is verified. Furthermore, the results of finite element analysis about the key components of the wheelchair shows that the structure strength meets the requirements and the safety is excellent.

Acknowledgments

Jin Rui would like to acknowledge assistance and encouragement from tutor Prof. Gao during my project and thesis.

References

[1] Coyle, Caitlin E., Putnam, Michelle. Identifying adults aging with disability using existing data: The case of the Health and Retirement Study. Disability and health journal, 2017, 10(4): 611-615.

[2] Xiaoning Hao, Juan Gu, Wei Fu. Social support and care needs of the disabled elderly population: An empirical study based on survey data from Beijing, China. Bioscience trends, 2017, 11(5): 507-515.

[3] Yuejun Jing, Yuan Li. Analysis on the Make-up of China's Disabled Seniors and the Demand for Long-term Care. China Population Today, 2014, 4(12): 55-63.

[4] Denny Brown, Noelle, Bonnie O'Day, Stephanie McLeod. Services and Supports People with Disabilities Need to Stay Employed. Mathematica Policy Research, 2013.

[5] Nobuhiro Nakata, Goro Obinata, Taira MakinoAssistance. System for Bedridden Patients to Reduce the Burden of Nursing Care (First report—Development of a multifunctional electric wheelchair, portable bath, lift, and mobile robot with portable toilet). International Symposium on System Integration, 2010: 132-139.

[6] Jong ChulRhee, NicolaeDone, Gerard F.Anderson. Considering long-term care insurance for middle-income countries: comparing South Korea with Japan and Germany. Health policy, 2015, 119(10):1319-1329.

[7] E. Fosch-Villaronga. Creation of a care robot impact assessment. International Conference on Social Robotics, 2015, 9(6): 1913-1917.

[8] Diansheng Chen, Qian Jiang, Jinghua Liu. The Function Orientation and Creative Ideas of Robotic Chair/Bed Based on Step-down User Research System. Journal of Graphics, 2013, 34(3): 85-89.

[9] Ciupitu Liviu, Simionescu Ion, Olaru, Adrian. Static Balancing of Mechanical Systems Used in Medical Engineering Field-Continuous Balancing. Advanced Materials Research, 2012, 463-464: 890-894.

[10] Roger Bostelman, James S Albus. A multipurpose robotic wheelchair and rehabilitation device for the home. International Conference on Intelligent Robots and Systems, 2007.

[11] Barbara O'Neill BA, BSN, GCNursED. Nursing home nurses' perceptions of emergency transfers from nursing homes to hospital: A review of qualitative studies using systematic methods. Geriatric Nursing, 2015, 36(6): 423-430.

[12] D. Liu, H.V. Atkinson, P. Kapranos. Microstructural evolution and tensile mechanical properties of thixoformed high performance aluminium alloys. Materials Science and Engineering A, 2003, 361: 213-224.

IOP Conf. Series: Materials Science and Engineering **538** (2019) 012046 doi:10.1088/1757-899X/538/1/012046

Unitized Association Design Based on Structural Feature Analysis

Shang Junzhi, Yang Bo[a] and Ma Ning

School of Mechanical Engineering, University of Jinan, 250022 Shandong Jinan, China

Corresponding author: 1603206175@qq.com

Abstract. Structural association design is an very important aspect in modern design. In order to realize change propagation effectively, the product association features and geometric features are analyzed, and the association rules are formulated according to the association strength. The associated directed network model as well as its simplifying rules is given. The termination rules are designed for the optimally selecting of the change propagation path. Finally, the feature segmentation change technology and the associated driving change design method are used to complete the product structural change propagation design.

1. Introduction

Structural association design is an efficient design method. When a structure changes or replaces, the structure associated with it changes correspondingly. many scholars have done a lot of research on it. Kima [1] used structural analysis techniques and a unified modeling language to complete the construction of structural association models in the product design process. Huang Boyuan [2] used the shape description and structural layout as the top-down design basic model, and established the product structure correlation model. Qian Qiankun [3] proposed the product form and structure correlation design, and the structure design aims to show the product form. VANBEET [4] establishes a method for calculating the correlation strength of matrix elements by modeling the functional behavior state of products. Ding Liping [5] based on product characteristics subdivide the related factors and attached weights, using the method of calculating the weighted average method to develop the product association design matrix. Yang Fan [6] analyzed the relationship between engineering changes and product characteristics, and developed a change propagation path search method based on the associated network model.

Since the structural variation design only involves changes in local structural parameters, it has great limitations [7]. In practical design, the change of topological relationship between structures is also required which has great influence on design result, so it needs to be studied in depth.

2. Associated directed graph basic information

2.1. Basic information of associate directed graph

Structural features consist of multiple geometric features with topological relationships. Due to factors such as the different of the functional unit or part model and the different process, the geometry of each part inside the unit is different, but the geometric features are similar. As shown in tab.1.

IOP Conf. Series: Materials Science and Engineering **538** (2019) 012046 doi:10.1088/1757-899X/538/1/012046

Table 1. Common geometric features.

Geometric Features	Subdivision geometry	Figure
plane		
	cylinder	
	cone	
curved surface	sphere	
	gear face	
composite face	thread surface	
Straight line segment		

To analyze an assembly model in a three-dimensional environment, it is first necessary to extract the constraint relationships in the model and the attribute information contained in the constraint relationship, based on which the associated geometry and feature information model can be built. Its relational expression is:

$$G=(H_k,Q_{(i,j)},W_{(i,j)},L_{(i,j)}) \tag{1}$$

Here, H is the assembly information matrix, and k is the functional unit number. Where $T_{(i,j)}=(Q_{(i,j)},W_{(i,j)},L_{(i,j)})$,T is the associated attribute, i,j are part numbers, $Q_{(i,j)}$ is a mating feature matrix between two part. W is a structural feature of participation constraints. L is the main geometric feature corresponding to the structural features.

2.2. Formulation of correlation strength

The correlation strength is the degree of interaction between features, which is affected by many factors. For the purpose of calculating the correlation strength, the correlation index and its application criterion are formulated. And fuzzy value of associated strength are given including associated strength based on number of matehes, and associated strength based on formula constraints as shown in tab.2. Here, x_1 means number of fits, the more the number of matches, the higher correlation strength is. x_2 means formula constraint, it is determined by whether relationship between dimensions is calculated by formulas. When there is a formula constraint, it can be represented by "have" or by the number "1"; if not, it is represented by "no" or "0".y means the associated strength level.

Table 2. Basic associated strength.

Number of matches x_1	Strength level y_{1i}	Formula constraints x_2	Strength level y_{2j}
1	0.5	have (1)	1
2	0.7	no (0)	0
3	1		

3. Construction of association model based on directed graph

3.1. Construction of association directed graph

The associated directed network model expresses the association between parts based on which the associated path can be built, as shown figure.1. The specific construction steps are as follows:

IOP Conf. Series: Materials Science and Engineering **538** (2019) 012046 doi:10.1088/1757-899X/538/1/012046

(1) Unit modeling of assembly model. The assembly is divided into functional units according to functions, and the directed graph of each unit is constructed.
(2) Construction of intra-unit associations according to the mating feature matrix Q. When there are matching features between two parts, there is a line connecting them. The direction of the directed line is determined according to the order of assembly. As shown in figure.1, the line between P_1 and P_2, means P_1 is assemble before P_2 during assembling process.
(3) According to the matching feature matrix Q, the parts with the matching features between the functional units are retrieved, the association between the functional units is constructed, and the directed line representation is used; Then the associated attribute $T_{(i,\ j)}$ and intensity level $y_{(i,\ j)}$ are added to the corresponding directed line. In this way, all functional units are traversed to complete the construction of the relationship between the units.

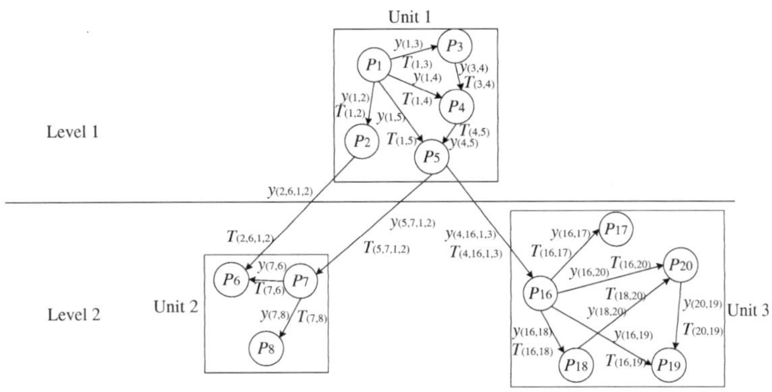

Figure 1. Extended associated directed network model.

3.2. Structure simplification of association directed graph

The above mentioned association directed graph is complex and hard trackle. So, structure is association network model is simplified step by step. The unit base component is used as the search starting point, and the one-way search is performed along the associated directed line until all levels of the unit are simplified. The unit 3 in the associated network model illustrated in figure. 1 is used as an example, the simplified result is shown in figure. 2.

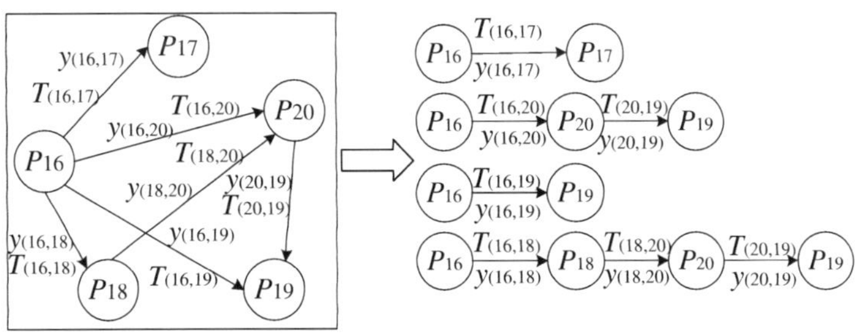

Figure 2. Simplification of the associated directed network model.

4. Structural association propagation analysis

4.1. Structure-related propagation rules

In order to analysis of the transmission more efficiently, the propagation termination rules and the transmission direction rules are given, based on which the direction of transmission and the propagation path have been established.

1.Termination rules. Termination rules are given based on association strength, association constraints and external cause. The rules are as follows:
(1) Correlation strength rule association constraints rule. If the association strength is less than 0.5 the association delivery is terminated.
(2) Association constraints rule .If the association strength is more than 0.5, the association strength transmission is satisfied. Then the associated attribute is verified. If the change of the structure does not cause a contradiction or conflict between the associated parts, the association is terminated.
(3) External cause rules. The external cause mentioned here are standard restrictions, space limitations, and user requirements. If those external cause are not conflict with each other. Then the delivery is not interrupted.
2.Determination of propagation direction. According to the location of the changed part in the simplified associated model, two types of propagation can be occurred.
(1) When the changed part or unit is at the start on end point of the directed graph model, the direction of the transfer is one-way.
(2) When the changed part or unit is in the middle of the simplified model, the direction of transmission is two-way.
3.Choose path. To reduce the complexity of the change propagation, the shortest principle is used to the selecting of the association path. When the associated attribute requirement can't be met, the other path is selected and transferd according to the delivery rule.
According to the above method, the change is made in the association model to complete the associated variant design of the product.

4.2. Association propagation analysis
Based on the simplified associated directed network model, association propagation can be performed as follows:
(1) Search for the in-cell associated path containing the changed part and sort it from small to large according to the number of included parts.
(2) The transmission is started from the shortest path, then the direction can be determined.
(3) Execute the transmission according to the direction of transmission and judge whether the transmission is terminated according to the termination rule.
(4) The transmission does not satisfy the termination condition and enters the structural association drive variant design.
According to the above structure association transfer method, the intra-unit association transfer process is established. As shown in figure 3.

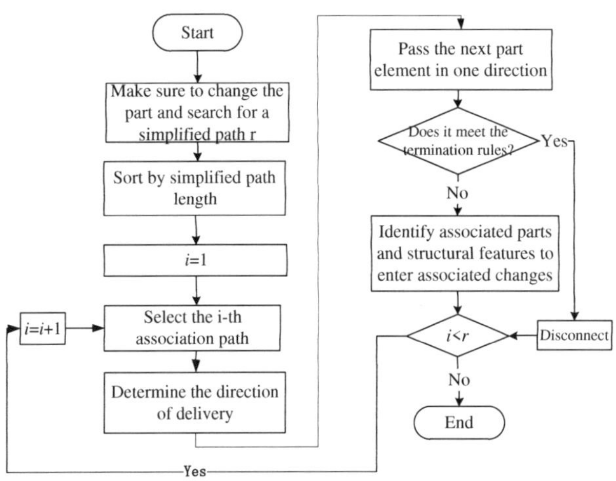

Figure 3. Structural association changes are passed in the simplified model within the unit.

5. Equations and mathematics

Additions or changes in performance, metrics and functionality sometimes cause structural changes, and due to the association between the parts, the structural changes cause a chain change in the associated structure in the associated part. When structural changes are propagated, it should be ensured that structural interferences are avoided while completing predetermined functions, performance and specifications.

The steps for structural association design based on change propagation are as follows:

1. Identify and delete the changed source structure feature. Determine the changed source feature, split and remove it from the source structure feature on the changed part. Then, build a new structure feature according to design requirement and save it.
2. Associate the transfer and determine the associated structural features. Find the associated part along the associated directed line; then search for the associated structural feature that matches the source geometry and the normal vector. Finally, the segmentation operation is performed on the associated structural features, and the conflicting features are deleted in the part model.
3. Build associated structural features. According to the geometric features of the new structural features and their normal vectors, the main shape of the new structure is judged, and based on the mapping relationship between the geometric features and the matching features, the geometric shape of the new structure on the associated parts is determined, thereby the structural features of the part are determined which associated with the new structure. Construct structural features and compose new part models with structural features of the part associated with the new structure.

6. Example

Taking the reducer shown in figure. 3a as an example, the meshing mode of the gears the design reguirement is that change from the ordinary key connection to the spline connection, figure.3b is the output shaft unit of the reducer.

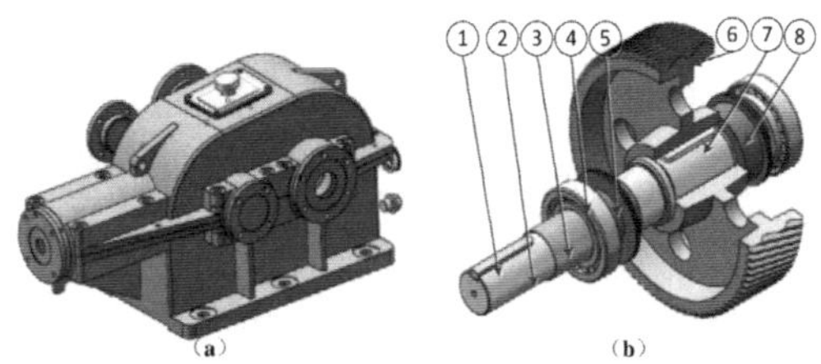

Figure 4. Reducer function unit (1.Key 2.Shaft 3.Sleeve 4.Bearing 5.Cast oil retaining plate 6.Large cylindrical gear 7.Key 8.Cast oil retaining plate).

Firstly, construct the associated directed network model as shown in figure.4, Then simplify the association model according to the simplified rule, and get 7 associated paths as shown in figure.5.

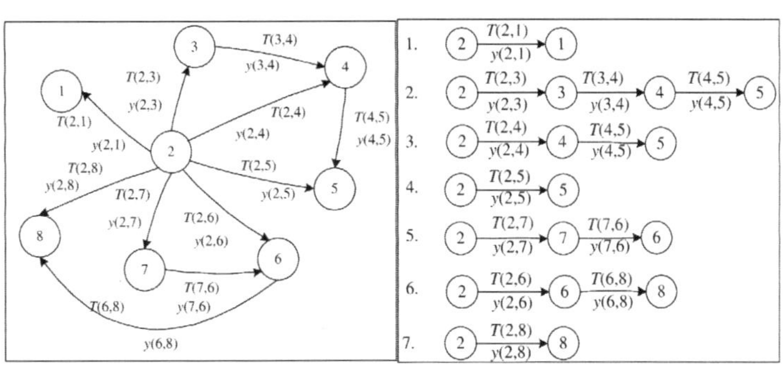

Figure 5. Simplification of the associated directed network model.

According to the above-mentioned association propagation rule, the path of 1, 4, 7, 3, 2 can be terminated; Path 5 is excluded because of the attribute constraints conflict between part 13 and 18, so the 5 path is excluded; Finally path 6 is selected, part 17 is changed until part 19 meets the termination principle and complete the modification. As shown in figure.6.
According to the structural association change design step, the drive shaft and other related parts can be changed as well. Finally, the input shaft system unit is changed to new structure to fit the design requirement of spline connection.

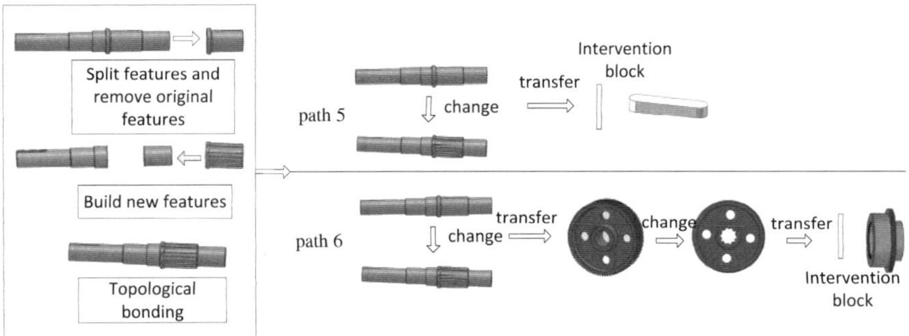

Figure 6. Structure Association Change Design.

7. Conclusions

In this paper, the basic correlation features are analyzed, the extended structure association model is established, and it is simplified, and the association rules including termination rule and determination of propagation direction are formulated. The method proposed in this paper can improve the accuracy and efficiency of change propagation design, which will be helpful in structural association design.

Acknowledgments

The study reported was supported by National Natural Science Foundation of China (51775239, 50975124)

References

[1] CH.Kima,RH.Weston,A.Hodgsonb.Comput.nd.50,(2003)
[2] H. Boyuan,X. Ping.Mech Eng & Auto. 06(2013)
[3] Q. Qiankun, Research on the correlation design of construction machinery product form and structure(Changsha: Central South University, 2013)
[4] D. Liping,T. Jian. Comput.Integr.Manuf.06,(2008)
[5] Y. Fan,T. Xiaoqing,D. Guijiang.Journal of Mech Eng, 47(2011)
[6] Y. Fan, T. Xiaoqing.Journal of Beijing University of Aeronautics and Astics, 38 (2012)
[7] H. Zhouyu.Research on Complex Modeling Technology and Its Application (Hangzhou: Zhejiang University, 2013)

IOP Conf. Series: Materials Science and Engineering **538** (2019) 012047 doi:10.1088/1757-899X/538/1/012047

Effect of Antioxidant on Storage Performance and Life Prediction of HTPB Propellant

DU Yongqiang[1], ZHENG Jian[1]*, ZHI Jianzhuang[1], TU Xin[2]

[1]Shijiazhuang Campus, Army Engineering University, Shijiazhuang, China
[2]Military Representative Office in Xiaogan District, Wuhan Regional Military Representative Bureau, Army Equipment Department, Xiaogan, China

*jzhengpaper@163.com

Abstract. In order to study the aging property of HTPB propellant in storage and the effect of antioxidant on the storage life of propellant, accelerated aging tests were carried out for two types of propellant. A non-linear aging model was introduced into the classical Arrhenius equation, and the starting point of aging was modified. With the maximum elongation as the evaluation parameter, the storage life at 25℃ is 9.6 years with the critical value is 20%, and the storage life is 17.5 years with the critical value is 15%. By controlling the content of antioxidant in propellant, the formula of type I propellant was modified, and the life of propellant after modification was predicted. The storage life is 12.4 years at 25℃ with the critical value is 20%, and the storage life of propellant is 21.3 years when the critical value is 15%. It was proved that the storage life of propellant could be effectively improved by controlling the content of antioxidant.

1. Introduction

HTPB composite solid propellant is an energetic composite material based on polymer and filled with metal powder and oxidant. It is an important power source of rocket engine. Its performance will have a direct impact on the combat capability of rocket [1]. The physical and chemical properties of propellant will change under the load/environment spectrum during storage, which will lead to the deterioration of the mechanical properties of propellant and the bonding properties of various bonding interfaces of rocket motor grain with the increase of storage time. Especially when the propellant of rocket engine is integrally adherent casting, the charge cannot be replaced during service [2]. The main cause of propellant deterioration is the slow chemical changes during storage. At present, it is generally believed that the main reaction is the oxidation, branching and cracking of polymer under the action of oxidant and air (oxygen and water vapor). For HTPB propellants, the aging reaction caused by double bond oxidation is more important [3, 4]. In order to prolong the life of HTPB propellant, many scholars have studied the effect of antioxidant on the performance of HTPB propellant. In order to avoid the waste caused by early retirement and the harm caused by overdue service, it is necessary to accurately predict the storage life of solid propellant for rocket motor.

To know exactly the storage life of a propellant, it is better to store propellant samples or propellant engines naturally. However, this method is not only expensive, but also time-consuming. In this paper, the change of mechanical properties of HTPB propellant under storage aging was studied by accelerated aging. A non-linear aging model was introduced into Arrhenius formula to predict the storage life of HTPB propellant. The effect of antioxidant on storage performance and life of HTPB propellant was also studied. The content of antioxidant in the formula was revised, and the life of the revised formula was estimated.

2. Teat section

2.1. Condition of Test

(1) Equipment: Type 402 and 802 thermal aging ovens with rotating speed of 12 min^{-1} and natural convection of air.
(2) Samples: Cut into 10×25×100mm specimens and hang vertically in the aging oven. Dumbbell-shaped specimens were made before the test. The state of specimens was adjusted for 16-46 hours at room temperature, and then the performance was tested.
(3) Temperature control accuracy: ±2℃.

2.2. Test Results

Tensile strength (σ_m) and maximum elongation (ε_m) were measured for aged samples. The aging results of type I propellant with different content of antioxidant at 80℃are shown in Figure 1. The aging results of type I propellant with 0.1% content of antioxidant at 35℃, 50℃ and 65℃are shown in Figure 2. For ease of comparison, the aging results of type II propellant at 80℃ are listed in Figure 3. The first point is the starting point of accelerated aging. The time of this point is the correction time considering the effect of curing process on aging. The correction method is shown in Table 1.

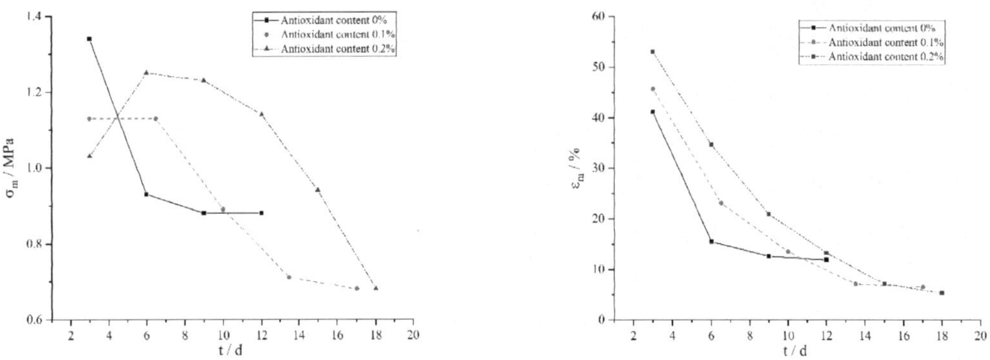

Figure 1. The aging results of type I propellant with different content of antioxidant at 80℃

Table 1. Revised Aging Starting Point

Aging Temperature (K)	298	308	323	338	343	353
K(343)/K(T)	3.67	2.58	1.64	1.12	1	0.8
Computation Time (d)	162	35.7	9.7	4.7	4	3
Approximate full time (d)	162	36	10	5	4	3

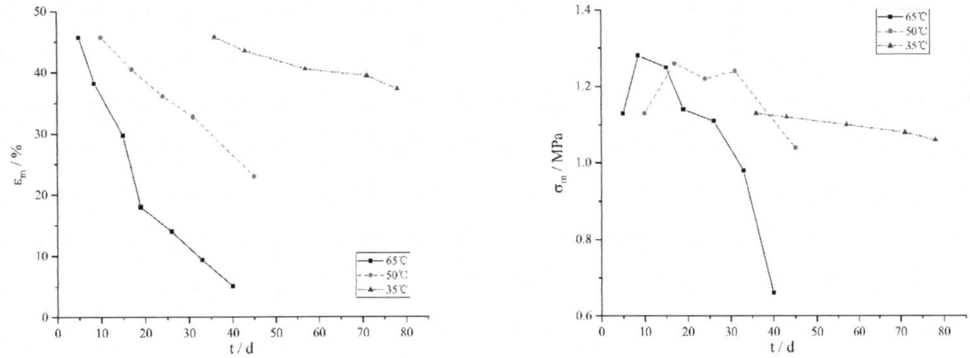

Figure 2. The aging results of type I propellant with 0.1% content of antioxidant at 35℃, 50℃ and 65℃

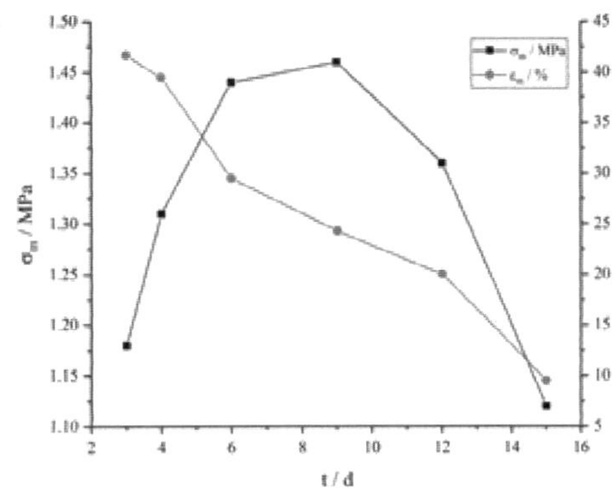

Figure 3. The aging results of type II propellant at 80℃

3. Selection of Aging Model

The basic teaching model of accelerated aging test is Arrhenius formula [6, 7]:

$$K = A\exp\left(-E / RT\right) \quad (1)$$

In formula: K is the rate constant of aging reaction at temperature T; A is the coefficient; E is the activation energy, J/mol; T is the absolute temperature, K; R is the general gas constant, 8.314 J/(mol·K).

According to the formula, the only effect of temperature is to increase the reaction rate, that is, K increases with the increase of temperature. Usually K is derived from the following formula [8]:

$$P = K_1 t + a \quad (2)$$

In the formula: K_l is a linear reaction rate constant; P is a physical property; a is a coefficient; t is time (d).

However, the test shows that the change of propellant performance cannot be described by formula (2) during the aging process of propellant. For example, by plotting time with the ε_m in Fig.1, it can be seen that the relationship between them is obviously not linear, that is, K is not a constant. However, if the logarithm of time is used to draw a graph, it is an ideal straight line, which can be expressed by the following formula:

$$P = K_2 \lg t + b \tag{3}$$

In the formula, b is the coefficient and K_2 is the non-linear reaction rate constant.
By replacing K in formula (1) with K_2 and taking logarithm of formula (1), formula (4) is obtained:

$$\lg K_2 = \lg A - \frac{E}{RT \ln 10} \tag{4}$$

The storage time at room temperature can be extrapolated by substituting the previous test results into the formulas (3) and (4).

4. Data Processing

4.1. Selection of Evaluation Parameters and Determination of Critical Value

In the aging process of propellant, all kinds of parameters are changing, so it is necessary to select a suitable parameter as the evaluation parameter. From the point of view of use, the important index to ensure the integrity of rocket engine is ε_m, so ε_m is taken as the evaluation parameter [9]. For different propellants and different engines, the critical value of marking the end point of aging should be different because of the different stress conditions. When the calculated value of polysulfide propellant charge is greater than 15% in an engine bonded with a certain shell, it is required that the ε_m be greater than 20%. In this paper, the ε_m=20% and 15% are selected for comparison.

4.2. Revision of Aging Starting Time

In the curing process, aging reaction is also taking place, so the curing end point is not the starting point of aging. The solidification time of propellant at 70℃ for 3 days can be calculated in 4 days, and the corresponding time at other temperatures can be modified by the following Arrhenius derivation [10]:

$$\ln \frac{K(T+\Delta T)}{K(T)} = \frac{E}{RT^2} \Delta T \tag{5}$$

By substituting the activation energy 21.338 kJ/mol into formula (5), the ratio of K value at different temperatures to K value at 70℃ can be obtained. Assuming that (P-b) in formula (3) after curing is a certain value, it can be derived:

$$\frac{K(343)}{K(T)} = \frac{\lg(t)}{\lg 4} \tag{6}$$

It can be obtained that the curing time at 70℃ for 4 days is equivalent to the aging time at different temperatures. The corrected results are shown in Table 1.

4.3. Result Calculation

Using the results of Figure 1 and 2, the regression equations corresponding to formula (3) at different temperatures containing 0.1% antioxidant:are calculated as follows:

35℃: $\varepsilon_m = 81.91 - 23.39 \lg t$ $r = 0.989$

50℃: $\varepsilon_m = 80.66 - 33.34 \lg t$ $r = 0.973$

65℃: $\varepsilon_m = 79.81 - 46.28 \lg t$ $r = 0.988$

80℃: $\varepsilon_m = 69.34 - 53.93 \lg t$ $r = 0.987$

According to the K_2 value of the above regression equation, the regression equation corresponding to (4) is obtained as follows:

$$\lg K_2 = 4.293 - 896.7\frac{1}{T} \quad r = 0.992 \tag{7}$$

K_2 = 19.21, E = 17.17 kJ/mol, b = 88.14 at 25℃, and ε_m is 45.7% at the beginning of aging, that is to say, the aging formula is as follows:

$$\varepsilon_m = 88.14 - 19.21\lg t \tag{8}$$

By substituting the above formula, the storage life at 25℃ is 3524 days, that is, 9.6 years when the critical value ε_m is 20%, and the storage life is 6417 days, that is, 17.5 years when the critical value ε_m is 15%.

4.4. Comparison with Type II Propellant and Revision of Formula

In the absence of actual storage results, it is very useful to compare with the accelerated aging results of similar products. According to the results in Figure 3, the regression equation of 80℃ aging of type II propellant can be obtained as follows:

$$\varepsilon_m = 56.36 - 37.33\lg t \text{ , r=0.970} \tag{9}$$

The critical aging time (ε_m =20%) at 80 C is 9.4 days, while the type I propellant containing 0.1% antioxidant is only 8.2 days. This shows that its aging performance is not as good as that of type II propellant. Therefore, the content of antioxidant was increased to 0.17% in the formulation.
The aging regression equation of type I propellant containing 0.2% antioxidant at 80℃ can be obtained by using the data in Figure 1 as follows:

$$\varepsilon_m = 85.11 - 66.44\lg t \text{ , r=0.999} \tag{10}$$

By interpolating it with the regression equation containing 0.1% antioxidant, the aging equation with 0.17% antioxidant can be obtained as follows:

$$\varepsilon_m\text{=80.38-62.69}\lg t \tag{11}$$

According to this equation, the critical aging time at 80℃ is 9.2 days, which is equivalent to the aging performance of type II propellant.

4.5. Storage Life Prediction of Modified Formula

Accelerated aging test was not carried out after formula modification. However, as can be seen from Figure 2, the influence of antioxidants on properties is regular, and there will not be much error in the calculation by interpolation method. Assuming that the activation energy of the projection remains unchanged, the Arrhenius formula of 0.17% antioxidant formulation can be obtained as follows:

$$\lg K_2 = 4.337 - 896.7\frac{1}{T} \tag{12}$$

The initial ε_m calculated by interpolation method is 50.0%, and then the aging equation at 25% can be obtained as follows:

$$\varepsilon_m = 97.82 - 21.28\lg t \tag{13}$$

According to the critical value of 20%, the storage life at 25℃ is 4539 days, that is 12.4 years. According to the critical value of 15%, the storage life is 7796 days, that is 21.3 years.

5. Conclusions

In this paper, a non-linear aging model was introduced into the classical Arrhenius equation, and the storage life of HTPB propellant was predicted by the accelerated aging equation. The storage life of type I propellant is 3524 days, that is, 9.6 years when the critical value is 20%, and the storage life is 6417 days, that is, 17.5 years when the critical value is 15%.

By controlling the content of antioxidant in propellant, the formula of type I propellant was modified, and the life of propellant after modification was predicted. When the critical value is 20%, the storage life is 12.4 years at 25℃. The storage life of propellant was 21.3 years when the critical value was 15%. It was proved that the storage life of propellant could be effectively improved by controlling the content of antioxidant.

From the regression equation, the correlation coefficients are all above 0.97, which shows that the mathematical model is in agreement with the experimental results, so the results are regrettable. However, there must be some differences between the hypothesis and the reality when establishing the digital model. In addition, interpolation and extrapolation will inevitably bring some errors. Its reliability needs to be further verified by actual storage.

References

[1] HOU Lin-fa.Composite solid propellant[M].Beijing:China Aerospace Publishing House, 1994.

[2] PANG Ai-min.Theory and engineering of solid rocket propellant [M].Beijing:China Aerospace Publishing House, 2014.

[3] ZHANG Xing-gao.Study on the aging properties and storage life prediction of HTPB propellant[D].Changsha: National University of Defense Technology, 2009 .

[4] WEI Xiao-qin, WU Hu-lin, ZHANG Lun-wu, et al. Natural Environmental Accelerated Aging Test Method of HTPB Propellant [J]. Equipment Environment Engineering, 2018, 15 (12): 104-108.

[5] Commission of Science, Technology and Industry for National Defense.Test method of propellant: GJB 770B-2005[S].Beijing: Standardization Administration of China, 2005.

[6] HONG Dong-pao,Wang Ying-hua,GUAN Fei,et al.Storage life assessment for solid propellant based on generalized linear model[J].Journal of Beijing University of Aeronautics and Astronautics, 2015,41(4):29-32.

[7] WANG Guo-qiang,SHI Ai-juan,DING Li,et al.Mechanical properties of HTPB propellant after thermal accelerated aging and its life prediction[J].Chinese Journal of Explosives & Propellants, 2015,38(1):47-50.

[8] CAO Fu-qi, LIU Zhi-cheng,LI Xiao-huan.Research on accelerated aging test and storage life prediction of solid rocket motor charge[J].Aero Weaponry, 2014 (4):59-61.

[9] ZOU Si-si, Yan Cong, MA Cen-rui, et al. The dualistic linear regression model of HTPB propellant's elongation under constant strain[J].Journal of Northeastern University: Natural Science,2010,31(2):261-264.

[10] GAO Da-yuan, He Bi, Liu Song-wei, et al. Discussion on limitations of the Arrhenius methodology [J]. Chinese Journal of Energetic Materials, 2006,14(2):132-135.

IOP Conf. Series: Materials Science and Engineering **538** (2019) 012048 doi:10.1088/1757-899X/538/1/012048

Aerodynamic Performance of Dragonfly Forewing–Hindwing Interaction in Gliding Flight

Yun Luo, Guoyi He[a], Huixiang Liu, Qi Wang and Hang Song

Nanchang Hangkong University School of Aircraft Engineering, China

[a]Corresponding author: hegy509@163.com

Abstract. In order to study the aerodynamic performance of dragonfly forewing–hindwing interaction in gliding flight, the three-dimensional (3D) model with corrugation of dragonfly forewing and hindwing are established. This paper studied the aerodynamic effects of dragonfly forewing–hindwing interaction at gliding motion using the method of computational fluid dynamics (CFD) for angles of attack changing from 0° to 25°(with an interval of 5°). Numerical simulation indicates that the interactions between forewing and hindwing enhance the aerodynamic performance of both forewing and hindwing to varying degrees compared to independent wings, especially the effect of forewing is more evident at small angle of attack.

1. Introduction

The micro-flapping-wing aircraft has the characteristics of small size, light weight, excellent stealth performance and strong operability. As a brand new aircraft design structure, it can not only take off and land vertically and accelerate in almost any direction like a micro-rotor aircraft, but also can maneuver precisely at high speed and long-distance cruising like the micro-fixed-wing aircraft. Dragonfly is the most suitable bionic object for the micro-flapping-wing aircraft because of its excellent flight ability.

Gliding flight is the most commonly used flight mode of the dragonflies, especially in the summer of high temperatures. On the one hand, the aspect ratio of a single wing is about five which is much larger than other insects, and can obtain better gliding ability by consuming very little energy; On the other hand, it can reduce the flapping frequency and energy consumption, and regulate body temperature by using the methods such as air convection[1]. Unlike other four-winged insects, dragonfly forewing and hindwing are independent and mutually influential. It is interest that dragonfly could spread a pair of hindwing but flapping a pair of forewing, or spread all wings stationary in gliding flight[2].

Due to the advances in CFD and experimental techniques, researchers are beginning to study the aerodynamic performance of the dragonfly wings. However, there are few studies about the aerodynamic performance of forewing–hindwing interaction in gliding flight. Sun and Lan[3] & Wang and Sun[4] studied the aerodynamic performance of forewing–hindwing interaction in hover flight and forward flight, using the method of CFD. Maybury and Lehmann[5] & Yamamoto and Isogai[6] conducted experimental studies on the forewing–hindwing interaction at hovering conditions. Zhang and Lu[7] studied the aerodynamic performance due to forewing and hindwing interaction with two-dimensional (2D) model of dragonfly wings in gliding flight. Huang and Sun[8] studied the aerodynamic performance of the dragonfly forewing–hindwing interactions by using 3D model wings, which are simplified to a flat model only with the outer contour of the dragonfly wings for both forewing and hindwing.

In fact, the structure of the dragonfly wings is very complicated. Although it seems to be a 2D planar structure, it is actually a complex 3D spatial structure with corrugation. Therefore, it is desirable to study the forewing–hindwing interaction by using 3D model with corrugation of the dragonfly forewing and hindwing.

In order to study the aerodynamic performance of dragonfly forewing–hindwing interaction in gliding flight, the 3D model with corrugation of the dragonfly forewing and hindwing are established. This paper studied aerodynamic effects of dragonfly forewing–hindwing interaction at gliding motion using the method of CFD for angles of attack changing from 0° to 25°(with an interval of 5°).

2. Numerical Method

2.1. Governing Equations

The 3D incompressible Navier-Stokes equations are as follows:

$$\frac{\partial u}{\partial x}+\frac{\partial v}{\partial y}+\frac{\partial w}{\partial z}=0 \tag{1}$$

$$\frac{\partial u}{\partial t}+u\frac{\partial u}{\partial x}+v\frac{\partial u}{\partial y}+w\frac{\partial u}{\partial z}=-\frac{\partial p}{\partial x}+\frac{1}{\mathrm{Re}}\left(\frac{\partial^2 u}{\partial x^2}+\frac{\partial^2 u}{\partial y^2}+\frac{\partial^2 u}{\partial z^2}\right) \tag{2}$$

$$\frac{\partial v}{\partial t}+u\frac{\partial v}{\partial x}+v\frac{\partial v}{\partial y}+w\frac{\partial v}{\partial z}=-\frac{\partial p}{\partial y}+\frac{1}{\mathrm{Re}}\left(\frac{\partial^2 v}{\partial x^2}+\frac{\partial^2 v}{\partial y^2}+\frac{\partial^2 v}{\partial z^2}\right) \tag{3}$$

$$\frac{\partial w}{\partial t}+u\frac{\partial w}{\partial x}+v\frac{\partial w}{\partial y}+w\frac{\partial w}{\partial z}=-\frac{\partial p}{\partial z}+\frac{1}{\mathrm{Re}}\left(\frac{\partial^2 w}{\partial x^2}+\frac{\partial^2 w}{\partial y^2}+\frac{\partial^2 w}{\partial z^2}\right) \tag{4}$$

where p and t is the pressure and time, and u,v,w are the components of the velocity along the x,y,z directions, and Re corresponds to the Reynolds number.

Lift coefficient and drag coefficient are the main basis for judging the aerodynamic performance, which are defined as:

$$\mathrm{Cl}=\frac{\mathrm{F_l}}{0.5\rho U^2 c} \tag{5}$$

$$\mathrm{Cd}=\frac{\mathrm{F_d}}{0.5\rho U^2 c} \tag{6}$$

where ρ, U and c are the density, velocity of the free stream and the average chord length of dragonfly wings, respectively, $\mathrm{F_l}$ and $\mathrm{F_d}$ are the lift and drag.

2.2. Geometric models

Dragonfly wings is mainly composed of thin cuticular membrane and veins. Veins can be divided into longitudinal veins and transverse veins, and the thickness of membrane is only 0.04 mm, moreover, the shape of the cross-section of the wing is pleated along the extension, and the thickness of the entire cross-section is also slightly different[9].

According to the characteristics of real dragonfly wings and the actual measured data by Okamoto[9], this paper have built the 3D model with corrugation of the dragonfly forewing and hindwing by using the 3D modeling software through reverse engineering. In order to simplify the model realistically, this paper ignored impact of the extremely tiny corrugations and some structures like pterostigma, nodus etc. In addition, the thickness of the cuticular membrane and veins, as well as the shape of the vein, are simplified to be consistent throughout the dragonfly wings[10].

Figure.1 shows the 3D model of dragonfly forewing and hindwing, respectively, Figure.1(a) is the 3D model of dragonfly forewing(where 1-5 is the corrugation structure of the dragonfly forewing to different positions), Figure.1(b) is the 3D model of dragonfly hindwing (where a-d is the corrugation structure of the dragonfly hindwing to different positions), and Figure.1(c) is the schematic diagram of wing membrane (where 1 is veins, 2 is cuticular membrane and 3 is the support structure composed of veins).

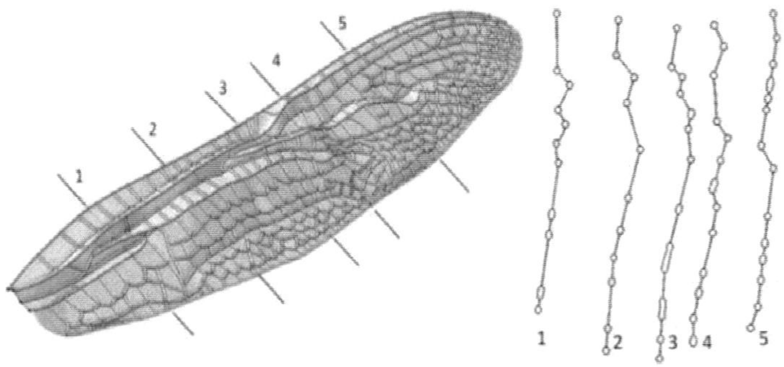

(a) The 3D model of the dragonfly forewing

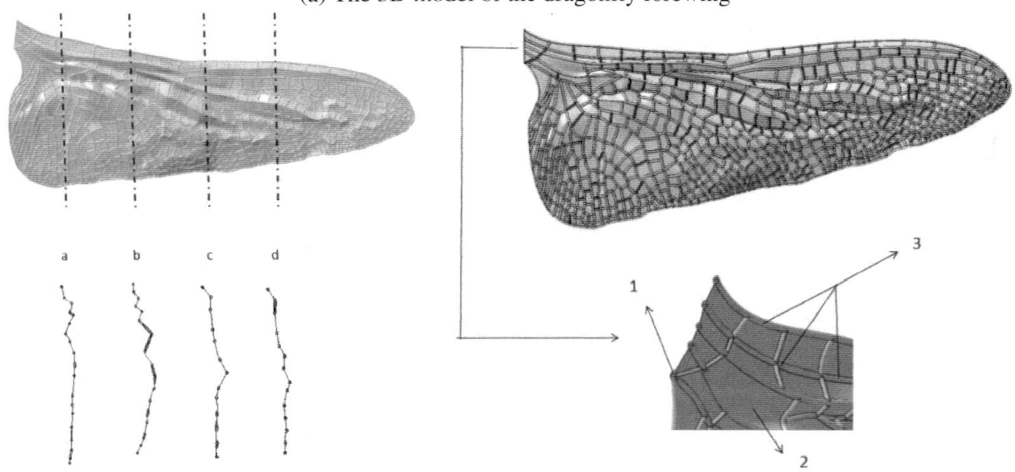

(b) The 3D model of the dragonfly hindwing (c) Schematic diagram of wing membrane

Figure 1. The 3D model of the dragonfly wings

2.3. Boundary Conditions

As shown in Figure 2, the computational domain is composed of cuboids with length, width and height of 15L, 10L and 10L respectively (where L is the length of the hindwing). The boundary conditions of left field and the dragonfly wings are symmetry and wall respectively.

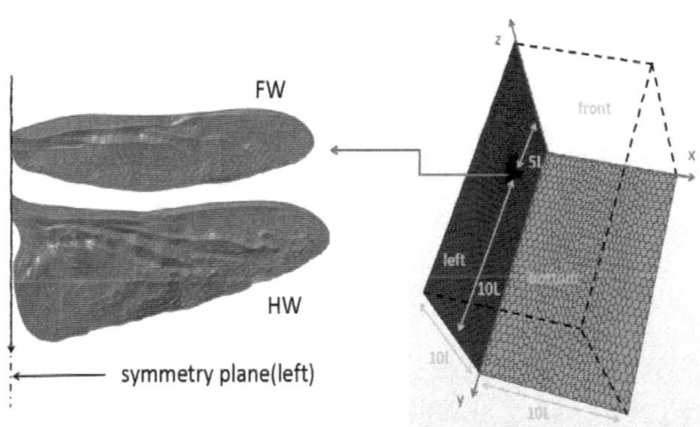

Figure 2. Computational domains

2.4 Grid Independence Verification

In order to eliminate the influence of grid size on the results, several different sizes mesh models of the dragonfly forewing and hindwing were calculated at gliding motion independently in the case of $\alpha = 0°, V = 4.5$ m/s.

The results show that when the grid sizes are 0.03mm, 0.04mm and 0.05mm, respectively, the lift coefficients of forewing are 0.119, 0.123, 0.129, and the lift coefficients of hindwing are 0.131, 0.134, 0.139. When the grid size is 0.04mm, the relative tolerance of two single wing are both less than 5%, so it can be considered that the result is independent of the grid size. Considering the balance between accuracy and efficiency of the numerical simulation, the mesh model with grid size of 0.04mm is selected as the solution model.

3. Results and Discussion

The flow fields around the aerodynamic effects of dragonfly forewing–hindwing interaction at gliding motion are simulated for V = 4.5 m/s and angles of attack changing from 0° to 25°(with an interval of 5°). To obtain quantitative data on the interactions between the forewing and hindwing, flows around independent forewing and hindwing that has the same condition are also computed.

The force coefficient of dragonfly forewing and hindwing obtained from numerical simulation are shown in Figure.3, where (a),(b) and (c) is the lift coefficient, drag coefficient and lift-drag ratio respectively.

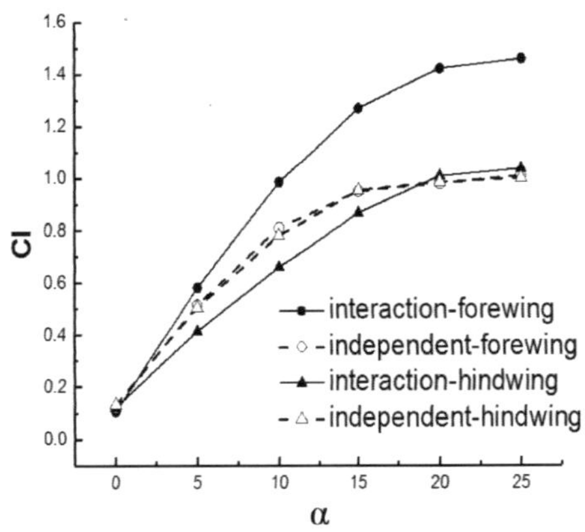

(a)lift coefficient

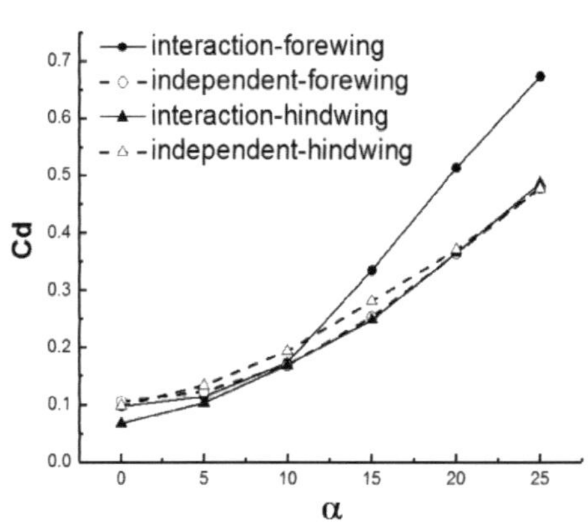

(b)drag coefficient

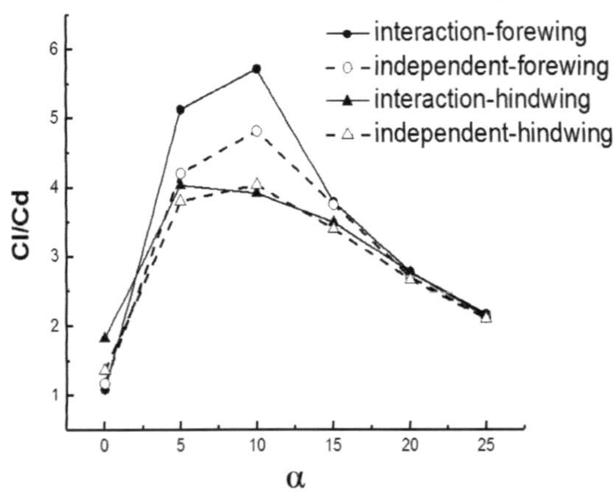

(c)lift-drag ratio

Figure 3. Force coefficient of forewing and hindwing

Only consider the case when the forewing and hindwing are gliding independently, the results are as follows: (1) The growth trend of both wing's lift and drag coefficient is consistent; (2) The lift coefficient of the independent forewing and independent hindwing is almost equal over the entire range of parameters tested, however the drag coefficient of hindwing is slightly larger than forewing when the angles of attack changing from 5° to 20°; (3) The aerodynamic performance of forewing is slightly better than hindwing, especially at small angle of attack.

To further understand the dragonfly forewing–hindwing interaction, the force coefficient in different conditions were compared. As shown in Figure.2, (1) The forewing–hindwing interaction increase the lift coefficient of forewing significantly, and the growth rate is −17.1%、11.4%、17.7%、25.0%、31.1%、30.7% compared to single forewing for angles of attack changing from 0° to 25°(with an interval of 5°); In the case of the smaller angle of attack, the drag coefficient of forewing is similar in both cases. However, it is evident larger than that of single forewing at larger angle of attack; (2)The forewing–hindwing interaction decrease both lift and drag coefficient of the hindwing to varying degrees almost over the entire range of parameters tested; (3)Compared to independent wings, the forewing–hindwing interaction enhances the aerodynamic performance of both forewing and hindwing to varying degrees, especially the effect of forewing is much more evident at small angle of attack.

Figure.4 shows the pressure contours of forewing–hindwing interaction at different angles of attack with profile cross sections shown below at 0.2L, in which the blue area represents the negative pressure, and the red or yellow areas represent the positive pressure. As shown in Figure.4, the magnitude and range of negative pressure on the upper surface and positive pressure on the lower surface of both forewing and hindwing are obviously enhanced with the increase of the angle of attack, especially the pressure changes on the upper and lower surfaces of forewing are significantly stronger than hindwing.

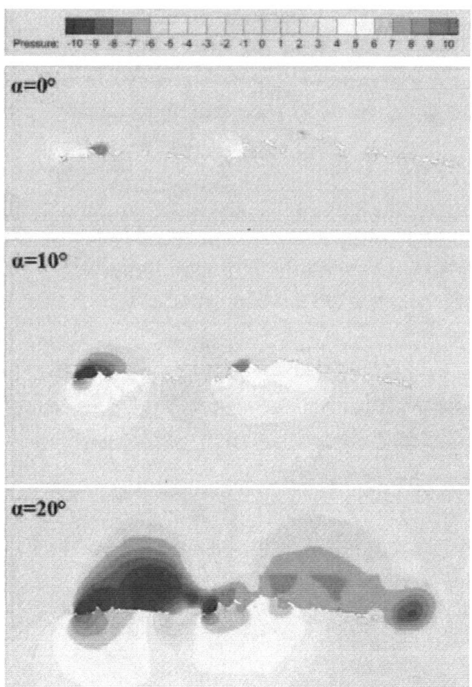

Figure 4. Pressure contours of forewing–hindwing interaction for different angles of attack with profile cross sections shown below at 0.2L

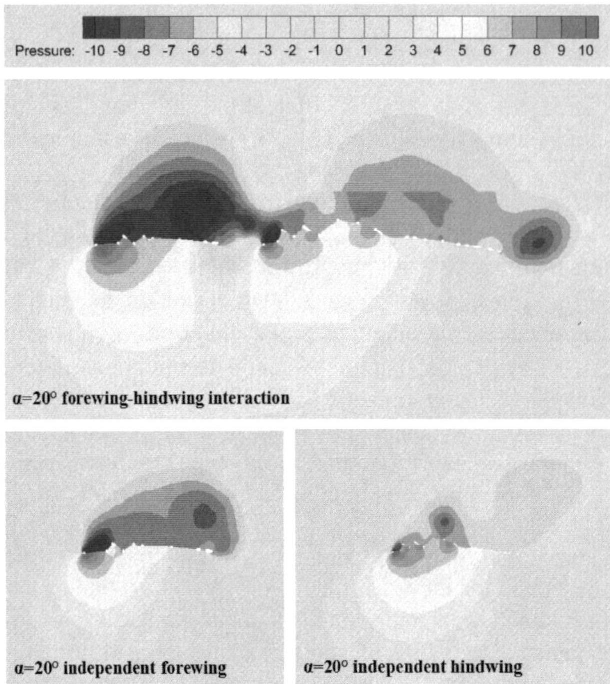

Figure 5. Pressure contours of dragonfly wings at angle of attack 20°shown below at 0.2L

Figure 5 shows the pressure contours of dragonfly wings in two different situations at angle of attack 20°. As shown in Figure 5, the forewing–hindwing interaction significantly enhance the magnitude and range of forewing's negative pressure on the upper surface and positive pressure on the lower surface compared to independent forewing; The positive pressure of the lower surface of the hindwing is not much different from independent hindwing, but the negative pressure on the upper surface is slightly larger.

Figure.6 shows the streamline of dragonfly hindwing in two different situations at angles of attack 10° and 20°. As shown in Figure 6, there is a trapped vortex inside each corrugation which rotates clockwise on the upper surface and counterclockwise on the lower surface, and it gradually moves out of the fold with the increase of the angle of attack. For the hindwing, the interactions only delayed the vortex shedding at large angle of attack, but had little effect on the aerodynamic performance of hindwing.

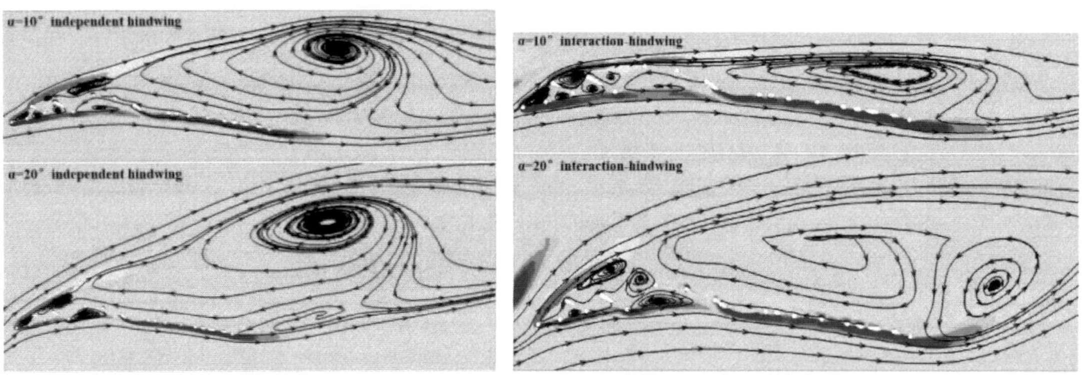

(a)independent hindwing (b)interactions-hindwing

Figure 6. The streamline of dragonfly hindwing in two different situations at angle of attack 10° and 20° with profile cross sections shown below at 0.2L

4. Conclusion

As the most suitable bionic object for the micro-flapping-wing aircraft, Dragonfly wings has always been a hot research topic. Numerical simulation indicates that the interactions between forewing and hindwing enhance the aerodynamic performance of both forewing and hindwing to varying degrees compared to independent wings, especially the effect of forewing is much more evident at small angle of attack.

The establishment process of 3D model of dragonfly wings can provide potential inspiration for other bionic calculation research. The results of the numerical simulation can also provide some reference for the design of bionic micro-flapping wing aircraft with different requirements.

References

[1] A.B. Kesel, J Exp Biol. J.203,31253-3135(2000)
[2] H.X Liu, G.Y.He,Y. Luo ,Sci. Tech.Engrg, 18 (2018).
[3] M. Sun, S. L. Lan, J Exp Biol. J.207,1887–1901(2004)
[4] J. K. Wang, M. Sun, J Exp Biol. J. 208, 3785–3804(2005)
[5] W. J. Maybury, F. Lehmann, J Exp Biol. J. 207, 4707–4726 (2004)
[6] M. Yamamoto, K. Isogai, AIAA Journal. J.43, 2475–2480(2006)
[7] J. Zhang, X.Y.Lu, Physical Review E Statistical Nonlinear & Soft Matter Physics, 80:017302(2009)
[8] M. Sun, H. Huang, AIAA Journal. J.45, 508-511(2007)
[9] M. Okamoto, K. Yasuda, A. Azuma, J Exp Biol. J.199, 281-294(1996)
[10] H.X Liu, G.Y He, C.Y Ma, et al. MATEC Web of Conferences. EDP Sciences, 151: 03001(2018)

Reliability Analysis of Certain Rocket Engine Based on Fault Tree

Wensheng Wang[1,3], Liying Jin[2], Xi Sun[3] and Jing Liu[3]

[1]Aviation Key Laboratory Science and Technology on Life-Support Technology, Hubei, Xiangyang, 441003, China
[2]School of Mechanical Engineering, Xi'an Jiaotong University, Shaanxi, Xi'an, 710049, China
[3]Aerospace Life-Support Industries, Ltd., Hubei, Xiangyang, 441003, China

Corresponding author: ggnjinliying@stu.xjtu.edu.cn

Abstract. In this paper, a fault tree model of certain rocket engine is established according to its structure and working principle. Through qualitative analysis and quantitative calculation, the factors affecting the reliability of a certain type of rocket engine are obtained, and the corresponding improvement measures are put forward to improve the reliability and safety of a certain type of rocket engine.

1. Introduction

A fault tree is a logical diagram used to mark out the failure of certain components of a product or external events or their combination that will cause a given fault of the product [1, 2]. It is a special inverted tree logic diagram of causal relationship, with events and logic gates are the compositional elements.

Fault Tree Analysis (FTA, for short) takes an undesired product failure event (or catastrophic product hazard) as the top objective for analysis [3]. By using rigorous top-down failure analysis with cause-and-effect logic [4], the necessary and sufficient direct causes of fault events based on deductive reasoning will be hierarchically found out, and all the causes and “cause packs” of the top event will be finally exposed. Calculate the occurrence probability of the top events, and try to reduce the occurrence rate through design improvement and implementation of effective fault detection, maintenance and other measures. Suggestions for product improvement will also be provided, which can be used for fault diagnosis and safety analysis. FTA will be helpful in analyzing the impact of the combined multiple fault factors on the product [5].

A certain type of rocket engine is of high reliability, which determines the life safety of the pilot in emergency ejection [6]. In this paper, FTA is used to analyze the reliability of a certain type of rocket engine, hoping to further confirm its reliability as well as the safety of pilot under emergency.

2. Establish the Fault Tree of A Certain Type of Rocket Engine

2.1. A Certain Type of Rocket Engine Works

When the ejection seat is preparing for ejection [7], the remote ignition system and the process control system will, at the same time, respectively gives an igniting signal to the gas ignition device and electric ignition device of a certain type of rocket engine. As soon as the black powder is ignited, the high temperature and high pressure gas will ignite the major fuel grain and start the rocket engine, promoting the pilot and the ejection seat rise continuously to the life-saving height.

2.2. FTA Procedure for A Certain Type of Rocket Engine

The FTA program of a certain type of rocket engine is shown in Fig. 1 which mainly includes relevant technical data and information prepared before FTA implementation [8, 9], such as project design report, failure modes and effects analysis (FMEA, for short) report, fault history data, etc; reasonable top events determined according to FMEA analysis results; the establishment of a fault tree and all the causing fault patterns; qualitative analysis through downlink method, up-link method or software instruments, obtaining all the minimum cut sets; the occurrence rate of top event and the importance of basic event according to the occurrence rate of basic event; and the design improvement to weak-links of the product, for improving the reliability of the certain type of rocket engine.

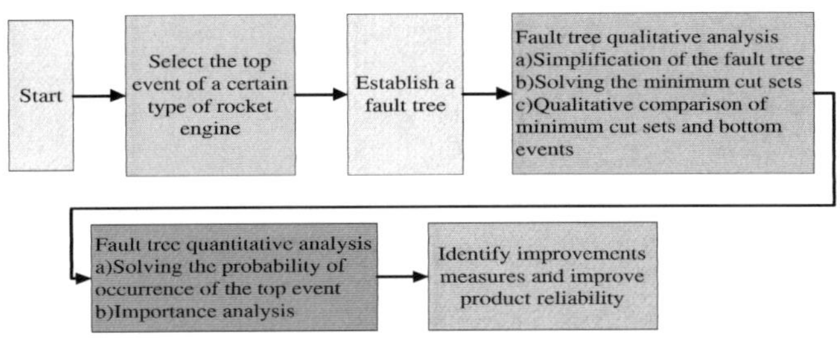

Figure 1. FTA procedure for a certain type of rocket engine.

2.3. Construction of A Certain Type of Rocket Engine Fault Tree

According to the FMEA carried out to a certain type of rocket engine, the critical fault modes (I severe degree of failure mode) that affecting safety and the success of tasks as well as the corresponding outcomes will be found out as top event, namely, the selected top event of a certain type of rocket engine is that the certain type of rocket engine is not able to provide thrust force. A top-down search will be carried out grade by grade according to the selected top event, various basic cause events will be found out and linked by appropriate logical gate symbols [10], so as to complete the establishment of the fault tree of a certain type of rocket engine top event, in which the Isograph was used, seeing Figs. 2 and 3.

3. FTA for A Certain Type of Rocket Engine

3.1. Qualitative Analysis

The task of qualitative analysis is to collect all the minimum cut set of fault tree. Minimum cut set has the minimum number of basic events, which means it will not be a cut set any more if any discard of the basic event happens. Many methods can be adopted to find the solution of the minimum cut set, among which upward method and downward method are most commonly used. However, for complex fault trees, commercial FTA software are often applied. As the fault tree of the certain type of rocket engine was built with Isograph software, this software will be directly used to calculate the minimum cut set of the top event of certain rocket engine (as shown in Table 1).

It can be seen from Table 1 that there are in total 19 top-event minimum cut sets of certain rocket engine, among which 9 are first-order minimum cut sets, 8 belong to second-order, and 2 are third-order.

IOP Conf. Series: Materials Science and Engineering **538** (2019) 012049 doi:10.1088/1757-899X/538/1/012049

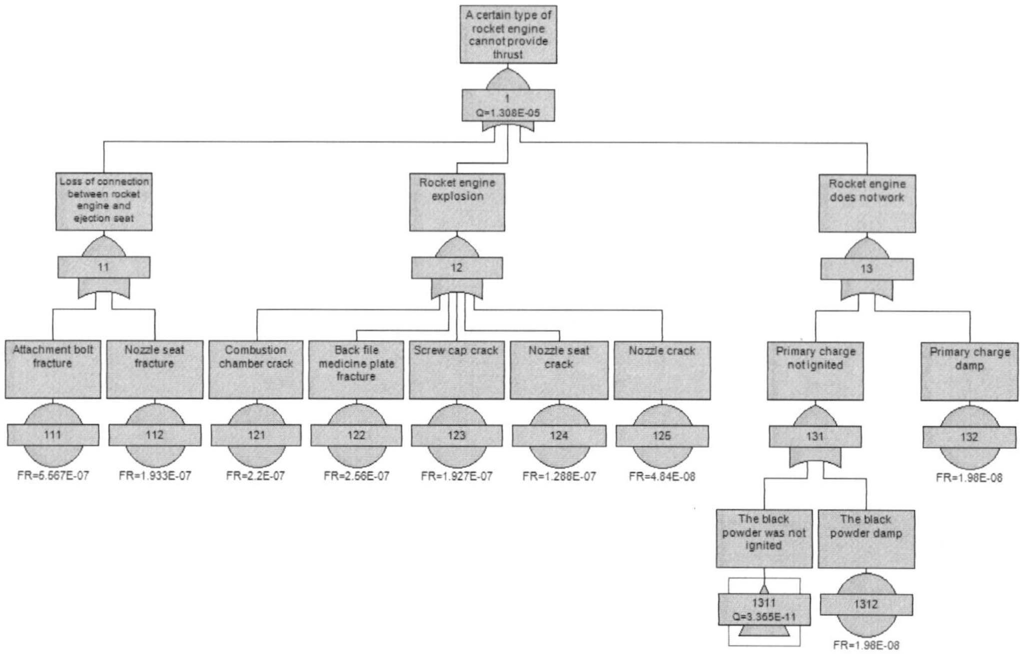

Figure 2. Fault tree main diagram of a certain type of rocket engine top event.

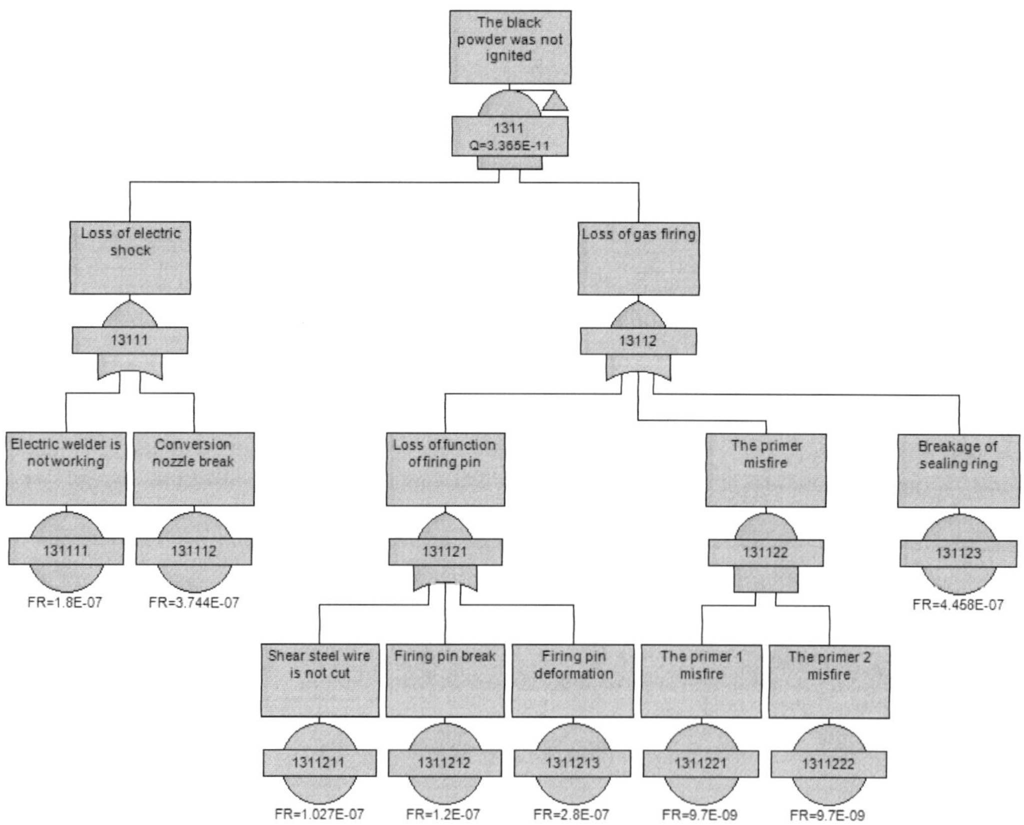

Figure 3. Fault tree diagram for percussion black powder function loss.

Table 1. The minimum cut set of the top event of a certain type of rocket engine.

Serial number	Cut set coding	Description of the minimum cut set event	Minimum cut set order
1	111	Attachment bolt fracture	1
2	112	Nozzle seat fracture	1
3	121	Combustion chamber crack	1
4	122	Back file medicine plate fracture	1
5	123	Screw cap crack	1
6	124	Nozzle seat crack	1
7	125	Nozzle crack	1
8	1312	The black powder damp	1
9	132	Main charge damp	1
10	131111、1311211	Electric detonator is not working、Shear copper wire is not cut	2
11	131111、1311212	Electric detonator is not working、Firing pin break	2
12	131111、1311213	Electric detonator is not working、Firing pin deformation	2
13	131111、131123	Electric detonator is not working、Breakage of sealing ring	2
14	131112、1311211	Conversion nozzle break、Shear copper wire is not cut	2
15	131112、1311212	Conversion nozzle break、Firing pin break	2
16	131112、1311213	Conversion nozzle break、Firing pin deformation	2
17	131112、131123	Conversion nozzle break、Breakage of sealing ring	2
18	131111、1311221、1311222	Electric detonator is not working、The primer 1 misfire、The primer 2 misfire	3
19	131112、1311221、1311222	Conversion nozzle break、The primer 1 misfire、The primer 2 misfire	3

Table 2. Failure rate of each bottom event of a certain type of rocket engine fault tree.

Serial number	Bottom event	Failure rate ($\times 10^{-6}$/h)
1	Attachment bolt fracture	0.5567
2	Nozzle seat fracture	0.1933
3	Combustion chamber crack	0.22
4	Back file medicine plate fracture	0.256
5	Screw cap crack	0.1927
6	Nozzle seat crack	0.1288
7	Nozzle crack	0.0484
8	The black powder damp	0.0198
9	Main charge damp	0.0198
10	Electric detonator is not working	0.18
11	Conversion nozzle break	0.3744
12	Shear copper wire is not cut	0.1027
13	Firing pin break	0.12
14	Firing pin deformation	0.28
15	The primer 1 misfire	0.0097
16	The primer 2 misfire	0.0097
17	Breakage of sealing ring	0.4458

IOP Conf. Series: Materials Science and Engineering **538** (2019) 012049 doi:10.1088/1757-899X/538/1/012049

3.2. Quantitative Analysis

Quantitative analysis refers to the calculation of the occurrence probability of the top event of the fault tree based on the occurrence probability of the basic event of the fault tree, as well as the calculation of the importance of the basic event. According to the "Non-electronic Parts Reliability Manual" and the accumulated data in the past engineering experience, the failure rate of each component is obtained. The failure rate of each basic event is also calculated by referring to the frequency ratio of each failure mode in the FMEA, seeing Table 2.

The probability significance of each basic event was calculated by Isograph software (as shown in Table 3).

Input the failure rate of each basic event into the software. The probability of occurrence of certain rocket engine top event is 1.308×10^{-5}/h.

Table 3. Probability importance of each bottom event of a certain type of rocket engine fault tree.

Serial number	Bottom event	Probability importance
1	Attachment bolt fracture	0.3404
2	Nozzle seat fracture	0.1182
3	Combustion chamber crack	0.1345
4	Back file medicine plate fracture	0.1565
5	Screw cap crack	0.1178
6	Nozzle seat crack	0.07875
7	Nozzle crack	0.02959
8	The black powder damp	0.01211
9	Main charge damp	0.01211
10	Electric detonator is not working	8.351×10^{-7}
11	Conversion nozzle break	1.737×10^{-6}
12	Shear copper wire is not cut	2.785×10^{-7}
13	Firing pin break	3.254×10^{-7}
14	Firing pin deformation	7.593×10^{-7}
15	The primer 1 misfire	0
16	The primer 2 misfire	0
17	Breakage of sealing ring	1.209×10^{-6}

3.3. Results Analysis

According to the calculation of the minimum cut set of certain rocket engine, nine single point failure of the top event of a certain type of rocket engine can be found, which will lead to the occurrence of the top event. Therefore, related design improvement or compensating measures must be carried out to reduce the occurrence rate of the single point failure. For example, use 30CrMnSiA materials to increase the fatigue strength and solve the cracks problem in the combustion chamber, carry out hydraulic pressure strength test, and perform magnetic inspection to the combustion chamber before assembly. The probability importance results of the basic event and the minimum cut set analysis results indicate that the weak link of the certain type of rocket engine is connecting bolt, so special attention should be paid to improve the strength of the connecting bolts. In addition, the rocket engine, as an initiating explosive device with limited service life, must be changed based on its life span (storage period or usage period), and the whole process must be performed strictly in accordance with the maintenance instructions for replacement. In Table 3, the probability significance of the primer 1 misfire and the primer 2 misfire is 0, which means that neither the primer 1 misfire nor primer 2 misfire make any contribution to the occurrence of the top event. It mainly because the top event can only be occurred through two redundancy design with primer 1 misfire and primer 2 misfire, which also proved that the reliability of the product can be improved with the addition of the redundancy design.

4. Conclusions

In this paper, a fault tree model is established for certain rocket engine that cannot provide thrust as a top event, and qualitative and quantitative analysis is carried out to find out the cause of the fault. By taking corresponding measures, the reliability of certain rocket engine is effectively improved. Sexuality provides a theoretical basis for further optimization of certain rocket engine.

Acknowledgments

We would like to thank the anonymous reviewers whose thoughtful comments improved the quality of this paper. This work was jointly supported by the Aerospace Life-Support Industries, Ltd. and the Aviation Key Laboratory Science and Technology on Life-Support Technology.

References

[1] Peichang Li, Hongjie Yuan, Jie Lan, Ming Cheng, Dynamic fault tree analysis using sequential binary decision diagrams [J], Journal of Beijing University of Aeronautics & Astronautics, (2017) 43(1):167-175.

[2] Shengkui Zeng, Reliability design and analysis [M], Beijing: national defense industry press, (2013).

[3] Jiming Guo, Jinping Qi, Xingyun Li, Status and development of fault tree analysis [J], Equipped with mechanical, (2018) 2, 61-66.

[4] Xiaoxing Yang, Logic Analysis of Electrical Machinery System Faults——Taking CNC Car as an Example [J], SHANDONG INDUSTRIAL TECHNOLOGY, (2018) 24.

[5] GJB/Z. 768A, Fault Tree Analysis Guide [S], National Military Standard of the People's Republic of China, (1998).

[6] Xinbo Li, He Liu, Chaoyun Wu, A Reliability Modeling Method of Liquid Rocket Engine Based on Task Process [J], Quality and Reliability, (2016) (6): 10-14.

[7] Yan Zhang, Tao Li, Xiangjun Duan, Failure Analysis and Research on Support Fracture of Detonation Transmission Mechanism of Certain Rocket Ejection Seat [J], Aviation Precision Manufacturing Technology, (2018) 54 (1): 35-39.

[8] Ruibing Du, Gang Wang, Analysis and Improvement of Ignition Failure of Solid Rocket Engine, Journal of Ordnance Equipment Engineering [J], SichuanOrdnanceJournal, (2008) 29 (4): 7-9.

[9] Wuyue Kuang, Songlin Tan, THE STUDY OF SYNTHETIC ANALYSIS ON DESIGN RELIABILITY OF LIQUID ROCKET ENGINE [J], JOURNAL OF PROPULSION TECHNOLOGY, (1997) 18(5): 9-12.

[10] Jinli Hou, Ping Jin, Guobiao Cai, Failure mode of reusable rocket engine based on fuzzy fault tree and factor analysis [J], (2014) 29(4): 987-992.

IOP Conf. Series: Materials Science and Engineering **538** (2019) 012050 doi:10.1088/1757-899X/538/1/012050

Correlation Analysis between Gross Parameters and Performance Parameters of the Light Aircraft

Fujia Liu[1,2,a], Qunfang Li[2] and Hao Geng[2]

[1]Shenyang Aircraft Design Institute, Shenyang 110035, China
[2]Liaoning General Aviation Academy, Shenyang 110136, China

Corresponding author: 718880677@qq.com

Abstract. Based on the elementary concepts of aircraft design, statistical analysis are made betweenthe gross parameters and performance parameters of modern two-seater and four-seater light aircraft. The relations among these parameters are found. 12 experiential expressions are obtained and some parameters' value scopes are determined. These expressions and data are proved to be useful for the conceptual design of two-seater and four-seater light aircraft.

1. Introduction

In conceptual design stage, aircraft design is largely dependent on previous design experience, so each aircraft design and research institutions typically has analyzed a large amount of statistical data, and has got a series of empirical formulas for overall parameters and performance parameters. Through these empirical formulas, the aircraft designers can select, analyze and evaluate the overall parameters and performance parameters of the scheme.

A large number of studies have been conducted on the light aircraft [1-3]. As the new technique, new materials, new technology are developed and applied, a lot of the new type light aircraft have been developed in recent years. Now the relationship between overall and performance parametersreflecting the current technical conditionshas not been set up. Based on these considerations above, the paper collects and sortsthe related parameters of light aircraft of the past and the recent at home and abroad, and the correlation of the overall and performance parameters of light aircraft is analyzed.

2. The source and classification of the overall and performance parameters

At present, there are many kinds of light aircraft. Due to the difference of mission requirements and design characteristics, there are different statistical rules between the overall and performance parameters of different types of light aircraft. Light aircraft are mainly used in tourism, pilot training and other aspects, and most of them are two-seater and four-seater in thc market. Therefore, this paper mainly studies two-seater and four-seater light aircraft. All the data cited in this paper are from the literature [4-6], with a total of 106 models, of which 64 are two-seater light aircraft and 42 are four-seater light aircraft.

The general and performance parameters of the aircraft are many, and the weight characteristic of the aircraft, the wing loading and the power to weight ratio are the three key parameters of the conceptual design phase. The wing area and the aspect ratio are important parameters affecting the aerodynamic characteristics of the aircraft. Endurance performance and flight speed are the main parameters reflecting the performance of light aircraft[7-8]. The following equation shows the weight characteristics of the aircraft:

$$W_{TO} = W_E + W_{PL} + W_F \tag{1}$$

In the formula, W_{TO} is the take-off weight, kg; W_E is the empty weight, kg; W_{PL} is the payload, kg; W_F is the fuel weight, kg.

Take-off weight is composed of empty weight, payload and fuel weight. Among them, the empty weight including structure weight, power system weight, avionics system weight and control system weight. The deformation of equation (1) can be obtained:

$$W_{TO} = \frac{W_{PL}}{1 - W_E/W_{TO} - W_F/W_{TO}} \tag{2}$$

where, W_E/W_{TO} is the coefficient of the empty weight, and W_F/W_{TO} is the coefficient of fuel weight.

Generally, W_{PL} is the data specified in the requirements of the aircraft design.Therefore, if the relationship between W_E/W_{TO}&W_F/W_{TO} and W_{TO} is known, or the relationship between W_E/W_{TO}&W_F/W_{TO} and W_{TO} is known, then the take-off weight can be obtained.

If we find out the mathematical relation between the wing loading& the power to weight ratio and the take-off weight, then the wing area and engine power can be obtained. The aircraft performance mainly depends on the overall parameters of the aircraft. After the overall parameters are determined, the aircraft performance will be roughly determined.

Based on the above considerations, the overall parameters and performance parameters studied in this paper are shown in table 1 and table 2.

Table 1. Overall parameters.

Parameter	Symbol	Unit
Empty weight	W_E	kg
Wing area	S	m^2
Aspect ratio	A	/
Engine power	P	kW
Take-off wing loading	$(W/S)_{TO}$	kg/m^2
Take-off power to Weight ratio	$((P/W)_{TO}$	kW/kg

Table 2. Performance parameters.

Parameter	Symbol	Unit
Take-off weight	W_{TO}	kg
Fuel weight	W_F	kg
Maximum level speed	V_{MAX}	m/s
Stall speed	V_S	m/s

3. Analysis and processing of statistical data

3.1. Statistical analysis of some parameters

Because different aircraft have different design requirements, for a particular parameter, the variation range of statistical data is generally large, which makes the inaccurate statistical relationship between the parameters. For this reason, the statistical values of all parameters are treated aslogarithm, and the processed values are relatively concentrated, so the mathematical relations between parameters can be established. Before establishing the mathematical relationship between the overall and performance parameters, this paper first gives the value range of some parameters determined according to the statistical results, which can provide useful data and reference for the same type of light aircraft in the conceptual design stage, as shown in table 3.

Table 3. Statistical analysis results of some parameters.

Parameter	Two-seater aircraft		Four-seater aircraft	
	Average	**Standard deviation**	**Average**	**Standard deviation**
Take-off wing load (kg/m^2)	58.075	23.143	94.258	25.566
Take-off power to weight ratio (kW/kg)	67.028	11.031	178.101	57.985
Engine power (kW)	62.266	14.422	134.692	34.982
Wing area (m^2)	10.947	1.855	13.824	2.812
Aspect	9.21	0.93	8.279	1.785
Maximum level speed (km/h)	219.514	41.215	295.628	44.825
Stall speed (km/h)	73.846	16.144	98.029	26.095

3.2. Empty weight and take-off weight

The magnitude of the take-off weight determines the value of the empty weight. Generally, the larger take-off weight will cause the larger empty weight. Statistics of empty weight and take-off weight of 64 two-seater light aircraft are shown in figure 1. The fitting equation is as follows:

$$W_E/W_{TO} = 0.04W_{TO}^{0.410} \tag{3}$$

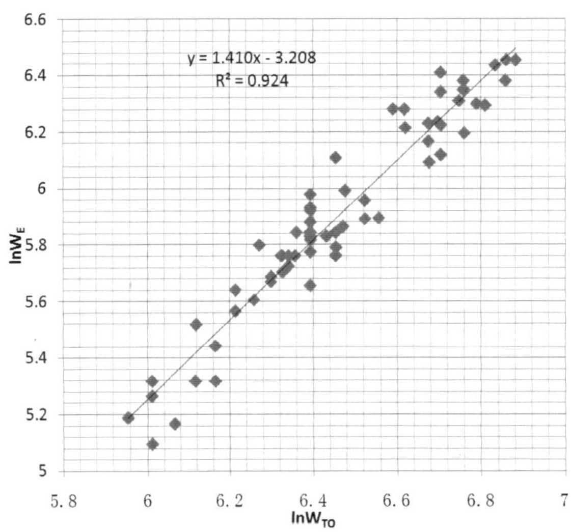

Figure 1. Empty weight and take-off weight of two-seater light aircraft.

Statistics of empty weight and take-off weight of 29four-seater light aircraft are shown in figure 2. The fitting equation is as follows:

$$W_E/W_{TO} = 0.056W_{TO}^{0.334} \tag{4}$$

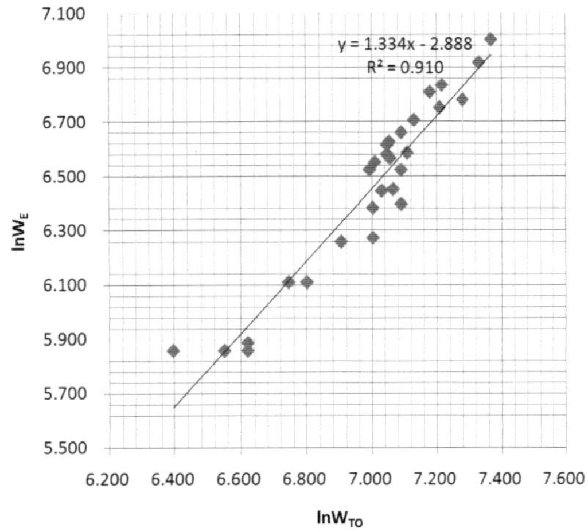

Figure 2. Empty weight and take-off weight of four-seater light aircraft.

It can be seen from Figure 1 and Figure 2 that there is a good linear relationship between the empty weight and the take-off weight, and the empty weight increases with the increase of the take-off weight. Equation (3) and equation (4) are the relation between the empty weight coefficient and the take-off weight. These formulas can provide a reference basis for the estimation of the take-off weight, and it can be seen that the empty weight coefficient increases with the increase of the take-off weight.

3.3. Fuel weight and take-off weight

The aircraft's fuel weight includes the mission fuel weight and the dead fuel weight, and the mission fuel weight depends on the flight mission, the aerodynamic characteristics of the aircraft and the fuel consumption characteristics of the engine. Statistical data of fuel weight and takeoff weight of 43 two-seater light aircraft are shown in figure 3, and the fitting equation is as follows:

$$W_F / W_{TO} = 0.00013 W_{TO}^{1.035} \tag{5}$$

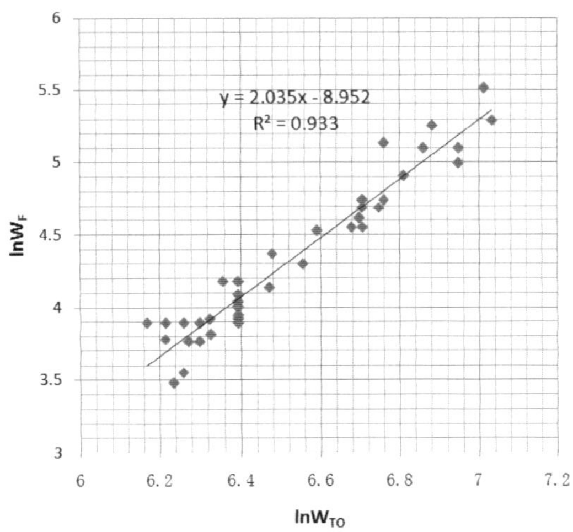

Figure 3. Fuel weight and takeoff weight of two-seater light aircraft.

Statistical data of fuel weight and takeoff weight of 21 four-seater light aircraft are shown in figure 4, and the fitting equation is as follows:

$$W_F/W_{TO} = 0.0124W_{TO}^{0.331} \tag{6}$$

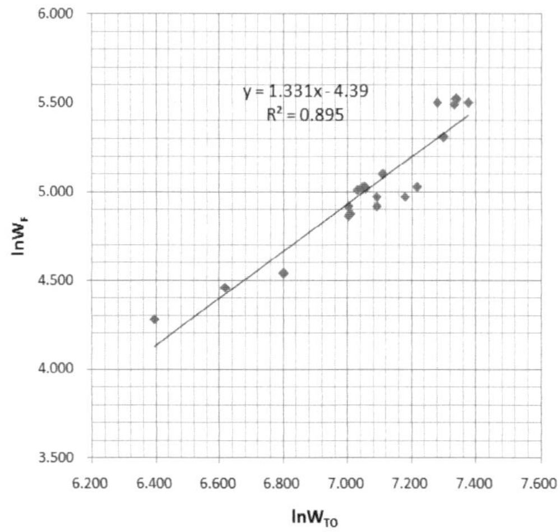

Figure 4. Fuel weight and take-off weight of four-seater light aircraft.

It can be seen from figure 3 and figure 4 that the fuel weight of most aircraft increases with the increase of the take-off weight, while a few aircraft do not show this feature due to mission requirements. Equation (5) and equation (6) are the relation between the fuel weight coefficient and the take-off weight. These formulas can provide a reference basis for the estimation of the take-off weight of light aircraft, and it can be seen that the fuel weight coefficient increases with the increase of the take-off weight.

3.4. Engine power and take-off weight

At the conceptual design stage, suitable engines are usually selected based on the required power of the aircraft. The engine power is higher, and the more fuel it takes to carry out the entire mission, thus the take-off weight of the aircraft will increase to complete the design mission. The statistical data of engine power and takeoff weight of 40 two-seater light aircraft are shown in figure 5. The fitting equation is as follows:

$$P/W_{TO} = 0.00118W_{TO}^{0.742} \tag{7}$$

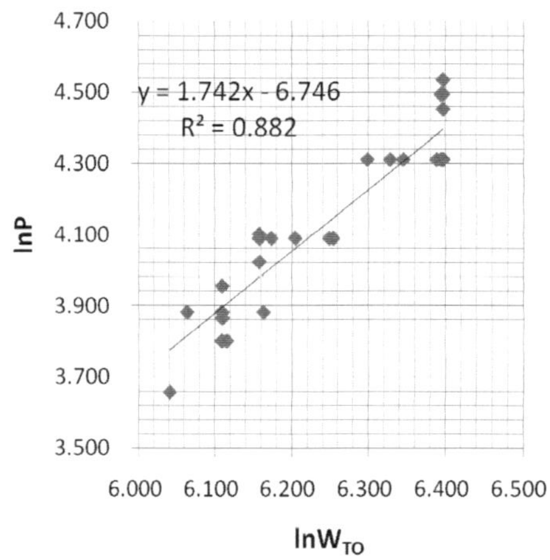

Figure 5. Engine power and take-off weight of two-seater light aircraft.

The statistical data of engine power and takeoff weight of 24 four-seater light aircraft are shown in figure 6. The fitting equation is as follows:

$$P/W_{TO} = 0.0429W_{TO}^{0.146} \tag{8}$$

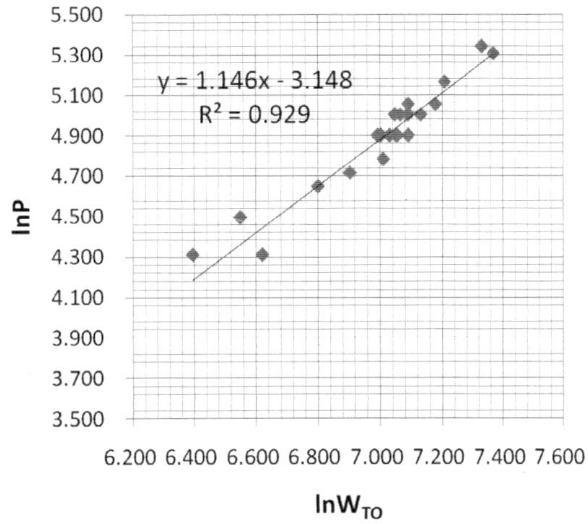

Figure 6. Engine power and take-off weight of four-seater light aircraft.

Equation (7) and equation (8) are the relation between the power to weight ratio and the take-off weight. If the take-off weight is known, the required engine power can be estimated.

3.5. *Wing area and take-off weight*

The take-off weight determines the maximum lift, which affects the wing area. Wing load is the weight of aircraft divided by wing area, which has a great influence on the determination of take-off weight. If the wing loading decreases, the wing area will be increased, which can improve the aircraft performance, however, additional drag and empty weight which are generated by the increase of wing area will result in the increase of take-off weight for the completion of the mission. The statistical data of wing area and take-off weight of 28 two-seater light aircraft are shown in figure 7. The fitting equation is as follows:

$$W_{TO}/S = 438.781W_{TO}^{-0.362} \tag{9}$$

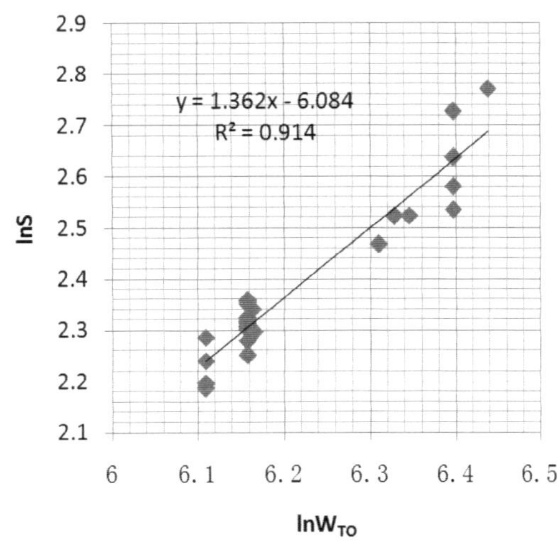

Figure 7. Wing area and take-off weight of two-seater light aircraft.

The statistical data of wing area and take-off weight of 32four-seater light aircraft are shown in figure 8. The fitting equation is as follows:

$$W_{TO}/S = 17.584W_{TO}^{0.222} \tag{10}$$

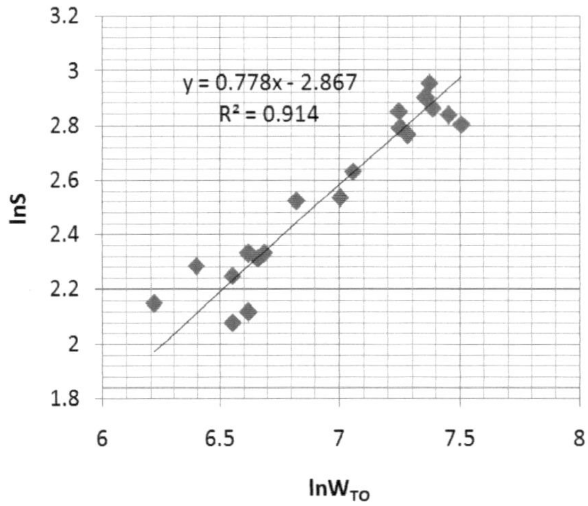

Figure 8. Wing area and take-off weight of four-seater light aircraft.

Equation (9) and equation (10) are the relation between the wing load and the take-off weight. If the take-off weight is known, the wing area can be estimated.

3.6. Stall speed and take-off wing loading

The aircraft stall speed is the main factor which affects the flight safety. It is directly determined by the aircraft wing loading and maximum lift coefficient. The statistical data of the stall speed and take-off wing loading of 26 two-seater light aircraft are shown in figure 9. The fitting equation is as follows:

$$V_S = 2.4303(W/S)_{TO}^{0.529} \tag{11}$$

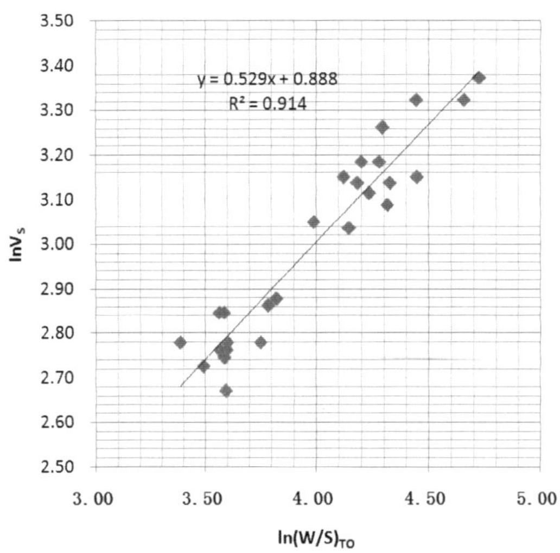

Figure 9. Stall speed and take-off wing loading of two-seater light aircraft.

IOP Conf. Series: Materials Science and Engineering **538** (2019) 012050 doi:10.1088/1757-899X/538/1/012050

The statistical data of stall speed and take-off wing loading of 35 four-seater light aircraft are shown in figure 10. The fitting equation is as follows:

$$V_S = 0.3379(W/S)_{TO}^{0.965} \tag{12}$$

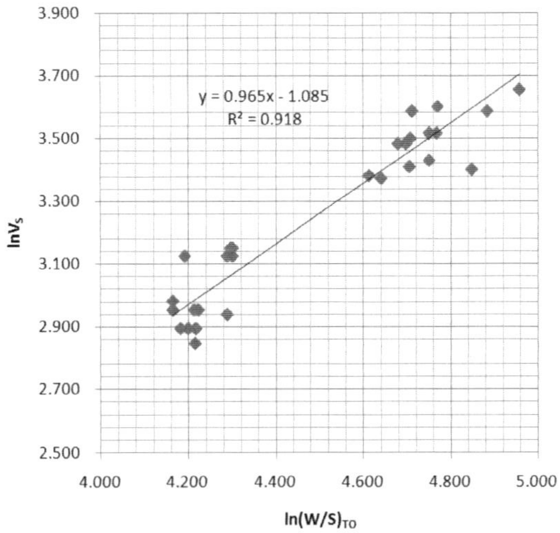

Figure 10. Stall speed and take-off wing loading of four-seater light aircraft.

Equation (11) and equation (12) are the relation between the stall speed and the take-off wing loading. If the take-off wing loading is known, the stall speed can be estimated.

3.7. Maximum level speed and aspect

In general, the drag of the high aspect ratio wing is less than the low aspect ratio wing. Therefore, the aspect ratio directly affects the maximum lift-to-drag ratio of aircraft. Statistics of maximum level speed and aspect ratio of 31 light aircraft with two seats are shown in figure 11. The fitting equation is as follows:

$$V_{MAX} = 0.0746A^{3.533} \tag{13}$$

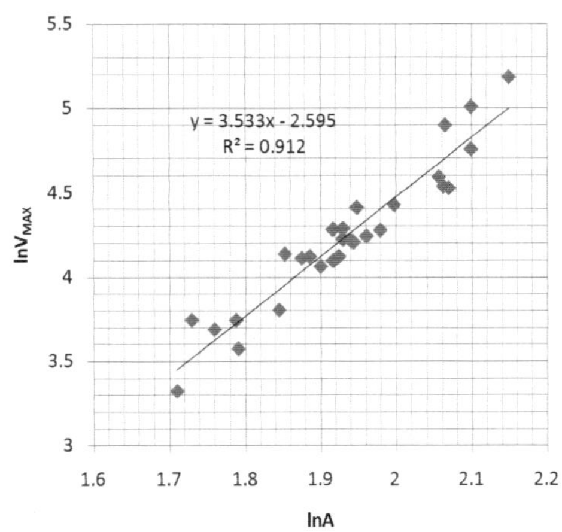

Figure 11. Aspect ratio and maximum level speed of two-seater light aircraft.

IOP Conf. Series: Materials Science and Engineering **538** (2019) 012050 doi:10.1088/1757-899X/538/1/012050

The statistical data of aspect ratio and maximumlevel speed of 42 four-seater light aircraft are shown in figure 12. The fitting equation is as follows:

$$V_{MAX} = 19.6288A^{0.678} \tag{14}$$

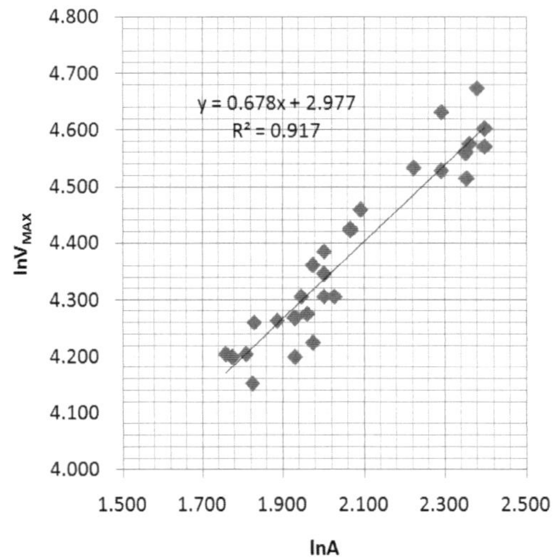

Figure 12. Aspect ratio and maximum level speed of four-seater light aircraft.

Equation (13) and equation (14) are the relation between the aspect ratio and the maximum level speed. If the aspect ratio is known, the maximum level speed can be estimated.

4. Analysis of example

Taking P2010 four-seater aircraft as an example, the statistical formula of four-seater light aircraft is used to calculate and compare with the real data, as shown in table 4. The results show that the satisfactory results can be obtained quickly by applying these formulas in the conceptual design stage.

Table 4. Comparison of estimated value and actual value of the P2010 four-seater aircraftparameters.

Parameter	Unit	Estimate value	Actual value
Empty weight	kg	681.937	710.000
Fuel weight	kg	148.676	151.200
Take-off wing loading	kg/m^2	84.222	79.452
Take-off power to weight ratio	kW/kg	0.120	0.116
Stall speed	km/h	87.726	88.000
Maximum level speed	km/h	278.290	280.000

5. Conclusion

Based on the statistical analysis of the overall and performance parameters of 106 kinds of light aircraft, this paper obtained 6 empirical formulas for two-seater aircraft and 6 empirical formulas for four-seater aircraft. The calculation example proved that the parameter estimation and selection in the conceptual design stage of light aircraft had certain guiding significance and practical value.

IOP Conf. Series: Materials Science and Engineering **538** (2019) 012050 doi:10.1088/1757-899X/538/1/012050

Because exhaust emissions of oil aircraft have an impact on air quality and climate, and with the constant improvement of the battery energy density, now the light electric aircraft begins to grow [9-10]. The empirical formula in this paper is given according to the statistical data of oil aircraft, but some empirical formula can also be applied to light electric aircraft conceptual design phase.

References

[1] Roskam J. Airplane design part I: Preliminary sizing of airplanes. Roskam Aviation and Engineering Corporation, Ottawa, Kansas, 1985.
[2] Raymer D P. Aircraft design: A conceptual app roach. American: Institute of Aeronautics and Astronautics, Inc, 1992.
[3] Aircraft Design Manual. Chapter 8: Weight and Balance Control. Beijing: Aviation Industry Press, 2000.
[4] International Aviation Magazine. World Aircraft Handbook. Beijing: Know l edge Press, 1982.
[5] China Aero-Information Center "World Aircraft Handbook" Edit OrialTeam. World Aircraft Handbook. Beijing: Aviation Industry Press, 2000.
[6] http://www.widola.com.
[7] Yu X Q, Xu H M, Ang H S. Aircraft Conceptual Design. Beijing: Aviation Industry Press, 2000.
[8] Gu S F, Xie S S. Aircraft Design. Beijing: BeijingUniversity of Aeronautics and Astronautics Pres s, 2001.
[9] ZHAO Changhui, CHEN Liwei, LU Libo, et al. Advances in Electric Aircraft Technology. Science&Technology Review, 2012, 30(12): 62-70.
[10] Huang Jun, Yang Fengtian. Development and challenges of electric aircraft with new energy. Acta Aeronautica et AstronauticaSinica, 2016, 37(1): 57-68.

IOP Conf. Series: Materials Science and Engineering **538** (2019) 012051 doi:10.1088/1757-899X/538/1/012051

Numerical Simulation of Deflagration to Detonation Transition in High-Energy Propellant

Zhuo-qun Wang[1], Li-qiong Wang[1,a] and Fei Zhen[1]

[1]Beijing Institute of Technology, Mechatronical Engineering Department, 100081 Beijing, China

[a]Corresponding author: wlq2018@sina.com

Abstract. The deflagration to detonation transition model was established based on AUTODYN/ANSYS to simulate the process of DDT and provide a basis for the safety analysis of high-energy propellant. The process of DDT was numerically simulated by the ignition and growth reactive flow model and ALE method. The detonation pressure, detonation velocity, the run distance and time to detonation were calculated. The simulating results indicate that the detonation velocity is about 2899m/s, the run distance to detonation is about 138.3mm when the density of propellant is 1860kg/m^3. The numerical results are in a good agreement with the experimental results.

1. Introduction

Solid propellant is the power source of weapon and rocket engine propulsion systems [1]. With the development of these systems, a large number of high-energy explosives, such as RDX and HMX, are added into solid propellants, which increase the energy and thrust of propellants, while increasing the hazard of propellants greatly at the same time.

The phenomenon of deflagration to detonation transition (DDT) is one of the important criteria for evaluating the safety of energetic materials [2]. Because DDT is a very complex, highly non-linear and scale-crossing transition process, and solid propellant is a composite energetic material based on polymer and filled with metal powder and oxidant, so it is extremely difficult to study DDT of propellant [3]. In recent decades, many scholars have studied DDT of solid propellants by means of numerical simulation method. At present, two phase mixed continuous flow models, including BN model proposed by Baer and Nunziato[4] and PSK model proposed by Powers[5] are widely used in the numerical simulation of DDT. Butler[6] and X. R. Jia[7] simulated DDT of solid propellant using two-phase reactive flow model. However, the actual propellant has low porosity, so it is not suitable to use two-phase flow model to study its DDT behavior[8]. Meantime, the three-dimensional numerical simulation of DDT needs further research and improvement. G. C. Qin[9] and N. Qin[10] studied the effects of different conditions on DDT for several propellants by experiments. Compared with the experimental method, the numerical simulation method not only has less cost and higher security, but also can obtain more data. Therefore, it is more conducive to the analysis of DDT process. From the existing experimental and numerical simulation studies[11], the researches on DDT of propellants mainly focuses on the low and medium density conditions, so it is necessary to carry out the research on the high density conditions.

In this paper, a three-dimensional numerical simulation model of DDT for high-energy propellant is established according to the actual experiment. The ignition and growth reactive flow model is used to describe the reaction process of high density propellant, and DDT process is analyzed. The detonation parameters, the run distance and time are calculated. It would provide a basis for analyzing and evaluating the safety of high-energy propellants.

2. Experiment and Numerical Models

The research model of DDT is mainly based on DDT tube. Based on the DDT tube experiment, the numerical model was established to solve DDT problem. The high-energy propellant in the tube is mainly composed of hydroxyl-terminated polybutadience (HTPB), ammonium perchlorate (AP), cyclotrimethylenetrinitramine (RDX) and aluminium (Al), and its density is 1860kg/m^3.

2.1. Experiment of DDT

The experimental device is shown in Figure 1. The high-energy propellant is loaded into a slightness steel tube(DDT tube), and the igniter made from black powder and RDX is placed at the end of the propellant column. The igniter ignites the ignition powder. Then, the propellant is ignited and combust, and transit to detonation under certain conditions.

The length of the DDT tube is 400mm, and the outer diameter and inner diameter of the shell are 28 mm and 10 mm. In order to fix the experimental device, the end caps are fixed on the thick steel plate and fastened by bolts. In order to measure the propagation velocity of detonation wave, some electrical pins are set along the axis of the tube wall. When the detonation wave reaches the positions of pins, the pins conduct electricity, and the signal reception device is triggered. The validation plate is placed under the experimental device, and DDT is judged by observing the deformation of the tube and validation plate.

Figure 1. The experimental device photo.

The DDT phenomenon of high-energy propellant was observed by the DDT tube experiment. By analyzing the fragments of tube and the indentation of validation plate as shown in Figure 2, the run distance to detonation was obtained about 135mm, and the stable detonation velocity is about 2940m/s measured by the electrical pins.

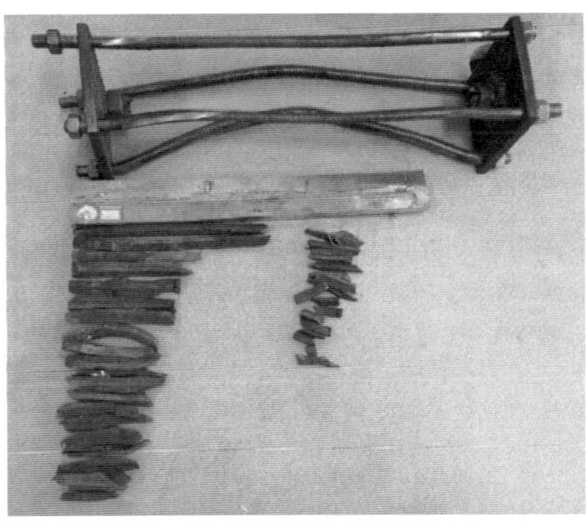

Figure 2. Experimental apparatus results photo of DDT.

2.2. Physical Model and Calculating Models

The physical model of DDT tube in this paper is shown in Figure 3 composed by end caps, shell, igniter and propellant. Compared with the experimental device, the validation plate and bolts are simplified.

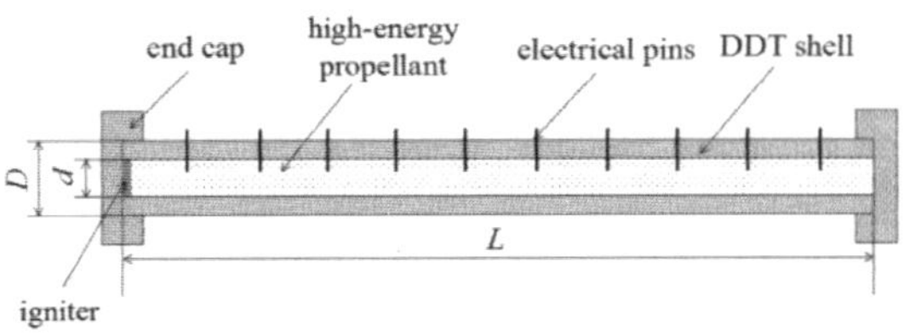

Figure 3. Physical model of the DDT tube.

ANSYS/AUTODYN is an explicit non-linear dynamic analysis software, which can analyze the dynamic characteristics of solids, fluids and gases. It is widely used in the field of explosion research. The calculating model is established based on the physical model shown in Figure 3. Because of the symmetry of the physical model, the 1/4 three-dimensional model is established as shown in Figure 4. The calculation model is composed of DDT tube, high-energy propellant and end caps, and their sizes are the same as the experimental model sizes. The igniter is simulated by applying the ignition boundary on the propellant end surface. Automatic contact is defined between parts of the model.

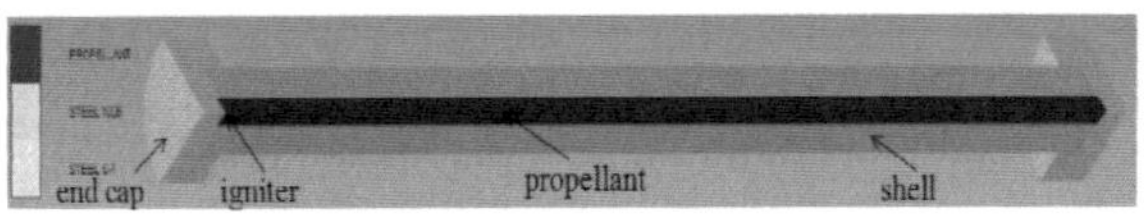

Figure 4. The calculating model.

Lagrange, Euler and Arbitrary Lagrangian Eulerian(ALE) methods are three methods widely used for dealing with continuum problems in numerical simulation. ALE method combines the advantages of Lagrange method and Euler method. On the one hand, it can effectively track the movement of material structure boundary. On the other hand, it can adjust the mesh position properly according to the defined parameters, so that the grid wouldn't be seriously distorted, and the flexibility of the grid would be increased. Therefore, ALE method was used to solve high-energy propellant, and Lagrange method was used to solve the shell and end caps.

The calculating model is regular in this paper and there is no difference in mesh density for the model, so uniform three-dimensional solid elements were used to mesh the calculating model. The gauges started with 0.4mm and were set up every 2mm as shown in Figure 5.

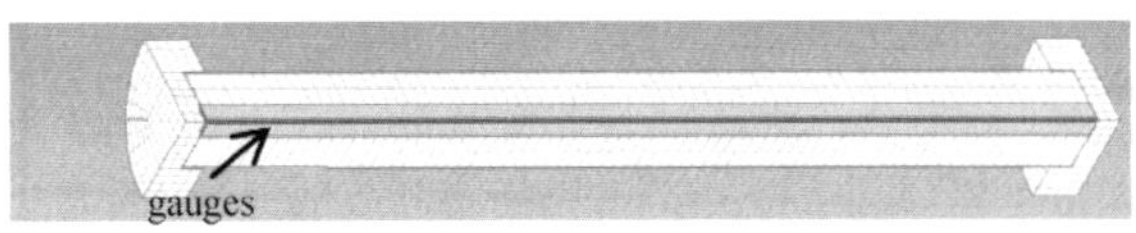

Figure 5. Mesh generations and gauges of the calculating model.

2.3. Physical Model and Calculating Models

The governing equations involved in the numerical simulation of DDT include conservation equations and material models, and material models include equations of state and strength models. The partial differential forms of the governing equations for the conservation of mass, momentum and energy are shown in equations (1)-(3).

$$\partial\rho/\partial t + u\cdot\nabla + \sigma\nabla\cdot u = 0 \tag{1}$$

$$\rho(\partial u/\partial t + u\cdot\nabla u) = \nabla\cdot\sigma \tag{2}$$

$$\rho(\partial e/\partial t + u\cdot\nabla e) = \sigma\cdot\psi \tag{3}$$

where ρ is density. u is the particle velocity. e is internal energy. t is time. σ is stress tensor. ψ is strain rate tensor.

The materials include shell, end caps and high-energy propellant in this model. The materials of shell and end caps are $45^{\#}$ steel and stainless steel. The shock equation of state and Johnson-Cook strength model were used in the material models as shown in equations (4)-(6). Some parameters of them used in calculation were listed in Table 1.

$$u_s = c_0 + Su_p \tag{4}$$

$$\sigma_y = (A_1 + B_1\varepsilon^n)(1 + C_1 \ln\dot{\varepsilon}^*)(1 - T^{*m}) \tag{5}$$

$$T^{*m} = (T - T_r)/(T_m - T_r) \tag{6}$$

In equations (4)-(6), u_s is stress wave velocity in solid medium and u_p is particle velocity of wave front. c_0 is sound velocity of the medium. σ_y is strain stress. ε is effective plastic strain. $\dot{\varepsilon}^* = \dot{\varepsilon}/\dot{\varepsilon}_0$ is dimensionless strain rate. T_r is reference temperature. T_m is material melting point. S, A_1, B_1, n, C_1, m are constants.

Table 1. The parameters of $45^{\#}$ steel and stainless steel.

Materials	Density /kg·m^{-3}	c_0 /cm·μs^{-1}	s	Gruneisen coefficient	Shear Modulus /GPa	Yield Stress /GPa
$45^{\#}$ steel	7900	0.457	1.49	2.17	81.8	0.35
Stainless steel	7750	0.457	1.49	2.17	81.8	1.54

It is very complex to establish the propellant material model. Not only the state and strength of propellant but also the process of DDT should be described. Ignition and growth reactive flow model has been widely used in shock ignition and DDT of explosives and propellants, which was used as the equation of state model of propellant materials. The model includes JWL equations describing the state of unreacted propellants and products, and Lee-Tarver model describing the reaction rate [12] as shown equations (7)-(8). The first item on the right-hand side of equation (8) is the ignition term representing the creation of hot spots. The second and third terms are the growth and completion terms. The second item of equation (8) represents the process with a lower reaction rate before the formation of the detonation. The third item of equation (8) represents the detonation transition process with rapid reaction. The parameters of ignition and growth reactive flow model was listed in Table 2[13].

$$p_i = A_2\left(1 - \frac{\omega}{R_1\bar{v}}\right)e^{-R_1\bar{v}} + B_2\left(1 - \frac{\omega}{R_2\bar{v}}\right)e^{-R_2\bar{v}} + \frac{\omega\rho_0 e}{\bar{v}} \tag{7}$$

$$\frac{dF}{dt} = \underbrace{I(1-F)^b(\rho/\rho_0 - 1 - a)^x}_{0<F<Figmax} + \underbrace{G_1(1-F)^c F^d P^y}_{0<F<F_{G_1max}} + \underbrace{G_2(1-F)^e F^g P^z}_{F_{G_2min}<F<1} \tag{8}$$

where P is pressure. i is 0 for unreacted propellant. i is 1 for reaction products. $\bar{v}$ is relative volume. A_2, B_2, R_1, R_2 are respectively constants. ω is Gruneisen coefficient. F is the degree of reaction. ρ and ρ_0 are current density and initial density respectively. I, G_1, G_2, a, b, c, d, e, g, x, y and z are respectively constants.

Table 2. The parameters of ignition and growth reactive flow model for high-energy propellant.

I	b	a	x	G_1	c	d	y	G_2	e	g	z
7.4×10^{11}	0.667	0	20	7	0.667	0.111	1.5	800	0.333	1	2

3. Simulation Results And Discussions

Based on the above calculation model, the appropriate mesh size was obtained by calculation, and the results of relevant parameters were obtained, and the DDT process of high-energy propellant was calculated and analyzed.

3.1. Verification of Grid Independence

The numerical simulation regarding grid independence is carried out in order to increase the accuracy of calculation. The sizes of grids in calculations are 5mm, 2mm, 1mm, 0.5mm and 0.25mm respectively.

Figure 6 shows the curves of velocity vs time calculated from different grid sizes. From the figure, it can be seen that the curves of velocity vs time are similar for different grid sizes, but the velocity values at the same time calculated by different grid sizes are different. When the grid sizes are 0.25mm and 5mm, the maximum deviation of velocity values is about 500m/s. It means that the difference of grid sizes could cause calculation deviation. The deviation under different grid sizes are shown in the Table 3. As the grid sizes decrease, the number of grids increase, and the calculation deviation decreases, but the calculation time increases. When the grid size is 0.5 mm, the deviation is 0.35%. The calculation time increases obviously, and the calculation speed is very slow when the grid size is 0.25 mm. Considering the calculation accuracy and cost, the grid size with 0.5mm is the most suitable determination to calculate the model.

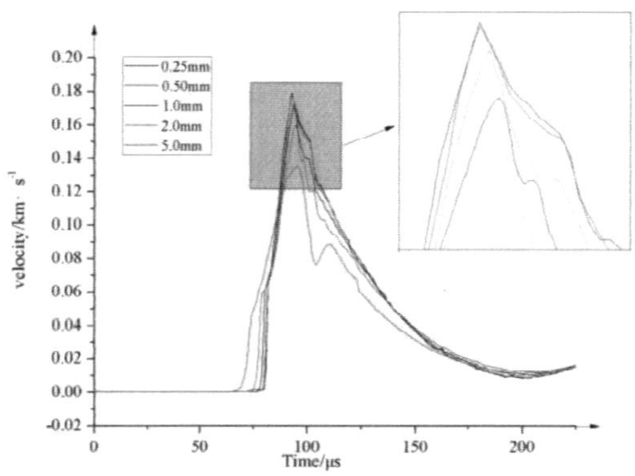

Figure 6. Variation of velocity values with different grid sizes.

Table 3. Deviation from different grid size calculations.

Grid Size /mm	5	2	1	0.5	0.25
Deviation	14.8%	5.2%	3.9%	0.35%	-

3.2. Simulation Results of DDT Parameters

The numerical simulation had been studied according to the experiment, and the results were compared with the above experimental results.

The reaction degree(F) of different reaction periods is different. It is an important parameter to describe DDT of propellant which can be obtained by calculation. When F=0, the propellant doesn't react and the propellant reacts completely when F=1. In order to analyze the change of the reaction degree during the DDT process of propellant and determine the run distance to detonation more accurately, the curves of the reaction degree-time at different positions were shown in Figure 7.

It can be seen that the reaction degree only reaches 40% when the distance is 18.4mm (Gauge#10), and the reaction is complete at 108mm (Gauge#55), but it still needs some reaction time, so the detonation doesn't occur yet. When the distance is 138.3mm (Gauge#70), F increases vertically to 1, and the reaction is complete quickly. The detonation occurs at this position, and the reaction time to the position is about 42.6μs. The phenomenon lasts until the whole column finishes the reaction. The stable detonation occurs. Therefore, the run distance and time to detonation of high-energy propellant are about 138.3mm and 42.7μs under this condition.

Figure 8 shows the peak values of pressure and velocity vs distance along the propellant column. It can be seen that the pressure and velocity linearly increase at first, and they suddenly rise at 118mm. When the detonation wave propagates to 138 mm after 42.7μs, the detonation pressure and velocity reach stable values of 35.6GPa and 2899m/s, respectively. Compared with the experimental results, the deviation of detonation velocity is about 1.4%, and the deviation of the run distance to detonation is about 2.4%. Therefore, the model is very effective to calculate DDT of high-energy propellant.

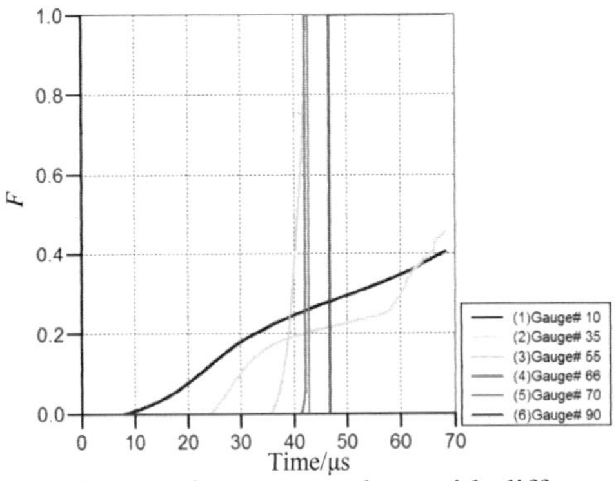

Figure 7. Variation of pressure values with different grid sizes.

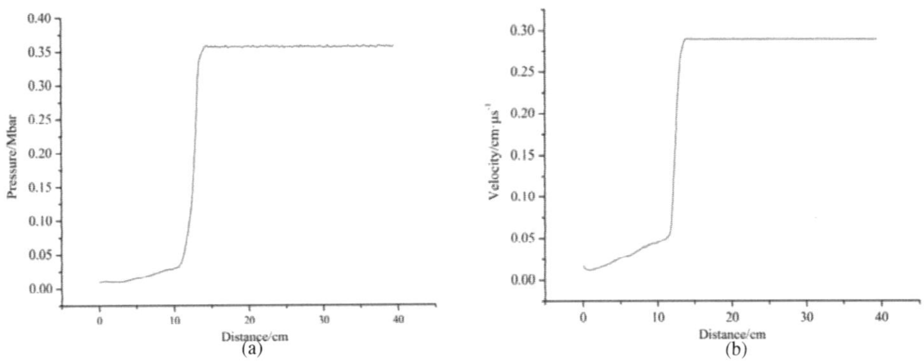

Figure 8. The peak values of pressure and velocity vs distance.

3.3. DDT Process of High-Energy Propellant

The DDT process of propellant was studied by observing the reaction process and analyzing the changes of parameters such as pressure and velocity in propellant.

Figure 9 shows the pressure nephogram at different times. As seen in Figure 9, the high-energy propellant is ignited and burned under the action of ignition energy at the boundary. As the reaction proceeds, the pressure increases, producing a large number of gas products and leading to ignition of more propellant and accelerating the reaction. The reaction is more and more intense, and the accelerated deflagration can eventually lead to shock compression of the unreacted propellant and finally transit to detonation. The width of the high-pressure reaction zone decreases with the increase of the reaction pressure, especially when the detonation occurs, the high-pressure reaction zone becomes very thin. Because when the detonation occurs, the pressure in the reaction zone is very high, and the reaction completes quickly and continue to react forward. Meanwhile, with increase of reaction time, pressure in the tube gradually increases beyond the allowable stress of the tube, the deformation of the tube occurs.

The pressure and velocity at different positions of the propellant varies with time as shown in Fig. 10. It can be seen from the curves that the pressure at the beginning is low and the reaction velocity is subsonic, showing combustion characteristics. As the increase of compressional wave strength, a strong shock wave forms, and the pressure and velocity rise sharply, showing the detonation characteristics. DDT occurs when the accelerating subsonic deflagration wave transmits to a high order detonation wave. The increase of pressure in the tube is not only due to the gas product released by the reaction, but also the restriction of the shell. Meantime, the reaction rate increases with the pressure that makes the velocity increase.

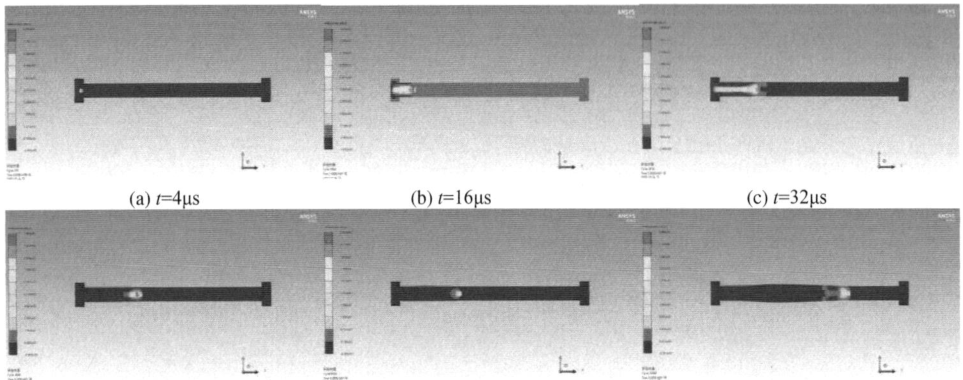

(a) t=4μs (b) t=16μs (c) t=32μs

(d) t=42μs (e) t=44μs (f) t=60μs

Figure 9. Pressure nephogram of propellant at different times

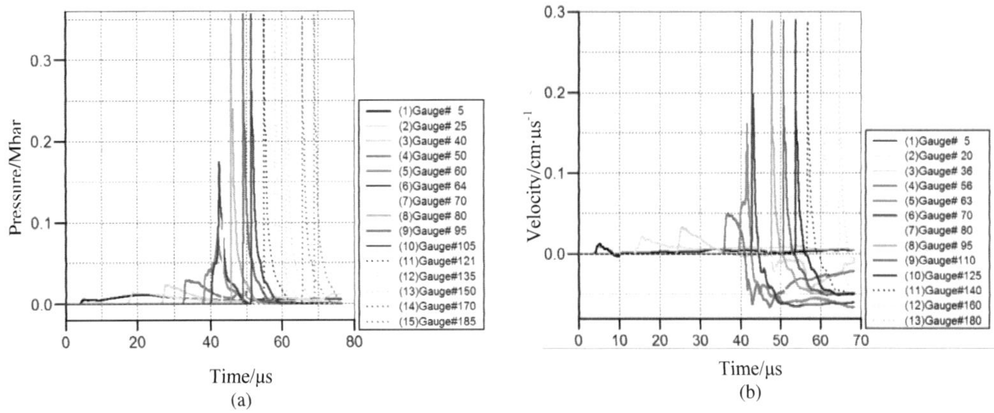

Figure 10. Curves of pressure and velocity vs time at different positions

4. Conclusions

According to the state of high density charge, the calculating model of DDT in high-energy propellant is established, and the process of DDT is analyzed. The results show that DDT could occur when the high-energy propellant ignited under strong confinement. The ignition and growth reactive flow model can describe the reaction process of propellant very well, and the ALE method can be used to solve the DDT problem. The detonation velocity and pressure are 2899m/s and 35.6GPa, and the run distance and time are about 138.3mm and 42.7μs, respectively. The numerical results are in a reasonable agreement with the experimental results. When the high-energy propellant transits from deflagration to detonation, a large amount of energy would be released, so it needs to pay attention to avoid DDT caused by accidental ignition of propellants.

Acknowledgments

Thanks for the support of National Security Project to this subject.

IOP Conf. Series: Materials Science and Engineering **538** (2019) 012051 doi:10.1088/1757-899X/538/1/012051

References

[1] A. Davenas, J Propul Power, 6, 1108 (2003)
[2] T. H. Zhang, T H Bian, Journal of ballistics, 2, (2001)
[3] S. Victor, N. A. Abrukov, O. C. Lukin, Atlanta, 53rd AIAA/SAE/ASEE J P C, (2017)
[4] M. R. Baer, J. W. Nunziato, Int J Multiphas Flow, 12, 861, (1986)
[5] J. M. Powers, D. S. Stewart, H. K. Krier, Combustion Flame, 80, 264, (1990)
[6] B. Narin, Y. Ozyoruk, A. Ulas, J Hazard Mater, 273, 44, (2014)
[7] X. R. Jia, D. X. Li, J. S. Sun, Acta Armamentarii, 1, 46, (1997)
[8] L Chen, F Wang, J. Y. Wu, J Energ Mater, 19, 697, (2011)
[9] G. C. Qin, X Hou, L. Q. Chen, J Solid Rocket Tech., 29, 186, (2006)
[10] N. Qin, L. Q. Liao, P G Jin, J Explo & Prop., 33, 86, (2010)
[11] C. Zhang, L Ma, F. Q. Zhao, J Explo & Prop, 23, 1028 (2015)
[12] F. Garcia, K. S. Vandersall, C. M. Tarver, J Phys CS, 500, 052048, (2014)
[13] J. Y Wu, L Chen, J Y Lu, Acta Armamentarii, 29, 1315, (2008)

IOP Conf. Series: Materials Science and Engineering **538** (2019) 012052 doi:10.1088/1757-899X/538/1/012052

Numerical Study on Tone Noise of Different Thickness Airfoils

Feng Huanhuan, Liu Yong[a], Wang Qi and Zou Sen

Nanchang Hang Kong University, School of Aircraft Engineering, 330063 Nanchang Jiangxi, China

[a]Corresponding author: liuyong@nchu.edu.cn

Abstract. Based on the Lattice Boltzmann Method(LBM)-Large Eddy Simulation(LES), a direct numerical calculation is presented of tone noise generated by flow past symmetric NACA airfoils with different thickness and at a Reynolds number based on chord of $Re = 2.0 \times 10^5$. The computation results indicate that the acoustic source of the airfoil mainly at region of trailing edge and the flow separation. For NACA series airfoils with different thicknesses of the same curvature, the increase of thickness could enlarge pressure fluctuation and produce larger scale of vortices. Noise sources are identified in the region of trailing edge and the flow separation. The sound pressure level spectrum analysis shows that the tone noise gradually disappears and the noise spectrum displays broadband features with the increases of airfoil thickness. The noise directivity shows that the sound pressure radiation around the airfoil characteristic the dipole acoustic field. In addition, data analysis shows that the generation of tone noise is closely related to the vortex shedding of the trailing edge of the airfoil.

1. Introduction

With the wide application of small or micro unmanned aerial vehicles, wind turbines and etal, the problem of airfoil flow around middle and low Reynolds number （$10^4 < Re < 10^6$）has received extensive attention in many military and civil flight engineering applications[1-2]. At these operating conditions, laminar boundary layer transition may occur on the airfoil due to an adverse pressure gradient, which may cause strong tone noise or broadband noise[3]. The increasingly serious noise problem, especially aerodynamic noise of rotating machinery and aircraft has gradually become an important factor affecting human health. Therefore, the research on the problem of airfoil tone noise is of great significance for engineering application problems such as low noise design of related mechanical equipment.

In recent years, many scholars have conducted a lot of repeated research on airfoil tone noise. Arbey & Bataille [4] conducted an experimental study on the noise generated by the NACA0012 airfoil in uniform laminar flow. It is believed that the generation of discrete tone noise at the trailing edge be ascribed to the scattering of instability waves, originating upstream and amplited through the separated shear layer, at the trailing edge. Atassi [5] found that the existence of boundary separation upstream of the trailing edge indicates that the Kelvin-Helmholtz instability as the cause for disturbance amplification. The resulting strong vortical structures convect past the trailing edge at a shedding frequency f_{sh} equal to that of the primary frequency $f_{n_{max}}$. S. Pröbsting & S. Yarusevych [6] studied

the feedback effect of tone noise generated by the airfoil suction surface separation bubble by PIV technique at the low Reynolds number. The results show that disturbance amplification on the suction and pressure surfaces causing roll-up vortices in separation bubbles and separated shear layers. When the vortex do not rupture upstream of the trailing edge, the passage of these structures over the trailing edge generates tonal noise. Huang Qian of Tsinghua University [7] used the large eddy simulation method to carry out the numerical simulation of the serrated trailing edge airfoil based on the NACA0018 airfoil. It was found that the airfoil pure tone noise is closely related to the state near the trailing edge, but did not make further studies.
Although some research has been done on tone noise of airfoils, many problems remain unclear due to the complexity of the process, especially the influence of airfoil geometry on tone noise. Therefore, the numerical calculations are conducted of NACA0008, NACA0012, NACA0015 and NACA0018 airfoils at zero incidence. The influence of airfoil thickness on tone noise was studied and research results have important reference significance for future low noise airfoil design.

2. Numerical Method

2.1. LBM
The Multiple-Relaxation Time (MRT) lattice Boltzmann equation (1) is given in discrete formulation in momentum space as:

$$\left|f_\alpha(x_i + e_\alpha\delta t, t + \delta t)\right\rangle - \left|f_\alpha(x_i, t)\right\rangle = M^{-1}\hat{S}\left[\left|m(x,t)\right\rangle - \left|m^{(eq)}(x_i,t)\right\rangle\right] \quad (1)$$

where δt is the time step; $\{f_\alpha|\alpha = 0,1,\ldots,N\}$ is the discretized distribution functions (DFs), which represents the probability to meet a particle at the position at the position x_i at the time t; $\{e_\alpha|\alpha = 0,1,\ldots,N\}$ is the discrete velocities, the superscript eq represents the equilibrium state; M is the transformation matrix; $\hat{\mathbf{S}}$ is the diagonal relaxation matrix, $m = M \bullet f$ [8].
The discrete velocity model D2Q9[9] is used for the two-dimensional, low Mach number flow. The discrete velocities are given by:

$$e_\alpha = \begin{cases} (0,0) & & \alpha = 0 \\ (\cos\theta_\alpha, \sin\theta_\alpha)c & \theta_\alpha = \dfrac{(\alpha-1)\pi}{2} & \alpha = 1,2,3,4 \\ \sqrt{2}(\cos\theta_\alpha, \sin\theta_\alpha)c & \theta_\alpha = \dfrac{(\alpha-5)\pi}{2} + \dfrac{\pi}{4} & \alpha = 5,6,7,8 \end{cases} \quad (2)$$

Where c is lattice velocity; the velocity distribution function for equilibrium particles is defined as:

$$f_\alpha^{(eq)} = \rho\omega_\alpha[1 + \frac{(e_\alpha \cdot u)^2}{c_s^4} - \frac{u^2}{2c_s^2}] \quad (3)$$

where c_s is the sound speed; ω_α is the weight parameters, $\omega_0 = \frac{4}{9}$, $\omega_{\alpha=1,2,3,4} = \frac{1}{9}$, $\omega_{\alpha=5,6,7,8} = \frac{1}{36}$; ρ is macroscopic density and u is macroscopic velocity. The macroscopic physical quantities can be obtained by the DFs: $\rho = \sum_{\alpha=0}^{8} f_\alpha$, $\rho u = \sum_{\alpha=0}^{8} f_\alpha e_\alpha$, $p = \rho c_s^2$. Assumping that the Knudsen number and Mach number are small enough, the continuum Navier-Stokes equation would be recovered by the Chapman–Enskog expansion.
For the D2Q9 model, the values of M, m, and m^{eq} can be found in[10]. The diagonal collision matrix $\hat{S}$ can be obtained by the linearization analysis:

$$\hat{\mathbf{S}} = \mathrm{diag}(1.0, 1.4, 1.4, 1.0, 1.2, 1.0, 1.2, s_7, s_8) \quad (4)$$

where S_i is collision frequency, and $S_7 = S_8$. The relaxation factors $\tau_i = \frac{1}{S_i}$. The shear viscosity υ can be expressed as:

$$\upsilon = \frac{1}{3}(\tau_7 - \frac{1}{2}) = \frac{1}{3}(\frac{1}{s_7} - \frac{1}{2}), \zeta = \frac{1}{3}(\frac{1}{s_1} - \frac{1}{2}) \tag{5}$$

2.2. LES

The large eddy simulation method assumes that the small scale vortices of the flow below the subgrid scale are similar and isotropic, and thus the subgrid scale model can be used to simulate the effects of the small scale on the large-scale vortices. The eddy viscosity model is the most widely used model for LES. It is only necessary to introduce the eddy viscosity υ_t into the Lattice Boltzmann equation(LBE), without changing the structure of LBE and the algorithm for solving LBE. The eddy viscosity υ_t can be expressed as:

$$\upsilon_t = C_x \Delta^2 |S| \tag{6}$$

where C_x is subgrid model constant; Δ is the grid filter width, which is proportional to the grid size; $|S| = \sqrt{2S_{ij}S_{ij}}$, $S_{ij} = \frac{1}{2}(\partial_i u_j + \partial_j u_i)$; S_{ij} can be obtained directly from m in moment space of the MRT-LBM method. However, in the process of solving N-S equations based on finite difference methods, etc, S_{ij} needs to be obtained by calculating the derivative of the physical quantities of adjacent points. The method of solution to S_{ij} is also one of the advantages of applying the LES method to the LB equation.

In the standard subgrid model, C_x is a constant. The dynamic subgrid model C_x can be obtained by filtering twice on a grid of different scales and then by the least squares method [11]. After obtaining the dynamic eddy viscosity υ_t, it is added to the fluid shear viscosity υ, ie $\upsilon = \upsilon_0 + \upsilon_t$, υ_0 is the hydrodynamic viscosity, and then the diagonal collision matrix $\hat{\mathbf{S}}$ can be determined according to the formula (5).

3. Grid and Acoustics Field Monitoring Point

3.1. Grid Design

The computational domain occupies a $40c \times 40c$ square, where c is airfoil chord. Duo to the multi-scale problem for flow over airfoil, it is not suitable using uniform meshes. The non-uniform quadtree grids with high retrieval efficiency is used to improve computation efficiency in this paper. The mesh in computational domain was shown in Fig.1, where the size of mesh is $5.2e^{-4}c$ near the wall and wake. The number of cell was 17 million, and the maximum of Δx^+ is 50 to ensure that the magnitude of the wall maximum Δy^+ [12]. Therefore, the wall function is not needed in the near wall area of the airfoil, and wall-resolved LES was used.

Acoustic applications are highly sensible to reflections of pressure waves that may occur at the inlet and outlet boundary conditions in compressible unsteady simulations. These reflections may interfere with the pressure measurements. To overcome this issue, the non reflecting Dirichlet boundary condition of velocity and pressure is derived from the local one-dimensional Viscous Free (LODI) equation[13]. Therefore, the inlet boundary is set as a velocity condition and the outlet boundary as a Non-reflective boundary condition. The top and bottom boundaries are set as periodic pairs.

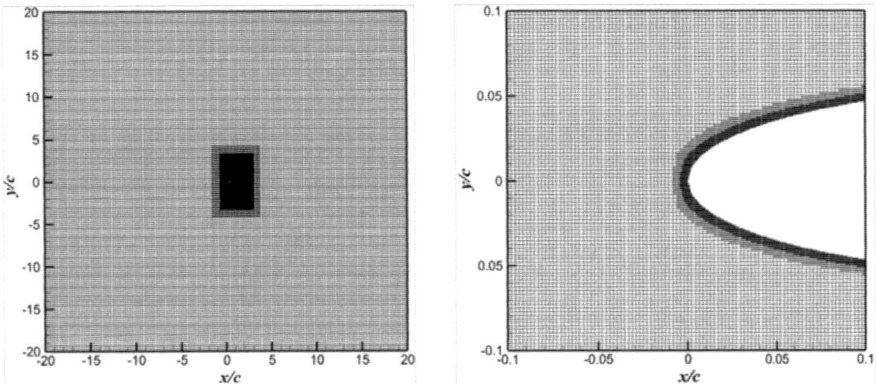

(a) whole domain (b) Grid details around airfoil leading edge

Figure 1. Grid in computational domain

3.2. Acoustic Field Monitoring Point Setup

The LBM-LES method can directly obtain the density fluctuation $\rho`$ of the fluid. The pressure fluctuation is similar to the sound pressure fluctuation far from the airfoil, and according to the formula $p = \rho c_s^2$ ($c_s$ is a constant), so the pressure fluctuation of the fluid is obtained $p` = \rho` c_s^2$, and the pressure fluctuation of the flow field and the sound field can be obtained by the formula[14]. Four monitoring points are placed vertically above and below the trailing edge of the airfoil from the trailing edge 0.5c and 1c, respectively. In addition, with the airfoil leading edge point as the center, 24 monitoring points are uniformly set in the near field of radius $r = 5c$ and the far field of $r = 10c$ as shown in figure 2:

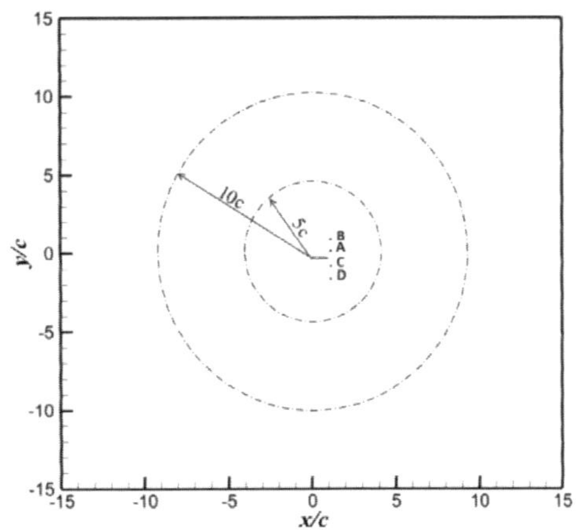

Figure 2. Acoustic field monitoring point distribution

4. Results and Discussion

In order to verify the reliability of the numerical method in this paper, the numerical calculations are performed of NACA0012 airfoil at 2° angle of attack and a Reynolds number based on chord of $Re = 2.0 \times 10^5$. And the calculation results of aerodynamic and noise performance are compared with the reference values in the literature [15-16]. It can be seen from Fig. 3 that the mean pressure coefficient distribution obtained by LBM-LES method agrees well with the experimental values of Mckee[15]. At the same time, it can be seen from Fig.4 that the sound pressure level spectrum of the monitoring point A obtained by LBM-LES method shows obvious spectral characteristics of tone noise, which is in good agreement with the trend of the results obtained by Desquesnes[16] using the DNS method and the maximum peak value of the sound pressure level is basically the same.

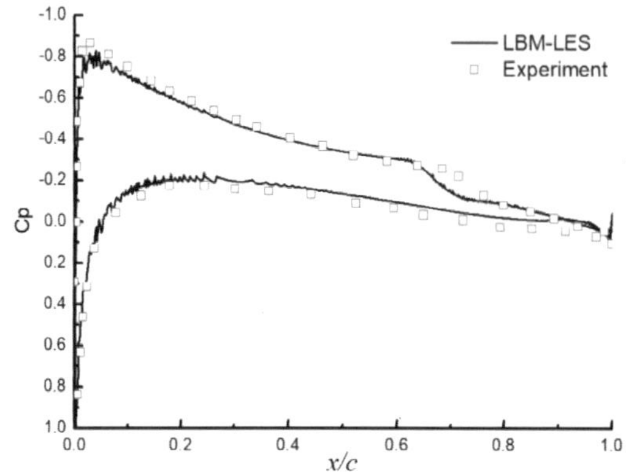

Figure 3. Mean pressure distribution on airfoil surface

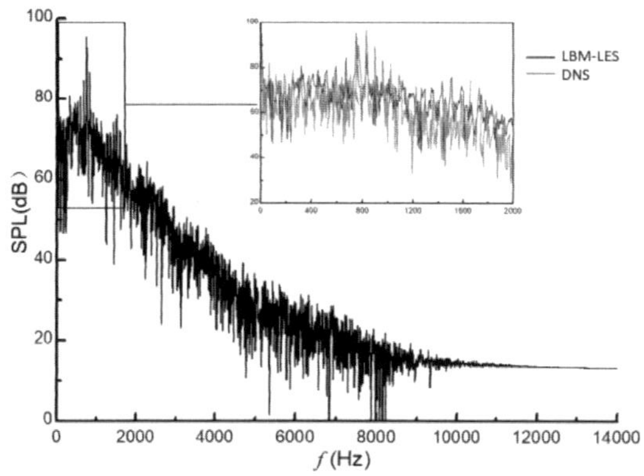

Figure 4. Sound pressure level spectra on point A

The aerodynamic performance and noise characteristics of four different thickness airfoils NACA0008, NACA0012, NACA0015 and NACA0018 were directly calculated by LBM-LES method at 0° angle of attack. The free stream Mach number is 0.1 and the Reynolds number based on chord was specified as $\mathrm{Re}=2.0\times10^{5}$.

Figure 5 shows the instantaneous vorticity and streamline for different thickness airfoils. For small thickness airfoils, the flow is mainly attached and there is no apparent vortex shedding is detected at the trailing edge of the airfoil. The local high curvature produces strong adverse pressure gradient with the increase of airfoil thickness, laminar boundary layer separation may occur on both the upper and lower surfaces of the airfoil. The free shear layer rolls up duo to Kelvin - Helmholtz instability and forms vortex structure. As for the thicker airfoil, more complex separation flow and vortex splitting and merging will be accompanied, resulting in a larger scale vortex structure, and the vortex structure will become more complex and strong.

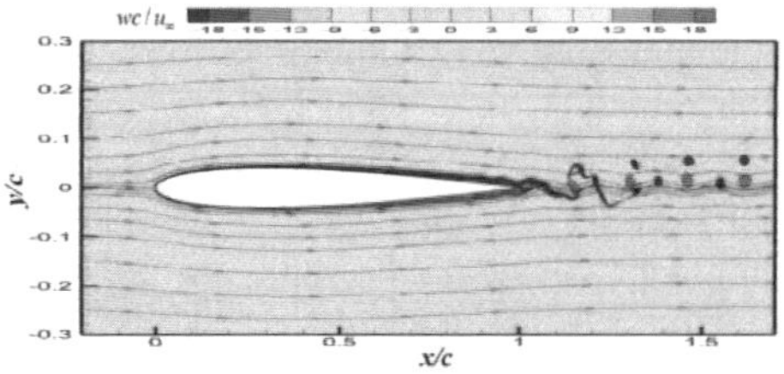

(a) NACA0008

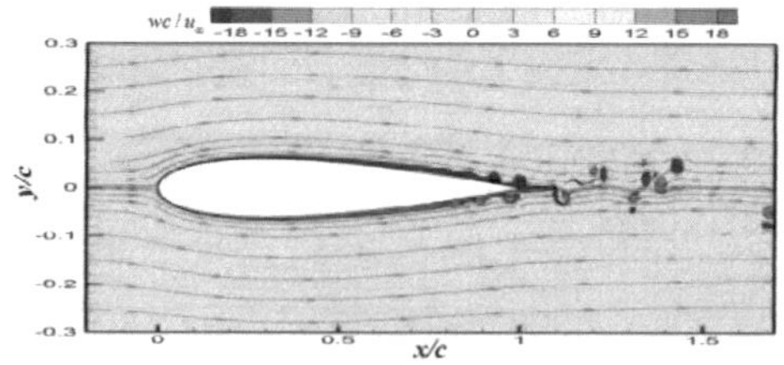

(b) NACA0012

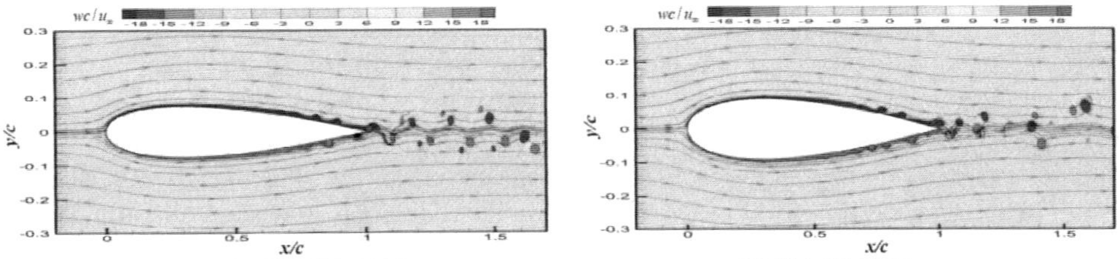

(c) NACA0015 (d) NACA0018

Figure 5. Instantaneous vorticity and streamline contours for different thickness airfoil

Figure 6 visualizes the near-field RMS pressure fluctuation for different thickness airfoils. The root mean squared pressure fluctuations p_{rms} is somewhat considered as a measure of the strength of sound sources. In this paper, p_{rms} is therefore defined by the following way:

$$p_{rms} = \sqrt{(\int_0^T (p - \overline{p})^2 dt / T}$$

Where, p denotes the instantaneous pressure and $\overline{p}$ is the time-averaged pressure. At 0° angle of attack, the RMS pressure distribution is symmetric about the centerline of the wake, and the main source is located in the trailing edge and wake of the airfoil, which corresponds well to the vorticity and streamline contours. This observation indicates that the region of trailing edge and the separation generation represent major sources of sound [17]. With the increase of airfoil thickness, the values of contour levels are also increased indicating that the sound sources are strengthened. For smaller thickness airfoils, the red area at the trailing edge of the NACA0008 airfoil is not obvious. For larger thickness airfoils, a large red area appears on the lower edge of the NACA0018 airfoil. This indicates that with the thickness of the airfoil increases, the pressure fluctuation at the trailing edge of the airfoil and turbulence intensity gradually increases. According to the definition of the root mean square, it is known that aerodynamic noise increases with the increase of airfoil thickness.

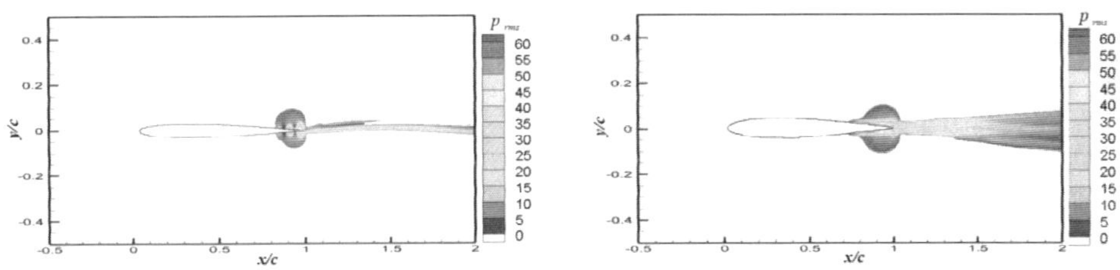

(a) NACA0008 (b) NACA0012

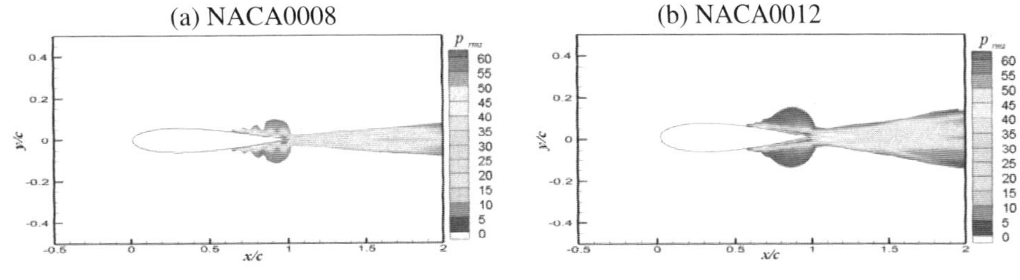

(c) NACA0015 (d) NACA0018

Figure 6. RMS Pressure contours for different thickness airfoil

Figure 7 shows the sound pressure level spectrum on point A for different thickness airfoil. It can be seen from Fig.7 that the first three airfoil sound pressure level spectrums shows obvious spectral characteristics of tone noise at low frequencies, the noise is observed to contain a superposition of discrete tones on a broadband hump [4]. And the spectrum tends to broadband features with the

IOP Conf. Series: Materials Science and Engineering **538** (2019) 012052 doi:10.1088/1757-899X/538/1/012052

frequency increases. For the small-thickness airfoil (such as the NACA0008 airfoil), the feature of tone noise spectrum is significantly higher than that of the medium-thickness airfoil (such as NACA0012). For thicker airfoils (such as the NACA0018 airfoil), there is no obvious primary frequency and sound pressure level peak at low frequency, the high frequency band also shows broadband features. With the thickness of the airfoil increases, the turbulence intensity gradually becomes stronger and produce larger scale of vortices. For larger thickness airfoils, the discrete noise will gradually disappear, and the noise spectrum reflects the feature of broadband noise. This is because the appearance of noise is accompanied by the separation and reattachment of the fluid. As the thickness increases, the fluid flow around the airfoil exhibits a high degree of instability, multi-scale vortex structures are produced in the separation and the trailing edges. The combination and splitting of these different scale vortex structures causes the noise to exhibit broadband features. Comparing the sound pressure level spectrum of different thickness airfoils, it can be known that with the thickness of the airfoil increases, the primary frequency $f_{n_{\max}}$ also changes greatly. The frequency of the primary frequency $f_{n_{\max}}$ is 619 Hz, 813 Hz, 649 Hz, respectively. And the spectral characteristics show a tendency to change from discrete to broadband features.

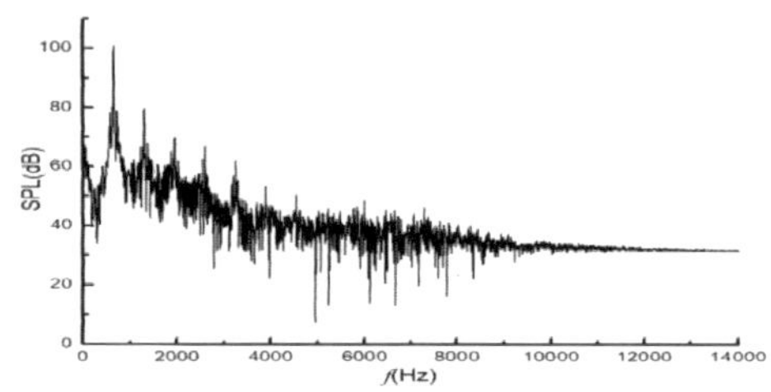

(a) NACA0008

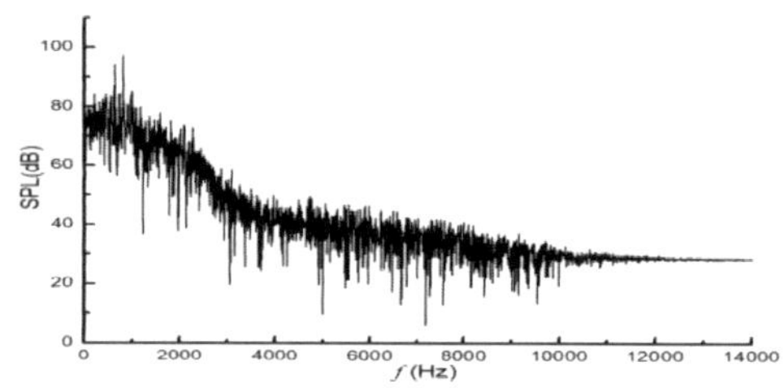

(b) NACA0012

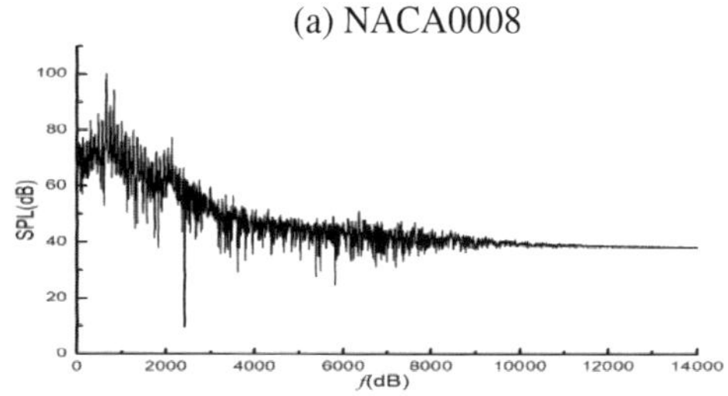

(c) NACA0015

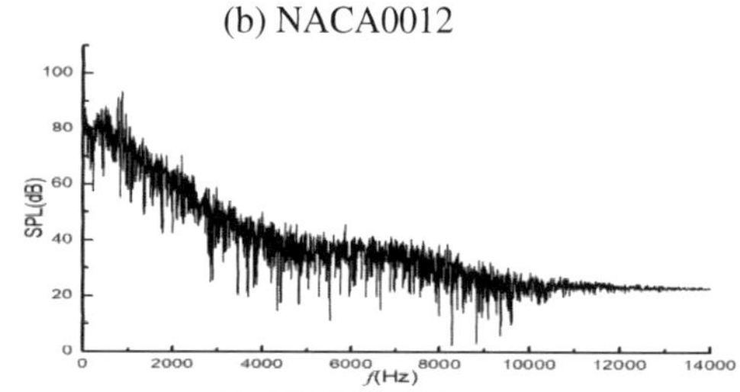

(d) NACA0018

Figure 7. Sound pressure level spectra on point A

Table 1 displays the total sound pressure levels of the four monitoring points. From this table, it can be conclude that the sound pressure level at the monitoring point A is higher than point B, and the sound pressure level at the monitoring point C is higher than point D. This indicates that the sound pressure level on the side of the suction surface is higher than that on the pressure side, but the difference is small. Meanwhile, it can be seen from the table that the sound pressure level increases with the thickness of the airfoil increases.

Table 1. Overall sound pressure level of 4 monitoring points

OASPL（dB）	NACA0008	NACA0012	NACA0015	NACA0018
A	78.2300	91.2709	92.9143	95.1686
B	77.9846	91.2201	92.7415	94.7964
C	75.9843	88.2316	89.7241	91.8628
D	75.6565	88.1651	89.4132	91.5371

The comparison of primary frequency $f_{n_{\max}}$ and vortex shedding frequency f_{sh} on point A are shown in Table 2. From this table, it can be conclude that primary frequency $f_{n_{\max}}$ is closely match the shedding frequency f_{sh}. This indicated that as for the condition of smaller airfoil thickness, vortex shedding at the trailing edge is one of the main factors causing pure tone noise. For larger thickness airfoils (such as NACA0018), the low frequency noise exhibits discrete features, but there is no obvious peak. After 3000Hz, noise spectrum displays broadband features. The main feature of broadband is that the energy at the high frequency position is similar, wide bandwidth and the amplitude is basically within a fixed range.

Table 2. Comparison of primary frequency and vortex shedding frequency on point A

	NACA0008	NACA0012	NACA0015	NACA0018
f_{sh}	619	813	649	_____
$f_{n_{\max}}$	612	802	641	_____

Figure 8 shows the airfoil directivity of OASPL for different thickness airfoil. It can be clearly seen from Fig. 8 that the sound pressure level around the airfoil increases rapidly with the thickness of the airfoil increases. The sound pressure level of the thicker airfoil is 8~12 dB higher than thinner thickness. The sound pressure distribution of the airfoil in the near and far fields is symmetrical. In addition, sound pressure radiation around the airfoil characteristic the dipole acoustic field.

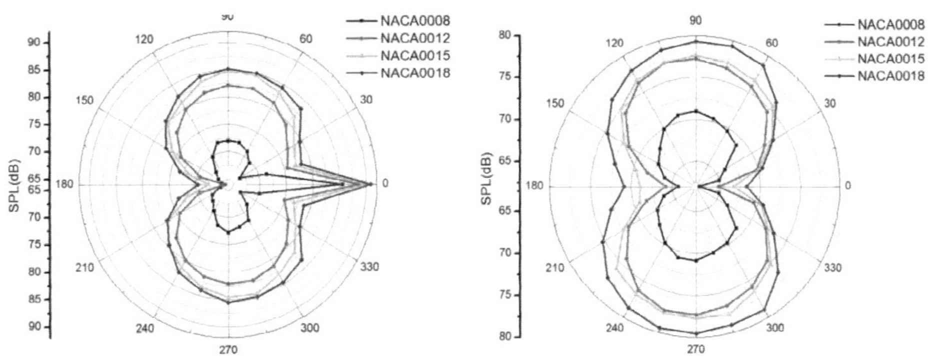

Figure 8. Airfoil directivity of OASPL for different thickness airfoil

5. Conclusion

In this paper, based on the LBM-LES method, the direct numerical calculation of four different thickness airfoil tone noise at low and medium Reynolds numbers is performed. The effects of airfoil thickness on tone noise are studied. The conclusion is as follows:

1. The aerodynamic performance and noise characteristics obtained by numerical simulation of the NACA0012 airfoil are in good agreement with the reference literature to verify the reliability of the LBM-LES method.
2. The RMS pressure contours shows that the main sound source of the airfoil mainly at region of trailing edge and the flow separation. The noise directivity shows that the sound pressure radiation around the airfoil characteristic the dipole acoustic field.
3. For NACA series airfoils with different thicknesses of the same curvature, the increase of thickness could enlarge pressure fluctuation and produce larger scale of vortices. The sound pressure level spectrum analysis shows that the tone noise gradually disappears and the noise spectrum displays broadband features with the increases of airfoil thickness. In addition, data analysis shows that the generation of tone noise is closely related to the vortex shedding of the trailing edge of the airfoil.

Acknowledgment
This work was financially supported by project founded by the Science and Technology Research Project of Jiangxi Education Department (GJJ180511), Jiangxi Province Graduate Innovation Fund Project (No. YC2017-S332).

References

[1] Selig M S, Guglielmo J J, "High-Lift Low Reynolds Number Airfoil Design", *Journal of Aircraft,* 34(1):72-79, (1997).
[2] Mutilin J C, Pauley L L, "Low-Reynolds-number separation on an airfoil", *AIAA J,* 34(8): 1570-1577, (1996).
[3] Amiet R K, "Noise due to turbulent flow past a trailing edge", *Journal of Sound & Vibration,* 47(3):387-393,(1976).
[4] Arbey H, Bataille J, "Noise generated by airfoil profiles placed in a uniform laminar flow", *Journal of Fluid Mechanics,* 134(134):33-47, (2006).
[5] Atassi H, "Feedback in separated flows over symmetric airfoils", *Aiaa Journal*, (1984).
[6] Probsting S, Yarusevych S, "Laminar separation bubble development on an airfoil emitting tonal noise", Journal of Fluid Mechanics, 780:167-191, (2015).
[7] Huang Qian, "A Numerical study of flow around airfoil with serrated trailing edges and the aerodynamic noise based on large eddy simulation [D]. Tsinghua University, (2015).
[8] D. d'Humi`eres, I Ginzburg, M. Krafczyk, et al, "Multiple Relaxation Time Lattice Boltzmann Models in Three Dimensions", Phil. Trans. R. Soc, 360:437-451,(2002).
[9] He Yaling, Wang Yong, Li Qing, "Theory and Application of Lattice Boltzmann Method", Science Press, (2009).
[10] Girimaji S, "Lattice Boltzmann Method: Fundamentals and Engineering Applications with Computer Codes", *Aiaa Journal,* 51(4):398–404, (2013).
[11] Georgiadis N J, Rizzetta DP, Fureby C, "Large-Eddy Simulation: current capabilities, recommended practices, and future research," *AIAA J,* 48(8):1772-1784, (2010).
[12] Chen Li, "Study of the methods of the numerical prediction of underwater flow noise based on the lattice Boltzmann method",Wuhan: Huazhong University of Science & Technology, 60-70, (2013).
[13] Brionnaud R, et al, "Direct Noise Computation with a Lattice-Boltzmann Method and Application to Industrial Test Cases", Aiaa/ceas Aeroacoustics Conference, (2015).
[14] Sandberg R D, Jones L E, Sandham N D, et al, "Direct numerical simulations of tonal noise generated by laminar flow past airfoils", *Journal of Sound & Vibration,* 320(4):838-858,(2009).
[15] Mckee M W, "An exploratory investigation of airfoil sections in low Reynolds number subsonic compressible flows", (1998).
[16] Desquesnes G, Terracol M, Sagaut P, "Numerical investigation of the tone noise mechanism over laminar airfoils," *Journal of Fluid Mechanics,* 591:155-182, (2007).
[17] Jiang M, Li X, Lin D, "Numerical Simulation on the Airfoil Self-Noise at Low Mach Number Flows", Aiaa Aerospace Sciences Meeting Including the New Horizons Forum and Aerospace Exposition, 28-39, (2013).
[18] Arcondoulis E J G, Doolan C J, Zander A C, "Airfoil noise measurements at various angles of attack and low Reynolds number", *Schools & Disciplines*, 1-8, (2009).

IOP Conf. Series: Materials Science and Engineering **538** (2019) 012053 doi:10.1088/1757-899X/538/1/012053

Reduction of Trailing-Edge Noise by Means of Brush-Serrated Coupling Bionic Structure

Yong Wang[a], Yu Zhao, Hao Tian, Kun Zhao and Yu-Bao Song

Key Laboratory of Aerodynamic Noise Control, China Aerodynamics Research and Development Center, Mianyang 621000, China

[a] Corresponding author: nudt604@aliyun.com

Abstract. Bio-inspired noise reduction based on the silent flight characteristics of the owl is an effective passive aerodynamic noise control method. Inspired by the silent flight of owls, this paper investigates trailing edge noise control of the NACA 0012 airfoil by using a brush-serrated coupling bionic structure at chord based Reynolds numbers ranging from Re_c=6×10^5 to Re_c=1.2×10^6. Preliminary experimental results showed that the proposed brush-serrated coupling bionic structure can not only avoid low-frequency vortex shedding tonal noise caused by single serrated trailing edges, but also reduce noise emission in a wider frequency range, which can be widely used in aerodynamic noise control.

1. Introduction

In the past few decades, bio-inspired aerodynamic noise control technique has attracted an increasing attention. It was originally inspired by three distinctive characteristics of the silent flying owls [1] (i.e., leading edge comb-like serrations, trailing edge fringes and velvet-like downy upper surface of wings and legs) and could be roughly classified into four main solutions: leading edge serrations [2-4], trailing edge serrations [5-12], fringe-type extensions (brushes or slits) [13-15], and porous material inspired noise reduction [16-20].

Though a large number of numerical and experimental studies have demonstrated that bio-inspired noise control technique can reduce medium frequency noise, several extra noise (e.g. increase in high-frequency or low-frequency tonal noise) was often observed, which may lead to a large penalty to the overall noise reduction [5, 6, 8, 17, 18]. Therefore, how to improve the design of existing bio-inspired devices to get better noise reduction goals is an interesting direction. In addition, existing bio-inspired devices learned only one of the three special noise reduction adaptations of the owl wings, how to combine or couple them together is still a meaningful direction. These two efforts motivate the experimental study presented in this paper, which coupled trailing edge serrations with brushes to reduce noise radiation. It is shown from the preliminary experimental results that the proposed coupling structure could suppress both high-frequency noise increase and extraneous low-frequency tonal noise caused by single trailing edge serrations.

2. Experimental Setup and Data Processing

All experimental tests were conducted in the 0.55 m × 0.4 m aeroacoustic wind tunnel at China Aerodynamics Research and Development Center (CARDC). The open test section of this wind tunnel, with dimension of 0.55 m (width) × 0.4 m (height) × 1.4 m (length), is surrounded by a 5.5 m × 3.7 m

× 4 m anechoic chamber. The walls of the anechoic chamber are acoustically treated with glass wool wedges, leading to a background noise of less than 76 dB(A) (measured at 80 m/s, 2 m away from the nozzle centerline) and cut-off frequency of 100 Hz. This wind tunnel can achieve a freestream turbulence intensity of about 0.2% and flow speeds of 8-100 m/s.

The experimental model under investigation is a symmetric 2D NACA 0012 airfoil, with chord length of c=300 mm and spanwise length of L=398 mm. As shown in Figure 1, the airfoil model is composed of a main steel body and three trailing edges: one with a straight, unserrated configuration (denoted as 'baseline') and two with directly cutting serrations. These cutting serrations offer better structural strength and integrity [7, 10-12], both of which have the same root-to-peak amplitude of H=30 mm. Two different serration wavelengths are compared in this study: one with λ=12 mm (termed as λ12H30 or wide serrations) and the other with λ=6 mm (termed as λ6H30 or narrow serrations).

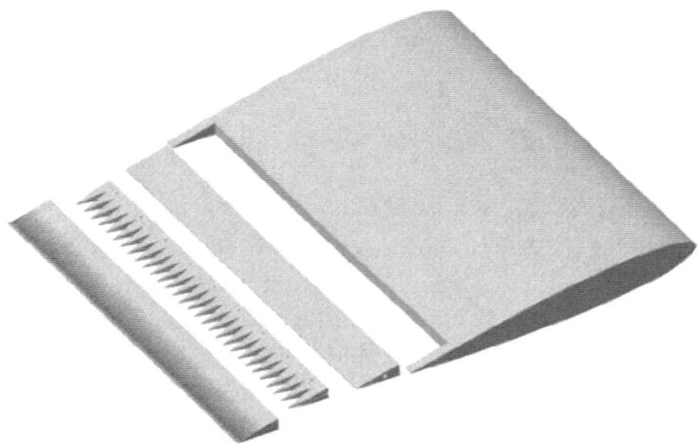

Figure 1. Illustrations of the NACA 0012 airfoil with three trailing edges (right to left: baseline, wide serrations and narrow serrations).

The proposed brush-serrated coupling bionic structure was achieved by inserting brushes into the trailing edge serrations, and thus a slot of 0.5 mm in thickness was made in the downstream face of the cutting serrations. As can be seen in Figure 2, the inserted brushes are a set of soft silicone rubbers.

Figure 2. Photograph of the tested brushes, separated (top) or coupled into the serrations (down).

Far-field noise measurements were made using a single G.R.A.S 46AE microphone, situated at a distance of 1.5 m from the airfoil trailing edge and mid-span. A detail sketch of the experimental setup is illustrated in Figure 3.

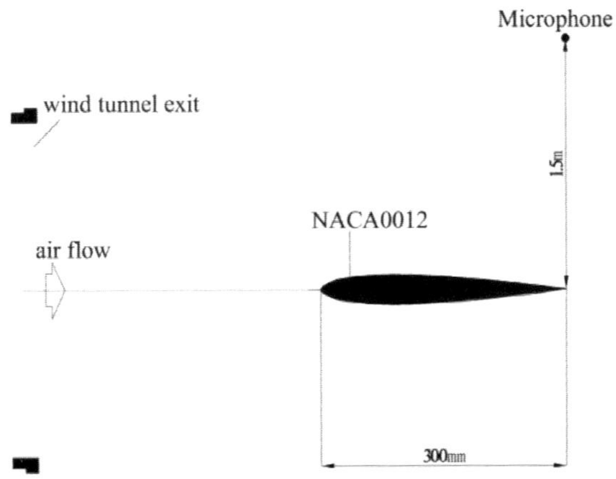

Figure 3. Sketch of the experimental setup (not in scale).

Noise data were acquired at a sampling frequency of 51.2 kHz and sampling time of 30 s. 1/3-octave Sound Pressure Level (SPL) was computed with hanning window function and a window size of 4,096 data points. Acoustic data were recorded at angle of attack α=0° and four mean flow speeds of V_∞=30, 40, 50 and 60 m/s, corresponding to the chord-based Reynolds numbers ranged from $Re_c=6\times10^5$ to $Re_c=1.2\times10^6$.

3. Experimental Results

3.1. Analysis of the Baseline Case

This section presents the baseline results, corresponding to the case when the main body of the NACA 0012 airfoil is equipped with the straight, unserrated trailing edge. The SPL plot of this case under different flow speeds is shown in Figure 4. From this figure, we can see that the NACA 0012 airfoil generates broadband self-noise at the investigated flow speeds or Reynolds numbers, and the noise generally increases with increasing flow speeds.

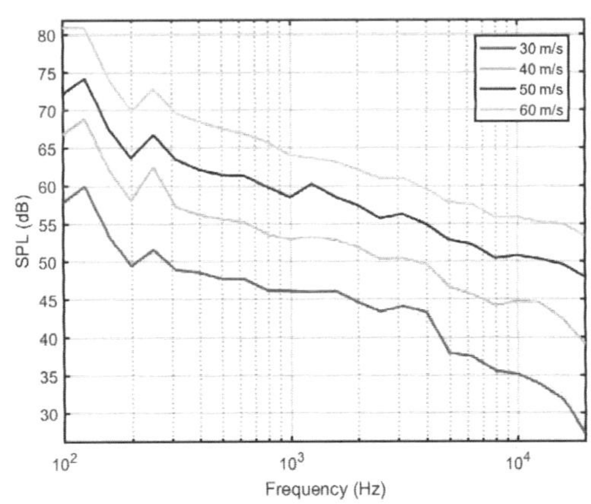

Figure 4. Acoustic spectra of the baseline NACA 0012 airfoil under different flow speeds.

3.2. Analysis of the Single Trailing Edge Serrations

Figure 5 shows the noise results produced by the two single serrated trailing edges under flow speeds of V_∞=30 m/s to V_∞=60 m/s, respectively. Several interesting observations can be seen from this figure: 1). Trailing edge serrations can reduce broadband noise at mid frequencies and the level of noise reduction decrease with the increasing of flow speeds. 2). Trailing edge serrations increase noise at high frequencies, similar as several previous observations [5, 6, 8]. 3). Single trailing edge serrations produce extraneous low-frequency narrow-band tonal noise (typically 10-25 dB above the baseline, similar as observations in [7, 10-12]), which seriously penalize the overall noise reduction. Moreover, low-frequency tonal noise increases with flow speed, while decreases with increasing wavelength of the serrations λ. To assess the source of the tonal noise, the difference of SPL between the trailing edge serrations and the baseline trailing edge Δ SPL are plotted in Figure 6 as a function of Strouhal number $St_\varepsilon = f\varepsilon / V_\infty$, based on the bluntness ε=8.5 mm of the trailing edge serrations at the root region. The positive value of Δ SPL represents noise increase, while negative value represents noise reduction. From this figure, we can see that the spectral hump collapses into nearly constant St_ε values under different flow speeds. Therefore, there is strong evidence that the tonal noise is generated from the vortex roll-up and shedding within each sawtooth gap of the serrations. 4). Wide serrations slightly outperforms narrow serrations, since the former has similar broadband noise reduce at mid frequencies while less noise increase at low frequencies and high frequencies.

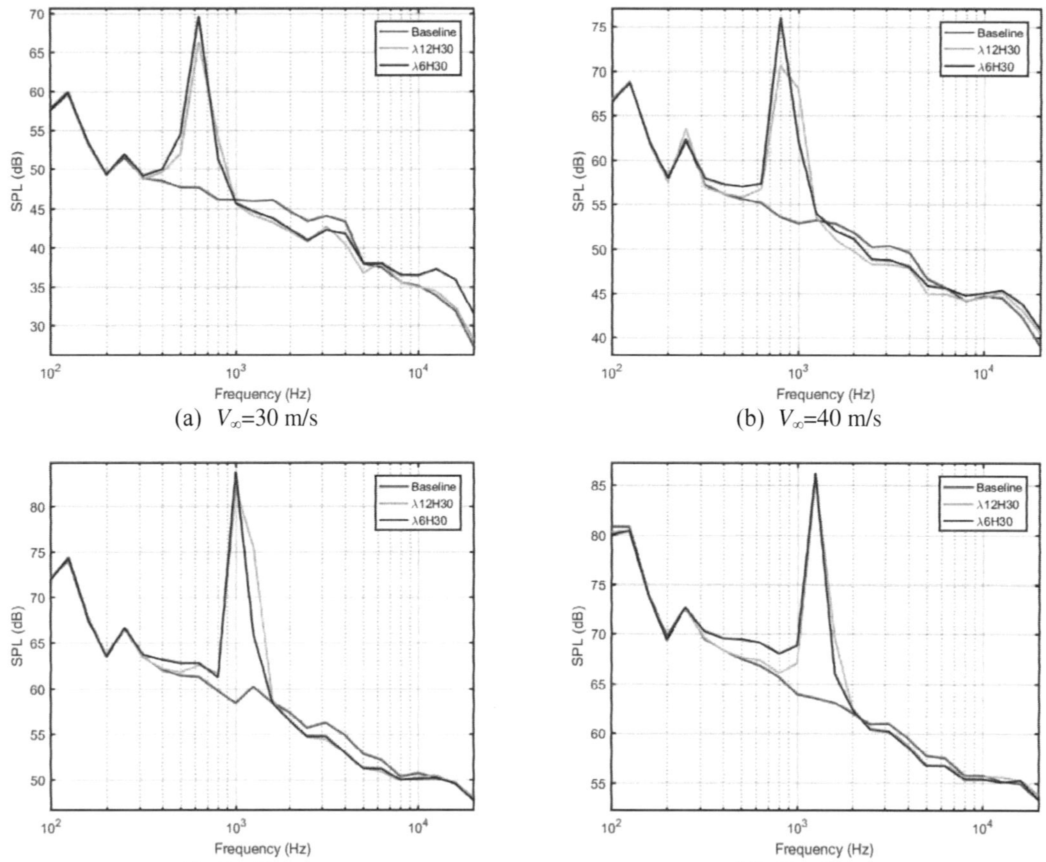

(a) V_∞=30 m/s (b) V_∞=40 m/s

(c) V_∞=50 m/s (d) V_∞=60 m/s

Figure 5. The comparisons of noise spectra between baseline trailing edge and single serrated trailing edges under different flow speeds.

3.3. Analysis of the Brush-Serrated Coupling Bionic Structure

Figure 7 shows the noise reduction of trailing edge sawtooth serrations coupled with brushes when used onto the wide serrations and narrow serrations, respectively. These experimental results reveal several interesting points: 1). Sawtooth serrations coupled with brushes maintain similar broadband noise reductions as the single trailing edge serrations at mid frequencies. 2). The brush-serrated coupling structure can also suppress noise increase at high frequencies. 3). The coupling structure can significantly suppress the amplitude and frequency of the narrow-band tonal noise components originating from the sawtooth gaps of the serrated trailing edges. The reason may be due to the fact that the coupled or inserted brushes inhibit the growth of the periodic oscillatory flow and vortical structures between each sawtooth gap and at the near wake. 4). Different to the case of single trailing edge serrations, narrow serrations when coupled with brushes obtain slightly better performance.

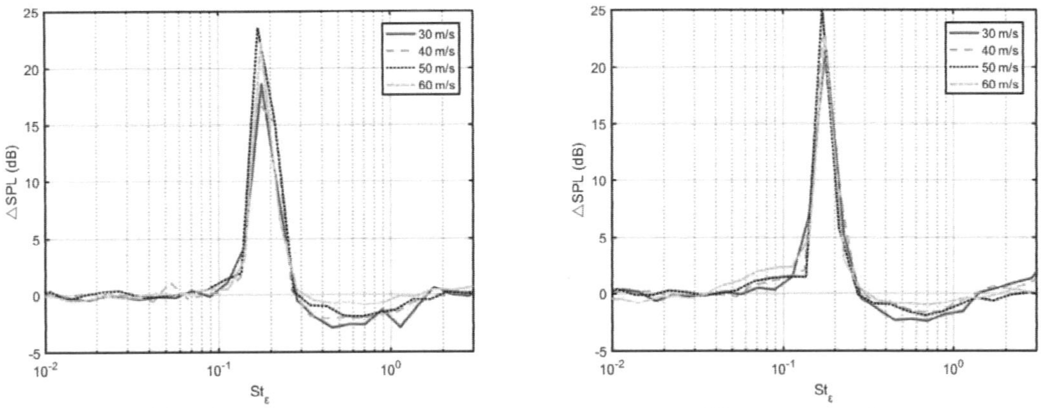

(a) wide serrations (b) narrow serrations

Figure 6. △SPL as a function of Strouhal number St_ε.

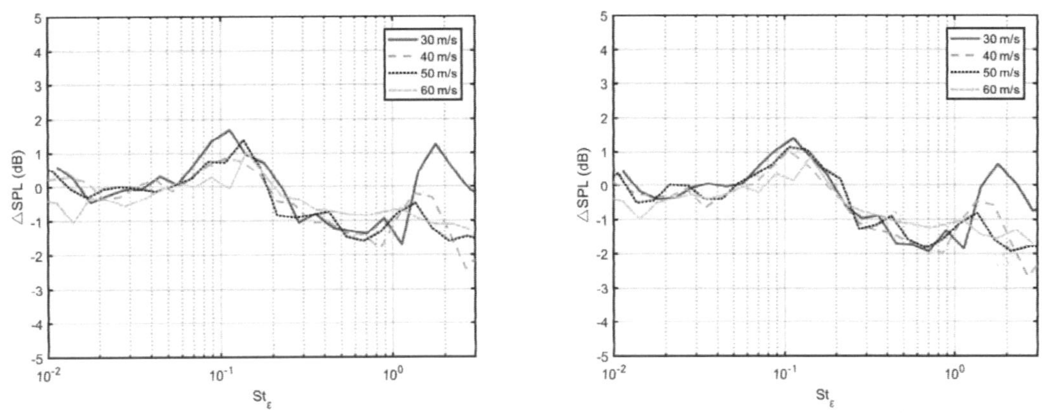

(a) wide serrations (b) narrow serrations

Figure 7. △SPL as a function of Strouhal number St_ε for the brush-serrated coupling bionic structure when used onto the different serrations.

4. Conclusions

In this paper, the preliminary results of an experimental study on noise reduction by using trailing edge serrations coupled with brushes have been described. Experimental results demonstrated that the proposed brush-serrated coupling structure could give better noise reduction, which not only obtain slightly wider broadband noise reduction, but also suppress extraneous increase of both high-frequency noise and low-frequency tonal noise. In the future, parameter influence of the brushes in detail, as well as the aerodynamic effects and underlying noise reduction mechanism will be further studied.

Acknowledgments

This work was sponsored by the National Science Foundation of China, under the grant number of 11602290.

IOP Conf. Series: Materials Science and Engineering **538** (2019) 012053 doi:10.1088/1757-899X/538/1/012053

References

[1] R.R. Graham, Journal of the Royal Aeronautical Society, 286, 837-843 (1934)
[2] K.L. Hansen, R.M. Kelso, C.J. Doolan, *16th AIAA/CEAS Aeroacoustics Conference*, 3700 (2010)
[3] S. Narayanan, P. Chaitanya, S. Haeri, P. Joseph, J.W. Kim, et al., Physics of Fluids, 27 (2), 25109 (2015)
[4] T.P. Chong, A. Vathylakis, A. McEwen, F. Kemsley, C. Muhammad, et al., *21st AIAA/CEAS Aeroacoustics Conference*, 2200 (2015)
[5] S. Oerlemans, M. Fisher, T. Maeder, K. Korler, AIAA Journal, 47, 1470-1481 (2009)
[6] M. Gruber, P.F. Joseph, T.P. Chong, 16th AIAA/CEAS Aeroacoustic Conference and Exhibit, 3803 (2010)
[7] T.P. Chong, P.F. Joseph, M. Gruber, *18th AIAA/CEAS Aeroacoustics Conference*, 2185 (2012)
[8] D.J. Moreau, C.J. Doolan, AIAA Journal, 51, 2513-2522 (2013)
[9] T.P. Chong, P.F. Joseph, Journal of Sound and Vibration, 332, 6335-6358 (2013)
[10] T.P. Chong, A. Vathylakis, P.F. Joseph, M. Gruber, AIAA Journal, 51, 2665-2677 (2013)
[11] T.P. Chong, P.F. Joseph, M. Gruber, Applied Acoustics, 74, 607-613 (2013)
[12] A. Vathylakis, T.P. Chong, P.F. Joseph, AIAA Journal, 53, 3379-3394 (2015)
[13] M. Herr, W. Dobrzynski, AIAA Journal, 43, 1167-1175 (2005)
[14] M. Herr, 13th AIAA/CEAS Aeroacoustics Conference, 3470 (2007)
[15] A. Finez, E. Jondeau, M. Roger, M.C. Jacob, *16th AIAA/CEAS Aeroacoustics Conference and Exhibit*, 3980 (2010)
[16] E. Sarradj, T. Geyer, 13th AIAA/CEAS Aeroacoustics Conference, 3719 (2007)
[17] T. Geyer, E. Sarradj, C. Fritzsche, *15th AIAA/CEAS Aeroacoustics Conference*, 3392 (2009)
[18] M. Herr, J. Reichenberger, *17th AIAA/CEAS Aeroacoustics Conference*, 2780 (2011)
[19] J.W. Jaworski, N. Peake, Journal of Fluid Mechanics, 723, 456-479 (2013)
[20] M. Herr, K.S. Rossignol, J. Delfs, M. Mößner, N. Lippitz, *20th AIAA/CEAS Aeroacoustic Conference*, 3041 (2014)

IOP Conf. Series: Materials Science and Engineering **538** (2019) 012054 doi:10.1088/1757-899X/538/1/012054

Research on Vulnerability and Damage Assessment of Tactical Ballistic Missile

Hongping Gu[1], Chun Cheng[2,a], Aixi Sun[2], Xiaoyun Lei[2], Yijian Shi[2] and Fenghua Huang[3]

[1]Xi'an Modern Chemistry Research Institute, Xi'an 710065, China
[2]School of Mechanical Engineering, Nanjing University of Science and Technology, Nanjing 210094, China
[3]Shanghai branch, No. 701 Research and Development Institute, China Ship Industry Corporation, Shanghai 201102, China

[a]Corresponding author: xiangchun893@163.com

Abstract. The threat of TBM (Tactical Ballistic Missile) on the battlefield makes it an important intercept target of anti missile system. In order to effectively evaluate the damage degree of TBM when intercepted, the system and function characteristics of TBM were analyzed with "Iskander" TBM as an example, the missile vulnerability model was constructed, the vulnerability and damage assessment program was developed, and the damage assessment test methods of TBM were summarized. The research results will help to carry out real-time evaluation of the damage effect of antimissile system towards TBM.

1. Introduction

TBM is mainly used to assault high value targets such as enemy groups, missile launching positions and air defense positions. It has become one of the most threatening weapons on the battlefield and the competing development of many countries because of its long range, high hit accuracy, great power and strong penetration ability[1]. Looking at the development of TBM on the world, U.S. Army Tactical Ballistic Missile System, the only active tactical ballistic missile system in service, can carry biochemical warheads, lethal explosive warheads, ground penetrating warheads to destroy hard and underground targets, cluster sub-carrier warheads to attack airport runways and precision terminal guided sub-carrier warheads, which launched using M270 Multiple-barrel rocket launcher. At present, the countries and regions that have purchased the missile system are the United States, Bahrain, France, Germany, Greece, South Korea, Turkey, Poland, Taiwan and so on. In the Gulf and Iraq wars, the United States used about 560 of these missiles to destroy enemy airports, air defense systems and command centers and other targets. The United Nations Department of Defense Strategic Capability Office (SCO) begun to implement the army's new long-range tactical ballistic missile project (the Army's Long-Range Precision Firepower project) to meet the demand for precision strike capability in future "multi-area warfare". The Russian "Iskander" TBM can carry a variety of warheads, such as cluster munitions, high explosive bombs, penetrating munitions, ground penetrating bombs, air combustion bombs and electromagnetic pulse bombs, which can strike command posts and communication hubs, large group armies, fire strike equipment, air defense and anti-missile facilities, berth at airports, aircraft and helicopters, important industrial and energy facilities and other targets.

The India TBM mainly includes the "fire" series and the "Earth" short range series. The ROK Army is mainly equipped with "Xuanwu-2A" and "Xuanwu-2B" ballistic missiles with missile bodies similar to the Russian-made "Iskander" tactical missiles. DPRK's active tactical ballistic missiles include short-range missiles Mars-5, Mars-6, Scud-D and KN-02, medium-range ballistic missiles Labor-1, Labor-2, medium and long-range Taepodong-1, Taepodong-2 and KN-08[3,4]. France is mainly equipped with Hades tactical ballistic missiles[5].
With the TBM playing a higher tactical strike capability on the battlefield, the anti-missile system of intercepting TBM has been gradually valued and developed by many countries[6-8]. Vulnerability analysis and damage assessment of TBM has gradually become a research hotpot in academic field[9-10]. In the present work Russian "Iskander" tactical ballistic missile was taken as an example to research the system and functional characteristics of TBM, develop vulnerability analysis and damage assessment software, investigate damage effect assessment test methods.

2. System and Function Characteristics

"Iskander" TBM is mainly composed of missile body system, guidance system, fuze and warhead system, control system, thrust system and power system. The system structure model is shown in Figure 1.

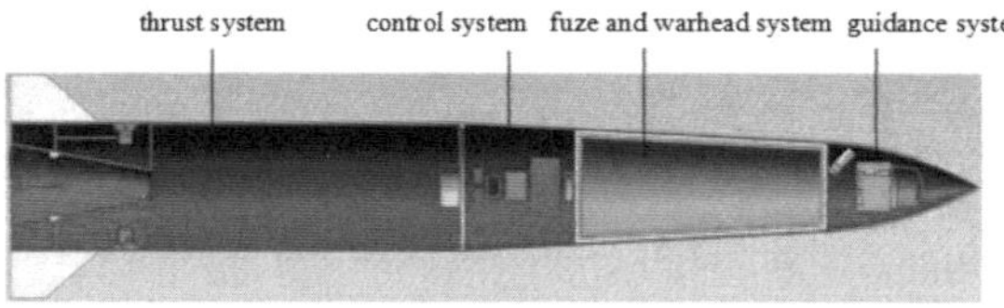

Figure 1. System structure model

As the carrier of warhead and missile equipment, the missile body system guarantees the corresponding working and storage environment, generates aerodynamic force, including maneuvering force and control force, and carries static and dynamic loads under ground working conditions and air flying conditions. The guidance system receives the matching position data from the real-time image processing system, forms the guidance command according to the flight deviation of the actual position and the pre-positioning position of the missile, and sends it to the control system. The control system responds to the command signal sent by the guidance system, and forces the missile to change the flight direction by changing the corresponding normal control force. The control system stabilizes the attitude or angular velocity of the missile and controls the missile according to the signal from the guidance system. Laser fuze is used in the fuze and warhead system. The fuze detects the target by receiving the laser beam which is reflected by the target. It can get rid of the dependence of the fuze on the target and realize the wide range detection of the target to meet the requirement of the missile attack. When the missile and target encounter, the fuze chooses the best time for the warhead to destroy the target effectively. Thrust system is a device that provides flight power for the missile to ensure that the missile obtains the acceleration, velocity and range required by tactical and technical specifications. In some cases, it can also be used to generate control force and maneuvering power.The power system provides electric source, gas and other power sources to the equipment on the missile during the whole process from takeoff to hitting the target.

3. Vulnerability Analysis and Damage Assessment Software

3.1. Software Design

The basic idea of missile vulnerability analysis and damage assessment is as follows. Firstly, read the original data according to the target design drawings and photographs, and draw the 3D graphics of the target. Then according to the principle of material equivalence or function equivalence, combined with the analysis of the key components of missile, the equivalent geometric model of missile vulnerability is given, which provides the target data platform for penetration paths of shooting lines.

Secondly, Monte Carlo method is used to simulate the intersection point coordinates and the penetration paths between the shooting lines and the geometric model of the target. Then, according to the characteristic parameters and damage mechanism of the projectile, the physical damage state of the components encountered by each shooting line in its penetrating path is calculated, and the approximate kill probability of each key component is given by statistical method combined with the damage criterion of the components. Finally, according to the requirement of target combat mission, the functional system is divided, and the functional damage level is defined in each functional system. The corresponding functional reduced-order logic damage tree graph is established. According to the connection relation of key components in the damage tree, the overall damage probability of each functional system is given by a certain logical operation. The vulnerability analysis and damage assessment process of the missile is shown in Figure2.

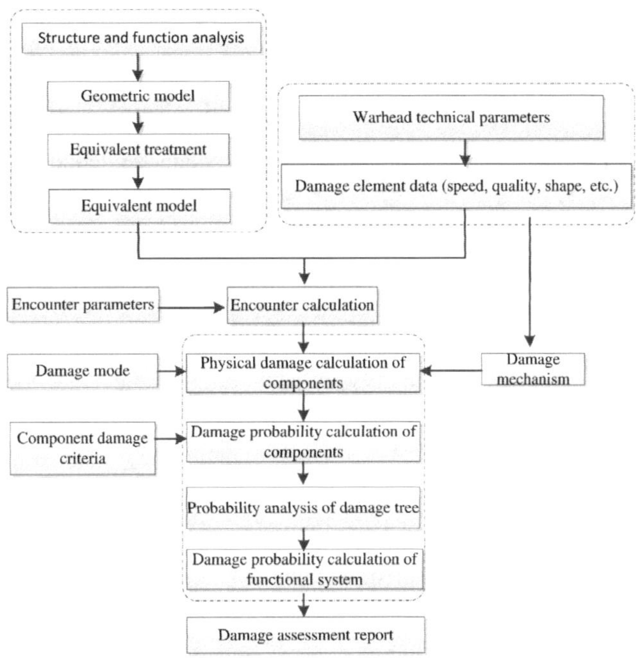

Figure 2. The vulnerability analysis and damage assessment process of the missile

The software system consists of eight function modules: user management module, warhead model parametric design and import module, target model parameter design and import module, damage criterion and criterion database, damage calculation module, missile target damage visualization module, missile target damage result and report output module and help document module. According to the relationship between functional modules, the logical relationship model of the system between them is constructed, as shown in Figure 3.

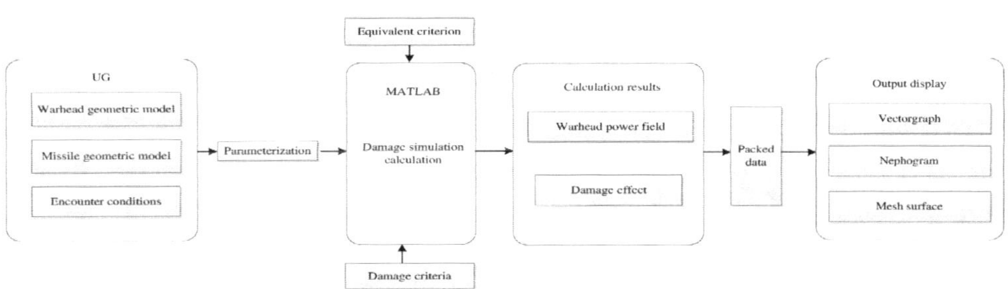

Figure 3. Logical relationship model of the system

IOP Conf. Series: Materials Science and Engineering **538** (2019) 012054 doi:10.1088/1757-899X/538/1/012054

3.2. Missile Vulnerability Analysis and Damage Probability Calculation

3.2.1. Damage Mode. The missile's battlefield mission is to accurately guide the missile warhead to the target area, timely and reliable detonates the warhead to damage the target. The possible damage modes of the missile are: (1) not accurately flying toward the target (yaw), (2) not detonating the warhead (dumb bomb), (3) catastrophic damage (disintegration). The mechanism leading to these damage modes is very complex. For example, under the action of fragments or shock waves, the deformation, bending of missile local body, breakage, deformation of the wing may cause aerodynamic asymmetry and make the missile yaw. The seeker and control system damage, fuel leakage, engine failure may cause the missile not to fly to the target accurately. Damage to the fuze and detonation sequence of a missile may lead to a dummy. If the inflammable and explosive components (such as warheads and fuel tanks) carried by the missile are impacted by high-speed fragments, combustion or explosion may occur, resulting in the disintegration of the whole missile.

3.2.2. Damage Rank. The classification and description of the damage level of the missile is shown in Table 1

Table 1. Classification and description of the damage level

Level	Description	Possible reasons
K	Missile disintegration immediately	The warhead or solid rocket motor is detonated, or the connecting mechanism is interrupted.
F	The missile failed to complete its scheduled task (the missile unable to fly to the target accurately or the warhead failed to detonate reliably).	Guidance system , Control system or fuze function is failure, thrust power of thrust system does not meet the requirement.
P1	The thrust system is slightly damaged, the thrust of the missile is lowered, the acceleration capability is weakened, or the thrust direction is out of control.	Breakdown of engine shell and nozzle leads to gas leakage.
P2	The thrust system is badly damaged, and the missile has no thrust action.	The solid engine is detonated.
C1	The control system is slightly damaged, the servo system is slightly distorted, and the rudder surface can not be accurately deflected.	The control box and signal amplifying circuit are damaged, and the servo mechanism is slightly damaged.
C2	The control system and servo system are seriously damaged, and the missile is out of control.	The control box and signal amplification circuit are damaged, and the servomechanism is seriously damaged.
N1	The guidance system is slightly damaged, and the error of data collection and receiving is increased.	Inertial guidance gyroscopes and accelerometers are slightly damaged, and optical guidance systems are slightly damaged.
N2	The guidance system is badly damaged, the guidance function fails, and the missile loses its target.	The inertial guidance system and optical guidance system are seriously damaged.
W1	Warhead can not be reliably detonated.	The fuze is damaged
W2	The warhead is detonated in advance.	Warhead charge is detonated.

3.2.3. Damage Tree. Damage tree is a functional reduced-order logic tree graph, which can reflect the logical connection between the key components in the system. The overall damage probability of each functional system can be calculated by a certain logical operation using the damage tree. K level damage tree and F level damage tree In the boost phase of TBM are shown in Figures 4 and 5.

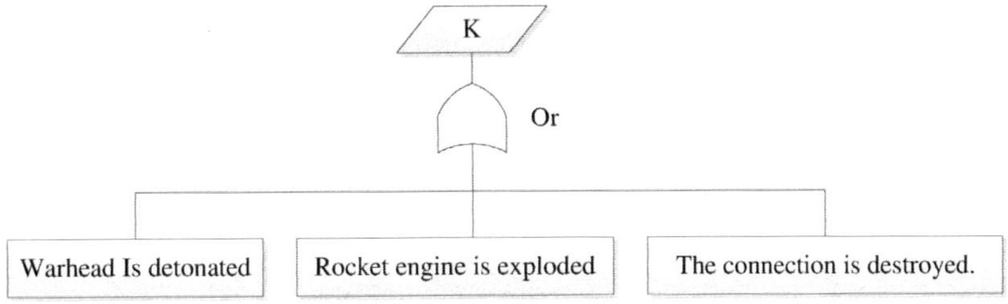

Figure 4. K level damage tree

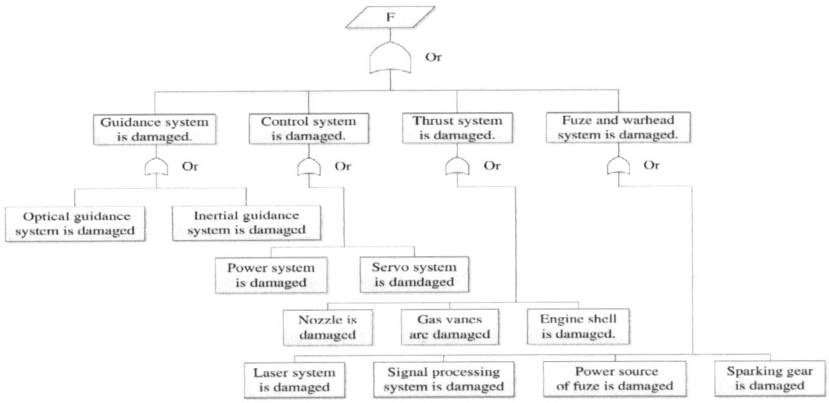

Figure 5. F level damage tree in the boost phase

3.2.4. Damage Probability Calculation. The kill probability of components $P_{k/h}$ can be calculated by the following formula.

$$P_{k/h} = \frac{A_v}{A_p} \tag{1}$$

Where, A_v represents the total vulnerable area of a component. A_p indicates the total area of presentation for a component.

The method of "shooting line scanning" can be used for calculating A_v and A_p when the component hit by a single fragment. In this method, a plane mesh (the length of the mesh is a and the width of the mesh is b) is laid flat on the quadrilateral finite element projection model of the component, and shooting lines (assuming a total of N lines) are generated in each element of the plane mesh. Scanning the target surface and accumulating the number of mesh elements on the surface of the component from shooting line 1 to shooting line N. The number of elements is multiplied by the area of the element (ab) to get the A_p.

A_v Can be calculated by the formula as follows.

$$A_v = \sum_{i=1}^{N} P_{k/hi} \tag{2}$$

where, $P_{k/hi}$ is the kill probability of the missile mesh unit hit by the firing line i and can be calculated according to the kill model. The killing mode (penetration, ignition) of each component and the killing probability of each component under a given attack should be considered in the calculation process. In the process of killing probability calculation, the parameters such as the incident velocity, the incident direction and the mass of the fragment should be considered. When calculating these parameters, the occlusion effect of key components and non-key components should be considered. Therefore, it is necessary to calculate the velocity, direction and mass change of fragments after striking the target skin or components.

According to the damage criterion of each component and the number of effective fragments hitting each component, the damage probability of each component can be calculated. Combined with the damage tree of the target, the damage probability of the whole target can be obtained.as follows.

$$P_t = 1 - \prod_{j=1}^{w}(1 - P^{(j)}) \tag{3}$$

where, P_t and $P^{(j)}$ denote the damage probability of the whole target and the damage probability of component j respectively. w denotes the number of components.

4. Damage Effect Assessment Test Method

Damage assessment of Intercepted TBM is an important link in ballistic missile defense system. The effectiveness of this assessment directly affects the re-interception capability of the system. Damage assessment is generally divided into two levels. The first level is to evaluate whether the interceptor hit the target, and the second level is to evaluate the damage degree of the target payload hit by the interceptor. At present, the damage effect evaluation methods mainly include the radar detection method which is widely used and the optical detection method which is not yet mature.

The use of radar detection for damage assessment mainly includes two aspects. (1) The damage of the missile is judged by judging the yaw trajectory. This judgment method is mainly aimed at the functional or structural damage of the warhead impacted by kinetic energy projectile or the impact of interceptor explosion shock wave, but no explosion occurs. When the warhead is impacted by external impact force, the flight trajectory changes greatly. Meanwhile, the influence of warhead maneuver avoidance trajectory on damage assessment should be considered. (2) When a warhead explodes under the direct impact of a kinetic energy projectile, the warhead and the shell break into debris clouds of different sizes and velocities. If the fragments are small enough, the target will disappear on the display of the tracking radar after the explosion. Thus, the complete damage of the target can be judged. However, if the intercepted target explodes and produces large debris, the radar will mistake it as the target and affect the damage assessment of the target [11].

At present, the main system of missile damage assessment by optical detection means is the space-based damage assessment system being developed by the United States, named SKA(Sky Kill Assessment)[12]. The SKA system mainly relies on a series of infrared sensor networks on commercial satellites as the implementation platform. In the experiment, infrared and visible images of the interceptor impacting the threatening target are obtained by the spaceborne infrared sensor. Based on real-time assessment algorithm, SKA quickly assesses whether interception projectile successfully intercept and destroy the threat targets. The key technologies of the evaluation system are the acquisition and processing of the collision image, and the processing of the spatial, spectral and luminance characteristics of the fireball based on the collision image.

5. Conclusion

In the present work, Russian "Iskander" tactical ballistic missile was taken as an example to research the System and functional characteristics of TBM, vulnerability analysis and damage assessment software was developed, damage effect assessment test methods were investigated. The research results will help to carry out real-time evaluation of the damage effect of antimissile system for tactical ballistic missile interception.

Acknowledgments

The work was supported by National Natural Science Foundation of China (Grant No. 11802141) and Postgraduate Research & Practice Innovation Program of Jiangsu Province (Grant No. KYCX18_0465).

IOP Conf. Series: Materials Science and Engineering **538** (2019) 012054 doi:10.1088/1757-899X/538/1/012054

References

[1] Wang S, Yang J J. Research on Interception Method of Tactical Ballistic Missile in Boost Phase[J]. Winged Missiles Journal, 2009(5):54-57.
[2] Zhang H. Research on the New Type of Tactical Ballistic Missile Development of U.S. Army Against the Background of Multi-domain Battle[J]. Tactical Missile Technology, 2018.
[3] Lv L L, Dai Y L. Overview of the Development of Foreign Tactical Ballistic Missiles in 2017[J]. Modern Military, 2017(z1):182-186.
[4] Lu Y Y, Tang S J. Analysis of the Development Status of Short Range Ground to Ground Tactical Ballistic Missiles Abroad[J]. Winged Missiles Journal, 2009(3):28-31.
[5] Hou B, He Z, Li D, et al. Maximum Correntropy Unscented Kalman Filter for Ballistic Missile Navigation System based on SINS/CNS Deeply Integrated Mode.[J]. Sensors, 2018, 18(6):1724.
[6] Shouqi Z, Yunbin Y, Lixin Q, et al. An Anti Tactical Ballistic Missile (ATBM) Warhead[J]. Propellants Explosives Pyrotechnics, 2015, 25(3):149-152.
[7] Zheng D U, Wang C, Astronautics S O, et al. Comprehensive Design Model of Anti-enhancer of Anti-TBM Interceptor Missile and Its Application[J]. Acta Aeronautica Et Astronautica Sinica, 2013.
[8] Li Z H, Liu C Y, Ni P, et al. Review on multisensor cooperative mission planning in anti-TBM system[J]. Journal of Astronautics, 2016.
[9] Li X. A Study on the Destruction Effect of the "ahead" Projectile to a Missile Target[j]. Acta Armamentarii, 2001.
[10] Li X. The Vulnerability Analysis and Computation for the Missile to the High Explosive Warhead[j]. Journal of Ballistics, 1998.
[11] Xiao-Wei H U, Guo-Ping H U, Wang Y C. Battle Damage Assessment Methods in TBM Reentry Interception[J]. Journal of Air Force Engineering University, 2014.
[12] Zhang Q, Guo J. The Development and Key Technologies of US Space Based Damage Assessment System[J]. Winged Missiles Journal, 2018(3).

IOP Conf. Series: Materials Science and Engineering **538** (2019) 012055 doi:10.1088/1757-899X/538/1/012055

Self-breathing Technology for Spacecraft Packaging Container Gas Drying Method

Yahong Zhu[1*], Xiaohuan Li[1], Shaohua Meng[1] and Chengli Zhang[1]
[1]NO.104, Youyi Road, Haidian, Beijing. PO BOX 5142-88 100094 China

Corresponding author: zhyhzymail@163.com

Abstract. The self-balancing method of packing container is to connect the packing container with the outside world to achieve pressure balance. The gas entering the packing container has been filtered and dried, and the technology of gas filtration has been mature. Under the open environment, the hygroscopicity of hygroscopicity agent needs to be tested. This paper introduces the requirement of hygroscopicant and the test of hygroscopicant under restraint condition, which is used in the design of self-balancing packaging equipment.

1. Introduction

The spacecraft is transported to the destination in the packing container, During the transportation ,the spacecraft must be kept in a clean, dry and pressurized environment[1-2].In this method, packing container pressure balance is proposed, the communication channel between the packing container and the external environment is a self-balance filter container. A moisture absorption agent and an air filter are installed in the filter box. The air entering the box must be filtered and dried by the filter container. In this method, air filtration is a relatively mature technology, but in the open environment, the moisture absorption capacity of the moisture absorbent needs further verification to assess whether it can meet the moisture absorption requirements. Therefore, this method focuses on the selection and verification of the moisture absorbent.

2. Calculation Condition

The requires are as follow:

(1) The volume of the purified packing box is about 240m^3;

(2) Purification time[3] is not less than 12 hours;

(3) In the external temperature for -10℃ ~ 40℃ environment (temperature range can be distinguished between summer and winter), Internal the container to maintain the temperature is 5 ~ 35℃;

(4) The cleanliness in the container: 100,000 grade[4];

(5) The humidity[5]of the the container: < 55%.

3. Principle and calculation

3.1. Principle

The amount of moisture absorbed should be calculated according to 95% of the maximum humidity of air. According to formula (1), the humidity of 12 hours is reduced from 95% to 55%, and the moisture absorbed by temperature change is calculated.

$$H=V_1 \times m_3 \times \rho \tag{1}$$

H——The humidity of 12 hours decreased from 95% to 55%, and the water content needed to be absorbed by temperature change;
V_1——Intake Volume of 240 m^3 container with 12 hours temperature Change;
m_3——Air humidity per cubic metre dropped from 95% to 55% of the water needed to be absorbed;
ρ—— Air density (1.2kg/ m^3);
Calculate the required moisture absorbent capacity as shown in Figure 1.

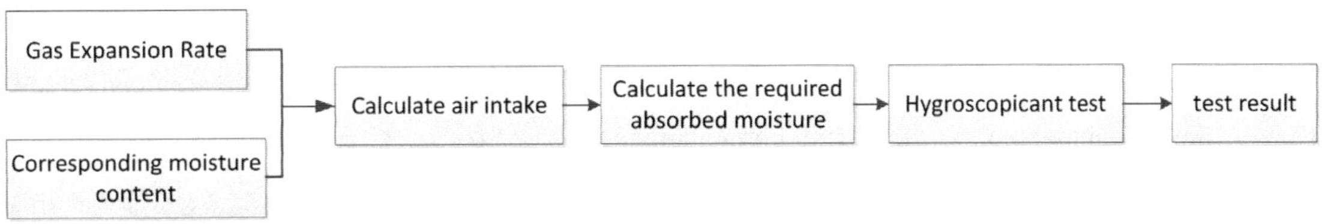

Figure 1. Computation flow

3.2. Gas Expansion Coefficient

According to Gay-Lusak's law (Uoseph Lollis Gay-lussas). When the pressure is constant, the volume of a certain mass of gas is proportional to the thermodynamic temperature. That is V1/T1=V2/T2=... =C constant. The expansion coefficient of the gas is 100/26666 (currently recognized as 1/273.15). When the altitude is fixed, the coefficient of thermal expansion of air volume can be found to be 1/273.15 (according to the law of thermal expansion of gas), that is, Air Volume (Abbreviation: AV) increases by 0.366% (av) for every degree of temperature rise.

3.3. Computation of Gas Volume Intake

The maximum range of temperature variation during transportation is 5 ~35 C, according to the record of temperature change curve in actual transportation, it is known that the temperature in the container rises at the fastest rate of 2℃ per hour and 24℃ at the most in 12 hours, so the air intake is calculated by the temperature difference of 30℃, Calculating gas volume according to formula (2)

$$V_1 = V2 \times (T1\text{-}T2) \times AV \tag{2}$$

V_2 —— cabin volume 240 m^3;
T_1=35℃;
T_2=5℃;
AV = 0.366%;
Data Substitution Formula (2);
V_1= 240 × 30 × 0.366% = 26.35 m^3.

3.4. Moisture Content

Packing container are transported between Beijing and Tianjin. The weather and climate conditions between Beijing and Tianjin are inquired. The highest temperature in summer is about 40 C, and the maximum humidity is 95%.

According to the dredging map, we can find that the moisture content of air is 46.26 g/kg when the temperature is 40℃ and the humidity is 95%. That is, the moisture content of air is 46.26 g/kg when the temperature is 40℃and the humidity is 95%.The moisture content of air is 46.26 g (h1).

According to the dredging map, we can find that the moisture content of air is 25.97 g/kg when the temperature is 40℃ and the humidity is 55% .That is, the moisture content of air is 25.97 g/kg when the temperature is 40℃and the humidity is 55%.The moisture content of air is 25.97g (h2). The weight of water absorbed per cubic meter of air from 95% The moisture content of air （Temperature 40℃ and humidity is 55% humidity to 55% humidity at 40℃ is calculated according to formula (3).

$$m_3=m_1-m_2 \tag{3}$$

m_1——Water content in air per cubic metre at 40℃ and 95% humidity (46.26g);
m_2—— Water content in air per cubic metre at 40℃ and 55% humidity (25.97g);
Data Substitution Formula
$m_3=m_1-m_2=20.29$(g).

3.5. Calculation of Moisture Absorption
Calculating Formula (1) Results Using Moisture Content and Intake data

$H=V_1\times m_3\times \rho=26.35\times 20.29\times 1.2=642$g

That is to say, during 12 hours of continuous transportation, under the most disadvantageous conditions, the moisture absorbed by the transport compartment is 642 g, which decreases from 95% to 55%.

4. Experiment

4.1 Experiment Principle
The air volume is regulated by a supply air equipment. The normal temperature and humidity air is fed into a sealed container to simulate the transportation environment of the equipment. The air volume is regulated at the inlet end and the temperature and humidity are measured. A certain amount of moisture absorbent is placed in the sealed container, and the temperature and humidity of dehumidified air are detected by another temperature and humidity detector at the outlet, thus the moisture absorption performance of raw materials is calculated. At the same time, a recorder, record the change of temperature and humidity time uninterrupted and after moisture absorption moisture. The test prototype is shown in Figure 2. The water content at the inlet minus the water content at the outlet is the water absorbed by the desiccant. The water content in each state is calculated according to formula (4).

$$H_n=V\times \rho\times dn \tag{4}$$

H_n——Absorbed water
V_2——Intake volume (7.2 m^3/h);
ρ——Density of air (1.2g/ m^3);
d——Moisture content;

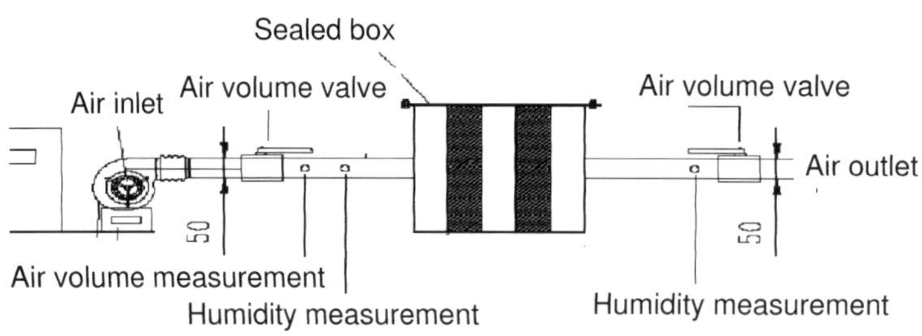

Figure 2. Test device

4.2. Test Conditions
Laboratory environment: Temperature: 5 - 35 C, humidity: 30% - 70%.
According to the conditions, the air intake of the packing container for 12 hours is v=26.35 m^3, while the air intake per hour is v=26.35 m^3.

$26.35\ m^3/12h = 2.2\ m^3/h$;

Input gas should be more than 2.2 m^3/h, At present, the input gas volume is 7.2m^3/h, which meets the requirements.

4.3. Experimental Data

Calculating data by taking hygroscopic material silica gel as an example,

W——Hygroscopic material weight (18.64kg);

T——Test time (m);

RH_1——Inlet relative humidity (59.6%);

d_1——Moisture content of inlet air (=9.77g/kg);

V_1——Inlet air volume (7.2 m3/h);

The above data substitution formula (4),

$H_1=V\times\rho\times d1$;

$H_1=7.2\times1.2\times9.77=84.41g$;

RH_2——Outlet relative humidity (34.3%);

d_2——Moisture content of outlet air (=5.87g/kg);

V_2——Outlet air volume (7.2 m3/h);

The above data substitution formula (4),

$H_2=V\times\rho\times d_2$;

$H_2=7.2\times1.2\times5.87=50.71g$;

According to the formula (5) calculation of water absorbent absorption,

$$H_3= H_1- H_2 \tag{5}$$

$H_3=84.41-50.71=33.7g$;

The maximum rheumatism rate was 34.3% within 35 minutes of the test and remained stable for several hours. The moisture absorption was calculated every 12 hours according to formula (6)

$$H_4=H/T\times60\times12 \tag{6}$$

H_4——12-hour water absorption (g);

H_4=694(g)

694>642, Therefore, the hygroscopicant meets the requirements.

The hygroscopicity of each substance is shown in Table 1.

Table 1. Water absorption capacity

	Silica gel	$CaCl_2$	montmorillonite	mineral
Weight (kg)	18.64	19.67	23.35	22.56
T (m)	35	60	85	48
RH_1 (%)	59.6%	66%	63.6%	34%
d_1 (g/kg)	9.77	9.99	9.21	4.53
H_1 (g)	84.4	86.3	79.6	39.1
RH_2 (%)	34.3%	34.6%	33.2%	16.5%
d_2 (g/kg)	5.87	5.46	5.08	2.25
H_2 (g)	50.7	47.2	43.9	19.4
H (g)	33.7	47.2	35.7	19.7
H_4 (g)	694	470	302	295
conclusion	Yes	No	No	No

The experimental results of several moisture absorbents show that silica gel can meet the requirements of open environment.

5. Concluding Remarks

The key technology of drying air in the self-balancing technology of packing container was tested through calculation and test analysis. The moisture absorbent and weight meeting the requirements were determined through the test results. However, this scheme only considers the gas exchange caused by temperature change, and the air pressure change transportation caused by altitude also needs to be considered. Therefore, this method is suitable for the transportation with little change in altitude. In the future, when this technology is widely used, the change of air pressure along the transport route should also be fully considered. The elevation of the transport route should be measured in detail to calculate the air exchange capacity. The weight of the hygroscopicant needed can be calculated by adding the air exchange capacity caused by the temperature change.

References

[1] GB/T 32301—2015, General requirements of packaging and transportation for spacecraft
[2] ISO14625, space systems-Ground support equipment for use at launch, landing or retrieval sites-General requirement
[3] Steven Buckley.Modular transport container for satellite:United States,Patent US 6237795 B1[P]. P105~106 (2001)
[4] Hu. Qizheng, Yang. Fang, Introduction to Astronautics, P376~379 (2010)
[5] G. Xiao, F.Guo, F Cong, The development of containers for air transport of spacecraft and an evaluation [J]. SEE, 27(6):P795-796.(2010)

IOP Conf. Series: Materials Science and Engineering **538** (2019) 012056 doi:10.1088/1757-899X/538/1/012056

CFD Simulation of Cavitation Bubble Collapse near a Rectangular Groove Wall

Yan Yu[1], Changguo Sun[1], Yazhe Liu[1], Yang Liu[2], Dongsheng Xia[2,a]

[1]Naval Architecture and Ocean Engineering College, Dalian Maritime University, 116026 Dalian, Liaoning, China
[2]Transportation Engineering College, Dalian Maritime University, 116026 Dalian, Liaoning, China

Corresponding author: xia@dlmu.edu.cn

Abstract. The collapse process of a vapour bubble near to a rectangular groove wall is numerically simulated based on the Volume of Fluid (VOF) method with considering bubble surface tension, liquid viscosity and gas compressibility. Under the conditions of different stand-off parameters, evolution of the bubble profile, the jet velocity and the shock pressure around the bubble are obtained. Variation of the maximum jet velocity and the shock pressure around the bubble with the stand-off parameters is analyzed. Moreover, the impact pressure on the groove wall induced by the jet and the shock wave is also discussed, respectively. Cavitation erosion of the groove wall is explored for different stand-off parameters.

1. Introduction

Cavitation as a complex hydrodynamic phenomenon has been a major concern in a variety of fields including hydraulic machinery [1], biomedical treatment [2-3], and ocean engineering. For the majority of the hydraulic machinery cases, cavitation usually resulting in vibration, noise, increase of hydrodynamic drag and damage of component surface is undesirable. On the other hand, cavitation could play a positive role in aspects such as degradation of organics [4], acceleration of chemical reactions [5], disruption of cells [5], and cleaning of surfaces [6]. For most of cavitation cases, the negative or positive effects of cavitation are closely associated with the destructive action of cavitation bubbles in the vicinity of a solid surface. Therefore, understanding the interaction between bubbles and walls is very important to inhibit or utilize cavitation.

Rayleigh firstly established the bubble dynamics equation for a bubble in an infinite liquid field [7], in which liquid viscosity and liquid compressibility was neglected. It is predicted that emission of shock waves as the bubble volume is compressed to the minimum at the final collapse stage. The shock wave propagates in the flow field, causing intense impulse pressure on the nearby wall. Mørch [8] reported the shock waves generated at the collapse of bubble clusters. Brujan [9] found that the shock wave pressure is 1.3±0.3GPa at a distance of 68μm from the bubble wall. The shockwave theory is recognized as one of main mechanisms for occurrence of cavitation erosion. In addition, the micro-jet theory is another important one, which was firstly proposed by Kornfeld and Suvorov [10] in 1944. It is believed that a bubble near a wall might behave an asymmetrical collapse characteristic, leading to generate a liquid jet towards to the wall. After the liquid jet penetrates the bubble, it finally impacts on the wall. The micro-jet theory was experimentally proved the earliest by Benjamin and Ellis in 1966

[11]. A lot of experimental and numerical research achievements relevant to the micro-jet phenomenon have been reported up to now, which provides information enriching the micro-jet theory [12-14].

Although a great number of researches concerning dynamics of cavitation bubbles close to walls have been performed experimentally and numerically for decades, few researches focusing on the interaction between bubbles and walls with special profiles are reported. In this paper, collapse of a cavitation bubble near to a rectangular groove wall is numerically investigated by the homogeneous multiphase model and volume of fluid (VOF) method. It is anticipated to obtain valuable information understanding the effect of wall profile on collapse behavior of a bubble adjacent to the wall.

2. Numerical Method

The physical description of the problem discussed in this paper is the dynamics of a spherical vapor bubble close to a rigid boundary with the micrometer-scale rectangular groove structure. It is assumed that only vapor, not other gas such as non-condensable gas, is contained in the single bubble. The vapor inside the bubble is compressible and assumed to behave like an ideal gas. The liquid surrounding the bubble (e.g. water) is considered as an incompressible newtonian fluid. The fluid flow state is laminar. The viscosity of water and vapor is considered in simulation. The effect of gravity is neglected. Moreover, mass transfer between vapor and water is not taken into account in the paper.

2.1. Governing Equations

The volume of fluid (VOF) method based on homogeneous flow theory was adopted to simulate collapse of the cavitation bubble close to the rigid boundary with micrometre-scale groove structure. In the model, the flow is assumed as a homogenous mixture fluid consisting of liquid and vapor, neglecting inter-phase relative motion. The two components share the same physical velocity and pressure.

The continuity, momentum, and energy equations for the mixture flow are

$$\frac{\partial \rho_{\mathrm{m}}}{\partial t}+\nabla\cdot\left(\rho_{\mathrm{m}}\vec{v}\right)=0 \tag{1}$$

$$\frac{\partial}{\partial t}\left(\rho_{\mathrm{m}}\vec{v}\right)+\nabla\cdot\left(\rho_{\mathrm{m}}\vec{v}\vec{v}\right)=-\nabla p+\nabla\cdot[\mu_{\mathrm{m}}(\nabla\vec{v}+\nabla\vec{v}^{\mathrm{T}})-\frac{2}{3}\mu_{\mathrm{m}}(\nabla\cdot\vec{v})I]+\vec{F}_{\sigma} \tag{2}$$

$$\frac{\partial}{\partial t}\left(\rho_{\mathrm{m}}E\right)+\nabla\cdot\left[\vec{v}\left(\rho_{\mathrm{m}}E+p\right)\right]=\nabla\cdot(k\nabla T) \tag{3}$$

where p is pressure, $\vec{v}$ is velocity vector, ρ_{m} and μ_{m} denotes the density and dynamic viscosity of the mixture, respectively, E is total energy, T is temperature and k is thermal conductivity. In equation (2), I represents the identity matrix.

The surface tension effect of bubble-water interface is included with the continuum surface force (CSF) model[15], which appends the surface tension term $F\sigma$ to the source term of the momentum equation (2) as a volume force.

$$\vec{F}_{\sigma}=\sigma\frac{\rho_{\mathrm{m}}\kappa\nabla\alpha_{\mathrm{l}}}{0.5(\rho_{\mathrm{g}}+\rho_{\mathrm{l}})} \tag{4}$$

where σ is the surface tension, κ is the curvature of the interface, and α_{l} is the volume fraction of the liquid phase in every cell. .

The transportation equation of the volume fraction of liquid phase α_{l} is given by

$$\frac{\partial \alpha_{\mathrm{l}}}{\partial t}+\nabla\cdot\left(\alpha_{\mathrm{l}}\vec{v}\right)=0 \tag{5}$$

As α_l=l or 0 for a grid cell at some time, it means that the grid cell is filled with pure liquid phase or pure gas phase. As $0<\alpha<1$, it represents that the interface between the two phase is located in this grid cell.
The density and the dynamic viscosity of the mixture are defined as

$$\rho_{\rm m} = \alpha_l \rho_l + (1-\alpha_l)\rho_{\rm g} \tag{6}$$

$$\mu_{\rm m} = \alpha_l \mu_l + (1-\alpha_l)\mu_{\rm g} \tag{7}$$

where the subscripts l and g correspond to liquid and vapor, respectively. The value of the liquid density is constant. Considering the compressibility of vapor, the density in the gas phase, ρ_g, is calculated by the ideal gas state equation.

$$\rho_{\rm g} = \frac{M_{\rm m}}{R_{\rm g} T} p \tag{8}$$

where R_g and M_m represent the universal gas constant and the gas molar mass, respectively.

2.2. Computational Domain, Mesh Generation and Boundary Conditions
To decrease computational cost, a 2D axisymmetric model is applied in the numerical simulation because the collapse process has the axial symmetric characteristics. The geometry of the computational model and the boundary condition is shown in figure 1. The rectangular groove shape is the commonly adopted one in a variety of researches concerned with configuration effect. The topography characteristics of rectangular groove wall may be represented by three geometric parameters, i.e. the width w, depth d and interval s of groove unit. In this paper, a distributed rectangular groove is arranged on the bottom rigid boundary. The width w, depth d and interval s of rectangular groove unit are designed to be 40μm (w=d=s=40μm), respectively. The rectangular groove boundary is set as wall with no-slip boundary condition. The left boundary is set as axisymmetric boundary. The pre-existing bubble is placed along the axisymmetric axis at a distance from the solid wall. Considering the profile characteristics of rectangular groove wall, two non-dimensional stand-off parameters γ_1 and γ_2 are defined as

$$\gamma_1 = h_1/r_{\rm max} \tag{9}$$

$$\gamma_2 = h_2/r_{\rm max} \tag{10}$$

where h_1 and h_2 are the distance of the initial bubble center to the wall and the bottom of groove, respectively, and r_{max} is the maximum bubble radius.
Both of the top and right boundaries are set as pressure outlet with the value being equal to 101kPa. The initial radius of cavitation bubble denoted by r_0 is chosen as 0.1mm. The initial pressure in the bubble is set as the saturated vapor pressure of 4247Pa at the ambient temperature of 300K. To decrease the effect of pressure boundary on the computational result, the distance between any two opposite boundaries is designed to be 150 times as the initial size of bubble.

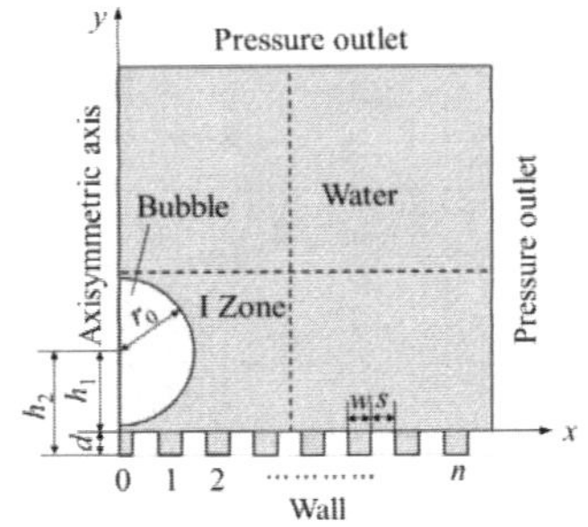

Figure 1. Computational model.

IOP Conf. Series: Materials Science and Engineering **538** (2019) 012056 doi:10.1088/1757-899X/538/1/012056

The whole domain is meshed with the structured quadrilateral cells by using ICEM CFD. The zone surrounding the bubble (I zone) is finely meshed in order to capture the details of the dynamics for bubble collapse. The minimum cell size in I zone is about $5\times5\mu m^2$. The total cell number of the whole domain is more than 240 thousand.

2.3. Discretization Scheme and Solution Algorithm
The finite volume method is adopted to discretize the governing equations. The PISO algorithm is employed for the pressure-velocity coupling in unsteady computation. The gradient term is discretized with the least squares cell based scheme. The liquid volume fraction at a cell interface is computed using Geo-Reconstruct scheme. The density term, the momentum term and the energy term are discretized by the second-order upwind scheme, respectively. The pressure term is calculated by the body-force-weighted scheme. The temporal term is discretized with the first-order implicit scheme.The time step was carefully adjusted between 10^{-9}s and 10^{-10}s during calculation to guarantee convergence.

3. Results and Discussion
In this paper, five cases with different stand-off parameters γ_1 and γ_2 as listed in Table 1 are investigated. Figure 2(a) depicts evolution of the bubble shape during collapse for γ_1=2.0 and γ_2=4.0. Initially, the bubble volume is compressed continuously due to pressure difference between the inside and outside of bubble. As t=9.96μs, the top part of bubble-liquid interface is completely flat and a high pressure zone with the maximum pressure of 1.41MPa exists above the bubble (figure 2(b)). Subsequently, liquid starts to penetrate inside the bubble forming a jet and the bubble takes on a heart-like shape (figure 2(a)). The jet passes through the bubble with the maximum jet velocity of 240m/s at t=10.2326μs, which results in forming a toroidal bubble (figure 2(c)). Then the toroidal bubble performs oscillation. A shock pressure of 5.23GPa is found around the bubble at t=10.2468μs (figure 2(d)). The groove wall below the bubble is subjected to an intensive impact pressure of 230MPa.
After the jet passes through the bubble, a high pressure appears ahead of the jet during the jet moves. The high pressure ahead of the jet continuously decreases as the jet velocity decays. At t=16.400μs, the jet starts to enter the rectangular groove with the jet velocity of 40m/s. As the jet moves downwards, the liquid fluid inside the groove flows outwards along the wall boundary, which leads to form two opposite-sense vortexes inside and outside the groove, respectively. At t=19-20μs, the jet reaches the groove bottom wall with the speed of 20-24m/s (figure 2(e)). The groove wall is subjected to a water-hammer pressure of 30MPa.

Table 1. Simulation conditions.

Case	Stand-off Parameter	Graphical Depiction
1	γ_1=1.0, γ_2=1.4	
2	γ_1=1.1, γ_2=1.5	
3	γ_1=1.5, γ_2=1.9	
4	γ_1=2.0, γ_2=2.4	
5	γ_1=3.0, γ_2=3.4	

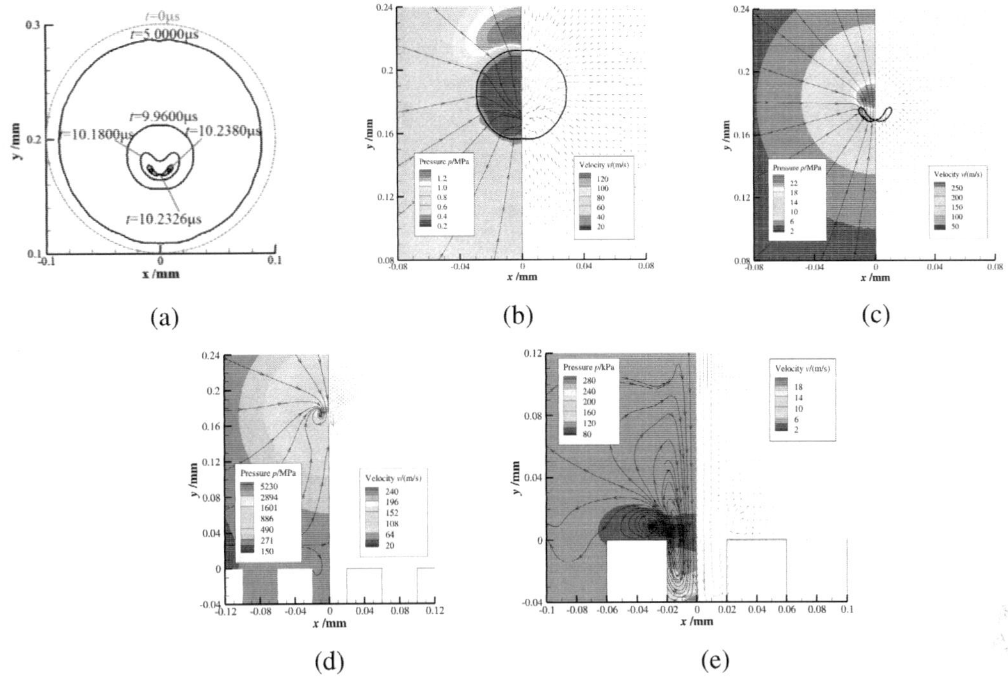

Figure 2. (a) Simulated bubble profile evolution; (b~e) Pressure field and velocity vector field around collapsing bubble at t=9.96μs, t=10.2326μs, t=10.2468μs and t=20μs, respectively, for γ_1=2.0 and γ_2=2.4. (the black solid line is the bubble-liquid interface).

For the case of γ_1=1.1 and γ_2=1.5, initially the deformation of the bottom part of bubble-liquid interface is significantly inhibited comparing with one of the upper part because motion of the liquid between the bubble and wall is impeded (figure 3(a)). The bubble center continuously approaches to the wall. At t=10.16μs, the top of bubble becomes flat meanwhile a high pressure zone with the maximum pressure of 0.89MPa appears above the bubble (figure 3(b)). Then a jet starts to penetrate inside the bubble. When the jet passes through the bubble at t=10.96μs, a toroidal bubble is formed and the jet velocity reaches the maximum value of 130m/s (figure 3(c)). Finally, the toroidal bubble behaves oscillation. The high pressure of 104MPa is formed around the bubble when the bubble rebounds at t=11.1244μs (figure 3(d)). The groove wall below the bubble is subjected to the impact pressure of 46MPa. As the jet reaches the groove wall at t=12.49μs, the jet velocity decays to be about 60m/s, which leads to a water-hammer pressure of 90MPa on the groove wall.

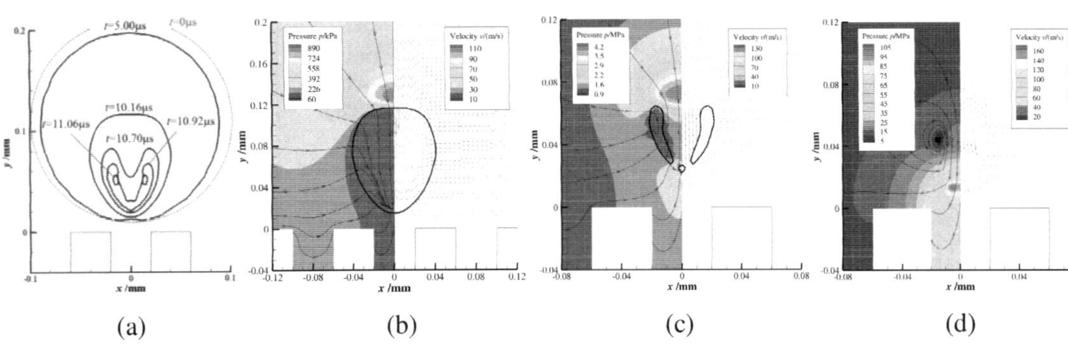

Figure 3. (a) Simulated bubble profile evolution; (b~d) Pressure field and velocity vector field around collapsing bubble at t=10.16μs, t=10.96μs and t=11.1244μs, respectively, for γ_1=1.1 and γ_2=1.5. (the black solid line is the bubble-liquid interface).

As the stand-off parameters γ_1 and γ_2 are further decreased to be 1.0 and 1.4, respectively, the bottom of bubble is very close to the top edge of groove wall at the initial condition as shown in Figure 4(a). At the beginning, the lower part of bubble contacts with the wall, which means that the bubble is absorbed to the wall. At t=10.16μs, the top of bubble becomes flat meanwhile a high pressure zone with the maximum pressure of 0.86MPa exists above the bubble (figure 4(b)). The lower part of bubble-liquid interface takes on a convex upward shape. Subsequently, a jet starts to penetrate through the bubble and the bubble appears to be the shape of saddle. At t=11.01μs, the jet passes through the bottom of bubble-liquid interface with the jet velocity of 135m/s (figure 4(c)). Compared to one in other cases, the jet is relatively wider. It is because the bubble is adsorbed to the wall, which makes the bottom of bubble keep a relatively large size in the jet forming process. Then an attached toroidal bubble is formed and continuously compressed. Finally, the attached bubble detaches from the wall to become a free one and behaves oscillation. It is observed that a pressure of 68.6MPa is formed around the bubble as the bubble rebounds at t=11.2857μs, which leads to the pressure impact of 46MPa on the groove wall (figure 4(d)). As the jet reaches the groove wall at t=12.608μs, the jet velocity decays to be 45-50m/s. It means that the groove wall is subjected to a water-hammer pressure of 67.5-75.0MPa.

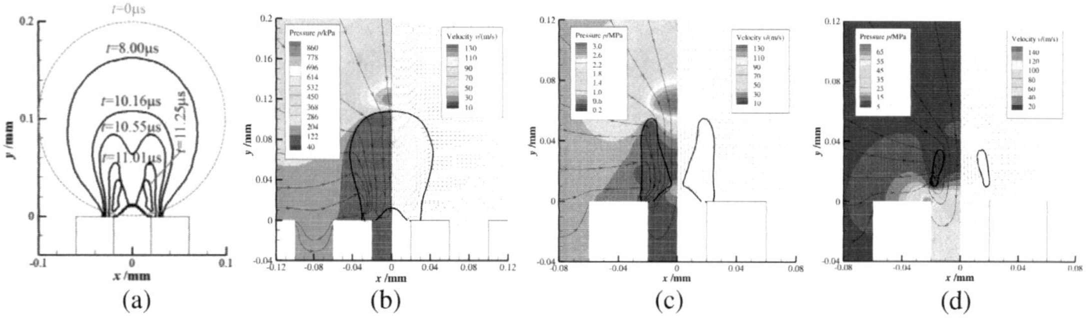

Figure 4. (a) Simulated bubble profile evolution; (b~d) Pressure field and velocity vector field around collapsing bubble at t=10.16μs, t=11.01μs and t=11.2857μs, respectively, for γ_1=1.0 and γ_2=1.4. (the black solid line is the bubble-liquid interface).

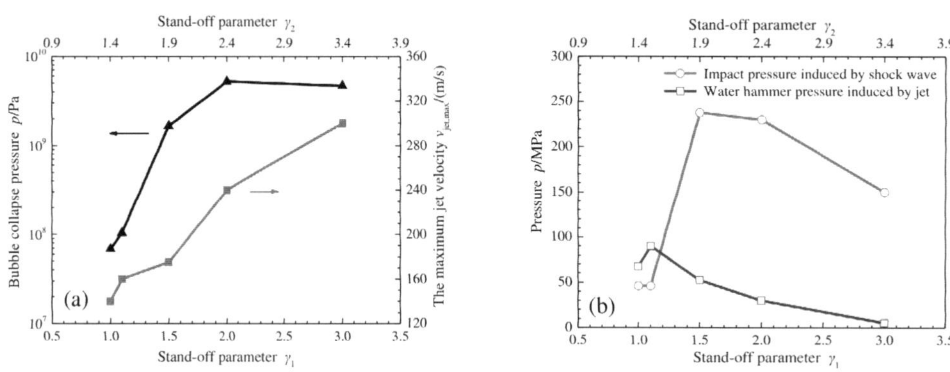

Figure 5. (a) The bubble collapse pressure and the maximum jet velocity, and (b) The impact pressure induced by shock wave and the water hammer pressure induced by jet on the groove bottom wall for different γ_1 and γ_2.

Figure 5(a) presents variation of the bubble collapse pressure and the maximum jet velocity with γ_1 and γ_2. It is indicated from figure 5(a) that the maximum jet velocity is promoted with increasing the stand-off parameters for $1.0<\gamma_1<3.0$ and $1.4<\gamma_2<3.4$. It is because that the wall effects on motion of the liquid below the bubble is weakened with increasing the distance between the bubble and the wall, which causes that the bubble tends to collapse symmetrically and the volume of bubble is compressed more before the jet is formed. The bubble collapse pressure is enhanced with increasing the stand-off parameters for $1.0<\gamma_1<2.0$ and $1.4<\gamma_2<2.4$. However, the bubble collapse pressure is found not to grow as γ_1 and γ_2 are further increased to be 3.0 and 3.4. Figure 5(b) shows variation of the impact pressure induced by shock wave and the water hammer pressure induced by jet on the groove bottom wall with γ_1 and γ_2. Even though the impact pressure on the groove wall induced by jet is higher than one induced by shock wave for $1.0<\gamma_1<1.1$ and $1.4<\gamma_2<1.5$, the shock wave still has a noticeable effect on the groove wall in this condition. Therefore, the cavitation erosion of groove wall should be resulted from both of jet and shock wave in the cases of 1 and 2. Moreover, the impact pressure on the groove wall induced by shock wave is considerably enhanced meanwhile one by jet is decreased as γ_1 and γ_2 are increased to be 1.5 and 1.9, respectively. For $1.5<\gamma_1<3.0$ and $1.9<\gamma_2<3.4$, the impact pressure induced by shock wave is far greater than one induced by jet on the groove wall. Therefore, the shock wave is mainly responsible for the cavitation erosion of groove wall for $1.5<\gamma_1<3.0$ and $1.9<\gamma_2<3.4$.

4. Conclusion

The dynamics of a micrometer-sized vapor bubble close to a rectangular groove wall is numerically simulated based on the volume of fluid method. Evolution of the bubble profile presents the distinctly different characteristics with changing the stand-off parameters γ_1 and γ_2. The maximum jet velocity is promoted with increasing γ_1 and γ_2. The bubble collapse pressure is enhanced with increasing γ_1 and γ_2 for $1.0<\gamma_1<2.0$ and $1.4<\gamma_2<2.4$. However, the bubble collapse pressure is not to grow as γ_1 and γ_2 are further increased. The cavitation erosion of groove wall should be caused by both of jet and shock wave for $1.0<\gamma_1<1.1$ and $1.4<\gamma_2<1.5$. The shock wave is mainly responsible for the cavitation erosion of groove wall for $1.5<\gamma_1<3.0$ and $1.9<\gamma_2<3.4$.

Acknowledgments

This work was financially supported by Natural Science Foundation of Liaoning Province of China (2015020136 and 201602090) and the Fundamental Research Funds for the Central Universities (3132019162).

References

[1] D. Suo, Z. Jin, X. Jiang, P.A. Dayton, Y. Jing, Appl. Phys. Lett., 110, 023703, (2017)
[2] C. G. Duan, V. Y. Karelin, *Abrasive Erosion and Corrosion of Hydraulic Machinery*. London: Imperial College Press, (2003)
[3] Y. Zhang, Y. Zhang, Y. L. Wu, Pro. Inst. Mech. Eng., Part C: J. Mech. Eng. Sci., 231, 1181-1204, (2017)
[4] R. Balachandran, Z. Patterson, P. Deymier, S. A. Snyder, M. Keswani, Chemosphere, 147, 52-59, (2016)
[5] K. Nakajima, D. Nishioka, M. Hirao, M. So, Y. Goto, H. Ogi, Ultrason. Sonochem., 36, 206-211, (2017)
[6] G. L. Chahine, A. Kapahi, J. K. Choi, C. T. Hsiao, Ultrason. Sonochem., 29, 528-549, (2016)
[7] L. Rayleigh, Philos. Mag. 34, 94-98, (1917)
[8] K. A. Mørch, *Erosion*. London: Academic Press, (1979)
[9] E. A. Brujan, T. Ikeda, Y. Matsumoto, Exp. Therm. Fluid Sci., 32, 1188-1191, (2008)
[10] M. Kornfeld, L. Suvorov, J. Appl. Phy., 15, 495-506, (1944)
[11] T. B. Benjamin, A. T. Ellis, Philos. Trans. R. Soc. London A Math. Phys. Eng. Sci., 260, 221-240, (1966)
[12] M. S. Plesset, R. B. Chapman, J. Fluid Mech., 47, 283-290, (1971).
[13] W. Lauterborn, H. Bolle, J. Fluid Mech., 72, 391-399, (1975)
[14] N. K. Bourne, J. E. Field, J. Fluid Mech., 244, 225-240, (1992)
[15] J. U. Brackbill, D. B. Kothe, C. Zemach, J. Comput. Phys., 100, 335-354, (1992)

Particle Concentration Distribution Control Based on Shear Rate Gradient and Marangoni Convection in Microchannel

Jinxin Shen, Xiangjun Zhang[a], Hao Zhong, Wenjun Cheng, Yong li

State Key Laboratory of Tribology, Tsinghua University, Beijing, 100084, China

Corresponding author: xjzhang@mail.tsinghua.edu.cn

Abstract. The application of the microfluidic chip to the immunoassay has many advantages such as high efficiency, low cost, and high integration. It is of great significance to understand the flow characteristics of fluids in microchannel, which can improve the efficiency and accuracy of detection. In this paper, the effects of microchannel structure on the particle concentration distribution in two-phase flow and the flow behaviour of near-wall layer are studied by microfluidic observation technique. Also, a kind of microchannel structure was designed to increase the particle concentration in the front-end of microfluidic.

1. Introduction

In recent years, with the rapid development of Point-of-care Testing (POCT), a variety of microfluidic chips have emerged. The microfluidic chip is based on microfluidic technology. The fluid handling, device manufacturing and sample detection concerns many basic disciplines such as fluid mechanics, surface interface mechanics, biological reaction and chemical modification. At present, microfluidic chips are mostly used for immunoassay. It is necessary to study the flow behaviour of complex fluids in microchannel.

The research on fluid flow in microfluidic chips is generally divided into three aspects: theoretical analysis, experimental exploration and numerical simulation. Theoretical analysis is based on the basic theory of fluid mechanics, in which appropriate theoretical models are established according to the flow characteristics of fluids. Experimental research focuses on relative parameters such as velocity, pressure and particle concentration distribution in microfluidic chip systems. Numerical simulation is based on commercial software of fluid mechanics.

In the application of immunoassay, increasing the detected particle concentration in the front-end of two-phase flow can improve the efficiency and accuracy effectively. This research studied the approaches to improve the particle concentration distribution within meniscus in a microfluidic chip by adjusting the two-phase behaviour of near-wall layer. Several kinds of microchannel structures were designed to control the particle concentration distribution of two-phase flow in microchips.

2. Materials and methods

2.1. Observation platform of microfluidic

The observation platform of microfluidic is consisted of inverted fluorescence microscope, CCD and automatic objective table (Figure 1). Nanoscale antigenic molecules modified with fluorescent particles are applied in POCT. Antigenic molecules react with antibody coating in the detection area.

The nanoscale particles are enriched in the detection area and emit a fluorescent signal. The particle concentration in the detection directly determines the intensity of fluorescent signal, which determines the lower limit of detection of microchip.

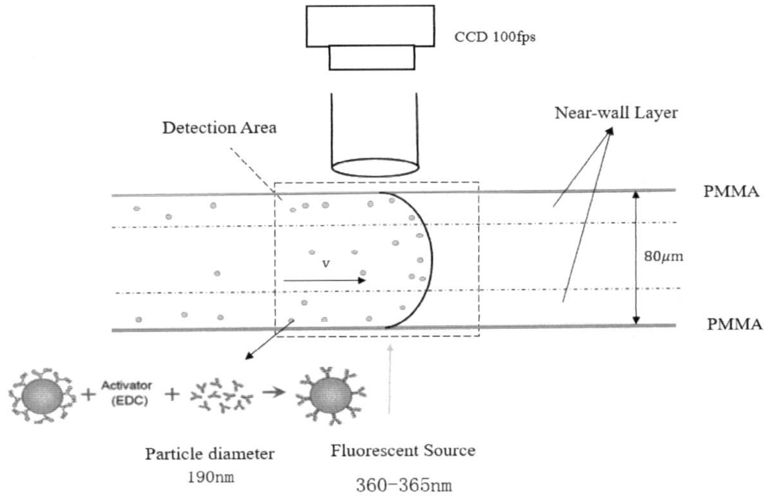

Figure 1. Observation platform of microfluidic.

2.2. Microfluidic chip

A new microfluidic chip structure is put forward in this research, which is low-cost and simple to manufacture. The chip consists of a three-layer structure (Figure 2). The material of the upper layer and the bottom layer is PMMA, which is hydrophilic material. The middle layer is a kind of double-sided tape with thickness of 80μm. It is hollowed out as the flow channel. So the height of channel is 80μm, which reaches the scale of microfluidic. The upper layer has an inlet and an outlet. The flow area of fluid can be limited by the micro-groove on the bottom layer, because the fluid in microchannel is driven by the capillary force.

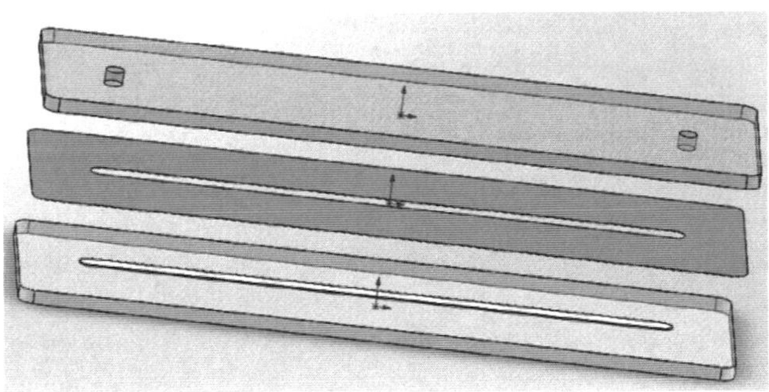

Figure 2. Structure of microfluidic chip.

2.3. Driving force

In the experiment, the fluid is driven by capillary force. According to the Laplace Capillary Pressure Formula:

$$p \sim 2\gamma/R$$

(γ: Surface Tension, R: Capillary Radius) Capillary pressure will decrease when the capillary radius increase. So we can limit the flow area of the fluid by the micro-grooves on the bottom layer. So the shape of micro-grooves is the shape of microchannel.

IOP Conf. Series: Materials Science and Engineering **538** (2019) 012057 doi:10.1088/1757-899X/538/1/012057

3. Results and discussion

Firstly, we set a lateral micro-protrusion on one side of the microchannel. The purpose is to speed up the flow of particles near the wall surface. The flow process is shown in Figure 3.

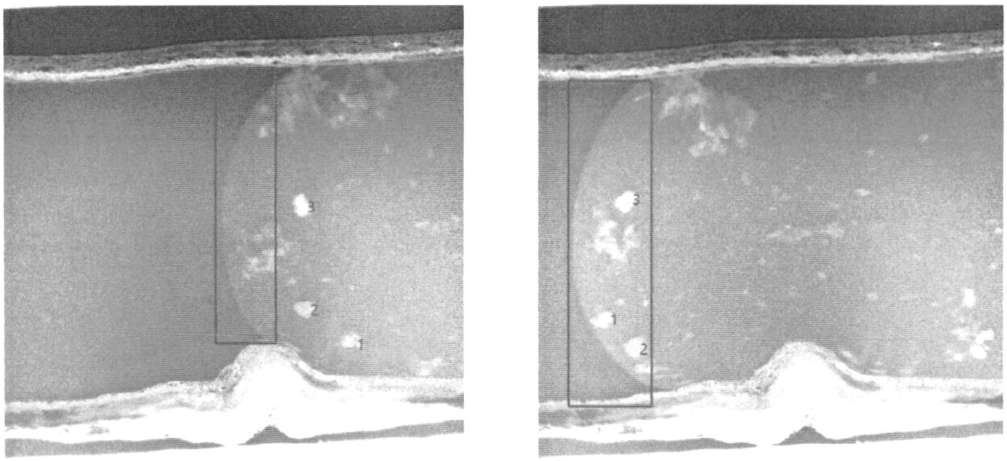

Figure 3. The effect of micro-protrusion on particle concentration distribution. (a) before the fluid passed by the micro-protrusion. (b) after the fluid passed by the micro-protrusion.

According to the result of observing, we find that the fluid maintained the characteristics of laminar flow very well. Before the fluid passed by the micro-protrusion, the front-end of meniscus had a low particle concentration, and no convergence occurred. When the fluid passed by the micro-protrusion, the fluid near the wall speeded up. Finally, lots of particles gathered at the front-end of the fluid meniscus. And the particle concentration of the front-end of fluid got to a high level.

Based on this phenomenon, we set the micro-protrusion on both sides of microchannel to make it more effective. With this structure, the particles near the wall on both sides of microchannel can speed up and gather at the front-end. The flow process is shown in Figure 4.

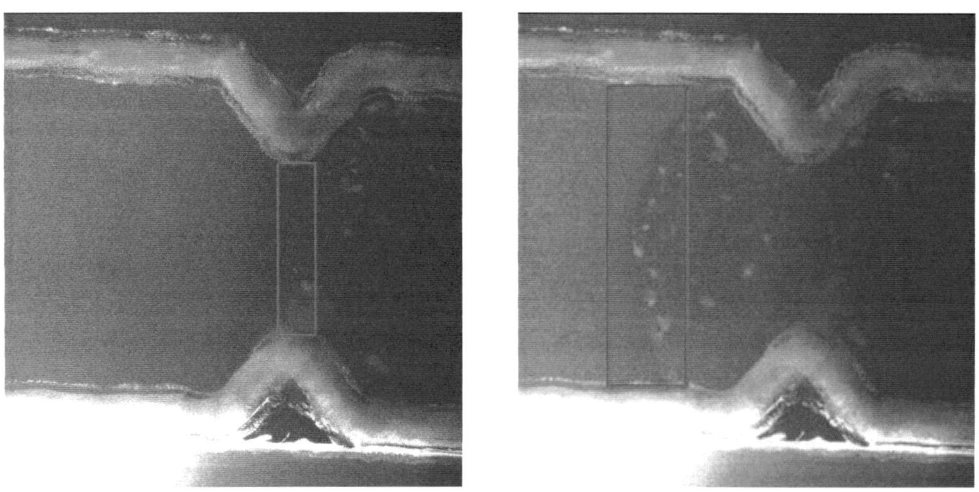

Figure 4. The Effect of Micro-protrusion on Both Sides of Microchannel. (a) Low concentration at the front-end of the fluid. (b) High concentration at the front-end of the fluid.

We explain this phenomenon as follows.

Under the condition of laminar flow characteristics, according to the Flow Continuity Equation, the fluid flow rate will be accelerated at smaller cross-sectional area to keep the fluid rate equivalent. The acceleration of the flow rate begins with the boundary layer, which increase the shear rate of boundary layer. Then it drives the fluid of the adjacent layer to accelerate under the action of shearing force. The shear rate between the apexes of the two micro-protrusions is shown in Figure 5.

The fluid in the channel forms a velocity gradient from the boundary layer to the central of microchannel. It caused that in the front-end of fluid, the particle concentration in boundary layer is higher than central. And the surface tension in central is higher than in boundary layer because of the concentration distribution. So, the surface tension gradient is formed in the front-end of fluid. And particles will flow to the central part under the surface tension gradient. This phenomenon is called Marangoni Convection. Therefore, increased shear rate of boundary layer and Marangoni Convection phenomenon makes more particles gather at the front-end of the fluid. The schematic diagram of fluid flow is shown in Figure 6.

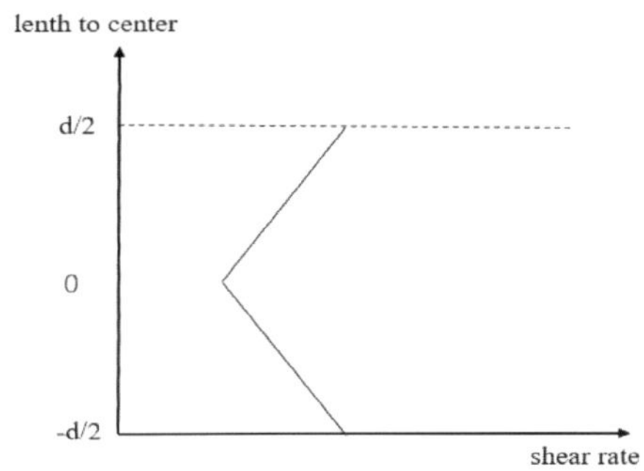

Figure 5. Shear rate schematic diagram.

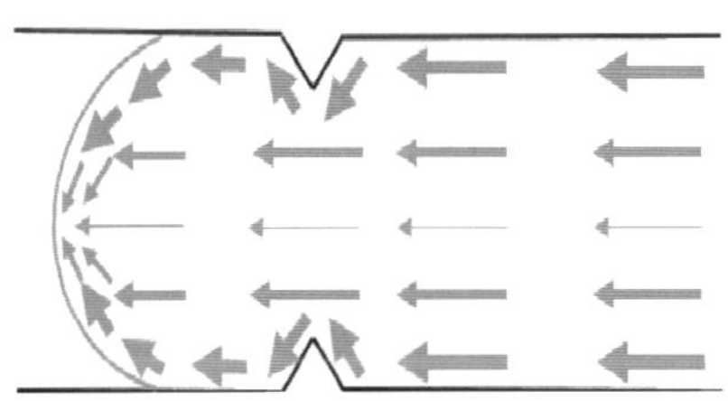

Figure 6. Flow process schematic diagram (Thicker streamlines represent higher speeds).

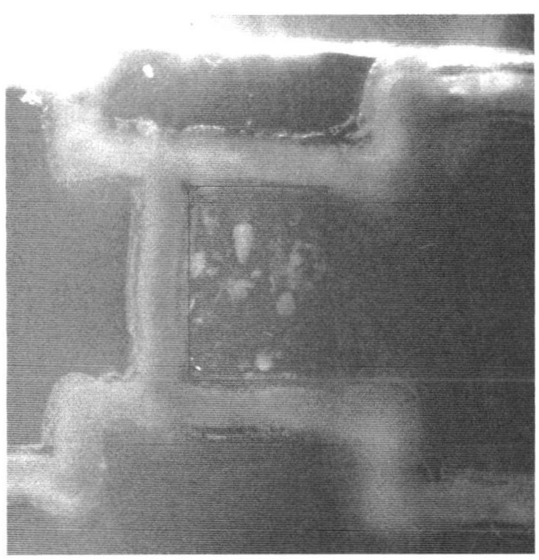

Figure 7. Detected particles gather at the detection area.

IOP Conf. Series: Materials Science and Engineering **538** (2019) 012057 doi:10.1088/1757-899X/538/1/012057

In immunoassay, it's hoped that the detected particles stay at the detection area for a period of time to improve the accuracy and efficiency. Therefore, the evolution of the above structure was made. The triangular micro-protrusions are changed to rectangular micro-protrusions, and a transverse groove is sculptured in the channel. So, the fluid is blocked by the change of capillary force temporarily. The particles are also blocked in this area temporarily because of the concentration phenomenon (Figure 7). And this area will be used as the detection area in immunoassay.

4. Conclusion

When microfluidic chips are applied to immunoassay, it's hoped that detected particles can gather at the front-end of sample and stay at the detection area for a short time. It can be achieved by designing a special microchannel structure. In this research, an observation platform of microfluidic was built to study the influence of microchannel structure on two-phase flow. And it is confirmed that the micro-protrusions with a transverse groove structure was effective.

References

[1] W. Chen, A. Y. Fadeey, M. C. Hsieh, D. Oner, J. Youngblood, and J. T. McCarthy, "Ultrahydrophobic and ultralyophobic surfaces: Some commentsand examples," Langmuir 15, 3395 (1995).

[2] K. K. S. Lau, J. Bico, K. B. K. Teo, M. Chhowalla, and G. A. J. Amaratunga, "Superhydrophobic carbon nanotube forests," Nano Lett. 3, 1701 (2003).

[3] J. Bico, C. Marzolin, and D. Quere, "Pearl drops," Europhys. Lett. 47, 220 (1999).

[4] T. J. McCarthy and J. P. Youngblood, "Ultrahydrophobic polymer surfaces prepared by simultaneous ablation of polypropylene and sputtering of poly(terafluoroethylene) using radio frequency plasma," Macromolecules 32, 6800 (1999).

[5] J. Bico, U. Thiele, and D. Quere, "Wetting of textured surfaces," Colloids Surf., A 206, 41 (2002).

IOP Conf. Series: Materials Science and Engineering **538** (2019) 012058 doi:10.1088/1757-899X/538/1/012058

Using Regularized Softmax Regression in the GNSS/INS Integrated Navigation System with Nonholonomic Constraints

Linlin Zhao, Haiyang Quan

Beijing Microelectronics Technology Institute, Beijing, China

Corresponding author's e-mail:elijaheslyz@buaa.edu.cn

Abstract. The integration of global navigation satellite system (GNSS) and Inertial navigation system (INS) is widely implemented in land-vehicle navigation applications. However, the satellite signal is vulnerable in some special urban scenarios, consequently errors in terms of position, velocity and attitude grow rapidly in stand-alone mode especially for low-cost MEMS-based INS. In the conventional tight combination navigation schemes, system works on predicting model during the GNSS signal outage and the positioning accuracy is determined by the precision of the inertial navigation. Besides the lack of observation makes the estimate of inertial navigation error with GNSS information less reliable due to the satellite signal loss. In this paper, an improved non-holonomic constraints (NHC) method based on regularized softmax regression is proposed to enhance navigation precision when the number of visible satellite is insufficient. The velocity constraint condition is applied to simplify the system calculating equations of MEMS-based INS. Furthermore, a regularization softmax regression model based on the collected data is trained to recognize the vehicle motion pattern so as to realize deeper constraints. Simulation and field-test results indicate that the method is beneficial to raise the precision of low-cost GNSS/INS integrated navigation receiver by efficiently reduce the navigation errors.

1. Introduction

Integrated navigation receiver-chips based on global navigation satellite system (GNSS) and Inertial navigation system (INS) have been widely applied in various fields such as land-vehicle navigation (LVN) [1]. The integration of GNSS and INS takes advantage of the both navigation methods since an INS is self-contained and able to provide full six degree-of-freedom solutions with high update rates which covers the shortage of satellite navigation on anti-jamming. Micro-electro-mechanical systems (MEMS) inertial sensors are predominant in power consumption, weight and cost which quitely meet the specifications and requirements for INS. However, the widespread uses of MEMS inertial systems that are of low-cost and low-grade inertial sensors outputs are characterized by high noise and large indeterminacy and consequently this reduction of cost has led to drop in accuracy as a whole [2-3]. Therefore, an INS that uses low-cost inertial sensors only can be used for standalone navigation for a very limited time period. Even though the system can estimate the sensor biases, the position error growth during the GNSS-signal outages is substantial and the navigation solution soon becomes useless.

Some research suggests that the navigation precision could be improved by raising hardware structure or importing other sensors data [4-6]. A preferable performance system can be realized, whereas the cost and systematic complexity increase as well. A short discussion about various information fusion

strategies is proposed in [7]. In most integrated schemes, the GNSS is considered to be an accurate reference with which to correct for the systematic errors of the inertial sensors. Admitting that an integrated navigation system can work in GNSS-denied environments, the lack of observation makes the estimate of inertial navigation error with GNSS information less reliable. It is valuable to introduce specific auxiliary information without adding extra hardware costs. For LVN applications, Non-Holonomic Constraints (NHC) is one of the most common types of auxiliary information. A detailed observability analysis of the contributions of the NHC from the perspective of observability is given in [8], which provides a deeper insight and shows how the NHC improves the navigation solutions in a loosely coupled GPS-aid INS. Niu [9] implemented the NHC in land-vehicle navigation which could significantly improve the heading accuracy. But it would barely contribute when the speed gets low. Shin [10] proposed a method to constrain the heading drift which is called Zero Integrate Heading Rate. The current approaches still need deeper constrains to cover the various situations.
In this paper, a tightly integrated navigation system with NHC based on low-cost INS and GNSS receiver is proposed. The equations of INS pseudo-range and pseudo-range rate are described in detail. The NHC condition is also imported to the combination filter for the LVN applications. Further, we implemented regularized softmax regression to recognize the vehicle motion mode so that deeper constrains could be realized. Simulation analysis verified the effectiveness of our method and field-test results showed a significant improvement on the performance during the satellite signal outage.

2. Method

2.1. Overall Design of the Tightly-Coupled Scheme
A succinct expression of the integrated navigation scheme is described in Fig.1. A relative distance to satellite that can be considered as the pseudo-range and pseudo-range rate of INS is calculated with the inertial navigation outcome and the satellite ephemeris information. The pseudo-range and pseudo-range rate offset between GNSS measurement and INS are implemented as the input factors of the extended Kalman filter [11]. It is mentioned that data pre-processing and initial alignment steps are not involved in the discussion of this paper which cannot be considered as vacancy.

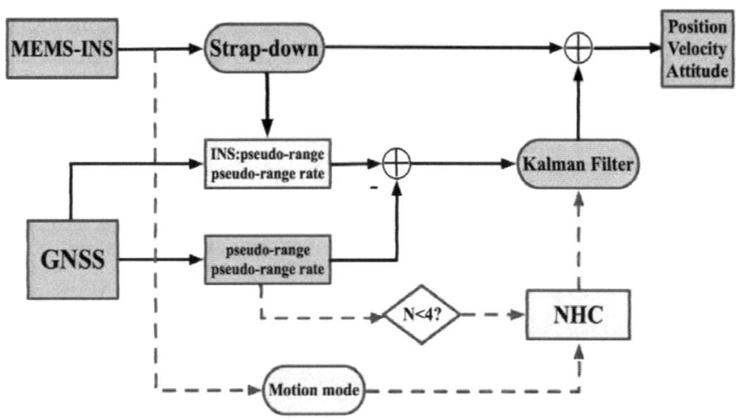

Figure1. The overall scheme of GNSS/INS

A 11-state vector which updates by the tightly-coupled INS/GNSS integrated filter based on the pseudo-range and pseudo-range rate is defined in (1).

$$X = [\phi_e\ \phi_n\ \phi_u\ \Delta\lambda\ \Delta l\ \Delta h\ \Delta v_e\ \Delta v_n\ \Delta v_u\ \delta t_u\ \delta t_{ru}]^T \tag{1}$$

Where the subscripts *e, n* and *u* represent the three axis of geocentric coordinate system and $\lambda,\ l\ and\ h$ represent the position of longitude, latitude and altitude. ϕ represents error-angle of platform and errors of velocity, clock and frequency are considered as Δv, δt_u and δt_{ru}.

2.2. Equations of pseudo-range and pseudo-range rate

The position measurement from INS can be assumed as $(\boldsymbol{x_I}\ \boldsymbol{y_I}\ \boldsymbol{z_I})^{\boldsymbol{T}}$ and accurate satellite position can be extracted from the ephemeris information as $(\boldsymbol{x_s}\ \boldsymbol{y_s}\ \boldsymbol{z_s})^{\boldsymbol{T}}$. The pseudo-range measurement from GNSS receiver is defined as $\boldsymbol{\rho_G}$. The real distance from INS to GNSS satellites can be described as (2).

$$r_j = \left[(x_{sj} - x)^2 + (y_{sj} - y)^2 + (z_{sj} - z)^2\right]^{\frac{1}{2}} \tag{2}$$

The factual coordinate orientation is set as $(\mathrm{x\ y\ z})^{\mathrm{T}}$ and Taylor series expansion in the measurement from INS can be defined as (3) and the first order is reserved. Therefore, we can express the pseudo-range with position error and $\mathrm{r_j}$ as (5).

$$r_j = \left[(x_{sj} - x_I)^2 + (y_{sj} - y_I)^2 + (z_{sj} - z_I)^2\right]^{\frac{1}{2}} - H \tag{3}$$

$$H = \frac{\partial r_j}{\partial x}\delta x + \frac{\partial r_j}{\partial y}\delta y + \frac{\partial r_j}{\partial z}\delta z \tag{4}$$

$$\rho_{Ij} = r_j - e_{j1}\delta x - e_{j2}\delta y - e_{j3}\delta z \tag{5}$$

$$\frac{\partial \rho_j}{\partial x} = -\frac{x_{sj} - x_I}{r_j} = -e_{j1} \tag{6}$$

$$\frac{\partial \rho_j}{\partial y} = -\frac{y_{sj} - y_I}{r_j} = -e_{j2} \tag{7}$$

$$\frac{\partial \rho_j}{\partial z} = -\frac{z_{sj} - y_I}{r_j} = -e_{j2} \tag{8}$$

Similarly, we can infer the expression of the pseudo-range rate as (11). Prior to that the expressions of pseudo-range and pseudo-range rate from GNSS receiver are given in (9) and (10). $v_{\rho j}$ stands for the ionospheric errors of satellite *j*.

$$\rho_{Gj} = r_j + \delta t_u + v_{\rho j} \tag{9}$$

$$\dot{\rho}_{Gj} = e_{j1}(\dot{x}_{sj} - \dot{x}) + e_{j2}(\dot{y}_{sj} - \dot{y}) + e_{j3}(\dot{z}_{sj} - \dot{z}) + \delta t_{ru} + \dot{v}_{\rho j} \tag{10}$$

$$\dot{\rho}_{Ij} = \dot{r}_j - e_{j1}\delta\dot{x} - e_{j2}\delta\dot{y} - e_{j3}\delta\dot{z} \tag{11}$$

The measurement equations of the pseudo-range and pseudo-range rate offset can eventually be expressed as follows.

$$\delta\rho_j = e_{j1}\delta x + e_{j2}\delta y + e_{j3}\delta z + \delta t_u + v_{\rho j} \tag{12}$$

$$\delta\dot{\rho}_j = e_{j1}\delta\dot{x} + e_{j2}\delta\dot{y} + e_{j3}\delta\dot{z} + \delta t_{ru} + \dot{v}_{\rho j} \tag{13}$$

The tightly-coupled integrated system then takes the pseudo-range and pseudo-range rate offset as observed quantity of the combination filter to update the navigation information.

2.3. The Non-Holonomic Costraints for LVN

When the number of visible satellites is less than four, although the combination filter still works, poor navigation results are given due to the lack of measurement. The NHC method is actually used commonly in the navigation of land vehicles [9]. For land vehicles in urban environment, it can be assumed that the height of carrier will not abruptly change in a short time. Thus, the height before it lost a moment can be used as a constraint in the process of GNSS disruption. Besides the NHC makes assumptions that in the case of LVN, the velocity of the vehicle in the plane perpendicular to the forward direction is almost zero unless the vehicle jumps off the ground or slides on the ground. The velocity constraint can be expressed as (14).

$$\left(v_x^b, v_z^b\right)^T \sim N(0, R_v) \tag{14}$$

where R_v stands for the perturbation error due to the angle misalignment. For the height constraint, a measurement equation is imported to the system as shown in (15).

$$h_{INS} - h_{const} = \delta h + v \tag{15}$$

Where h_{INS} and h_{const} represent the current solution of INS and previous solution of GNSS in height. If h_{const} is of highly authenticity, the navigation solution is equal to the normal mode. Otherwise the measured values have a deviation of the system. In accordance with the standard Kalman filtering equation, the measurement deviation will be imported the state estimation equation to influence the filtering solution, the height constraint can cause larger error when h_{const} is unreliable. However, for land vehicle system, height will not cause large-scale fluctuations and even assuming that the height is not accurate anyway, the positioning solutions are still better than that of without constraints. When it comes to the velocity constraints, it is equivalent to add another two assumed truth values. Similar to the height constraint, if the vehicle is in high dynamic circumstances, such as making a turn or greatly changing directions, the estimation error could be worse. Therefore, deeper constraints are still in need for the NHC method.

3. Improved Method on Deeper Constraints

3.1. Reguarized Softmax Regression in Motion Mode Recognition

As is mentioned that constraints of position and velocity are not reliable when the vehicle changes attitude or directions. In other word, the performance of NHC is affected by the motion state. Inertial sensors can reflect the movement of the carrier which can determine the current motion state to import different constraint conditions according to the output of INS.

Softmax regression is an extension for logistic regression on the multiple classification problems [12] which is a kind of real-time supervised learning model. The softmax function and loss function are defined as follows.

$$h_\theta\left(x^{(i)}\right) = \frac{1}{\sum_{j=1}^{3} exp\left(\boldsymbol{\theta}_j^T \boldsymbol{x}^{(i)}\right)} \begin{bmatrix} exp\left(\boldsymbol{\theta}_1^T \boldsymbol{x}^{(i)}\right) \\ exp\left(\boldsymbol{\theta}_2^T \boldsymbol{x}^{(i)}\right) \\ exp\left(\boldsymbol{\theta}_3^T \boldsymbol{x}^{(i)}\right) \end{bmatrix} \tag{16}$$

$$\boldsymbol{J}(\boldsymbol{\theta}) = -\frac{1}{m}\left[\sum_{i=1}^{m}\sum_{j=1}^{3}\left\{\boldsymbol{y}^{(i)} = \mathbf{j}\right\}\boldsymbol{ln}\left(\frac{exp(\boldsymbol{\theta}_j^T x^{(i)})}{\sum_{n=1}^{3} exp(\boldsymbol{\theta}_n^T x^{(i)})}\right)\right] + \boldsymbol{\lambda}\|\boldsymbol{\theta}\|^2 \tag{17}$$

Where m stands for the quality of data samples and $\boldsymbol{\theta}$ is the weighting parameter vector. Besides we imported $\lambda\|\boldsymbol{\theta}\|^2$as the regulation term to avoid parameter redundancy and overfitting. Therefore, the loss function is composed of cross entropy and regularization and iteration methods [12] can be applied to obtain the optimal solution. For vehicles, general motion states can be summarized as static, going straight and turning. A training method based on the actual data is given as Fig.2.

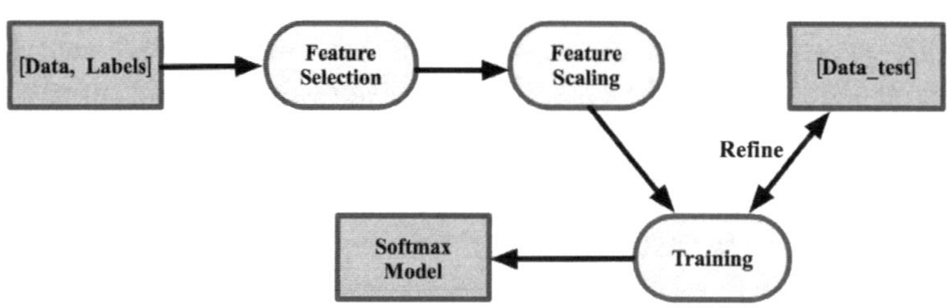

Figure 2. Training method

The data samples and calibration are generated from real road collections. Since MEMS-INS is fixed on the vehicles, the perturbation could form acceleration in the vertical direction in the process of driving. Considering the motion characteristics of land vehicles, a practical method of feature selection is proposed here. The expressions of feature are shown as follows.

$$x_1 = \sum_{t_i=t_k-n}^{t_k} \omega_{norm}(t_i) \tag{18}$$

$$x_2 = \sum_{t_i=t_k-n}^{t_k} |a_{norm}(t_i) - a_{norm}(t_i - i)| \tag{19}$$

$$x_3 = \frac{1}{m}\sum_{t_i=t_k-m}^{t_k} \omega_z(t_i) \tag{20}$$

where $\omega_{norm} = \sqrt{{\omega_x}^2 + {\omega_y}^2}$ represents the horizontal angular velocity and $a_{norm} = \sqrt{{a_x^b}^2 + {a_y^b}^2 + {a_z^b}^2}$ stands for the total acceleration. It is noted that scale normalization is essential in the feature scaling. The range of features is unified to [0, 1]. We can get the motion state probability by softmax regression based on the current feature vector as shown in Fig.3.

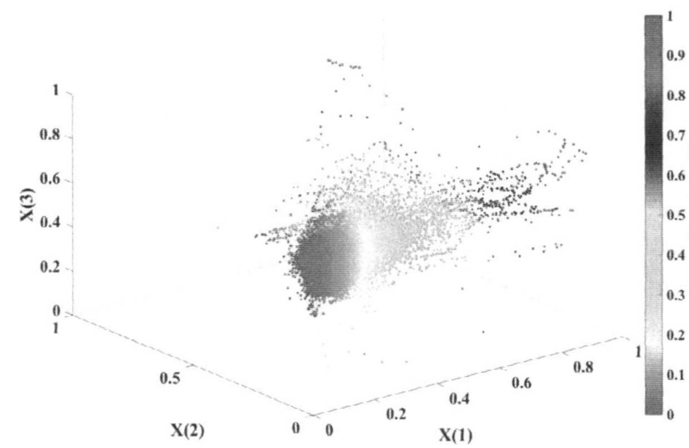

Figure 3. The probability distribution by regularized softmax regression

The motion state can be determined by the optimal probability estimate. The recognition performance in a real-road test is shown in Fig.4. The outcomes are consistent with the real motion state in most cases.

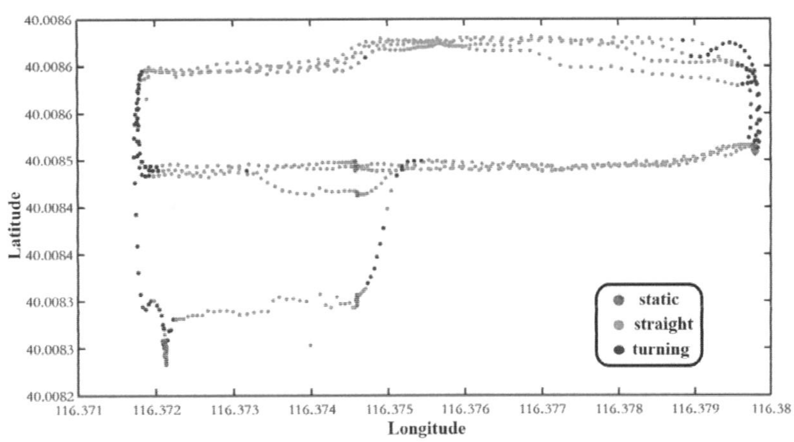

Figure 4. The motion mode solution by regularized softmax regression

3.2. Deeper Constraints for NHC

After motion mode judgement, different conditions can be derived as deeper constraints. First of all, for the static state, we need to make zero velocity update and modify the pitch and the roll by the accelerometer calibration, besides the course angle should remain and the gyroscope zero bias need recalculated. The constraints can be described as follows.

$$\boldsymbol{v} = 0 \tag{21}$$

$$\theta = \sin^{-1}\left(\frac{a_y}{g}\right) \tag{22}$$

$$\emptyset = \sin^{-1}\left(\frac{-a_x}{\sqrt{g^2 - a_y^2}}\right) \tag{23}$$

$$B_\omega = \omega \tag{24}$$

For a linear motion, the roll can be corrected by the accelerometer of x-axis which is shown in (25). As for turning mode, the velocity can be obtained by the relationship between the centripetal acceleration and the angular velocity as (26).

$$\emptyset = \sin^{-1}\left(\frac{-a_x}{\sqrt{g\cos\theta}}\right) \tag{25}$$

$$v = \frac{a_x + g\sin\phi\cos\theta}{\omega_z} \tag{26}$$

4. Experiments and Discuss

4.1. Simulation of the GNSS/INS with Improved NHC

We have showed the performance of our method on the vehicle movement state recognition. Further, we imported the NHC model with regularized softmax regression to a self-developed GNSS/INS integrated navigation receiver as shown in fig.5.

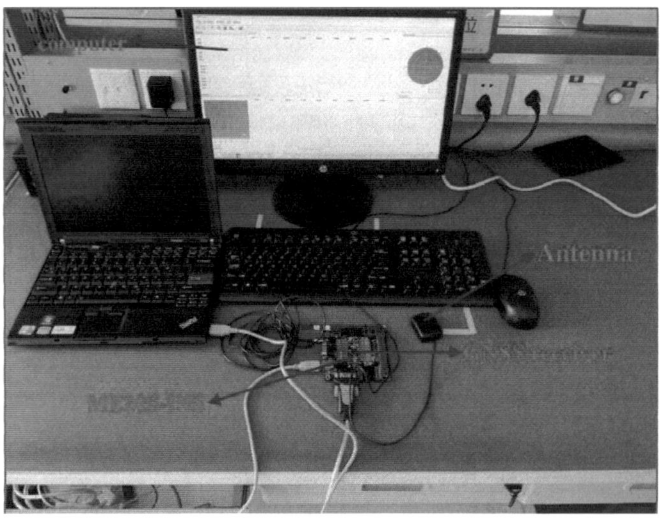

Figure 5. The self-developed GNSS/INS

The simulated environment of insufficient visible satellites is generated by the signal simulator. When the number of visible satellites is less than four, the system can still work at an excellent precision based on the improved NHC. Fig.7 shows the result error of static simulation.

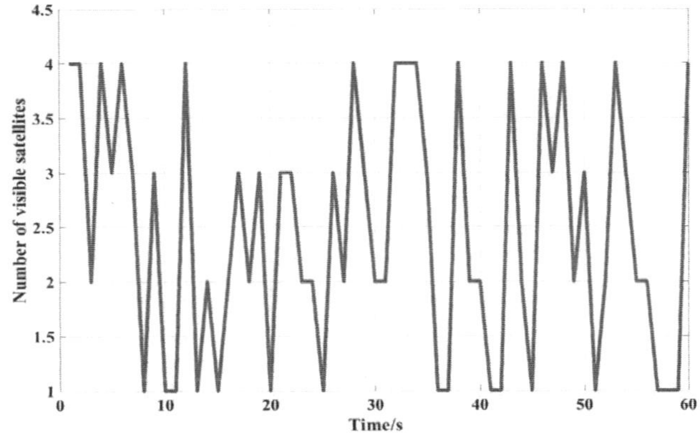

Figure 6. The number of visible satellites

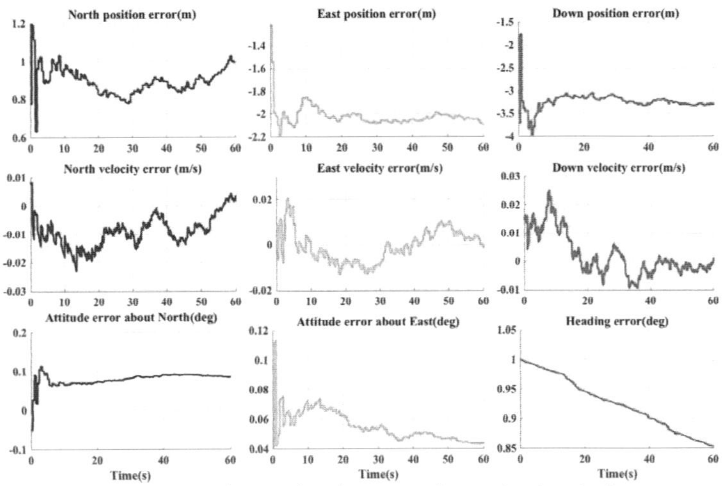

Figure 7. The navigation error in static simulation

Table1. RMSE of location error

RMSE of location error	Static Simulation
East position (m)	2.87
North position (m)	0.95
Down position (m)	3.37
East velocity (m/s)	0.02
North velocity (m/s)	0.01
Down velocity (m/s)	0.12
Pitch (°)	0.18
Roll (°)	0.12

Splendid performance of error limitation could be realized by our method in the static simulation as shown in table 1. It's understandable that the improvement on precision is produced by NHC. The system will work at zero-velocity updating mode once it is recognized as static.

4.2. Field-test Experiment

A field-test is conducted along the route shown in Fig.8 which is composed of general urban scenario. We measured the navigation errors as well and performance of the method can be seen in Fig.9.

Figure 8. The road-test route

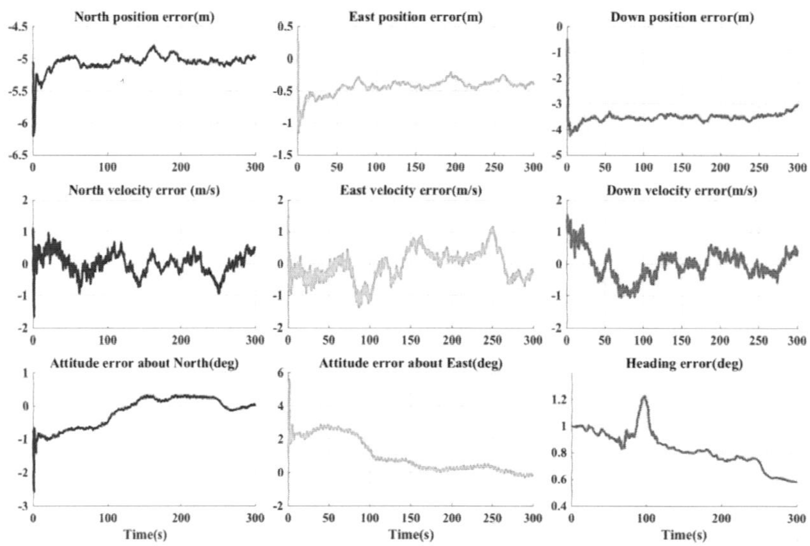

Figure 9. The navigation error of road test

Table2. RMSE of location error

RMSE of location error	Static Simulation
East position (m)	5.47
North position (m)	0.75
Down position (m)	3.89
East velocity (m/s)	0.98
North velocity (m/s)	1.12
Down velocity (m/s)	0.87
Pitch (°)	1.12
Roll (°)	1.84

As is shown in Fig.9 and table 2, since the practical test is complex, the navigation error is worse than static simulation but still proves an acceptable precision. Besides, the down position error is caused by the initial alignment. For low-end MEMS based navigation system in LVN applications, we improved the navigation performance by using the regularized softmax regression in NHC to realize deeper constraints and the effectiveness has been verified.

5. Conclusion

In this paper, we successfully implemented regularized softmax regression into a self-developed tightly integrated navigation system with NHC based on a low-cost INS and GNSS receiver. The regularized softmax regression is applied to recognize the vehicle motion mode so that deeper constrains could be realized. Detailed analysis and field-test results inferred that our method could significantly raise the performance of the system especially during the GNSS outage.

Acknowledgments

This study was partially supported by the College specialized research project from Beihang University and the navigation research center of Beijing Microelectronics Technology Institute.

References

[1] Aboelmagd Noureldin, Tashfeen B Karamat, and Jacques Georgy. Funda- mentals of Inertial Navigation, Satellite-Based Positioning and Their Inte- gration. Springer, 2013.

[2] E. H. Shin and N. El-Sheimy, "Accuracy improvement of low cost INS/GPS for land applications," Department of Geomatics Engineering, University of Calgary, 2001.

[3] E. H. Shin, "Estimation techniques for low-cost inertial navigation," UCGE report, 2005.

[4] Y. Tawk, P. Tome, and C. Botteron, "Implementation and Performance of a GPS/INS Tightly Coupled Assisted PLL Architecture Using MEMS Inertial Sensors," *Sensors* 2014, 14, 3768-3796.

[5] Yang, Y. "Tightly Coupled MEMS INS/GPS Integration with INS Aided Receiver Tracking Loops," University of Calgary: Calgary, AB, Canada, 2008.

[6] Niu, X., Hassan, T., Ellum C., and El-Sheimy, N., 2006. Directly Georeferencing Terrestrial Imagery using MEMS- based INS/GNSS Integrated Systems. In: *XXIII FIG (International Federation of Surveyors) Congress*, Munich, Germany, 8-13 October 2006. .

[7] Isaac Skog and Peter Handel, "In-car Positioning and Navigation Technologies—A Survey," IEEE transactions on Intelligent Transportation systems, VOL. 10, NO. 1, March 2009.

[8] X. Niu, Y. Li, Q. Zhang, Y. Cheng and Ch. Shi, "Observability Analysis of Non-Holonomic Constraints for Land-Vehicle Navigation Systems," Journal of Global Positioning Systems, Vol.11, No.1 : 80-88, 2012.

[9] X. Niu, S. Nassar, and N. El-Sheimy, "An accurate land-vehicle MEMS IMU/GPS navigation system using 3D auxiliary velocity updates," *Navigation, Journal of the Institute of Navigation*, vol. 54, no. 3, pp. 177–188, 2007.

[10] Shin, E. H., 2005. Estimation Techniques for Low-Cost Inertial Navigation. PhD Thesis, MMSS Research Group, Department of Geomatics Engineering, University of Calgary, Calgary, AB, Canada, UCGE Report No. 20219.

[11] Shin, E. H., 2005. Estimation Techniques for Low-Cost Inertial Navigation. PhD Thesis, MMSS Research Group, Department of Geomatics Engineering, University of Calgary, Calgary, AB, Canada, UCGE Report No. 20219.

[12] Pereira F., Mitchell T. and Botvinick M., Machine Learning Classifiers and fMRI: A Tutorial Overview. NeuroImage, 2009, 45(1): S199-S209

Study on Assessment Method of Satellite Damaged by Space Debris

Hongping Gu[1], Yijian Shi[2,a], Chun Cheng[2] ,Quan Wang[2], Jie Fu[2], Hongxia Sui[3] and Tongkun Xu[3]

[1] Xi'an Modern Chemistry Research Institute, Xi'an 710065, China
[2] School of Mechanical Engineering, Nanjing University of Science and Technology, Nanjing 210094, China
[3] Heilongjiang North Tools Co. Ltd, Mudanjiang, Heilongjiang 157013, China

[a] Corresponding author: shiyijian@163.com

Abstract. The damage assessment model of space debris was established to satellite, included the description of space debris distribution field, the analysis of function and structure of satellite components, the rendezvous analysis between space debris and satellite, the damage criterion and so on. NASA ORDEM model and Shot-line method were used to describe the probability distribution field of millimeter and centimeter debris in space. The equivalent model of satellite was established by energy conservation principle and basic body method. The failure probability of satellite system was calculated under the impact of space debris. Establishing the vulnerability analysis method of satellite target is great significance to the risk assessment of space debris impacting satellite, and provides reference for the optimization design of satellite protection structure.

1. Introduction

Increasingly frequent space activities, the existence of space debris seriously threatens the safety of spacecraft on orbit. Shielded spacecraft can protect debris less than 1.0mm. Space debris with an average size of more than 10 cm can be monitored and warned by ground-based radar and telescope, and the spacecraft can be avoided collision by orbital maneuver avoidance strategy [1]. However, the debris with a diameter of 0.1-10 cm is beyond the observation and tracking capability, and the debris can't be positioned. The spacecraft has no effective maneuver to avoid these debris, which is extremely dangerous to the spacecraft.

For the description of space debris, The NASA's ORDEM database was generated and the space area was meshed. The distribution of debris was calculated by finite element method, and the debris environment was formed according with the observed data [2, 3]. N. Welty and others combined environmental model, ballistic limit equation and spacecraft exposure area occlusion processing algorithm to calculate the structure damage probability of spacecraft under space debris impact.

The basic body method [4] was applied to satellite target description. The basic body method is to simulate the geometric shape of the target with regular basic shapes. Based on the damage tree, the relationship between target damage and component damage is established.

The paper described the space debris based on ORDEM model, established the satellite target with the basic body method, and found the relationship between component damage and target damage with the damage tree [5, 6]. A damage assessment model of space debris to satellite target was established, which provides a new solution for the impact risk assessment of unmanned spacecraft such as satellites.

2. Satellite Equivalent Model

2.1. Satellite Description

Top-down analysis was used to determine the key components of a satellite, also known as damage tree method. The physical and functional characteristics [7,8] of the target satellite were acquired. The target functional block diagram and finally the damage tree diagram was made. Satellite public system is a system that any type and purpose of satellite must be equipped with. The structural system is set as an ellipsis event in Figure 1.

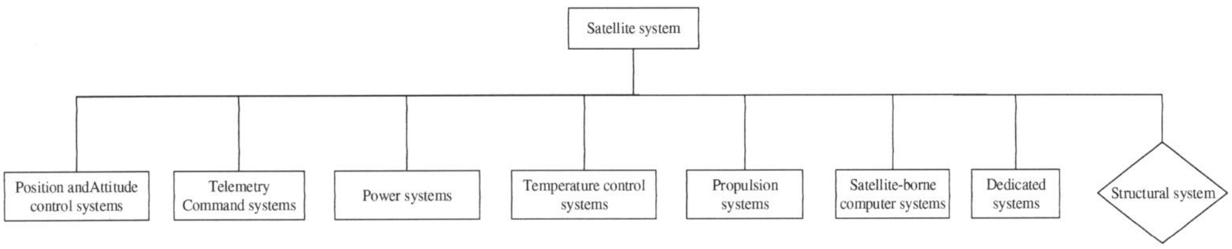

Figure 1. Satellite system structure diagram

2.2. Equivalent Model

The residual velocity of debris penetrating any material target and the equivalent material of the same specification is equal at the same speed and rendezvous mode. In this paper, 2A12 aluminum alloy was used as the equivalent material. The model was established with LS-DYNA finite element software to acquire the residual velocity after the ball penetrated honeycomb aluminum plate.
Three different positions including the hexagonal cell center, the midpoint of the cell wall and the rendezvous of the cell wall, were simulated. The honeycomb aluminum plate was meshed in HyperMesh as shown in Figure 2.

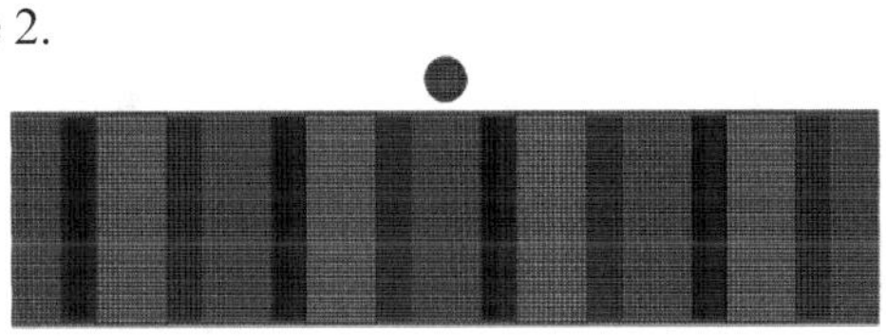

Figure 2. Finite element model

Three groups of velocity curves varying with time were obtained by numerical simulation of ball penetration into honeycomb aluminum at three different positions as shown in Figure 3.

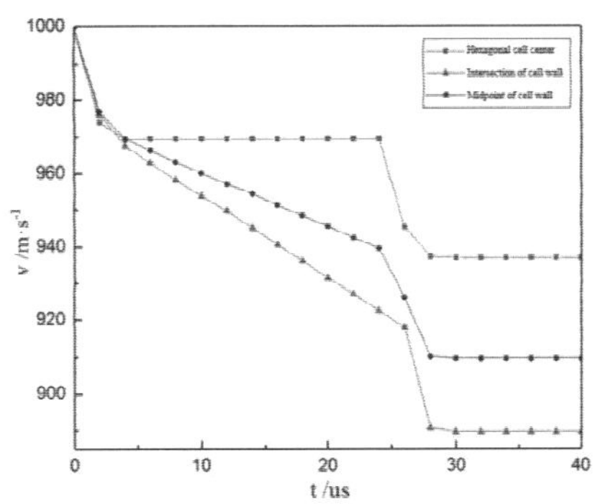

Figure 3. Penetration velocity curve

The average residual velocity was taken as the residual velocity of the ball penetrating honeycomb aluminum. The Von Mises stress of 2A12 aluminum alloy is shown in Figure 4, The velocity of equivalent 2A12 aluminum alloy is obtained by LS-DYNA simulation.

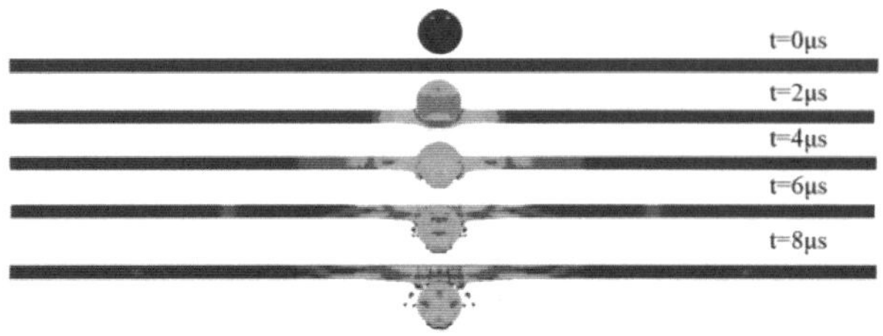

Figure 4. Von Mises stress of 2A12 aluminum alloy

3. Damage Criteria

The interior of these components is made up of many fatal parts, which are destroyed and invalidated only when debris strikes the fatal parts or areas within them. Otherwise, the components will not be damaged, and the key parts in the components are not necessarily tightly arranged. The damage probability of such components can be calculated by the following formula:

$$P_{k/h} = \frac{A_v}{A_p} = \sum_{i=1}^{n} \frac{A_V^{(i)}}{A_P^{(i)}} \tag{1}$$

Where A_v is the total vulnerable area of components, A_p is the total rendering area of components, n is to consider the incident direction of debris. For a two-dimensional region, the ratio of the vulnerable area to the rendered area; for a three-dimensional component, such as a satellite computer, the ratio of the total rendered area to the total rendered area in each direction, or the volume of the vulnerable part to the total volume of the component as shown in Figure 5.

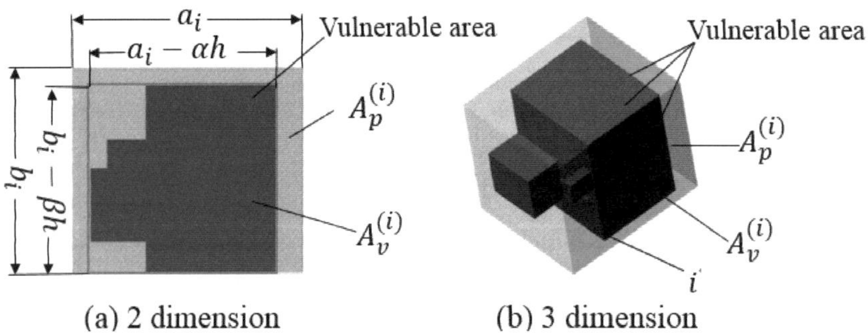

Figure 5. Parts vulnerable area and presentation area

The following formula can be used to calculate the damage probability of components when the internal precise structural parameters can't be determined:

$$P_{k/h} = \frac{\sum_{n=1}^{n}(a_i - \alpha h)\times(b_i - \beta h)\times\zeta_i + \sum_{m=1}^{m}\pi\times(r_m - \lambda h)^2\times\xi_m}{\sum_{n=1}^{n}a_i\times b_i + \sum_{m=1}^{m}\pi\times r_m^2}, \quad (m+n=6, 0\le n\le 6, 0\le m\le 6) \tag{2}$$

where a_i and b_i are the length of area of presentation from the incident direction, ξ and ζ are shape coefficients, r_m is the radius of area of presentation from the j incident direction. α, β and λ are coefficients, h is equivalent thickness.

4. Damage Assessment

4.1. Rendezvous Analysis of Space Debris And Satellite

Whether the debris breaks through the target satellite, the number of debris hit the satellite target (based on breakdown) and the target damage probability. The Shot-line method is simple and image, which can well handle the multilayer penetration of debris and the problem of mutual shielding between components as shown in Figure 6.

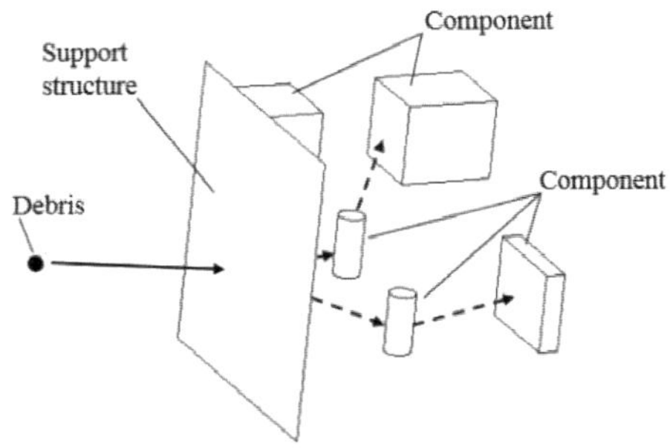

Figure 6. The diagrammatic sketch of Shot-line method

The impact process is Calculated between debris and satellite target, and which parts may be hit by the debris. The coordinates of the impact point and the impact angle can be obtained by calculating the rendezvous of the debris and the components Whether the component structure of satellite target can be broken down or not, the single-layer trajectory limit equation [9] is adopted.

$$t = 4.082 \times d\left(\sigma / \sigma_t C^2\right)^{-0.227}\left(\rho_p / \rho_t\right)^{0.503}(V / C)^{0.56} \tag{3}$$

where d is debris diameter, V is debris velocity, ρ_P is debris velocity, ρ_t is equivalent aluminum plate density, σ is equivalent aluminum plate yield limit, t is the thickness of the equivalent aluminum plate.

The debris penetrates the component structure of the satellite target, assumed that the motion direction of the debris remains unchanged during the rendezvous process. Characteristic parameters such as residual velocity, residual mass and deflection angle of debris are calculated by THOR equation [6,10] after the rendezvous of debris and satellite components.

$$\begin{aligned}
V_{ijr} &= V_{ijs} - 0.3048 \times 10^{c11} \times \left(61023.75 h A_{ij}\right)^{c12}\left(15432.1 m_{ijs}\right)^{c13}\left(\sec\theta_{ij}\right)^{c14}\left(3.28084 V_{ijs}\right)^{c15} \\
V_{ijo} &= 0.3048 \times 10^{c21} \times \left(61023.75 h A_{ij}\right)^{c12}\left(15432.1 m_{ijs}\right)^{c23}\left(\sec\theta_{ij}\right)^{c24}\left(3.28084 V_{ijs}\right)^{c25} \\
m_{ijr} &= m_{ijs} - 0.3048 \times 10^{c31} \times \left(61023.75 h A_{ij}\right)^{c32}\left(15432.1 m_{ijs}\right)^{c33}\left(\sec\theta_{ij}\right)^{c34}\left(3.28084 V_{ijs}\right)^{c35}
\end{aligned} \tag{4}$$

where V_{ijr} is debris residual velocity, V_{ijs} is debris impact velocity, V_{ijo} is ballistic limit velocity, h is target material thickness, A_{ij} is debris impact area, m_{ijs} is debris initial mass, m_{ijs} is debris residual mass, θ_{ij} is the angle between the velocity direction and the target normal, c11 ~ c35 is material constant or coefficient [10].

4.2. Damage Calculation of Components

During the rendezvous operation between debris and target satellite, each impact between debris and target is considered as an independent event. During each rendezvous, each event uses the initial state of the satellite to calculate the path. It's ignored the damage effect of previous debris on satellite targets, in the rendezvous operation of debris and target satellite. The damage to the satellite target caused by debris and debris is independent of each other. The damage probability of the satellite target is calculated by n hits in Eq. (5).

$$P_{K/H}^{(i)} = 1 - \left(1 - P_{k/h}^{(i)}\right)^n \quad (5)$$

where $P_{k/h}^{(i)}$ is the kill probability of a single debris hitting the component, n is Times of being impacted by the debris.

4.3. Target Damage Calculation

The damage to the satellite was divided into five levels: Damage of position and attitude control system (W level); Damage of communication system (T level); Damage of power system (D level); Damage of functional system (G level) and satellite damage (K level). The damage of the target can be calculated by component damage as shown in Figure7.

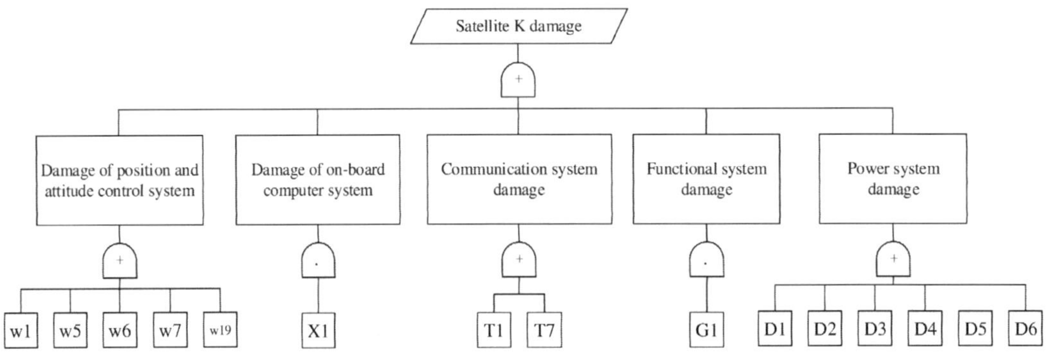

Figure 7. K damage tree for satellite (part)

where W1 is Earth sensor damage, W5 is Sun sensor damage, W6 is Gyroscope damage, W7 is Flywheel damage, W19 is Propulsion system damage, X1 is On-board computer damage, T1 is Damage of Telemetry Command System, T7 is Communication subsystem damage, G1 is Functional damage, D1 is Solar wing A damage, D2 is Solar wing B damage, D3 is Power control equipment damage, D4 is Battery damage, D5 is Power converter damage, D6 is Power distribution and cable net damage.

The relationship between components in damaged tree is "or", and the formula for calculating combined damage probability is shown in Eq. (6).

$$P_{K/H} = 1 - \prod_{i=1}^{n}\left(1 - P_{K/H}^{(i)}\right) \quad (6)$$

The relationship between components in damaged tree is "and", and the formula for calculating combined damage probability is shown in Eq. (7).

$$P_{K/H} = \prod_{i=1}^{n} P_{K/H}^{(i)} \quad (7)$$

where $P_{K/H}$ is component damage probability, n is component number under the damage tree, $P_{K/H}^{(i)}$ is component damage probability under the damage tree.

5. Summary

The damage assessment method of space debris to satellite considers the distribution of space debris, the structure of satellite key components and satellite components, and the equivalent of satellite components with finite element method proposed in this paper. It has important reference value for the impact risk assessment and the satellite protection optimization design of the current spacecraft. So the damage of satellite temperature control system is ignored in debris impact satellites. Further research is needed to quantify the damage criterion of temperature control system.

Acknowledgments

The work was supported by National Natural Science Foundation of China (Grant No. 11802141) and Postgraduate Research & Practice Innovation Program of Jiangsu Province (Grant No. KYCX18_0465).

References

[1] Rbital debris quarterly news [EB/OI]. (2011-01).http://orbital debris. jsc. nasa, gov/newsletter/ newsletter, html.

[2] C. Liou. NASA's ORDEM2010 Status[R].28th Inter-agency Debris Corditn Committee Meeting (IADC), Trivandrum, India, 9-12 March 2010.

[3] H. Krisko, S. Flegel, M.J. Matney, D. R. Jarkey, V. Braun. ORDEM 3.0 and MASTER-2009 modeled debris population comparison[J]. Acta Astronautica, 2015,113.

[4] CIILMAC111; K K N. A combinatorial geometry computer description of the M578 light recovery vehicle:AD-A1411945[R].USA: US Army Ballistic Research Laboratory,1984.

[5] Ao L I, Xiang-Dong L I, Xian-Kun G E, et al. Assessment of Fragmentation Warhead Damaging Typical Phased Array Radar[J]. Journal of Ballistics, 2015.

[6] Ueblood J W. A case study of a combat helicopter's single hit vulnerability. [J]. A Case Study of A Combat Helicopters Single Unit Vulnerability, 1987.

[7] Annides R T, Pany T, Gibbons G. Known Vulnerabilities of Global Navigation Satellite Systems, Status, and Potential Mitigation Techniques[J]. Proceedings of the IEEE, 2016, 104(6):1174-1194.

[8] J, Cui W, Man X, et al. New satellite structure design technology[J]. Chinese Space Science & Technology, 2010.

[9] Courpalais B G. Hypervelocity Impact Investigations and Meteoroid Shielding Experience Related to Apollo and Skylab[J]. 1985.

[10] G-Gang L U, Yang S Q. The Analytical Model Of KE-rod Piercing Armor Based on THOR Formula[J]. Journal of Projectiles. rockets. missiles & Guidance, 2005.

IOP Conf. Series: Materials Science and Engineering **538** (2019) 012060 doi:10.1088/1757-899X/538/1/012060

Base on PBN's General Aviation Airspace Planning Research

Jing Mu[1, 2]

[1]Chengdu Technician College, Chengdu City, Sichuan Province, China
[2]Chengdu Industry and Trade College, Chengdu City, Sichuan Province, China

Corresponding author: Xianhao Zhang , cafczxh@126.com

Abstract: At present, all kinds of general aviation in China developing fast. The market of general aviation facing the never seen before appeared challenges. Improve the utilization of airspace resources and expand operational capability of general aviation have become very urgent task. Nowadays PBN Navigation can be used for the operation of general aviation. Improving the utilization of airspace resources, promoting the quality of the flight operation and ensuring flight safety under the same condition. This paper put forward the basic procedure method and operational recommendations for PBN based airspace planning by studying the PBN based airspace planning method.

1. Introduction

With the continuous development of China's low altitude opening policy, China's general aviation is developing rapidly. At present, the total number of general aviation aircraft operated by our country has exceed 2000. General aviation has been widely used in pilot training, emergency rescue, line pipe patrol, forest-fire prevention, marine aviation commute, tourism and sightseeing, aerial photography, entertainment aviation and many other service field. General aviation has become an important part of improving China's civil aviation service level and promoting economic development as same as transport aviation.

The slow advance and slow development of China's low altitude opening policy of air traffic have much to do with the unclear usage method, characteristics of China's air traffic in the area of air traffic and the imperfect way of supervision over air traffic. This resulted in extensive development of general aviation airspace, inadequate use of supervision, and ineffective protection of flight safety. These objective factors are also the important reasons why the low altitude opening of China's general aviation has not yet achieved a major breakthrough and the general aviation can not improve the quality of service. At present, China's general aviation development policy signal is strong, industrial capital and investment funds are highly concerned. China's general aviation will usher in a new round of development boom, the general airport serving general aviation flight will also usher in a new construction climax. With the support of the policy, our country should apply the new PBN technology to general aviation to realize the healthy and rapid development of general aviation in China.

The superiority of PBN enables it to better integrate with general aviation. Higher requirements for track control, safety and flexibility are put forward by the greater differences between general airports and operating areas. The complexity of fleet composition and the flexibility of business content urgently need a unified national and even global operation mode. The implementation of PBN in

A Project Supported by:
1. Scientific Reserch Fund of SiChuan Provincial Education Department: "Research on Key Technologies of general aviation airspace planning based on PBN"(18ZA0012);
2. Civil Aviation Safety Capacity Building Project: "Research on PBN-based ADS-B IN Operation Approval and Flight Test Verification and Demonstration Application" and"Research on Safety Management Model and Policy of General Aviation Operation"

general airport can no longer rely entirely on ground navigation facilities as traditional flight procedures, such as VOR, DME, NDB, ILS and so on. Thus reducing the construction and maintenance of ground equipment and significantly reducing airport construction. The PBN program is implemented for general aircraft. The optimal flight path can be implemented horizontally and the sustainable descent can be implemented on the vertical section, which can reduce flight time, increase traffic flow and airspace capacity, reduce fuel consumption and greenhouse gas emission. Not only reduce operating costs, but also green environmental protection.

2. Research Status and Development Trend at China and Abroad

Currently, navigation based on PBN has been widely used in transport aviation, and can also be applied to general aviation. With the wide application of PBN airborne equipment and the reduction of cost, it is possible to install various types of PBN airborne equipment in general aviation aircraft. At the same time, transport the airport PBN flight programming capabilities can also be easily transferred to general aviation airports on the basis of VFR and IFR flight procedures. By applying the latest PBN technology to general aviation, it can improve the flight safety of general aviation, use airspace more efficiently and conduct flight supervision more conveniently. Therefore, it can effectively solve the problems of high difficulty of general aviation airspace delimitation, high efficiency in airspace use, flight safety and low altitude supervision. Break through the technical bottleneck of general aviation in the use of airspace, and promote the rapid development of general aviation services.

China has signed a letter of intent with the International Civil Aviation Organization （ICAO） for the Asia Pacific Regional PBN flight procedure office, which is in line with the international practice in terms of technology application, program design concepts and personnel training. China's civil aviation PBN implementation will be divided into three stages, the recent implementation of PBN key applications, the medium-term implementation of PBN comprehensive applications, and long-term implementation of PBN and CNS/ATM system integrations. At present, Lhasa, Linzhi, Lijiang, Xiamen and other airports PBN program has completed the design and flight test validation, accumulated a certain amount of PBN operation experience.

3. General Aviation Airspace Planning Method Based on PBN

According to the characteristics of general aviation development in China and the experience of applying PB technology to general aviation development in Europe and America, the general aviation airspace planning based on PBN mainly includes the following aspects:

3.1. Application Scope and Capability of PBN Navigation in General Aviation Airspace

Currently, China's general aviation low altitude airspace is classified by controlled airspace, surveillance airspace, reporting airspace and visual flight routes. Instrumented flights are allowed in controlled airspace and visual flights are allowed in surveillance, reporting airspace and visual flight routes. PBN navigation can be used not only for instrument flight but also for visual flight guidance. By studying the characteristics of general aviation for various types of flights, the range and capability requirements of PBN navigation are determined. For example, the feasibility of applying PBN technology in flight transit training, oil pipeline patrol, highway rescue, City patrol and other specific flights, the requirements for airborne equipment, the use of navigation equipment, and the determination of possible navigation specifications.

3.2. Characteristics and Requirements of General Aviation in Various Airspace Uses

The general aviation flight area can be divided into airport area, flight area and operation area. But general aviation has many kinds of flight, wide range and low altitude. The requirements of different general aviation flights to airspace have their own characteristics and require more complicated. This project will further refine the characteristics of the airspace requirements for different types of general aviation flights, more clearly define their spatial scales and the relationship with other airspace, and provide a technical basis for the effective designation and efficient use of airspace.

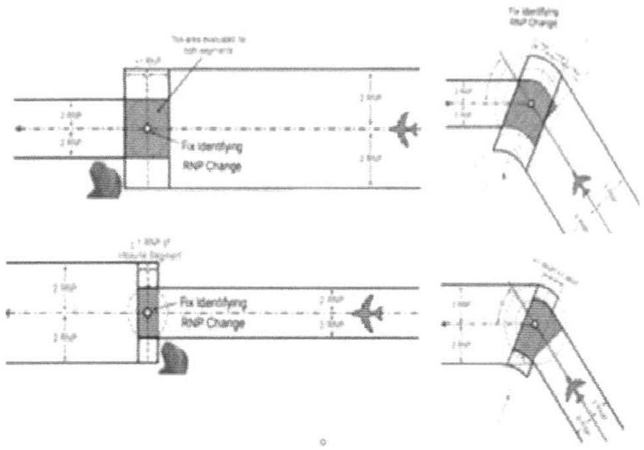

Figure 1. Characteristics of airspace usage based on PBN

3.3. Define the PBN Planning and Design Methods and Recommendations for General Aviation Airspace

According to the characteristics of different types of general aviation, refine the navigation performance and accuracy requirements of general aviation in different flight types, flight stages and different airspace environments. To determine the space scale and related performance requirements of the PBN program, and set a alarm threshold in the PBN airspace to ensure the flight in the designated area. And then formulate specifications and standards for all kinds of general aviation mm airspace planning and design. Define programming requirements and navigation equipment requirements.

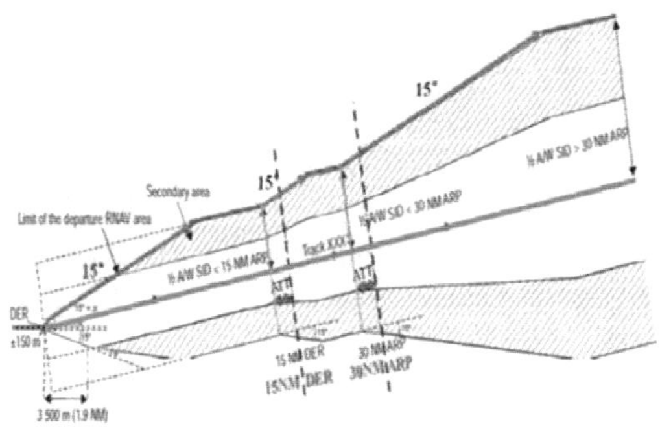

Figure 2. Design of general aviation PBN instrument and visual program

4. The Next Stage Needs to Be Completed

Research on the characteristics of various types of general aviation flight, define the feasibility of applying PBN technology in various specific flights, the scope and capability requirements of PBN navigation, and determine the navigation modes that can be used in various general aviation flights.

According to the area and type of general aviation flight and from the relationship between the space scale and the width of the protection area of all kinds of generally aviation flight with the flight safety guarantee such as take-off, climb, cruise, operation and descent of aircraft. Determine the relevant technical parameters required for general aviation airspace planning based on PBN to provides technical support for effective delimit and efficient utilization of airspace.

Define the width of protective areas for all kinds of general aviation flight, the connection of protective areas and turning areas in different flight phases. And propose the general aviation airspace planning method and recommended standard specification base on PBN.

5. Concluding Remarks

The application of PBN in general aviation planning can provide the corresponding technical reference for general aviation airspace planning, improve the efficiency and rationality of general aviation airspace planning. Fully tap the airspace resources of our country and expand the airspace of general aviation flight. Effectively solve the difficulties of general aviation airspace delimitation, airspace efficient use, flight safety and low altitude supervision. To break through the technical bottleneck of low altitude airspace development and promote the rapid development of general aviation services.

At present, general aviation has begun to widely use airborne GAMIN navigation and display equipment such as GPS. Through this project, the airspace of general aviation flight planning can be integrated into navigation system, which can be widely used in general aviation, improve the flight safety of general aviation, and can use airspace more effectively.

Using PBN program for airspace planning can ensure the guidance of general aviation in the flight route and area, enhance the flight safety under visual conditions. It can ensure the interval between general aviation aircraft and other airspace aircraft and enhance the ability of general aviation aircraft to use airspace dynamically. Through the combination of geographical information system to enhance the situational awareness of general aviation aircraft, improve the safety interval between aircraft and aircraft, between aircraft and terrain, improve the planning and management of airspace based on PBN, and promote the rapid development of general aviation.

Acknowledgments

A Project Supported by Scientific Reserch Fund of SiChuan Provincial Education Department: "Research on Key Technologies of general aviation airspace planning based on PBN" (18ZA0012);
Civil Aviation Safety Capacity Building Project: "Research on PBN-based ADS-B IN Operation Approval and Flight Test Verification and Demonstration Application" and "Research on Safety Management Model and Policy of General Aviation Operation"

References

[1] Analysis of key technologies of airspace planning in modern air traffic management [J]. Chen Xu. Silicon Valley. 2014 (03)
[2] Terminal Airspace Planning and Design Method Based on Flexible Airspace Use [J]. Zhang Ming, Han Songchen. Traffic Information and Safety. 2013 (05)
[3] Research on Key Technologies of air traffic planning in air traffic management [J]. Deng Qi Wen. China new communications. 2018 (03)
[4] Realize the "double flying" of transport aviation and general aviation [N]. Xu Yegang, Zhao Yu, CAAC Daily. 2016-01-22 (001)
[5] Volume 2 Construction of Visual and Instrument Flight Procedures (Volume 2)Aircraft Operations. International Civil Aviation Organization. Doc 8168-OPS/611 . 2014
[6] Performance-based Navigation (PBN)Manual (DOC.9613). . 2008

IOP Conf. Series: Materials Science and Engineering **538** (2019) 012061 doi:10.1088/1757-899X/538/1/012061

Joint Optimization of Level of Repair Analysis and Civil Aircraft Inventory System Based on PSO Algorithm

Yuchang Liu[a], Yunwen Feng, Xiaofeng Xue and Cheng Lu

Northwestern Polytechnical University, School of Aeronautics, Xi'an, China

Corresponding author: lyc93424@126.com

Abstract. The traditional optimization approach for the level of repair analysis (LORA) of civil aircraft and inventory system allocation is to optimize repair level and spare parts allocation separately. But the joint cost of maintenance and inventory allocation policy will not be the necessarily optimal, due to the fact that the direct effect on the inventory system is neglected in the LORA process. Thus, in this paper, a joint optimization model for LORA and civil aircraft inventory system is constructed with the system total cost as the objective function and the fleet availability as the constraint. Next, the particle swarm optimization (PSO) algorithm is applied to solve the proposed multivariable nonlinear optimization model. Finally, some components of the landing gear system are selected as the examples, and through the comparison with the results from traditional sequence method and iterative algorithm, the proposed method is validated to be feasible and effective.

1. Introduction

Level of repair analysis (LORA) has a significant impact on civil aircraft supportability and maintainability, it generally starts from the aircraft design and development stage to determine which failed components to repair or discard and which echelon to conduct the repair or discard decision. Especially for the expensive products like the engine, landing gear and cabin door systems, LORA can formulate more economical maintenance strategies to allocate the logistics support resources by identifying the best maintenance options and the best repair locations.

Although an estimated cost for spare parts stocking can be considered in the LORA process, this cost is too rough to mimic the accurate stocking policy. Therefore, in the traditional optimization method for LORA and inventory system allocation, the repair level and spare parts stocking are optimized separately. So, the first step in the traditional sequence method is to find the best maintenance strategy, and then according to the repair level decided from the previous step to get the most economical inventory configuration policy. But the joint cost will not be the necessarily optimal, due to the fact that repair or discard option is only decided by the repair or discard cost. The backorder levels, transportation time between echelons as well as the difference of repair time at different locations are neglected in the LORA process. For example, if there is a component decided to be repaired at the central depot in LORA, but actually, it will take more time for transportation and repair when comparing the base repair option, which means it will result in higher number of backorders and inventory quantity. So, in order to acquire the same fleet availability and the minimum system total cost, repair at the base level may be the best choice. Therefore, it is necessary to establish the joint optimization model for LORA and inventory system.

To the best of our knowledge, the first model for solving LORA problem is proposed by Barros, and the problem is considered as an integer linear programming model [1]. Saranga and Kumar develop a different model with Barros, and the genetic algorithm is applied to solve this new nonlinear problem [2]. Wu et al. use an improved particle swarm optimization (PSO) algorithm to solve a three-echelon three-indenture LORA model for civil aircraft [3]. Basten et al. establish a minimum cost flow model with side constraints, and remarkably decrease the computation time [4]. Xue et al. propose a concept of LORA decision flow, and take into consideration the maintenance characteristics of the aircraft in the economic analysis model [5]. Jia et al. combine the analytical hierarchy process (AHP) and set pair analysis (SPA) to present a comprehensive analysis model of AHP-SPA for the LORA problem [6]. Besides, a large number of scholars have conducted researches on spare parts stocking problem. Sherbrooke develops the METRIC model (Multi-Echelon Technique for Recoverable Item Control), which is considered as the seminal work in this field [7]. Sun et al. establish a multi-echelon inventory optimization model based on VARI-METRIC model, and use the marginal analysis to optimize the inventory of civil aircraft components [8]. Francesco et al. combine the Italian Air Force data to research the multi-echelon multi-indenture inventory configuration of spare parts based on VARI-METRIC model and marginal analysis method [9]. Basten et al. study the existing single-echelon and multi-echelon inventory configuration models and further explore the scope of application of these classic models [10]. Feng et al. integrate the maintenance rate into the METRIC model, and accomplish the multi-echelon inventory allocation of civil aircraft spare parts [11]. What's more, in order to deal with the defect of the sequence method, Basten et al. develop a joint optimization model and an integrated algorithm, but this method is restricted to two-echelon single-indenture problems [12]. Both Basten and Xue propose to use the iterative algorithm to solve the generalized multi-echelon multi-indenture problem [13, 14]. However, the iterative algorithm is only used for solving symmetrical network problems, it assumes each component owns the same configuration quantity in every base, which is insufficient in practice. Although it can be extended to the asymmetric problems, the extended iterative algorithm is quite complex and difficult to program due to the frequent iterative process for inventory cost.
Therefore, in order to solve the problems above, a joint optimization model of LORA and inventory system is established in this paper with the system total cost as the objective function and the fleet availability as the constraint. Then, the PSO algorithm is applied to solve the proposed asymmetric model due to its characteristic of easy programming and the superior ability for direct search and global optimization. Finally, some spare parts from the civil aircraft landing gear system are chosen as an example, and the model proposed in this paper is validated to be feasible and effective through comparing with the traditional sequence method and iterative algorithm.

2. Assumptions and mathematical model

2.1. Assumptions
In the field of civil aviation, the airlines prefer to use the two-echelon maintenance support system with a number of bases and one central depot as the Fig.1 shows below.

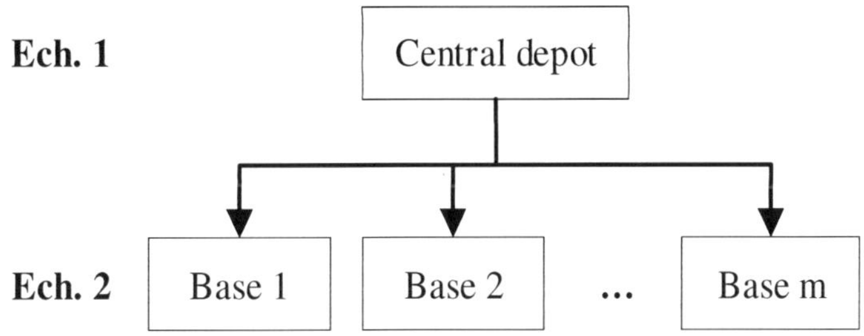

Figure 1. Two-echelon inventory system.

When spare parts demands occur at the bases, the central depot will provide direct replenishment to the base stocks. And most capital goods in the civil aircraft consist of several line replaceable units (LRUs), which can be replaced by functioning spare parts at the base echelon. Besides, the LRUs are made up of a number of subcomponents, shop replaceable units (SRUs), which are generally repaired at the central depot as shown in the Fig. 2.

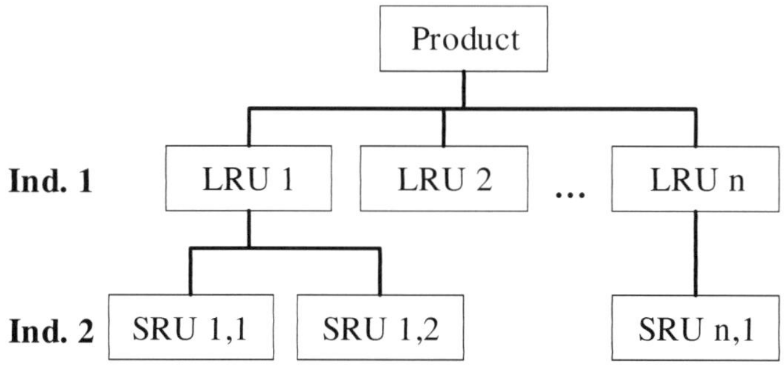

Figure 2. Two-indenture component system.

In this paper, combined the stocking theory of the METRIC model, this paper develops a joint optimization model with two-echelon and two-indenture. In order to simplify the modeling process and improve the engineering applicability, some basic assumptions are proposed here:

- The LRUs can be repaired at the base echelon, or transferred to the central depot for repair.
- The SRUs only can be repaired or discarded in the depot.
- If an LRU is chosen to be discarded, it means all the subcomponents will be discarded together.
- All the discard options will be performed at the central depot only, since the newly purchased components in practice will be sent to the depot generally.
- The (S-1, S) stocking control policy (one for one replenishment) is applied to each location.
- Component faults yield to the Poisson distribution.
- Replacement of a defective component takes zero time, but the repair time, transportation time, as well as the reorder time will be taken into consideration.
- The repairs are considered always successful.
- Spare parts are generally allocated in both levels unless the failed components are discarded or repaired in the central depot.
- There are no lateral transshipments allowed among the bases.

2.2. *Joint optimization model*

Let the maintenance support system consist of two levels with m bases (J_1, ..., J_m) and one central depot (J_0). The product is made up of some LRUs and several SRU subcomponents, the LRU total number I is set to n. And the maintenance option P includes $\{r, d, t\}$, represent repair at base, discard and transfer respectively. Then define the nonlinear joint optimization model for two-echelon two-indenture LORA and spare part stocks as follows:

$$\min \sum_{j=0}^{m}\sum_{p\in P}\sum_{i\in I} vc_{j,p,i}\cdot\beta_i\cdot x_{j,p,i} + \sum_{j=0}^{m}\sum_{i\in I} fc_{j,r,i}\cdot x_{j,r,i} + \sum_{j=0}^{m}\sum_{i=1} hc_i\cdot s_{j,i} \tag{1}$$

Subject to:

$$\sum_{p\in P} x_{j,p,i} = 1,\ \forall j\in\{J_1,\ldots,J_m\},\forall i\in I \tag{2}$$

$$x_{j,t,i} \le \sum_{p\in P} x_{J_0,p,i},\ \forall e \in \{J_1,\ldots,J_m\},\ \forall i \in I \tag{3}$$

$$x_{j,d,i} \le x_{j,d,i1},\ldots,x_{j,d,i\theta},\ \forall j \in \{J_0,\ldots,J_m\}, \forall i \in I \tag{4}$$

$$x_{j,p,i} \in \{0,1\},\ \forall j \in \{J_0,\ldots,J_m\}, \forall i \in I, \forall p \in P \tag{5}$$

$$A \ge A_{\min} \tag{6}$$

in which, $vc_{j,p,i}$ is defined as the variable cost for component i (i=1,...,n) at location $j \in \{J_1,\ldots,J_m\}$ to perform the maintenance option $p \in P$. Besides, $fc_{j,p,i}$ is the fixed cost for component i when the repair option is chosen at location j. Generally, the variable costs consist of labor costs, transportation costs and storage costs. And the fixed costs may include the maintenance tools, equipments as well as technical documents. Additionally, β_i represents the annual average number of failures for each component. $s_{j,i}$ is defined as the spare parts quantity of every component at each location, and hc_i is the unit price for each component.
What's more, x is the decision variable defined in this paper:

$$x_{j,p,i} = \begin{cases} 1, & \text{if the component } i \text{ chooses the maintenance option } p \text{ at the location } j \\ 0, & \text{else} \end{cases} \tag{7}$$

There are three parts in the objective function, the maintenance variable costs, repair fixed costs and the spare parts costs. The constraint (2) assures that for each component a decision will be made at the base level. Constraint (3) makes sure that there will be an option made at the central depot, if a component decides to transfer. Constraint (4) guarantees that when an LRU is discarded, all the subcomponents will be discarded together. Constraint (5) is value range for the decision variable, and Constraint (6) is the minimum requirement for the fleet availability A.

2.3. Inventory allocation model

In the joint optimization model, the spare parts cost is a significant part, so it is necessary to elaborate the inventory allocation model and theory.

2.3.1. Annual average demand. Component faults may occur at any time during the operation of the aircraft, and the average annual demand for each component is equal to the average replacement in one year.

$$\lambda_i = \frac{FH \cdot QPA_i \cdot N}{MTBUR_i} \tag{8}$$

where λ_i is the annual average demand for each component, the aircraft annual flight time is FH, and QPA_i is the installation number for each plane. N is the number of fleet size and $MTBUR_i$ is the mean time between unscheduled removals for every component.
The plane may pass through multiple bases in one route, so the demand of each base can be calculated based on the flight distance.

$$\lambda_{j,i} = \frac{D_j}{\sum D} \lambda_i \tag{9}$$

where $\lambda_{j,i}$ is considered as the annual demand of base j for each component, and D_j is the distance between base j and the former base.

And the average annual demand of central depot is:

$$\lambda_{0,i} = \sum_{j=1}^{m} \lambda_{j,t,i} \cdot x_{j,t,i} \tag{10}$$

in which $\lambda_{0,i}$ is the average annual demand of depot for each component.

2.3.2. Annual average number of supply channel. μ0,i is defined as the number of supply channel for the central depot, $RT_{0,i}$ is the expected repair time in depot maintenance spot, and $Ts_{0,i}$ is the expected reorder time for every component. According to the assumption, component faults obey Poisson distribution. So, the equation below can be derived by Palm theorem.

$$\mu_{0,i} = \begin{cases} \lambda_{0,i} \cdot RT_{0,i}, & x_{0,r,i}=1 \\ \lambda_{0,i} \cdot Ts_{0,i}, & x_{0,d,i}=1 \end{cases} \tag{11}$$

And according to Palm theorem, the number of supply channel for each base is:

$$\mu_{j,i} = \begin{cases} \lambda_{j,i} \cdot RT_{j,i}, & x_{j,r,i}=1 \\ \lambda_{j,i} \cdot (TRT_{j,i} + MWT_{0,t,i}), & x_{0,r,i}=1 \\ \lambda_{j,i} \cdot (TRT_{j,i} + MWT_{0,d,i}), & x_{0,d,i}=1 \end{cases} \tag{12}$$

where $RT_{j,i}$ is the expected repair time in each base, and $TRT_{j,i}$ is the expected transportation time between the base and depot. Besides, according to the Little formula, the delay time due to depot backorder $MWT_{0,i}$ can be calculated by:

$$MWT_{0,i} = \frac{EBO(s)_{0,i}}{\lambda_{0,i}} \tag{13}$$

2.3.3. Expected number of backorders and fleet availability. The expected number of backorders in the central depot and bases are:

$$\begin{cases} EBO(s)_{0,i} = \sum_{k=s+1}^{+\infty} (k-s) \cdot \frac{(\mu_{0,i})^k e^{-\mu_{0,i}}}{k!} \\ EBO(s)_{j,i} = \sum_{k=s+1}^{+\infty} (k-s) \cdot \frac{(\mu_{j,i})^k e^{-\mu_{j,i}}}{k!} \end{cases} \tag{14}$$

And the fleet availability is calculated as below:

$$A_j = \prod_{i=1}^{n} \left(1 - \frac{EBO(s)_{j,i}}{N \cdot QPA_i} \right)^{QPA_i} \tag{15}$$

3. Algorithm

It is clear that the joint optimization model established here is a multivariable nonlinear problem, which is quite hard to solve. Although Basten develops an iterative algorithm, the method for adding spare parts costs into the variable costs in every iteration is complex and hard to program. So, it is important to find a new algorithm for the proposed model. It is noticed that the PSO algorithm has a great potential to solve such problems, since it is simple to program and has a strong generality.

Additionally, it also owns a superior ability for direct search and global optimization. Thus, in this paper, the PSO algorithm is applied for the joint optimization of LORA and inventory system.
The PSO algorithm steps are summarized as the following:
Step 1: Set the particle quantity Q, and initialize the velocity and position for each particle.
Step 2: Calculate the fitness value for each particle, and the fitness value y is as the same as the objective function (1):

$$y=\sum_{j=0}^{m}\sum_{p\in P}\sum_{i\in I} vc_{j,p,i}\cdot\beta_i\cdot x_{j,p,i}+\sum_{j=0}^{m}\sum_{i\in I} fc_{j,r,i}\cdot x_{j,r,i}+\sum_{j=0}^{m}\sum_{i=I} hc_i\cdot s_{j,i} \tag{16}$$

Step 3: For each particle, compare the fitness value with the best fitness value (P_{best}) in history, and update the fitness value and position of P_{best}.
Step 4: Choose the particle with the best fitness value in history from all the particles, and assign the best as the global best (G_{best}).
Step 5: Update the speed and position values according to equation (17) and (18).

$$V_i^{k+1}=w\cdot V_i^k+c_1\cdot r_1\cdot(Pb_i^k-X_i^k)+c_2\cdot r_2\cdot(Gb_i^k-X_i^k) \tag{17}$$

$$X_i^{k+1}=X_i^k+V_i^{k+1} \tag{18}$$

where V represents the particle vector, X represents the particle position, and k is the number of iterations. w is the inertia weight factor, c_1 and c_2 are the acceleration constants. Besides, r_1 and r_2 are random numbers over the range [0,1]. Pb and Gb are defined as the particle best position and global best position separately.
Step 6: Once the search criteria are attained, the global best fitness value is the searched solution. Otherwise, return to Step 2.
What's more, the marginal analysis method is used in this algorithm for calculating the spare parts quantity $s_{j,i}$. The process of marginal analysis method is elaborated a lot in the papers introduced above, so it won't be repeated here.

4. Computational experiment

In this section, a computational experiment is used here to compare solving the joint optimisation problem using PSO algorithm with using the traditional sequence method and iterative algorithm. 10 components of the civil aircraft landing gear system are selected as the research objects. The fleet is made up of 10 Boeing 787 airplanes with an average flight time of 2950 hours per year as well as the minimum fleet availability at 0.97. And the inventory system consists of 1 central depot and 3 bases as shown in Fig. 3.

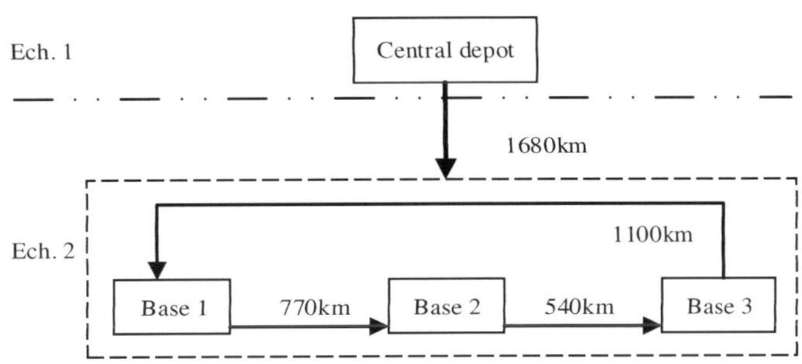

Figure 3. Inventory allocation system for landing gear system.

And the parameters for inventory allocation are given in Table 1. And the input parameters relating to LORA, such as variable costs and fixed costs, are shown in Table 2. Then, the PSO algorithm is applied to solve the example, the computational experiment is carried out with 50 particles with the iteration criterion of 50 steps. Besides, the inertia value w and the acceleration factors c_1 and c_2 are set as 0.7, 0.8 and 0.8 respectively. It is noticed that the PSO algorithm stops at the 4th step, and acquires the new maintenance options as well as the minimum joint cost at 5.011 million dollars.

Table 1. Parameters for inventory allocation of each component.

No.	QPA_i	$MTBUR_i/h$	$TRT_i/year$	$RT_{j,i}/year$	$RT_{0,i}/year$	$Ts_{0,i}/year$	$hc_i/\$$
LRU$_1$	32.0	3975	0.0029	0.012	0.012	0.014	9510
LRU$_2$	3.0	48133	0.0055	0.052	0.028	0.038	149905
LRU$_3$	8.0	1383	0.0035	0.02	0.0195	0.017	35984
LRU$_4$	8.0	4564	0.006	0.081	0.039	0.041	176815
LRU$_5$	2.0	1191	0.003	0.017	0.0165	0.016	15359
LRU$_6$	1.0	4020	0.0042	0.018	0.0175	0.02	65868
SRU$_{6,1}$	2.0	53606	0.00275	0.021	0.012	0.013	7599.7
SRU$_{6,2}$	2.0	4057	0.0027	0.017	0.01	0.013	3374
LRU$_7$	4.0	4639	0.0033	0.013	0.012	0.017	25752
LRU$_8$	1.0	13392	0.006	0.082	0.04	0.042	179161

Table 2. Variable and fixed costs of each component.

No.	β_i	$fc_i/\$$	$VC_{j,r,i}/\$$	$VC_{j,t,i}/\$$	$VC_{0,r,i}/\$$	$VC_{0,d,i}/\$$
LRU$_1$	4	800	3600	360	3500	9510
LRU$_2$	2	5000	35000	480	25000	149905
LRU$_3$	9	1300	9670	420	6100	35984
LRU$_4$	7	5000	40000	500	26000	176815
LRU$_5$	3	900	4995	400	4900	15359
LRU$_6$	1	1500	16200	430	13200	65868
SRU$_{6,1}$	1	500	/	360	2500	7599.7
SRU$_{6,2}$	4	500	/	360	970	3374
LRU$_7$	2	1000	6150	410	6090	25752
LRU$_8$	2	5000	42000	500	25000	179161

In order to verify the feasibility and effectiveness of the proposed algorithm, this example is also solved by the iterative algorithm and sequence method separately. By using the iterative algorithm, the obtained optimal cost is as the same as the result of PSO algorithm, which is 5.011 million dollars, but it is found that the solving process iterates 6 times before stopping. However, the result of the traditional sequence method reaches 5.375 million dollars, which is 6.76% higher than the PSO algorithm and the iteration algorithm. What's more, it can be found that all the three methods have achieved the minimum fleet availability requirement, and the fleet availabilities of PSO and iterative algorithm are slightly higher than the sequence method. Through comparing with the results of iterative algorithm and sequence method, it is cleared that the PSO algorithm can solve the joint optimization problem accurately and effectively. The comparison for results of three methods is shown in Table 3.

Table3. Results comparison for three methods.

	Iteration times	Fleet availability	Joint cost/$	Cost reduction
Sequence method	/	0.9714	5.375×10^6	/
PSO algorithm	4	0.9719	5.011×10^6	6.76%
Iteration algorithm	6	0.9719	5.011×10^6	6.76%

Table 4. Maintenance option and inventory quantity by PSO algorithm.

No.	Base repair cost/$	Depot repair cost/$	Discard cost/$	Inventory quantity
LRU_1	45600	/	/	14
LRU_2	/	157880	/	1
LRU_3	/	177340	/	23
LRU_4	/	561500	/	12
LRU_5	47655	/	/	7
LRU_6	/	42390	/	4
$SRU_{6,1}$	/	9080	/	4
$SRU_{6,2}$	/	16460	/	7
LRU_7	39900	/	/	4
LRU_8	/	158000	/	2
LORA cost/$	1255805			
Inventory cost/$	4118788			
Joint cost/$	5374593			

Table 5. Maintenance option and inventory quantity by sequence method.

No.	Base repair cost/$	Depot repair cost/$	Discard cost/$	Inventory quantity
LRU_1	45600	/	/	14
LRU_2	/	157880	/	1
LRU_3	264990	/	/	12
LRU_4	/	561500	/	12
LRU_5	47655	/	/	7
LRU_6	53100	/	/	3
$SRU_{6,1}$	/	9080	/	4
$SRU_{6,2}$	/	16460	/	7
LRU_7	39900	/	/	4
LRU_8	/	158000	/	2
LORA cost/$	1354165			
Inventory cost/$	3657096			
Joint cost/$	5011261			

What's more, Table 4 and Table 5 show the new maintenance options after the joint optimization by PSO algorithm and the original options decided according to the sequence method. It can be seen that LRU3 and LRU6 change the maintenance option from depot repair to base repair due to the reduction of spare parts allocation quantity by 11 and 1 respectively. Although the LORA cost increase slightly, the inventory cost shows a remarkable decrease because of the reduction of inventory quantity. The reason can be explained by the theory from METRIC model, because the repairs are considered always successful, so the base repair option will lead to zero spare parts demand in the central depot, but all the other options will cause a high demand instead. Thus, once the base repair option is decided, the expected number of backorders in the depot is 0, which will lead the depot inventory allocation quantity to 0, too.

Take LRU3 as an example, the average backorder numbers for all bases by base repair, depot repair and discard option are shown in Fig. 4, and the average backorder numbers of the whole inventory system with the different options are presented in Fig. 5. It can be seen that the bases average backorder numbers share the similar level, although the backorder numbers for base repair and discard are slightly less than the depot repair. But it is clear that when considering the depot backorder numbers, the system average backorder numbers for depot repair option and discard option increase dramatically. Therefore, for LRU3, the base repair option will lead less spare parts quantity than the depot repair or discard, which is the reason for its maintenance option changing.

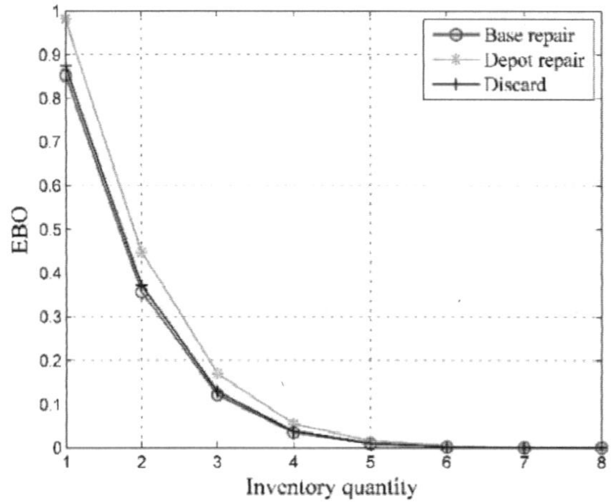

Figure 4. Average expected number of backorders in all bases by different maintenance option.

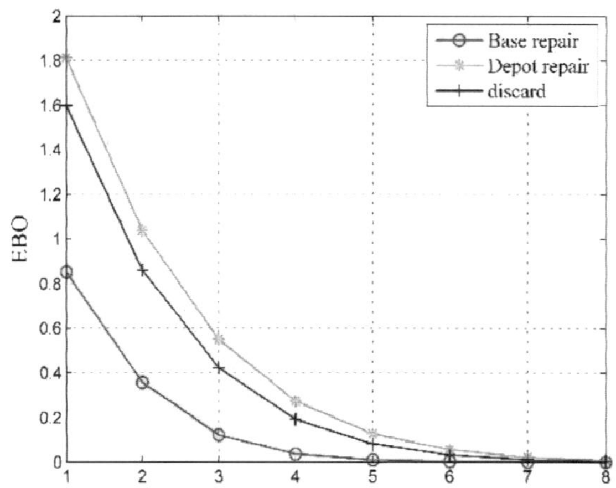

Figure 5. Average expected number of backorders for whole system by different maintenance option.

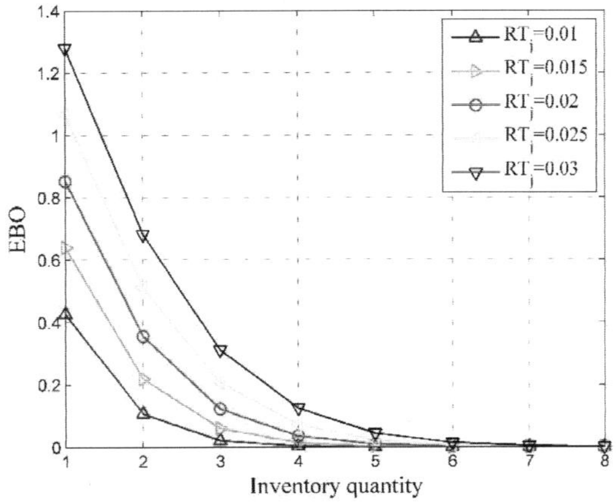

Figure 6. Influence for the inventory system backorder number by the change of base repair time when repair at the base.

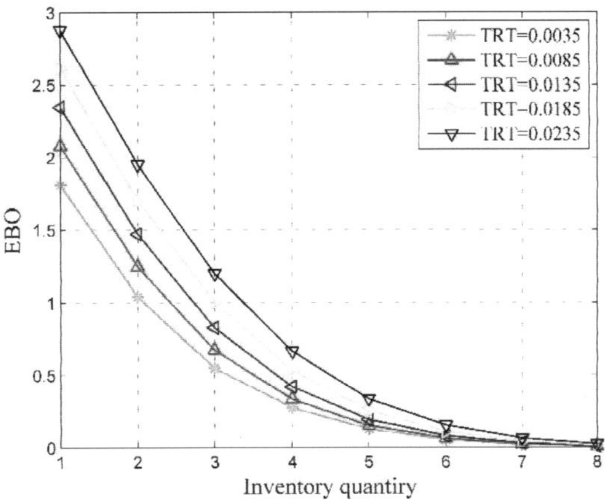

Figure 7. Influence for the inventory system backorder number by the change of transportation time when repair at the depot.

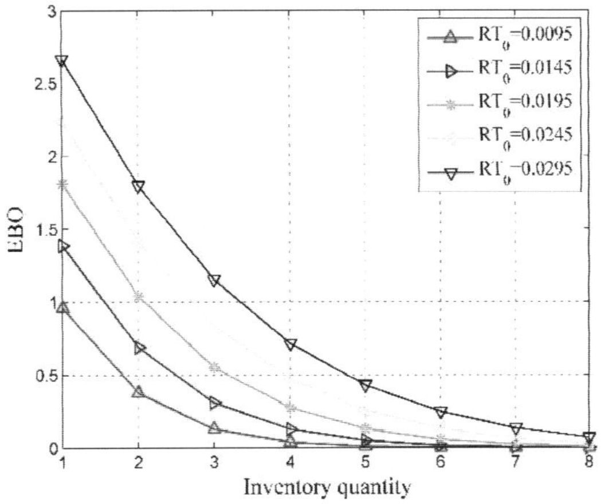

Figure 8. Influence for the inventory system backorder number by the change of depot repair time when repair at the depot.

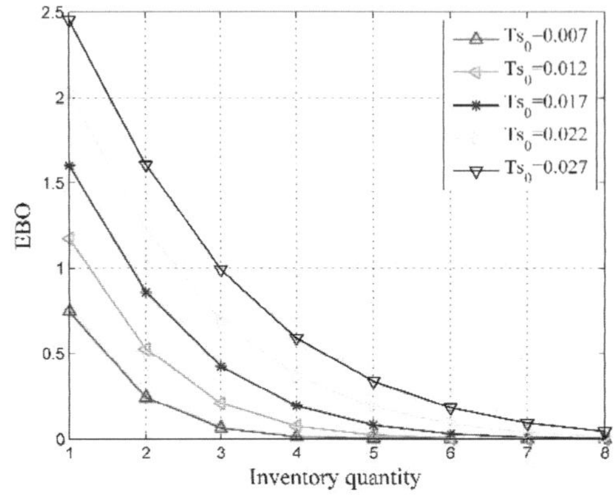

Figure 9. Influence for the inventory system backorder number by the change of reorder time when discard at the depot.

However, the great difference among central depot backorder numbers caused by the METRIC model is only one reason for the repair location change. The maintenance time for repair, reorder and transportation also has an important impact on the backorder numbers. Still take LRU3 as an example, the influences for the inventory system backorder numbers by the changes of maintenance time are shown in the following figures. In Fig. 6 and Fig. 7, it is noticed that the base repair time and transportation time show a slight effect on the backorder numbers. However, the depot repair time and reorder time present a large impact on the backorder numbers in Fig. 8 and Fig. 9. Especially, if the depot repair time drops to 0.0095 and the base repair time rises to 0.025, the system backorder number of base repair will exceed that of the depot repair, which means repair at bases will cause more spare parts stocks and the depot repair option will be the better choice.

5. Conclusion

In this paper, the following conclusions can be obtained:

(1) Based on the existing research on LORA and inventory allocation, a joint optimization model for LORA and inventory system is established with the system total cost as the objective function and the fleet availability as the constraint.

(2) The PSO algorithm is applied to solve the joint optimization problem, and the solution steps are elaborated in detail.

(3) 10 components of the aircraft landing gear system are selected as the research objects, and the PSO algorithm is validated to be feasible and effective for solving this joint optimization model through comparing with the traditional sequence method and iterative algorithm.

(4) The joint optimization method leads to a joint cost reduction of 6.76%, and in this paper, the reasons for causing maintenance option change and cost decrease are discussed in detail.

Acknowledgments

The authors wish to thank the anonymous referee for the helpful comments. The authors gratefully acknowledge the support of the National Natural Science Foundation of China (Grant No. 5187051969). Besides, the research work has also been supported by the Civil Aircraft Special Scientific Research Technology Research Project (Grant No. MJZ-2016-Y-84).

IOP Conf. Series: Materials Science and Engineering **538** (2019) 012061 doi:10.1088/1757-899X/538/1/012061

References

[1] L. L. Barros, The optimization of repair decisions using life-cycle cost parameters. IMA. J. APPL. MATH., **9**, 430-413 (1998)
[2] H. Saranga, U. D. Kumar, Optimization of aircraft maintenance/support infrastructure using genetic algorithms - level of repair analysis. ANN. OPER. RES., **143**, 91-106 (2006)
[3] H. Wu, H. F. Zuo, W. Sun, A new level of repair analysis optimization model for civil aircraft maintenance. Acta Aeronautica ET Astronautica Sinica, **30**, 247-293 (2009)
[4] R. J. I. Basten, M. C. Heijden, J. M. J. Schutten, A minimum cost flow model for level of repair analysis. INT. J. PROD. ECON., **133**, 233-242 (2011)
[5] T. Xue, Y.W. Feng, X.F. Xue, Y.S. Feng, An aircraft economic evaluation model for level of repair analysis. Acta Aeronautica ET Astronautica Sinica, **34**, 97-103 (2013)
[6] B. H. Jia, L J. Yu, Y. G. Lin, X. Lu, Comprehensive analysis model for determination of civil aircraft repair level based on AHP-SPA. Acta Aeronautica ET Astronautica Sinica, **38**, 178-186, (2017)
[7] Sherbrooke, C. Craig, METRIC: a multi-echelon technique for recoverable item control. OPER. RES., **16**, 122-141 (1968)
[8] L. Sun, H. F. Zuo, W. Liu, J. Xu, LRU multi-echelon inventory optimal model for aircraft parts based on VARI- METRIC model. Journal of Nanjing University of Aeronautics & Astronautics, **45**, 532-537 (2013)
[9] C. Francesco, D. G. Giulio, T. Massimo, Multi-echelon, multi-indenture spare parts inventory control subject to system availability and budget constraints. RELIAB. ENG. SYST. SAFE., **119**, 95-101 (2013)
[10] R. J. I Basten, G. J. Houtum, System-oriented inventory models for spare parts. Surveys in Operations Research and Management Science, **19**, 34-55 (2014)
[11] Y. W. Feng, C. Lu, X. F. Xue, Y. C. Liu, Multi-echelon inventory allocation of civil aircraft spare parts considering maintenance ratio. Journal of Northwestern Polytechnical University, **36,** 582-589 (2018)
[12] R. J. I. Basten, M. C. Heijden, J. M. J. Schutten, Joint optimization of level of repair analysis and spare parts stocks. EUR. J. OPER. RES., **222**, 474-483 (2012)
[13] R. J. I. Basten, M. C. Heijden, J. M. J. Schutten, E. Kutanoglu, An approximate approach for the joint problem of level of repair analysis and spare parts stocking. ANN. OPER. RES., **224**, 121-145 (2015)
[14] T. Xue, Y. W. Feng, X. F. Xue, Research on Joint Optimizing Model of Equipment LORA and Spare Parts Inventory. Mechanical Science and Technology for Aerospace Engineering, **34**, 963-968 (2015)

IOP Conf. Series: Materials Science and Engineering **538** (2019) 012062 doi:10.1088/1757-899X/538/1/012062

Fatigue Analysis of Suspension Control Arm Based on Road Spectrum

Kang Yuan-chun

Hubei University of Automotive Technology, Key Laboratory of Automotive Power Train and Electronics (Hubei University of Automotive Technology), HuBei, ShiYan, 442002

Corresponding author: cookie518@163.com

Abstract. Aiming at precisely predicting fatigue life for suspension control arm, a fatigue life analysis method is presented. First of all, 2D road spectrum is structured using Sayers. Then, a rigid-flexible coupling multi-body model is built for vehicle dynamics simulation based on ADAMS software. Automobile test is simulated and the time-load histories of each connecting point of the suspension control arm are obtained. By means of the rain flow counting method, fatigue analysis load spectrums are obtained. Finally, the fatigue life of the control arm is analyzed. The result of fatigue life analysis shows that the control arm is designed to meet the requirements of durability.

1. Introduction

Suspension system is an important part of automobile, whose performance is an important factor which affects the ride comfort, handling stability and speed of the vehicle. Mcpherson suspension has a simple structure and occupies a small space, so it has been widely used in the car. The control arm is an important part of the McPherson suspension, whose fatigue life has a great influence on the performance of the vehicle.

At present, the fatigue life of the control arm has been studied extensively. The load spectrum of the control arm is obtained through road test and bench test [1-2].

The fatigue life of the control arm is analyzed by finite element method and multi-body dynamic analysis method. First, based on ISO/TC108/SC2N67, GB7031-86 and Sayers empirical model, two dimensional road spectral curves are constructed by applying Sayers theory. Then, the rigid flexible coupling multi-body dynamics simulation model is established in Adams. The motion of the vehicle in the road spectrum is simulated, and the time load history of each connection point of the suspension control arm is obtained. The load spectrum of fatigue analysis is obtained by rain flow counting method. Finally, the fatigue life of the control arm is analyzed.

2. Basic Theory

2.1. Power Spectrum Density of Road Surface

The expression of road power spectral density is recommended by ISO/TC108/SC2N67 and GB7031-86. As follows:

$$G_q(n) = G_q(n_0)(\frac{n}{n_0})^{-w} \tag{1}$$

Where n is spatial frequency (m-1); n0 is reference spatial frequency, n0 = 0.1 m-1; Gq(n) is power spectrum density of spatial frequency; Gq(n0) is road roughness coefficient(m3). According to the above two documents road roughness grading standards, at different levels, Gq(n0) have different values; W is frequency index, W=2. [3]
Taking into account the speed, the spatial frequency power spectral density Gq (n) can be converted to the time frequency power spectral density Gq (f).

$$G_q(f) = G_q(n_0)n_0^2 \frac{u}{f^2} \tag{2}$$

2.2. Road Spectrum in Adams/Car

In Adams/Car, Sayers empirical model shows that there is a relationship between spatial power spectral density and spatial frequency of road profile:

$$G_d(n) = G_e + \frac{G_s}{(2\pi n)^2} + \frac{G_a}{(2\pi n)^4} \tag{3}$$

Where, n is spatial frequency, G_e is the amplitude of white noise of spatial power spectral density. G_s is the amplitude of white noise of velocity power spectrum density. G_a is the amplitude of white noise of acceleration power spectrum density.
The mean value of each white noise is 0, the standard deviation is

$$\sigma = \sqrt{G/2\Delta} \tag{4}$$

where G is G_e, G_s or G_a;
△ is sampling interval. [4]
Road of automobile reliability test include Smooth asphalt pavement, asphalt pavement, Rough asphalt pavement, Smooth cement pavement, cement pavement, Rough cement pavement and so on. Parameters of these road surfaces are shown in Table 1.

Table 1. Parameters of sayers model

Type of road surface	Ge *10-6m3/cycle	Gs *10-6m3/cycle	Ga *10-6m3/cycle
Smooth asphalt pavement	0	6	0
asphalt pavement	0	12	0.17
Rough asphalt pavement	0.003	20	0.20
Smooth cement pavement	0	1	0
cement pavement	0.1	20	0.25
Rough cement pavement	0.1	35	0.3

Based on Sayers methods, the model of 2D road roughness is built in Adams/Car. The result is shown in Figure 1.

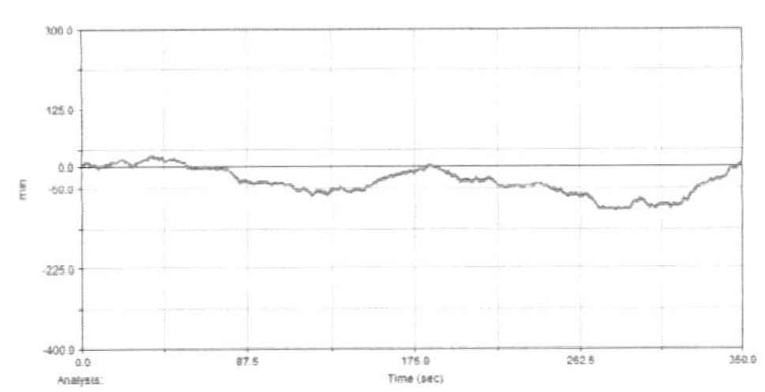

Figure 1. 2D road roughness

3. Multi-body Dynamics Simulation

3.1. Vehicle Road Coupling Model

The front suspension of the vehicle is Macpherson suspension, and the rear suspension is a multi-link suspension. In order to get high precision stress of Macpherson suspension control arm, Macpherson suspension rigid flexible coupling dynamic model is built by finite element method and multi-body dynamics method, in which the control arm is as flexible body. The control arm of the Macpherson suspension is connected to the steering knuckle by a ball joint and is connected to the body by front and rear rubber bushing. According to the coordinates of suspension hard points, a rigid flexible coupling Macpherson suspension model is built based on Adams. As shown in Figure 2.

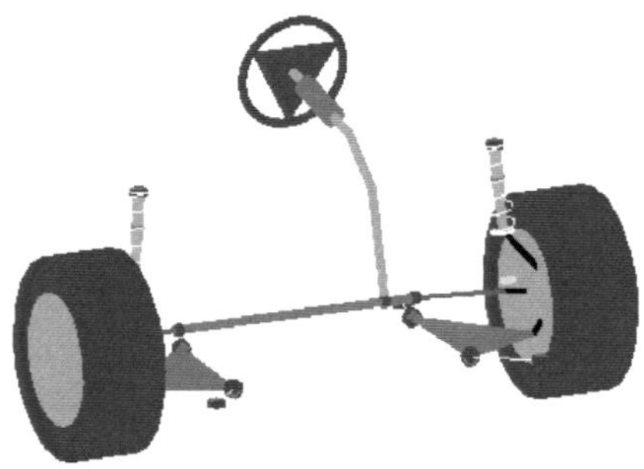

Figure 2. Macpherson front suspension model

Other subsystems, including rear suspension, wheel, steering system, brake, power system, body, etc., are modeled separately and finally these systems are assembled. Simulate the movement of vehicles on the road on the four column test stand in Adams, and the simulate model is shown in Figure 3.

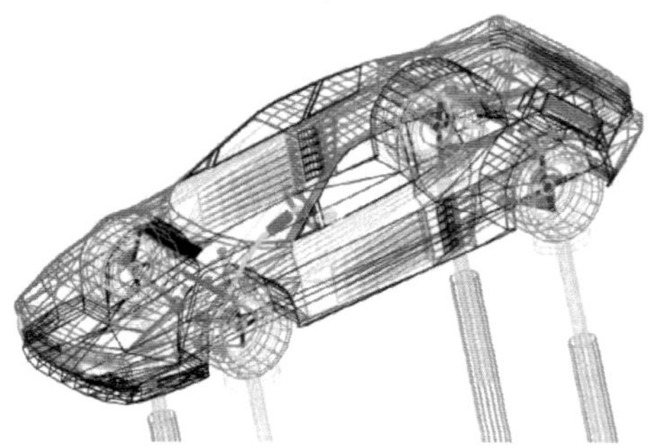

Figure 3. Simulate model

3.2. Multi-body Dynamics Simulation

The result is that the load spectrum of each connection point of the left control arm is obtained. The load spectrums of left control arm rear bushing are shown in Figure 4.

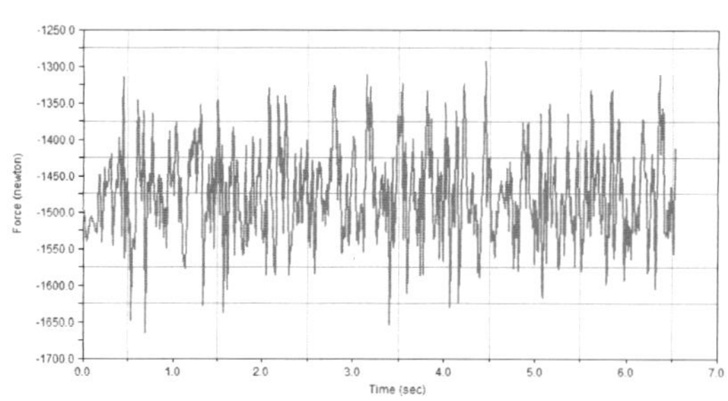

a)X direction

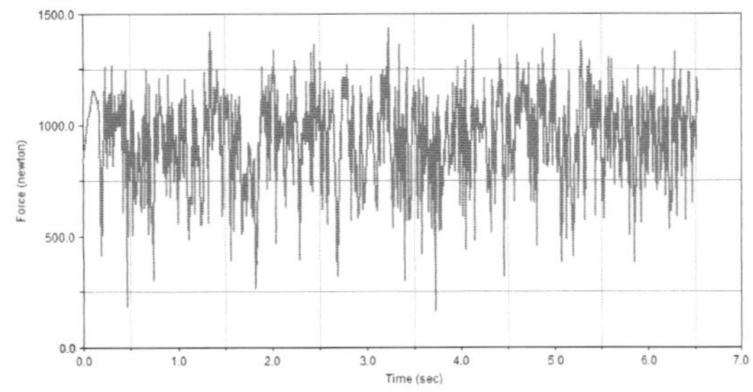

b)Y direction

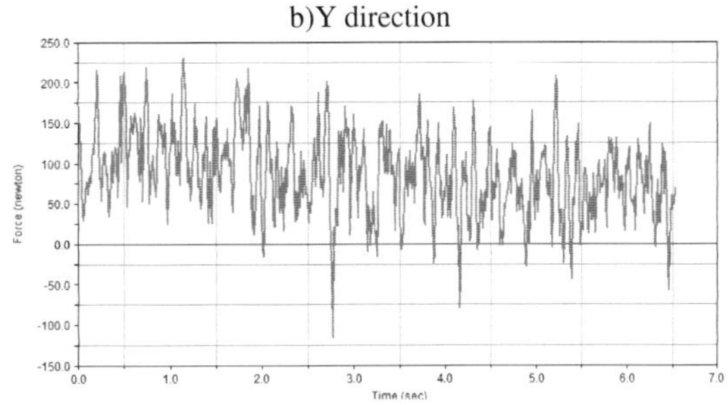

c)Z direction

Figure 4. Load spectrums of left control arm rear bushing

4. Control Arm Fatigue Analysis

4.1. The Finite Element Model of the Control Arm

The CAD 3D model of the Macpherson suspension control arm is imported into the CAE software. At the same time, the control arm model geometry clean and proper simplification is carried out. Then, the control arm is meshed with tetrahedron elements as shown in Figure 5. The material of the control arm is 40cr, whose parameters are shown in Table 2. S-N curve of fatigue analysis can be obtained by Yield strength and Ultimate strength.

According to the traditional method, the static analysis of the control arm is carried out, the front and back bushes of the control arms are fully restrained to prevent rotation of the control arm. This is not in conformity with the actual movement of control arm, and causes the control arm stress is greater than the actual stress. The static analysis and fatigue analysis of the control arm are carried out by using the inertia release.

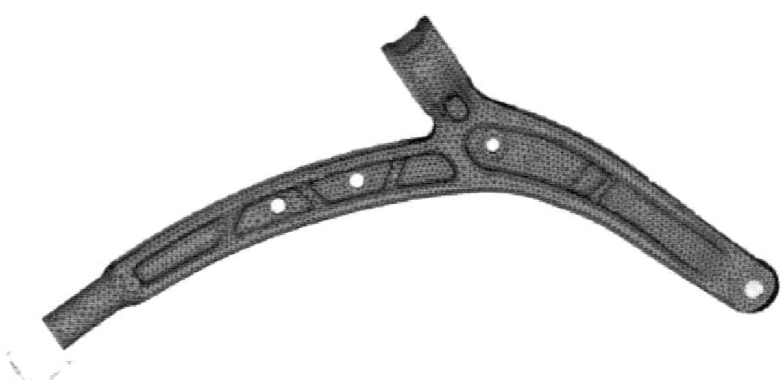

Figure 5. FEA model of control arm

Table 2. Physical property

material	Elastic modulus	Poisson's ratio	Density	Yield strength	Ultimate strength
40cr	2.06 E5MPa	0.3	7.82E-9 t/mm3	780MPa	910MPa

IOP Conf. Series: Materials Science and Engineering **538** (2019) 012062 doi:10.1088/1757-899X/538/1/012062

4.2. Control Arm Fatigue Analysis

According to the results of the dynamic analysis, the load time history of each connection point of the control arm is obtained. These load time histories are simplified as a series of full or half cycles of different amplitudes, which is called the counting method. Rain flow counting method is the most widely used in the study of fatigue.

For the same stress amplitude, the higher the average stress, the shorter the fatigue life. Therefore, the mean stress must be corrected in fatigue analysis. Mean stress correction include Goodman and Gerber. Here, GERBER is used. Give the material S-N fatigue property, specify various parameters, and get the cumulative damage coefficient directly.

The fatigue life of the control arm is shown in Figure 6.

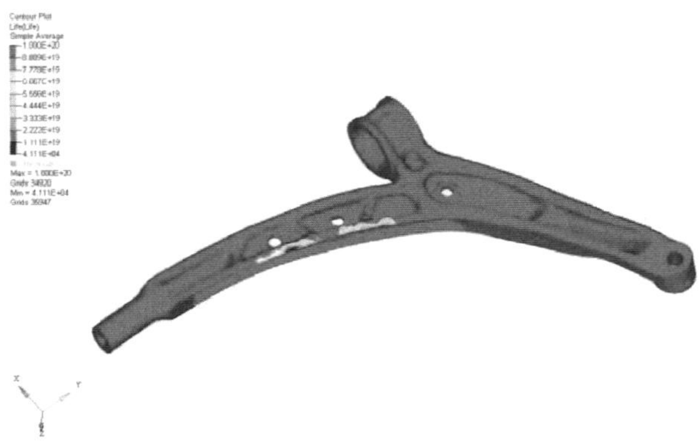

Figure 6. Fatigue life of the control arm

The results show that the minimum fatigue life of control arm is 41110 load cycle. The life of the control arm is 17 years, according to a vehicle running 8 hours a day and 300 days a year. In general, the service life of a vehicle is 15 years, so the control arm is designed to meet the requirements of durability.

5. Conclusion

(1) According to Sayers model, 2D road spectrum is structured. Then the road is structured in Adams, which is used as road model in the vehicle road coupling multi-body dynamic analysis.

(2) The vehicle dynamics model of rigid flexible coupling is constructed. The load time history of the connecting point of the control arm is obtained;

(3) The fatigue life of the control arm was predicted by the method of CAE fatigue analysis.

(4) The fatigue analysis process provides a reference for the fatigue analysis of other automobile parts.

Acknowledgements

The authors gratefully acknowledge the financial support of the Foundation of Key Laboratory of Automotive Power Train and Electronics (Hubei University of Automotive Technology) (Project No. 2015xtzx0407).

References

[1] Qian Lijun, Wu Daojun, Yang Nianjiong, Zhu Anding. Wang Ting. Multi-axial Fatigue Analysis of Vehicle Lower Control Arm Based on Simulated Road Excitation [J]. Automotive Engineering. 2012.3:249-254 (in Chinese).

[2] Liu Yongchen, Wang Guolin, Sun Li. Fatigue Damage Analysis and Life Prediction for Vehicle Control Arm [J]. Transactions of the Chinese Society of Agricultural Engineering. 2013.8:83-91(in Chinese).

[3] Yu Zhisheng. Automobile Theory. China Machine PRESS.2009.3:

[4] Wang Yong-tao, Li Cheng, Ma Zh-lei. Simulation on vehicle ride comfort based on

ADAMS/Car ride[J]. Machinery Design & Manufacture. 2010.4
[5] Hu Zhigang. Research on pavement model system based on VPG virtual test field. Huazhong University of Science and Technology 2011.
[6] Chinese Standard. Reliability running test method for automobiles. GB/T12678-90.1990:1-27
[7] Wu Daojun. Research on the key technologies of vehicle fatigue durability analysis, test and optimization. HeFei University of Technology. 2012.

IOP Conf. Series: Materials Science and Engineering **538** (2019) 012063 doi:10.1088/1757-899X/538/1/012063

Optimization of A Novel Retractable Suspension Structure

Kun Yin[a], Zhongkai Chen, Ye Sun, Xiaodong Liang and Erkang Li

The Northwest Institute of Nuclear Technology, Xi'an Shaanxi 710024, China

[a]Corresponding author: yinkun5991@foxmail.com

Abstract. This paper designs a novel retractable suspension structure for amphibious equipment, combined with the principle and characteristics of the retractable suspension mechanism, and uses ADAMS/Insight to develop a set of optimization strategies that meet the characteristics of its dual-function process. Then based on this optimization strategy, the retractable index and the wheel alignment parameters of the retractable suspension are optimized respectively, and significant results are obtained, which verifies the feasibility of the optimization strategy.

1. Introduction

Amphibious platforms are important for water transport, marine development and military activities in a country. [1] The walking system of amphibious platform is one of the most critical factors to realize the high speed in water and high adaptability in land. Based on this, we propose a walking system with "deformable track + retractable suspension".

When the platform is traveling on land, it can adjust the touchdown area of the track according to the road features automatically. When the platform is running in water, the retractable suspension can pull up the deformable track and reduce the water resistance. The amphibious platform using this new type of walking system has the function of high mobility and high efficiency, and the speed is fast in water. The structure of the walking system is shown in Figure 1.

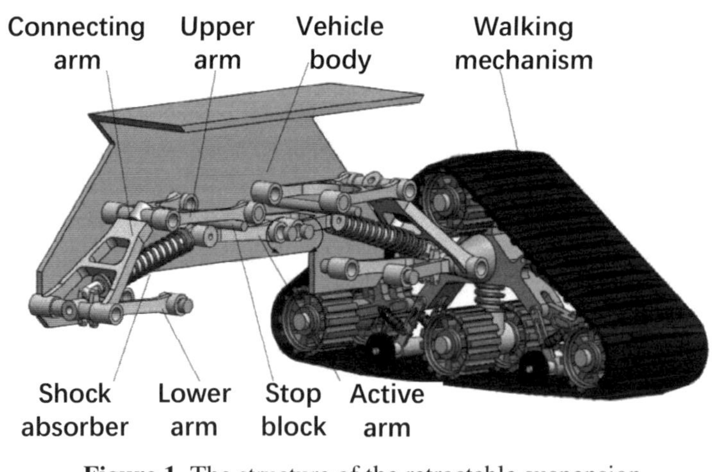

Figure 1. The structure of the retractable suspension.

The kinematics and dynamics models of the proposed retractable suspension can be established by using the predicted parameters. However, in order to obtain the maximum retraction of the walking mechanism and make the suspension have reasonable kinematics characteristics when traveling on land, it is necessary to optimize the predicted parameters. [2, 4]

2. Optimization Strategy

The designed mechanism realizes two motion processes, and the two processes are coupled to each other. And it is impossible to obtain a clear objective function and constraint function, this paper proposes a new optimization method for two mutually coupled motion processes, whose specific process is shown in Figure 2.

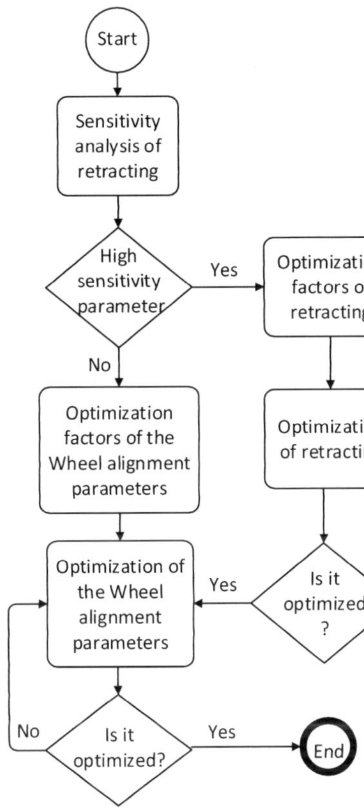

Figure 2. Optimal flowchart of retractable suspension system.

Considering that the absolute value of the wheel alignment parameters is small and sensitive to the change of system parameters, sensitivity analysis of the two motion processes is carried out first. Then the parameters with high sensitivity to one motion process and low sensitivity to another are selected as the optimization variables. Finally, Adams/insight was used for experimental design and repeated optimization to determine the final system parameters.

3. Optimization

3.1. Sensitivity Analysis

According to the structure of retractable suspension, the mechanism can be simplified into a plane figure, as shown in Figure 3. And then we use D as the origin of the frame. At the same time, it is assumed that the lower arm remains horizontal at the design position, so the coordinate parameters of C_1 and C_2 only need to take the x coordinates as variables. In addition, as the connecting arm remained relatively fixed with the walking mechanism during the whole process, the lifting, turning and positioning parameters of the walking mechanism were all dependent on point E and C_2, rather than the connection point C_3.

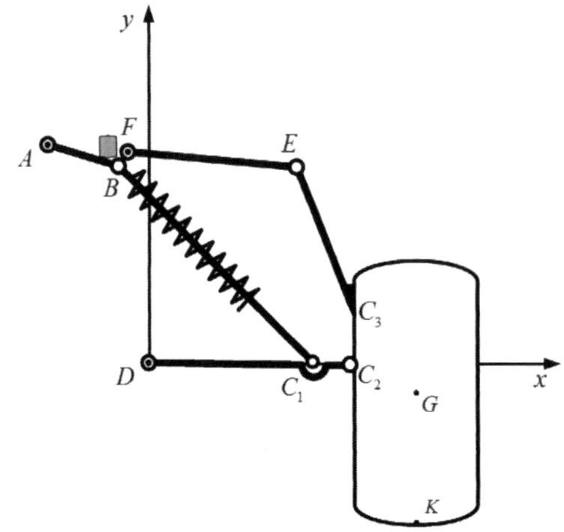

Figure 3. Schematic diagram of retractable suspension system.

Therefore, the factors actually used for sensitivity analysis are shown in Table 1.

Table1. Factors of Sensitivity analysis.

Factor	Meaning	Original value	variation range
DV_1	coorX of C_2	300	-5.0, 5.0
DV_2	coorX of C_1	225	-4.5, 4.5
DV_3	coorX of B	-50	-2.0, 2.0
DV_4	coorY of B	275	-4.5, 4.5
DV_5	coorX of A	-150	-2.5, 2.5
DV_6	coorY of A	300	-5.0, 5.0
DV_7	coorX of F	-25	-1.0, 1.0
DV_8	coorY of F	275	-4.5, 4.5
DV_9	coorX of E	175	-2.5, 2.5
DV_10	coorY of E	325	-5.0, 5.0

Screening studies (Screening) of linear Fractional Factorial types were used in the design of experiments. A total of 16 simulation experiments were conducted on the lifting height of the walking mechanism during the retracting and releasing process. The analysis results are shown in Figure 4.

Factor	From	To	Effect	Effect %
DV_4	2.6263e+02	2.8738e+02	-6.0743e+01	-725.92
DV_6	2.8500e+02	3.1500e+02	5.4447e+01	650.68
DV_10	3.0875e+02	3.4125e+02	8.4855	101.41
DV_5	-1.5375e+02	-1.4625e+02	-7.7809	-92.99
DV_1	2.8500e+02	3.1500e+02	7.5916	90.72
DV_8	2.6263e+02	2.8738e+02	-4.3529	-52.02
DV_9	1.7063e+02	1.7938e+02	-1.4534	-17.37
DV_3	-5.1000e+01	-4.9000e+01	1.3405	16.02
DV_2	2.1488e+02	2.3512e+02	6.5279e-01	7.8
DV_7	-2.5250e+01	-2.4750e+01	-3.0169e-01	-3.61

Figure 4. The results of sensitivity analysis of lifting height.

Then, by using the same experimental design method, 16 simulation experiments were conducted on the turning angle of the walking mechanism during the retracting and releasing process. The result is shown in Figure 5.

Factor	From	To	Effect	Effect %
DV_4	2.6263e+02	2.8738e+02	-5.6413	-66.36
DV_6	2.8500e+02	3.1500e+02	4.7544	55.92
DV_1	2.8500e+02	3.1500e+02	-3.4459	-40.53
DV_5	-1.5375e+02	-1.4625e+02	-1.4296	-16.82
DV_9	1.7063e+02	1.7938e+02	-1.3831	-16.27
DV_10	3.0875e+02	3.4125e+02	6.7771e-01	7.97
DV_3	-5.1000e+01	-4.9000e+01	4.4533e-01	5.24
DV_2	2.1488e+02	2.3512e+02	-3.6775e-01	-4.33
DV_7	-2.5250e+01	-2.4750e+01	-1.5283e-01	-1.8
DV_8	2.6263e+02	2.8738e+02	1.4102e-01	1.66

Figure 5. The result of sensitivity analysis of turning angle.

The results of sensitivity analysis showed that the design variables with high sensitivity for both targets are DV_4, DV_6, DV_5, DV_1, DV_10, and DV_9. DV_3, DV_2 and DV_7 are less sensitive to both targets. Therefore, DV_4, DV_6, DV_10, DV_5 and DV_1 are selected as the optimization variables of the retracting process after comprehensive consideration. DV_3, DV_2 and DV_7 were selected as the optimization variables of the wheel alignment parameters.

3.2. Optimization of Retractable Index

According to the Fig. 3, it is easy to know that retractable suspension can correctly realize the retraction of the walking mechanism, under the following conditions:

$$AB+AD<BC_1+DC_1 \tag{1}$$

$$|BC_1-DC_1|<AD-AB \tag{2}$$

$$AB<AD\ ;AB<BC_1\ ;AB<DC_1 \tag{3}$$

At the same time, when the retracting process reaches the extreme position of the four-bar mechanism ABC_1D, the mechanism needs to satisfy:

$$C_2F \leqslant EF+EC_2 \tag{4}$$

and:

$$x_{C1}<x_{C2} \tag{5}$$

Formulas (1) to (5) are connected as the constraint conditions for the optimization of the retracting process, and the maximum value of the coorY of G is used as the optimization target. The values of the optimized design variables after multiple iterations are shown in Table 2.

Table2. Variables before and after the first optimization.

Factor	Before optimization	After optimization	Change
DV_4	275.00 mm	262.63 mm	-0.045%
DV_6	300.00 mm	315.00 mm	0.05%
DV_10	325.00 mm	341.25 mm	0.05%
DV_5	-150.00 mm	-153.75 mm	0.025%
DV_1	300.00 mm	315.00 mm	0.05%

The lifting height and the turning angle before and after optimization are compared respectively, as shown in Figures 6 and 7.

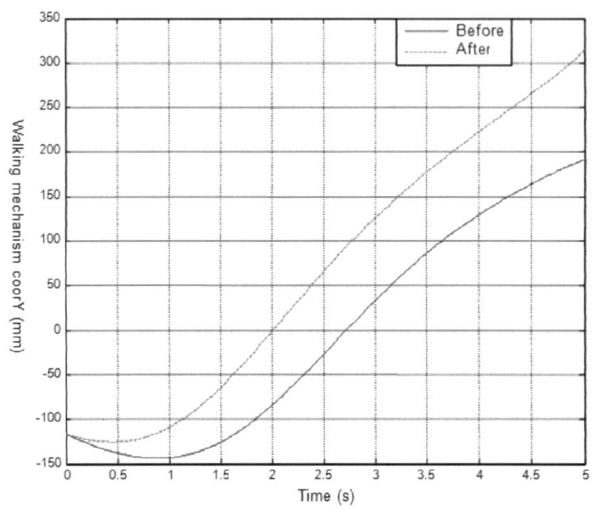

Figure 6. Optimization results of lifting height.

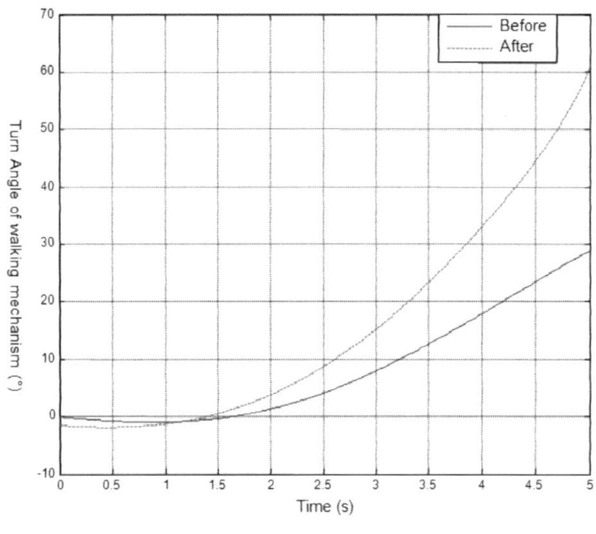

Figure 7. Optimization results of turning angle.

According to Fig. 6 and Fig. 7, the maximum lifting height and maximum turning Angle of the walking mechanism were significantly increased, with 64.8% and 112.2% respectively. At the same time, the optimized lifting speed has been increased. On the other hand, in the process of lowering the walking mechanism before lifting, its lowering maximum value was reduced by 12.9% after optimization.

3.3. Optimization of Wheel Alignment Parameters

The novel amphibious platform is driven by four identical tracks independently and differential steering. Considering the different concepts and uses of wheel alignment parameters, the kinematics characteristics of retractable suspension described in this paper need to study the wheel camber angle, kingpin inclination angle and wheel-base of the walking mechanism, as shown in Figure 8. [3] In particular, if the amphibious platform adopts wheeled walking mechanism and steering knuckle, the optimization strategy described in this paper can also be applied to other wheel alignment parameters optimization.

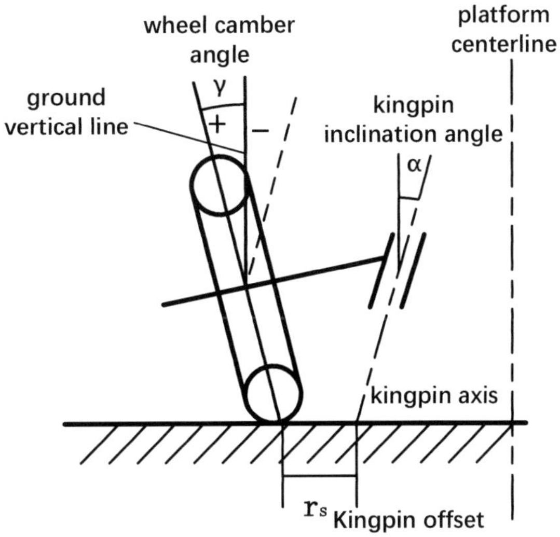

Figure 8. Schematic diagram of wheel alignment parameters.

According to Figure 8, the formulas for calculating the wheel camber angle γ, the kingpin inclination angle α and wheel-base L are as follows:

$$\gamma = arctan\ [(x_{C3}\text{-}x_{C2}) \ / \ (y_{C3}\text{-}y_{C2})] \quad (6)$$

$$\alpha = arctan\ [(x_{C2}\text{-}x_{E}) \ / \ (y_{E}\text{-}y_{C2})] \quad (7)$$

$$L = 2x_K \quad (8)$$

Based on the ADAMS model, we add a wheel-jump experiment table, as shown in Figure 9. At the same time, three measurements in Eq. (6), (7) and (8) are established, and then DV_2, DV_3, DV_7 and DV_9 are selected as optimization variables of suspension kinematics characteristics.
With the objective of minimizing the maximum kingpin inclination angle, the optimized variables are obtained by many iterations in the [-50,50] mm wheel-jump experiment as shown in Table 3.

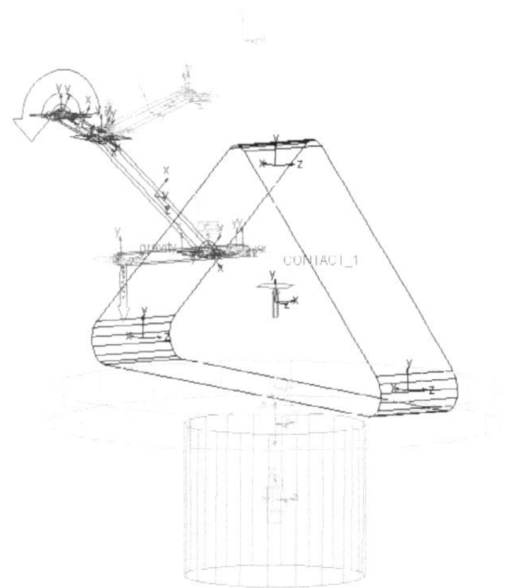

Figure 9. Retractable suspension wheel-jump test table.

Table3. Variables before and after the second optimization.

Factor	Before optimization	After optimization	Change
DV_2	225.00 mm	225.00 mm	0.00%
DV_3	-50.00 mm	-50.00 mm	0.00%
DV_7	-25.00 mm	-25.25 mm	-0.01%
DV_9	175.00 mm	179.38 mm	0.25%

The comparison of the kingpin inclination angle, wheel camber angle and wheel-base before and after optimization is made. The result is shown in Figure 10~12.

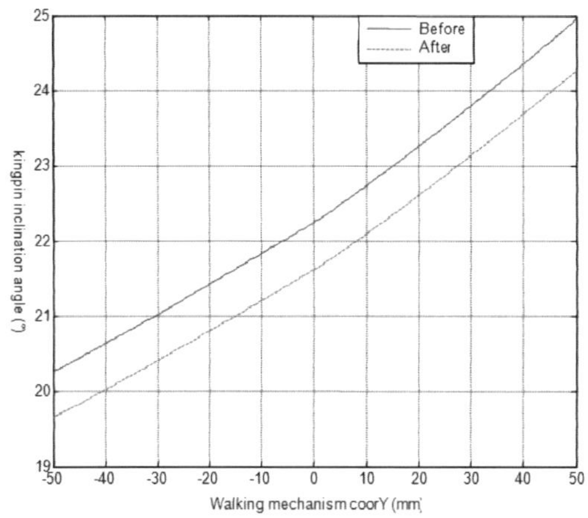

Figure 10. Optimization results of kingpin inclination angle.

According to the optimization results, the maximum and minimum pin inclination angles of [-50,50] mm wheel-jump test are reduced by 0.75 degrees, and the corresponding wheel inclination angle is increased by about 1 degree. At the same time, the balance position inclination angle is increased from the original approximate 0 degree to 1 degree, which reduces the wheel inclination angle to zero when the platform is fully loaded. It is beneficial to reduce the abnormal wear of the walking mechanism. And the wheelbase basically did not change.

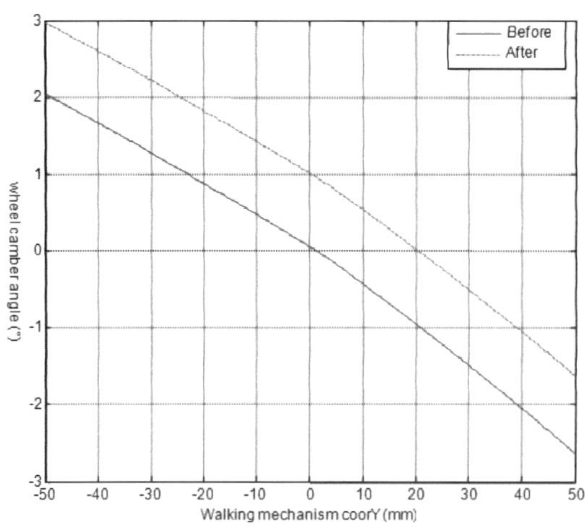

Figure 11. Optimization results of wheel camber angle.

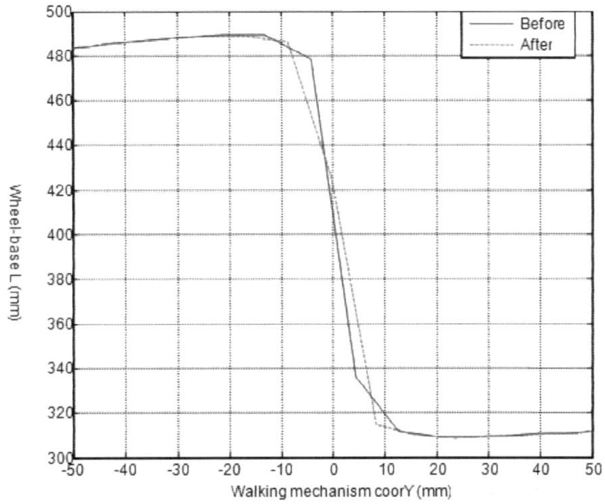

Figure 12. Optimization results of wheel-base.

4. Conclusion

Through the analysis above, we can draw a conclusion: the optimization strategy described in this paper has obvious effect on retractable index of the retractable suspension for greatly improving the retractable height and the turning angle; at the same time, the kinematics characteristics of the suspension are improved by optimizing the alignment parameters of the walking mechanism.

References

[1] Song Guixia, Key issues of drag reduction and growth of amphibious vehicles. Nanjing University of Aeronautics and Astronautics. D. (2008)

[2] Yi-Xin X U, Shu-Wei L I, et al. Optimization the Body of Amphibious Vehicle via Compositional Modeling. Computer Simulation. J. 30(3):173-147(2013)

[3] Zhang wenchun. Automotive theory. 2nd edition [M]. Mechanical industry press, (2010)

[4] Guo Z X, Kui-Wu L I, Pan Y T, et al. Research on Calculation and Prediction of Amphibious Vehicle Resistance Based on CFD Theory. Journal of Gun Launch & Control, (2012)

IOP Conf. Series: Materials Science and Engineering **538** (2019) 012064 doi:10.1088/1757-899X/538/1/012064

The Design of A Motorcycle Featuring Fully Independent Adjustability for Front Suspension and Steering Geometry

Marco Gadola, Daniel Chindamo[a], Federico Vitale and Emanuele Bonera

Department of Industrial and Mechanical Engineering, Automotive Group, University of Brescia, I-25123 Brescia, Italy.

[a] Corresponding author: daniel.chindamo@unibs.it

Abstract. The paper presents a feasibility study for a prototype motorbike (bike in short) with an alternative front suspension. Novel features include the adjustability of steering axis inclination, front trail, and anti-dive effect i.e. the main parameters affecting rider feedback and perception as well as motorcycle behaviour. Although being based on the well-known double-wishbone layout, the system can be adjusted to replicate the geometry of a conventional fork as a baseline for experimental testing. Unlike similar concepts that can be found in literature the kinematics can be modified by means of straightforward adjustments that do not require complex disassembly. Independent adjustability is provided for each parameter. These innovative features enable a back-to-back comparison between significantly different settings, in order to explore a wide range of characteristics in terms of dynamic response, handling properties and rider's perception with a single vehicle. The project is aimed at fostering knowledge on motorcycle dynamics within the research group and especially on rider feedback and subjective perception, somehow a neglected topic in current literature. It was undertaken under the form of a student educational project in mechanical engineering and named "F.A.B." (Fully Adjustable Bike). The paper is mainly focused on front suspension kinematics; chassis design and related structural aspects are outlined as well.

1. Introduction

The dynamics of single-track vehicles like motorcycles and bicycles is a consolidated subject. Authors like Cossalter [1] and Sharp ([2-11] among many others) explain the dynamic equilibrium equations and the importance of steering geometry. [12] describes the development of motorcycle suspension over the years and explains pros and cons of the fork as a front suspension system vs alternative layout e.g. the Hossack system [13].

The need for an adjustable front-end geometry to cater for different requirements on a motorcycle has long been felt. [14] for instance is a well-known patent owned by Bombardier regarding a fork head with adjustable steering axis inclination. A similar system is currently the standard in racing, where bike set-up is often adjusted to suit riding style, circuit characteristics and weather conditions. [15] is again a patent where the fork inclination (aka rake) can be modified in order to adjust the trail without affecting steering axis inclination. Such concepts require front suspension dismantling and re-assembly in the workshop, and do not provide independent adjustability for rake and trail. [16] is one of many patents owned by motorcycle manufacturers (Honda in this case) and dealing with the tendency to nose dive under braking, a typical consequence of conventional telescopic fork kinematics, as described in [17, 18].

In Europe, USA and Japan the motorcycle is mainly seen as a leisure vehicle. These markets are notoriously very demanding, leading to an offer made of particularly refined, high-end products, where the emotional aspects play an extremely important role [19]. As a matter of fact riding a bike on a twisty road can be a deeply involving experience. It is well known that riding position, feedback of interface commands such as the front brake and gearbox lever, throttle response and engine sound [20], and even the quality of powerplant vibration all add to the "character" and "personality" of a motorbike and should be carefully designed to suit a specific rider "family". In other words, in a market made of discerning customers the quality of emotional feedback is considered at least as important as objective dynamic characteristics like outright engine power, overall performance, stability, and active safety for instance.
Steering feedback and response in particular are among the main factors affecting the riding experience in terms of subjective perception. They involve turn-in, bike roll transient and all those subtle interactions between rider's body and bike balance. However, it seems that the impact of objective design factors like motorcycle geometry on rider perceptions is still a fairly unexplored zone, at least outside of the manufacturers' design and R&D departments. Apparently, road and/or track testing performed by experienced riders is still the most straightforward route to an effective correlation between subjective "feelings" and engineering parameters.
Automobile steering feeling is often considered to be the most important among driver feedbacks, see [21-25] among many others. This is a largely explored topic, especially with the advent of Electric Power Steering (EPS), driving simulators and Advanced Driving Assistance Systems (ADAS). On the other side it is true that various papers deal with rider/motorcycle interaction from the control theory point of view (from "historical" works like [26-29] to the renowned research of R. Sharp, again [2-11], to recent contributions [30]). However, there is not much evidence of similar studies concerning rider feedback, steering feeling and subjective-objective correlation assessment, although the role of perception is acknowledged as the core of riding a motorbike.
The authors felt the need for an attempt at bridging this gap and building their own experience on the subject, also in order to guide the students in motorcycle-related projects such as future Motostudent competitions. These are the reasons why the project of a concept motorcycle with fully adjustable front end was born, where set-up changes could be easily made "on the go" in order to enable a back-to-back comparison between different steering geometries hence between different characteristics in terms of dynamic response. The design enables independent adjustment of each relevant parameter without taking the bike apart:

- Steering axis inclination
- Front trail arm
- Front anti-dive percentage

The prototype will undergo manufacturing and assembly in the next months and road test results will be addressed in a future publication.

2. Material and Methods

The Automotive Group of the University of Brescia has been active in the fields of vehicle dynamics and design, research on high-performance vehicles and sustainable mobility for many years. With regard to racing cars for instance, a work dealing with lap time simulation of a professional single-seater racecar is [31], including optimal control application. The interaction of ground-effect aerodynamics with non-linear suspension dynamics is dealt with in [32], while a thorough revision of the working principles of the limited-slip differential and its impact on vehicle behaviour are described in [33]. Recent involvement in student design competitions like the Formula SAE and Motostudent contests has boosted applied research in the automotive sector along with the build and development of some one-off vehicles, as shown by papers like [34] (on the development of a peculiar front suspension and steering geometry for a FSAE racing car), [35] (describing four-poster vibration testing on a FSAE car aimed at the development of a suspension friction model, to be incorporated into vehicle dynamics simulations), and [36, 37] (describing estimation methods for the so-called

Vehicle Sideslip Angle). Research on the impact of intensive use of light alloys for chassis substructures is described in [38-43]. Experimental vibration testing on cars and motorbikes is also dealt with in [44-46]. Some papers are related to human-vehicle interaction like [21], describing how friction affects driver's feedback from the steering system, as well as the results of a related experimental campaign. A peculiar approach to energy management for hybrid vehicles is presented in [47].

The vital role of student design competitions in engineering education through experiential learning and "learning by doing" is described in [48, 49] according to the theory outlined by authors like Wankat and Oreovicz [50], Bandura [51, 52], Siddique at al. [53], Kolb [54, 55], and a world-famous article written by cognitive psychologist Miller [56]. Among others, [57] states the specific value of motorsport-related design competitions in engineering schools.

Along with the guidelines above, the design of a motorbike with a fully adjustable front end was undertaken as a final project in mechanical engineering [58]. It is based on the experience recently gained by entering the 2016 international Motostudent event with BRX 250, a small racing bike fully designed, built and developed by students (again [48], and Fig. 1). So far, this project resulted in extended knowledge within students and university staff and is aimed at generating guidelines for future designs.

Figure 1. BRX 250, the 2016 Motostudent project of the University of Brescia.

2.1. Fundamentals of Bike Geometry

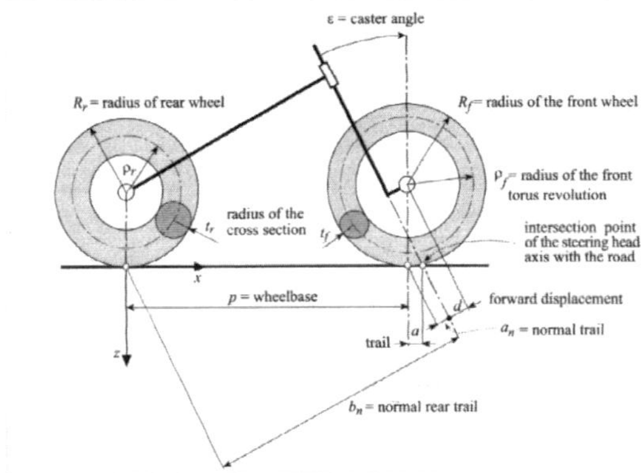

Figure 2. Motorcycle front end geometry (courtesy [1]: Cossalter, V. Motorcycle dynamics, 2nd edition).

With reference to [1], a simple bike model with no suspension is composed of four rigid bodies: the front wheel is connected to the front end i.e. the steering system, and the rear wheel is connected to the chassis by means of revolute joints allowing for wheel rotation. A third revolute joint connects the front end to the chassis; the relative rotation between these two parts and around the steering axis is the steering angle, to be directly controlled by the rider. Additional degrees of freedom are the longitudinal motion and the roll angle. The basic geometry can be described by means of the following three values, all defined in the bike vertical plane (Fig. 2) and related to the basic handling characteristics:

1. Steering axis inclination (aka caster angle)
2. Normal front trail i.e. the distance between the steering axis and the front wheel contact patch centre
3. Normal rear trail i.e. the distance between the steering axis and the rear wheel contact patch centre

Table 1. Motorcycle front end geometry: Reference range for the main parameters.

Motorbike type	**Racing** *track racing only*	**Sports** *road-going motorbike, high power-to-weight ratio*	**Touring** *suitable for long-distance travelling*
Handling requirements	← *agile*		*stable* →
Caster angle	20°-25°		25°-30°
Normal trail ratio (front/rear)	5-7%	6-8%	7-9%
Front trail (on ground plane)	75-100 mm		90-130 mm
Wheelbase	1300-1500 mm		1400-1600 mm

Reference is usually made to the so-called front trail (namely the projection of the front normal trail on the ground plane when the bike is in the upright position), and to the wheelbase, as these dimensions are easily understood. Table 1 shows a typical range for these parameters that play a fundamental role in determining the bike handling "character" in terms of dynamic response. A large front trail in particular makes for a large self-aligning moment on the front wheel and results in a very stable bike with heavy steering, at the expense of agility on a twisty route.

Nearly all the models on sale feature the conventional fork as front suspension/steering system. In general, the fork is deemed to ensure a good compromise in terms of handling, comfort, reliability, and production costs, and offers the traditional look that is normally appreciated in such a largely conservative market. Apart from inherent friction between the sliding components, the main drawback of the fork is related to the tyre contact patch trajectory which is inevitably aligned with the fork axis. As the suspension is compressed the wheel moves backwards resulting into a so-called "pro-dive" effect. Such a geometry gives favourable ride comfort on bumps but amplifies the effect of the longitudinal load transfer in braking, resulting in larger pitch angles (up to a 100% increase and beyond, see Table 2) with the related "lazy" feeling in transients plus a reduction of the front trail, at the expense of stability [1, 58, 59]. A modern, non-kinematic solution to this typical behaviour is active damping control, see [60].

Currently BMW is one of the few high-volume manufacturers to offer alternatives to the fork: the well-known Telelever® (Fig. 3; corresponding to a McPherson system in automotive terms) and the Duolever®, originally known as the Hossack system [12, 13] (Fig. 4), corresponding to a double-wishbone layout in automotive terms. The handlebar pivots directly on the chassis and it is connected to the front wheel carrier by means of a scissor link. In the BMW case these bikes usually combine a mass well beyond 200 kg with a fairly high power-to-weight ratio, long-distance touring capabilities and a high-tech look as well.

Table 2. Comparison of the F.A.B. concept with a reference bike model.

	KTM Duke 690	**F.A.B. concept**
Engine	KTM "mono"	Aprilia V-twin
Capacity	690 cc	550 cc
Power	54 kW	50 kW
Dry weight	150 kg	120 kg (estimate)
Wheelbase	1467 mm	1450 mm
Front trail	115 mm	110 mm
Caster angle	26.5°	21°

Such systems decouple steering geometry parameters from the front wheel contact patch trajectory in bump, enabling reduced pro-dive or even anti-dive properties to be built in the front suspension, hence reducing pitch angle in braking and shortening the related transient [1, 12, 17, 18]. From the subjective point of view this is usually perceived as a noticeable advantage, especially when riding a heavy touring bike with passenger and luggage.

Figure 3. The BMW Telelever® front suspension (source: www.bmw-motorrad.de).

Figure 4. The BMW Duolever®/Hossack front suspension (source: www.bmw-motorrad.de).

2.2. The F.A.B. (Fully Adjustable Bike) Concept

The F.A.B. has been conceived as a road-going, single-seater, naked-style motorcycle (Fig. 5). It features a spaceframe chassis made of 25CrMo4 tubes accommodating a stock Aprilia SXV V-twin, 550 cc engine, which is extremely light and compact. This unit was readily available as it was previously employed on the FSAE racing cars. The rear suspension is as simple as possible, featuring a Honda CRF 450 swing-arm in cantilever configuration and a 4.5" wheel fitted with a 160/60-17 tyre. A progressive linkage is deemed redundant as there is no passenger seat.

With regard to the front end, suspension geometry is usually fixed on customer motorbikes, whereas the F.A.B. concept provides quick, easy and independent adjustability of the main parameters including the amount of anti-dive effect. This led to the design of an unconventional layout akin to a double-wishbone or SLA (Short-Long Arm) system like the Hossack and BMW Duolever®, where the hardpoints have been purposely located to guide the front wheel centre/contact patch along the desired

IOP Conf. Series: Materials Science and Engineering **538** (2019) 012064 doi:10.1088/1757-899X/538/1/012064

trajectory. A proprietary tool based on a spreadsheet has been used to define the in-plane kinematics (Fig. 6) along wheel travel, and to automatically define the geometry in the CAD software accordingly by means of a parametric baseline. This model, also used by the authors for race car suspension design, can be applied to any combination of the adjustments outlined in the following sections in order to plan the testing activity, and will be described in a future publication. Low-friction, motorsport-type ball joints connect the suspension arms to the chassis (side joints) and each wishbone to the wheel carrier (main joints): the main joints are of the wide-angle type to allow for the required steering angle and their centres define the actual steering axis, hence the nominal caster angle and front trail values.

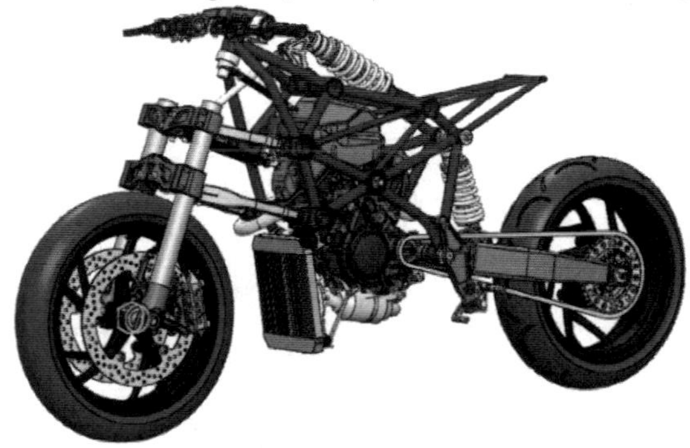

Figure 5. The F.A.B. project assembly.

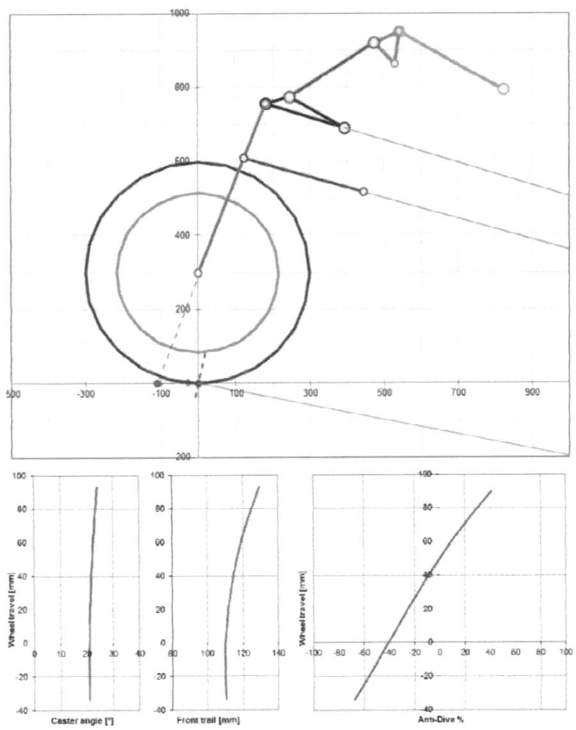

Figure 6. Spreadsheet-based model for kinematics design.

Table 2 compares the F.A.B. design configuration with a roughly similar bike available on the market. The anti-dive percentage has been computed statically i.e. at the beginning of braking, before any pitch occurs, assuming brake bias is 100% on the front wheel. The Centre of Gravity height was estimated at 745 mm with the help of the CAD solid model comprehensive of a 95-percentile rider equipped with full-face helmet.

IOP Conf. Series: Materials Science and Engineering **538** (2019) 012064 doi:10.1088/1757-899X/538/1/012064

Unlike the Duolever® the handlebar is connected to the wheel carrier by means of a rocker coupled with a pair of conrods in parallel, allowing for kinematic compatibility within the adjustability range while also reducing the required working space ahead of the handlebar in combined bump and steer (Fig. 7).

Springing is provided through a Formula 1-style pushrod system acting on a coaxial spring-damper unit via a rocker located just behind the handlebar (Fig. 8). The design configuration is based on a motion ratio equal to 1.4 with a progressive trend in bump. The motion ratio characteristic could be easily modified by changing the shape of the rocker. The spring-damper unit is located on hand above the engine/airbox for easy adjustments, and eventually hidden under a dummy fuel tank cover, while the actual fuel tank is located under the saddle.

Figure 7. Steering assembly.

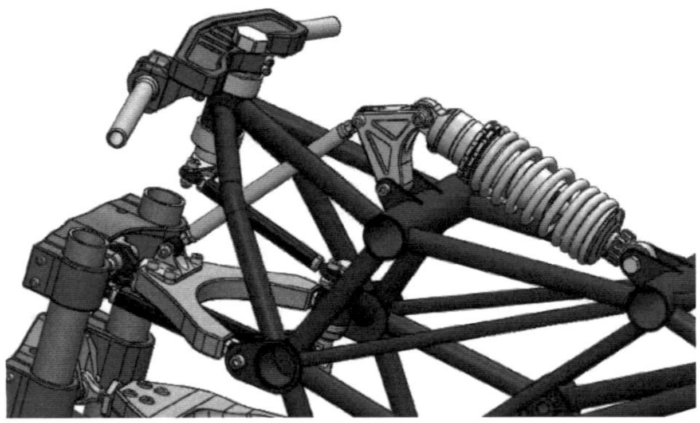

Figure 8. Pushrod/rocker/spring-damper unit assembly.

3. Results and Discussion

The bike in itself is a stripped-down, all-essential design hence potentially it is extremely light and sensitive to setup changes: 140 kg in running order, according to an estimate based on the CAD solid model. On top of this, removing the fork provides further potential in terms of weight reduction - unsprung mass reduction in particular- and yaw moment of inertia optimization.

While similar prototype bikes feature machined plate-style wheel carriers, currently the F.A.B. wheel carrier has been designed with light-alloy plates accommodating steel tubes running down to the wheel pivot, a layout that should ensure a torsional stiffness comparable to an upside-down fork (Fig. 7). It is possible to replace steel with lighter materials like CFRP or even titanium as a cost-no-object option. Eventually fatigue should also be taken into account carefully for these advanced solutions.

3.1. The Baseline Configuration

The baseline or design configuration is positioned in the middle of the range available for all the three adjustable parameters. In kinematic terms it is equivalent to a geometry suitable for an agile bike with prompt response to rider inputs. The anti-dive amount and its variation along wheel travel are similar to a Duolever® layout. A fork-equivalent geometry is available at the lower end of the anti-dive adjustability range (see section 3.4).

3.2. Front Trail Adjustment

With regard to caster angle and trail it is possible to apply significant set-up changes to the front geometry with simple tools and some training, even during a short stop along the road. There is no need to take the motorbike apart in a workshop, and this means a back-to-back comparison between different configurations is made possible.
Suspension arm ball joints are connected to the wheel carrier by means of machined blocks, bolted to the fork-style plates and conceptually similar to the so-called camber blocks on single-seater racecars. The front wheel carrier position relative to the chassis can be offset by changing the thickness of the shims located between each block and the top and bottom plates (Fig. 9). Therefore, the front trail can be modified by simply releasing a few Allen screws while keeping the steering axis angle and position as well as the whole suspension kinematics characteristics unaltered. As shown in Table 3, front trail adjustability is very wide and covers nearly the whole range specified in Table 1. The original wheelbase length and ride heights can be eventually restored by adjusting the longitudinal rear wheel position relative to the swingarm and the front spring-damper unit length respectively. The latter modification might require action on the spring platform like changing the rear preload, an ordinary adjustment often made when preparing a touring bike for a trip with passenger and luggage.

Table 3. Front trail adjustability range.

Offset modification		Front	Front trail	Wheelbase &	Comments
Top Shim	**Bottom Shim**	**trail**	**variation**	**variation**	
-5 mm	+5 mm	81 mm	-29 mm	+27 mm	Minimum trail available
0	+5 mm	93 mm	-17 mm	+16 mm	Bottom shim mod only
-5 mm	0	98 mm	-12 mm	+11 mm	Top shim mod only
+5 mm	+5 mm	104 mm	-6 mm	+6 mm	
0	**0**	**110 mm**	**0**	**1450 mm**	**Design configuration**
-5 mm	-5 mm	115 mm	+5 mm	-5 mm	
+5 mm	0	123 mm	+13 mm	-10 mm	Top shim mod only
0	-5 mm	126 mm	+16 mm	-15 mm	Bottom shim mod only
+5 mm	-5 mm	138 mm	+28 mm	-26 mm	Maximum trail available

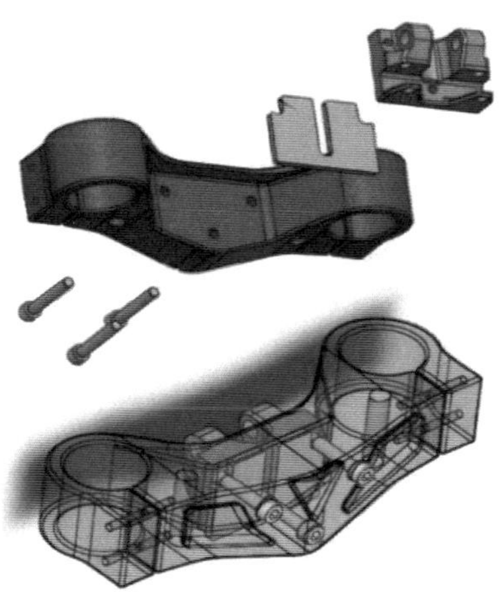

Figure 9. Shim adjustment for front trail variation.

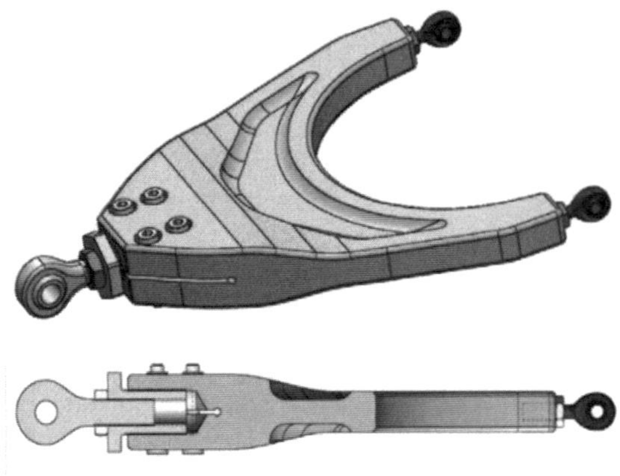

Figure 10. Variable-length lower wishbone. A threaded insert and a clamp hold the main ball joint in place and allow for continuous adjustability.

3.3. Steering Axis Inclination Adjustment

The lower wishbone has been conceived for quick length adjustability as the ball joint is installed via a threaded bushing clamped in the suspension arm. By releasing the clamp and a locknut and rotating the bushing with the help of a spanner the caster angle can be adjusted continuously. Alternatively, the top wishbone length can also be modified by adjusting the ball joint length. In this case the adjustment is discrete as the ball joint can be rotated by 180° at a time. This modification requires lifting the bike off the ground. Once again, the adjustability range is very wide. Figures 11 and 12 show the combined effect of wishbone length variation on the caster angle, as well as the consequent trail variation, to be eventually restored by means of shim replacement with the help of the spreadsheet-based model. A ~10% caster increase for instance (from 21° to 23°) can be achieved with a length adjustment of 5.5 mm on the bottom wishbone only. This will result in a 11 mm front trail increase approx, to be corrected by removing a 5 mm shim from the top only. These figures show that large geometry modifications can be easily performed in order to achieve significant changes in terms of handling properties and rider's perception.

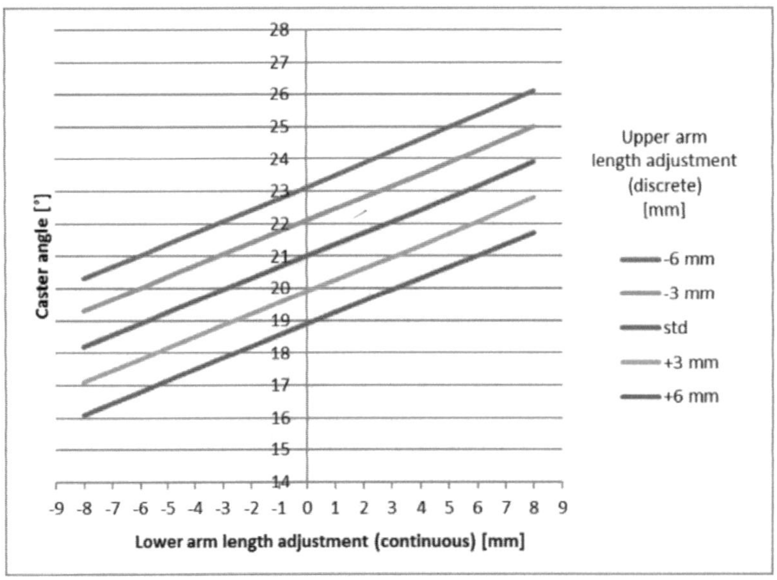

Figure 11. Caster angle adjustment with suspension arm length variation.

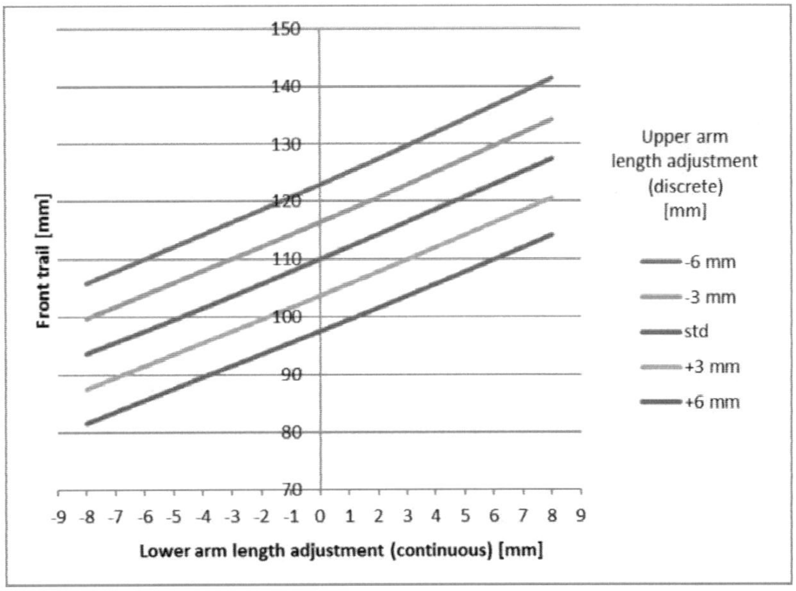

Figure 12. Caster angle and front trail adjustments with suspension arm length variation.

3.4. Anti-dive Adjustment

Adjustment of the anti-dive characteristics is more complex since it inevitably requires re-locating the suspension arm mounting points: this modification cannot be made "on the go". Nevertheless, it is quick and relatively easy: three different positions have been designed for each wishbone on the chassis side. They are located 12 mm apart in the vertical direction and along an arc (Fig. 13), hence no suspension arm length modification is virtually required Again, a ride height compensation can be eventually made by acting on the front spring-damper unit length. Small corrections to caster angle and front trail might be required as well and can be computed by means of the kinematics model.

It should be stated that unlike a fork, the Hossack geometry produces variable anti-dive properties along suspension displacement. The BMW Duolever® for instance features a very progressive trend from mild pro-dive in static configuration to anti-dive in full bump. That's how the F.A.B. front end was designed as well, although the remarkable arm length will generate a more consistent characteristic in this case. Table 4 lists the pick-up point combinations available with the related anti-dive percentage from static to full bump.

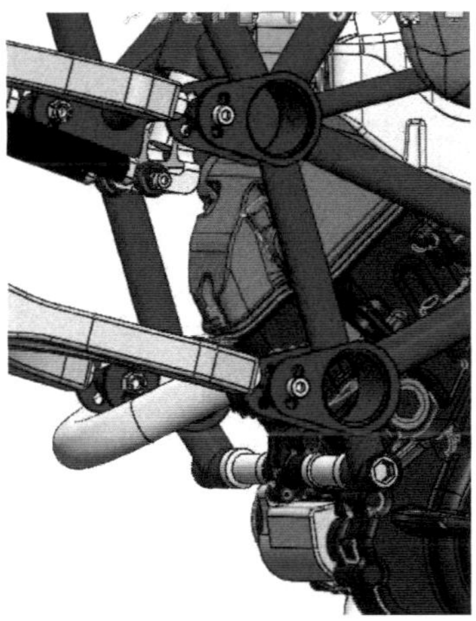

Figure 13. Adjustable suspension arm pick-up points for anti-dive modification.

Table 4. Anti-dive variation in bump and adjustability range.

Pick-up points vertical position chassis side		% anti-dive range (pro-dive if <0)		Comments
Upper arm	Lower arm	Static	Full bump (90 mm)	
+12 mm	-12 mm	-110%	-36%	Full pro-dive (equivalent to fork)
+12 mm	0	-79%	-1.5%	
0	-12 mm	-47%	+11%	
+12 mm	+12 mm	-47%	+30%	
0	**0**	**-40%**	**+41%**	**Design configuration**
-12 mm	-12 mm	-34%	+52%	Similar to BMW Duolever®
0	+12 mm	-8%	+67%	
-12 mm	0	-2%	+77%	
-12 mm	+12 mm	+28%	+95%	From mild to full anti-dive

3.5. Structural Requirements

FEM analysis was carried out on the chassis with main focus on stiffness. A rigid chassis is vital to guarantee a responsive behaviour and to stay away from dangerous dynamic effects like wobble oscillations [10]. The F.A.B. engine is semi-stressed as it is bolted directly on the chassis, therefore the FEM model takes the engine block into account as a dummy solid structure (Fig. 14) and shows that it contributes heavily to overall structural performance. With regard to torsion stiffness, simulation results are beyond 200 kNm/rad. This is probably overestimated: welding is not yet taken into account for instance, however the value is well within the desirable range suggested by [1] (100-300 kNm/rad). Moreover, it is in line with the results of a previous study on the Aprilia RS 250 chassis [61], still considered a benchmark structure for compact, light and sporty road bikes. The FEM model was composed of the chassis alone in this case (Fig. 15), as the engine is not stressed. In any case the construction of a stiffness test rig is under way within the Motostudent project: it should allow for FEM validation.

Chassis pick-up points, suspension arms and ball joints were also verified for strength under the main load cases found in the literature [1]. Future experimental tests on the above-mentioned rig will be required to evaluate longitudinal, side and torsion stiffness for the front suspension assembly.

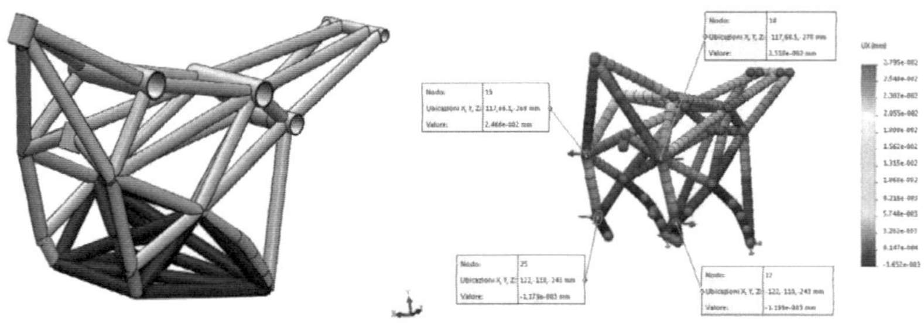

Figure 14. Chassis FEM model with and without engine contribution.

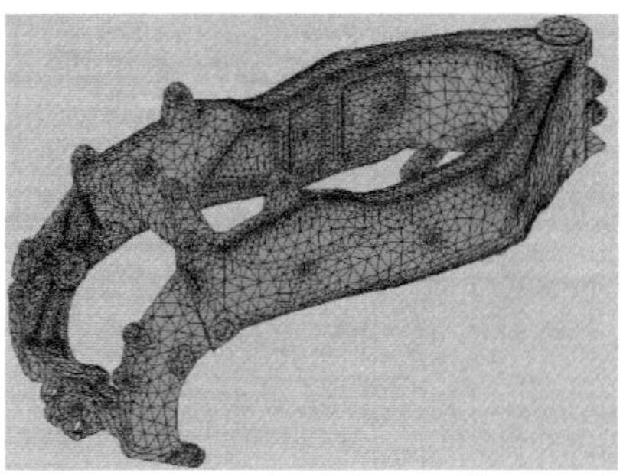

Figure 15. FEM mesh of the aprilia RS chassis.

4. Conclusion

Although it was fairly successful, the recent design and construction of a Motostudent racing bike highlighted the need to gain knowledge and experience on the effects of steering geometry on the motorbike dynamic behaviour. Even more important, it was also found that correlation between rider perception and bike geometry seems to be a neglected topic in literature. As an attempt to fill such a gap the design of an experimental motorcycle with a fully adjustable front end was undertaken from scratch under the form of an educational project.

The paper presents the first stage of the project i.e. a feasibility study including overall motorbike layout, a spaceframe chassis, and a traditional rear suspension. The focus is placed where rider interaction with the bike is most significant: the front end, featuring a peculiar layout for a Hossack-style front suspension with built-in adjustability of the most significant parameters influencing rider feedback as well as motorcycle dynamics.

The overall concept and methodology were inspired by typical racing car suspension system design: steering axis inclination, front trail arm and anti-dive amount are adjustable separately. A wide range is available for each parameter, thus offering the opportunity to change the bike "character" noticeably. Geometry modifications can be performed easily, even by the rider alone with the bike resting on the centre stand. There is no need to take the front end apart.

The following F.A.B. project stages will be addressed in a future publication: the final design, manufacturing and construction, and the results of a back-to-back testing campaign focused on subjective-objective correlation, which should give more insight and issue guidelines for future projects. The overall longitudinal, lateral and torsion stiffness of the front end will also require careful assessment in order to avoid dynamic effects that might jeopardize rider perception and motorcycle performance. The reduction of both steering and suspension friction is also a challenge to be tackled during the manufacturing process and might require further developments in the concept design as well.

References

[1] Cossalter, V. (2006). Motorcycle dynamics, 2nd edition. ISBN 978-1430308614.
[2] Sharp, R.S. (1976). The Dynamics of Single-Track Vehicles. Vehicle System Dynamics. 5 (1-2): 67-77.
[3] Sharp, R.S., Alstead, C.J. (1983). Frequency responses of motorcycles to steering torque inputs and to front wheel and tyre imperfections. I Mech E Conference Publications (Institution of Mechanical Engineers). 193-200.
[4] Sharp, R.S. (1985). The Lateral Dynamics of Motorcycles and Bicycles. Vehicle System Dynamics. 14 (4-6): 265-283.
[5] Sharp, R.S. (2001). Stability, control and steering responses of motorcycles. Vehicle System Dynamics. 35 (4-5): 291-318.
[6] Sharp, R.S. (2004). Advances in the modelling of motorcycle dynamics. Multibody System Dynamics. 12 (3): 251-283.
[7] Sharp, R.S. (2006). Single-track vehicle modeling and control: Bicycles, motorcycles, and models. IEEE Control Systems. 26 (5): 34-61.
[8] Sharp, R.S. (2008). On the stability and control of the bicycle. Applied Mechanics Reviews. 61 (1-6): 0608031-06080324.
[9] Sharp, R.S. (2012). Rider control of a motorcycle near to its cornering limits. Vehicle System Dynamics. 50 (8): 1193-1208.
[10] Sharp, R.S. (2013). Chatter vibrations of high-performance motorcycles. Vehicle System Dynamics. 51 (3): 393-404.
[11] Sharp, R.S. (2014). A method for predicting minimum-time capability of a motorcycle on a racing circuit. Journal of Dynamic Systems, Measurement and Control, Transactions of the ASME. 136 (4), 041007.
[12] Foale, T. (2003). Motorcycle handling and chassis design: the art and science. ISBN 978-

8493328634.
[13] Hossack, N.H. Motorcycle suspension. GB Patent No. 2121364A. Published 21 Dec. 1983.
[14] Robison, G.A. Motorcycle steering head angle adjustment. U.S. Patent No. 3866946A. Published 18 Feb. 1975.
[15] Callaluca, P.A.; Fuller, J.F.III; Conn, D.W. Adjustable motorcycle triple tree. U.S. Patent No. 5967538A. Published 19 Oct. 1999.
[16] Kawaguchi, T. Motorcycle front-wheel suspension with means for restraint of compression of a telescopic front upon braking. U.S. Patent No. 3989261A. Published 2 Nov. 1976.
[17] Mavroudakis, B.; Eberhard, P. (2007). Analysis of alternative front suspension systems for motorcycles. Vehicle System Dynamics. 44 (Sup1): 679-689.
[18] Cossalter, V.; Doria, A.; Lot, R. (2000). Optimum Suspension Design for Motorcycle Braking. Vehicle System Dynamics. 34 (3): 175-198.
[19] Steinberg, F.; Woll, R. (2015). Evaluation of product design analyses for the identification of perception-defining characteristics. Management and production engineering review, 6 (2), 45-55.
[20] Lu, M.H.; Jen, M.U. (2016). Acoustic simulation and validation to realize sound quality refinement of a motorcycle. Proceedings of the INTER-NOISE 2016 - 45th International Congress and Exposition on Noise Control Engineering. 5286-5295.
[21] Gritti, G., Peverada, F., Orlandi, S., Gadola, M., Uberti, S., Chindamo, D., Romano, M., Olivi, A. (2017). Mechanical steering gear internal friction: Effects on the drive feel and development of an analytic experimental model for its prediction. Proceedings of the International Joint Conference on Mechanics, Design Engineering & Advanced Manufacturing (JCM 2016), 14-16 September 2016, Catania, Italy.
[22] Kim, N.; Cole, D.J. (2011). A model of driver steering control incorporating the driver's sensing of steering torque. Vehicle System Dynamics, 49 (10), 1575-1596.
[23] Sharp, R.S.: Vehicle dynamics and the judgement of quality. 87-96 in Pauwelussen, J.P. (1999). Vehicle performance – understanding human monitoring and assessment. ISBN 978-9026515422.
[24] Nybacka, M.; He, X.; Su, Z.; Drugge, L.; Bakker, E. (2014). Links between subjective assessments and objective metrics for steering, and evaluation of driver ratings. Vehicle System Dynamics, 52 (Sup1), 31-50.
[25] Gomez, G.L.G.; Nybacka, M.; Bakker, E.; Drugge, L. (2015) Findings from subjective evaluations and driver ratings of vehicle dynamics: steering and handling. Vehicle System Dynamics, 53 (10), 1416-1438.
[26] Rice, R.S. (1978). Rider skill influences on motorcycle maneuvering. SAE paper 780307.
[27] Weir, H.D. (1997). Motorcycle Dynamics and Rider Control. Vehicle System Dynamics, 6 (2-3), 187-190.
[28] Katayama, T.; Aoki, A.; Nishimi, T. (1998). Control Behaviour of Motorcycle Riders. Vehicle System Dynamics, 17 (4), 211-229.
[29] Cossalter, V.; Da Lio, M.; Biral, F.; Fabbri, L. (1998). Evaluation of motorcycle manoeuvrability with the optimal manoeuvre method. SAE paper 983022.
[30] Kooijman, J.D.G.; Schwab, A.L. (2013). A review on bicycle and motorcycle rider control with a perspective on handling qualities. Vehicle System Dynamics, 51 (11), 1722-1764.
[31] Dal Bianco, N.; Lot, R.; Gadola, M. (2018). Minimum time optimal control of a GP2 race car. Proceedings of the Institution of Mechanical Engineers, Part D: Journal of Automobile Engineering. 232(9), 1180-1195.
[32] Marchesin, F.P.; Barbosa, R.S.; Gadola, M.; Chindamo, D. (2018). High downforce racecar vertical dynamics: aerodynamic index. Vehicle System Dynamics, 56(8), 1269-1288.
[33] Gadola, M.; Chindamo, D.; Lenzo, B. (2018). On the Passive Limited Slip Differential for High Performance Vehicle Applications. 14th International Symposium on Advanced Vehicle Control AVEC 18, Beijing, July 16-20, 2018.

[34] Marchesin, F.P.; Barbosa, R.S.; Alves, M.A.L.; Gadola, M.; Chindamo, D.; Benini, C. (2016). Upright mounted pushrod: the effects on racecar handling dynamics. The Dynamics of Vehicles on Roads and Tracks. Proceedings of the 24th Symposium of the International Association for Vehicle System Dynamics, IAVSD 2015. 543-552.
[35] Benini, C; Gadola, M; Chindamo, D; Uberti, S; Marchesin, F.P.; Barbosa, R.S. (2017). The influence of suspension components friction on race car vertical dynamics. Vehicle System Dynamics. 55 (3): 338-350.
[36] Chindamo, D.; Lenzo, B.; Gadola, M. (2018). On the vehicle sideslip angle estimation: a literature review of methods, models and innovations. Appl. Sci. 2018, 8(3), 355; DOI:10.3390/app8030355.
[37] Gadola, M.; Chindamo, D. (2018). Estimation of vehicle side-slip angle using an artificial neural network. MATEC Web of Conferences 166, 02001 - The 2nd International Conference on Mechanical, Aeronautical and Automotive Engineering (ICMAA 2018). Singapore, Feb 2018. DOI: 10.1051/matecconf/201816602001.
[38] Cecchel, S.; Chindamo, D.; Turrini, E.; Carnevale, C.; Cornacchia, G.; Gadola, M.; Panvini, A.; Volta, M.; Ferrario, D.; Golimbioschi, R. (2018). Impact of reduced mass of light commercial vehicles on fuel consumption, CO2 emissions, air quality, and socio-economic costs. Science of the total environment. (613-614), 409-417.
[39] Cecchel, S.; Chindamo, D.; Collotta, M.; Cornacchia, G.; Gadola, M.; Panvini, A.; Tomasoni, G. (2018). Lightweighting in automotive: cradle- to- grave life cycle assessment of a safety relevant component. International Journal of Life Cycle Assessment. 23(10), 2043-2054.
[40] S. Cecchel, G. Cornacchia, M. Gelfi; Corrosion behavior of primary and secondary AlSi High Pressure Die Casting alloys, Materials and corrosion, 2017, vol. 68, issue 9, pp. 961-969, https://doi:10.1002/maco.201709526Xxx
[41] D. Dioni, S.Cecchel, G.Cornacchia, M. Faccoli, A. Panvini; Effects of artificial aging conditions on mechanical properties of gravity cast B356 aluminum alloy, Transactions of Nonferrous Metals Society of China, 2015, vol. 25, issue 4, pp. 1035-1042
[42] S. Cecchel, D. Ferrario, A. Panvini, G. Cornacchia, Lightweight of a cross beam for commercial vehicles: development, testing and validation; Materials and Design, 2018, 149, p. 122–134, https://doi.org/10.1016/j.matdes.2018.04.021
[43] Borboni, A., Amici, C., Cappellini, V., Faglia, R., Vibration modes of piezoelectric bimorphs: A sensitivity analysis, 1st International Conference of IFToMM ITALY, IFIT 2016; Vicenza; Italy; 1 December 2016 through 2 December 2016; Code 186419, Volume 47, 2017, Pages 467-474.
[44] Chindamo, D.; Gadola, M.; Armellin, D.; Marchesin, F.P. (2017). BikeShake: the design of an indoor simulator dedicated to motorcycle ride testing as an interactive project. International Journal of Interactive Design and Manufacturing. Article in press. https://doi.org/10.1007/s12008-017-0435-7.
[45] Chindamo, D.; Gadola, M.; Armellin, D.; Marchesin, F.P. (2017). Design of an indoor simulator for motorcycle applications. Applied Sciences. 7 (12), 1220.
[46] Chindamo, D., Gadola, M. (2017). Reproduction of real-world road profiles on a four-poster rig for indoor vehicle chassis and suspension durability testing. Advances in Mechanical Engineering. 9 (8): 1-10.
[47] Chindamo, D.; Gadola, M.; Romano, M. (2014). Simulation tool for optimization and performance prediction of a generic hybrid electric series powertrain. International Journal of Automotive Technology, 15:135–144.
[48] Gadola, M., Chindamo, D. (2017). Experiential learning in engineering education: the role of Student Design Competitions and a case study. International Journal of Mechanical Engineering Education. Article in press. DOI: 10.1177/0306419017749580.
[49] Gadola, M.; Chindamo, D.; Legnani, G; Comini, M. (2018). Teaching automotive suspension design to engineering students: Bridging the gap between CAD and CAE tools through an

integrated approach, Int. Journal of Mechanical Engineering Education. Article in press. DOI: 10.1177/0306419018762803.

[50] Wankat, P.C.; Oreovicz, F.S. (2015). Teaching engineering, second edition. Purdue University press. ISBN 978-1557537003.

[51] Bandura, A. (1986). Social foundations of thought and action: a social cognitive theory. Englewood Cliffs, N.J.: Prentice-Hall.

[52] Bandura, A. (1997). Self-efficacy: the exercise of control. New York: W.H. Freeman.

[53] Siddique, Z.; Hardré P.L.; Bradshaw, A.C.; Mistree, F. (2010). Fostering innovation through experiential learning. Proceedings of the ASME 2010 International Design Engineering Technical Conferences & Computers and Information in Engineering Conference IDETC/CIE 2010 August 15-18, 2010, Montreal, Quebec, Canada.

[54] Kolb, D. (1984). Experiential learning: experience as the source of learning and development. Englewood Cliffs, NJ: Prentice Hall.

[55] Kolb, A.Y. and Kolb, D.A. (2005). Learning styles and learning spaces: enhancing experiential learning in higher education. Academy of Management Learning and Education. 4 (2): 193-212.

[56] Miller, G.A. (1956). The magical number seven, plus or minus two: some limits on our capacity for processing information. Psychological Review. 63 (2): 81–97.

[57] Wickenden, P.W.; Stobart, R.K. (2007). An evaluation of formula SAE in the context of the UK engineering education: culture, money and space. In SAE World Congress & Exhibition, Detroit, MI, USA. SAE paper 2007-01-1051.

[58] Vitale, F. (2017). Studio di fattibilità per sospensione anteriore alternativa di motocicletta con relativo telaio. MSc degree thesis in Mechanical Engineering, University of Brescia, Italy.

[59] Gadola, M. (2013). Vehicle Dynamics lecture notes on suspension kinematics: the "anti" effects.

[60] Savaresi, S.M.; Tanelli, M.; Corno, M. (2014). Modelling, Simulation and Control of Two-Wheeled Vehicles. John Wiley & Sons.

[61] Muffolini, D. (2011). Reverse engineering e successiva progettazione di un telaio motociclistico in lega leggera a carattere sportivo. MSc degree thesis in Mechanical Engineering, University of Brescia, Italy.

IOP Conf. Series: Materials Science and Engineering **538** (2019) 012065 doi:10.1088/1757-899X/538/1/012065

Thermal Analysis and Simulation of Cryopump Based on CFD Method

Yueshuai Zhao[a], Lichen Sun, Wei Sun and Rongping Shao

Beijing Institute of Spacecraft Environment Engineering, Beijing 100094, China

[a] Corresponding author: zhaoyueshuai@163.com

Abstract. Based on CFD method, thermal design and numerical simulation of cryopump first array and second array was performed during the cool-down period and the steady state. The analysis of the temperature distribution of the cryopump was made under various conditions, such as operating conditions, the setting of the thermal rods and the thickness of the radiation shield. The const thermal property and variable thermal property was used in this paper to calculate the cool-down process of the cryopump, the effect of the first array structure was also taken into account. Some important results with typical parameters are: The temperature difference of the first array of the cryopump is greatly affected by the working conditions, especially when facing the room temperature vacuum system, the main heat load is concentrated on the baffle; A variable thickness bottom plate can be used to reduce the overall temperature change of the cryopump; The setting of the thermal rods is very useful to reduce the temperature change of the cryopump; The setting of the thermal rods is preferably close to the center of the baffle; If the variable property was not considered, the deviation of the cool-down time is about 24%.

1. Introduction

In large vacuum system, cryopump was usually used as the main pump for obtaining clean vacuum. The cryopump remove gases and vapors from the vacuum chamber by means of condensation and adsorption on the cold faces under 15K. In order to reduce the heat load of the second array, the baffle and radiation shield under 80K (first array) were used[1].

The first array and second array were cooled by Gifford-McMahon cryocooler. For cryopumps, The normal operating temperature of the second array was between 10K and 20K. In this case, the thermal conductivity of oxygen-free copper at 15K can be 15 times higher than at the room temperature[2]. The maximum temperature difference of the second array is usually less than 1K. However, the maximum temperature difference of the first array is much larger, and can reach about 10-20K [3, 4].

In this condition, detail thermal analysis of the cryopump is necessary, especially for the first array. By optimizing the structure of the cryopump, it is possible to reduce the temperature difference of the second array without changing the cryocooler, and obtain a lower average temperature. On the other hand, after optimizing the structure, when designing the cryopump with the same pumping speed, it is possible to select a cryocooler with a smaller cooling power, which is also advantageous for improving economic efficiency.

The simplest method to improve the temperature distribution of the cryopump, especially for the radiation shield and the baffle is increase the thickness. However, this method also greatly increases the mass of the first array and so as the cool-down time of the cryopump, which needs to be balanced from these two aspects.

In this paper, the steady-state temperature distribution and cooling process of the first array were analyzed for the DN400 cryopump. The computational fluid dynamics (CFD) numerical simulation was used to analyze the heat transfer of the cryopump. Different thickness of the radiation shield and the heat conducting plate was considered, the difference of the thermal conductivity of copper and the different cooling power of the refrigerator at different temperatures were also calculated. A design method was proposed to improve the temperature uniformity of the cryopump.

2. Numerical Simulation Method and Mathematical Model

2.1. Structure and Composition of the Cryopump

The cryopump is mainly composed of G-M cryocooler, pump housing, first array (radiation shield, baffle, thermal rods, thermal plate), second array (including condensation surface and adsorption surface), etc. Shown as Fig. 1. The radiation shield and the baffle mainly remove water vapor; the condensation surface of the second array is mainly used for removing other condensable gases except He, Ne, H_2; the adsorption surface of the second array is adhered with activated carbon, to adsorb the remaining non-condensable gas. The radiation shield, baffle and second array of the cryopump were cooled by the cryocooler. The radiation shield, baffle and second array of the cryopump were made of oxygen-free copper.

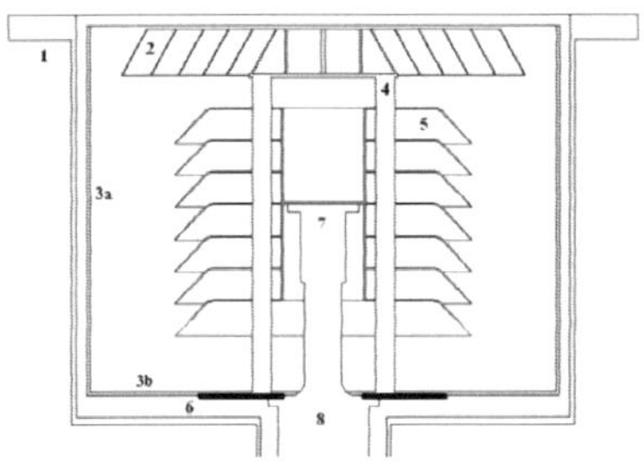

Figure 1. Stucture of DN400 cryopump.

1.Pump housing 2.baffle 3.Radiation shield (3a Shield wall 3b Shield bottom) 4.Thermal rods 5.Second array 6.Thermal plate 7.2nd stage of the cryocooler 8.1st stage of the cryocooler

2.2. Governing Equations

The governing equations for the solid three-dimensional unsteady heat conduction problem is [5]:

$$\frac{\partial(\rho h)}{\partial t} = \nabla \cdot (k \nabla T) \tag{1}$$

where: ρ is density, h is entropy, t is time, k is thermal conductivity, T is temperature.

To simplify the calculation, the radiant heat of the first and second array is calculated directly by the formula (2), and the calculation result of each unit on the wall is connected to the calculation model and substituted as a boundary condition by UDF(User-defined functions):

$$q = \varepsilon_s \sigma \left(T_1^4 - T_2^4\right) \tag{2}$$

where: q is the heat flux per unit area, ε_s is the system emissivity, σ is the Stefan-Boltzmann constant.

All the surfaces in the cryopump are assumed to be gray surface, and the surface emissivity of each surface is set as follows:

Baffle: 0.9; pump housing (inner surface): 0.1; radiation shield (outer surface): 0.1; radiation shield (inner surface): 0.9; second array condensation surface: 0.1; second array adsorption surface: 0.9. The specific heat capacity and thermal conductivity of copper are temperature dependent[6-7].
The cooling power of the G-M cryocooler was setting as Fig. 2.

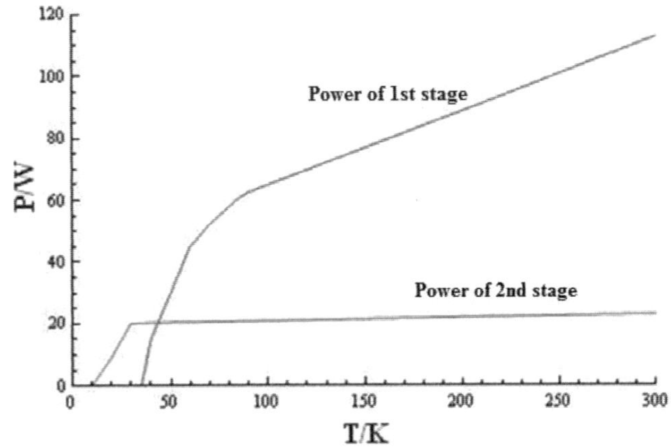

Figure 2. The power of the 1st and 2nd stage of the G-M cryocooler

The physical properties of the material and the cooling parameters of the cryocooler are also connected to the calculation model by means of UDF.

3. Results and Analysis

3.1. The steady Temperature Distribution of the Cryopump

3.1.1. The Steady Temperature Distribution of the Cryopump. Generally, there are mainly three operating conditions when the cryopump was used in a high-vacuum system.
1) Pre-cooling condition: the cryopump and the vacuum system are isolated by the valve;
2) The liquid nitrogen condition: The cryopump operates in a vacuum system with liquid nitrogen heat sink;
3) Room temperature condition: the cryopump operates in a vacuum system at room temperature.
Under case 2) and case 3), The vacuum vessel is usually much larger than the cryopump, the system emissivity can directly adopt the surface emissivity of the baffle.
The three-dimensional steady-state temperature distribution of the first array of the cryopump calculated under different working conditions is shown in Fig. 3.

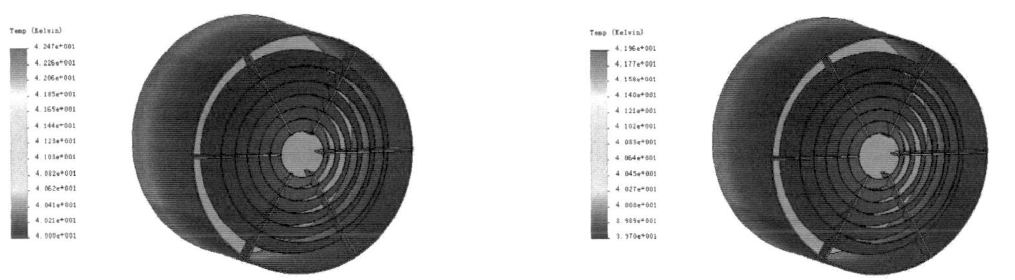

(a)Case 1 (b)Case 2

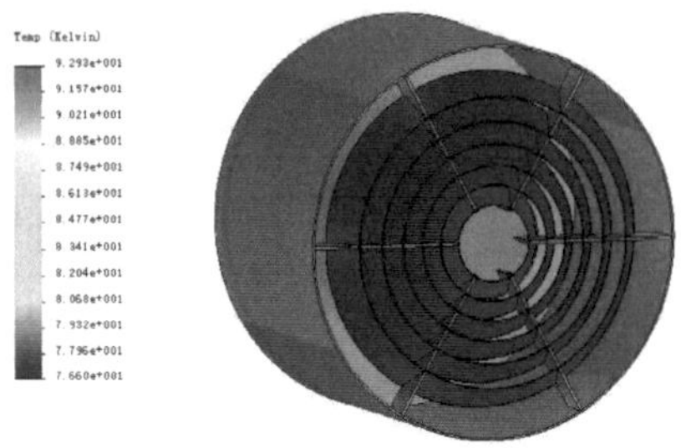

(c)Case 3

Figure 3. The three-dimensional steady-state temperature distribution of the first array of the cryopump calculated under different working conditions

In the pre-cooling condition, the average temperature of the first array is about 41K, and the maximum temperature rise is about 2.5K, as shown in Fig. 3a). In the liquid nitrogen condition, the average temperature of the first array is about 40K, the maximum temperature rise is about 2K, as shown in Fig. 3b). In room temperature condition, the average temperature of the first array is about 85K, and the maximum temperature rise is about 16.3K, as shown in Fig. 3c).

It can be seen from the figure that different operating conditions have a great influence on the temperature distribution of the cryopump. In room temperature condition, the temperature rise of the cryopump is the largest, which is the key point of the optimal design of the cryopump. The hottest part of the cold array takes on the main heat load of the cryopump.

Under room temperature condition, the temperature rise of each part of the cryopump's first array is shown in Fig.4 (the abscissa indicates the distance from the primary refrigerator), and the maximum temperature difference of the shield bottom plate is about 5.6K, The maximum temperature difference of the shield wall is about 6.5K, and the maximum temperature difference of the baffle is about 4.2K.

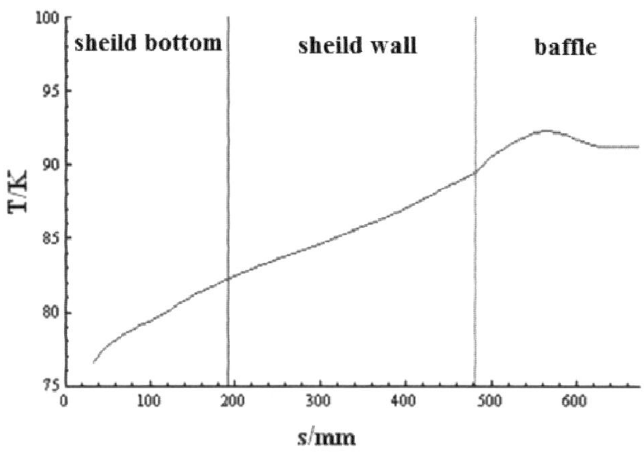

Figure 4. The temperature distribution of each part of the cryopump's first array

3.1.2. Effect of Thermal Rods Setting. In first array of the cryopump, the lowest and the highest temperature parts were connected by several thermal rods.

In order to investigate the effect of the thermal rods on the temperature distribution of the cryopump's first array, three different thermal rods settings were simulated:

Case1: no thermal rod was set;

Case2: 4 thermal rods (10 mm diameter) were set;

Case3: 4 thermal rods (15 mm diameter) were set.

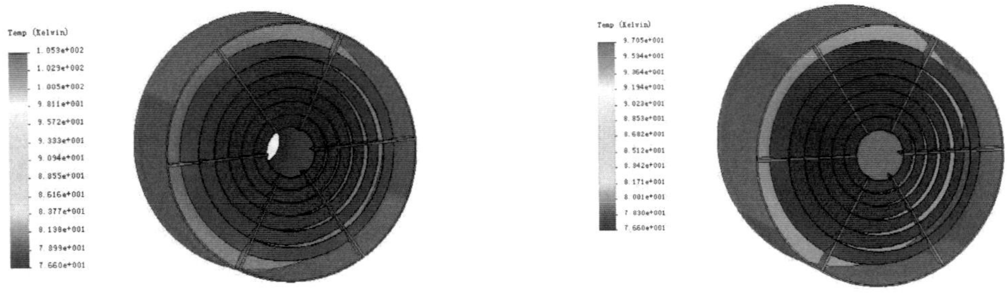

a) Without thermal rods b)With 4-φ10mm thermal rods

Figure 5. The temperature distribution of the first array under different thermal rods setting

In case1, the average temperature of the first array is about 91K, and the maximum temperature rise is about 28.9K, as shown in Fig. 5a); In case2, the average temperature of the first array is about 86.8K, and the maximum temperature rise is about 20.5K, as shown in Fig. 5b); In case3, the temperature distribution is shown in Fig. 3c).
Fig. 6 shows the temperature distribution of the baffle in case1 and case3:

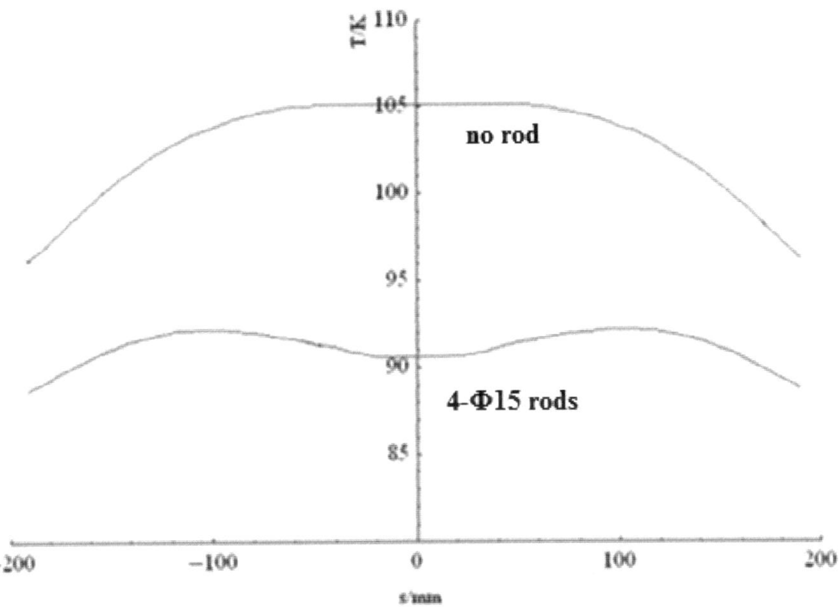

Figure 6. The temperature distribution of the baffle under different thermal rods setting

It can be seen from the figure that the arrangement of the thermal rods greatly reduces the thermal resistance of the baffle to the cryocooler, reduces the overall temperature difference of the cryopump, making the baffle temperature more uniform.

3.1.3. Effect of Thickness of the Radiation Shield. In the design of the cryopump first array, the temperature rise on the radiation shield can account for more than 70% of the total temperature rise, and the thickness of the radiation shield has a great influence on the total temperature rise control of the cryopump. Fig. 7 shows the temperature distribution of the first array with the radiation shield thickness of 1.0 mm and 1.5 mm. The temperature distribution with the radiation shield thickness of 2 mm is shown in Fig.3c). Under the three conditions, the maximum temperature rise of the first array of the cryopump is 21.5K, 18.4K and 16.3K.

IOP Conf. Series: Materials Science and Engineering **538** (2019) 012065 doi:10.1088/1757-899X/538/1/012065

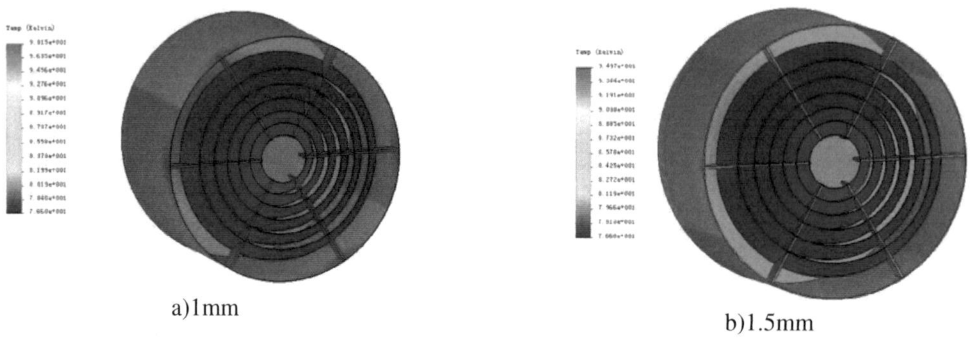

Figure 7. The temperature distribution of cryopump under different thickness of the radiation shield

Fig. 8 shows the temperature distribution of the radiation shield.

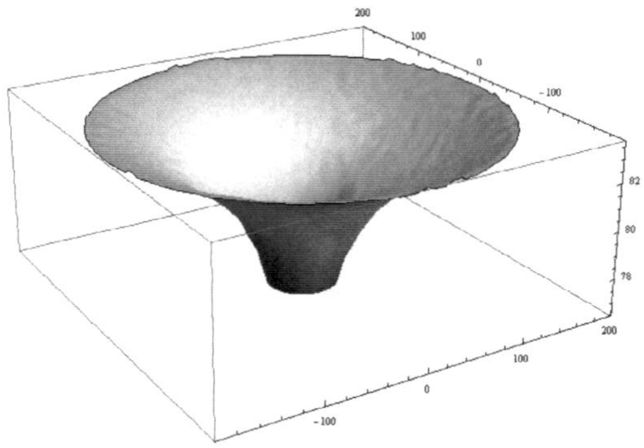

Figure 8. The temperature distribution of the bottom of the radiation shield (Max. 7K)

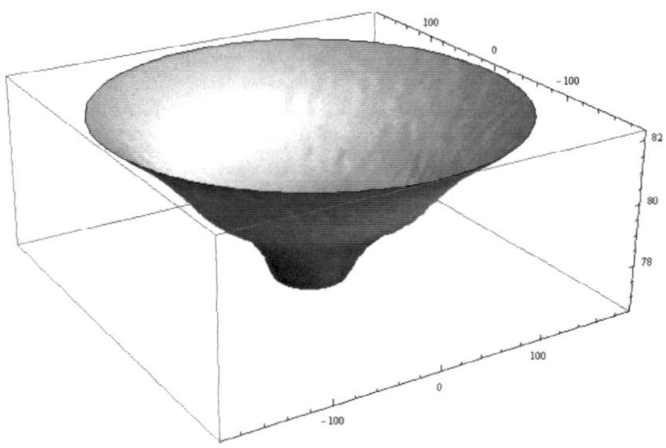

Figure 9. The temperature distribution of the bottom of the radiation shield (Max. 5.6K)

Combined with Fig. 4, it can be seen that the temperature rise of the shield bottom of the radiation shield exhibits a tendency of being fast and slow first. In order to control the temperature rise of the shield bottom plate, thickening the central portion of the bottom plate of the cold screen helps to increase the temperature uniformity of the bottom plate. . As shown in Fig. 9, a thermal plate（2 mm thick） is added to the central portion of the shield bottom plate to reduce the maximum temperature difference of the bottom plate by 1.4K.

3.2. The Temperature of the Cryopump During the Cool-Down Period

3.2.1. Model and Parameter Settings. The cool-down time of the first array of the cryopump refers to the time when the first array is cool-down from room temperature to 100K. The cool-down time of the first array has a great relationship with the thermal load of the cryopump [8-9].

Generally, the finial temperature of the cool-down time of the cryopump is set 20K(which is 10K higher than the equilibrium temperature of the second array), the finial temperature of the cool-down time for the first array is set 10K higher than the balance temperature.

3.2.2. Effects of Constant and Varying Physical Properties. Considering that during the cool-down period of the cryopump, the first array of the cryopump is cooled from 300K to about 40K, and the second array is cooled from 300K to about 10K, the temperature gradient is very large. The selection of the qualitative temperature of the first array and the second array has a great influence on the cooling performance of the cryopump. If the selection is improper, it will generate a large calculation error.

Under the pre-cooling condition, The change of the temperature during the cool-down calculated by the const thermal properties (first array qualitative temperature 170K, second array qualitative temperature 155K) and variable thermal properties was showed as Fig.10.

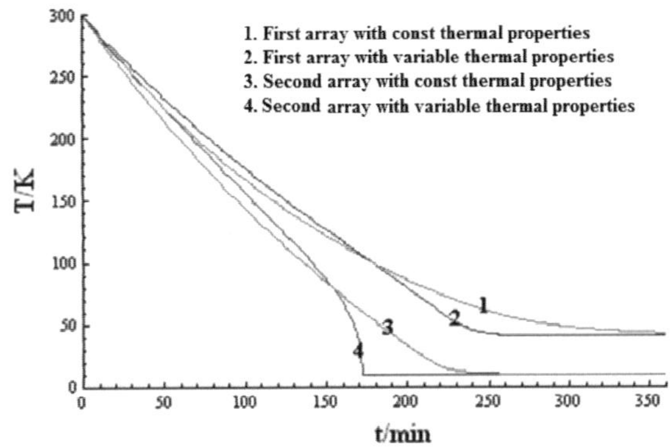

Figure 10. The change of the temperature during the cool-down calculated by the const thermal properties and variable thermal properties

It can be seen from Fig. 12 that the variation in the result calculated by the const thermal properties and variable thermal properties is about 24%. Obviously, if the temperature at the room temperature or the equilibrium temperature was used as the qualitative temperature, the calculation error will become larger.

3.2.3. The Effect of the First Array Structure. Fig.11 shows the cooling curves in three different structural states:

a) no thermal rods, shield wall thickness 1mm;

b) 4 Φ10 thermal rods, shield wall thickness 1.5mm;

c) 4 Φ15mm thermal rods, shield wall thickness 2mm.

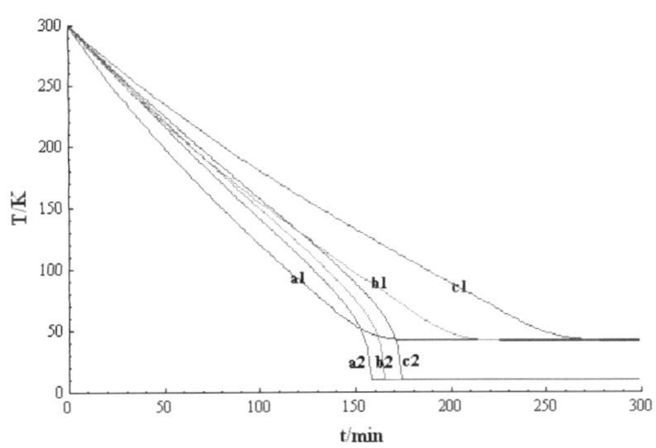

Figure 11. The change of the temperature during the cool-down under different structure of first array

It can be seen from the figure that the cool-down time of the cryopump is prolonged with the increase of the quality of the first array. The cool-down time of the second array of the DN400 cryopump is usually about 100-180 min, within a reasonable range of the cool-down time. It's reasonable to increase the quality of the first array, in order to reduce the thermal resistance and reduce the temperature rise.

4. Conclusion

In this paper, the temperature distribution and temperature drop characteristics of cryopump first and second arrays are studied by numerical simulation. In this study, the thermal conductivity and heat capacity of oxygen-free copper is considered by means of a user defined functions (UDF).
The conclusion can be made as follows:
1) The temperature difference of the first array of the cryopump is much larger than the temperature difference of the second array. The temperature difference of the first array is about 16K, and the temperature difference of the second array is about 1K.
2) The temperature difference of the first array of the cryopump is greatly affected by the working conditions, especially when facing the room temperature vacuum system, the main heat load is concentrated on the baffle;
3) For the bottom plate of the DN400 cryopump, a variable thickness bottom plate or a thermal plate can be used to reduce the overall temperature change of the cryopump;
4) The setting of the thermal rods is very useful to reduce the temperature change of the cryopump. At the same time, according to the temperature distribution of the baffle without the thermal rods, it can be considered that the setting of the thermal rods is preferably close to the center of the baffle;
5) If the variable property was not considered, the deviation of the cool-down time is about 24%。

Acknowledgments
Financial grants from the National Natural Science Foundation of China (NSFC Grant No. U1537109) is gratefully acknowledged.

References
[1] Rogalski A. Progress in focal plane array technologies[C]. Progress in Quantum Electronics. 2012;36(2):342-473.
[2] Kimo M Welch. Recommended practices for measuring the performance and characteristics of closed-loop gaseous helium cryopumps[J]. Journal of Vacuum Science and Technology A, 1999, 17(5):3081-3095.
[3] Baechler, W.G. Cryopumps for research and industry[J]. Vacuum, 1987, 37(1): 21-29.
[4] Johnson V J.Properties of Materials at Low Temperature[M],Pergamon Press,1961.
[5] Baechler WG. Cryopumps for research and industry. Vacuum. 1987;37(1–2):21-9.
[6] Dermois OC, Schmidt PW. A 20 K cryopump with a high conductance baffle design[J]. Vacuum. 1980;30(10):411-4.
[7] Hands BA. Introduction to cryopump design[J]. Vacuum. 1976;26(1):11-6.
[8] Iwasa Y, Ito S. A new type of cryopump with a metal cryopanel cooled below 3.6 K by a two-stage GM refrigerator[J]. Vacuum. 1996;47(6–8):675-8.
[9] Meng D, Sun L, Yan R, Shao R. et al. Effects of cryopump on vacuum helium leak detection system[J]. Vacuum. 2017;143:316-9.

IOP Conf. Series: Materials Science and Engineering **538** (2019) 012066 doi:10.1088/1757-899X/538/1/012066

Double Fuzzy Logical Control in Hybrid Power Supply Electric Vehicle

Meng-Qiang Wu[1,a], Hao Guo[1], Zi-Qiang Xu[1], Yun-Long Yang[1], Zhi-Mei Yuan[2],Chao-Qun Xiao[3] and Jian Liu[4]

[1] School of Materials and Energy, University of Electronic Science and Technology of China, Chengdu, 611731, China
[2] Sichuan Jia Yi Co. Ltd. Guang'an, 638500, China
[3] Sichuan Southwest Vocational College of Civil Aviation, Chengdu, 610400, China
[4] Chengdu Hongtao new material co. Ltd. Chengdu, 611432, China

[a] Corresponding author: mwu@uestc.edu.cn

Abstract. The hybrid power supply of fuel cell and super capacitor combines the advantages of high energy density of fuel cell with high power density of super capacitor. Based on the fuzzy logical control strategy, two fuzzy logical controllers are set up to participate in power distribution, and the simulation verification is carried out under the standard working Condition experiment, which effectively reduces the driving energy consumption of pure electric vehicles. This scheme control strategy is simple and easy to engineer.

1. Introduction
Hybrid power supply is combined with the high energy density characteristics of the fuel cell and the high power density characteristics of the super capacitor, which can recover the braking energy of the vehicle under the power distribution scheme of the fuzzy logical control, and give full play to the advantages of the fuel cell and the super capacitor[1] [2].

2. Parameter Matching
Vehicle parameters as shown in table 1:

Table 1. Vehicle parameters

Items	*Parameters*
Mass of the empty vehicle	1160kg
Mass of the whole vehicle	1750kg
Reduction ratio	7.93
Frontal area	$2.4m^2$
Coefficient of rolling friction	0.137
Drag coefficient	0.284
Transmission efficiency	0.96

2.1. Motor Parameter Matching

P_1^{MAX} and P_2^{MAX} are obtained based on the maximum climbing degree and the parameter matching of the motor with the maximum driving speed.

1) According to the maximum climbing degree to match the motor power:

$$P^{MAX}{}_1 = \frac{1}{3600\eta_T}(\mathrm{mgf} \bullet \cos\alpha_{MAX} + \frac{C_D A V_1^2}{21.15} + mg \bullet \sin\alpha_{MAX}) \bullet V_1 \tag{1}$$

Where P_1^{MAX} indicates vehicles at the 15° power requirements when uphill, η_T indicates that the total efficiency of vehicle power system transmission is 96%, m indicates vehicle quality 1750kg, g indicates that the gravitational acceleration which equals 9.8m/s2, α_{MAX} indicates the maximum climbing angle of 20° , C_D indicates that the air resistance coefficient which equals 0.284, A indicates windward area 2.4m2, V1 indicates the speed at which driving on the maximum slope which equals 20km/h, f indicates that the rolling friction coefficient which equals 0.0137

2) According to the maximum driving speed to match the motor power:

$$P_2^{MAX} = \frac{1}{3600\eta_T}(mgf + \frac{C_D A V^2{}_{MAX}}{21.15}) \bullet V_{MAX} \tag{2}$$

where P_2^{MAX} indicates the power requirements of the vehicle when driving at maximum speed 120km/h, V_{MAX} indicates the maximum speed which equals 120km/h.

The maximum power requirement of the motor P_E meets the following formula:

$$P_E \geq MAX(P_1^{MAX}, P_2^{MAX}) \tag{3}$$

According to the formula (1) formula (2) we could calculate the P_1^{MAX}=35.30kw, P_2^{MAX}=24.27kw. By formula (3) know P_E equals the maximum value in P_1^{MAX} and P_2^{MAX}. In view of the normal operation of the vehicle will encounter temperature and specific road conditions and other factors of the affection. Therefore, the peak power of the motor should be greater than the calculated value, take P_E=40kw to ensure the normal driving of the vehicle [3].

2.2. Fuel Cell Parameter Matching

In this paper, the parameters of the fuel cell are matched according to the maximum mileage of the vehicle.

$$P_{FC} = \frac{1}{3600\eta_T}(mgf + \frac{C_D A V_2^2}{21.15}) V_2 \tag{4}$$

In the formula:

P_{FC} is the power that fuel cells need to provide when cruising at V_2 speed *of* 60km/h, V_2 is the vehicle's cruising speed which equals 60km/h.

Calculate the capacity of the fuel cell based on the maximum mileage of the vehicle:

$$W = \frac{SP_{FC}}{V_2} \tag{5}$$

where S indicates the maximum mileage to which equals 800km, W indicates the energy required for the vehicle to travel the maximum mileage.

After calculating get the P_{FC}=6.09kw, the energy required when the vehicle travels at a distance of 800km W=81.2kw·h.

3. Power Distribution Scheme

As shown in Figure 1, fuzzy logical control, as a new control strategy, can imitate the precise input blur processing, inference and the output of reasonable results through the solution fuzziness, and has a greater advantage over the logical threshold control in terms of reliability[4]-[6].

IOP Conf. Series: Materials Science and Engineering **538** (2019) 012066 doi:10.1088/1757-899X/538/1/012066

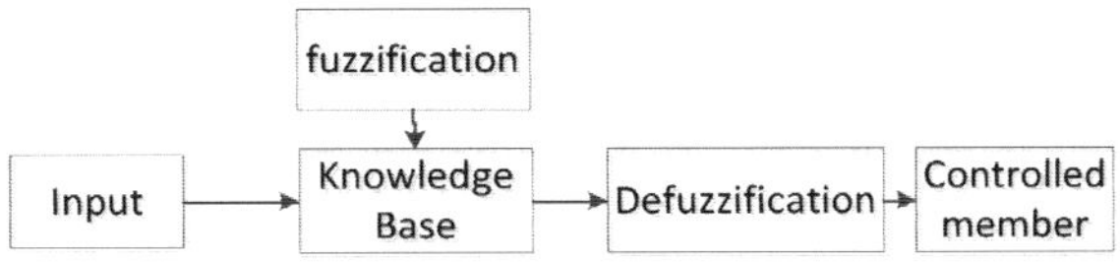

Figure 1. Process of fuzzy logical control

3.1. Selection of Control Quantities

As shown in Figure 2.When driving normally, the fuzzy logical controller FLC1 is involved in the work, with the input of LoadSignal, the charge state of the fuel cell (FC_SOC) and the Charge State (UC_SOC) of the super capacitor, and the output is the power demand for the fuel cell (FC_req) And the power demand for supercapacitors (UC_req). When the vehicle is in a braking state, the fuzzy logical controller FLC2 is involved in the input of the braking pressure and the Charge State (UC_SOC) of the supercapacitor, and the output is the power requirement for the supercapacitor (UC_req).

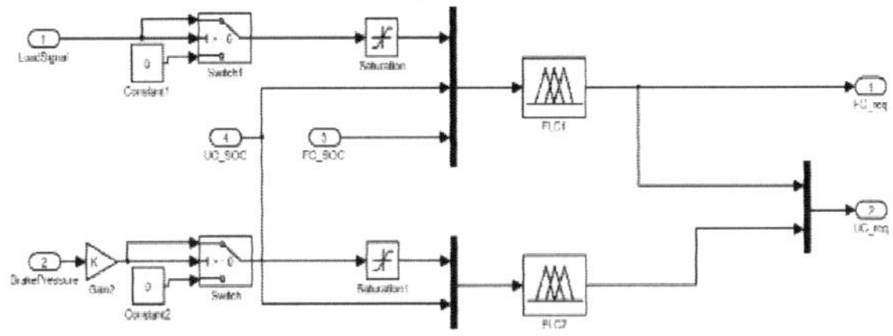

Figure 2. Power allocation strategy

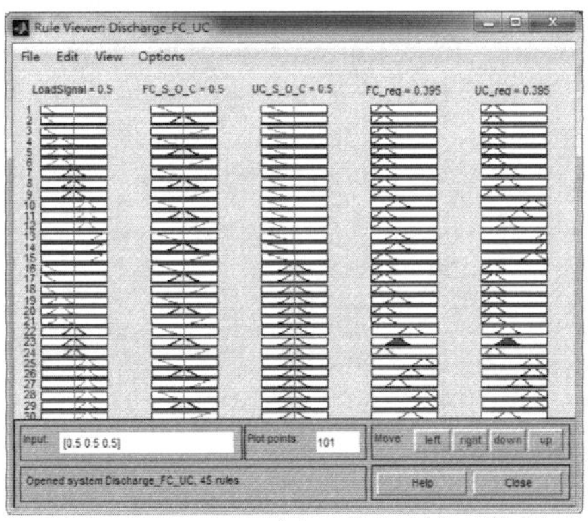

(a)

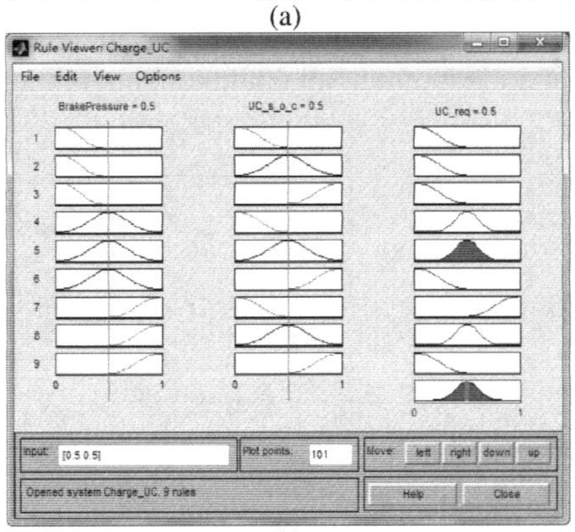

(b)

Figure 3. Rule watcher when driving normally and braking

3.2. Building of Knowledge Base

As shown in Figure 3 when the electric vehicle is in normal driving, the LoadSignal is at a low or medium level, and when the FC_SOC is large, the fuel cell's separate output power meets the power demand of the vehicle, and when the LoadSignal of the electric vehicle is at a higher level, the individual output of the fuel cell is no longer sufficient to meet the power demand of the vehicle , will be output by supercapacitors and fuel cells.at the same time to meet vehicle power requirements.
When the electric vehicle is in the braking state and the input BrakePressure and UC_SOC is small, the output UC_req is larger; When the UC_SOC is large, the output UC_req is smaller.

3.3. Defuzzification Method

In this paper, the center of gravity method is used as the solution defuzzification method, that is, by calculating the center of gravity of the graph formed by the membership function and the axis as the precise output, the calculation method is as follows:

$$V_0 = \frac{\int V \mu_V(v) dv}{\int \mu_V(v) dv} \tag{6}$$

4. Verification of Hybrid Power Supply

Figure 4 is a comparison diagram of the speed/acceleration - time between the single-power ev and the hybrid power source ev. It can be seen that from 0 to 100km/h, the acceleration time of single power source electric vehicles is 13.76 seconds, while that of pure electric vehicles with hybrid power source is 12.95 seconds.
Figure 5 is a single power supply of pure electric vehicle climbing capacity and composite power supply pure electric vehicle climbing-speed comparison chart, you can see that the composite power supply in the same slope when driving, faster; at the same speed, composite power electric vehicle is afraid of greater slope. At 60km/h, the climbing degree of single power source electric vehicles is 14.23%, while that of hybrid power source pure electric vehicles is 23.39%.
To verify the vehicle's cruising ability, the vehicle is tested in a fixed speed, cruising at 60km/h speed, and the SOC range for the fuel cell is set at 90%~10% when cruising. According to the figure 5 we can know 17.79% of the fuel cell's SOC is left when the vehicle reaches 800km in travel distance.
As shown in figure 7,in WLTP conditions under the fuzzy logical controlled power distribution measurement, the fuel cell always discharges at a small current, while the supercapacitor carries out a high current discharge according to the power demand of the vehicle, receiving the braking current during braking to enhance the vehicle's life capacity.

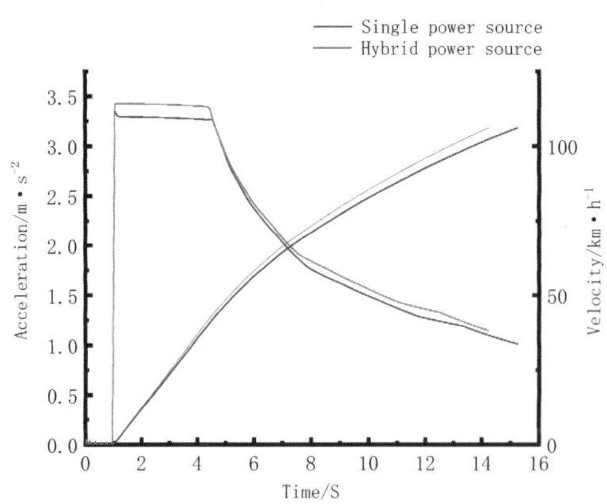

Figure 4. Comparison of velocity/acceleration - time

IOP Conf. Series: Materials Science and Engineering **538** (2019) 012066 doi:10.1088/1757-899X/538/1/012066

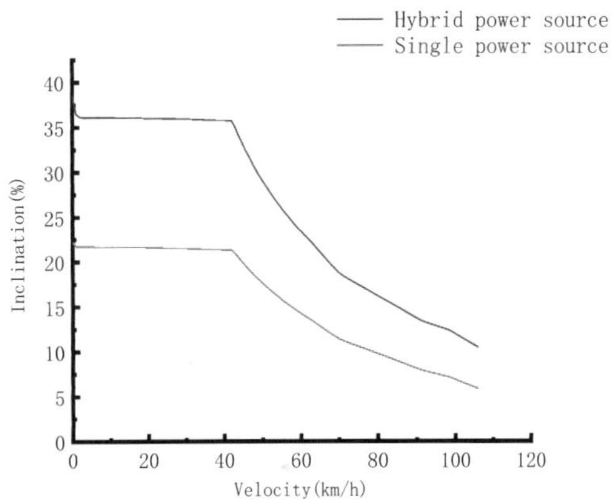

Figure 5. Slope-speed comparison chart

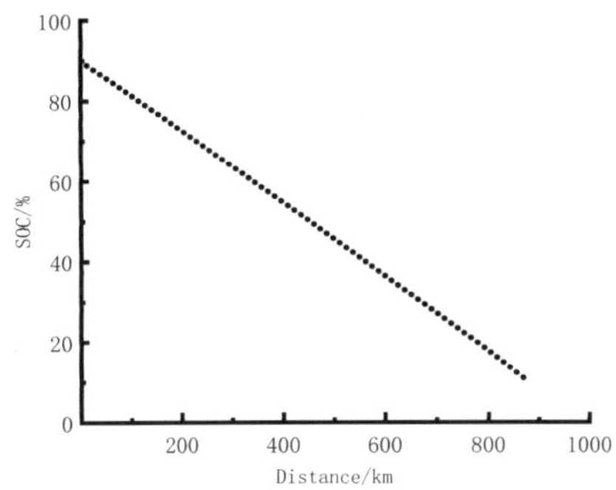

Figure 6. Change of SOC with driving distance

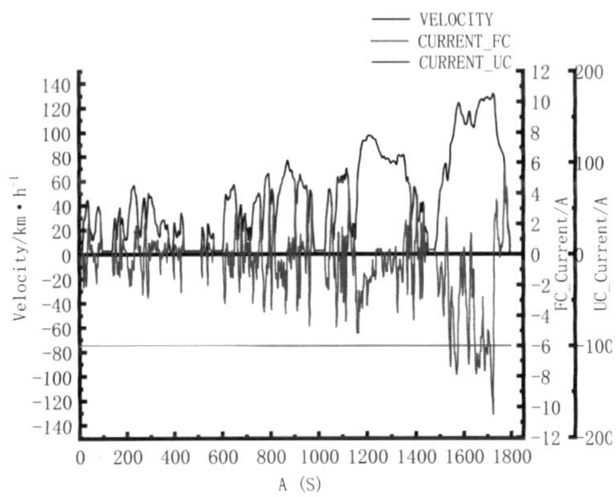

Figure 7. Current flow under WLTP conditions

5. Conclusions

In this paper, a hybrid power supply composed of fuel cell and super capacitor is studied, which improves the recovery efficiency of braking energy under the power distribution strategy of double fuzzy control, avoids the too frequent charging and discharging of the battery, and realizes the long mileage cruising and extends the service life of the cell. At the same time improve the climbing ability and acceleration ability of electric vehicles. To a certain extent, the problem of insufficient endurance and service life of pure electric vehicles is improved.

Acknowledgments
This work was supported by the Sichuan Science and Technology Program (2017GZ0102, 2018GZ0010), and Chengdu Electric Vehicle Industry Cluster Collaborative Innovation Project (2017-XT00-00002-GX).

References
[1] Choi M E, Kim S W, and Seo S W. Energy Management Optimization in a Battery/Supercapacitor Hybrid Energy Storage System. IEEE Transactions on Smart Grid, Vol. 3, No. 1, pp.463-472. 2012
[2] Song Z, Hofmann H, Li J, et al. A comparison study of different semi-active hybrid energy storage system topologies for electric vehicles. Journal of Power Sources, pp. 400-411. 2015
[3] Georgakellos D A. A polygon-based environmental appraisal of new vehicle technologies combined with renewable energy sources. Transportation Research Part D Transport & Environment, Vol 13,No. 4, pp. 283-288. 2008
[4] Wang J, Wang Q, Liu J, et al. Forward simulation model precision study for hybrid electric vehicle International Conference on Mechatronics and Automation. IEEE pp. 2457-2461, 2009
[5] Wang G, Yang P, and Zhang J. Fuzzy optimal control and simulation of battery-ultracapacitor dual-energy source storage system for pure electric vehicle[C]// International Conference on Intelligent Control and Information Processing. IEEE, ,pp. 555-560, 2010
[6] Yin H, Zhou W, Li M, et al. An Adaptive Fuzzy Logic-Based Energy Management Strategy on Battery/Ultracapacitor Hybrid Electric Vehicles. IEEE Transactions on Transportation Electrification, Vol 2, No.3, pp. 300-311. 2016

IOP Conf. Series: Materials Science and Engineering **538** (2019) 012067 doi:10.1088/1757-899X/538/1/012067

Sliding Mode Control Algorithm for Regenerative Braking of an Electric Bus with a Pneumatic Anti-lock Braking System

Yiming Zhao[1], Junzhi Zhang[1,2], Chao Li[1] and Chengkun He[1]

[1]State Key Laboratory of Automotive Safety and Energy, Tsinghua University, Beijing, People's Republic of China

[2]Innovation Center of Electric Vehicles, Beijing, People's Republic of China

Corresponding author: jzhzhang@tsinghua.edu.cn

Abstract. Equipped with regenerative braking system, electric vehicles have the function of coordinating regenerative braking and friction braking simultaneously in normal braking situations. However, the usual operation is withdrawing the regenerative braking when the anti-lock braking system (ABS) is triggered. In order to make full use of regenerative braking, this article focuses on the cooperative control between the regenerative braking and the pneumatic friction braking during emergency stops. A sliding mode control (SMC) algorithm for regenerative braking is proposed to control the electric motor and help improve the braking performance and braking comfort. The proposed control algorithm is validated in the MATLAB/Simulink environment. A maximum mean deceleration of 25.3% is achieved and the stopping distance is shortened by 13.2m comparing to the pure pneumatic ABS on the wet gravel road, which verifies the feasibility and effectiveness of the developed control algorithm greatly.

1. Introduction

Recent years, with the development of new energy technologies and the necessity of environmental protection, electric vehicles come into our lives gradually [1]. Compared to conventional vehicles, the electric motor of an electrified vehicle can work as a generator and capture the kinetic energy when the vehicle is braking [2]. However, due to the restrictions of an electric motor and the batteries, the friction braking system is also indispensable. Thus, it is essential to achieve the coordinated control between the regenerative braking system and the friction braking system.

Meanwhile, in order to guarantee the safety under emergency conditions, ABS is mounted on most of the vehicles. It is also an aspect which should be noticed of the cooperative control between the regenerative braking and the friction braking when ABS is activated.

By far, the cooperative control between the regenerative braking and the friction braking is researched worldwide [3-6], but rarely related to the participant of electric motor in a panic stop. For electric vehicles, when the vehicle takes an emergency braking, the regenerative braking force is withdrawn immediately [7], [8], leaving the friction braking to take over the ABS function only. It is a simple and convenient way, but the potential advantages of the electric motor are not fully utilized.

Based on these facts, there are also some researchers to carry out innovations beyond the simplification. A pure electric motor regenerative braking method was developed in [9]. Test results indicate the advantages of the method, but the braking capability is limited. This in turn validates that the friction braking is indispensable to guarantee the safety. According to this thought, reference [10] recommended three control modes to realize the cooperative control, namely PQ-method, frequency selection by filter and model following control. The effectiveness is confirmed by simulations and road tests. Reference [11] developed a strategy in which the regenerative braking is prior to load to respond the braking command. Once the regenerative braking force can't meet the total braking force, the hydraulic braking force will compensate immediately. A phase-theory method, dividing the total braking torque to a stable one and a dynamic one, was proposed by reference [12]. The friction braking and regenerative braking take responsible for the two torques respectively. Reference [13] introduced the optimal control theory and thus obtained maximal available adhesion. A robust wheel slip ratio controller was designed for in-wheel-motors-driven vehicles in [14] and reached a good compromise between stopping distance and regenerative braking torque. Other control methods could also be found in references [7], [15], [16].

Nevertheless, most of the strategies are not fully using the logic threshold control method and are about the hydraulic braking system of passenger cars. Due to the widespread utilization of the pneumatic logic threshold control on buses and the significant dynamic differences between the hydraulic braking system and the pneumatic braking system [17-19], it is essential to explore the regenerative braking control method on the basis of the pneumatic logic threshold ABS of a bus.

In this paper, a SMC algorithm of regenerative braking is proposed to help control the slip ratio on an electric bus with the pneumatic ABS. A system model is built in the MATLAB/Simulink environment and simulations are carried out in four typical road surfaces. The rest of this paper is organized as follows. In section 2, the logic threshold ABS is introduced. In Section 3, the SMC algorithm of regenerative braking is designed. Simulation results and conclusions are displayed and concluded respectively in section 4 and section 5.

2. Logic threshold ABS and the ideal slip ratio section of your paper

2.1. Logic threshold ABS

The logic threshold control of ABS is the most widely used control method nowadays. The control core is to calculate the real slip ratio and judge the state of the wheel according to the wheel angular speed and the wheel angular deceleration. Once the wheel has the tendency of locking up, the ABS is activated and the system switches the pneumatic proportional valve among increasing pressure, holding pressure and decreasing pressure to prevent the lock-up of the wheel. The five threshold values, namely three angular decelerations and two slip ratios are the triggered values and used to switch the control zones.

In practical usage, the threshold values are obtained through a number of tests and need to be validated for a long time. Because of the control logic, the complexities and the uncertainties of road conditions, the slip ratio is controlled in a range but not an ideal value during the procedure. Thus, although the wheels will not be locked during the emergencies, it is believed that the braking performance will be worse than the situation in which the slip ratio is stabilized at the ideal one.

Table 1. Constants of C_1, C_2 and C_3 on different roads.

Road surface	C_1	C_2	C_3
Ice	0.05	306.39	0.001
Snow	0.1946	94.129	0.0646
Wet gravel	0.4404	33.708	0.1204
Wet bituminous	0.857	33.822	0.347

2.2. The relations between the road adhesion and the slip ratio

The relations between the road adhesion and the slip ratio are complicated. This article adopts the following equation to calculate the ideal slip ratio on different roads [20].

$$\mu(\lambda) = C_1\left(1 - e^{-C_2\lambda}\right) - C_3\lambda \tag{1}$$

where μ is the road adhesion, λ is the slip ratio, C_1, C_2, C_3 are constants on different roads. According to Table 1, the ideal slip ratio could be gotten in usual road surfaces.

3. The SMC algorithm of regenerative braking

The SMC algorithm has the advantages of good control accuracy and strong robustness. Once the controller is well designed and the control input is able to be guaranteed, the control performance is only affected by the convergence rate. It is especially proper for the control of a nonlinear system.

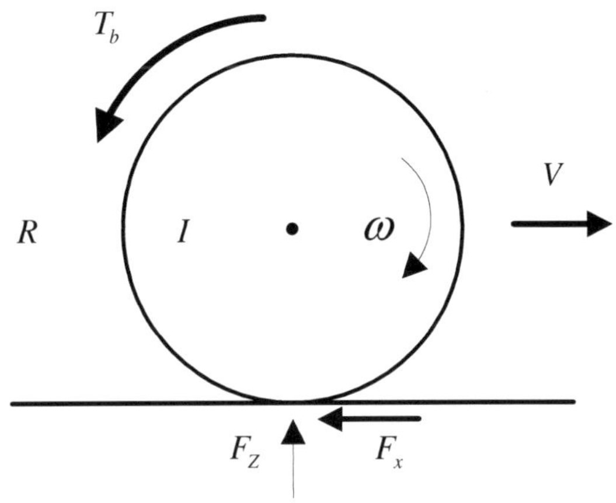

Figure 1. Longitudinal dynamical model of the wheel.

The motion equations of the researched electric bus can be described as (The wind resistance is neglected.):

$$m\dot{V} = -F_x \tag{2}$$

$$I\dot{\omega} = F_x R - T_{b\_total} \tag{3}$$

where m is the quarter mass of the vehicle, V is the vehicle velocity, F_x is the longitudinal force, R is the wheel radius, ω is the wheel's angular speed, I is the inertia of the wheel and T_{b_total} is the total braking torque.
As for the rear-left wheel,

$$T_{b\_total} = T_b + 0.5 i_0 i_g T_m + T_{wheel\_drag} \tag{4}$$

where T_b is the pneumatic braking torque, T_m is the motor torque, T_{wheel_drag} is the wheel's drag torque, i_g is the transmission ratio of the gearbox and i_0 is the final drive ratio.
The slip ratio of the wheel is defined as

$$\lambda = \frac{V - R\omega}{V} \tag{5}$$

Conduct the derivation of equation (5),

$$\dot{\lambda} = \frac{\dot{V}(1-\lambda) - R\dot{\omega}}{V} = \frac{-\dot{V}}{V}\lambda + \frac{\dot{V}}{V} - \frac{R\dot{\omega}}{V} \tag{6}$$

Let

$$\frac{-\dot{V}}{V}\lambda = f(\lambda) \tag{7}$$

$$\frac{\dot{V}}{V} - \frac{R\dot{\omega}}{V} = u \tag{8}$$

where $|f(\lambda)| < \rho * \lambda$, $\left|\frac{-\dot{V}}{V}\right| < \rho$ and ρ is one of the upper limits of $\left|\frac{-\dot{V}}{V}\right|$, then we have

$$\dot{\lambda} = f(\lambda) + u \tag{9}$$

Define the tracking error of the slip ratio as:

$$e = \lambda_d - \lambda \tag{10}$$

where λ_d is the ideal slip ratio of the corresponding road. Perform derivation of equation (10) and equation (11) is obtained

$$\dot{e} = \dot{\lambda}_d - \dot{\lambda} \tag{11}$$

Let

$$u_1 = \dot{\lambda}_d + ke + \rho\frac{|e|}{e} \tag{12}$$

$$u_1 = u \tag{13}$$

where $k > 0$, thus,

$$\frac{\dot{V}}{V} - \frac{R\dot{\omega}}{V} = \dot{\lambda}_d + ke + \rho\frac{|e|}{e} \tag{14}$$

Define the Lyapunov function as follows.

$$V(e) = \frac{1}{2}e^2 \tag{15}$$

Then

$$V(\dot{e}) = e\dot{e} = -ke^2 - f(\lambda)e - \rho|e| < -ke^2 < 0$$

From $V(\dot{e}) < 0$, it can be concluded that the system is stable. It means that given the input of u, the slip ratio of λ will convergence to λ_d in a limited time and the convergence rate is decided by the value of k.
Combine equations of (2) (3) (4) (14), the motor torque command can be expressed as follows:

$$T_m = \frac{2}{i_0 i_g}\left[\frac{I\dot{V}}{R} - \frac{kVI}{R}(\lambda_d - \lambda) - \frac{VI}{R}\dot{\lambda}_d - \frac{VI\rho}{R}\mathrm{sgn}(e) + F_x R + T_b + T_{wheel\_drag}\right] \tag{16}$$

where $\mathrm{sgn}(e)$ is the sign function and equation (16) is the motor torque command during the braking. In practical usage, in order to decrease the chattering of the motor command, the sign function is usually replaced by saturation function. After replacement, the final motor torque command is shown in (17).

$$T_m = \frac{2}{i_0 i_g}\left[\frac{I\dot{V}}{R} - \frac{kVI}{R}(\lambda_d - \lambda) - \frac{VI}{R}\dot{\lambda}_d - \frac{VI\rho}{R}\mathrm{sat}(e) + F_x R + T_b + T_{wheel\_drag}\right] \tag{17}$$

Thus, the controller design of the SMC is completed.

4. Simulation and analysis

In order to evaluate the proposed control algorithm, the simulation model is built in a Matlab/Simulink environment.

4.1. Simulation scenarios

Because the maximum possible deceleration of the researched bus is about 5 m/s^2 (absolute value) and skid is much easier to occur on slippery roads. The simulations are carried out on the snow, ice and wet gravel roads, of which the maximum adhesion coefficients are 0.2, 0.1 and 0.5 respectively. And a transitional road (from 0.8 to 0.2) is also conducted. The details of simulation are: set the initial velocity of 60 km/h and press the brake pedal fully to trigger the ABS. The simulation results are displayed as follows.

4.2. Analysis of simulation results

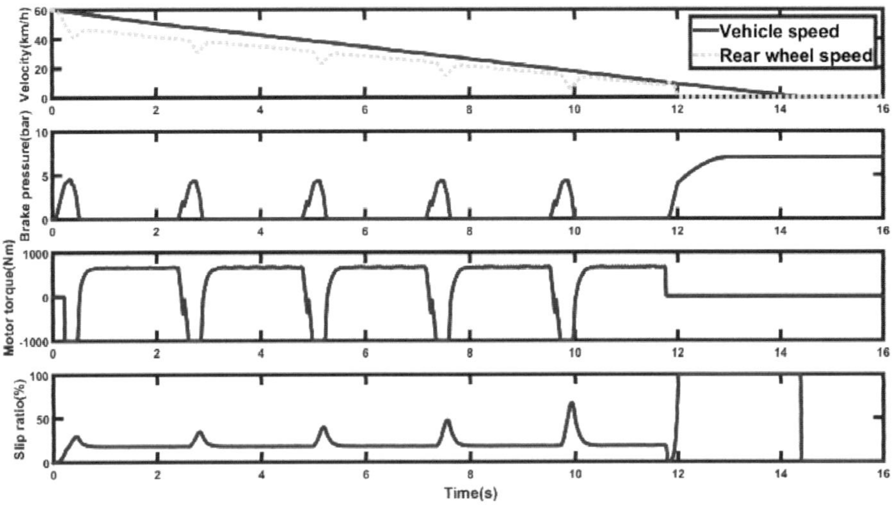

Figure 2. Simulation results of the SMC algorithm on a snow road.

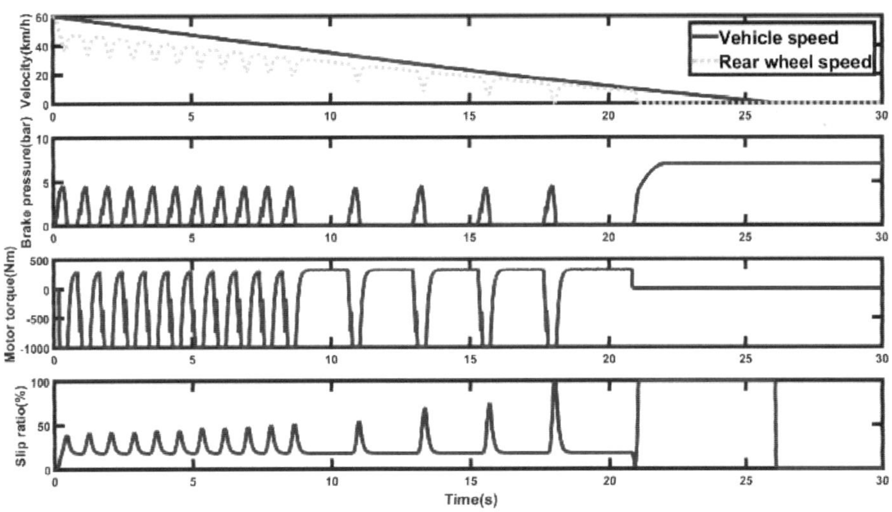

Figure 3. Simulation results of the SMC algorithm on an ice road.

Figure 2 is conducted on the snow road. When the brake pedal is fully depressed, the pressure of pneumatic braking increases greatly, following the increase of the slip ratio. The system detects the tendency of lock-up and prevents the slip ratio from increasing any further. Because of the slow response of pneumatic system, the proposed control strategy sets the motor torque immediately to negative value (drive torque) to resist the oversize slip ratio. When the pneumatic pressure decreases to a low value, the slip ratio has the tendency of less than the ideal value. The proposed control strategy sets the motor torque from negative to positive to help maintain the slip ratio to the ideal value.

When the pneumatic pressure is high enough again, the motor torque is set to negative value once more. These cycles come into play again and again until the vehicle speed is below 10 km/h, at which the motor torque is removed and only the pneumatic braking takes over the rest braking process.
Viewed from the full range, the proposed strategy is able to control the braking smoothly and no skid occurs during the procedure. Meanwhile, the slip ratio is stabilized at the ideal value most of the time, which achieves the expected control performance.
Then the ice road is conducted following the snow road. Similar to the snow road, no skid appears in Figure 3. The real slip ratio is also kept at the ideal one at some time. The difference is that the adjustment cycles are more frequently than the snow road. This is understandable. Due to the lower adhesion of ice road, the wheel is easier to lock up, so the control logic is triggered frequently to perform more control cycles. But the proposed strategy is able to handle them appropriately.

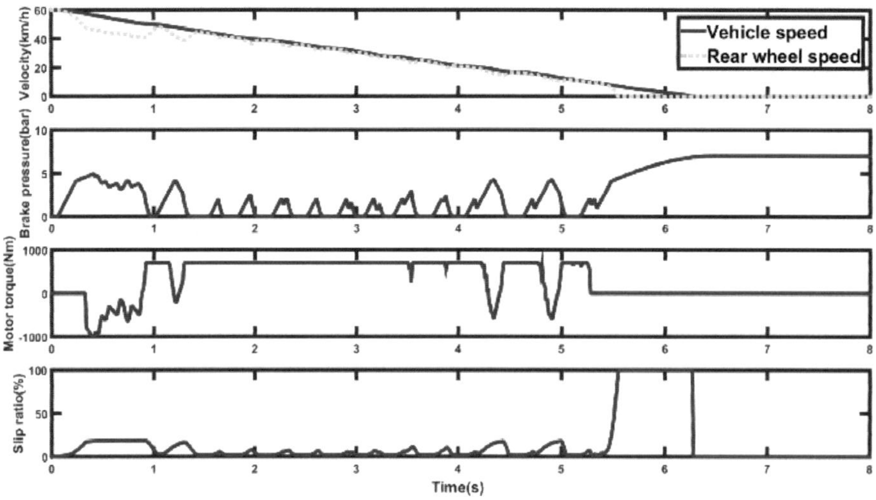

Figure 4. Simulation results of the SMC algorithm on a wet gravel road.

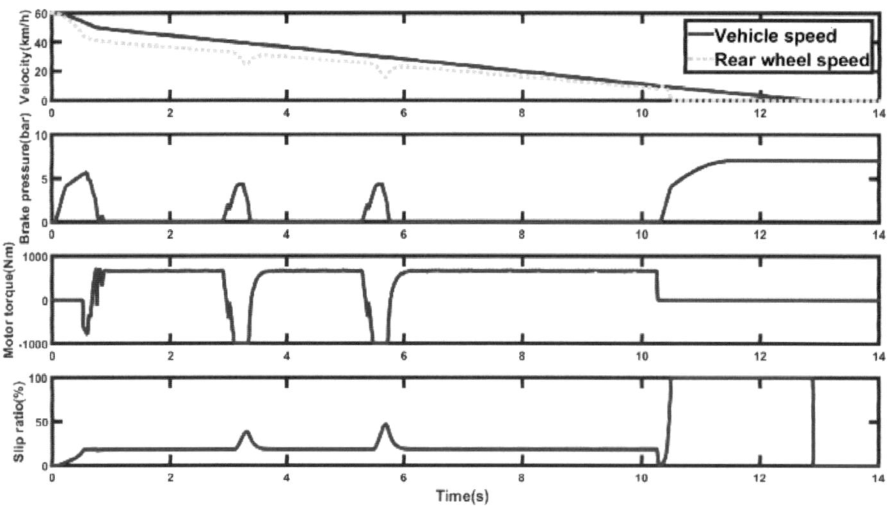

Figure 5. Simulation results of the SMC algorithm on a transitional road.

Besides the above two situations, the simulations are also performed on a wet gravel road and a transitional road. The Figure 4 and 5 show that the control results are desirable and no locking up occurs on both roads. However, owning to the motor's capability, the slip ratio can't always adjust to the ideal value but close enough to the ideal value.

From the above results and analyses, it can be concluded that the braking performance and braking stability are guaranteed, demonstrating the feasibility and effectiveness of the SMC strategy.

4.3. Comparison to the other two strategies

In order to assess the SMC strategy further, the simulations of a proportional-integral-differential (PID) control strategy and the pure pneumatic control strategy are carried out in the same scenarios. The difference between the SMC strategy and the PID control strategy is replacing the SMC algorithm to a PID control algorithm.

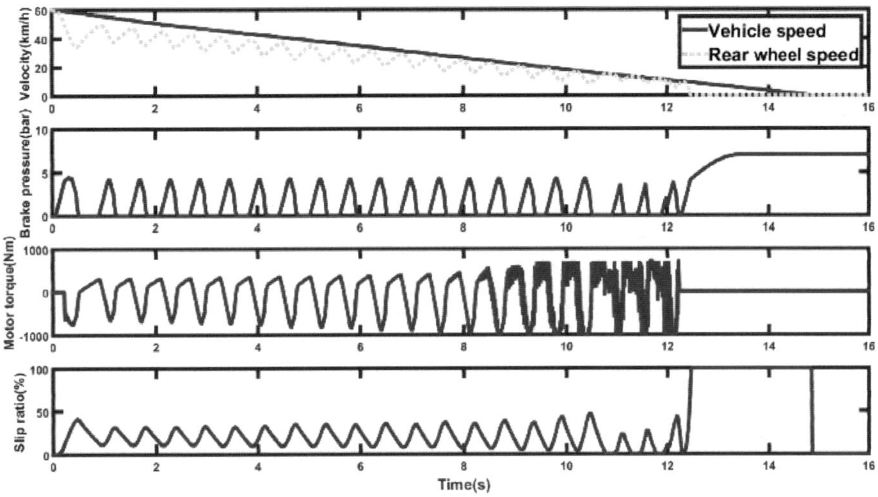

Figure 6. Simulation results of the PID control strategy on a snow road.

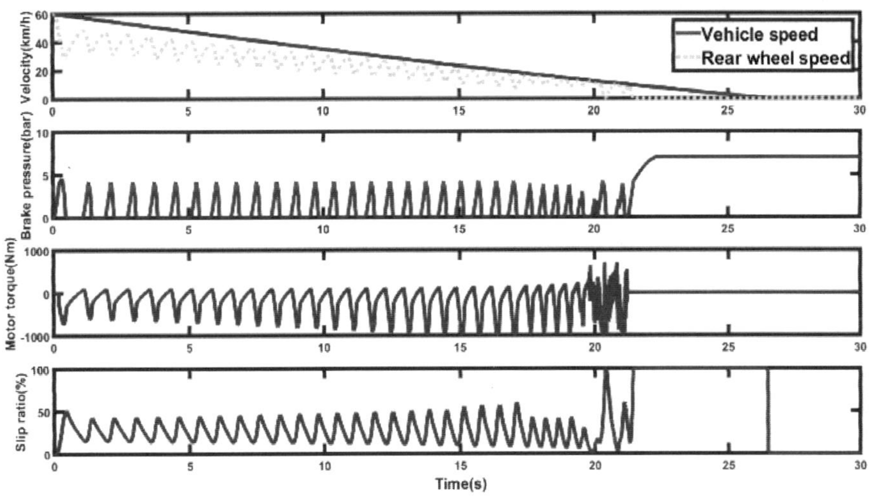

Figure 7. Simulation results of the PID control strategy on an ice road.

Figure 6 and Figure 7 are the simulation results on the snow and ice road. Results indicate that the PID control strategy can also realize the anti-locking of the wheels. But being restricted to the control characteristics of the PID controller, the slip ratio is hard to stabilized to the ideal value but a smaller range around the ideal value. This can be seen from the variation trend of the motor torque.

Meanwhile, in order to correct the error of the slip ratio ahead, the differential control will cause the strong chattering of the motor torque, which will increase the jerk and worsen the braking comfort during the braking. The chattering can be seen from Figure 8 and Figure 9 as well.

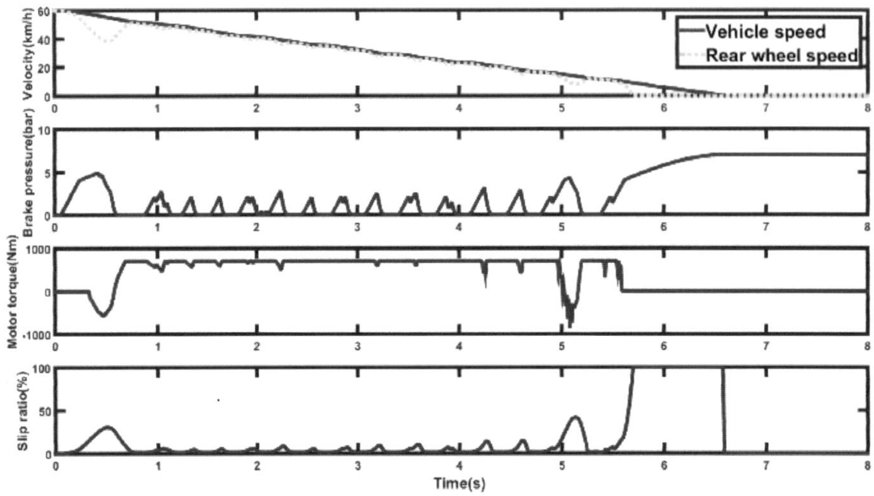

Figure 8. Simulation results of the PID control strategy on a wet gravel road.

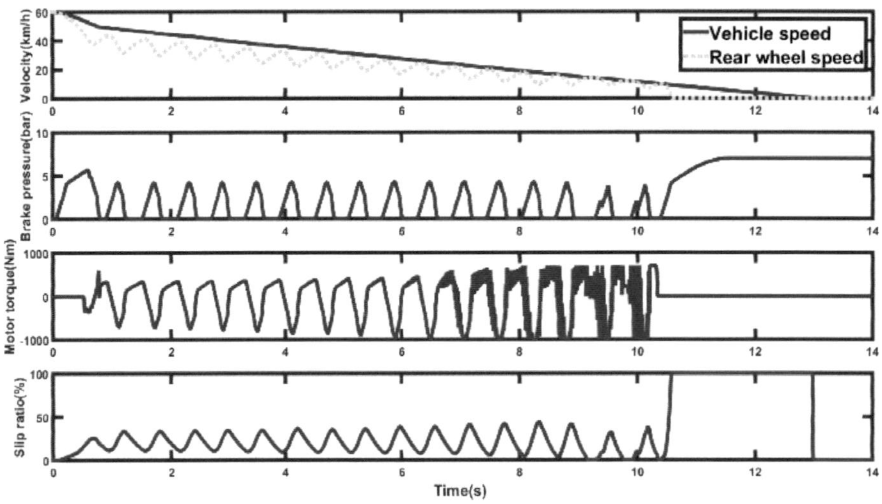

Figure 9. Simulation results of the PID control strategy on a transitional road.

In order to evaluate the control strategy more comprehensively, the root mean square of jerk (RMSJ) which can be used to weigh the braking comfort is introduced.
The RMSJ is defined as：

$$RMSJ = \sqrt{\frac{1}{n}\sum_{k=1}^{n} J^2} \tag{18}$$

where

$$J = \frac{d^2 V}{dt^2} \tag{19}$$

Thus, the Table 2 is obtained.

From the table, it can be concluded that both the PID control strategy and the SMC strategy can achieve a better braking performance regarding of the stopping distance and the mean braking deceleration. However, on the one hand, due to the control characteristics of the PID controller, it need a lot of work to adjust the control parameters but not SMC. On the other hand, the SMC strategy can greatly improve the braking comfort (Except the wet gravel road which is affected by the control strategy of the original pneumatic ABS.), but the PID control strategy deteriorates the braking comfort to a great extent. The comparisons demonstrate the advantages of the SMC strategy greatly.

Table 2. Comparison of the three control strategies.

Road condition	Strategy	Stopping distance (m)	Mean deceleration (m/s^2)	Deceleration improvement (%)	RMSJ	RMSJ improvement (%)
Ice	Pneumatic	217.30	0.639	——	0.288	——
	PID	207.60	0.669	4.68	0.308	-6.76
	SMC	204.90	0.678	6.10	0.258	10.37
Snow	Pneumatic	130.00	1.069	——	1.035	——
	PID	120.10	1.156	8.14	1.212	-17.10
	SMC	117.70	1.180	10.40	0.928	23.43
Wet gravel	Pneumatic	65.40	2.124	——	5.566	——
	PID	54.82	2.534	19.30	6.777	-21.76
	SMC	52.15	2.663	25.30	6.143	-10.37
Transitional	Pneumatic	109.30	1.270	——	1.529	——
	PID	93.70	1.482	16.69	1.953	-27.73
	SMC	94.95	1.463	15.20	1.048	31.46

Therefore, comparing to the other two control strategies, it is encouraging that the SMC strategy is much better.

5. Conclusion

A SMC strategy for regenerative braking of an electric bus equipped with an anti-lock braking system is proposed in this paper. The SMC strategy is able to coordinate the pneumatic ABS strategy and to help maintain the slip ratio at the ideal value or a more precise range. Meanwhile, the PID control strategy and the pure pneumatic strategy are taken as benchmarks to compare the SMC strategy.
Simulation results indicate that owing to the strong robustness and low parameter adjustment workload of SMC, not only better braking performance can be achieved, but better braking comfort. The mean deceleration and braking comfort are improved by 15.2% and 31.46% on the transitional road respectively comparing to the pure pneumatic ABS control strategy. Simulation results in four typical roads fully demonstrate the adaptability and effectiveness of the SMC strategy.
Future work will focus on the hardware-in-loop tests to verify the strategy further. Meanwhile, the impact on energy usage of overall mileage will also take into consideration.

Acknowledgments

This article was supported by the National Key Research and Development Program of China [project no.2018YFB0106103].

References

[1] A S. Tie, C. Tan, "A review of energy sources and energy management system in electric vehicles", *Renew. and Sustain. Energ. Rev.*, **20**, 82-102 (2013)

[2] G. Sovran, D. Blaser, "Quantifying the potential impacts of regenerative braking on a vehicle's tractive-fuel consumption for the U.S., European, and Japanese driving schedules", *SAE paper*, 01-0664 (2006)
[3] C. Kumar, S. Subramanian, "Cooperative control of regenerative braking and friction braking for a hybrid electric vehicle", Jour. of Auto. Eng., **230**, 103-116 (2016)
[4] J. Ko, S. Ko, H. Son, er al., "Development of brake system and regenerative braking cooperative control algorithm for automatic-transmission-based hybrid electric vehicles", IEEE Trans.on Veh. Tec., **64**, 431-440 (2015)
[5] P. Fajri, S. Lee, V. Prabhala, et al, "Modeling and integration of electric vehicle regenerative and friction braking for motor/dynamometer test bench emulation", IEEE Trans. On Veh. Tec., **65**, 4264-4273 (2016)
[6] R. Bravo, V.Negri, A. Oliveria, "Design and analysis of a parallel hydraulic–pneumatic regenerative braking system for heavy–duty hybrid vehicles", App. Energ., **225**,60-77 (2018)
[7] D. Savitski, V. Ivanov, K. Augsburg, et al, "The new paradigm of an anti-lock braking system for a full electric vehicle: experiment investigation and benchmarking", Jour. of Auto. Eng., **230**, 1364-1377 (2016)
[8] L. Zhang, L. Yu, N. Pan, et al, "Cooperative control of regenerative braking and friction braking in the transient process of anti-lock braking activation in electric vehicles", Jour. of Auto. Eng., **230**, 1459-1476 (2016)
[9] O. Tur, O. Ustun, R. Tuncay, "An introduction to regenerative braking of electric vehicles as anti-lock braking system", IEEE Intelli. Veh. Sym., 944-948 (2007)
[10] T. Okano, S. Sakai, T. Uchida, "Braking performance improvement for hybrid electric vehicle based on electric motor's quick torque response", Proc.19th Inter. Ele. Veh. Sym., (1999)
[11] Z. Zhang, J. Zhang, D. Sun, et al.,"Research on control strategy of electric –hydralic hybrid anti-lock braking system of an electric passenger car", IEEE Intelli. Veh. Sym., 419-424 (2015)
[12] C. Lv, J. Zhang, Y. Li, et al., "Novel control algorithm of braking energy regeneration system for an elcctric vehicle during safety–critical driving maneuvers", Energ. Conver. and Manage., **106**, 520-529 (2015)
[13] J. Zhang, D. Kong, L. Chen, et al,"Optimization of control strategy for regenerative braking of an electrified bus equipped with an anti-lock braking system", Jour. of Auto. Eng., 1-13 (2011)
[14] B. Wang, X. Huang, J. Wang, et al, "A robust wheel slip ratio control design combing hydraulic and regenerative braking systems for in-wheel-motors-driven electric vehicles", Jour. of the Fra Instit. **352**, 277-602, (2015)
[15] L. Li, X. Li, X. Wang, et al, "Transient switching control strategy from regenerative braking to anti-lock braking with a semi-brake-by-wire system", Veh. Sys. Dyna., **54**, 231-257 (2016)
[16] T. Sadarmehni, A. Heydari, "Optimal switching in anti-lock brake systems of ground vehicles based on approximate dynamic programming", Proceed.of the ASME 2015 Dyna. Sys.and Cont. Confer. (2015)
[17] S. Subramanian, S. Darbha, K. Rajagopal, "Modeling the pneumatic subsystem of an S-cam air brake system", Jour. of Dyna. Sys. Mea. and Cont. **126**, 36-46 (2004)
[18] J. Zhang, Y. Li, C. Lv, et al., "New regenerative braking control strategy for rear-driven electrified minivans", Energ. Conver. and Manage., **82**,135-145 (2014)
[19] K. Huang, K. Quang, K. Tseng, "Study of recycling exhaust gas energy of hybrid pneumatic power system with CFD" , Energ. Conver. and Manage., **50**, 1271-1278 (2009)
[20] M. Burckhardt, J. Reimpell, "Fahrwerktechnik, Radschiupf-Regelsysteme", Vogel Fachbuch (1993)

IOP Conf. Series: Materials Science and Engineering **538** (2019) 012068 doi:10.1088/1757-899X/538/1/012068

Method for Determining Accurate Initial Air Pressure Range of Air Suspension Airbag Based on Vehicle Ride Comfort Simulation Analysis

Quan Zhou[1,a], Puguang Yuan[1], Tim Tudor[2], Xudong Wang[1], Zhangwei Guo[1], Yihui Huang[1], Jingfeng Sun[1], Boyun Wang[3]

[1]Institute of vehicle suspension, Automotive Engineering Department, Wuhan University of Technology, China
[2]School of Engineering, University of Wales Trinity Saint David, Swansea, UK
[3]Bellevue College, Bellevue, Washington, US

Corresponding author: 1805150@student.uwtsd.ac.uk

Abstract. With the rapid development of automobile technology and the improvement of people's quality of life, people put forward higher requirements for vehicle ride comfort. Shock absorption has become an important direction for automobile research, manufacturing and optimization. The traditional air suspension airbag delimits the initial air pressure range based on the bearing capacity of the airbag material, so the range of the pressure is inaccurate. Considering optimizing ride comfort of the vehicle, this paper proposes a method to determine the accurate initial air pressure range. In order to determine the accurate initial air pressure range of the airbag, this paper needs to know the vehicle ride comfort under any air pressure. The airbag spring stiffness is calculated using the airbag model in MATLAB environment. A shock absorber with appropriate damping coefficient is selected by using the single mass system model and the airbag spring stiffness. Suspension stiffness is calculated by constructing a single wishbone beam independent suspension model and using the airbag spring stiffness. The damping coefficient and suspension stiffness are introduced into the Simulink five-degree-freedom vehicle model to obtain the acceleration signals of every part including the seat whose RMS value of the acceleration signal is used to judge the ride comfort of the vehicle. Under the specific simulation conditions in this paper, there is an accurate range (14~16kPa) in the rough range (12~20kPa) of the initial air pressure of the airbag to make the vehicle ride smoothly.

1. Introduction

Vehicle ride comfort is a significant performance indicator to compete for advantages in market competition. With the improvement of people's living standards, the problem of structural vibration comfort has attracted more and more researchers' attention. The vehicle ride comfort directly affect the experience of the occupants, and indirectly affect the vehicle's power and economy. The suspension system plays a significant role in the shock absorption performance of the vehicle. In practical applications, the air suspension system can meet the various requirements of the suspension design and save energy and material, thus driving many scholars and engineers extensively research and gradually promote it. Air suspension has been widely used in various vehicles. However, for the vast majority of air suspension airbags on the market, only a rough initial air pressure range is given, while

IOP Conf. Series: Materials Science and Engineering **538** (2019) 012068 doi:10.1088/1757-899X/538/1/012068

the accurate initial air pressure range of various airbags are not given. This paper predicts that there is a specific pressure value or accurate pressure range in the rough initial air pressure range to maximize the vehicle ride comfort. This paper intends to find the accurate pressure value or an accurate range of the initial air pressure of the airbag by simulation.

The initial air pressure of the airbag is the main factor determining the natural frequency of the air spring, which determines the air spring stiffness, the air suspension stiffness and the damping coefficient of the shock absorber. Xiong established the vibration model of the air suspension system, and proposed the method to calculate the parameters of the air spring and air suspension and the method to select shock absorber with appropriate parameters [1]. Li proposed the formula to calculate the stiffness of three different shaped air springs and the stiffness of air suspensions matching different shaped springs [2].

Meanwhile, the rough road surface is the root cause of the vibration of the vehicle, so it is necessary to establish a mathematical model to describe road roughness. Xie used computer simulation of random road surface excitation as input, and established the road surface roughness model by MATLAB/Simulink simulation software for computer simulation [3]. Liu built a Simulink simulation model for Class B Road based on pseudo white noise method [4]. Zheng used the probability and statistics theory and the central limit theorem to establish a mathematical model of the uneven road surface spectrum based on the power spectral density of the principle of harmonic superposition [5]. Zhang's research summarizes the triangular series method and the Fourier inverse transformation method for road roughness [6].

In general, whole body vibration is the main form of vibration that causes physical injury to the driver so a standard should be established to evaluate the ride comfort. This paper decides to use the impact of the body's vibration to evaluate the ride comfort. In order to evaluate the impact of the body's vibration, the International Organization for Standardization (ISO) introduced the international standard "Guidelines for the Evaluation of Human Body's Resistance to Whole Body Vibration" (ISO 2631-1978) [9] in 1978. In order to accurately evaluate the ride comfort of the vehicle, Zhang uses the RMS value of the vertical weighted acceleration of the passenger to evaluate the index of the flatness. [7] Zhang summarized two internationally accepted evaluation criteria for ride comfort and three quantitative calculation methods for passenger comfort [6].

Table 1. Relationship between weighted RMS acceleration and riding comfort.

Weighted Acceleration RMS Value/ ($m \bullet s^{-2}$)	People's feeling
0.315	Comfortable
0.315 ~ 0.63	Slightly Uncomfortable
0.5 ~ 1.0	A Little uncomfortable
0.8 ~ 1.6	Uncomfortable
1.25 ~ 2.5	Very uncomfortable
2.0	Particularly Uncomfortable

The weighted RMS acceleration is related to seat acceleration signal. According to the relationship between weighted RMS acceleration and riding comfort (Table 1), it is necessary to establish a suitable mathematical model to analyze the vibration of every part of the vehicle to get seat acceleration signal. In order to accurately evaluate the ride comfort of the vehicle, Zhang constructed a five-degree-freedom vehicle model of human-vehicle-road, and analyzed the interaction between human and vehicle roads by transfer matrix method [7]. Xie performed mechanical analysis to obtain the system vibration differential equation and built the four-degree-freedom vehicle model by Simulink. [3] Liu, through the dynamic analysis of vehicles, established the eight-degree-freedom vehicle model, calculated the eight-degree- freedom coupled vibration equation of the vehicle four-point random excitation, and used the block matrix method to derive the transfer function of the vehicle vibration system. [8] Zhang's research summarizes the two-degree-freedom, the four-degree-freedom, five-degree-freedom, the seven-degree-freedom, seventeen-degree-freedom vehicle model for vibration analysis of each part [6].

The above literatures provide a theoretical basis and modeling basis for studying the relationship between initial air pressure and vehicle ride comfort of air suspension airbags, but none of these literatures study the initial air pressure of air suspension airbags for the purpose of maximizing vehicle ride comfort. Therefore, the purpose of this paper is to find a method for determining accurate initial air pressure range of air suspension airbag based on vehicle ride comfort simulation analysis

In order to ensure the best ride comfort of a vehicle under a specific road roughness level, this paper proposes a method to determine the accurate initial air pressure range of the air suspension airbag. For a certain type of vehicle using air suspension, the control variable method is used to change the initial air pressure of the air suspension airbag, thereby changing the suspension stiffness and suspension damping coefficient. Vehicle parameters and the road roughness excitation derived from the Simulink road roughness model are introduced to five-degree-freedom vehicle Simulink model. Using the transfer matrix method to solve the vibration equation, the acceleration signal of the vehicle seat under each initial air pressure of airbag can be obtained. Then use the acceleration signal of the vehicle seat to find the seat acceleration RMS value. The RMS value of the seat acceleration at each initial air pressure is compared to find an accurate range of the initial airbag pressure that optimizes ride comfort. The method can be used to determine the optimal initial air pressure range of the air suspension airbag of the vehicle under various road conditions and the vehicle can adjust the initial air pressure of the air suspension in real time according to the road condition; The method can also be used to initialize the air suspension airbag gas pressure during maintenance.

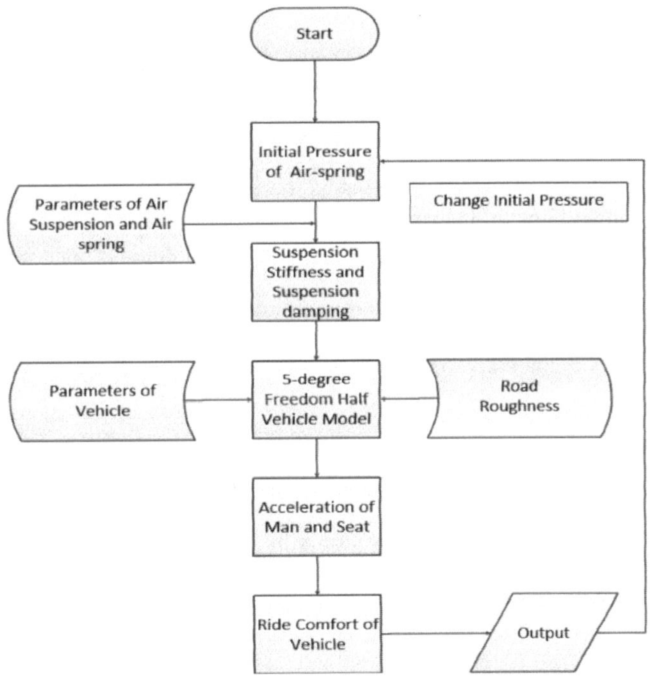

Figure 1. System model of optimal initial air pressure optimization method for airbags.

2. System description

The working process of the optimal initial air pressure optimization system for airbags proposed in this paper is shown in the figure 1. First, an initial air pressure is required to set the air suspension airbag. Next, the suspension stiffness and suspension damping coefficient can be calculated by inputting the initial air pressure of the airbag and the parameters related to airbag and air suspension into the single mass system model and the airbag model. Then, the suspension stiffness damping coefficient, vehicle parameters and road roughness excitation obtained from the road model are input into the five-degree-freedom vehicle Simulink model to obtain the signals of the vibration displacement, velocity and acceleration of each part of the vehicle. Then, the RMS value of the existing seat acceleration signal is obtained to evaluate the ride comfort of the vehicle as an output and recorded. Finally, change the initial air pressure of the airbag and repeat the above steps to find the initial air pressure of the airbag with the optimal ride comfort.

3. Mathematical model

In this paper, a five-degree-freedom vehicle model is built in the Simulink environment, and the ride comfort simulation analysis of the model's vibration differential equation is carried out. First, in order to obtain the input parameters of the model, the road roughness excitation $q(t)$ is obtained through the simulation of the road roughness model. The second is to obtain the front air suspension stiffness and rear air suspension stiffness k_{cf}, k_{cr} and front suspension damping coefficient and rear suspension damping coefficient c_{cf}, c_{cr} through the airbag model, the single mass system model and the single wishbone beam independent suspension model. The road roughness excitation $q(t)$, the suspension stiffness k_{cf}, k_{cr}, the suspension damping coefficient c_{cf}, c_{cr} and the vehicle parameters are input to the five-degree-freedom vehicle model to obtain the seat acceleration $a_w(t)$.

3.1. Road roughness model

The input of the five-degree-freedom vehicle model is mainly the road roughness excitation $q(t)$and parameters of the vehicle, so establishing a reasonable road model helps to obtain accurate simulation results. Through a large number of literatures [3, 4], the white noise method has the advantages of small calculation, fast running speed and high calculation accuracy. Therefore, this paper chooses the white noise method to simulate the road model.

Road roughness refers to the deviation of the road surface from the ideal plane. It affects the power, comfort and safety of the vehicle. The power spectral density of the road is often used to describe the statistical characteristics of road roughness. The international standard ISO8608[9] and the national standard GB7031-86[5] suggest that the fitting expression of the road power spectral density $G_q(n)$ is:

$$G_q(n) = G_q(n_0)\left(\frac{n}{n_0}\right)^{-W} \tag{1}$$

where: $G_q(n_0)$ is the road surface roughness coefficient, which is the value of the road surface power spectral density at the reference spatial frequency n_0 , in m^3; n represents the spatial frequency of several wavelengths per m length m^{-1}; n_0 is the reference spatial frequency.

According to the road power spectrum density, the road roughness is divided into 8 levels, as shown in the following table, and the geometry mean of the corresponding root mean square value q_{rms} (σ_q) of the road roughness in the range of $0.011m^{-1}$<n< $2.83m^{-1}$ is listed.

Under normal circumstances, the numerical range of road roughness is: wavelength $\lambda = 0.1{\sim}100$m , amplitude A $= 1{\sim}200$mm road roughness classification as shown in the table 2.

Table 2. 8-level-classification of road roughness.

Road Level	Roughness Level $n_0 = 0.1m^{-1}, 0.011m^{-1} < n < 2.83m^{-1}$ $G_q(n_0)/10^{-6}m^3$			
	Lower Limit	Geometric Mean	Upper Limit	Geometric Mean
A	8	16	32	3.81
B	32	64	128	7.61
C	128	256	512	15.23
D	512	1024	2048	30.45
E	2048	4096	8192	60.9
F	8192	16384	32768	121.8
G	32768	65536	131072	243.61
H	131072	262144	524288	487.22

The method proposed in this paper takes B-level road as an example. For reference to the literature [4], using the pseudo white noise method, the mathematical model of a single wheel subjected to road roughness excitation in the time domain can be described by (2) :

$$\dot{q}(t) = 2\pi f_0\sqrt{G_q(n_0)u}W(t) - 2\pi f_0 q(t) \tag{2}$$

In equation (2), $q(t)$ is the road roughness excitation; $G_q(n_0)$ is the road surface roughness coefficient; $W(t)$ is the random simulated white noise; f_0 is the cut-off frequency; u is the vehicle speed.
Because the probability of a passenger car driving on a highway or a road with good road conditions is high, here is a B-level road as an example:

$$\begin{cases} G_q(n_0) = 64 \times 10^{-6} m^{-3} \\ f_0 = 0.1Hz \end{cases}$$

According to the road roughness model and the initially set parameters, the road roughness excitation $q(t)$ can be obtained in the Simulink environment.

3.2. Airbag model

Filled with compressed gas (this paper takes the rough pressure range 12kpa ~ 20kpa), the airbag (figure 2 and figure 3) uses the compressibility of the gas to achieve the spring action.

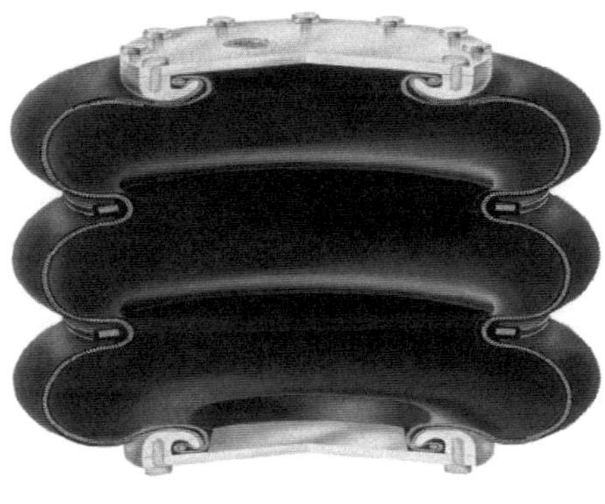

Figure 2. Airbag physical map.

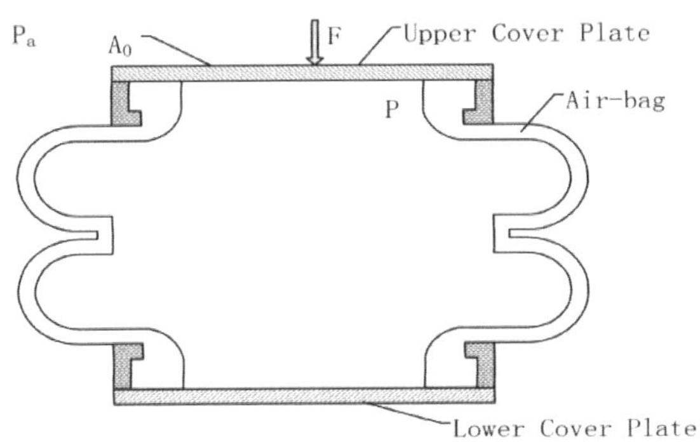

Figure 3. Airbag structure diagram.

The relationship between the load F on the air spring, the absolute pressure P of the air inside the airbag, the atmospheric pressure P_a , and the effective area A_0 is:

$$F = (P - P_a) \times A_0 \tag{3}$$

When the load causes the height of the airbag to change, the volume and internal pressure of the airbag also change, and the variation law satisfies the gas state equation:

$$\mathrm{P} = P_0\left(\frac{V_0}{V}\right)^n \tag{4}$$

where: V represents the volume within the airbag at any position; P_0 represents the absolute pressure of the volume within the airbag at the static equilibrium position; V_0 represents the volume within the airbag at the static equilibrium position; n represents the polytropic index. When the vehicle speed is slow, $\mathrm{n} = 1$; when the road condition is poor, the gas state changes close to the adiabatic process, then $\mathrm{n} = 1.4$; in general, take $\mathrm{n} = 1.33$, so in this paper $\mathrm{n} = 1.33$. The vehicle simulation speed is $41\mathrm{km/h}$.

The vertical stiffness k of the air spring is:

$$\begin{aligned} \mathrm{k} &= \frac{dF}{dx} = A_0\frac{d(P-P_a)}{dx} + (P-P_a)\frac{dA_0}{dx} \\ &= (P-P_a)\frac{dA_0}{dx} - A_0 n P_0 \frac{V_0^n dV}{V^{n+1}dx} \end{aligned} \tag{5}$$

where:

$$\frac{dA_0}{dx} = 0, \frac{dV}{dx} = -A_0$$

Consider $\mathrm{V} = \mathrm{V}_0$ when the airbag is in static equilibrium position. Spring stiffness at static equilibrium position can be obtained:

$$k_0 = \frac{nP_0A_0^2}{V_0} \tag{6}$$

From equation (6), the static equilibrium spring stiffness of the front suspension airbag is:

$$k_{0f} = \frac{nP_0A_{0f}^2}{V_{0f}} \tag{7}$$

The static equilibrium spring stiffness of the rear suspension airbag is:

$$k_{0r} = \frac{nP_0A_{0r}^2}{V_{0r}} \tag{8}$$

The static equilibrium spring stiffness k_{0f} of the front suspension airbag and the static equilibrium spring stiffness k_{0r} of the rear suspension airbag can be obtained by the airbag model. With the single mass system model, the front suspension damping coefficient c_{cf} and the rear suspension damping coefficient c_{cr} can be calculated from the values of k_{0f}, k_{0r}. The front suspension stiffness k_{cf} and the rear suspension stiffness k_{cr} can be obtained by the single wishbone beam independent suspension model.

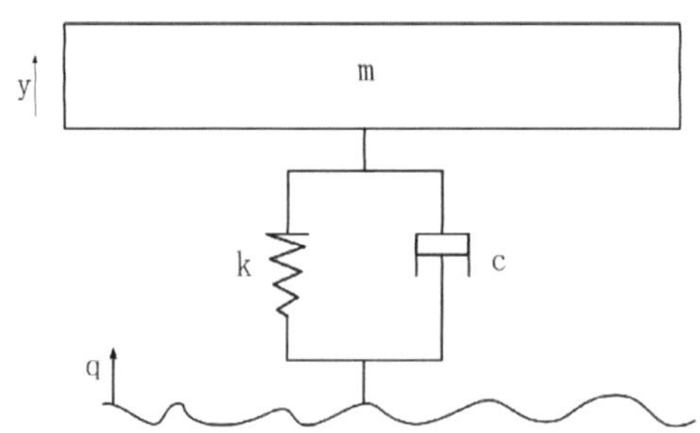

Figure 4. Single mass system model.

3.3. Single mass system model

In general, the suspension damping coefficient matching the stiffness of airbag can be obtained from the single mass system model shown in Figure 4. **m** is the suspension mass of the vehicle. It consists of the body, the frame and the assembly on it. **k** is the stiffness of the airbag, **c** is the damping coefficient of the shock absorber, and the three constitute the suspension system of the car. **q** is the input road roughness excitation.
Apply Newton's second law to the entire system:

$$m\ddot{y} + c(\dot{y} - \dot{q}) + k(y - q) = 0 \tag{9}$$

The solution of this equation consists of the sum of the solution of the free vibration homogeneous equation and the special solution of the nonhomogeneous equation.

$$\text{Let } 2\mathrm{a} = \mathrm{c}/\mathrm{m},\ w_0^2 = k/m$$

where: w_0 is the natural circular frequency of the system. The influence of damping on the motion depends on the ratio ζ of a and w_0.ζ is called the damping ratio.

$$\zeta = \frac{\mathrm{a}}{w_0} = \frac{c}{2\sqrt{km}} \tag{10}$$

So:

$$\mathrm{c} = 2\zeta\ \sqrt{km} \tag{11}$$

According to the equation (11), suspension damping coefficient c can be obtained.
The front suspension damping coefficient c_{cf} is:

$$c_{cf} = 2\zeta\ \sqrt{k_{0f}M_f} \tag{12}$$

The rear suspension damping coefficient c_{cr} is:

$$c_{cr} = 2\zeta\ \sqrt{k_{0r}M_r} \tag{13}$$

where: M_f is the front suspension mass. M_r is the rear suspension mass. k_{0f} is the stiffness of front airbag. k_{0r} is the stiffness of rear airbag.

3.4. Single wishbone beam independent suspension model

If the car does not move, the car is in a static stress state, that is, the ground normal reaction force on a tire is $\frac{(G_s+G_u)}{2}$, $\boldsymbol{G_s}$ is the suspension mass gravity, $\boldsymbol{G_u}$ is Non-suspension mass gravity; then add an upward micro-element force $\Delta \boldsymbol{F'_z}$ on the tire, thereby causing the micro-component displacement $\Delta \boldsymbol{s_t}$ of the wheel in the vertical direction and the micro-element displacement $\Delta \boldsymbol{s_s}$ of the spring along its center. The spring force also increases $\Delta\mathbf{Q}$, $\Delta\mathbf{Q} = \boldsymbol{k_s}\Delta\boldsymbol{s_s}$, $\boldsymbol{k_s}$ is the spring stiffness.

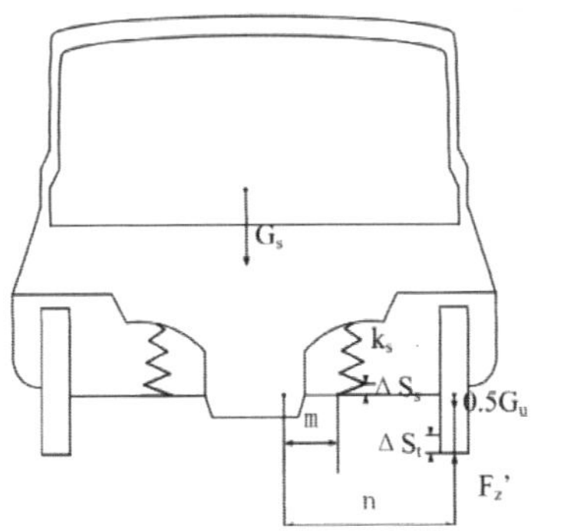

Figure 5. Single wishbone beam independent suspension model.

As can be seen in figure 5:

$$\frac{\Delta s_s}{m} = \frac{\Delta s_t}{n} \tag{14}$$

Where m is the distance from the center of the spring to the hinge point of the wishbone beam; n is the length of the wishbone beam.
In addition, according to the torque balance:

$$\Delta F'_z n = \Delta Q m = k_s \Delta s_s m \tag{15}$$

According to (14)(15):

$$\Delta F'_Z = k_s \frac{m}{n} \Delta s_s = k_s \left(\frac{m}{n}\right)^2 \Delta s_t \tag{16}$$

One side suspension stiffness is:

$$\frac{\Delta F'_Z}{\Delta s_t} = k_s \left(\frac{m}{n}\right)^2 \tag{17}$$

The whole suspension stiffness is:

$$k_c = 2k_s \left(\frac{m}{n}\right)^2 \tag{18}$$

According to (18), the front suspension stiffness is:

$$k_{cf} = 2k_{0f} \left(\frac{m_f}{n_f}\right)^2 \tag{19}$$

The rear suspension stiffness is:

$$k_{cr} = 2k_{0r} \left(\frac{m_r}{n_r}\right)^2 \tag{20}$$

Road roughness excitation (t), the stiffness of vehicle front suspension and rear suspension k_{cf}, k_{cr}, the damping coefficient of front and rear suspension c_{cf}, c_{cr} and vehicle parameters are available. These parameters are input together to the five-degree-freedom vehicle model and the seat acceleration signal $a_w(t)$ is obtained.

3.5. Road roughness model

As is shown in figure 6, the displacements of the seat, the front suspension system and the rear suspension system are $y - a\Phi$, $y - d\Phi$ and $y + b\Phi$. The vibration equation is established by using Newton's second law.

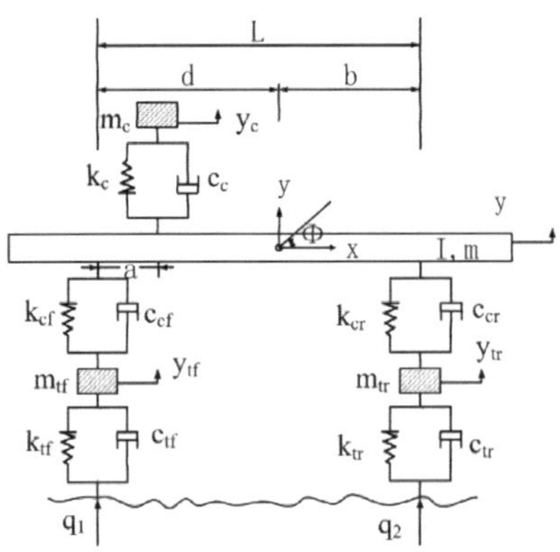

Figure 6. Five-degree-freedom vehicle model.

For seat:

$$m_c\ddot{y}_c + c_c[\dot{y}_c - (\dot{y} - a\dot{\Phi})] + k_c[y_c - (y - a\Phi)] = 0 \tag{21}$$

For front wheel:

$$m_{tf}\ddot{y_{tf}} + c_{cf}[\dot{y_{tf}} - (\dot{y} - d\dot{\Phi})] + c_{tf}(\dot{y_{tf}} - \dot{q_1}) + k_{cf}[y_{tf} - (y - d\Phi)] + k_{tf}(y_{tf} - q_1) = 0 \tag{22}$$

For rear wheel:

$$m_{tr}\ddot{y_{tr}} + c_{cr}[\dot{y_{tr}} - (\dot{y} + b\dot{\Phi})] + c_{tr}(\dot{y_{tr}} - \dot{q_2}) + k_{cr}[y_{tr} - (y + b\Phi)] + k_{tr}(y_{tr} - q_2) = 0 \tag{23}$$

For the vehicle body:

$$m\ddot{y} - c_c[\dot{y}_c - (\dot{y} - a\dot{\Phi})] - c_{cf}[\dot{y_{tf}} - (\dot{y} - d\dot{\Phi})] - c_{cr}[\dot{y_{tr}} - (\dot{y} + b\dot{\Phi})] - k_c[y_c - (y - a\Phi)] - k_{cf}[y_{tf} - (y - d\Phi)] - k_{cr}[y_{tr} - (y + b\Phi)] = 0 \tag{24}$$

For the torque balance of vehicle body:

$$I\ddot{\Phi} + ac_c[\dot{y}_c - (\dot{y} - a\dot{\Phi})] + dc_{cf}[\dot{y_{tf}} - (\dot{y} - d\dot{\Phi})] - bc_{cr}[\dot{y_{tr}} - (\dot{y} + b\dot{\Phi})] + ak_c[y_c - (y - a\Phi)] + dk_{cf}[y_{tf} - (y - d\Phi)] - bk_{cr}[y_{tr} - (y + b\Phi)] = 0 \tag{25}$$

Using the transfer matrix method to solve the vibration equation of a five-degree-freedom vehicle model, (21)~(25) can be written as:

$$\frac{dQ}{dt} = XQ + Y_1q_1 + Y_2q_2 + Y_3\dot{q_1} + Y_4\dot{q_2} \tag{26}$$

where:

$$Q = [y_c \quad \dot{y}_c \quad y_{tf} \quad \dot{y_{tf}} \quad y_{tr} \quad \dot{y_{tr}} \quad y \quad \dot{y} \quad \Phi \quad \dot{\Phi}]^T$$

$$Y_1 = \begin{bmatrix} 0 & 0 & 0 & \frac{k_{tf}}{m_{tf}} & 0 & 0 & 0 & 0 & 0 & 0 \end{bmatrix}^T$$

$$Y_2 = \begin{bmatrix} 0 & 0 & 0 & 0 & 0 & \frac{k_{tr}}{m_{tr}} & 0 & 0 & 0 & 0 \end{bmatrix}^T$$

$$Y_3 = \begin{bmatrix} 0 & 0 & 0 & \frac{c_{tf}}{m_{tr}} & 0 & 0 & 0 & 0 & 0 & 0 \end{bmatrix}^T$$

$$Y_4 = \begin{bmatrix} 0 & 0 & 0 & 0 & 0 & \frac{c_{tr}}{m_{tr}} & 0 & 0 & 0 & 0 \end{bmatrix}^T$$

$$X = \begin{bmatrix}
0 & x_{1.2} & 0 & 0 & 0 & 0 & 0 & 0 & 0 & 0 \\
x_{2.1} & x_{2.2} & 0 & 0 & 0 & 0 & x_{2.7} & x_{2.8} & x_{2.9} & x_{2.10} \\
0 & 0 & 0 & x_{3.4} & 0 & 0 & 0 & 0 & 0 & 0 \\
0 & 0 & x_{4.3} & x_{4.4} & 0 & 0 & x_{4.7} & x_{4.8} & x_{4.9} & x_{4.10} \\
0 & 0 & 0 & 0 & 0 & x_{5.6} & 0 & 0 & 0 & 0 \\
0 & 0 & 0 & 0 & x_{6.5} & x_{6.6} & x_{6.7} & x_{6.8} & x_{6.9} & x_{6.10} \\
0 & 0 & 0 & 0 & 0 & 0 & 0 & x_{7.8} & 0 & 0 \\
x_{8.1} & x_{8.2} & x_{8.3} & x_{8.4} & x_{8.5} & x_{8.6} & x_{8.7} & x_{8.8} & x_{8.9} & x_{8.10} \\
0 & 0 & 0 & 0 & 0 & 0 & 0 & 0 & 0 & x_{9.10} \\
x_{10.1} & x_{10.2} & x_{10.3} & x_{10.4} & x_{10.5} & x_{10.6} & x_{10.7} & x_{10.8} & x_{10.9} & x_{10.10}
\end{bmatrix}$$

The elements in X that are not 0 are as follows:

$$x_{1.2} = 1, x_{2.1} = -\frac{k_c}{m_c}, x_{2.2} = -\frac{c_c}{m_c}, x_{2.7} = \frac{k_c}{m_c}$$

$$x_{2.8} = \frac{c_c}{m_c}, x_{2.9} = -\frac{k_c a}{m_c}, x_{2.10} = -\frac{c_c a}{m_c}, x_{3.4} = 1$$

$$x_{4.3} = -\frac{k_{cf} + k_{tf}}{m_{tf}}, x_{4.4} = -\frac{c_{cf} + c_{tf}}{m_{tf}}, x_{4.7} = \frac{k_{cf}}{m_{tf}}$$

$$x_{4.8} = \frac{c_{cf}}{m_{tf}}, x_{4.9} = -\frac{k_{cf} d}{m_{tf}}, x_{4.10} = -\frac{c_{cf} d}{m_{tf}}, x_{5.6} = 1$$

$$x_{6.5} = -\frac{k_{cr} + k_{tr}}{m_{tr}}, x_{6.6} = -\frac{c_{cr} + c_{tr}}{m_{tr}}, x_{6.7} = \frac{k_{cr}}{m_{tr}}$$

$$x_{6.8} = \frac{c_{cr}}{m_{tr}}, x_{6.9} = \frac{k_{cr} b}{m_{tr}}, x_{6.10} = -\frac{c_{cr} b}{m_{tr}}, x_{7.8} = 1$$

$$x_{8.1} = \frac{k_c}{m}, x_{8.2} = \frac{c_c}{m}, x_{8.3} = \frac{k_{cf}}{m}, x_{8.4} = \frac{c_{cf}}{m}$$

$$x_{8.5} = \frac{k_{cr}}{m}, x_{8.6} = \frac{c_{cr}}{m}, x_{8.7} = -\frac{k_c + k_{cf} + k_{cr}}{m}$$

$$x_{8.8} = -\frac{c_c + c_{cf} + c_{cr}}{m}, x_{8.9} = \frac{k_c a + k_{cf} d + k_{cr} b}{m}$$

$$x_{8.10} = \frac{c_c a + c_{cf} d - c_{cr} b}{m}, x_{9.10} = 1, x_{10.1} = -\frac{k_c a}{I}$$

$$x_{10.2} = -\frac{c_c a}{I}, x_{10.3} = -\frac{k_{cf} d}{I}, x_{10.4} = -\frac{c_{cf} d}{I}$$

$$x_{10.5} = -\frac{k_{cr} b}{I}, x_{10.6} = -\frac{c_{cr} b}{I}$$

$$x_{10.7} = \frac{k_c a - k_{cf} d - k_{cr} b}{I}$$

$$x_{10.8} = \frac{c_c a - c_{cf} d - c_{cr} b}{I}$$

$$x_{10.9} = -\frac{k_c a^2 + k_{cf} d^2 + k_{cr} b^2}{I}$$

$$x_{10.10} = -\frac{c_c a^2 + c_{cf} d^2 + c_{cr} b^2}{I}$$

4. Simulation model

In order to determine the optimal initial air pressure of the suspension airbag under the B-level road surface, the paper first gives an arbitrary initial air pressure, and calculates the suspension stiffness and suspension damping coefficient through the initial air pressure in the MATLAB environment. Then, the vehicle parameters and the road roughness excitation are introduced to five-degree-freedom vehicle model to obtain the acceleration signal of seat whose RMS value a_w can be an evaluation index of the ride comfort of the vehicle. The simulation of this paper consists of two parts. The first part is the road surface roughness excitation measured by the road roughness Simulink simulation model. The second part is the five-degree-freedom Simulink model to measure the acceleration signal of the seat and find its RMS value. The vehicle parameters are derived from the bus of the literature [7]. The airbag parameters are the airbag parameters that are reasonably matched according to the vehicle parameters of the bus. The most widely used vibration comfort standards in the world are BS 6841-1987 and ISO 2631-1-1997 [9], both of which use the weighted acceleration of seat RMS as the basic evaluation index. [7] At present, when studying the ride comfort of a vehicle, the human comfort index is generally based on the weighted acceleration RMS a_w, and the calculation formula is as follows:

$$a_w = \left[\frac{\int_0^T a_w^2(t)dt}{T}\right]^{\frac{1}{2}} \tag{27}$$

4.1. Road roughness Simulink model

According to (2), this paper chooses to simulate road roughness excitation based on Simulink environment. There are two main outputs for road roughness, the front wheel road roughness excitation q(t) and the rear wheel road roughness excitation q(t + T). Rear wheel output lags front wheel output for T. Assuming that the front and rear wheel axles are on the same rut, the front and rear road roughness displacement outputs q_1 and q_2 are only one time lag in time Δt, where $\Delta t = L/v$, v is the vehicle speed. The hysteresis function can be implemented using the Transport Delay module.

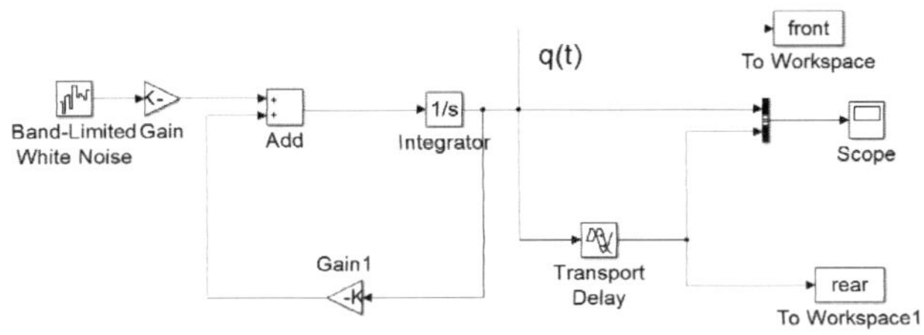

Figure 7. Road roughness Simulink model.

4.2. Five-degree-freedom vehicle Simulink model

According to (26), a five-degree-freedom vehicle model can be constructed in the Simulink environment. The q_1 and q_2 obtained from the road roughness Simulink simulation model and the X, Y_1, Y_2, Y_3, Y_4matrices composed of vehicle parameters are introduced into the five-degree-freedom vehicle Simulink simulation model to obtain the output matrix Q.

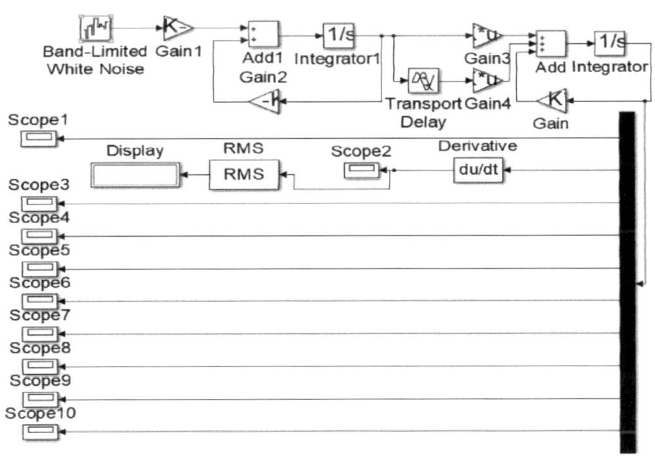

Figure 8. Road roughness Simulink model.

5. Simulation results

The parameters introduced in this paper are shown in the table 3.

Table 3. The table of vehicle parameters.

Front suspension airbag static balance volume V_{0f}/cm^3	13650
Effective area of front suspension airbag A_{0f}/m^2	0.286
Rear suspension airbag static balance volume V_{0r}/cm^3	13650
Effective area of rear suspension airbag A_{0r}/m^2	0.0905
Polytropic index n	1.33
Vehicle speed $v/km \bullet h^{-1}$	41
Front suspension mass M_f/kg	594.8
Rear suspension mass M_r/kg	290.2
Relative damping ratioζ	0.3
People and seat mass m_C/kg	53
Stiffness of the seat $k_c/N \bullet m^{-1}$	19030
Damping coefficient of the seat $c_c/N \bullet s \bullet m^{-1}$	170
Rear wheel mass m_{tr}/kg	210
Stiffness of rear wheel $k_{tr}/N \bullet m^{-1}$	571420
Rear wheel damping coefficient $c_{tf}/N \bullet s \bullet m^{-1}$	0
Front wheel mass m_{tf}/kg	180
Stiffness of front wheel $k_{tf}/N \bullet m^{-1}$	571420
Front wheel damping coefficient $c_{tf}/N \bullet s \bullet m^{-1}$	0
Seat center and centroid distance a/m	0.183
Front wheel center and centroid distance d/m	1.353
Distance from the center of the front suspension spring to the hinge point of the wishbone beam m_f/m	1.2
Distance from the center of the rear suspension spring to the hinge point of the wishbone beam m_r/m	1.59
Front suspension wishbone beam length n_f/m	0.5
Rear suspension wishbone beam length n_f/m	0.5
Vehicle body mass m/kg	1770
Moment of inertia of the frame around the center of mass $I/kg \bullet m^2$	1988
Rear wheel center and centroid distance b/m	0.947
Front and rear wheel center distance L/m	2.3

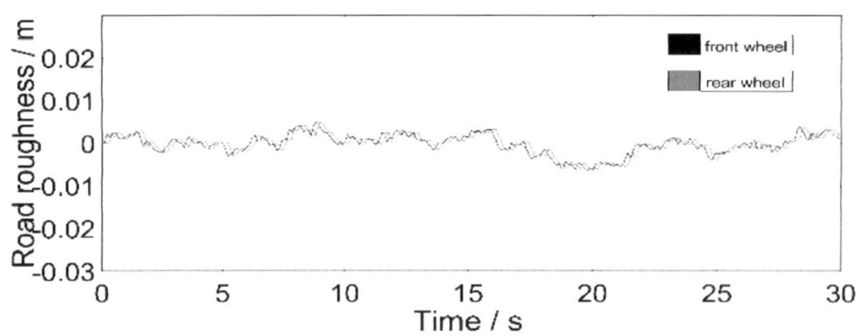

Figure 9. Front wheel and rear wheel road roughness excitation q_1, q_2.

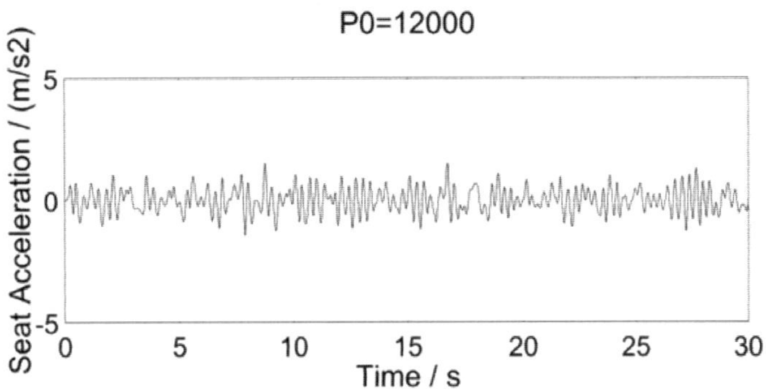

Figure 10. Seat acceleration signal when $P_0 = 12000$.

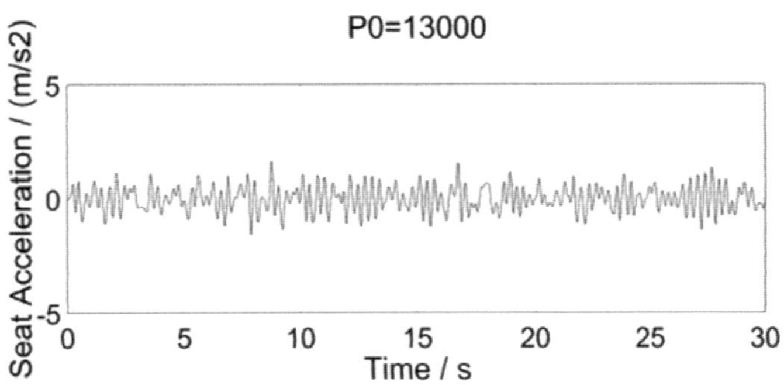

Figure 11. Seat acceleration signal when $P_0 = 13000$.

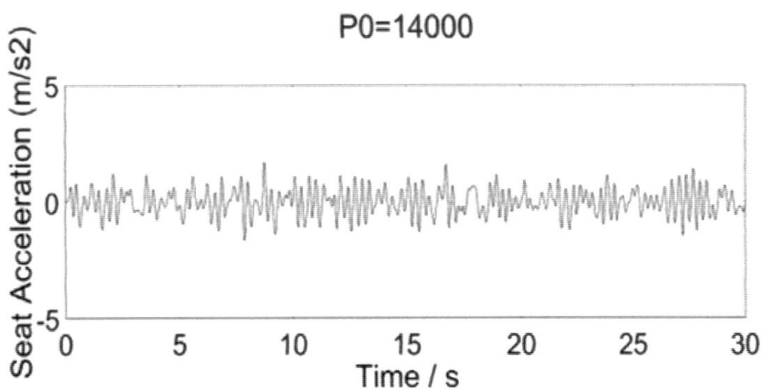

Figure 12. Seat acceleration signal when $P_0 = 14000$.

IOP Conf. Series: Materials Science and Engineering **538** (2019) 012068 doi:10.1088/1757-899X/538/1/012068

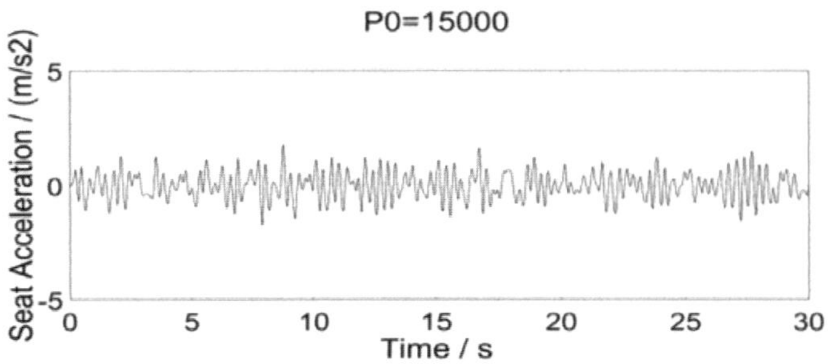

Figure 13. Seat acceleration signal when $P_0 = 15000$.

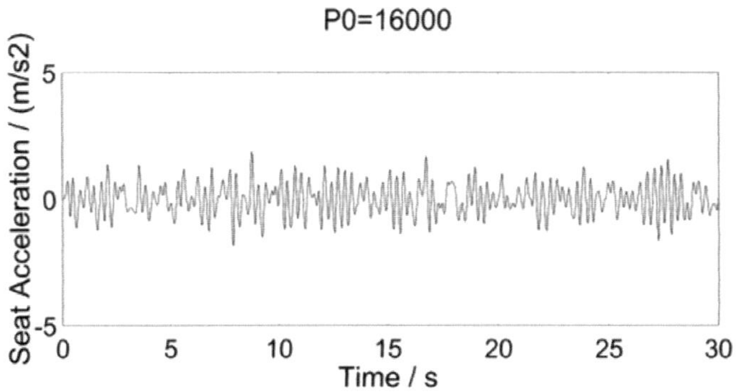

Figure 14. Seat acceleration signal when $P_0 = 16000$.

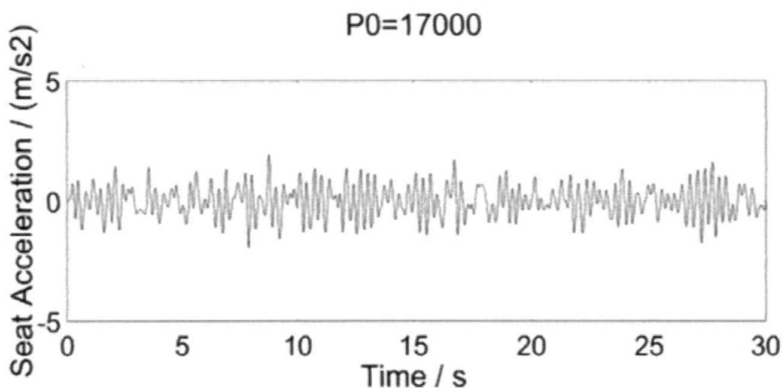

Figure 15. Seat acceleration signal when $P_0 = 17000$.

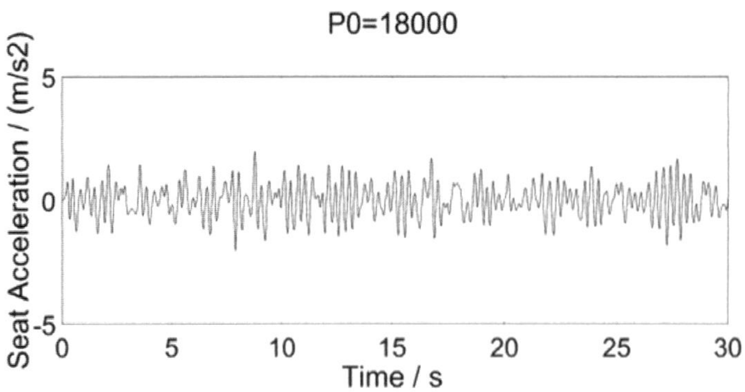

Figure 16. Seat acceleration signal when $P_0 = 18000$.

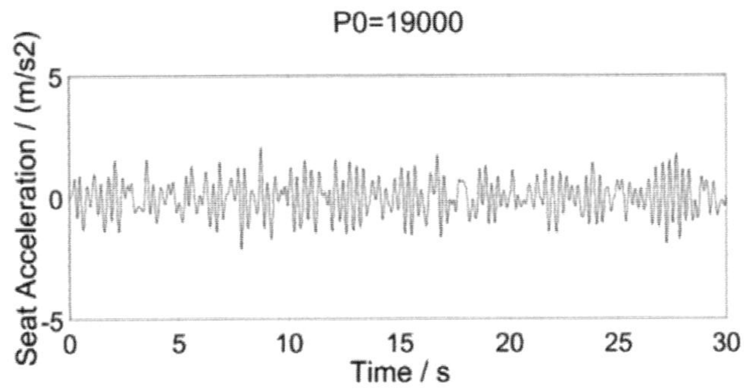

Figure 17. Seat acceleration signal when $P_0 = 19000$.

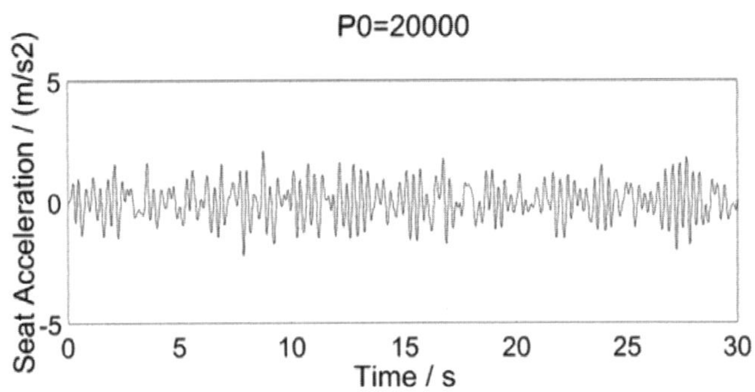

Figure 18. Seat acceleration signal when $P_0 = 20000$.

Figure 10 to Figure 18 show the relationship between the acceleration of the seat $a_w(t)$ with time t when P_0 takes different values. It can be seen that the difference between the maximum and minimum values of the acceleration of the person and the seat increases as the initial air pressure of the airbag increases.

Table 4. The table of a_w value corresponding to any P_0.

P_0 / Pa	$a_w = \left[\frac{\int_0^T a_w^2(t)dt}{T}\right]^{\frac{1}{2}}$ / (m/s^2)
12000	0.08882
13000	0.06459
14000	0.03773
15000	0.009946
16000	0.01768
17000	0.04326
18000	0.06132
19000	0.0707
20000	0.09029

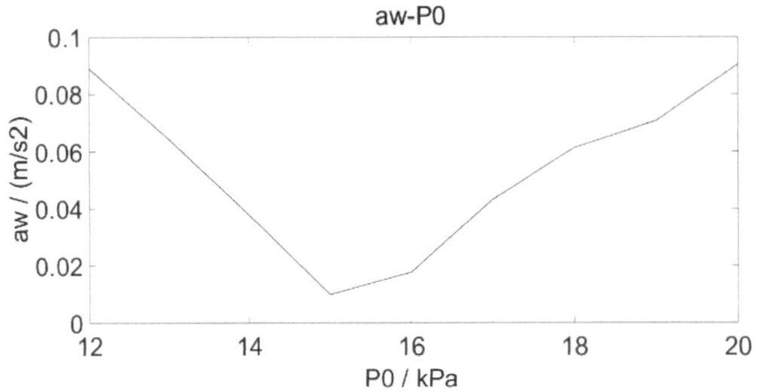

Figure 19. The line diagram of a_w value corresponding to any P_0.

Table 4 reflects the relationship between the seat acceleration RMS value a_w and the initial air pressure P_0 of airbag. Figure 19 is a line diagram drawn based on the data of Table 4. It can be seen from the line diagram that as the initial air pressure P_0 of the airbag increases, the acceleration RMS value a_w of the seat decreases first and then increases while the ride comfort of the vehicle increases first and then decreases. This reflects the existence of an accurate initial air pressure of airbag with optimum vehicle ride comfort and the initial air pressure ranges between 14kPa and 16kPa.

6. Conclusion

Taking the B-level road as an example, in order to optimize the ride comfort of the vehicle, this paper proposes a Method for determining accurate initial air pressure range of air suspension airbag based on vehicle ride comfort simulation analysis. Through simulation and calculation, the main conclusions are as follows:

The five-degree-freedom vehicle model established in this paper can fully reflect the vibration characteristics of the vehicle. Using random road excitation, air suspension parameters and other relevant vehicle parameters as inputs, the five-degree-freedom vehicle Simulink model can simulate the road roughness excitation (Figure 9) and the vibration displacement signal, speed signal and acceleration signal of each part of the vehicle. As is shown in Figure 8, the Simulink model can still output displacement, velocity and acceleration signals of other parts such as body, wheel and centroid. This paper only selects the person and the seat acceleration signal to analyse and research. Displacement signal, velocity signal and acceleration signal of other parts can also be used as research objects in other fields.

There is an optimal initial air pressure of the airbag to maximize the vehicle ride comfort. It can be seen from Figure 19 that under the B-level road surface, as the initial air pressure of the airbag increases, the ride comfort of the vehicle increases first and then decreases. Obviously, there is an accurate initial air pressure maximizing the vehicle ride comfort, which is consistent with the expected result of this paper.

The method proposed in this paper can obtain a more accurate initial air pressure range of the airbag. The optimal initial air pressure range of the vehicle airbag is between 14~16kPa (Figure 19). The rough range given in this paper is 12~20kPa. The length of the range is reduced from 8kPa to 2kPa, and the length of the accurate range is reduced to 1/4.

Here are a few innovative points:

(1) Using the airbag model, the single mass system model and the single wishbone beam independent suspension model can determine the specific values of the suspension stiffness and suspension damping coefficient corresponding to any initial air pressure of airbag.

(2) The five-degree-freedom vehicle model established in this paper can comprehensively reflect the vibration characteristics of every part of the vehicle under the condition of rough road surface. This model is the basis for studying the ride comfort of vehicles and analysing the vibration state of vehicles.

(3) When analysing the ride comfort of a rough road surface, it is not appropriate to use the vehicle vibration acceleration to evaluate the ride comfort of the human body. Therefore, the vibration condition of the seat is particularly important and can be directly used to measure the ride comfort of the vehicle.
(4) The initial air pressure of the air suspension airbag has a great influence on the ride comfort of the vehicle. Under the B-level road surface, as the initial air pressure of the airbag increases, the ride comfort of the vehicle increases first and then decreases, so there is an initial air pressure. Therefore there is an accurate initial air pressure makes the vehicle ride smoother. In this paper, the method of finding the accurate range of the initial air pressure of the airbag can be used as the initial air pressure reference value for air suspension production and can also be used as an important reference for airbag maintenance.

7. Prospective

The five-degree-freedom vehicle model constructed in this paper can reflect the vibration characteristics of every part of the vehicle more comprehensively, so it can be used to study the influence of other vehicle parameter changes on vehicle ride comfort.
This paper provides a method to determine the optimal initial air pressure range of air suspension airbags. It can be used as a reference for the study of shock absorption. It can also be used as a method to determine the precise initial air pressure range of airbags during airbag maintenance. It can also be used as the basis for producing variable stiffness air suspension.
The rough range given in this paper is 12~20kPa, and the accurate range found in this paper is 14~16kPa. In theory, if the selected pressure gradient is smaller, a more accurate initial pressure range or even an accurate initial pressure value can be obtained. However, if the selected pressure gradient is too small, the workload simulated and calculated using the method proposed in this paper will be too large. So the models and method of this paper need to be optimized.

References

[1] S Xiong, W Bao, *Calculation and Selection of Air Suspension System Parameters*. Mechanical Manufacturing, Page 8-10, **07**(2008).
[2] X Li, *Research and Simulation Analysis of Dynamic Characteristics of Single Air Chamber Air Spring*, Northeastern University, Page 89(2013).
[3] J Xie, Q Zhang, *Simulation Analysis of Vehicle Ride Comfort Based on Simulink*, Electromechanical Technology, Page 14-18, **01**(2013).
[4] H Liu, W Zhou, J Chen, Vehicle Ride Simulation and Optimization Based on Five-degree-freedom Vehicle Model, 33rd China Control Conference, Page 5(2014)
[5] H Zheng, Standard Road Surface Reconstruction Based on Harmonic Superposition Method, Mechanical Research and Application, **06**(2014).
[6] B Zhang, Vehicle-Road System Coupling Vibration Analysis and Ride Comfort Evaluation, Central South University, Page 156(2013).
[7] H Zhang, W Yang, Evaluation Method of Road Flatness Based on Human-vehicle-road Five-degree-freedom Vibration Model, Journal of Transportation Engineering, Page 16-22, **04**(2010).
[8] C Liu, *Study on The Transfer Function of Eight-degree-freedom Vehicle Model*, Journal of Gansu Agricultural University, Page 113-117, **03**(2006).
[9] Z Yu, *Automotive Theory*, Mechanical Industry Press(2000).

IOP Conf. Series: Materials Science and Engineering **538** (2019) 012069 doi:10.1088/1757-899X/538/1/012069

A Road-Holding Index Based on Ride Dynamics for High-Downforce Racing Cars

Felipe Marchesin[1], Roberto Spinola Barbosa[1], Marco Gadola[2,a] and Daniel Chindamo[2]

[1]Department of Mechanical Engineering, University of Sao Paulo, Sao Paulo, Brazil.
[2]Department of Industrial and Mechanical Engineering, Automotive Group, University of Brescia, I-25123 Brescia, Italy.

[a] Corresponding author: marco.gadola@unibs.it

Abstract. This work builds on some of the current industry techniques used in racing to evaluate vertical dynamics performance and propose a new methodology to evaluate vehicle performance. The proposed method creates a quantitative numerical index from the classic tyre vertical load variation frequency response function with some novelties that cover all peculiarities of high downforce race cars. In this method the aerodynamic forces are included as non-linear functions vs ride height and it is shown that they affect system stiffness and damping. As a result, the system response changes as a function of vehicle longitudinal speed. The importance of non-linear suspension rates in this kind of vehicle is also highlighted. The proposed index can also be customized for a certain vehicle speed range. A Formula 3 racecar model has been used as an application example.

1. Introduction

Since the early era of motor racing, race car dynamics has been studied by adopting a similar approach used for road cars. In the past race cars were usually derived from road cars, but since the early sixties they followed different design and development paths quite often. Road cars are designed for everyday use, customer comfort, fuel economy, reduced environmental impact etc. Race cars are designed to be as fast as possible under a set of given regulations. As a matter of fact, aerodynamics became a key performance factor in the late sixties: downforce can increase cornering speed and reduce lap time, while road car aerodynamics, in most cases, is aimed at decreasing drag to improve fuel efficiency [1]. The aerodynamic devices used in race cars, such as wings and diffusers, are not applied on road cars (except for special, high-performance road cars) as the applications have different aims, as highlighted above. Therefore, the impact of aerodynamics in terms of maximum lateral acceleration developed by a road car is usually small (values are around 0.9g) while a race car can even reach 5g [2].

With regard to vehicle mass, the minimum weight is usually given by the regulatory structure in motorsport, while weight reduction is seen as a key to sustainable mobility as far as road vehicles are concerned [3-5].

From the point of view of vertical dynamics, road car designers must achieve a compromise between comfort, road holding and stability, while race car engineers mainly focus on road holding alone. Apart from the different goals, some other characteristics are very different in magnitude. According to the authors experience and public data [6], the typical race car weight can be around 50% to 70% of a typical road car, the centre of gravity height is around 50% lower, the ground clearance is around 25mm, and the tyres feature completely different functional characteristics. Race car tyre vertical stiffness is about the same of the suspension system (or even softer), and the presence of wings and diffusers generates high values of aerodynamic downforce. Of all differences, the most important one is aerodynamics: race car dynamics is heavily speed dependent as this work will show and discuss.

In terms of vertical dynamics, road holding can be translated into tyre contact patch load variation. The longitudinal and lateral forces available from the tyres are proportional to vertical load. Any suspension oscillation will result into fluctuation of the contact patch load that will affect vehicle performance [7, 8].

As far as vertical dynamics is concerned, several works can be found in literature on both road and racing vehicles or motorcycles. Many of them are focused on the vehicle response as function of the road input [9, 10] or on the calibration of active suspension systems [11, 12]. Benini et al. [13] investigated the influence of the suspension components friction on vertical dynamics of a race car. Studies based on the search for natural frequencies and analysis for noise, vibration and harshness (NVH) purposes can also be found [14]. Vertical dynamics can also affect cornering response in general [15, 16], torque side distribution and yaw moment if the vehicle is fitted with a limited-slip differential or a torque vectoring system [17].

A research was carried out at Clemson University coupling race car vertical dynamics and aerodynamics using a quasi-steady state approach for aerodynamics [18, 19]. The researchers' work on race car damper settings was based on frequency response and damping for different vehicle speed values. They showed that indoor testing can be replicated by simulation, thus accelerating the optimization process. They also showed that aerodynamic forces can cause instability as they affect overall stiffness and damping.

The authors developed a novel method to evaluate aerodynamic performance in high downforce race cars [20]. The impact of aerodynamic forces as non-linear functions of ride heights on lap time performance is dealt with in [21]. A work analysing aerodynamic effects on ride comfort and road holding was also published, the aerodynamic forces were non-linear functions of the ride height, but the suspension stiffness was linear [22].

1.1. Seven-post rig testing: Time domain and frequency domain approach

The four-post rig is the main indoor testing facility used to tune vertical dynamics. Basically, the four wheels are resting each on a hydraulic actuator. The system replicates the road profile applying force or position inputs to each tyre contact patch. Lateral and longitudinal accelerations are neglected in this case. However, three extra actuators can also be attached to the sprung mass to replicate aerodynamic loads and lateral load transfer. In this case the system is called seven-post rig [2] and it is mainly used for high-downforce racecars. Some works were published with time and frequency domain procedures applied to four/seven-post rigs [23, 24]. The influence of aerodynamic stability on vertical dynamics was also studied by plotting the system roots of a linearized vehicle model, trying to replicate seven-post rig testing [25].

The four- and seven-post rigs allow for time domain simulations to be performed on the vehicle using real-world road profiles. In this case, to create a suitable signal, the so-called Iterative Control System (ICS) can be used: it is an energy-based iterative process that computes an equivalent road input time history starting from suspension displacement and/or unsprung mass acceleration data recorded on a real vehicle running on a relevant road surface [9, 26].

The replication process for a high-performance racing car normally requires at least eight ICS iterations to achieve proper fitting of a full race course lap [26]. A similar method is also used by MTS [27], although [28] shows that even if the four unsprung mass vertical accelerations from the uprights can be replicated, it doesn't mean that the road profile is accurately replicated.

Several racing series are limited in terms of the maximum number of on-board data acquisition channels during official events to control costs, hence it is difficult to have access to a car with all channels required to replicate the track profile. In such a case a frequency domain-based approach can be adopted, trying to optimize vehicle performance by means of stochastic theory. An artificial profile signal like a step or a constant velocity sine sweep can be used for basic response analysis. The vertical displacement inputs are applied in phase on the four wheels. According to some works, performance improvements can be achieved when pitch response is minimized for a certain bounce input, regardless of the type of racing car: from high-downforce formula cars to Nascar-type cars [24, 28].

In these tests, steady-state aerodynamic forces can be simulated by means of low-pressure air spring actuators (featuring near-zero stiffness and damping) applying equivalent downforce to the sprung mass. The correct car attitude is achieved by keeping suspension displacement at the correct position. Therefore, suspension kinematic properties like a variable motion ratio are considered automatically. Usually additional components giving strongly non-linear suspension rates (bump stops for instance) are removed for this kind of test [28]. It makes sense as the frequency domain approach is a linear system tool. However, the test result is often an "optimum" setup that cannot be used for impracticality, as a highly non-linear suspension rate is necessary to prevent vehicle bottoming at high speed and/or to improve aerodynamic efficiency.

The interaction between aerodynamic forces and the suspension system can be studied in two different situations: quasi-static and transient. The quasi-static analysis is mainly used to set spring stiffness, static ride heights and bump stop setup; the target is full control of ground clearance, vehicle attitude and aero balance (i.e. percentage of total downforce on the front axle) keeping them within the desired range along the whole circuit. The transient analysis is mainly used for damper setting [27].

Some authors [24, 25, 28, 29] prefer frequency domain to time domain tests. According to them the time domain approach is situation based and the track profile generated is only compatible with the related system settings. If it is modified, the system response also changes: that input profile is not realistic anymore, and it cannot be used for system optimization.

In order to make indoor testing efficient, a performance index can be adopted as a reference for each parameter to be evaluated. Typical performance indexes are listed below for a frequency domain analysis based on a sine sweep input [30]:

- Bounce and pitch coupling factor to be minimized, targeting a pitch gain reduction to improve both mechanical and aerodynamic balance;
- Front and rear ride height stability, translating into aerodynamic stability;
- Wheel load variation for both front and rear axles, to be minimized for maximum grip;
- Ratio between tyre and suspension deflection as an indicator for tyre temperature gain due to hysteresis;
- Heat energy absorbed by tyres, again as an indicator for tyre temperature.

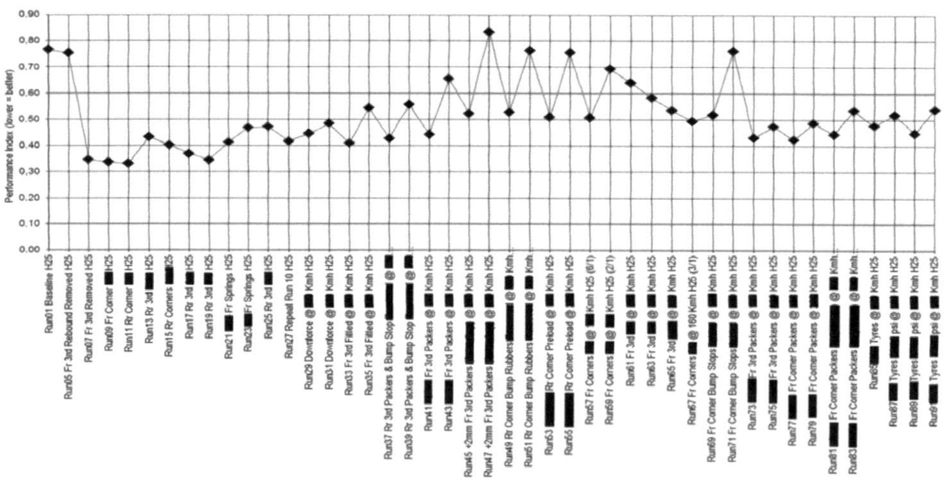

Figure 1. Performance index evolution in a series of frequency domain tests (courtesy of Fortec Motorsports).

The effect of a kerb impact on handling and vehicle stability can also be simulated by means of a step input on one of the front wheels followed by the same input on the same side rear wheel, after a time delay inversely proportional to the vehicle speed considered for the test. An overall performance index is generated by combining the 5 indexes above plus the kerb strike results, each one with a specific weight. Figure 1 shows a series of runs conducted in sequence and the overall performance index for each one of them. Each run (showed in the x axis) is performed with a different car configuration. The lower the index the higher the performance on track. Some comments were removed for confidentiality.

Both proposals require non-linearities to be removed from suspension rates, either on the seven-post rig and eventually in a software simulation environment, thus resulting in a theoretically optimum setup which is scarcely applicable to a real-world, race track setup, as already stated.

During a full championship season, as a matter of fact a race car is used on many different circuits, each one requiring a specific car configuration to improve lap time: both suspension and aerodynamics settings change form circuit to circuit and even from session to session. Table 1 below shows some circuit data and highlights their peculiarities.

Table 1. Circuit data.

Circuit	Aerodynmic configuration	Track roughness level	# turns	Speed (Km/h)			
				min	med	max	avg min at apex
circuit 1	High downforce	high	6	56	154	216	108
circuit 2	High downforce	low	11	80	158	224	122
circuit 3	High downforce	low	9	66	166	238	93
circuit 4	High downforce	low	11	80	165	225	123
circuit 5	Medium downforce	medium	12	62	170	245	115
circuit 6	Low downforce	medium	13	63	184	241	136

1.2. Proposal of a novel performance index

While the current industry approach tries to find a generic optimum for the suspension configuration without

any use of circuit-specific information, the index proposed in this work is aimed at representing race car road holding performance in terms of vertical dynamics in the frequency domain, including some of the non-linear effects that are usually neglected.

Downforce is proportional to vehicle speed squared and generates suspension deflection. Two kinds of non-linearity should be considered: the suspension stiffness is a function of suspension deflection because of the above-mentioned non-linear suspension rates and on top of this, downforce is a function of front and rear ride heights according to the so-called aeromaps, as explained in Section 2.2. This means in turn that suspension stiffness is related to vehicle speed through suspension deflection and aerodynamic properties.

The aerodynamic forces will be modelled using the induced angle of attack approach [18, 25, 34], where the angle of attack is a function of the vehicle body attitude and also of the bounce and pitch velocity. When the vehicle is bouncing, it creates an induced angle of attack proportional to bounce velocity. With pitch oscillation, the same happens. Using the induced angle of attack in the calculation of aerodynamic forces and moments, they become proportional to bounce and pitch velocities resulting in aerodynamic damping in bounce and pitch oscillations.

To calculate this novel index, the race car model is linearized for several values of longitudinal speed within a specific range, and the frequency response is plotted for each one of these speed values. The frequency response will change because the system stiffness and damping will change with vehicle longitudinal speed. This paper aims at showing that by means of these curves it is possible to evaluate a performance index representing the system road holding performance, thus improving the current approach results to achieve a circuit dependent optimum setup.

IOP Conf. Series: Materials Science and Engineering **538** (2019) 012069 doi:10.1088/1757-899X/538/1/012069

In this work the word "race car" refers to single seaters (from F4 up to F1). The results are based on a Formula 3 car. All vehicle data sets came from measurements and are courtesy of Fortec motorsports. They were validated for application purposes and also used in a previous work [20].

2. Methods

2.1. Vehicle model

The half car model is a classic model used in vehicle dynamics to evaluate vehicle comfort (through the estimate of vehicle body accelerations) and road holding (through evaluation of tyre contact patch load variations). The model features four degrees of freedom: sprung mass pitch and bounce, and front and rear unsprung mass vertical displacements, see Figure 2.

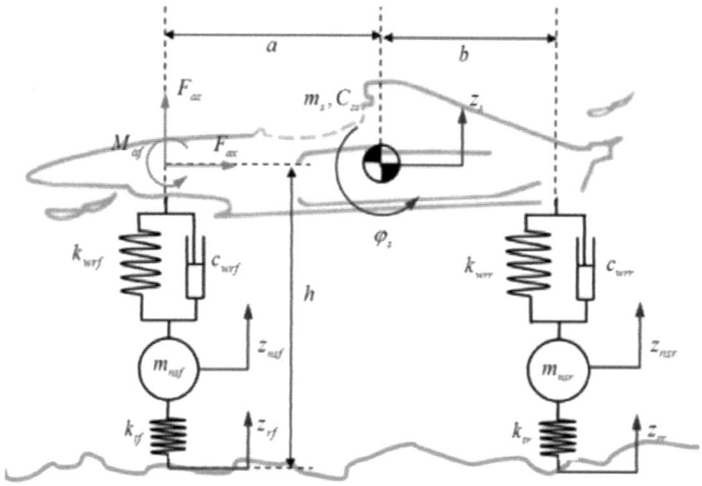

Figure 2. Half car model.

The aerodynamic actions are divided into downforce and pitch moment on the vehicle body. The reference for the aerodynamic forces is lying on the ground and aligned with the front axle. Equations (1), (2), (3) and (4) are the equations of motion. Equations (5) and (6) compute the tyre contact load while equation (7) compute the time delay between front and rear axle inputs.

$$I_y.\ddot{\alpha}_{ns} = F_a + a.F_a + F_{spring\,f}.a + F_{spring\,r}.b + c_{wf}(-\dot{z}_s + \dot{\varphi}_s.a + \dot{z}_{nsf}).a + \\ + c_{wr}(\dot{z}_{nsr} - \dot{z}_s - \dot{\varphi}_s.b).b \quad (1)$$

$$m_s.\ddot{z}_s = -F_a + F_{spring\,f} + F_{spring\,r} + c_{wf}(\dot{z}_s - \dot{\varphi}_s.a - \dot{z}_{nsf}) + c_{wr}(\dot{z}_{nsr} - \dot{z}_s - \dot{\varphi}_s.b) \quad (2)$$

$$m_{nsf}.\ddot{z}_{nsf} = F_{spring\,f} + c_{wf}(\dot{z}_s - \dot{\varphi}_s.a - \dot{z}_{nsf}) + k_{tf}(z_{rf} - z_{nsf}) \quad (3)$$

$$m_{nsr}.\ddot{z}_{nsr} = F_{spring\,r} + c_{wr}(\dot{z}_{nsr} - \dot{z}_s - \dot{\varphi}_s.b) + k_{tr}(z_{rr} - z_{nsr}) \quad (4)$$

$$F_{cf} = k_{tf}(z_{rf} - z_{nsf}) \quad (5)$$

$$F_{cr} = k_{tr}(z_{rr} - z_{nsr}) \quad (6)$$

$$z_{rr}(t) = z_{rf}\left(t - \frac{L}{V}\right) \quad (7)$$

Aerodynamic forces as well as wheel rates (i.e. suspension stiffness curves) are non-linear in the model. Downforce is non-linear mainly because of the interaction with the ground, the so-called ground effect. The wheel rates are non-linear in order to resist the high aerodynamic loads at high vehicle speed. Formulations for both actions are presented below.

2.2. Aerodynamic forces

In order to include the ground effect in the aerodynamic forces and moments, the vehicle ride height is included in the formulation. The non-linear relationships between forces and front and rear ride heights are expressed by the so-called aeromap surfaces. Figure 3 presents typical aeromap shapes for overall downforce (left) and aerobalance (right) derived from wind tunnel tests (values are related to a reference position due to a specific non-disclosure agreement). Both are functions of either front and rear ride height. The non-linear aerodynamic forces and coefficients are presented in equations (8) and (9) and are based on formulations used in other works [32] and by the industry [33].

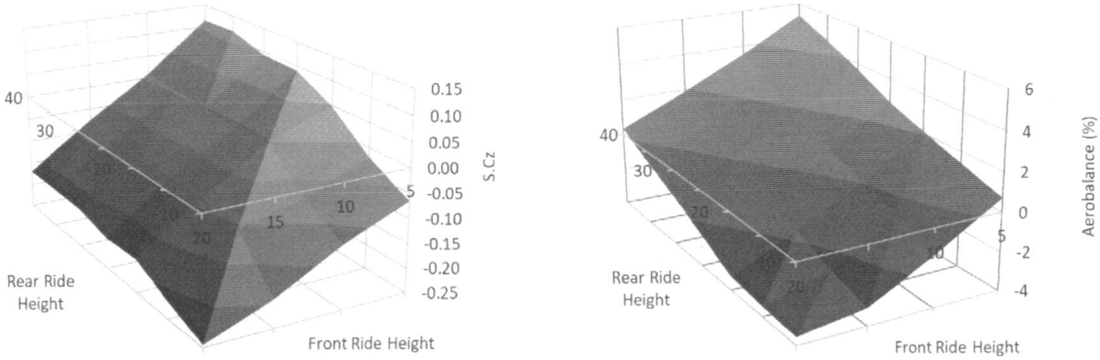

Figure 3. Typical aeromaps shape.

$$F_a = \frac{1}{2}\rho S C_z V^2 = pSC_z = pS\left(C_{z0} + \frac{\partial C_z}{\partial h_f}h_f + \frac{\partial C_z}{\partial h_r}h_r + \frac{\partial^2 C_z}{\partial h_f^2}h_f^2 + \frac{\partial^2 C_z}{\partial h_r^2}h_r^2 + \frac{\partial^2 C_z}{\partial h_f \partial h_r}h_f h_r\right) \quad (8)$$

$$M_a = \frac{1}{2}\rho S C_m L V^2 = pL\left(C_{m0} + \frac{\partial C_m}{\partial h_f}h_f + \frac{\partial C_m}{\partial h_r}h_r + \frac{\partial^2 C_m}{\partial h_f^2}h_f^2 + \frac{\partial^2 C_m}{\partial h_r^2}h_r^2 + \frac{\partial^2 C_m}{\partial h_f \partial h_r}h_f h_r\right) \quad (9)$$

The front and rear ride heights in equations (8) and (9) are replaced by equations (10) and (11), as both are a function of centre of mass displacement and pitch angle.

$$h_f = z_s + \varphi_s a \quad (10)$$

$$h_r = z_s - \varphi_s b \quad (11)$$

Replacing equations (10) and (11) into equation (8), the downforce can be calculated as presented in equation (12) and the pitch moment coefficient can be calculated in the same way.

$$C_z = C_{z0} + \frac{\partial C_z}{\partial z}z + \frac{\partial C_z}{\partial \varphi}\varphi + \frac{\partial^2 C_z}{\partial z^2}z^2 + \frac{\partial^2 C_z}{\partial \varphi^2}\varphi^2 + \frac{\partial^2 C_z}{\partial z \partial \varphi}z\varphi \quad (12)$$

In the transient calculations the model is linearized for a specific value of vehicle longitudinal speed. Equations (13) and (14) present the linearized aerodynamic coefficients.

$$C_{z(v)} = C_{z\,base} + \frac{\partial C_z}{\partial \varphi}_{(v)}\varphi + \frac{\partial C_z}{\partial z}_{(v)}z \quad (13)$$

$$\frac{\partial C_z}{\partial \varphi}_{(v)} = \frac{\partial C_z}{\partial \varphi} + 2\frac{\partial^2 C_z}{\partial \varphi^2}\varphi + \frac{\partial^2 C_z}{\partial z \partial \varphi}z \quad (14)$$

The aerodynamic model formulation follows the approach adopted by other authors that work with unsteady aerodynamics in ground effect [34, 35]. They showed that three main effects are important for unsteady aerodynamics: the ground clearance, the induced angle of attack and the added mass. These effects are related to bounce displacement, velocity and acceleration respectively. Each one of these effects works in a specific range of frequency [36]. In this work the effect of added mass will be neglected as it affects high frequencies only and in the proposed work the model will be limited to

inputs up to 60 Hz. This limit assures that the sprung and unsprung mass natural frequencies will be covered.
The pitch angle used in the computation of aerodynamic forces and moments will be a function of pitch displacement and velocity, and bounce velocity, as presented in equation (15). In the aerospace industry this approach is known as induced angle of attack.

$$\varphi = \varphi_S + \dot{\varphi}_S \frac{L}{V} + \frac{\dot{z}_S}{V} \tag{15}$$

Equation (16) presents the aerodynamic downforce coefficients linearized for a fixed vehicle longitudinal speed and with the effect of the induced angle of attack as shown in equation (15).

$$C_Z = C_{Z\,0} + \left(\frac{\partial C_Z}{\partial z}_{(v)} a + \frac{\partial C_Z}{\partial \varphi}_{(v)}\right)\varphi_S + \left(\frac{\partial C_Z}{\partial z}_{(v)}\right) z_S + \left(\frac{\partial C_Z}{\partial \varphi}_{(v)} \frac{1}{V}\right)\dot{z}_S + \left(\frac{\partial C_Z}{\partial \varphi}_{(v)} \frac{L}{V}\right)\dot{\varphi}_S \tag{16}$$

2.3. Spring and bump stop forces
In order to include the non-linear effects of the bump stop, a non-linear stiffness curve was used. This force can be modelled using a lookup table. In the linearized model, the stiffness was calculated in the operation point. The bump stop hysteresis has been neglected as it is smaller compared to damper forces.
The bump stop is not always being compressed: the static gap between the bump stop itself and the damper body is used to set its initial position, as showed in Figure 4. Equation (17) presents how the spring and bumpstop forces were implemented.

$$\mathrm{F} = \mathrm{K.x} + f(\mathrm{x} - \mathrm{gap}) \tag{17}$$

3. Results and discussion

3.1. Quasi-static
The quasi-static calculations were conducted by applying a vehicle longitudinal speed sweep from 0 to 230 km/h, therefore neglecting any longitudinal acceleration. This is a common procedure to understand how ride heights evolve as a function of vehicle speed and to set static ride heights and bump stop gaps for a specific circuit. This procedure is also used to validate the suspension deflection with aerodynamic load. Figure 4 shows a comparison between the model and track data for circuit 2.

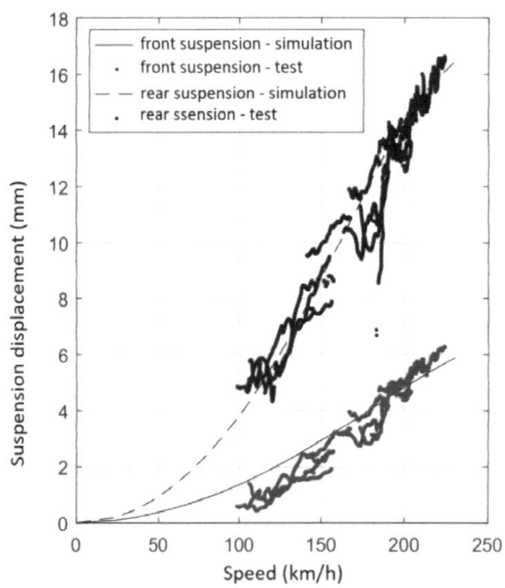

Figure 4. Suspension deflection due to aerodynamic forces.

Race cars must have their suspension set to maximize vehicle performance on track. This implies that the car ground clearance should be as low as possible. As vehicle longitudinal speed increases suspensions and tyres are compressed by increasing downforce, and ground clearance is therefore reduced. At first glance, the key point is to set the race car not to touch the ground and because of that the suspension stiffness should be high enough. Therefore, bump stops are normally used to achieve a rising rate. The key variables in quasi-static vertical dynamics are called dynamic ride heights: they are the front and rear ride heights as a function of vehicle longitudinal speed, as shown in Figure 5.

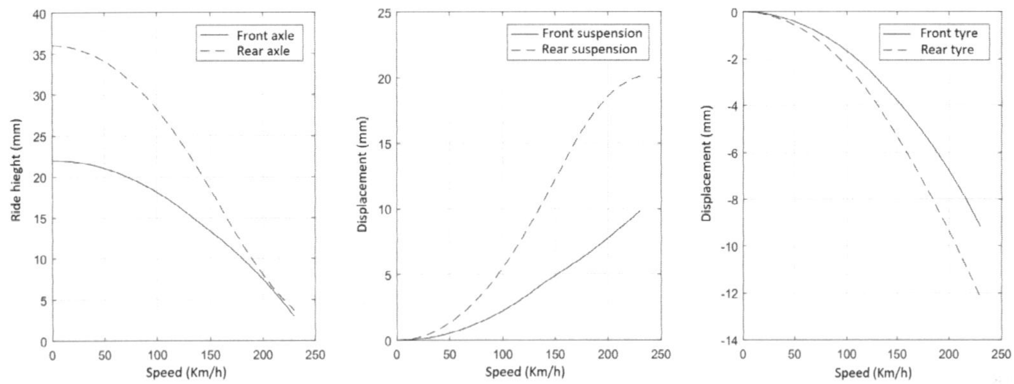

Figure 5. Dynamic ride heights, suspension displacement and tyre deflection against vehicle longitudinal speed.

On the left of Figure 5 the dynamic ride height vs vehicle speed curves show that the rear suspension features a visible change in stiffness while the front is smoother. This happens because the rear bump stop stiffness is much higher than the spring stiffness. The front bump stop gap is 3 mm while the rear is 6 mm. The front bump stop is engaged at around 130 km/h and the rear one around 160 km/h.
Formula cars generally use fairly wide tyres with a high aspect ratio due to the mandatory use of 13” rims. Because of that, their radial stiffness is relatively low with respect to a road car tyre. As a matter of fact, the suspension/tyre stiffness ratio can be as low as 1:1 (a typical road car stiffness ratio is about 1:10). When the bump stops come into play at high vehicle longitudinal speed the suspension can get nearly solid, creating a system with awkward dynamics when compared to a road car. Whenever tyre stiffness is smaller than suspension stiffness it dominates the dynamic response, leaving the system almost undamped. In Figure 5 at maximum vehicle speed, the overall ride height drop at the front is about 19 mm; tyre radial deflection accounts for about 50%. The rear ride height drop is about 32 mm, of which 37.5% is due to tyre deflection. This is common practice for a typical single seater: the front suspension is set to be much stiffer than the rear.

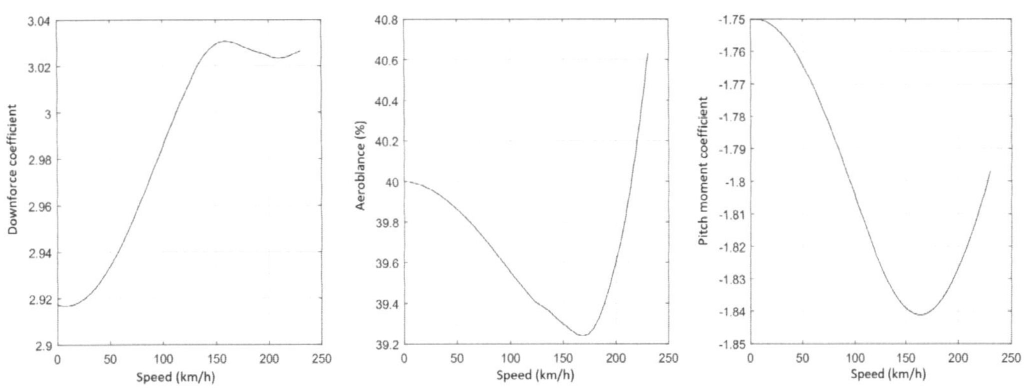

Figure 6. Aero parameters as a function of vehicle speed.

As stated above the dynamic ride heights have a direct impact on the so-called ground effect and heavily affect aerodynamic performance. As the aerodynamic coefficients are a function of both front and rear ride heights, setting the correct dynamic ride heights for a given circuit is vital for performance. Figure 6 is based on the so-called vehicle aeromaps and shows the downforce, aerobalance and pitch moment coefficients as a function of vehicle longitudinal speed. The drop of downforce above 150 km/h is peculiar of a single seater aerodynamic behaviour and it is due to diffuser stall: it happens when the rear ride height becomes low and it is mainly due to the boundary layer. This drop of downforce also results in a drop of aerodynamic drag; hence it is desirable in order to increase vehicle top speed at the end of a long straight.

3.2. Transient

Race car vertical dynamics is a compromise between aerodynamic stability and road holding. The first one is about the longitudinal movement of the centre of pressure and the second one has been explained above. When vehicle speed increases, chassis ground clearance and suspension stiffness change, and vehicle frequency response changes as a consequence.

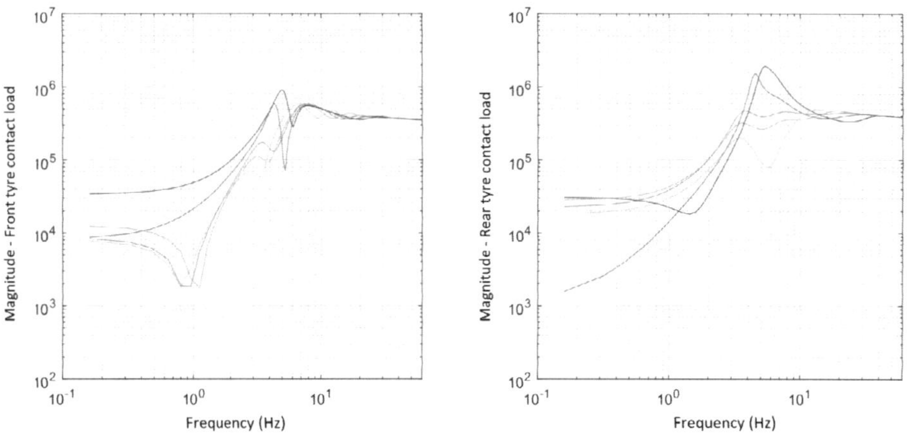

Figure 7. Tyre contact load frequency response (higher vehicle speeds in darker colours).

The equations of motion presented before were linearized for a range of vehicle speed values (from 0 up to 230 km/h) and transformed into a system of first-order differential equations presented in equation (18). Contact patch load variation for front and rear axles are calculated as outputs as shown in equation (19). As matrix A is not diagonal the pitch and bounce vibration modes are coupled [37]. The rear tyre input was modelled with a time delay as a function of vehicle speed and wheelbase length, see last item of equation (18) and equation (19). The frequency response from 0 up to 230 km/h for a Formula 3 car was obtained by deriving the equations above. Figure 7 shows the tyre contact load frequency response when the model is linearized at 5 different vehicle speed values: its shape is strongly related to vehicle speed because of the wheelbase filtering effect, the suspension rising rate, the relationship between aerodynamic actions and ride heights and the induced angle of attack.

$$\begin{bmatrix} \dot{z}_s \\ \dot{\alpha}_s \\ \dot{z}_{nsf} \\ \dot{z}_{nsr} \\ \dot{\omega}_s \\ \dot{v}_s \\ \dot{v}_{nsf} \\ \dot{v}_{nsr} \end{bmatrix} = \mathbf{A}. \begin{bmatrix} z_s \\ \alpha_s \\ z_{nsf} \\ z_{nsr} \\ \omega_s \\ v_s \\ v_{nsf} \\ v_{nsr} \end{bmatrix} + \begin{bmatrix} 0 \\ 0 \\ 0 \\ 0 \\ 0 \\ 0 \\ 1 \\ 0 \end{bmatrix} . u(t) + \begin{bmatrix} 0 \\ 0 \\ 0 \\ 0 \\ 0 \\ 0 \\ 0 \\ 1 \end{bmatrix} . u(t-\tau) \tag{18}$$

$$\begin{bmatrix} Fc_f \\ Fc_r \end{bmatrix} = \begin{bmatrix} 0 & 0 & -K_{tf} & 0 & 0 & 0 & 0 & 0 \\ 0 & 0 & 0 & -K_{tr} & 0 & 0 & 0 & 0 \end{bmatrix} \boldsymbol{x} + \begin{bmatrix} K_{tf} \\ 0 \end{bmatrix} \mathrm{u}(t) + \begin{bmatrix} 0 \\ K_{tr} \end{bmatrix} \mathrm{u}(t-\tau) \quad (19)$$

4. The Road Holding Performance Index (RHPI)

The proposed performance index is calculated using the contact load frequency response, a track profile density function and a vehicle speed range (from 0 up to 230 km/h in steps of 5 km/h in the present case). The contact load frequency response is multiplied by a track profile density function, and the results (inertance) are integrated over a frequency range of 0.1 Hz up to 60 Hz, generating the contact load RMS value for each vehicle speed value. The integral of the RMS values over a vehicle speed range (circuit dependent) generates the proposed index. The calculation process is explained in the chart below (Figure 8).

To evaluate the vehicle vertical time response attributable to excitation by pavement irregularities, the convolution of the vehicle function with the surface vertical displacement profile can be used [37]. However, in the frequency domain, the spectral response may be obtained directly only by multiplying the spectrum of the two systems. The resulting function is the inertance magnification of vehicle contact load attributable to spectral pavement roughness.

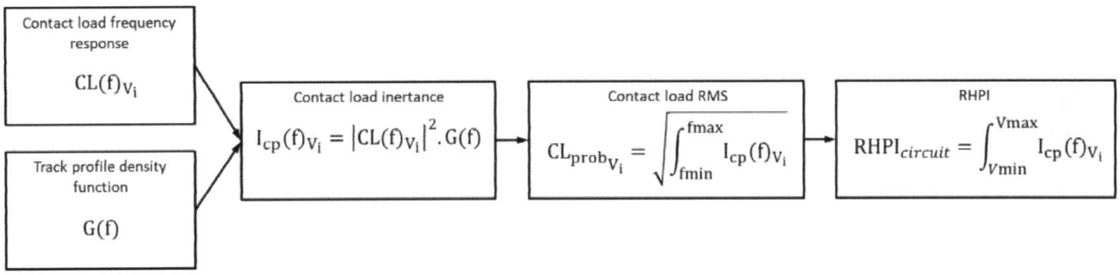

Figure 8. Block diagram representing the index calculation steps.

In this work, the proposed performance index was calculated for a mathematical road profile compatible to standard asphalt type "C" [38]. The Figure 9 presents the contact load inertance. By integrating the inertance function in the frequency range it is possible to plot the contact load RMS as a function of vehicle speed.

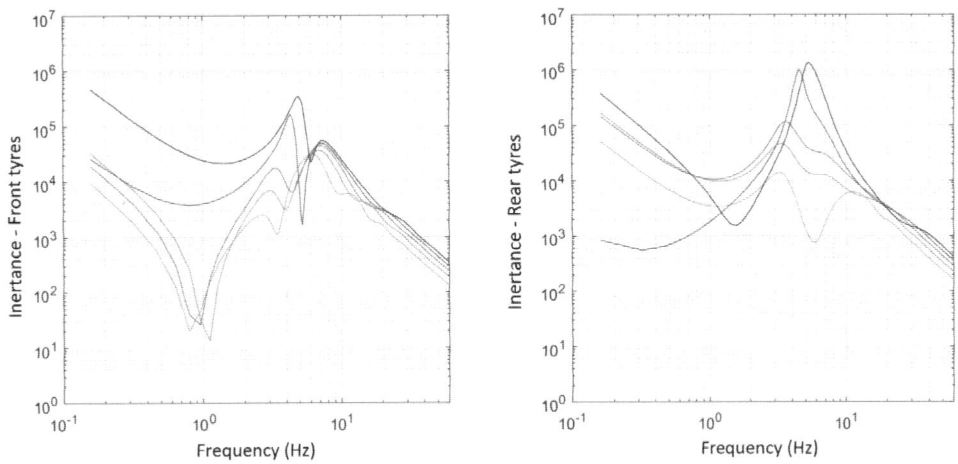

Figure 9. Contact load inertance function for front and rear tyres (higher vehicle speeds in darker colours).

Figure 9 shows the contact load inertance for the front and rear tyres. The track profile amplifies the low frequency and reduces the high frequency gains from the contact load frequency response (Figure 7).

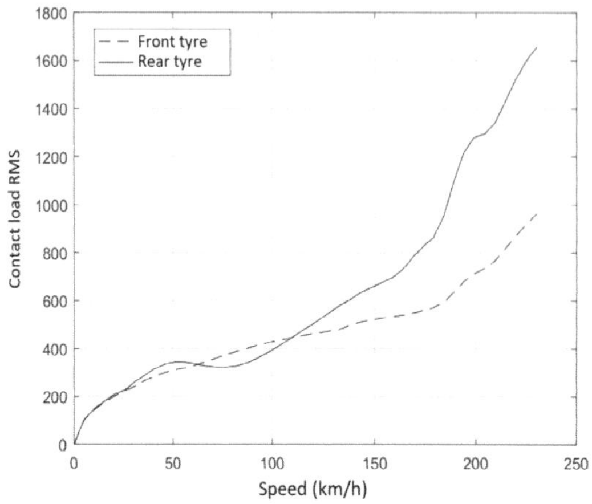

Figure 10. Tyre contact patch load RMS vs. vehicle speed.

The integral of the contact load inertance (contact load RMS) at each vehicle longitudinal speed value is presented in Figure 11: low values correspond to better road holding behaviour. Any change in car settings (spring stiffness, etc.) will also change this curve. Each circuit presented in Table 1 force the vehicle to work in a specific vehicle speed range. This means that the values in Figure 10 can be windowed for each circuit speed range. The non-linearities are due to the changes in system stiffness and damping. The front bump stop starts to be compressed at around 130 km/h and the rear around 160 km/h.

Figure 11 shows a sensitivity check of the contact load RMS for different race car settings. The influence of front and rear static ride heights is presented in the first column. The second column shows the influence of front and rear bump stop gaps. The third column presents the effects of a change in terms of aerobalance (i.e. the position of the centre of pressure, on top) and a of a suspension rate variation (front and rear). The latter basically compares a "stiff" setup for a smooth track surface with a "soft" setup designed to cope with the bumps of a street circuit.

Although small, a change of the performance index is clearly visible in all cases. It is important to remember that race car performance is often measured in tenths or even hundredths of a second.

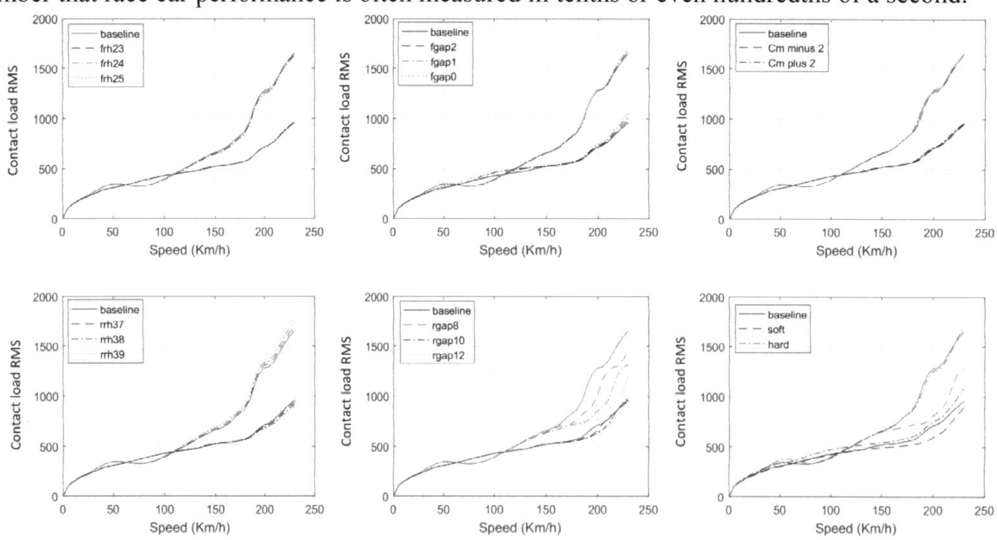

Figure 11. Sensitivity of the contact load RMS: from left to right, effects of front and rear static ride heights (frh and rrh respectively); effects of front and rear static bump stop gaps (fgap and rgap); effects of aerobalance variation and suspension rate (softer vs harder).

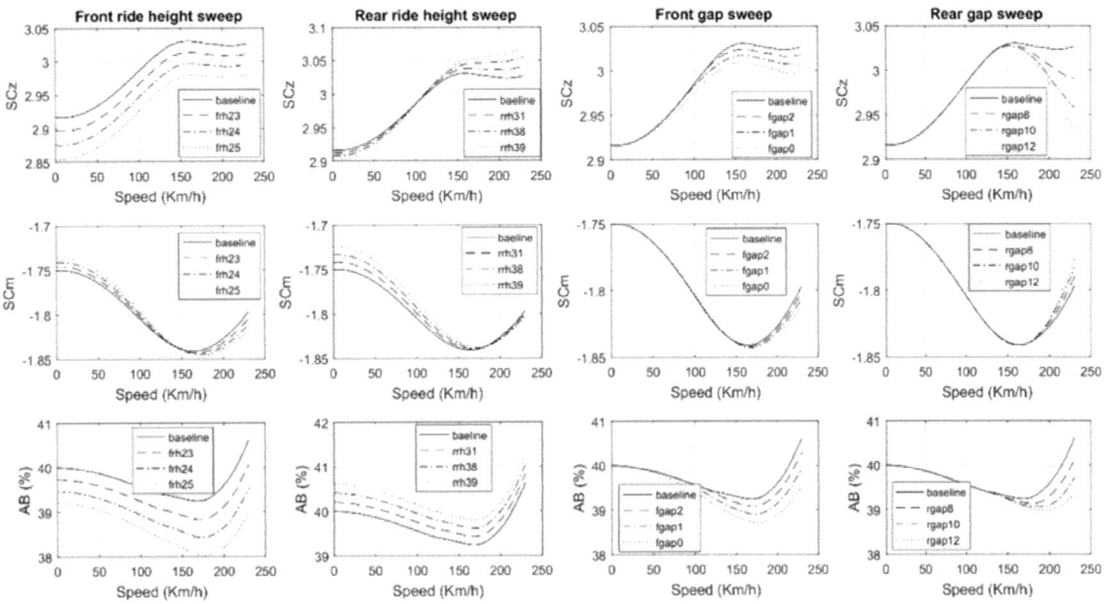

Figure 12. Influence of the simulated suspension settings on aerodynamic coefficients.

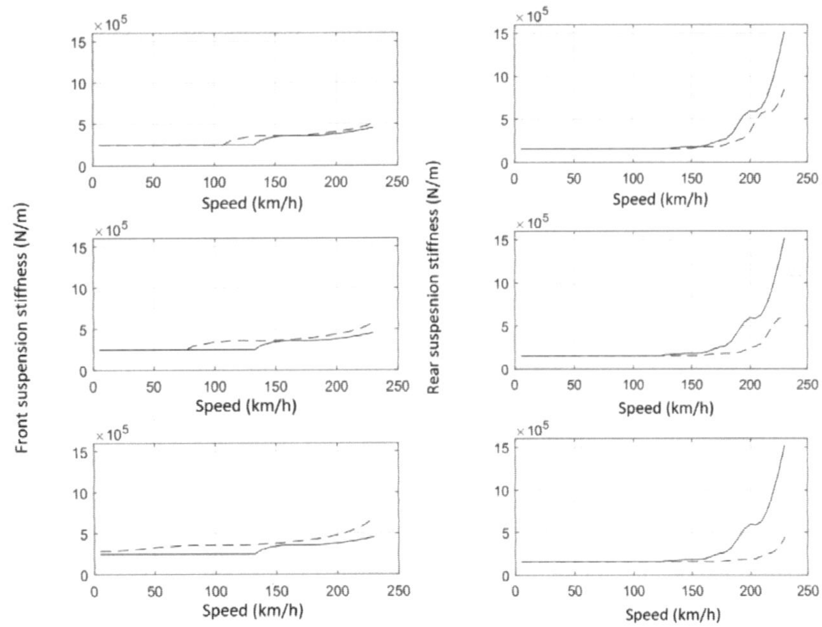

Figure 13. Suspension stiffness vs vehicle longitudinal speed. Bump stop gap changes as in third and fourth columns of Figure 12.

Figure 11 suggests that sensitivity of the rear axle is usually more visible than the front one for all changes made. The main factor is the rear bump-stop gap. It also shows how different setup settings in terms of suspension rate can have a big impact on the curves.

Figure 12 shows how the aerodynamic coefficients are influenced by the suspension changes presented in the first and second columns of Figure 11 (front and rear static ride heights and bump stop gaps). Any variation of the rear bump stop gap has the largest effect on aerodynamic forces, and consequently on the rear tyre contact patch loads (Figure 11). This is an effect of the change in terms of dynamic ride heights due to the bump stops and to the non-linear relationship between aerodynamic forces and ride heights.

Figure 13 shows how the static bump stop gaps affect suspension stiffness: front bump stop gap in the left column, rear gap in the right. The dashed line is the baseline while the solid line is the modified setup. Although the gap steps in millimetres are the same, the rear is more sensitive to changes as the bump stop/spring stiffness ratio is higher than the front.

Finally, the integral of the tyre contact patch load RMS versus vehicle longitudinal speed (Figure 10), gated for a speed range of a target circuit (the example shows circuit 1 and circuit 2) represents the proposed Road Holding Performance Index (RHPI), summarizing the road holding behaviour in a single, RMS-based number. Higher values represent high tyre contact patch load variation, meaning less performance. Figure 14 shows the effect of various setup changes on the RHPI for the front and rear axles as well as the front/rear average RHPI as a representative index for the complete vehicle. These plots show that a different vehicle speed range (different circuits) will excite the system in a different way, consequently unsettling the contact patch load in different magnitudes. Once more, the biggest changes are related to the rear suspension gap, as this will change rear suspension stiffness and move the onset of rear diffuser stall to a different vehicle speed as well (Figure 12).

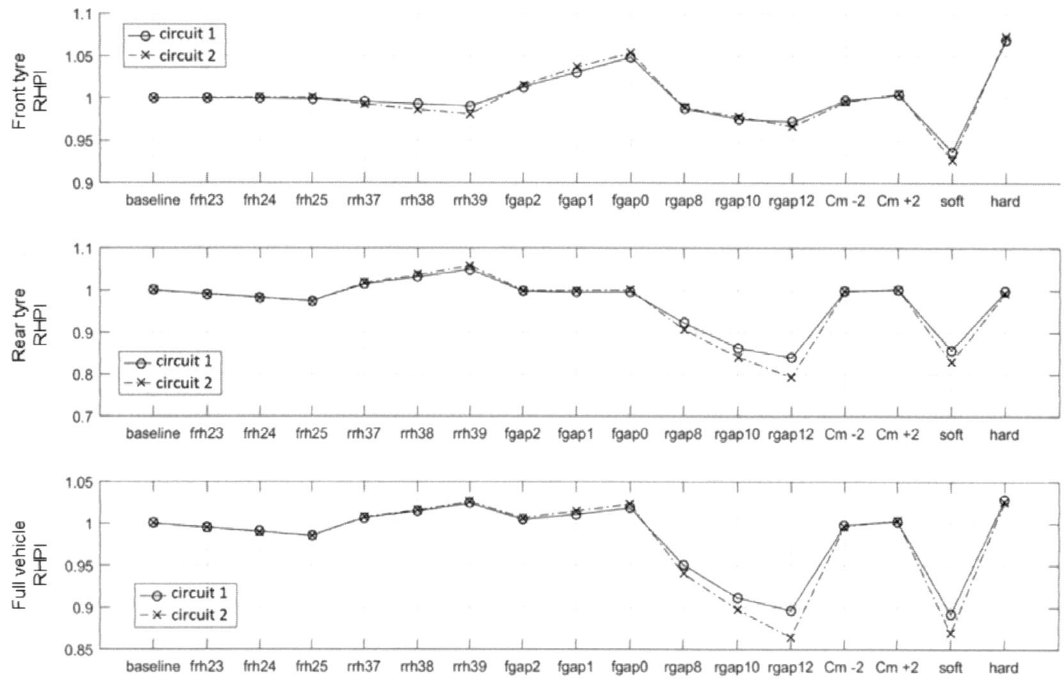

Figure 14. Road Holding Performance index for different car settings and for circuits 1 and 2.

5. Conclusion

The paper proposes a Road Holding Performance Index including several racecar parameters that are not commonly used in the literature and by the industry.

Aerodynamic forces change vehicle stiffness progressively as speed increases. Suspension stiffness non-linearities are important to control ground clearance and to keep the car within a proper ride height range (as shown in the quasi-static simulation); they also have a significant impact on vehicle system dynamics. The analysis of the system behaviour without these non-linearities would generate a vehicle setup that is impractical for circuit use or at least not fully realistic. The linearization as a function of vehicle longitudinal speed showed a very good correlation with typical racecar dynamics. The non-linear effects are clearly visible by means of the proposed performance index (Figure 14).

The performance index was calculated using numerical simulation, but it is also possible to generate it by means of the so-called four- or seven-post rig testing. This unprecedented approach would require further research.
Although the optimization of the transient response is important to improve racecar performance, setting suspension stiffness and static ride heights is a priority. The RHPI offers the opportunity to optimize the design of damping and spring rate characteristics by means of the traditional quasi-static analysis (Figure 5 and 6).
Further developments are possible by taking bump stop hysteresis into account and also by creating a new performance index coupling aerodynamic stability with vertical dynamics.

Symbols

a	distance from sprung mass center of gravity to front axle
AB	aerobalance
b	distance from sprung mass center of gravity to rear axle
c_i	weight factor
c_f	contact force index
C_m	pitch moment coefficient
C_{wr}	front damping coefficient at wheel
C_z	downforce coefficient
F_a	aerodynamic force
$F_{a\,mass}$	aerodynamic force due to added mass effect
F_{spring}	force at suspension spring
F_{wheel}	spring force at wheel
gap	spring deflection required to engage bump stop
h	ride height
K_{wr}	spring rate at wheel
L	wheelbase length
M_a	aerodynamic pitch moment
m_{ns}	unsprung mass
m_s	sprung mass
p	dynamic pressure
rh	ride height
S	vehicle reference frontal area
t	time
V	vehicle longitudinal speed
x_{spring}	spring displacement
x_{wheel}	relative displacement between wheel and sprung mass
z_{ns}	unsprung mass vertical displacement
z_r	road displacement input to the tyre
z_s	sprung mass vertical displacement
φ_s	sprung mass pitch angle
ρ	air density
ϑ	air volume displaced

Subscripts

0	static value
f	front
r	rear

Acknowledgments
The authors would like to thank Fortec Motorsports UK for all data supplied and their support.

References

[1] Chindamo, D.; Gadola, M.; Romano, M. (2014). Simulation tool for optimization and performance prediction of a generic hybrid electric series powertrain. International Journal of Automotive Technology, 15:135–144.
[2] Wright, P. Formula 1 Technology. 1st edition. SAE, 2001.
[3] Cecchel, S.; et al. (2018). Impact of reduced mass of light commercial vehicles on fuel consumption, CO2 emissions, air quality, and socio-economic costs. Science of the total environment. (613-614), 409-417.
[4] Cecchel, S.; et al. (2018). Lightweighting in automotive: cradle- to- grave life cycle assessment of a safety relevant component. International Journal of Life Cycle Assessment. 23(10), 2043-2054.
[5] Chindamo, D.; Gadola, M. (2018). What is the Most Representative Standard Driving Cycle to Estimate Diesel Emissions of a Light Commercial Vehicle? Proceedings of the 1st IFAC Workshop on Integrated Assessment Modelling for Environmental Systems, (IAMES 2018), May 2018, Brescia, Italy. IFAC-PapersOnLine, 51(5), 73-78.
[6] Reimpell, J.; Stoll, H.; Betzler, J. The Automotive Chassis: Engineering Principles. 2001, Second Edition, Butterworth Heinemann.
[7] Barcelo, J. D. P. (2012). Optimization of Racing Car Suspensions featuring Inerters. (PhD Thesis). Department of Mechanical Engineering and Mathematical Sciences, Oxford Brookes University.
[8] New, T. M. Random Road Analysis and Improved Gear Ratio Selection of a Front Wheel Drive Drag Racing Car. 2008. (MSc Thesis). Clemson University.
[9] Chindamo, D.; et al. (2017). Design of an indoor simulator for motorcycle applications. Applied Sciences. 7(12), 1220.
[10] Chindamo, D.; et al. (2017). BikeShake: the design of an indoor simulator dedicated to motorcycle ride testing as an interactive project. International Journal of Interactive Design and Manufacturing. (IJIDeM). 12(4), 1371-1383.
[11] Zhao, J.; et al.; Chassis integrated control for active suspension, active front steering and direct yaw moment systems using hierarchical strategy, Vehicle System Dynamics, 2017, 55(1), 72-103.
[12] Coric, M. et al.; Optimisation of active suspension control inputs for improved vehicle handling performance, Vehicle System Dynamics, 2017, 54 (1), 1574-1600.
[13] Benini, C.; et al. The influence of suspension components friction on race car vertical dynamics, Vehicle System Dynamics, 2017, 55 (3), 338-350.
[14] Matsubara, M. et al.; Natural frequency analysis of tyre vibration using a thin cylindrical shell model, SAE Technical Paper, SAE Noise and Vibration Conference and Exhibition; DeVos Place Convention CenterGrand Rapids; United States; 2015.
[15] Chindamo, D.; et al. (2018). On the vehicle sideslip angle estimation: a literature review of methods, models and innovations. Appl. Sci. 2018, 8(3), 355.
[16] Gadola, M.; Chindamo, D. (2018). Estimation of vehicle side-slip angle using an artificial neural network. MATEC Web of Conferences 166, 02001 - The 2nd International Conference on Mechanical, Aeronautical and Automotive Engineering (ICMAA 2018). Singapore, Feb 2018.
[17] Gadola, M.; et al. (2018). On the Passive Limited Slip Differential for High Performance Vehicle Applications. 14th International Symposium on Advanced Vehicle Control AVEC 18, Beijing, July 16-20 2018.
[18] Floyd, R.; Law, E. Simulation and Analysis of Suspension and Aerodynamic Interactions of Race Cars. Motorsport Engineering Conference & Exposition: SAE. SAE Paper 942537 1994.

[19] Kasprzak, JL.; Floyd, RS. Use of simulation to tune race car dampers. SAE Paper 942504, 1994.
[20] Marchesin, F.P: et al. (2017). High downforce racecar vertical dynamics: aerodynamic index. Vehicle System Dynamics, 56(8), 1269-1288.
[21] Dal Bianco, N.; et al. Minimum time optimal control of a GP2 race car. Proceedings of the Institution of Mechanical Engineers, Part D: Journal of Automobile Engineering. 2018, 232(9), 1180-1195.
[22] Doniselli, C.; et al. Aerodynamic Effects on Ride Comfort and Road Holding of Automobiles. Vehicle System Dynamics, v. Supplement 25, 1996.
[23] Kelly, J.; Kowalczyk, H.; Oral, H. Track Simulation and Vehicle Characterization with 7 Post Testing. Motorsport Engineering Conference & Exhibition, 2002, SAE.
[24] Kowalczyk, H. Damper tuning with use of a seven-post shaker rig. SAE 2002 World Congress. Detroit Michigan.
[25] Cambiaghi, D.; Gadola, M.; Vetturi, D. Suspension System Testing and Tuning with the Use of a Four-Post Rig. SAE 1998.
[26] Chindamo D.; et al. Reproduction of real-world road profiles on a four-poster rig for indoor vehicle chassis and suspension durability testing. Advances in Mechanical Engineering, Vol (9)8, 2017.
[27] MTS, RPC Pro Software, 2014 MTS Corporation
[28] Marchesin, et al. (2016). Upright mounted pushrod: the effects on racecar handling dynamics. The Dynamics of Vehicles on Roads and Tracks. Proceedings of the 24th Symposium of the International Association for Vehicle System Dynamics, IAVSD 2015. (10), 543-552.
[29] Crema, C.; et al. (2015). Smartphone-based system for vital parameters and stress conditions monitoring for non-professional racecar drivers. Proceedings of the 2015 IEEE SENSORS. 7370521.
[30] Multimatic. A note in the use of four post test rigs. p.20. 1998
[31] Fung,Y.C., An Introduction to the Theory of Aeroelasticity, Dover Publications, 2008
[32] Mavroudakis, B. About the simulations of Formula 1 Racing Cars. 2010. Institute of Engineering and Computational Mechanics, University of Stuttgart
[33] Dallara, Formula Renault 3.5 2012 Aerobook, release 1.1, Dallara Automobili 2012
[34] Fuller, J., et al., The importance of unsteady aerodynamics to road vehicle dynamics, Journal of Wind Engineering and Industrial Aerodynamics, 2013, vol. 117, 1-10.
[35] Molina J.; Zhang, Z.; Arodynamics of a heaving airfoil in ground effect, AIAA Journal, Vol. 49, No. 6, 2011
[36] Barbosa, RS.; Vehicle Dynamic Safety in Measured Rough Pavement. Journal of Transportation, 2011
[37] Newland, D. E. (1984). An introduction to random vibrations and spectral analysis, 2nd Ed., Longman Scientific & Technical, New York, 377.
[38] AASHTO (2001). Standard Specifications for Transportation Materials and Methods of Sampling and Testing. Standard TP 62-03, Standard Method of Test for Determining Dynamic Modulus of Hot Mix Asphalt Concrete Mixtures, Washington, DC.

IOP Conf. Series: Materials Science and Engineering **538** (2019) 012070 doi:10.1088/1757-899X/538/1/012070

Effects of Riblet Shape of Non-smooth Surface on Aerodynamic Characteristics of Vehicle

Liu Rong[1,a] and Wu Yiwan[2]

[1]College of Computer and Information Sciences, Fujian Agricultural and Forestry University, 350002 Fuzhou, China
[2]School of Mechanical Engineering and Automation, Fuzhou University, 350116 Fuzhou, China

Corresponding author: chensi1983@163.com

Abstract. To reduce the aerodynamic drag of the vehicle, three kinds of riblet shape non-smooth surfaces were arranged in the tail of the Motor Industry Research Association (MIRA) stepped-back vehicle mode. The shapes of the non-smooth riblet elements are triangular, circular and rectangular. Computational fluid dynamics (CFD) method was used for numerical analysis. To verify the effectiveness of the analysis method, the analysis result of the standard MIRA stepped-back vehicle mode was compared with wind tunnel test data. The tail-flow, pressure and turbulent kinetic energy of the MIRA stepped back vehicle modes, which were arranged with riblet elements, were analyzed. The effects of riblet shape of non-smooth surfaces on the reduction of vehicle' aerodynamic drag were compared with each other. The simulation results show that the three kinds of non-smooth surfaces have a good effect on drag reduction by reducing the negative pressure in the tail zone of the model, reduce the turbulent kinetic energy and reduce energy dissipation. The circular non-smooth surface is better than the other, with drag reduction effect of 5.53%.

1. Introduction

Global oil consumption has been increasing in recent decades. But oil is a non-renewable resources. How to effectively reduce energy consumption is a worldwide problem to be solved urgently. The total number of vehicles in the world has exceeded 1 billion in 2011. The oil is consumed by vehicles accounts for a large proportion of the world's oil consumption. When a vehicle is driving at high speed, the output power of the engine is mainly used to overcome the air drag. Therefore, reducing the aerodynamic drag of the vehicle is beneficial to improve vehicle power and fuel economy, thus reducing oil consumption. Bionics studies have found that the surface of sharks is covered with shield scales with riblets. These shield scales can inhibit and delay the occurrence of turbulence in the fluid boundary layer of shark body surface and effectively reduce the flow drag [1]. How to apply the research results of bionics to the aerodynamic drag reduction of the vehicle body is a very interesting topic.

Walsh et al. [2] carried out turbulent drag reduction studies on fluted flat plates. Their results show that the drag reduction effect of non-smooth riblet of shield scale can reach 8% under the condition of high-speed fluid flow. Bechert et al. [3-4] proposed the calculation theory and optimization method of riblet section, which laid the theoretical foundation for the design and manufacture of riblet drag reduction structure.

To reduce the aerodynamic drag of the vehicle, researchers have conducted a lot of researches on aerodynamic drag reduction effect and mechanism of pit type non-smooth element [5-8, 10, 12-13]. Previous studies have shown that the arrangement of non-smooth elements in vehicle tail can achieve better drag reduction effect than other surfaces.

Scholars have conducted extensive research on the plate riblet model and achieved beneficial results. Although riblet drag reduction technology has been applied in the shipbuilding industry [9], there are few studies on aerodynamic drag reduction of vehicles with non-smooth riblet structures. In this paper, three different non-smooth riblet structures are arranged at the vehicle tail. And then the influence of the shape of non-smooth elements on the aerodynamic characteristics of the vehicle is compared and analysed by Computational Fluid Dynamics (CFD, Fluent).

2. Vehicle Model And Validation

The stepped-back vehicle model given by the Motor Industry Research Association (MIRA) was selected as the research target. The MIRA model group is an internationally recognized standard model [10] and is widely used in basic research of vehicle aerodynamic characteristics. As shown in figure 1, the 3D model of MIRA stepped-back vehicle was built by UG based on the data in reference [11].

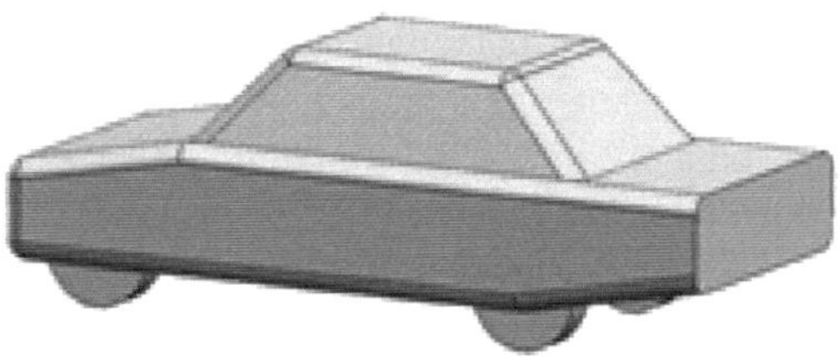

Figure 1. MIRA stepped-back vehicle model.

To ensure the accuracy of the numerical calculation, the parameters of the computational domain of Fluent are shown in table 1.

Table 1. Computational domain (unit: mm).

Parameter	Length	Width	Hight	The distance from inlet to vehicle head	The distance from vehicle tail to outlet
Value	10×4165	9×1625	5×1421	3×4165	4×4165

The unstructured tetrahedral mesh was used to divide the mesh. The triangular prism mesh was added to the curved surface of the vehicle body as the boundary layer for eliminating the influence of the wall function. The maximum mesh size of each part in the simulation is shown in table 2. Settings of Fluent are shown in table 3.

Table 2. Maximum mesh size of each part (unit: mm).

Part	Flat surface of vehicle body	Curved surface of vehicle body	The overall basin	The basin around vehicle body	The basin around vehicle tail
Value	32	16	8	512	64

Table 3. Settings of Fluent (unit: mm).

Object	Setting value	Object	Setting value
Solver	Pressure-Based	Ground	Slip wall
Turbulence model	Realizable k-ε	Other wall surface	No-slip wall
Inlet	Velocity inlet	Temperature	15℃
Out let	Pressure outlet	Residuals	0.000001
Vehicle body	No-slip wall	Iteration steps	3000

The aerodynamic drag coefficient of the stepped-back vehicle model (smooth body) of MIRA is 0.3257 through numerical calculation.

To verify the accuracy and reliability of the simulation model and method, the simulation data of MIRA stepped-back vehicle model was compared with the measured wind drag data of HD-2 wind tunnel in Hunan University and IVK wind tunnel in Stuttgart [12].

As shown in table 4, the error between the numerical simulation value and the typical wind tunnel test value is small. It means that the numerical simulation method adopted in this paper has high reliability and accuracy.

Table 4. Comparison between simulation and test.

Wind tunnel	Test data	error
HD-2	0.3225	0.99
IVK	0.3204	1.65

3. Riblet Shapes And Parameters

As shown in figure 2, non-smooth riblet structures such as triangle, circular arc and rectangle were selected to be arranged at the tail of MIRA stepped-back vehicle model.

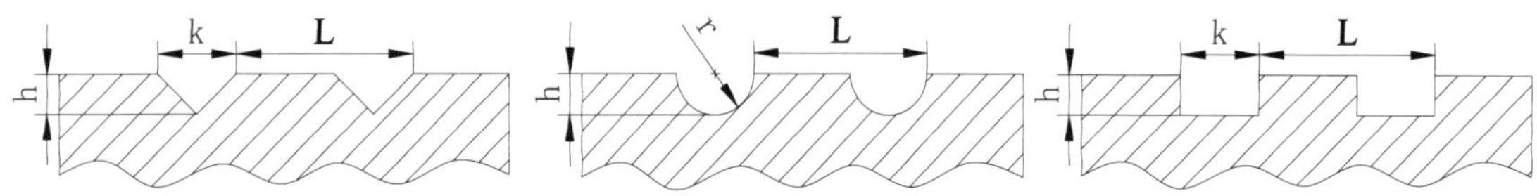

(a)Triangular riblet (b)Circular riblet (c)Rectangular riblet

Figure 2. MIRA stepped-back vehicle model.

Previous studies have shown that the non-smooth surface morphology can change the boundary layer flow field structure, control the surface flow field and reduce aerodynamic drag. Therefore, the dimension of the non-smooth riblet can be chosen according to the thickness of the boundary layer at the tail of the vehicle. The selection of riblet element size should consider making the riblet depth less than the thickness of the boundary layer. Reference [13] shows that the thickness of the tail boundary layer of MIRA stepped back model is 4.13mm. Therefore, the non-smooth riblet structure depth h is set as 4mm. In this study, L=30mm, k=8mm, r=k/2.

Bionic studies have shown that the parallel arrangement of the riblets and the incoming flow is better than the vertical arrangement of the drag reduction. In this study, the riblets were arranged along the flow direction, as shown in figure 3.

IOP Conf. Series: Materials Science and Engineering **538** (2019) 012070 doi:10.1088/1757-899X/538/1/012070

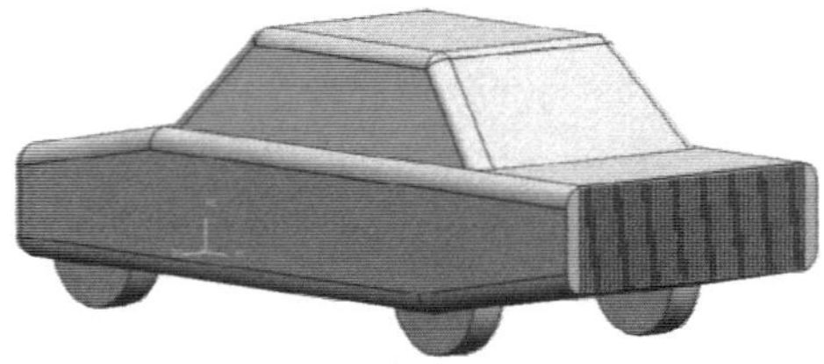

Figure 3. Layout of riblet structure.

4. Effect of Riblet Shape on Drag Reduction Performance

The MIRA stepped-back vehicle model with three types of riblets at the tail was analysed by the method given in section 2. Due to the size of the tablet is too small, the riblet surface is encrypted (maximum size 2mm) in the meshing process to obtain more accurate flow field information.

4.1. Effect of Riblet Shape on Wind Drag Coefficient

Drag reduction rate (k) was used as the evaluation index of drag reduction effect.

$$k = \frac{C_{ds} - C_{dn}}{C_{ds}} \tag{1}$$

where C_{ds} is the aerodynamic drag coefficient of smooth vehicle body, and C_{dn} is the aerodynamic drag coefficient of non-smooth vehicle body.

From table 5, it can be seen that the non-smooth riblet structure at the tail of the MIRA stepped-back vehicle model can effectively reduce the wind drag coefficient of the vehicle, and the circular riblet has the most obvious drag reduction effect.

Table 5. Drag coefficient and k of different shapes.

Shape	Drag coefficient	*k*
Triangular	0.3119	4.24%
Circular	0.3077	5.53%
Rectangular	0.3099	4.85%

4.2. Effect of Riblet Shape on Wake Flow

It can be seen from figure 4 that after the non-smooth riblet elements are arranged at the tail of the MIRA model, the shape of the wake flow becomes narrow and the center of the wake vortex is farther from the tail. It indicates that the tail of the vehicle is less affected by the low-pressure region of the wake flow after the non-smooth riblet elements are arranged, and the effect of differential pressure drag is reduced.

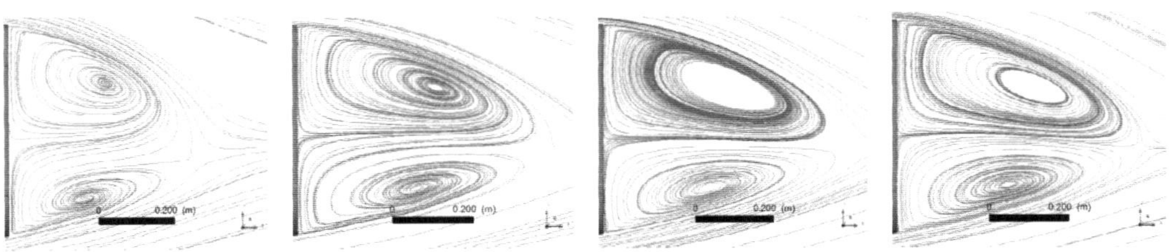

(a) Smooth model (b) Triangular riblet (c) Circular riblet (d) Rectangular riblet

Figure 4. Wake flow field maps of longitudinal symmetry plane.

IOP Conf. Series: Materials Science and Engineering **538** (2019) 012070 doi:10.1088/1757-899X/538/1/012070

4.3. Effect of Riblet Shape on Pressure

The pressure drag accounts for about 50%~60% of the total aerodynamic drag of the vehicle. To reduce the pressure drag, the positive pressure area at the front of the vehicle and the negative pressure area at the tail of the vehicle should be reduced [10]. This paper mainly focusses on the influence of the riblet shape of non-smooth surface on the vehicle tail on the aerodynamic characteristics of the vehicle. Therefore, the pressure distribution in the front area of each model is the same, and the pressure drag of the vehicle is determined in the tail wake flow region.

It can be seen from figure. 5 that the negative pressure region of the vehicle tail was reduced after the non-smooth riblet element was arranged at the tail. It indicates that the pressure at the tail of the model was increased, and the pressure drag was reduced. It also can be seen from figure 5 that the surface of the circular riblet have the most obvious influence on the pressure drag.

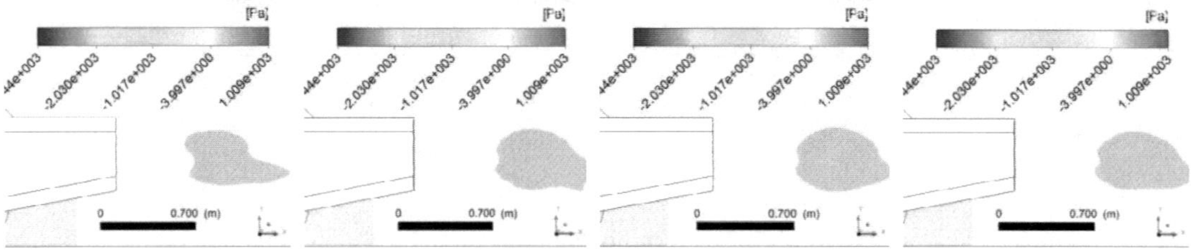

(a) Smooth model (b) Triangular riblet (c) Circular riblet (d) Rectangular riblet

Figure 5. Pressure maps of longitudinal symmetry plane.

4.4. Effect of Riblet Shape on Turbulent Kinetic Energy

Figure 6 shows the comparison of turbulent kinetic energy in the 400mm plane at the rear of each model. As shown in figure 6, the turbulent kinetic energy in the tail region of the model can be effectively reduced by adding non-smooth riblet elements in the vehicle tail. The smaller the turbulent kinetic energy is, the less energy will be dissipated when the air flows over the vehicle surface. The circular riblet element has the best effect among several shapes.

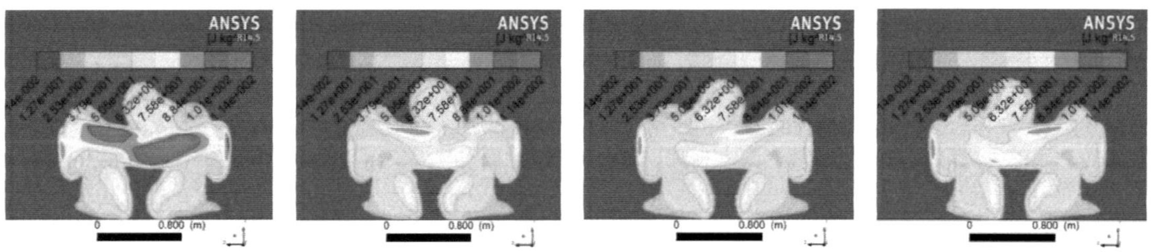

(a) Smooth model (b) Triangular riblet (c) Circular riblet (d) Rectangular riblet

Figure 6. Cloud of turbulent kinetic energy.

5. Equations and Mathematics

In this paper, the computational fluid dynamics method was used to evaluate the influence of riblet shape on the aerodynamic characteristics of the vehicle body. The main conclusions, which can be drawn from the conducted simulations, are as follows:

1) By adding non-smooth riblet elements on the vehicle tail, the aerodynamic drag of the vehicle can be reduced significantly. The circular riblet has the largest amount of drag reduction, which is up to 5.53%.

2) Three kinds of riblet shape can effectively reduce the negative pressure at the vehicle tail, and reduce the pressure drag, turbulent kinetic energy and energy dissipation.

In this paper, non-smooth riblet element surface is only arranged on the vehicle tail, and no other surface is involved. At the same time, the relationship between the geometry of non-smooth riblet and the aerodynamic drag reduction of the vehicle body needs to be further studied.

Acknowledgments
This work was supported by the National Natural Science Foundation of China (No.11626060) and by Scientific Research Fund of Fujian Provincial Education Department(No.JAT160172).

References
[1] B. Qu, J. Wang, C. Pan, et al. Preparation of shark skin surface micro-structure materials. Journal of Dalian Ocean University, 26, 2: 173-175 (2011).
[2] M. J. Walsh. Riblets as a viscous drag reduction technique. AIAA Journal (S0001-1452), 21(4): 485-486 (1983).
[3] D. W. Bechert, M. Bartenwerfer. The viscous flow on surfaces with longitudinal ribs. Journal of Fluid Mechanics, 206(206): 105-129 (1989).
[4] D. W. Bechert, M. Bruse, W. Hage. Experiments with three-dimensional riblets as an idealized model of shark skin. Experiments in Fluids, 28(5): 403-412 (2000).
[5] Y. Yang, J. F. Huang, G. H. Fan. Research on Non-smooth Surfaces for Control of the Automobile Trailing Vortex Structure. Journal of Mechanical Engineering, 52(8): 133-140 (2016).
[6] K. Chen, Y. Wang, K. Y. Cal, et al. Analysis of aerodynamic drag reduction on automobile by using bionic non-smooth surface. China Mechanical Engineering, 23(8): 1001-1006 (2012).
[7] F. Xie, Y. M. Ding, L. Qin, et al. Effects of non-smooth surface dimple configuration on vehicle body aerodynamic characteristics. Applied Mathematics and Mechanics, (5): 505-514 (2015).
[8] J. F. Xie, J. L. Zhang. Influence analysis of pit-type non-smooth elements arrangement on aerodynamic drag of vehicle. Journal of Zhengzhou University (Engineering Science), 37(5): 86-90 (2016)
[9] L. S. Zhao, X. Q. Bai, F. L. Fu, et al. Simulation analysis and experimental validation of drag reduction performance of marine micro-trapezoidal groove surfaces. Ship Engineering, (10): 15-20 (2015).
[10] Y. Yang, Z. Liu, Z. Q. Gu, et al. Optimization and design mnethod of non-smooth surface at stepped back MIRA mode's tail. Journal of Chongqing University, 38(4): 10-17 (2015).
[11] Aerodynamic Testing of Road Vehicles--Open Throat Wind Tunnel Adjustment Standard, SAE J2071 Revised JUN94. 1994, 06.
[12] S. Wang. Experimental investigation on aerodynamic characteristics of MIRA model group in wind tunnel. Hunan University (2011).
[13] Y. H. Wu. The analysis of non-smooth unit structure effects on vehicle aerodynamic characteristic. Hunan University (2014).

IOP Conf. Series: Materials Science and Engineering **538** (2019) 012071 doi:10.1088/1757-899X/538/1/012071

Dynamic Characteristics Analysis of Cutterhead System with Uncertain Loads

Bao Youneng[1], Huo Junzhou[1] and Xu zhaohui[1,a]

[1]School of Mechanical Engineering,Dalian University of Technology, 116023 Ganjingzi District, Dalian City, Liaoning Province, China

[a] Corresponding author:baoyu_1994@qq.com

Abstract. Tunnel boring machine (TBM) is a large and complex tunnel tunnelling equipment, which is widely used in underground works such as water conservancy, transportation, national defense and energy. In the construction of TBM, it is very difficult to predict the dynamic characteristics of the cutterhead system, because of its complexity and the uncertainty of the service environment, geological uncertainties and manufacturing process. To solve the above problems, first of all, this paper applies concentrated mass method, establishing a multi degree of freedom coupling dynamic model of TBM cutterhead system under uncertain excitation. Considering the changes between hob location and establishing the uncertain geological condition model, the load boundary with uncertainties is obtained and replaced by the load boundary. The dynamic characteristics of TBM cutter head system with uncertain load are obtained, which provides important basis for TBM vibration reduction and optimization design.

1. Introduction

Tunnel boring machine (TBM) is a complex large tunnelling equipment integrated with the functions of tunneling, ballast and lining. It is also a landmark product in the equipment manufacturing industry. Cutterhead is the most important core component of TBM. Because the cutterhead system has complex driving environment, the rock has the characteristics of high hardness, high wear resistance, high temperature and high confining pressure. In addition, the TBM hob has the characteristics of multi-point impact when rock breaking and there will be strong impact load when the hob is cutting rock. These loads will be transferred to the cutter head, which will cause severe vibration of the cutterhead during the working process, resulting in fatigue damage of the cutter head. Therefore, it is very necessary to analyze the dynamic characteristics of the cutter head system. However, due to the uncertainties in the geological distribution of tunneling, uncertainties in the tunneling process and manufacturing and assembly process, such uncertainties will result in uncertainties in the load of TBM, and consequently the dynamic characteristics can not be accurately predicted. Therefore, the dynamic characteristics analysis of the cutter system under the uncertainty factors is more close to the actual working environment, which is of great engineering significance for the structural design and vibration reduction optimization of the TBM cutter head system.

At present, domestic and foreign scholars have done a lot of analysis on the dynamic characteristics of TBM key components from theoretical modeling and experimental aspects. Tang Guowen and Xie Qijiang[1,2] set up the dynamic equations of TBM propulsion system, considering the load of the cutterhead and the supporting interface stiffness under variable constraints. At the same time, they also use interval numbers to describe the uncertainty of geological parameters, and analyze the dynamic characteristics of TBM propulsion system. The dynamic characteristics of TBM under the condition of uncertain geological distribution are evaluated according to the speed and speed of driving angle. Zhang Kaizhi [3] sets up a shield experimental platform in proportion to 10:1. By changing the position of the driving motor, he can get the torque and speed of each drive motor under all loads, and study the load sharing and stability of the driving system. Liu Ran [4] applies the simulated load to build the test bench of the cutter driving system. The validity of the adaptive load sharing control strategy is verified based on the testbed. Zhu Zhenhuan [5] studies the contact characteristics of disc hob during rock breaking process based on the finite element method, and analyzes the kinematics characteristics of disc hob and the three directional force characteristics of disc hob. Huo Junzhou [6] carries out a dynamic simulation of the TBM main engine system, and studies the influence of cutter speed, pinion speed fluctuation and other factors on the dynamic characteristics of the cutter head. Considering the nonlinearity of hob bearing, Huo [7] sets up the dynamic model of hob cutter holder system with multiple coupling surfaces, and modifies the model with experiments, and explores the difference of dynamic characteristics between different hob cutter holder systems. Li [8] established the shield dynamic model of the complex geological conditions, calculated the dynamic response of the cutter rotary system, and explored the influence of the key parameters. Sun [9] proposes a hierarchical modeling method for TBM complex cutter driving system, and deduces the dynamic equation of the cutter driving system with multi-source nonlinear factors, and verifies the model with the test bench.

In summary, scholars at home and abroad have made some researches on the dynamic characteristics of TBM, and also achieved some results, however, they all have their own limitations. There are few studies on the dynamics of the TBM cutter system under the condition of working conditions, and the uncertainties in the TBM tunneling process are often neglected. Therefore, this paper will take the typical TBM cutter head system as the research object, consider the influence of uncertainty factors and establish a dynamic model of TBM cutter head system under uncertain excitation, so as to explore the dynamic behavior of TBM cutterhead system and make the theory more realistic.

2. Model Establishment

2.1. Geological Uncertainty Model

In the process of driving, there are many strata in the excavation interface of TBM. In the TBM excavation interface, the continuous change curve is used to represent the geological change of the excavation interface, thus forming a generalized geological feature model in the TBM cutterhead driving interface.

Assuming that the dividing line between strata is a continuous curve, the curve equation is $f(x, y)$, and xy as the end face of cutter head. For the sake of generality, if there are n strata distributed in the whole excavation interface, the composite geological model can be described by the $n-1$ divider line, and the combined geological model of the excavation interface can be obtained, as shown in Figure 1.

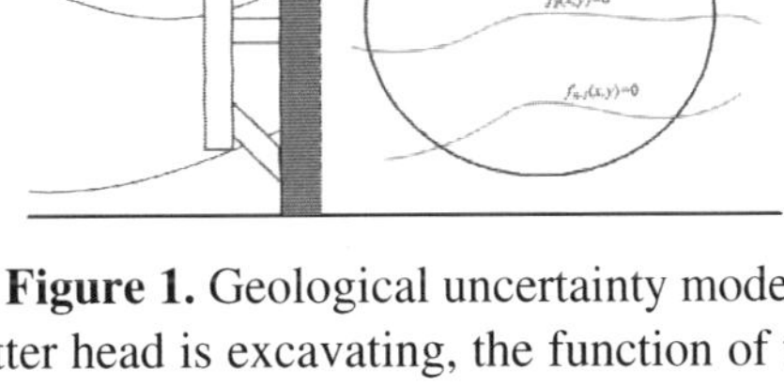

Figure 1. Geological uncertainty model

The t at any time when the cutter head is excavating, the function of the dividing line is $f_{i(t)}(x, y)$, that is, the stratum dividing line equation for the interface of the cutter head at the time. According to the dividing line equation, the stratum of each hob at the corresponding time can be determined, and the driving load of each hob can be obtained and the load of the cutter can be calculated by substituting the calculation formula of the load of the cutter.

IOP Conf. Series: Materials Science and Engineering **538** (2019) 012071 doi:10.1088/1757-899X/538/1/012071

According to the engineering statistics, the composite geology in the cutterhead excavation is usually a combination of soft rock and hard rock. Therefore, a composite stratum model consisting of hard rock and soft rock is established in this paper, that is, only one stratigraphic separation line is adopted, and two kinds of geotechnical models with the largest proportion are combined to model the strata.

2.2. Method for obtaining Interval Boundary of Uncertain Load Under Different Working Conditions

In the process of TBM tunneling, there will be different driving conditions. The external load of each cutter is different, and the vibration of cutter head will be very different. Therefore, it is necessary to calculate the vibration response of the cutterhead under different working conditions.According to the engineering experience and the design conditions of several TBM design companies, the typical driving conditions include 4 main types: maximum thrust condition, maximum overturning condition, correcting turning deviation condition and out of difficulty working condition. According to the analysis of different working conditions, and considering the uncertainty of tunnel geological distribution and uncertainty of hob driving process, the calculation flow of each load boundary of cutter head is as follows:

(1) According to the geological survey report and the measured data, the time domain diagram of the single hob is obtained;

(2) According to the driving conditions, the number and location of the hobs involved in the excavation are determined;

(3) We need to determine the formation of each hob at each time in the process of tunneling, and bring the position coordinates (xi, yi) of the i hob at any time to the surface equation of the hierarchical function, and calculate the y coordinates of the corresponding hob points at the current boundary surface. Compared with the yi of hob, the stratum of the hob is judged, and the load of the corresponding hob is extracted to calculate the load of the cutter head;

(4) The time history of the load of the cutter head under each working condition is obtained, and the minimum value of each load of the cutter head is obtained;

(5) In order to ensure the safety design, the maximum load of the cutterhead is generated under the nominal load of each hob, and the boundary of the cutter load can be obtained under different working conditions.

2.3. Equivalent Mechanical Model

The TBM cutter system mainly includes cutter head, big gear ring, drive gear, shield body and motor. The schematic diagram of the cutter head system is shown in Figure 2.

This paper adopts to TBM to analyze the model of the fractional cutterhead system. When establishing the dynamic model, the cutter disc body, the big gear ring, the driving gear, the shield body, the motor and the reducer are equivalent to the mass point, and the motor and the reducer are equivalent to a mass point. By analyzing the vibration of each mass point in radial, axial, overturning and torsional degrees of freedom, the equivalent mechanical model of typical cutterhead is derived by lumped mass method. Among them, X represents the horizontal and radial direction of the cutter head, Y represents the vertical direction of the cutter head, and Z is the axial direction, that is, the heading direction of TBM. The equivalent mechanical model is shown in Figure 3.

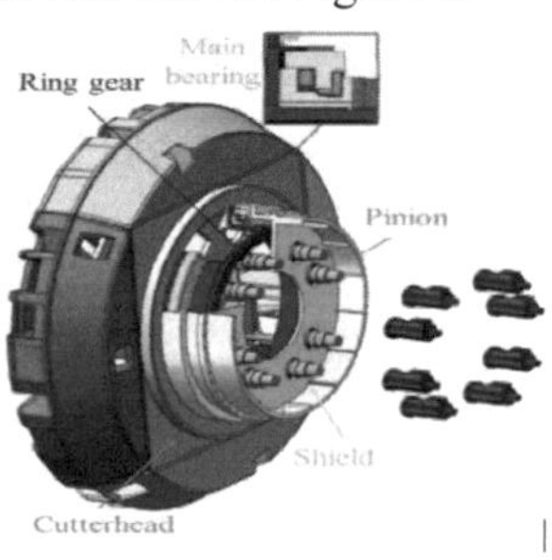

Figure 2. Schematic diagram of cutterhead system structure

This paper adopts to TBM to analyze the model of the fractional cutterhead system. When establishing the dynamic model, the cutter disc body, the big gear ring, the driving gear, the shield body, the motor and the reducer are equivalent to the mass point, and the motor and the reducer are equivalent to a mass point. By analyzing the vibration of each mass point in radial, axial, overturning and torsional degrees of freedom, the equivalent mechanical model of typical cutterhead is derived by lumped mass method. Among them, X represents the horizontal and radial direction of the cutter head, Y represents the vertical direction of the cutter head, and Z is the axial direction, that is, the heading direction of TBM. The equivalent mechanical model is shown in Figure 3.

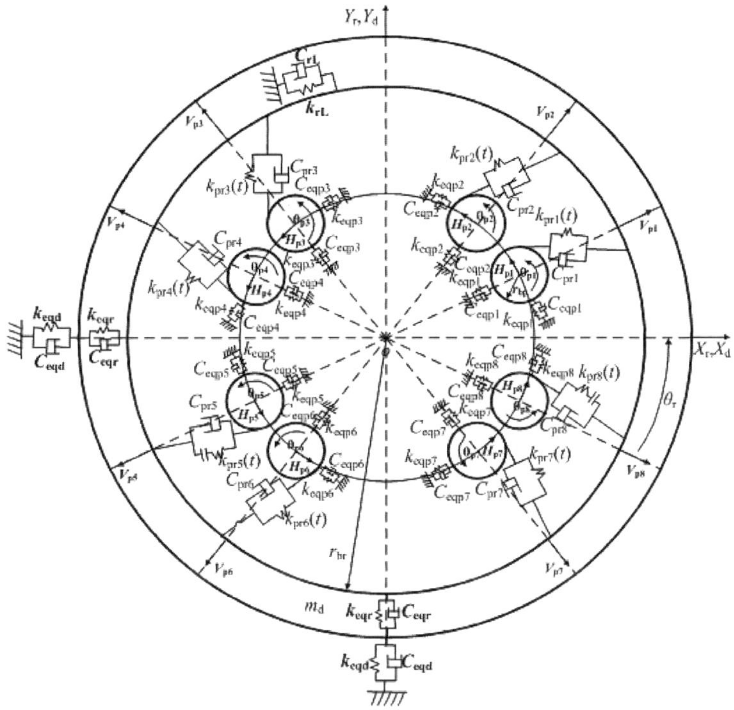

(a) Torsional mechanical model of rotating system

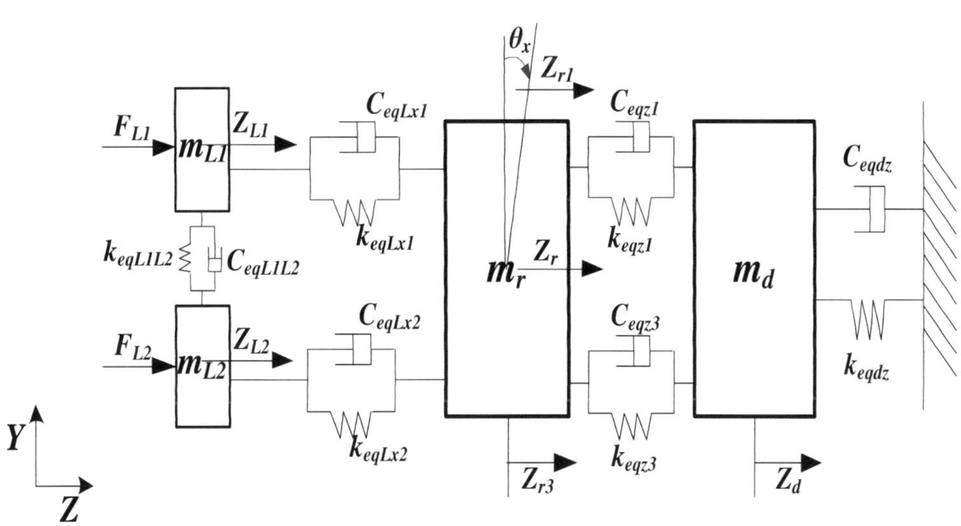

(b) Axial mechanical model and capsizing

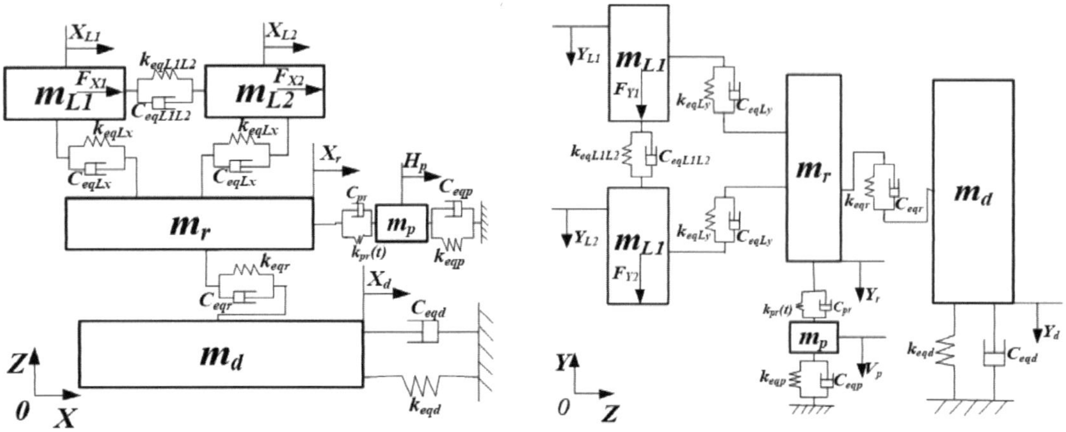

(c-1) Radial mechanical model (c-2) Radial mechanical model

Figure3. Equivalent mechanical model

2.4. Kinetic Equation

According to the established equivalent mechanical model and Newton's second law, this paper takes into account the load balance relationship of various components, derives the differential equations of cutter body split, cutter center block, big gear ring, shield body, pinion and driving motor, and then obtains the multi degree of freedom differential equations of cutter head system.

(1)Cutterhead

The multi degree of freedom dynamics model of the Chinese five-fraction cutterhead is shown in Figure 3.4. The center of the cutterhead, the big gear ring and the shield body are in the generalized coordinate system. And the cutter coordinate system ζi, ηi are relative coordinate systems. With the rotation of the generalized coordinates, coordinate transformation is needed.

$$\begin{cases}
m_{L1}\ddot{X}_{L1}+C_{eqLx}(\dot{X}_{L1}-\dot{X}_r)+C_{eqL1L2}(\dot{X}_{L1}-\dot{X}_{L2})+k_{eqLx}(X_{L1}-X_r)+k_{eqL1L2}(X_{L1}-X_{L2})=F^I_{X1}\\
m_{L1}\ddot{Y}_{L1}+C_{eqLy}(\dot{Y}_{L1}-\dot{Y}_r)+C_{eqL1L2}(\dot{Y}_{L1}-\dot{Y}_{L2})+k_{eqLy}(Y_{L1}-Y_r)+k_{eqL1L2}(Y_{L1}-Y_{L2})=F^I_{Y1}\\
m_{L2}\ddot{X}_{L2}+C_{eqLx}(\dot{X}_{L2}-\dot{X}_r)+C_{eqL1L2}(\dot{X}_{L2}-\dot{X}_{L1})+k_{eqLx}(X_{L2}-X_r)+k_{eqL1L2}(X_{L2}-X_{L1})=F^I_{X2}\\
m_{L2}\ddot{Y}_{L2}+C_{eqLy}(\dot{Y}_{L2}-\dot{Y}_r)+C_{eqL1L2}(\dot{Y}_{L2}-\dot{Y}_{L1})+k_{eqLy}(Y_{L2}-Y_r)+k_{eqL1L2}(Y_{L2}-Y_{L1})=F^I_{Y2}\\
m_{L1}\ddot{Z}_{L1}+C_{eqLx1}(\dot{Z}_{L1}-\dot{Z}_{r1})+C_{eqL1L2}(\dot{Z}_{L1}-\dot{Z}_{L2})+C_{eqLy1}(\dot{Z}_{L1}-\dot{Z}_{r2})+C_{eqL1L2}(\dot{Z}_{L1}-\dot{Z}_{L2})\\
\quad +k_{eqLx1}(Z_{L1}-Z_{r1})+k_{eqL1L2}(Z_{L1}-Z_{L2})+k_{eqLy1}(Z_{L1}-Z_{r2})+k_{eqL1L2}(Z_{L1}-Z_{L2})=F^I_{L1}\\
m_{L2}\ddot{Z}_{L2}+C_{eqLx2}(\dot{Z}_{L2}-\dot{Z}_{r3})+C_{eqL1L2}(\dot{Z}_{L2}-\dot{Z}_{L1})+C_{eqLy2}(\dot{Z}_{L2}-\dot{Z}_{r4})+C_{eqL1L2}(\dot{Z}_{L2}-\dot{Z}_{L1})\\
\quad +k_{eqLx2}(Z_{L2}-Z_{r3})+k_{eqL1L2}(Z_{L2}-Z_{L1})+k_{eqLy1}(Z_{L2}-Z_{r4})+k_{eqL1L2}(Z_{L2}-Z_{L1})=F^I_{L2}\\
I_{Lx}\ddot{\theta}_{Lx}+r_{br}[C_{eqLz1}(\dot{Z}_{L11}-\dot{Z}_{r1})-C_{eqLz3}(\dot{Z}_{L33}-\dot{Z}_{r3})+k_{eqL1}(Z_{L11}-Z_{r1})-k_{eqLz3}(Z_{L33}-Z_{r3})]=M^I_X\\
I_{Ly}\ddot{\theta}_{Ly}+r_{br}[C_{eqLz2}(\dot{Z}_{L22}-\dot{Z}_{r2})-C_{eqLz4}(\dot{Z}_{L44}-\dot{Z}_{r4})+k_{eqLz2}(Z_{L22}-Z_{r2})-k_{eqLz4}(Z_{L44}-Z_{r4})]=M^I_Y\\
I_L\ddot{\theta}_L+C_{rLQ}(\dot{\theta}_L-\dot{\theta}_r)+k_{rLQ}(\theta_L-\theta_r)=-T^I_L
\end{cases} \tag{1}$$

The ζi is the tangential ηi of each component of the cutterhead, which is normal; the pinion coordinate system is Hp, Vp, Hp is the pinion tangent, and Vp is the radial of the pinion. In the formula $I(k = L, r, Lx, Ly, rx, ry, pj, mj)$, they respectively represent the torsion of the center block, the twist of the big gear ring, the cutter coiling around the x axis and the cutter coiling around the y axis, the large gear ringing around the x axis and the big gear ringing around the y axis, the pinion twisting, and the inertia moment of the motor torsion. In the formula $m_{\Theta}(\Theta = i, L, r, d, pj)$, they respectively represent each component of cutter head, the center of the cutterhead, the big gear ring, the shield body and the equivalent mass of pinion. In the formula $k_{L\varsigma i}, k_{L\eta i}, k_{Li}$, they represent the tangential equivalent supporting stiffness of the cutter body, the normal equivalent supporting stiffness and the axial equivalent supporting stiffness respectively. In the formula $k_{eq\vartheta}(\vartheta = x, y, L, r, z, d, dz, p)$, they represent the lateral equivalent supporting stiffness of the cutter head, the longitudinal equivalent supporting stiffness of the cutter head, the radial equivalent supporting stiffness of the gear ring, the axial stiffness of the gear ring, the lateral equivalent supporting stiffness of the shield body, the longitudinal equivalent supporting stiffness, the axial equivalent supporting stiffness, and the equivalent supporting stiffness of the pinion. In the formula $k_{rLQ}, k_{mpQ}, k_{prt(t)}$, they represent the torsional stiffness of the cutterhead, the torsional stiffness of the pinion connecting shaft and the mean mesh stiffness of the gear respectively. In the formula $C_{eq\vartheta}(\vartheta = x, y, L, r, z, d, dz, p)$, they represent the lateral damping coefficient of the cutter center block, the longitudinal damping coefficient of the cutter center block, the radial damping coefficient of the big gear ring, the axial damping coefficient of the big gear ring, the transverse damping coefficient of the shield body, the longitudinal damping coefficient, the axial damping coefficient, and the damping coefficient of the pinion. In the formula $C_{rLQ}, C_{mpQ}, C_{prt(t)}$, they represent cutterhead damping, torsional damping coefficient of pinion connecting shaft and gear mesh damping respectively. In the formula $T_L, T_{mj}, F_X, F_Y, M_X, M_Y, F_L$, they represent cutter torque, motor torque, lateral load of center block, longitudinal load of center block, cutterhead overturning moment and axial load of cutterhead. In the formula $F_{\varsigma i}, F_{\eta i}, F_{Li}$, they separately represent the tangential load of the cutterhead body, the normal load of the cutterhead body and the axial load of the cutterhead body. In the formula b_{prj}, e_{prj}, they represent the meshing error of pinion and large gear tooth side clearance as well as pinion and large gear tooth clearance. In the formula r_{br}, r_{bp}, they represent the base circle radius of the big gear ring and the radius of the pinion base circle respectively.

According to the equivalent mechanics model, the generalized displacement matrix of cutter head system $\{\delta\}$ is:

$$\begin{aligned}\{\delta\} = \{&\zeta_i, \eta_i, Z_i, X_L, Y_L, Z_L, \theta_{Lx}, \theta_{Ly}, \theta_L, X_r, Y_r, \\ &Z_r, \theta_x, \theta_y, \theta_r, X_d, Y_d, Z_d, H_{pj}, V_{pj}, \theta_{pj}, \theta_{mj}\}^T\end{aligned} \tag{2}$$

(i=1~4, j=1~number of pinion) Matrix in 3.1, the formula ς_i, η_i, Z_i, they represent the tangential displacement, normal displacement and axial displacement respectively. In the formula $X_L, Y_L, Z_L, \theta_{Lx}, \theta_{Ly}, \theta_L$, they represent the lateral displacement, longitudinal displacement, axial displacement, overturning vibration displacement and torsional vibration displacement of cutter head respectively. In the formula $X_r, Y_r, Z_r, \theta_x, \theta_y, \theta_r$, they represent the lateral displacement, longitudinal displacement, axial displacement, overturning vibration displacement and torsional vibration displacement of the gear ring respectively. In the formula X_d, Y_d, Z_d, they represent lateral displacement, longitudinal displacement and axial displacement respectively. In the formula $H_{pj}, V_{pj}, \theta_{pj}$, they represent the tangential displacement, radial displacement and torsional vibration displacement of each pinion, respectively. The symbol θ_{mj} represents the angular displacement of each motor's torsional vibration.

3. Interpretation of Results

Through the detailed analysis of many uncertain complex influence relationships such as geological factors, this paper establishes the uncertain geological condition model and takes account of the relationship between the hob position, and finally obtains the cutterhead load boundary under different conditions by interval method, and introduces it into the dynamic equations established and calculates the dynamic response boundaries of the five-fraction cutterhead in each working condition of each part of the project, respectively, as shown in Figure 4. In this paper, the response is analyzed, and then the vibration condition of the cutter head system during tunneling is theoretically understood.

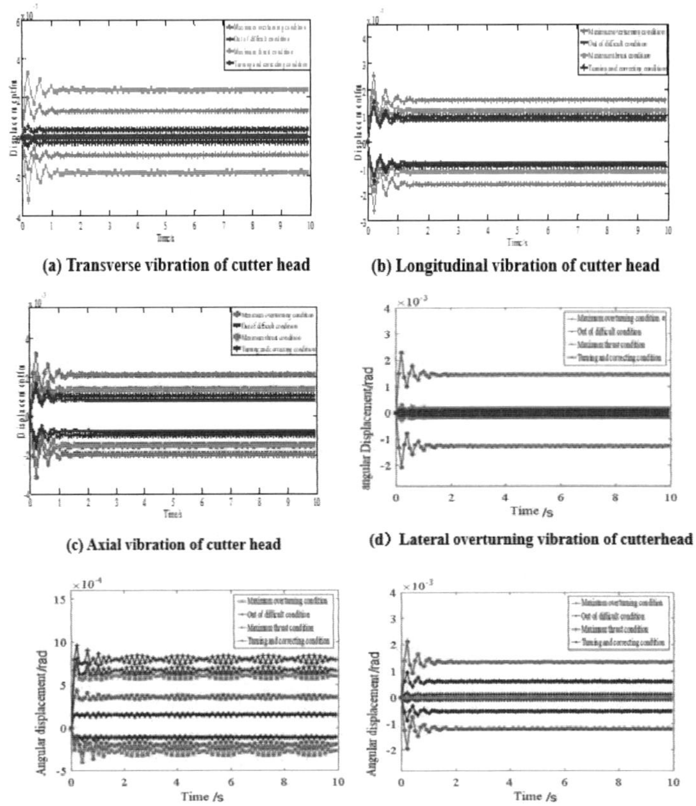

(a) Transverse vibration of cutter head (b) Longitudinal vibration of cutter head

(c) Axial vibration of cutter head (d) Lateral overturning vibration of cutterhead

(e) Longitudinal overturning vibration of cutterhead (f) Torsional vibration of cutterhead

Figure 4. Dynamic response boundary of various degrees of freedom of cutterhead system under different working conditions

As shown in the figure above, the free vibration response of cutterhead can be divided into two stages, namely, non-stationary stage and stationary stage. At the initial stage of cutterhead excavation, the vibration response boundary is non-stationary, oscillating with time, and the vibration response interval enters a stationary stage and oscillates with time. By analyzing the maximum response interval and the steady-state response interval of one circle cutterhead under different conditions, the stationary response intervals of the degrees of freedom of the cutterhead are all narrower than those of the maximum response interval, and the vibration response interval of the cutter body translation is wider than that of the center block translational vibration response area.

Through the dynamic analysis of different degrees of freedom on different degrees of operation, it can be obtained that under different degrees of freedom, which kind of working condition has greater response, therefore, designers can comprehensively consider the vibration response of different working conditions and different forces when designing the cutter head, and aim at the vibration reduction design of the cutter head structure.

4. Conclusion

This paper first considers the geological uncertainty and establishes the geological uncertainty model, and analyzes the uncertainty factors of the cutter head system, and solves the external load interval boundary of the TBM cutter head system. Then, based on multi-body dynamics, this paper establishes a typical TBM cutter system dynamics model under uncertain excitation, the dynamic equation, and obtains the dynamic characteristics. The analysis results can provide a theoretical reference for the structural design and vibration reduction optimization of the cutterhead, and further improve the damage of the cutterhead and ensure the quality of the engineering excavation.

Acknowledgments
This work is supported by the National Natural Science Foundation of China(Grant No.51875076), NSFC-Liaoning United Key fund(Grant No.U1708255).

References

[1] Tang G, Haidong Y U, Xie Q. Evaluation of Dynamic Characteristics of Tunnel Boring Machine Excavating in Uncertain Geological Parameters[J]. Journal of Shanghai Jiaotong University, 2016.
[2] Xie Q J. Research on Modeling and Dynamic Behavior of TBM with Variable Constraints [D]. Journal of Shanghai Jiaotong University, 2015.
[3] ZHANG Kaizhi. Study on dynamic characteristics of redundantly driven revolving system of shield TBM[D]. Shanghai:Shanghai Jiao Tong University,2011.
[4] LIU R.Research on Adaptive Load Sharing Control for synchronous Driving System of cutterhead in shield machines [D].Dalian University of Tecnology, 2013.
[5] Zhu Z H. Numerical Simulation of Wear Evolution of disc cutter of TBM [D]. Central South University, 2013
[6] Huo J, Ouyang X, Wang Y, et al. Analysis of influencing factors of vibration behaviors of TBM cutterhead under heavy impact loads[J]. Journal of Harbin Engineering University, 2015, 36(4):555-559.
[7] Huo J, Hou N, Sun W, et al. Analyses of dynamic characteristics and structure optimization of tunnel boring machine cutter system with multi-joint surface[J]. Nonlinear Dynamics, 2017,87(1):237-254.
[8] Xian H L, Hai B Y, Ming Z Y, et al. Dynamic Modeling and Analysis of Shield TBM Cutterhead Driving System[J]. JOURNAL OF DYNAMIC SYSTEMS MEASUREMENT AND CONTROL-TRANSACTIONS OF THE ASME, 2010, 132(4):1-14.
[9] Sun W, Ma H, Song X, et al. Modeling and Dynamic Analysis of Cutterhead Driving System in Tunnel Boring Machine[J]. Shock and Vibration, 2017:1-12.